Diesel Technology
Fundamentals / Service / Repair

by

Andrew Norman
ASE Certified Master Technician/Machinist
Member, North American Council of Automotive
Teachers

John "Drew" Corinchock
Automotive Author
Technical Writer

Robert Scharff
Automotive Author

Publisher
THE GOODHEART-WILLCOX COMPANY, INC.
Tinley Park, Illinois

Library of Congress Catalog Card Number 00-028759

International Standard Book Number 1-56637-733-1

1 2 3 4 5 6 7 8 9 10   01   04 03 02 01 00

Important Safety Notice

Proper service and repair is important to the safe, reliable operation of motor vehicles. Procedures recommended and described in this book are effective methods of performing service operations. Some require the use of tools specially designed for this purpose and should be used as recommended. Note that this book also contains various safety procedures and cautions, which should be carefully followed to minimize the risk of personal injury or the possibility that improper service methods may damage the engine or render the vehicle unsafe. It is also important to understand that these notices and cautions are not exhaustive. Those performing a given service procedure or using a particular tool must first satisfy themselves that neither their safety nor engine or vehicle safety will be jeopardized by the service method selected.

This book contains the most complete and accurate information that could be obtained from various authoritative sources at the time of publication. Goodheart-Willcox cannot assume any responsibility for any changes, errors, or omissions.

Library of Congress Catalog-in-Publication Data

Norman, Andrew.
    Diesel Technology: fundamentals, service, repair/by Andrew Norman, John "Drew" Corinchock, Robert Scharff.

    p.   cm.
    ISBN 1-56637-733-1
    1. Diesel motor. I. Corinchock, John A. II. Scharff, Robert. III. Title.
TJ795 .N67 2001
621.43'6--dc21   00-028759
                                                              cip

Introduction

Diesel Technology is designed to provide students in the rapidly changing diesel engine field with up-to-date information on the construction, operation, service, and repair of diesel engines. Diesel engines have been the "workhorses of industry" for many years, powering ships, agricultural and construction equipment, generators, pumps, and a host of other applications. They also power the over-the-road tractor-trailers and trucks that carry goods.

In addition to detailing the fundamentals of operation, **Diesel Technology** contains information on the latest developments in the diesel engine field, including engine control computers, fuel management, and emissions control systems. Today's diesel technician must understand the operation of these components and systems in order to service them properly. For many years, the basic design and operation of the diesel engine remained unchanged. Manufacturers concentrated on increasing engine power output and improving reliability. However, the need to meet strict fuel economy and exhaust emissions standards has prompted manufacturers to add complex emissions control devices and electronic control systems to their engines.

Diesel Technology is written in a clear, logical, and interesting manner. Each chapter begins with learning objectives and ends with a summary, a list of important terms, Review, and ASE-type questions. Numerous illustrations are used throughout the text to show and reinforce important concepts. Safety is stressed throughout the text to help ensure safe work practices. Proper hazardous chemical and waste handling and disposal are encouraged throughout the text. A Glossary that contains definitions to many of the terms found in the diesel field is located in the back of the text.

This edition of **Diesel Technology** has been updated to include the latest in electronic fuel injection and computer control technology. The semiconductor information has been included with the basic electronic information. The lubrication chapter has been updated to include the latest standards.

While being an ideal text for the beginning diesel student, **Diesel Technology** is also an invaluable resource for technicians who are currently working in the service and repair of diesel engines. **Diesel Technology** is designed to help students and working technicians prepare to take the ASE Medium/Heavy Duty Truck Diesel Engines, Electricity and Electronics, and PMI tests.

The diesel service and repair industry offers many challenges and rewards for the well-trained service technician. **Diesel Technology** will help you gain the skills necessary to succeed in this dynamic industry.

Andrew Norman
John "Drew" Corinchock
Robert Scharff

Table of Contents

8

Chapter 1

Introduction to Diesel Engines

After studying this chapter, you will be able to:
- ❏ List five reasons diesel engines are more efficient than gasoline engines.
- ❏ Define and explain how compression ignition is used in a diesel engine.
- ❏ Name two major advantages and disadvantages of diesel fuel.
- ❏ Explain in simple terms how electronic engine controls work and why they have been introduced on all modern mobile diesel engines.
- ❏ Name the major applications for diesel engines in modern industry and transportation.
- ❏ Name several key milestones in the development of the diesel engine.

This chapter will introduce you to the history behind the diesel engine and the man that invented it, Rudolf Diesel. It also covers the basic operating principles of diesel engines and lays the foundation for the material that will follow in the later chapters. You will also see how the diesel has become the "workhorse" engine of industry.

Diesel Versus Gasoline Engines

The **diesel engine** is an internal combustion engine that uses the heat of compression to ignite a fuel charge, **Figure 1-1.** Mechanical ignition components used in gasoline engines, such as spark plugs, coils, and distributor assemblies, are not required for ignition. Instead, as the piston in a diesel engine cylinder moves upward on its compression stroke, it compresses the air in the cylinder. The air temperature in the cylinder increases to the point that the diesel fuel ignites as it is injected into the cylinder.

Despite the fundamental difference in the way ignition takes place, the major components of a diesel engine are similar to those found in a gasoline engine. Both engines have a cylinder block, crankshaft, valve train, camshaft, pistons, and connecting rods. They both require a lubrication system and a cooling system.

Diesel Engine Stroke Cycle

Diesel and gasoline engines can be designed to operate on a **four-stroke cycle** or a **two-stroke cycle**. Each **stroke** in the cycle corresponds to the up or down movement of the piston within the cylinder. Four-cycle gasoline and diesel engines use four piston strokes to complete one operating cycle—one stroke each for **intake, compression, power,** and **exhaust.** See **Figure 1-2.** Two-stroke cycle engines accomplish intake, compression, power, and exhaust using only two piston strokes, one upward and one downward.

Figure 1-1. *Modern diesel engines are available in a wide range of sizes and configurations. All use the heat of compression to ignite internal combustion. (Caterpiller)*

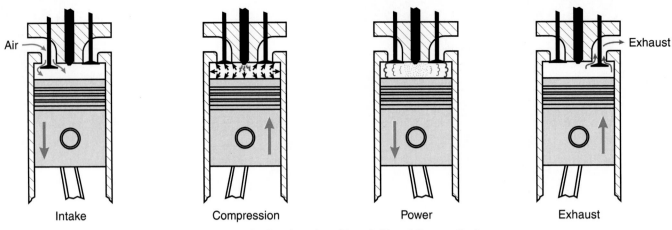

Figure 1-2. *The four piston strokes of a four-cycle diesel engine. (Detroit Diesel Corporation)*

Virtually all high horsepower gasoline engines are four-cycle engines. Two-cycle gasoline engines are used primarily for power tools, lawn and garden equipment, chain saws, outboard boat motors, and other relatively light-duty applications.

In contrast, both two- and four-cycle diesel engines have been used successfully in high horsepower applications. While four-cycle diesel engines are the most common, certain manufacturers have had great success with two-cycle designs. In a two-cycle diesel engine, intake and compression occur on the upward piston stroke, while power and exhaust occur during the downward piston stroke. See **Figure 1-3.**

Diesel engines are also more efficient than gasoline engines. This is due to several factors, including:

- ❑ The method of supplying fuel to the combustion chamber.
- ❑ The high compression ratio used in diesel engines.
- ❑ The method of igniting the fuel in the combustion chamber.
- ❑ The grade and type of fuel used.
- ❑ The greater integral strength of diesel engine components.

Supplying Air and Fuel

Diesel engines are always fuel injected. Unlike gasoline engines, however, the liquid diesel fuel is not mixed with air before it enters the combustion chamber. Instead, air is drawn into the diesel engine cylinder through the intake manifold and compressed by the piston. As the air is compressed, its temperature increases. An atomized mist of liquid fuel is then sprayed into the cylinder at the top of the piston's compression stroke, **Figure 1-4.** The fuel instantly ignites and burns with the high temperature air in the cylinder, forcing the piston down on its power stroke.

In a carbureted or fuel injected gasoline engine, speed is controlled by regulating the amount of air-fuel mixture that is delivered to the cylinders. However, a diesel engine has no throttling valve in its intake manifold. Air pressure in a diesel engine's intake manifold remains constant at all loads. This results in high efficiency at light loads and idle speeds because there is always an excess of combustion air delivered to the cylinders.

Because diesel engines do not have a throttling valve, another method must be used to control engine speed. Diesel engine speed is controlled by varying the amount of fuel injected into each cylinder. The two most common

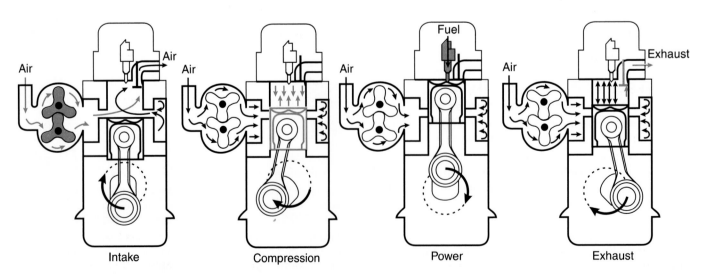

Figure 1-3. *Operating principle of a two-cycle diesel engine. (Detroit Diesel Corporation)*

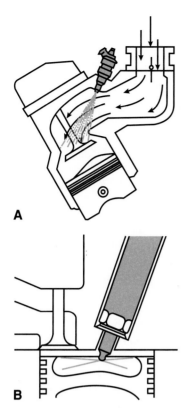

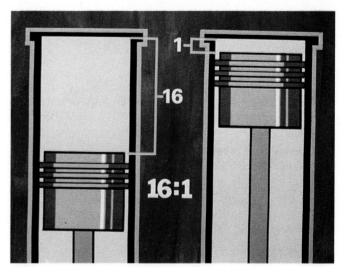

Figure 1-5. *Compression ratios in a diesel engine are high. The compression ratio is determined by comparing the amount of space in the cylinder when the piston is at the top of its travel to the space available when the piston is at its lowest point of travel.*

Figure 1-4. *A–In a gasoline engine, air and fuel are mixed and then introduced into the combustion chamber through the intake valve. B–In a diesel engine, air and fuel are not premixed. The fuel is injected directly into the compressed air inside the combustion chamber.*

ways of controlling the amount of fuel is to either change the fuel injection pump stroke length or to vary the fuel pressure to the injectors based on operating conditions.

Electronic controls have increased the precision and speed at which fuel delivery can be altered. In all cases, a governor plays a key role in the fuel metering process. A **governor** is a device capable of sensing engine speed and load and changing fuel delivery accordingly. Diesel engine governors can be mechanically, servo-mechanically, electronically, hydraulically, or pneumatically controlled.

Compression Ratios

The **compression ratio** is a comparison of the volume of air in a cylinder before compression with its volume after compression. A 16:1 compression ratio means that at the top of the compression stroke, the air in the cylinder takes up 1/16 the volume it did when the piston was at the bottom of the compression stroke, **Figure 1-5.**

Diesel engines operate at higher compression ratios than gasoline engines. Depending on whether the engine is naturally aspirated, turbocharged, or supercharged, typical diesel compression ratios range from 13.5:1 to 22:1. In contrast, gasoline engines typically use compression ratios ranging from 7.5:1 to 9.5:1.

The higher compression ratios used in diesel engines result in greater thermal expansion of gases in the cylinder following combustion. The end result is a higher percentage

of fuel energy being converted into useful power. If the high compression ratios used in diesel engines were used in gasoline engines, preignition or detonation would occur.

Compression Ignition

Because compression raises the air temperature above the fuel's ignition temperature, combustion occurs instantly as the fuel is injected, eliminating the need for a mechanical ignition system. This process is called **compression ignition.** The elimination of spark plugs, coils, ignition wiring, distributors, and transistorized ignition controls is a major factor in the diesel's simplicity and maintenance economy. It also eliminates systems that are the cause of many performance problems in gasoline engines.

Diesel Fuel

Diesel fuel contains more heat energy (BTUs or joules) than does gasoline. However, diesel is much less volatile than gasoline. The diesel engine's design, including its high compression ratio, is intended to extract the maximum amount of power from its fuel. No gasoline or other internal combustion, reciprocating piston-driven engine can match the diesel engine's ability to get the most power out of its fuel. Fuel efficiency in diesel engines can be as high as 40%. In contrast the gasoline engine, at peak efficiency, converts only about 25% of its fuel energy into usable power. The rest is lost as heat.

Durability

Diesel engine cylinder blocks, heads, crankshafts, camshafts, and other major components are designed to withstand operation at high compression ratios and high air temperatures. A typical heavy-duty diesel truck engine will run for 250,000-500,000 miles (400 000-800 000 km) before it must be rebuilt. Also, the same engine can be successfully rebuilt many times. Gasoline engine components are not as durable as diesel engine components.

Diesel Drawbacks

Despite their many advantages, diesel engines are not perfect. The characteristics that make them more durable than gasoline engines also make them more expensive to build. Diesel fuel injection pumps and injectors must be built to precise tolerances, a requirement that also adds to the initial cost of the engine, **Figure 1-6.**

When compared to a diesel engine, a gasoline engine delivers usable torque through a much wider rpm range. To compensate for this, a heavy-duty truck is equipped with a 10-, 18-, or even 20-speed transmission. Gasoline engines also run quieter and offer faster acceleration than diesel engines.

Diesel fuel also presents some problems. Its sulfur content is higher than that of gasoline. A poorly-adjusted diesel engine can produce soot-laden, foul-smelling exhaust. In comparison, gasoline engines rarely emit solid pollutants. Also, diesel fuel has a tendency to attract water, which can cause bacteria to form in the fuel system.

The power output of the diesel engine can also be affected by the temperature of the fuel. At lower temperatures, wax in diesel fuel begins to solidify into crystals that can block fuel flow through filters, lines, and injectors. This condition can lead to fuel starvation or no-start conditions. Many diesel engines are equipped with fuel heaters to help prevent this problem. High fuel temperatures above 100°F (38°C) can also thin out the diesel fuel and reduce power output. At temperatures above 150°F (66°C), diesel fuel loses much of its lubricating ability. When this occurs, damage to the injectors and other parts can result.

Diesel Engine History

The steam engine was the driving force behind the Industrial Revolution in the United States and Europe during most of the nineteenth century. Toward the end of the nine-

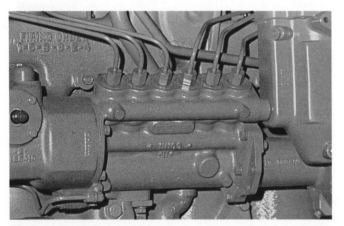

Figure 1-6. *Diesel injection pumps are quite complex and must be built to exacting tolerances. (Caterpillar)*

teenth century, however, it could no longer meet the needs of all industries, particularly smaller businesses. Steam engines were very large, making them best suited for heavy industry and transportation. They required an external firebox to burn fuel, a boiler to store the steam that drove the piston, and a condenser to turn the steam back into water. Steam engines required constant tending by an operator and were highly inefficient. The average steam engine converted only 6% of its fuel heat into usable power.

Internal combustion engines filled the need for a small, efficient power source. In an **internal combustion engine**, fuel burns inside the engine itself, rather than in a separate furnace or firebox. Higher working pressures are also possible when air and combustion gases are used instead of steam to propel the piston inside the cylinder.

The gasoline engine and the diesel engine are the two most widely used internal combustion engines in the world. Credit for their invention is historically given to two men—**Nikolaus Otto** for the gasoline engine and **Rudolf Diesel** for the engine that bears his name. However, the triumphs of Otto and Diesel would not have been possible without the theories and experiments of earlier inventors and scholars. The fact that both Diesel and Otto faced challenges to their patents (Otto's patent was eventually broken) proves that many people were working toward the same goal.

Early Theories and Successes

Nicholas Sadi Carnot, a French scientist, formulated theories on the thermodynamics of engine operation that would influence both Otto, and to a greater extent, Diesel. In papers published in 1824, Carnot introduced the idea of a cycle in heat engine operation. The cycle consisted of a series of steps that would both produce power and return the engine to its original position. Carnot theorized that a piston could be driven back and forth in a cylinder by first heating and then cooling a gas trapped inside the cylinder. The controlled expansion of the gas as it is heated and the contraction of the gas as it cooled would drive the piston.

In theory, Carnot's engine was 100% efficient. All heat generated would be turned into useful power. Carnot admitted that a 100% efficient engine was an impossibility, but it was an idea that would intrigue a young Rudolf Diesel over fifty years later. Although Carnot never attempted to build a working model based on his theory, his ideas on engine cycling and the use of an expanding gas to move a piston were major contributions to small engine design.

In 1860, Etienne Lenoir demonstrated a double-acting internal combustion engine fueled with street lamp gas. It used many of the principles found in modern engines. A mixture of air and lamp gas was drawn into one end of a long cylinder that housed a piston connected to a crankshaft. The mixture was ignited by an electric spark. The resulting explosion drove the piston along the cylinder to the end of its stroke. The process was then repeated at the other end of the cylinder with the explosion occurring on the opposite face of the piston. This drove the piston back to its original position. On its return trip, the piston helped clear the cylinder of the waste gases remaining from the previous explosion. The two firings per cycle were timed to occur at 180° of crankshaft rotation apart.

The Silent Otto Engine

Nikolaus Otto, a German traveling salesman, heard of the Lenoir engine during one of his many trips throughout Europe. Although he lacked an engineering background, Otto became interested in building a small engine that could successfully control an explosive fuel. Otto began work on his engine in the 1860s. He received financial backing from the German industrialist Eugen Langen, and by 1876, their company, Deutz Motors, was producing an internal combustion engine that operated at atmospheric pressures. The engine, called the Silent Otto, used a four-stroke combustion cycle. The Silent Otto drew in a mixture of gasoline and air, compressed it in the cylinder to a compression ratio of about 2.5:1, ignited it with a flame, and then expelled the exhaust gases. The thermal efficiency of the Silent Otto was about 14%, more than twice that of the steam engine.

A reliable ignition system was a problem for early gasoline engines. The first solution used a *hot-bulb chamber*, located next to the main combustion chamber. This small chamber was continually heated by the burning fuel and stayed hot enough to ignite each subsequent fuel charge. Because it remained hot all the time, the hot-bulb chamber tended to preignite the air-fuel mixture, especially at low engine speeds when the charge could not be compressed fast enough by the piston. Another problem was initial engine start-up. The hot-bulb chamber had to be preheated with a blow torch before it could successfully fire the first charge.

The introduction of an **electric ignition** system that used spark plugs solved many of the gasoline engine's ignition problems. With electric ignition, the widespread acceptance and success of the gasoline engine was assured. Gasoline engines proved to be compact and lightweight for the power they produced. They continue to be the engine of choice in most passenger vehicles and serve in countless light manufacturing and portable power applications.

The Development of the Diesel Engine

The success of the Silent Otto did not stop others from pursuing a better engine design. Otto's 14% fuel efficiency offered plenty of room for improvement and the dangers involved in igniting and containing an explosive charge convinced many engineers that safer, more efficient engines were possible. One of these engineers was Rudolf Diesel, **Figure 1-7**.

Rudolf Diesel was an extremely brilliant man. He was educated in Germany, attending the Commercial School at Augsburg and then the Technical University at Munich. One of Diesel's university instructors was Carl Linde, the founder of the modern refrigeration business. Linde introduced Diesel to the heat engine theories of Carnot and planted the seed of an idea that eventually grew into the diesel engine.

After graduating with the highest test scores in the university's history, Diesel went to work for Linde. He traveled throughout Europe and North Africa repairing refrigeration systems and acting as a company representative. Diesel began his inventive career while working for Linde. His first invention was a refrigeration system that made clear ice instead of crystalline ice. However, a clause in his employment contract prevented him from marketing the system, and Diesel's interests soon turned to engine design.

Diesel's First Engine

Diesel's first engine was an *external combustion engine* that used ammonia vapors instead of steam to drive the piston. The low boiling point of ammonia meant that less heat was required for vaporization. Diesel also devised a system in which glycerin absorbed the ammonia vapor as it was expelled from the cylinder. The glycerin held the vapor's heat so that it could be recycled to heat more ammonia. The goal was a super-efficient engine that could run for hours after a short initial warm-up period. Although

Figure 1-7. *Rudolf Diesel, the father of the diesel engine. (MAN Historical Archives)*

his model worked, problems with heat exchange, leaks, and power output led Diesel to abandon the design in April of 1889.

Diesel's experiments led him to two important conclusions. The first was that the difference between the pressure of the compressed gas at the start of the power stroke and the pressure of the expanded gas at the end of the power stroke should be as great as possible. This would provide an equally wide heat range difference, with the maximum amount of heat being converted to usable power. To accomplish this, Diesel estimated that pressures of fifty to sixty atmospheres would be needed inside the cylinder.

Diesel's second conclusion was that both ammonia vapor and steam were too difficult to handle at such high pressures. The best gas for the job was simply air. In the ammonia vapor engine, both the ammonia and glycerin had to be heated and cooled from external sources. For a short time, Diesel approached his high compression air engine in the same way, considering the use of an external source to heat the air.

The Beginning of the First Diesel Engine

Diesel was driven in his work by Carnot's vision of a perfectly efficient heat engine. He had not set out to build an internal combustion engine. Finally, in his own words, Diesel formulated "the idea of using air not only as a working medium, but also as a chemical medium for combustion." He considered his work of the previous five years, "a long detour to an idea that has been used for a long time in gas and hot air engines: combustion in the cylinder itself."

Diesel also realized that igniting the air-fuel charge using a hot bulb or external spark would not be practical at the high temperatures and pressures his engine would generate. Instead, Diesel decided on "undertaking combustion in the highly compressed air itself." In other words, the high heat of the compressed air would be used to ignite the fuel.

The ideas of internal combustion air as a working medium, and compression ignition were not new. The first two were being successfully used in the Silent Otto engine and many other engines that were produced before 1892. The principles of compression ignition were well documented by several scholars. Diesel's originality came in the way he proposed to control combustion in the cylinder.

Refining the Engine

Diesel believed the gasoline engine operated on a poor working principle. Since an almost equal mixture of gasoline and air was used, ignition inside the combustion chamber resembled more of an explosion than a controlled burning of fuel. The result was a major increase in temperature. Much of the fuel heat was lost through the cylinder walls or in the exhaust gases.

To correct this inefficiency, Diesel proposed drawing in about nine times more air than was actually needed to burn the fuel. The air would be compressed until it reached a temperature that was much higher than the temperature

at which the fuel burned. At the point of maximum air compression, a small amount of fuel would be introduced into the cylinder. The high heat of the compressed air would immediately ignite the fuel. As the fuel burned, its expanding combustion gases would drive the piston down on its power stroke. In Diesel's engine, these expanding combustion gases would theoretically perform another equally important job. They would absorb the heat generated by the burning fuel so the temperature inside the cylinder would remain constant.

With little or no excess heat generated, Diesel calculated that his engine would be 72% efficient. In his 1892 patent draft, Diesel proposed coal dust as the engine's probable fuel. Diesel imagined a system in which coal dust would be stored in hoppers and gravity fed into the cylinders through rotating disks. In theory, the job of introducing fuel into the cylinder appeared rather simple. In practice, the task would prove much more difficult.

Early Engine Models

Diesel made a mistake by filing his patent before building a working model of his new engine. This would eventually cause him numerous problems, because the engine he would ultimately produce differed significantly from the one he proposed, **Figure 1-8A**. Diesel pursued thermal efficiency and the Carnot cycle at the expense of mechanical efficiency. It was soon clear that the power needed to compress the air inside the cylinder could not be generated using the amount of fuel Diesel proposed. Simply put, Diesel's engine would not run.

Diesel corrected this problem by decreasing compression levels inside the cylinder dramatically and increasing the amount of fuel used by a factor of eight. Both steps decreased operating efficiency. Diesel's engine now had to be liquid cooled to remove excess heat. He also abandoned the theory of constant temperature operation in favor of constant pressure operation. In other words, enough fuel would now be introduced into the cylinder so that expanding combustion gases could maintain a constant pushing pressure on the piston through its entire power stroke. These changes brought Diesel's engine closer to other engine designs of the day, **Figure 1-8B**.

With the exception of compression ignition, Diesel's engine now closely resembled an engine proposed by Otto Kohler five years earlier. Diesel steadfastly insisted that his ideas were not influenced by others. His cold dismissal of such charges was viewed by many peers as a sign of his guilt. Rather than refile a patent that reflected the changes in his design, Diesel foolishly attempted to defend his 1892 patent. Much of Diesel's later life was spent defending his early ideas rather than formulating new ones.

The years between 1892 and 1897 were Diesel's most productive. During this period, he supervised the building of the first diesel engines at the Augsburg Engine Works. However, all did not go smoothly. Leaking valves and gaskets made reaching the high air pressures needed to start combustion nearly impossible. When these problems were solved, the engine exploded violently when the first fuel

A   **B**

Figure 1-8. A–Diesel's first experimental engine was built in 1893. B–Modified diesel engine. Following modification, including enlargement of the piston bore, the first power tests were performed on the diesel engine in 1895. The tests showed that the new engine had an efficiency of 26.2%.
(Diesel & Gas Turbine Publications)

charge was injected into the cylinder. Numerous liquid fuels were tried, with kerosene producing the best results.

Vaporizing the fuel proved especially troublesome. An air-blast system using an air compressor from a refrigerator was the first workable solution to this problem. Vaporization was later refined with the development a sieve-type atomizer. The piston was also enlarged and the combustion chamber redesigned. A better fuel pump was installed and a crude fuel distributor was devised. By February of 1897, after almost five years of extensive trial-and-error testing, Diesel's engine was ready for sale to business and industry. Although a far cry from Carnot's perfect heat engine, the diesel engine offered a thermal efficiency of over 30%. Fuel consumption was only one-half that of the gasoline engine.

The First Diesel Engines in Industry

The first commercial installation of a diesel engine was a two-cylinder, 60 horsepower engine installed in a Bavarian match factory. At an exhibition for engines and driven machines in Munich in 1898, the diesel engine was officially presented to the public for the first time, **Figure 1-9.** Interest in the diesel engine was high, and Diesel quickly sold licensing rights to build the engine to companies in France, England, Germany, Russia, and the United States. Diesel was now a rich man and was riding a wave of success.

Nevertheless, production was slow and problems persisted. The sheer size and weight of the diesel engine limited its use. The bulky air compressors, air tanks, valves, and lines needed to run the fuel injection system prevented mobile applications, and the fledgling automobile industry

quickly adopted the gasoline engine as its power source. By 1902, only 350 diesel engines were in operation. Problems with fuel vaporization and distribution, fuel quality, and air compression were common in the early diesel engines.

Diesel's attempt at operating his own production plant was a failure. Also, personal clashes with his industrial backers resulted in Diesel having little input in the technical improvements made to his original design after 1900.

The greatest area of promise for the diesel engine appeared to be marine applications. The first diesel-powered, ocean-going ship was the 7000 ton *Seelandia,* a Danish vessel powered by two 1000 horsepower, eight-cylinder, four-stroke engines. Its maiden voyage to Bangkok in 1912 sparked a great deal of interest. As Europe and the world moved steadily toward war, interest in diesel-powered warships and submarines increased.

The Demise of Rudolf Diesel

Despite the increasing application of his engine design, Rudolf Diesel faced impending bankruptcy by the autumn of 1913. He complained of poor health and was beset by critics eager to discredit his work and challenge his patents. In late September, Diesel boarded the steamship *Dresden* for a trip to England. He planned to attend the ground breaking for the British Diesel Company's new production plant. Diesel would never arrive for the groundbreaking. Sometime during the night of September 30th, Diesel disappeared from the ship. In early October, a Dutch pilot boat recovered a body from the sea. As was the custom, the captain of the pilot boat removed certain personal effects from the body and then returned it to the sea. These personal effects were later identified as Diesel's.

Figure 1-9. At an exhibition in 1898, Diesel's engine was officially shown to the public for the first time.
(Diesel & Gas Turbine Publications)

The exact circumstances of Diesel's death were never fully determined. Many factors point toward suicide, but some suspect Diesel possibly met an untimely death at the hands of German military agents, anxious to keep diesel engine technology from the British.

Had Diesel lived, he would have seen a rapid succession of technological advances that finally unleashed the diesel engine's full potential. After World War I, marine diesels were rapidly refined. The 1920s also saw the perfection of **solid fuel injection**, which allowed diesel fuel to be injected directly into the combustion chamber without using an air compressor.

Another major milestone occurred when the Robert Bosch firm of Stuttgart, Germany developed a compact, reliable fuel injection pump. These advances opened the door for smaller, lighter, faster diesel engines that could power trucks and automobiles. The Daimler-Benz Company introduced the first production model diesel automobile in 1936.

Continued Development of the Diesel Engine

In America, interest in the diesel engine was spearheaded by such men as Clessie Cummins, owner of the Cummins Engine Company in Columbus, Indiana. In 1930, Cummins outfitted a Packard automobile with a modified diesel marine engine and drove it from Indianapolis to New York City—over 800 miles at a cost of $1.38. Cummins also drove the only diesel-powered vehicle ever entered in the Indianapolis 500, completing the race without a fuel stop.

Several companies, including General Motors (Detroit Diesel), Caterpillar, Allis-Chalmers, and Worthington were building and marketing diesel engines by the late 1930s. The diesel proved its worth in World War II so well that Detroit Diesel's Gray Marine earned the nickname "the engine that won the war of the seas." Land-based diesels proved their reliability in thousands of installations.

Diesel-Electric locomotives rapidly replaced less efficient steam engines in railroad applications, **Figure 1-10.** By 1941, almost 1400 diesel locomotives were operating on American rail systems, and by the 1950s, the transition to diesel locomotives was complete.

Diesels on the Road

By the 1950s, diesel engines had also become dominant in heavy duty trucks and construction equipment. The development of the compact Roosa Master fuel distributor paved the way for diesel-powered farm tractors and agricultural equipment. Buses and other public transportation vehicles were converted to fuel-efficient, durable diesel engines. Electric power generation, mining, and irrigation are all areas that came to rely heavily on diesel power.

Interest in fast, lightweight diesels for automobiles and light trucks peaked in the early 1980s, as the auto industry searched for a solution to gasoline shortages and increased air pollution. However, stabile gasoline prices and improved

Figure 1-10. *By the 1950s, diesel locomotives replaced all of the inefficient steam engines on railroads.*

computer, fuel, and emission control systems dampened this enthusiasm and the automotive market remains only a small portion of the total number of diesel applications in the United States. In Great Britain, however, more than 25% of the cars on the road are powered by diesel engines.

Modern Diesel Applications

The many applications of the modern diesel engine can be grouped into two broad categories:

❑ Mobile applications.
❑ Stationary applications.

Mobile Applications

In **mobile applications**, the engine propels the vehicle or machine upon which it is mounted from one location to another. In most cases, the diesel engine pushes or pulls a load while also driving external accessories. For example, the engine may propel its machine via the powertrain to the wheels, while another train of gears from the engine drives a power takeoff. A **power takeoff (PTO)** is essentially a supplemental drive. The PTO may drive another machine that is pulled behind the first one. A good example of this is a farm tractor pulling and powering a hay baler. All mobile applications require a power train to convert engine speed to the desired machine speed. Examples of mobile diesel engine applications include:

❑ Cars, trucks, and buses, **Figure 1-11.**
❑ Ships and boats, **Figure 1-12.**
❑ Locomotives.
❑ Railcars.
❑ Construction equipment, **Figure 1-13.**
❑ Forestry equipment.
❑ Harvesting and farm equipment.
❑ Wheel and crawler tractors.

Stationary Applications

Diesel engines used in **stationary applications** produce power from a fixed location for industrial use, **Figure 1-14.**

A

B

Figure 1-11. Various applications of mobile diesel engines. A—Cars and light trucks. B—Medium and heavy-duty trucks and buses. (Mercedes-Benz)

Figure 1-12. Applications of marine diesel engines vary from tankers to cutters to tugboats to yachts. (US Coast Guard)

Figure 1-14. One typical application of stationary diesel engines. While these engines are permanently fixed, smaller diesel engines are transportable to any place where a powerful, efficient engine is needed.

Figure 1-13. As the workhorses of industry, diesel engines power off-highway and construction equipment as well as over-the-road medium and heavy-duty trucks. (Jack Klasey)

Figure 1-15. Typical gen-set. Diesel power is used to generate electricity, pump fluids, and hundreds of other industrial applications.

They are often called **power units** and are mounted on a stand, transmitting their power to the load through a coupler or driveline. Because no mechanism is needed to propel the engine and machinery, the distribution of power is simplified. Modern stationary applications for diesel engines include:

❑ Electric power generators (gen-sets), **Figure 1-15.**
❑ Industrial PTO units.
❑ Motorized pumps.
❑ Cranes and power shovels.
❑ Diesel starting units.

Modifications to Increase Diesel Engine Efficiency

For many years, the basic design and operation of diesel engines remained unchanged. Manufacturers concentrated on increasing engine power output and overall reliability. However, concerns over exhaust emissions and the need to conform to strict emission standards have led diesel manufacturers to design **electronically controlled fuel injection systems** and **emission controls.** The electronic controls are similar to those that have been adopted by the automotive industry for gasoline engines. They use **sensors** to monitor operating conditions, a **computer** to calculate the ideal fuel and engine settings for these conditions, and electromagnetic **actuators** to make the necessary adjustments. Diesel engine technicians must understand how these electronic controls operate in order to service them properly. Mastering new troubleshooting techniques and test equipment is essential for success.

In addition, many heavy-duty truck diesel engines are now equipped with **turbochargers** and **superchargers** to increase power output and engine responsiveness under load. Changes in design and the materials used to construct engine components are occurring at an ever increasing speed. Other significant developments in the past ten years include:

❑ The use of high-efficiency air-to-air **aftercoolers** that cool the air supplied to the engine. Cool air is denser and holds more oxygen than warm air. The result is a 3%-5% improvement in fuel mileage and a reduction in exhaust emissions.

❑ Increased use of **unit fuel injectors,** which are capable of attaining higher injection pressures than inline pump fuel injection systems. Again, the result is reduced emissions.

❑ Matched intake and exhaust flow systems.

❑ Redesigned intake and exhaust valves.

❑ Low-flow cooling systems.

❑ Computerized fuel management systems.

The new generation of diesel engines operates on the basic principles set forth by Rudolf Diesel over a century ago, but they offer today's diesel service technician many challenges and opportunities to learn new skills.

Summary

Diesel engines are internal combustion engines that use compression for ignition. Air and fuel are not premixed before entering the combustion chamber. Liquid diesel fuel is injected into the highly compressed, high-temperature air inside the cylinder, where it ignites and burns.

Since there is always an excess of combustion air in the cylinder, engine speed is controlled by metering the amount of fuel injected into the cylinder. Electronic controls have increased the precision and speed at which this can be done.

When compared to gasoline engines, diesel engines operate at high compression ratios. Diesels are more durable than gasoline engines, but are more expensive to manufacture. With proper care, diesel engines can be rebuilt several times and will provide many years of service.

The diesel engine was invented by Rudolf Diesel, an engineer who never lived to see his engine fully appreciated or used to its potential.

Robert Bosch developed the first compact, reliable fuel injection pump during the late 1920s. It opened the door for smaller, faster diesel engines in trucks, farm equipment, boats, and other mobile applications.

Modern diesel engines are used in many mobile and stationary applications, such as trucking, construction, mining, agriculture, rail, marine, and power generation.

Important Terms

Diesel engine	Electric ignition
Four-stroke cycle	Solid fuel injection
Two-stroke cycle	Mobile applications
Stroke	Power takeoff (PTO)
Intake	Stationary applications
Compression	Power units
Power	Electronically controlled fuel injection systems
Exhaust	
Governor	Emission controls
Compression ratio	Sensors
Compression ignition	Computer
Diesel fuel	Actuators
Internal combustion engine	Turbochargers
	Superchargers
Nikolaus Otto	Aftercoolers
Rudolf Diesel	Unit fuel injectors

Review Questions–Chapter 1

Do not write in this text. Place your answers on a separate sheet of paper.

1. What is a diesel engine?

2. List at least four reasons why diesel engines are more efficient than gasoline engines.

3. In a diesel engine, the _____ is not mixed with air prior to entering the combustion chamber.

4. In a diesel engine, speed is controlled by regulating the amount of _____ delivered to the cylinders.

5. There is always an _____ of combustion air in a diesel engine.

6. What is the function of the governor in a diesel engine?

7. Diesel fuel contains more _____ than gasoline.

8. What problems are associated with diesel fuel at both low temperatures and at very high temperatures?

9. Who was the first engineer to develop a compact, reliable diesel fuel injection pump?
   (A) Nikolaus Otto.
   (B) Rudolf Diesel.
   (C) Robert Bosch.
   (D) Clessie Cummins.

10. Define the terms *mobile* and *stationary* as they pertain to diesel engine applications.

ASE-Type Questions

1. Typical compression ratios used in heavy-duty, high-speed diesel engines range between _____.
   (A) 8.5 and 10:1
   (B) 10 and 15:1
   (C) 13.5 and 22:1
   (D) 17.5 and 25:1

2. Technician A says that when compared to a diesel, a gasoline engine delivers usable torque through a much wider rpm range. Technician B says gasoline engines also run quieter and offer faster acceleration. Who is correct?
   (A) A only.
   (B) B only.
   (C) Both A & B.
   (D) Neither A nor B.

3. Technician A says unit fuel injectors are being used on diesel engines because they are capable of attaining higher injection pressures that lead to reduced emissions. Technician B says that inline injection pumps are capable of attaining higher injection pressures than unit fuel injectors. Who is correct?
   (A) A only.
   (B) B only.
   (C) Both A & B.
   (D) Neither A nor B.

4. Technician A says that aftercoolers are used to cool the diesel fuel that returns to the fuel tank from the injectors. Technician B says that aftercoolers are used to cool the air supplied to the engine so it is denser and contains more oxygen. Who is correct?
   (A) A only.
   (B) B only.
   (C) Both A & B.
   (D) Neither A nor B.

5. Technician A states that both four-stroke and two-stroke diesel engines are used for high-horsepower diesel engines. Technician B states that only four-cycle engines are used for high-horsepower diesels. Who is correct?
   (A) A only.
   (B) B only.
   (C) Both A & B.
   (D) Neither A nor B.

6. All of the following are characteristics of diesel engines EXCEPT:
   (A) they are more durable than gasoline engines.
   (B) diesel engines operate at lower compression ratios than gasoline engines.
   (C) all diesel engines are fuel injected.
   (D) diesel engines convert a higher percentage of fuel into useful power.

7. All of the following are drawbacks to the diesel engine EXCEPT:
   (A) does not produce usable torque through a wide rpm range.
   (B) power output can be affected by the temperature of the diesel fuel.
   (C) they offer faster acceleration than gasoline engines.
   (D) they are more expensive to build.

8. Technician A states that solidified wax can lead to fuel starvation or a no-start condition. Technician B states that high temperatures can thin out diesel fuel and increase wear. Who is correct?
   (A) A only.
   (B) B only.
   (C) Both A & B.
   (D) Neither A nor B.

9. All of the following are changes or systems designed in the last 10 years to improve the efficiency of diesel engines EXCEPT:
   (A) inline fuel injection pumps.
   (B) computer-controlled fuel management systems.
   (C) high-efficiency aftercoolers.
   (D) redesigned intake and exhaust valves.

10. Stationary applications for diesel engines include:
    (A) motorized pumps.
    (B) starting units.
    (C) industrial power takeoff units.
    (D) All of the above.

Diesel technicians sometimes have to go out to a worksite to maintain or repair engines. (John Deere)

Shop Safety

After studying this chapter, you will be able to:
- ❏ Explain the importance of shop safety in a repair shop.
- ❏ Describe the personal safety precautions that a technician must follow.
- ❏ Outline the general shop housekeeping procedures that must be maintained.
- ❏ Explain the safety rules that must be practiced when working with fuel injectors.
- ❏ Describe the three basic types of fires.
- ❏ Describe the proper use of the different types of fire extinguishers.
- ❏ Know your rights under the right-to-know laws.

The most important consideration in any shop should be accident prevention and safety. Safety is everybody's business. Most accidents result from one or more broken safety rules. The injured person learns to respect safety rules the hard way—by experiencing a painful injury. You must learn to respect safety rules by studying and following the rules in this book and in manufacturers' service manuals.

Safety Notices

The chapters in this text contain special notes labeled **note, caution,** and **warning.** Notes contain information that can help you complete a particular task or make a job easier. Cautions are placed to warn you not to make a mistake that could damage the engine or vehicle. Warnings are a reminder of those areas that could cause personal injury or death. An example of a typical notice is shown in **Figure 2-1.**

In addition, most manufacturers' service manuals also contain safety symbols that provide an additional signal for the need to use caution. Always pay attention to these symbols. Personal safety is always the *number one* consideration when working with diesel engines. Safety is not based simply on do's and don'ts—it is a matter of understanding the job at hand and using common sense.

Personal Safety

The following are general guidelines for personal safety, as well as accident prevention. Knowing and following these safety rules is very important. Keep yourself neat and clean. This means a clean, pressed uniform—not clothing that has oil or grease stains and keeping yourself well groomed. Not only does this promote health and safety, it shows professionalism.

You should dress comfortably, but safely. Avoid wearing dangerous clothing, such as ties, open jackets, sweaters, or long sleeve shirts. Remove any rings, watches, or other jewelry. Never wear jewelry or any other item that can become caught in moving machinery. Loose clothing or jewelry can become caught in the moving parts of an engine or a rotating shaft, pulling you in with it. If your job position requires you to wear a tie, tuck it inside your shirt when around engines or other machinery. If you wear your hair long, it should be tied up or kept under a hat.

Footwear is the most important part of the diesel technician's attire. Proper footwear for the shop is a pair of steel toe safety shoes with slip-resistant soles, as they can protect against flying sparks, heavy falling objects, and chemicals. Most safety shoes are constructed using materials that will

⚠️ **Warning: Do not create sparks or flames around diesel fuel. While diesel fuel is not as volatile as gasoline, it will flash and burn if exposed to a heat source.**

Figure 2-1. *This a typical warning used in this textbook. Cautions and notes are styled in a similar manner.*

give the shoe a long life expectancy in a harsh shop environment. These shoes also have additional support for standing on the concrete floors in most shops. Protective head covering such as a hard hat is advised when working in a pit or under an overhead hoist.

Personal Protective Equipment

Personal protection is vital when performing many jobs in the shop. Failure to wear the proper eye, respiratory, skin, or hearing protection can result in a permanent injury. In some cases, the damage does not appear until it is too late to correct, often occurring slowly over a period of months or years.

Wear eye protection at all times. When you enter a shop, you are entering an area that at any time, may result in a situation where dirt, metal, or liquid is splashed in your face. You should wear special safety glasses, goggles, or a face shield whenever a particular job requires it, such as welding, chiseling, or grinding, **Figure 2-2.**

Respiratory protection, such as face masks, should be worn whenever you are working on systems that can produce dust and dirt. While most dust is not potentially harmful, some dust may contain **asbestos**, which is a known carcinogen (can cause cancer). Respiratory protection is also a good idea when working around any equipment that gives off fumes, such as a hot tank or steam cleaner.

Protective gloves are necessary whenever you are working with chemicals such as parts cleaning solvents. If you spill oil, diesel fuel, cleaning solvents, or any other substance on your skin, clean it off immediately. Prolonged exposure to even mild substances may cause severe rashes or chemical burns.

Hearing protection is important when working around noisy engines and machinery. Largely ignored for many years by technicians, shops are beginning to supply ear plugs and other hearing protection, **Figure 2-3.** It is important to wear hearing protection when working in a noisy area or when using certain tools, such as pneumatic impact wrenches and chisels.

Figure 2-3. Hearing protection is an often overlooked safety device. However, it is a very important one.

Shop Safety Rules

It is not only important to dress safely, but you must also work safely as well. These general shop rules should also be kept in mind while you are working. Following them can prevent a painful injury or death. Horseplay is never permitted in the shop area. Report all accidents promptly so that proper first aid or medical assistance can be provided. Always lift heavy objects by bending down and using your leg muscles so as not to strain your back, **Figure 2-4.**

Shop Cleanliness

Keep all workbenches clean. This reduces the chance of tools or parts falling from the bench, where they could be lost, damaged, or cause injury. A clean workbench also reduces the possibility that critical parts will be lost in the clutter. It also reduces the chance of a fire from oily debris.

Figure 2-2. Various types of eye protection. You should always wear some sort of eye protection whenever you are in the shop.

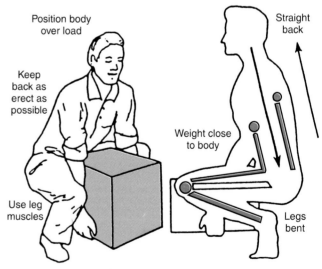

Figure 2-4. Using the correct lifting technique will prevent a painful back injury. Never lift more than you are capable of handling safely.

Do not allow oily rags to accumulate. Place them in a secure, self-closing fireproof safety container. Place all trash, defective parts, and paper debris in a waste can. Recyclable core parts should be kept in a location away from the primary work areas.

Return all tools and equipment to their proper storage places. This saves time in the long run, as well as reducing the chance of accidents, damage, and theft. Do not leave any pieces of equipment where others could trip on them. If tools are dirty or oily, clean them before you put them away. This will not only extend the life of the tool, but is also a courtesy to others who use them. Never use a damaged tool or piece of equipment—report any damaged shop equipment immediately. Remove creepers from the floor area when not in use. Always be aware of and follow all safety rules when using any piece of equipment. If you do not know how to use a piece of equipment, read the manufacturer's instruction manual before use. Use all tools properly; more accidents are caused by the improper use of hand tools than the improper use of power tools.

Clean up any spilled oil, grease, diesel fuel, coolant, or other fluids from the floor immediately. Many people are injured when they slip on floors coated with oil, fuel, antifreeze, or water. Do not leave open containers of any chemical in the shop or outside.

Vehicle Safety Rules

Move vehicles very slowly (about 2 mph) in and out of the shop, checking to see that no one is in the way. Get someone to help you guide the vehicle in and out of the shop. Do not step in the path of a moving vehicle. Do not work beneath a vehicle that does not have properly installed jack stands, **Figure 2-5.** Do not stand under a rack or lift unless safety pins are in place. When lifting the hood of a heavy-duty truck, be sure that it is in a locked position and is well supported, **Figure 2-6.**

When lifting an engine from a vehicle, be sure that the lift or crane has the proper lift capability to do the job, **Figure 2-7.** Be certain all throttle linkage connections are intact and all fuel lines are connected before cranking the engine. When running an engine in the shop, it must have an exhaust hose connected to the outside for proper ventilation, **Figure 2-8.** If the shop is not properly ventilated, *carbon monoxide* will quickly build up, possibly causing death. Never leave a running vehicle unattended.

Figure 2-6. *When working on an engine, be sure that the hood is in the locked position and is well supported.*

Figure 2-7. *When removing an engine, use a lift that is rated to do the job. This lift is sometimes referred to as a "cherry picker." (Snap-On Tools)*

Diesel Fuel Safety Precautions

There are additional safety rules when working with diesel fuel, fuel injectors, and injector test equipment. While diesel fuel is not as volatile as gasoline, it will still burn. Avoid sparks, smoking, or open flame around diesel fuel.

Use caution when testing or working on the diesel injection system. Diesel fuel, under high pressure, can easily penetrate the skin, causing infection or a severe reaction. Seek medical attention immediately if you are ever injured by a stream of pressurized diesel fuel. Never compress a fuel injector while the tip is touching your skin.

Figure 2-5. *Jackstands should be used whenever you must work under a vehicle. Be sure to locate them under the vehicle's frame, not under a drivetrain or suspension part. (Snap-On Tools)*

Figure 2-8. *Carbon monoxide can quickly build up in a shop, causing asphyxiation and possibly death. Always use exhaust hoses to vent these gases, even if the engine will be running for only a few seconds. (Nederman, Inc.)*

When using a starting aid in cold weather, keep it away from all sparks and flames. Never accelerate a diesel engine to high speed under no-load conditions. Never use glow plugs and a starting aid simultaneously. Never prime an engine while it is being cranked.

Cleaning Equipment

When working with cleaning equipment, the following safety factors must be kept in mind. Use extreme caution when using an acid cleaning solvent—a rash or other irritation can result. Do not allow any cleaning solution to contact your skin as it could irritate or cause injury. Be careful when using solvents as most are toxic, caustic, and/or flammable. Read all manufacturer's precautions and instructions, **Figure 2-9,** and *Material Safety Data Sheets (MSDS)* before using any chemical. Do not use gasoline or kerosene to clean components.

Do not mix solvents or any chemical. They could react violently and/or act as a catalyst, causing a reaction with other chemicals. Wipe up spilled solvents promptly with a clean cloth and dispose of all used rags in marked metal containers. Store all solvents either in their original containers or in approved, properly labeled containers. When using a commercial parts washer, **Figure 2-10,** be sure to close the lid when you are finished. Always have a fire extinguisher close to the cleaning area. In case of accidental poisoning, seek medical attention immediately and take the suspect container so the chemical can be quickly identified.

Figure 2-9. *Warning labels are placed on containers that hold materials that are potentially harmful. Be sure to read any and all warning labels.*

Fire Prevention

While diesel fuel does not ignite as easily as gasoline, it is flammable. Care must be taken when handling or storing diesel fuel and other combustibles such as cleaning

Figure 2-10. *A closed parts cleaner is the best way to clean any part. (Am/Pro Machinery)*

fluids and paints. The following three conditions must be present at the same place and time in order for combustion or a fire to occur:

❑ Oxygen must be present.
❑ There must be a supply of burnable material, or fuel, available.
❑ Temperature must be above the fuel's flash point.

There are three basic types of fires that can occur in the typical diesel shop. Each has their own characteristics and subsequently, a different type of extinguishing medium must be used for each. Most shops now have multipurpose ABC dry chemical **fire extinguishers** that are capable of fighting all three types of fires.

❑ **Class A**—A fire where the combustible material consists of paper, wood, cloth, or trash. Extinguishing this type of fire involves using lots of water, a solution containing a high percentage of water and/or foam, or a multi-purpose dry chemical extinguisher.

❑ **Class B**—A fire where the combustible material is a liquid such as gasoline, diesel fuel, oil, grease, or solvents. This is the most common type of fire that can occur in the shop. Extinguish this fire with smothering from either foam, carbon dioxide, or dry chemical. Using water on this type of fire will only cause it to spread.

❑ **Class C**—In this case, the burning material consists of live electrical equipment—motors, switches, generators, transformers, or general wiring. To extinguish this type of fire, use a non-conductive smothering agent such as carbon dioxide or dry chemical.

A fourth type of fire is a metal-based fire, however, the chance of this type of fire occurring in a diesel shop is small.

General Fire Safety Rules

Become familiar with the location and operation of all firefighting equipment in the shop area. Make sure that any fire extinguisher that has been used is recharged immediately, **Figure 2-11.** Here are some fire safety rules that should be followed in any diesel engine shop.

❑ Never create flames or sparks near fuel or any other flammable liquid.
❑ Store volatile liquids in properly labeled containers and keep them stored safely in an isolated area.
❑ Keep batteries away from sparks or flame to avoid possible explosions.

An important fact to remember is that you can only hope to contain a small fire with a fire extinguisher. If the fire extinguisher cannot contain the fire after a short period of time, the fire begins to spread, or creates great quantities of smoke, get out of the shop immediately and call the fire department.

Using a Fire Extinguisher

Here are some suggestions for the proper use and handling of different types of fire extinguishers. More extensive

Figure 2-11. Most shops are equipped with dry chemical fire extinguishers. A dry chemical extinguisher is also referred to as an ABC fire extinguisher, since it can be used to extinguish most common fires. (Courtesy of Lab Safety Supply Inc, Janesville, WI)

on-site training can be scheduled by contacting your local fire department. Since most shops have foam, carbon dioxide, or dry chemical fire extinguishers, only these will be discussed.

When using a foam extinguisher, do not aim the stream directly into the burning liquid. Instead, allow the foam to fall lightly on the fire. As the fire diminishes, concentrate it more directly on the fire.

Direct a carbon dioxide extinguisher's stream as close to the fire as possible. Start at the edge of the flames and gradually move forward and upward. Do not use a carbon dioxide fire extinguisher in an enclosed space, as the gas will quickly displace any oxygen in the area, possibly causing asphyxiation.

The use of a dry chemical extinguisher depends on the type of fire. In most cases, direct the stream at the base of the flames. With a Class A fire, continue to direct the chemical stream at any remaining burning materials.

Right-to-Know Laws

Every employee of a diesel shop is protected by **right-to-know laws.** Right-to-know laws were first outlined in 1983 in the Occupational Safety and Health Administration (OSHA) Hazard Communication Standard. This publication was originally created for chemical companies and manufacturers who require their employees to handle potentially hazardous materials in the workplace.

The law is intended to ensure that employees are provided with a safe working place in regard to hazardous materials. There are three major areas of employer responsibility:

❑ *Employee Training and Education.* Under this legislation, all employees must know about their rights, as well as the type of hazardous chemicals in their workplace, the labeling of these chemicals, and the information about each chemical as posted on Material Safety Data Sheets (MSDS). MSDS sheets give product composition and precautionary information for any product that could present a health or safety hazard, and are prepared by the material's supplier. It is important that all employees understand the proper use, major characteristics, protective equipment needed, and accident or spill procedures associated with all major chemical groups. New employees should receive this training as part of their orientation, and they should be updated annually.

❑ *Labels and/or Information about Potentially Hazardous Chemicals.* All hazardous materials should be properly labeled to show what health, fire, or reactivity hazard they may have and to indicate what type of productive equipment to use when handling each chemical. The material's manufacturer must also provide all warnings and precautionary information that is to be read by the user before application.

❑ *Record Keeping.* A company must keep records of the hazardous materials in the work area, proof of training programs, and records of accidents and/or spill incidents. The company must also keep a list of employee requests for Material Safety Data Sheets, as well as a Right-to-Know procedure manual containing company policies.

 Note: When handling any hazardous material, be sure to wear the safety equipment listed in the MSDS and follow all recommended procedures correctly.

Hazardous Materials

Hazardous materials and wastes of most concern to the diesel technician are organic solvents, flammable, corrosive, and/or toxic materials, and wastes that contain heavy metals, including lead. Some common materials include:

❑ Used engine oil

❑ Contaminated fuel

❑ Antifreeze

❑ Solvents

❑ Coolants used in grinders and cutters.

The areas of most concern to the diesel engine shop will be used chemicals and any cleaning solutions. Most shops generate a large quantity of used oil, fuel, and antifreeze, which must be stored properly for pickup. Spray cabinets and dip tanks that use caustic chemicals produce high alkaline solutions and contain heavy metals. Thermal baking ovens generate ash containing heavy metals. Small parts washers generally use solvents that are classified as hazardous materials. Testing for hazardous wastes can be done by any qualified laboratory that performs tests on drinking water.

It should be noted that no material is considered hazardous waste until the shop is finished using it and is ready

to dispose of it. When the shop is ready to dispose of hazardous waste, it must be handled accordingly. For instance, a caustic cleaning solution with a heavy concentration of lead in the cleaning tank is not considered hazardous waste until it is ready to be replaced.

Handling Hazardous Waste

Equipment and services are available to help the diesel engine shop cope with hazardous waste disposal. There are now parts cleaners that use nonevaporative, recyclable cleaning solvent. There are also machines that will now recycle antifreeze (and other material considered hazardous), **Figure 2-12,** as well as companies that pick up used oil and recycle it. These recycling companies pick up the hazardous material at a shop and dispose of it in accordance with environmental regulations. It is important that any recycling equipment or recycling company is EPA approved.

Summary

Most manufacturers' engine service manuals contain safety symbols that signal a need for the use of caution or safety. Always pay attention to these symbols. Always wear safety glasses, goggles, or a safety face shield whenever the job requires it. Long hair should be tied up or kept under a hat to ensure it doesn't get caught in moving parts of machinery. Avoid wearing loose clothing, ties, or jewelry that can become caught in moving machinery. Proper

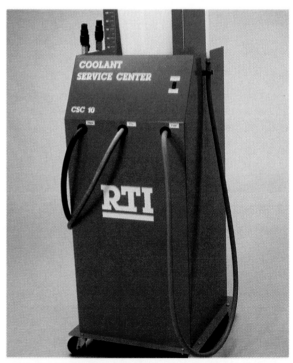

Figure 2-12. *Chemicals such as oil, refrigerant, and antifreeze should be recovered and recycled. This coolant recycling unit can remove the old antifreeze, flush the cooling system, and refill with new antifreeze. (RTI Technologies Inc.)*

footwear such as heavy-duty work shoes or steel toe safety shoes are recommended, as they can protect against flying sparks, heavy falling objects, and even battery acid. Protective head covering such as a hard hat is advised when working in a pit or under an overhead hoist.

Always use any extra safety devices or protective equipment provided for special jobs such as welding or grinding. Never get underneath a vehicle that does not have properly installed safety jack stands. Report all accidents right away so that the proper first aid or medical assistance can be given. Always be aware of and follow machine safety rules. Do not use any damaged tool or piece of equipment—report any damage to the shop supervisor. Lift heavy objects by bending at the knees in order not to strain the back muscles.

When lifting an engine from a vehicle be sure that the lift or crane has the capability to do the job. Move vehicles very slowly (about 2 mph) in and out of the shop, checking to see that no one is in the way. A running vehicle in the shop must have an exhaust hose connected outside to properly ventilate the area. Never prime an engine while it is being cranked. Be certain all throttle linkage connections are made and all fuel lines are connected before cranking the engine. Never accelerate a diesel engine to high speed under no load. Clean up any spilled oil, grease, or other fluids from the floor area. Never leave a running vehicle unattended. Store any oily rags in self-closing metal safety containers. Never use glow plugs and a starting aid simultaneously.

There are three types of fires that can occur in the diesel shop. Use the right type of fire extinguisher for the fire. Diesel shops generate large quantities of hazardous waste. This waste must be dealt with in accordance with environmental regulations. There are many companies that will pick up the hazardous material and recycle or dispose of it properly

Important Terms

Note	Fire extinguishers
Caution	Class A
Warning	Class B
Asbestos	Class C
Carbon monoxide	Right-to-know laws
Material Safety Data Sheets (MSDS)	

Review Questions–Chapter 2

Do not write in this text. Place your answers on a separate sheet of paper.

1. List at least four steps to ensure your personal safety when working in the shop.

2. Which of the following tool operations does not require the use of safety goggles, glasses, or shields?
   (A) Chiseling.
   (B) Grinding.
   (C) Welding.
   (D) All of the above.

3. Why should you wear hearing protection?

4. List three safety precautions when working with diesel fuel.

5. Why should you *not* mix solvents or any chemicals?

6. List the three common classes of fires.

7. The most common fire extinguisher in the diesel shop is the _____.

8. What are the three areas of employer responsibility under the Right-to-Know laws?

9. The initials for the organization that enforces regulations that protect the environment are:
   (A) PTA.
   (B) ASE.
   (C) EPA.
   (D) ERA.

10. The proper procedure for disposing of hazardous waste involves:
    (A) dumping it down a storm drain.
    (B) contacting a company that specializes in handling hazardous waste.
    (C) storing it on the property in plastic drums.
    (D) Both B & C.

ASE-Type Questions

1. All of the following are considered inappropriate shop attire EXCEPT:
   (A) steel toe shoes.
   (B) jewelry.
   (C) ties.
   (D) loose fitting jackets.

2. Technician A says most shop accidents involve power tools. Technician B says that most accidents are caused by the improper use of simple hand tools. Who is right?
   (A) A only.
   (B) B only.
   (C) Both A & B.
   (D) Neither A nor B.

3. Technician A says keeping the work area and components clean is vital to a safe work area. Technician B says that some debris and parts on a workbench will cause no major problem. Who is right?
   (A) A only.
   (B) B only.
   (C) Both A & B.
   (D) Neither A nor B.

4. When using a parts cleaner, you should wear _____.
   (A) protective gloves
   (B) face shield
   (C) jewelry
   (D) Both A & B.

5. Technician A says you should store soiled and oily rags in a ventilated container so fumes will not become concentrated. Technician B says you should store soiled and oily rags in a closed properly marked metal container. Who is right?
   (A) A only.
   (B) B only.
   (C) Both A & B.
   (D) Neither A nor B.

6. All of the following are unacceptable uses for a carbon dioxide fire extinguisher EXCEPT:
   (A) cooling your drinks.
   (B) putting out a fuel-based fire.
   (C) putting out a burning metal fire.
   (D) putting out a wood or paper fire.

7. Technician A says you should aim a foam-type fire extinguisher directly into a burning liquid. Technician B says you should allow the foam to fall lightly on the fire. Who is right?
   (A) A only.
   (B) B only.
   (C) Both A & B.
   (D) Neither A nor B.

8. The form that describes a substance is called a _____.
   (A) Master Safety Data Sheet
   (B) Material Substance Data Sheet
   (C) Material Safety Data Sheet
   (D) Master Substance Data Sheet

9. All of the following are safe substances EXCEPT:
   (A) water.
   (B) oxygen.
   (C) carbon monoxide.
   (D) nitrogen.

10. All of the following places are acceptable for hazardous waste storage EXCEPT:
    (A) sealed drum.
    (B) lined storage tank.
    (C) labeled container.
    (D) storm drain.

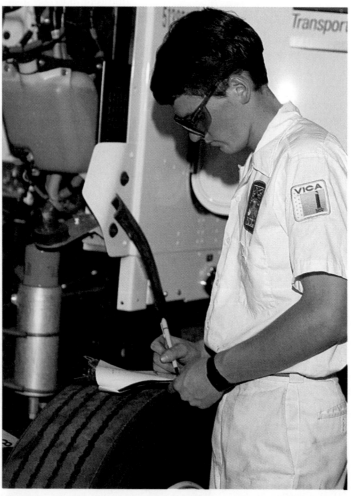

Proper documentation is extremely important in diesel engine repair. (Lloyd Wolf, VICA)

Chapter 3

Tools, Precision Tools, and Fasteners

After studying this chapter, you will be able to:
❑ Describe the proper use of hand tools.
❑ Describe how to operate portable and stationary tools safely.
❑ Explain how to use a torque wrench.
❑ Explain how to use a micrometer.
❑ Name the different types of gauges used to take precision measurements.
❑ Understand the purpose of the various diesel injection system tools and measuring devices.
❑ Define the difference between U.S. customary and metric fasteners.

The tools used for testing and servicing diesel engines all have one purpose—to help the technician do the job better and faster. It would be almost impossible to describe all of the specialized tools found in a diesel repair shop. However, the following are some of the more common ones used to take measurements, check the engine and injection system, and those for electronic service. The latter portion of this chapter covers the important subject of fasteners used with diesel engines.

Hand Tools

Many of the tools a diesel technician uses every day are general purpose hand tools. For example, a complete collection of wrenches and sockets is indispensable. This is also true of power tools, both electric and pneumatic.

While **hand tools** are simple to handle and operate, it must be remembered that more accidents are caused by hand tools than power tools. Always select the correct size and type of hand tool for the job and only use the tool for the job it was designed to do. Even though several different tools may be used to loosen a bolt, usually one will do a better job

than the others. It may be faster, grip the bolt better, or be less likely to break, or require less physical effort.

Always keep hand tools in good working condition and store them safely when not in use. A technician will have hundreds of different tools. For each tool to be quickly accessible, they must be neatly arranged. A **toolbox** will help keep your tools well organized, **Figure 3-1.** Remember: there should be a place for every tool and every tool should be in its place.

Never put tools in your pocket. A sharp or pointed tool can penetrate a vehicle's seat or your skin if you sit on it. Any tools with loose or cracked handles should be

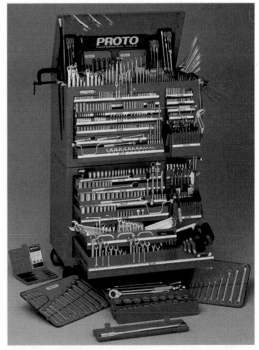

Figure 3-1. *Store all tools in a toolbox. The drawers of this toolbox are open for display purposes only. Leaving the drawers open in this manner may cause the box to fall over. (Stanley Proto)*

repaired, replaced, or discarded. Wipe all tools clean and dry after each use. A greasy or oily tool is not only unprofessional, it is dangerous and can contribute to dirt contamination. It is very easy for your hand to slide off a dirty wrench, cutting or even breaking a finger or hand. When using any tool, always remember to use all safety equipment provided with it.

Always purchase quality tools. Most quality hand tools come with a lifetime warranty from the manufacturer. While they may be more expensive, a quality tool will last longer than a less expensive tool. Quality tool vendors will come to your shop, allowing you to stay on the job. Often, an inexpensive tool does not have the same warranty as the more expensive quality tool. If the less expensive tool breaks, you will have to buy a new one rather than simply turning it in for a replacement covered under a tool manufacturer's warranty.

Chisels and Punches

Chisels are used for cutting off damaged or badly rusted nuts, bolts, and rivet heads. **Figure 3-2** shows some various chisel shapes. Use common sense when selecting a chisel shape. *Punches* also come in several configurations. See **Figure 3-3**.

Always wear safety glasses when working with chisels and punches. The best way to hold a chisel or punch is steadily, but loosely, just below the head. Never use a chisel or punch for prying as they are too hard and brittle. The excessive force can damage or break the tool.

When chiseling, select a chisel with a blade at least as large as the cut to be made. To chisel a small piece, clamp the piece tightly in a vise and chip toward the stationary jaw.

Be sure to use the correct punch for the job. If the punch is too small, it could break and cause an injury. Using a punch that is too large can result in the corners of

the cutting edge breaking. Often, more than one punch is needed for a particular job. For example, a starter punch can easily withstand the heavy blows needed to begin driving out rivets or pins, while a pin punch has a long straight shank that can finish the job.

Be sure to grind off any mushroom heads, **Figure 3-4**, as the sharp edges could cause personal injury. Keep a smooth bevel ground on the heads of all punches and chisels. Do not use chisels or punches on hardened materials like bearing races or locating pins. Keep chisels properly sharpened.

Saws and Knives

Sharp objects such as *saws* and *knives* should be stored safely out of the way. Knives should be sheathed when not in use. Never pry with a knife, as the blade could break and cause an injury.

A *hacksaw* is the most frequently used saw in the diesel shop. Various blade lengths and teeth sets can be mounted in its adjustable frame. The blade teeth should point away from the handle, **Figure 3-5**. Keep all blades sharp as a sharp blade requires less force to use, and allows more control over the cutting action. The blade should be fastened tightly in the frame.

Always cut away from your body, keeping your hands and fingers behind the cutting edge. Keep all handles clean and dry to help you keep a firm grip. Always complete a saw cut slowly to avoid any sudden movement that could result in injury.

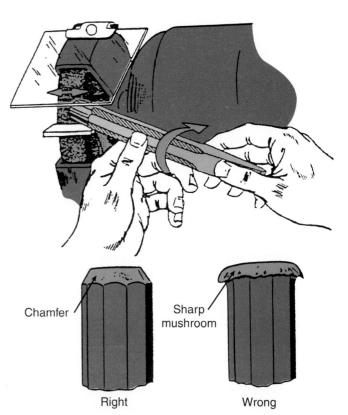

Figure 3-2. A technician should own a variety of chisels.

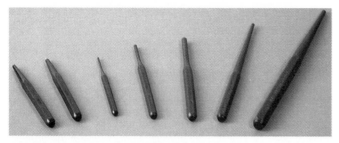

Figure 3-3. Punches are available in a variety of shapes and lengths and are best purchased as a set.

Figure 3-4. Chisel heads and blades can be brought back to sharpness and shape using a grinder. Make sure the head has a good chamfer after grinding. Always wear suitable eye protection.

A   **B**

Figure 3-5. *A—When installing a hacksaw blade, make sure the teeth point away from the handle. B—The blade should be tight, but not too tight.*

Files

Files are used to remove sharp edges, burrs, nicks, and perform other smoothing operations. They are useful when only a small amount of metal must be removed. The basic parts of a file and file patterns are shown in **Figure 3-6.** Keep in mind the following safety rules when using a file:

❑ A tang-end file should have a handle that protects your hand from injury if the file slips.

❑ Never use a hammer on a file, or use a file as a prying tool. The tool could shatter or break and cause injury.

❑ Always stroke away from yourself, **Figure 3-7,** and use a file card to keep the file clean.

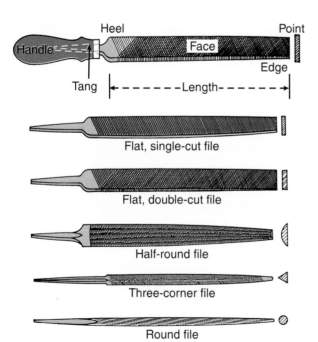

Figure 3-7. *When filing a part, stroke the file away from you. Move the part as necessary to provide the proper file angle.*

Screwdrivers

The *screwdriver* is one of the most basic and essential of all the technician's hand tools, **Figure 3-8.** It is also the most frequently misused hand tool. Use a screwdriver to drive screws only. Using this tool as a chisel, punch, or prybar will make it unfit to properly tighten or loosen screws.

Make sure that the screwdriver tip fits properly into the screw slot. An improper fit could damage both the screw and the screwdriver tip. Keep the screwdriver tip perpendicular to the screw to minimize any slippage. When applying pressure to a screwdriver, keep your other hand away from the area to avoid injury if the tool slips.

Flat, single-cut file

Flat, double-cut file

Half-round file

Three-corner file

Round file

Figure 3-6. *Various file types used in automotive and diesel engine repair.*

Figure 3-8. *Screwdrivers are essential tools in any toolbox. Shown here are flat edge, Phillips, Reed Prince and Torx® screwdrivers. (Thorsen)*

Screwdrivers designed for use with wrenches have either a square shank or bolster at the handle to withstand the extra force. For electrical work, minimize the chance of shock by using a screwdriver with an insulated handle.

Hammers

While there are various types of **hammers** used by the technician, the most common are the hard-face ball peen and sledge, **Figure 3-9,** and soft-face plastic, brass, or copper. The second type is used to prevent part breakage or damage to surfaces. Another type that is commonly used is a plastic hammer called a *dead-blow hammer*. This type of hammer is filled with metal or lead shot that reduces rebound when the hammer strikes an object or tool.

When using a hammer, always wear the proper eye protection to protect your eyes from flying chips. Select the right type of hammer for the job. A hammer that is too light will not do the job, while one that is too heavy can be hard to control or can cause damage to the part. Check to be sure handles are tightly fastened to the hammer heads. Replace cracked or splintered handles. Do not use the handles for prying or bumping. Never strike one hammer with another. The hammer could be damaged or chipped and the flying chips could cause injury. Never swing a hammer near anyone.

Grip the handle close to the end to increase leverage for more productive blows. This also reduces the chance of your fingers becoming caught between the hammer head and the work surface. Keep the hammer parallel to the

work. This will distribute the striking force over the whole hammer face, reducing the chances of the hammer chipping or slipping off the work area. Keep the handle free from grease and oil.

Wrenches

Wrenches are used to install and remove nuts and bolts. The four basic wrenches used by diesel technicians are *combination, open-end, box-end,* and *flare-nut,* **Figure 3-10.** Their size is determined by measuring across the wrench jaws. Wrenches come in U.S. conventional (inch) and metric (millimeter) sizes. The size is usually stamped on the side of the wrench.

To use any wrench safely, use the proper size wrench for the job. If the wrench does not fit properly, it can damage the bolt head or nut, result in skinned knuckles, and even lead to falls. Do not use shims to make a wrench fit. Use a wrench that gives a straight, clean pull. Do not cock the wrench, as this creates stress that can lead to fastener or tool failure. Use *angle head, offset,* and *socket* wrenches as needed to achieve a straight, clean pull in tight places. Never use a pipe to extend the length of a wrench. Under excessive force, the wrench or bolt can slip or break, causing damage and/or injury.

Pull a wrench toward you whenever possible. If you must push the wrench, use the open palm of your hand to push it in order to reduce the chance of skinned knuckles. Do not use a hammer with a wrench unless it is a *striking wrench.*

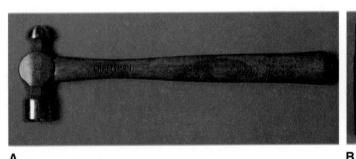

A   **B**

Figure 3-9. *Hard- and soft-faced hammers are used throughout engine service. A—Ball peen hammer. B—Sledgehammer.*

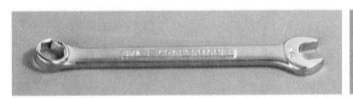

A   **B**

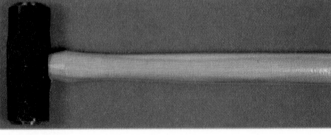

C   **D**

Figure 3-10. *Various wrenches used in diesel engine work. A—Combination wrench. B—Open-end wrench. C—Box-end wrench. D—Flare-nut wrench.*

While an *adjustable wrench* can be used on a job, a properly fitting box-end wrench is preferred. When using an adjustable wrench, position the open jaws toward you and pull on the wrench, **Figure 3-11**. This technique will place most of the pressure on the solid, stronger jaw and will reduce the chances of slippage.

Use *pipe wrenches* only for turning pipe or round stock. The teeth of a pipe wrench can easily damage bolt heads. If this happens, wrenches and sockets will no longer be able to turn the bolt. Replace all cracked or worn wrenches. Do not attempt to straighten a bent wrench as this will only weaken it further.

Sockets and Socket Drivers

A *socket* is basically a box-end wrench designed to be used on a socket handle. See **Figure 3-12**. The socket has a square opening that fits onto the driver handle. Sockets come in various sizes and lengths. *Socket drivers* (ratchets, breaker bars) provide a fast means of loosening and tightening fasteners. A *ratchet* is the most commonly used type of socket driver. A ratchet allows quick selection of turning direction for either loosening or tightening, **Figure 3-13**. *Breaker bars* are used to loosen tight nuts and bolts.

There are several safety rules to remember when working with sockets and socket drivers. Always use the proper size socket and ratchet or breaker bar for the application. Keep the ratchet as close as possible to the socket.

When loosening or tightening fasteners with a ratchet or breaker bar, always pull the tool toward you. When space permits, use a socket rather than an open-end wrench as it grips the fastener better. Use *extensions* and *deep-well sockets* only when necessary, to obtain the necessary reach, **Figure 3-14**.

When using extensions, support the ratchet or breaker bar and head to prevent applying side load on the socket and fastener. Never hammer on ratchets or breaker bars to increase loosening or tightening torque. Never use a pipe as an extension to improve leverage on a ratchet or breaker bar. Damage to the tool and fastener may result. Always replace damaged sockets, ratchets, and breaker bars.

Allen Wrenches

Allen wrenches are used to loosen and tighten socket Allen head bolts, **Figure 3-15**. Allen wrenches come in metric and U.S. conventional sizes. They are also available as *Allen sockets*. Allen sockets fit on a ratchet or breaker

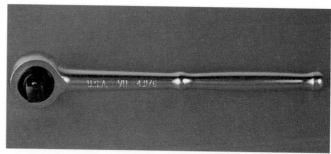

Figure 3-13. *Ratchets give you the ability to change turning directions by switching the knob on the head.*

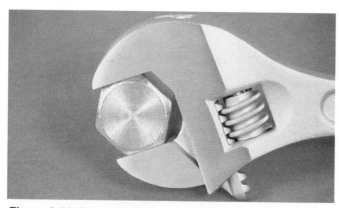

Figure 3-11. *If an adjustable wrench must be used, make sure the force is concentrated on the solid jaw.*

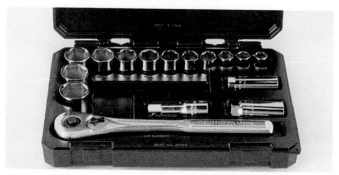

Figure 3-12. *A good set of hand sockets are a necessity for any engine work.*

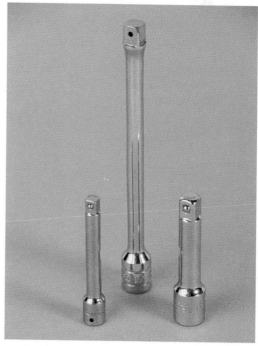

Figure 3-14. *Socket extensions give you the ability to reach recessed fasteners while giving the socket driver room to turn.*

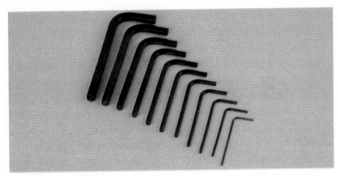

Figure 3-15. *Allen wrenches come in a variety of sizes and are important in diesel engine work.*

bar for high torque applications. Remember to follow these safety rules when using Allen wrenches:

- ❏ Always select the correct size Allen wrench for the job.
- ❏ Always replace worn Allen wrenches.
- ❏ Never use an impact wrench on Allen sockets.

Pliers and Cutters

Pliers and *cutters* of various types, styles, and sizes are used by every technician to perform many different operations, **Figure 3-16.** Pliers are used for cutting as well as for holding and gripping small parts in situations where it may be inconvenient or impossible to use hands. However, do not use pliers when another tool such as a wrench or socket will work. Pliers can easily nick and scar an object.

Pliers are made for holding, pinching, squeezing, and cutting, but not for turning. When using pliers, keep the following safety rules in mind. Do not substitute pliers for a wrench or socket, as the pliers can damage bolt heads and nuts. *Always* wear eye protection when using pliers or cutters. There are several special pliers available, such as *snapring pliers* that are used to install and remove snap rings.

There are additional precautions to observe when using *wire cutters*. Select a wire cutter that is the right size for the job. Keep the blades at right angles to the work. Do not rock the cutter to make a faster cut. Keep a slight clearance when adjusting cutter blades. This prevents the blades from striking each other as they close.

Vise and C-Clamps

A *vise* is used to hold parts during cutting, filing, drilling, hammering, and pressing operations, **Figure 3-17.** Often called a *bench vise,* it has scored jaws that prevent the work from slipping. Most have a swivel base that allows the vise to be rotated.

Because of its scored jaws, avoid clamping a smooth, machined part surface in the uncovered jaws of a vise. If a machined surface is scratched, the part may be ruined. *Vise caps* (plastic or lead jaw covers) or wood blocks should be used when mounting precision parts in a vise.

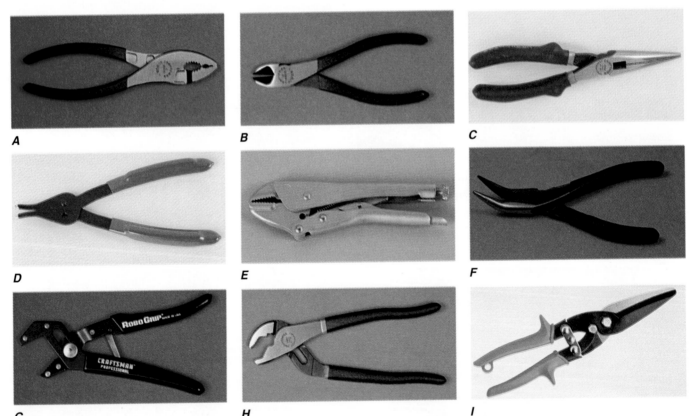

Figure 3-16. *Technicians own a wide assortment of pliers and cutters for various jobs. A—Slip joint pliers. B—Diagonal, side cutting pliers. C—Needle nose pliers. D—Snap-ring pliers. E—Vise grip pliers. F—Bent needle nose pliers. G—Spring-loaded rib joint pliers. H—Rib joint pliers. I—Metal snips.*

Figure 3-17. *The parts of a shop vise. Most shops provide bench vises for a technician's use.*

They will not only protect the part, but provide a more secure grip. There are a few vise safety rules that should always be followed. These include:

☐ Wear safety glasses when using a vise. A vise can exert tremendous clamping force and the parts may break and fly out with great force.

☐ Never hammer on a vise handle to tighten or loosen the vise. Use the weight of your body.

☐ Keep the moving parts of the vise clean and oiled.

C-clamps are often called portable vises because they can hold parts on the work surface when filing, cutting, drilling, and other similar operations, **Figure 3-18.** Being portable, they can be taken to the job.

Torque Wrench

There are several designs of **torque wrenches**, **Figure 3-19.** These wrenches allow the diesel engine technician

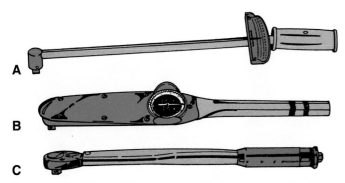

Figure 3-18. *C-clamps come in various sizes. However, a C-clamp cannot provide the compression force possible in a vise.*

Figure 3-19. *Three different styles of torque wrench. A—Beam torque wrench. B—Dial torque wrench. C—Snap torque wrench.*

to duplicate the pressure used when the engine was originally assembled, thereby duplicating conditions that meet manufacturer's specifications.

The *flex bar torque wrench* bends a metal beam with a pointer across a scale, which indicates the torque. The *dial indicator torque wrench* is very accurate and can be used even when an extension is added to the handle to gain more leverage. Consult the manufacturer's instructions for proper use of hand extensions.

The snap or *click torque wrench* is easy to use. Set the torque valve by turning the handle and stop when a click or pop sound is felt or heard. With most click torque wrenches, however, accurate torque values can be found only if the point of pull remains at the center of the hand grip area. This is why no handle extension should be used with a click torque wrench unless the tool is specifically designed and calibrated for this purpose.

Using a Torque Wrench

When torquing a fastener with a torque wrench, it is important to use steady pressure. Apply torque smoothly and evenly to achieve an accurate torque value. When torquing a nut and bolt assembly, wherever possible, hold the bolt stationary and apply the torque to the nut. If the torque value required cannot be obtained, a longer wrench may be used.

> **Note: Make sure the torque wrench is able to obtain the specified torque value.**

When an adapter is attached to a dial torque wrench, the dial reading (when compared with the actual torque applied) will be affected if the torque wrench length is increased. A correction factor is needed in this case to achieve an accurate dial reading. To calculate this correction factor, proceed as follows:

1. Divide total length **A** by torque wrench length **B**, **Figure 3-20.**
   Example: Total length A = 24″
   Torque wrench length B = 12″
   $$\frac{A}{B} = \frac{24″}{12″} = 2 \text{ (quotient)}$$

2. Multiply the dial reading times the quotient to determine the torque at the end of the adapter.
   Example:
   Dial reading = 50 ft.-lbs.
   Quotient = 2
   Torque at end of the adapter = 50 × 2 = 100 ft.-lbs.

3. If the required torque is known, divide by the quotient to determine what the torque reading should be.
   Example:
   Required torque = 100 ft.-lbs.
   Quotient = 2
   $$\frac{100}{2} = 50 \text{ ft.-lbs.}$$

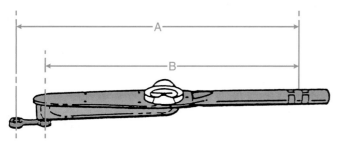

Figure 3-20. When an adapter is used with a torque wrench, it is necessary to compute the actual amount of torque applied. **A** indicates the total length of the wrench/adapter assembly, while **B** is the actual torque wrench length.

Figure 3-21. Conventional gear puller. Many other general and specialized pullers are used in diesel engine work.

The torque wrench has been used for years to tighten bolts and nuts that clamp parts together. Up to 90% of fastener torque is used up by friction. The remaining 10% is used to clamp parts together.

Because late-model engines are designed with extremely close tolerances, torque alone is not adequate for proper fastener installation. Torque/angle measurement ensures like fasteners will exert the same amount of force without deviation from one fastener to the next. Without proper clamping forces, leaking head gaskets, as well as cylinder and bearing bore distortion can occur.

The torque/angle method first specifies a low threshold torque on the fastener to get components to touch. Torque is least affected by fastener friction at these low values. The fasteners are then turned through a specified angle. The force of each fastener will then be consistent.

Other Hand Tools

Other hand tools are used by the diesel technician. For instance, when it is necessary to remove a gear, wheel, pulley, or bearing from a shaft, a puller is used. There are two basic types: the conventional *gear puller,* **Figure 3-21,** and the *T-bar puller.* Other hand tools such as tap and dies, bolt removal devices, and various engine hand tools, including gauges, are covered later in this chapter.

Power Tools

Power tools fall into two basic categories: portable (hand-held) and stationary (located in a fixed location). Both can make many jobs easier but can also cause serious injury if they are used improperly or safety rules are not closely followed.

Power tools can be driven by either air pressure or electricity. Power tools can be either hand-held or stationary. Although electric drills, wrenches, grinders, chisels, drill presses, and various other tools are found in shops, pneumatic (air) tools are used more frequently. Pneumatic tools have four major advantages over electrically powered equipment:

- ❑ Air tools run cooler and have the advantage of variable speed and torque; potential for damage from overload or stalling is reduced.
- ❑ The air tool is lighter in weight and has a better power-to-weight ratio than electrical tools. For instance, a typical 3/8″ air drill weighs 2.5 pounds (1.12 kg), while the same size electric drill weighs 4.5 pounds (2.02 kg).
- ❑ Air equipment reduces the danger of fire and shock hazards in some environments where the sparking of electric tools can be dangerous.
- ❑ Air tools require fewer repairs and less preventive maintenance. Also, the initial cost of air-driven tools is usually less than an equivalent electric tool.

Three of the most common air-powered tools used by diesel engine technicians are shown in **Figure 3-22.** Follow all applicable safety rules when using either portable air or electric tools. Always wear the proper eye protection. Noise can be a hazard with some portable power tools, especially pneumatic tools. Wear the proper hearing protection when working with these tools. Always wear gloves and a face shield when operating air chisels or air hammers.

Never try to make adjustments, lubricate, or clean a power tool while it is running. Do not operate a power tool without being properly instructed on how to use it cor-

A

B

C

Figure 3-22. Common air tools used by the diesel technician A—Impact wrench. Impact wrench sizes up to 1″ are used in diesel work. B—Air ratchet. C—Air drill.

rectly. The ***air gun*** is good for cleaning parts when properly used, **Figure 3-23,** but like all power tools, use it only for its intended purpose. Never clean yourself or others with compressed air.

When connecting pneumatic tools and lines, check to make sure they are attached properly. Turn off all machinery and power tools when not in use, and return all equipment to its proper place. Never use a hand socket with an impact wrench.

Stationary Power Tools

When operating any stationary power tools, always wear the proper eye protection and follow the safety rules when operating or observing any machinery. Inspect all equipment for any defects before using it.

Do not use a two-prong electrical adapter to plug in a three-prong, grounded piece of equipment. Do not use electrical equipment that has the ground prong removed. Replace all frayed or damaged cords on electrical equipment. When using electrical power tools, never stand on a wet or damp floor. Disconnect electrical power before performing any service on the machine or tool. Never abuse an electrical cord by yanking it from its receptacle. Whenever safety devices are removed to make adjustments or repairs, the equipment's main switch should be locked and tagged that it is out of service.

All electrical equipment should be well grounded. Before plugging in any electric tool, make sure the switch is off to prevent serious injury. When you are through using it, turn it off and unplug it. Do not attempt to use the machine or tool beyond its stated capacity or for operations requiring more than the rated horsepower of the motor. Never use a tool or machine for operations it was not designed to perform.

Check to see that any keys or adjusting wrenches are removed from the machinery before turning it on. Give the machinery your undivided attention when using it—do not look away or talk to others. Be sure to keep the work area clean and well lighted. Do not leave any loose rags around operating equipment. Never lean against any machinery. Maintain a balanced stance to avoid slipping.

Do not operate any machinery without receiving instructions on the correct operating procedures. Read the owner's manual to learn the proper applications for a tool as well as any hazards related to its use. Always allow the tool or machine to reach full operating speed. Stay with the machine from the start and remain with it until it has come to a complete stop. When using stationary power equipment on small parts, never hold the part in your hand as it could slip. Always use a vise or other clamp instead.

Keep hands away from moving parts when the machine or tool is under power. Never clear chips or debris when the machine is under power. Never use your hands to clear chips; use a brush or chip rake. Also, do not try to strip broken belts or debris from a pulley while it is in motion, or reach between belts and pulleys. Allow a minimum of 4″ (101 mm) between your hands and any blades, cutters, or moving parts.

In **Chapters 4, 6,** and **8,** specialized stationary machining equipment such as lathes, boring bars, and honing machines are described. Two of the most used general shop tools are the drill press, **Figure 3-24,** and the bench grinder, **Figure 3-25.**

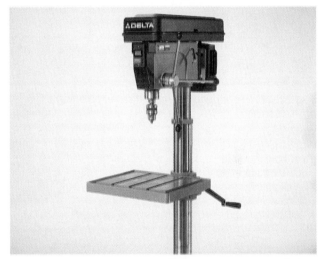

Figure 3-24. *A drill press is a powerful tool that can easily injure or kill if you are careless.*

Figure 3-25. *Bench grinders should not be used without the proper guards in place.*

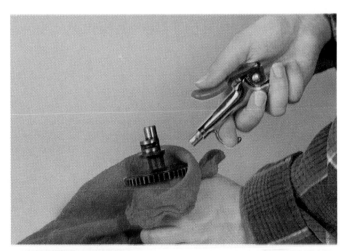

Figure 3-23. *An air gun is useful for cleaning and blow drying parts. Do not use an air gun to blow dirt or debris off yourself.*

Drills

When using a **drill press** or any other drill for machining a hole, remember the following rules. Keep all guards in place and in working order. Always use a center punch to start the drill bit, **Figure 3-26.** Drill a pilot hole if a large drill bit is to be used, **Figure 3-27.** Make sure the drill bit is tight in the drill chuck, **Figure 3-28.** Always remove the chuck key before starting the drill. The workpiece should be held firmly. Use cutting fluid (oil) on thick or hard metal to cool the drill bit. Do not spin the drill bit too fast or press too hard. Overheating will quickly soften and ruin the bit. Always let the drill bit do the work. Drilling usually causes a butt (rough edge) on both sides. A burr should be removed with a deburring tool or countersink.

Grinders

Grinders are used for rough metal removal or cleaning parts, depending on the wheel or stone. While there are several grinders in a typical diesel engine shop, (surface grinders, piston grinders, flywheel grinders, and so on), the *bench grinder* is the one that is most frequently used.

Improper use of a bench grinder can result in serious injury. For this reason, be sure to follow these general grinder safety rules. Wear eye protection. Never use a grinder without the proper guards in place. When grinding small parts, hold them with locking pliers. This can help prevent serious hand injuries. Use light pressure to avoid uneven stone wear, overheating, and catching. Quench metal parts in water to cool them during the grinding operation. Check the area for flammable materials before grinding; do not grind near combustible materials. If possible, avoid grinding aluminum or brass with abrasive stones. These soft materials will clog the stone.

Pneumatic Tool Safety Rules

Pneumatic tools and the **air compressor,** **Figure 3-29,** must be operated at the pressure recommended by their manufacturer. Make sure the compressor has a pressure relief valve and that it works properly. Do not use compressed air to clean clothes. Even at low cleaning pressure, compressed air can cause dirt particles to become embedded in the skin, which can result in infection.

When working with a **hydraulic press,** **Figure 3-30,** make sure that hydraulic pressure is safely applied. It is generally wise to stand to the side when operating the press. If the shop has a hydraulic **vehicle lift,** be sure to read the instruction manual before using it. It is also important before using any engine lift to check the owner's manual.

Figure 3-26. Use a punch to mark the piece so that the drill bit can start easily.

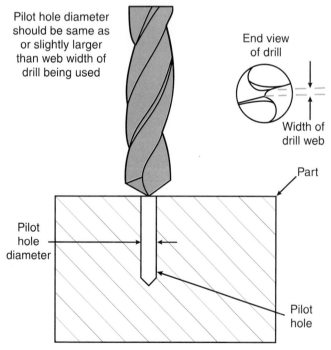

Pilot hole diameter should be same as or slightly larger than web width of drill being used

End view of drill

Width of drill web

Part

Pilot hole diameter

Pilot hole

Figure 3-27. If a large hole has to be drilled, drill a pilot hole first. It may be necessary to drill more than one pilot hole if the final hole is very large or the metal is very thick or hard.

Figure 3-28. Make sure the drill bit is tight in the chuck. If a drill bit slips in the chuck, it will catch and possibly break.

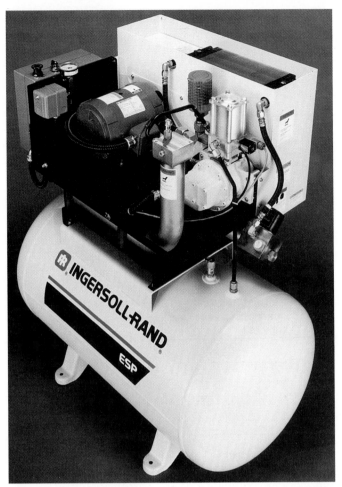

Figure 3-29. An air compressor supplies the air pressure that powers hand and stationary tools and in some cases, hydraulic lifts. (Ingersoll-Rand)

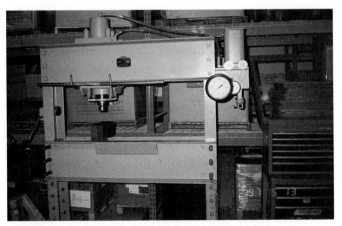

Figure 3-30. Hydraulic presses come in various sizes and can be either pneumatic or hydraulic powered.

Welding Equipment

In the course of engine or vehicle repair, welding equipment is sometimes needed to join pieces of metal, seal cracks, cut a frozen bolt or nut, apply heat for part removal, and hundreds of others jobs. The most frequently used welders in the diesel shop are *oxyacetylene, shielded metal arc* (referred to as MIG or TIG), and *arc welding* systems.

Wear the appropriate safety goggles or welding helmet, depending on the type of welding job. The lenses in the goggles or helmet should be of the correct shade for the particular welding process. Also wear leather gloves, apron, and clothing that will not catch sparks or hot pieces of metal, **Figure 3-31.**

Engine Measuring Tools

In diesel engine repair, the technician is frequently required to make very close measurements, often in ten thousandths of an inch or millimeter. Accurate measurements to this high standard of precision can only be made through the use of measuring devices especially designed to show these very small differences in size.

Measuring tools are precise and delicate instruments. The more precise the tool, the more susceptible to damage it is. When using any measuring device, always place it so it cannot fall or strike other tools. Never pry, strike, drop, or force these instruments as they could be damaged beyond repair.

Precision measuring instruments, especially micrometers, are extremely sensitive to rough handling. Clean them after every use. Never touch the measuring surfaces of a micrometer. Fingerprints can promote rust formation. Even body heat can affect the micrometer's accuracy. All measuring operations should be performed on parts at room temperature. Never measure a part that is still warm from machining operations.

 Note: Check measuring instruments regularly against known good instruments to ensure they are operating properly and are capable of accurate measurement. All measuring tools should be calibrated regularly.

Figure 3-31. Welding is a vital skill in any shop. This technician is using a shielded metal inert gas (MIG) welder. (Miller)

Micrometers

Examples of measurements required in diesel engine service involve the size of the outside and/or inside diameter of a shaft or a cylinder bore. The micrometer is the common instrument used in taking these measurements. Both *outside micrometers,* **Figure 3-32,** and *inside micrometers,* are calibrated and read in the same manner. They both operate so that the measuring points exactly contact the surfaces being measured.

The major components and markings of a micrometer are the *frames, anvil, locknut, spindle, sleeve numbers, sleeve long line, thimble marks, thimble,* and *ratchet.* Micrometers may be calibrated in either inch or metric graduations. The thimble on both outside and inside micrometers is revolved between the thumb and forefinger. Very light pressure is required when bringing the measuring points into contact with the surfaces to be measured. It is important to remember that the micrometer is a delicate instrument and that excessive pressure will result in an incorrect reading.

A standard micrometer is made so that each turn of the thimble moves the spindle .025″. This is accomplished by using forty threads per inch on the thimble. The sleeve long line is marked with sleeve numbers 1, 2, 3, etc. These sleeve numbers represent .200″, .300″, and so on. The sleeve of the micrometer contains sleeve marks that represent 1″ in .025″ increments. Each of the thimble marks represents .001″. In one complete turn, the spindle will move 25 marks or .025″. Inch-graduated micrometers come in a range of sizes—zero to 1″, 1″ to 2″, 2″ to 3″, 3″ to 4″, etc. The most commonly used micrometers are calibrated in thousandth of a inch increments.

Reading a Micrometer

To read a micrometer, first read the last whole number visible on the sleeve long line. Then count the number of full sleeve marks past the number. Finally, count the number of thimble marks past the sleeve marks. Add these measurements together for the total measurements. These three readings indicate tenths, hundredths, and thousandths of an inch respectively. For example, a 2″–3″ micrometer that has taken a measurement is described as follows, **Figure 3-33.**

1. The largest sleeve number visible is 4, indicating .400″.

2. The thimble is three full sleeve marks past the sleeve number. Each sleeve mark indicates .025″, so this indicates .075″.

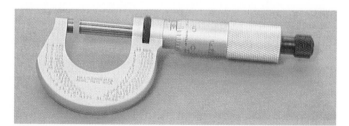

Figure 3-32. *An outside micrometer. Inside and outside micrometers come in various sizes.*

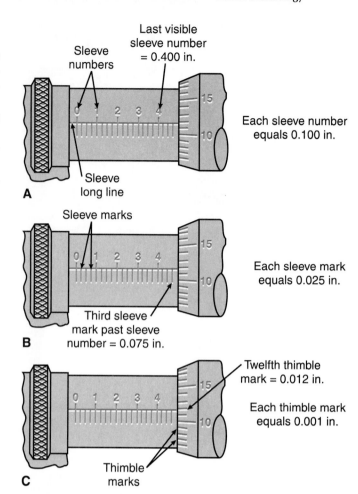

Figure 3-33. *The three steps in reading a micrometer. A—Measuring tenths of an inch. B —Measuring hundredths of an inch. C—Measuring thousandths of an inch.*

3. The number 12 thimble mark is lined up with the sleeve long line. This indicates .012″.

4. Add the readings from steps 1, 2, and 3. The total of the three is the correct reading. In our example:

Sleeve	.400″
Sleeve marks	.075″
Thimble marks	.012″
Total =	.487″

5. Now add 2″ to the measurement, since this is a 2″–3″ micrometer. The final reading is 2.487″. Even with frequent use, misreadings often occur. Digital micrometers that give an exact number readout are now available. **Figure 3-34** shows a typical direct reading electronic outside micrometer.

Depth Gauge

A *depth gauge* or depth micrometer, **Figure 3-35,** is used to measure the distance between two parallel surfaces. Depth micrometers are available in inch-graduated and metric models. Operation and measurement readings are similar to the inside and outside micrometers described earlier.

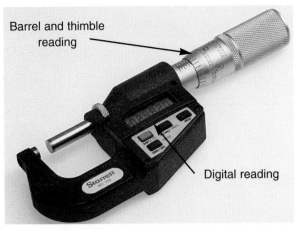

Barrel and thimble reading

Digital reading

Figure 3-34. Typical direct reading electronic outside micrometer.

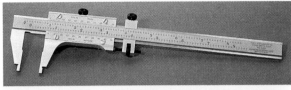

A

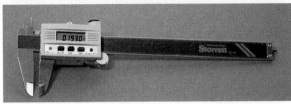

B

Figure 3-36. Dial or vernier calipers can be used to take inside and outside measurements. A—Vernier caliper. B—Direct reading digital caliper.

Figure 3-35. A depth gauge is used to measure the difference between the parallel surfaces of a part and a recessed space in the part.

Vernier Calipers

The *dial* or *vernier caliper,* **Figure 3-36A,** is a versatile measuring instrument because it can take inside, outside, and depth measurements. It can measure these dimensions from 0–6". Metric vernier calipers typically measure from 0–150 mm in increments of .02 mm.

The vernier caliper features a *depth scale, bar scale, indicator, inside measurement jaws,* and *outside measurement jaws.* The bar scale is divided into one-tenth of an inch graduations. The indicator is divided into .001" graduations.

The metric vernier caliper is similar in appearance to the U.S. (inch reading) model. However, the scale is divided into .02 mm increments. Direct electronic digital reading calipers are also available, **Figure 3-36B.**

Telescoping and Small Hole Gauges

Telescoping gauges or snap gauges, **Figure 3-37,** are used for measuring bore diameters and other clearances. They normally come in a range of sizes. This type of gauge consists of two telescoping plungers, a handle, and a lock

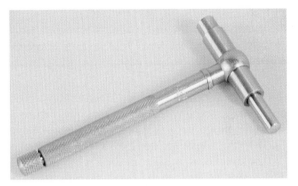

Figure 3-37. A telescoping snap gauge.

screw. Telescoping gauges are normally used in combination with an outside micrometer.

The *small hole gauge,* **Figure 3-38,** functions in the same manner as the telescoping gauge. The gauge is put into a hole and expanded until it fits snugly, but not too tightly to be removed easily. Measuring the gauge with a micrometer gives the diameter of the hole, **Figure 3-39.**

Feeler Gauges

The *feeler gauge* consists of strips of known and closely controlled thicknesses. Several of these metal strips

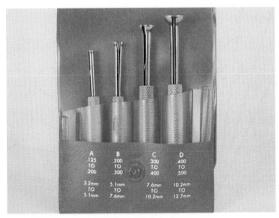

Figure 3-38. The small hole gauge is used to measure holes too small for snap gauges.

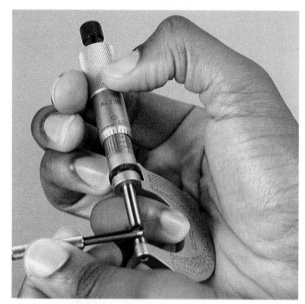

Figure 3-39. Checking the small hole gauge's measurement with a micrometer.

Figure 3-41. A dial indicator can be set up to measure a part's radial and lateral movement.

may be combined into a multiple measuring instrument that pivots in the same way as a pocket knife, **Figure 3-40.** A steel feeler gauge pack will normally contain leaves of .002″–.010″ thickness (in steps of .001″) and leaves of .012″–.024″ thickness (in steps of .002″).

The feeler gauge can be used by itself to measure piston ring side clearance, piston ring end gap, connecting rod side clearance, crankshaft end play, and other clearances. The feeler gauge can also be used with a straightedge to measure main bearing bore alignment and cylinder head/block warpage.

Dial Indicator

The **dial indicator** is calibrated in .001″ increments, **Figure 3-41.** Metric dial indicators are also available. Both types are used to measure movement. Common uses of the dial indicator include measuring valve lift, journal concen-

tricity, flywheel runout, gear backlash, and crankshaft end play. Dial indicators are designed with face markings and ranges to suit many different measuring jobs.

Cylinder Bore Dial Gauge

The **cylinder bore dial gauge** is used to determine cylinder bore size, out-of-round, and taper. These three measurements, in addition to main bearing saddle alignment, provide basic information about the condition of the cylinder block. A block outside specification in any of these areas requires machining to restore it to specified tolerances. Cylinder bore dial gauges, **Figure 3-42,** are read in the same way that a dial indicator is read.

Digital Tachometer

A **digital tachometer** provides the technician with an accurate and safe method of determining a diesel engine's speed. It operates by directing a visible light source onto a reflective tape attached to a rotating object. It then counts the light pulses reflected by each revolution and compares them to a highly accurate crystal time base.

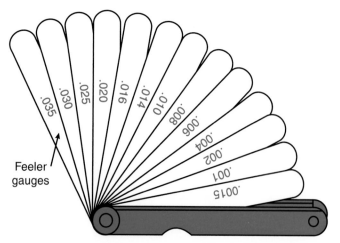

Feeler gauges

Figure 3-40. A feeler gauge along with a straightedge work hand-in-hand to check surface flatness.

Figure 3-42. This technician is measuring a cylinder's bore size with a bore dial gauge. (Andrew Norman)

Infrared Thermometer

An *infrared thermometer* allows non-contact temperature measurement. Aim the unit, pull the trigger, and read the temperature. Since there is no need to touch what you are measuring, temperatures of hazardous, hard-to-reach, or moving materials can be taken without the risk of injury. Some multimeters have a feature that allows them to be used as thermometers with an infrared probe, **Figure 3-43.**

Cylinder Compression Gauge

Engines operate efficiently when each cylinder carries an equal portion of the total load. The most reliable method of measuring cylinder load is to measure the firing pressure of each cylinder. A cylinder *compression gauge,* **Figure 3-44,** allows you to check the compression level in

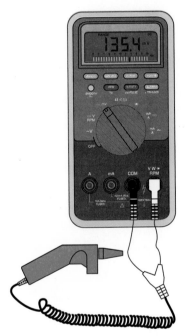

Figure 3-43. An infrared thermometer allows non-contact temperature measurements. This multimeter has a feature that allows it to be used as a thermometer when attached to an infrared probe. (Fluke)

Figure 3-44. Checking a cylinder's ability to produce power with a compression gauge. (Kubota)

each engine cylinder. Poor compression may indicate faulty piston rings or valves. If an engine has poor compression, it will be hard to start and will have low power.

Manometer

Either a water or mercury *manometer,* **Figure 3-45,** can be used to check engine blowby and restriction in both the intake and exhaust systems. See **Chapters 12** and **13** on how it is used.

Specialized Diesel Tools and Equipment

Diesel engine work requires the use of many specialized tools to perform particular tasks or procedures. Examples of such tools include cylinder liner pullers/installers, cylinder reboring bars, cylinder ridge reamers, cylinder deglazers, piston ring groove cutting tools, piston ring compressors, and piston pin bore gauges. Special tools are also available to test, service, clean, and calibrate fuel injection pumps, nozzles, and injectors.

Injection Systems

The following are the more common "special" diesel fuel injection system tools that a technician may come across while servicing an engine.

Injection Puller/Extractor

One of the most common tasks a diesel engine technician performs is to remove the injectors. The *injector puller*

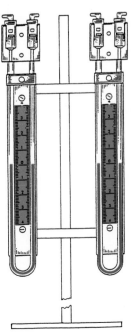

Figure 3-45. A water and mercury manometer are used to check air and exhaust restrictions.

provides a grip on the injector body when clearance between the valve springs is limited. The heavy-duty puller legs will not deform when using a slide hammer to remove the injector.

An *injector extractor* is a slide-hammer puller designed to remove press-fit diesel injectors. Several truck diesel injectors require this tool, **Figure 3-46.** However, most automotive diesel injectors screw into the cylinder head.

Diesel Injector Service Set

The *diesel injector service set,* shown in **Figure 3-47,** includes brass brushes and special devices for cleaning and repairing injectors and other fuel system components. Because the clearances in a diesel injector or injection pump are very precise, care must be taken not to scratch or mar components. Never use a steel brush to clean diesel injection parts. Use an ultrasonic cleaner to remove carbon from the fuel system components, energy cells, precombustion chambers, exhaust valves, injector tips, and cups.

Diesel Injection Pump Tester

The *diesel injection pump tester* or test bench, contains a set of pressure gauges and valves that can be used

to measure injector pump operation, **Figure 3-48.** If the diesel injection pump pressures are not near or within specifications, repairs or adjustments can be performed.

Nozzle Pop Tester

Nozzle pop testers are designed for checking the condition of a diesel injector nozzle, **Figure 3-49.** The injector is mounted on the tester. By pumping the tester handle and watching the pressure gauge, you can check opening pressure, spray pattern, and injector leakage. Checking the condition of injection nozzles is a basic and essential step in diagnosing diesel engine performance. Peak engine performance, maximum fuel economy, and minimum exhaust smoke require injector nozzles that are functioning properly.

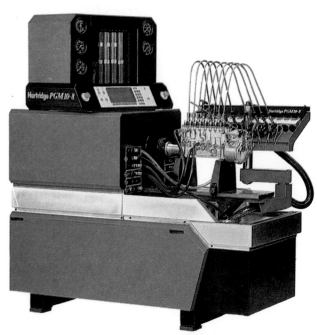

Figure 3-48. *This test bench is used for checking diesel injection pump operation. (Lucas)*

Figure 3-46. *Special extractor removes injectors which are pressed into the cylinder head. It is commonly needed on larger diesel engines. (Cummins)*

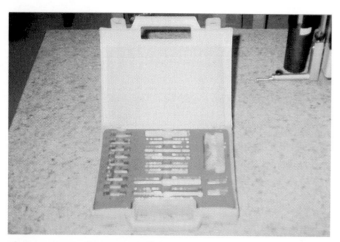

Figure 3-47. *Diesel injector service kit includes soft brass brushes and other special tools.*

Figure 3-49. *A nozzle pop tester is used to test diesel injection nozzle. (Lucas)*

An injector *pop-n-fixture* is shown in **Figure 3-50.** The diagnostic test benefits of this tool are:

- ❑ Checks the spray pattern to determine if the tip holes are plugged or excessively worn.
- ❑ Checks injector valve operation to ensure clean fuel atomization.
- ❑ Allows a visual test for external leaks that can cause fuel dilution or hydrostatic lock.
- ❑ Determines if internal sealing areas are leaking, which can cause fuel dilution of the engine oil.

> ⚠ **Warning: Extremely high pressures are developed when pop testing a diesel injector nozzle. Wear eye protection and keep your hands away from the fuel spraying out of the nozzle.**

Plunger and Bushing/Tip Flow Gauge

The **tip flow gauge** provides a means for classifying injector tips based on relative restriction of air flow through the tip holes. Injector tips are compared to a built-in "master" orifice, that has been calibrated to an exact air flow value. Three master orifices are incorporated to cover the full range of injector tip flows.

The companion function of this unit, utilizing the same air circuit and controls, provides the capability of measuring the plunger's effective stroke within the bushing. The beginning and end of the effective stroke is indicated by air pressure gauge readings and the plunger travel is read directly from the digital micrometer head. By matching plunger and bushing assemblies with proper tips, predictable injector outputs can be maintained. High flow tips should be assembled with short stroke, and mean flow with mean stroke.

Glow Plug Resistance Tester

A **glow plug resistance tester** combines a special wiring harness with an ohmmeter, **Figure 3-51.** It may be used to assist in troubleshooting rough idle problems. The tester harness has leads that connect to each glow plug. After a period of engine operation, combustion will increase the temperature of the glow plugs, affecting their resistance. A comparison of glow plug resistance may indicate that one or more cylinders are not firing properly. The tester helps the technician find a "dead cylinder" (cylinder not burning fuel charge).

Diesel Injection Timer

A **diesel injection timer** is a device for adjusting injection timing (when fuel is injected in relation to piston position in the cylinder), **Figure 3-52.** The timer can detect the pulse produced by the injector on the combustion stroke. This will let the technician adjust the injection pump to set timing properly.

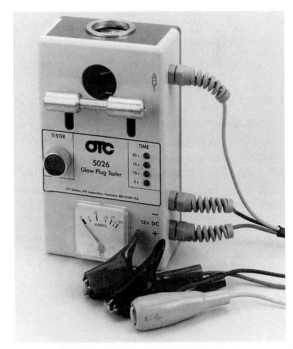

Figure 3-51. Glow plug testing is easy with this device. An ohmmeter can also be used to check glow plugs. (OTC Division)

Figure 3-52. Diesel injector timer reacts to the firing or burning of diesel fuel in the cylinder by sending a pulse the instant the fuel charge ignites. (Andrew Norman)

Figure 3-50. Injector pop-n-fixture tester. (Lucas)

Opacity Smoke Meter

Federal regulations have established quality levels for diesel exhaust emissions. State and local agencies enforce these emission control standards while the vehicle is on the road. Similar standards are enforced for stationary diesel engines.

Diesel exhaust quality is usually expressed in terms of the exhaust smoke's opacity. The **opacity meter** provides an inexpensive, reliable, and accurate instrument for diesel exhaust monitoring. If the exhaust smoke blocks too much light, engine or injection system repairs or adjustments are needed. The opacity meter is used to check for economical engine performance, because excessive smoke usually indicates a rich fuel condition. Another smoke testing device is shown in **Figure 3-53.**

Electrical Test Equipment

It is possible to locate an electrical fault without much trouble using the following diagnostic tools: test meters, test lights, and jumper wires. But keep in mind that electrical systems can be damaged if the wrong type of test equipment is used or if it is used improperly.

Multimeters

While there are two basic types of **multimeters,** digital and analog, the digital type is the one recommended by most diesel engine manufacturers, **Figure 3-54.** In order to avoid damage when diagnosing and servicing today's sensitive electronic systems, a high impedance (minimum 10 megohms) multimeter is recommended. Digital readouts

give precise measurements and can react with the speed needed for proper diagnosis.

Test Lights

A **non-powered test light** is often used to determine if there is current flowing through the circuit. One lead of the test lamp is connected to a good ground, the other lead to a point in the circuit, **Figure 3-55.** If the lamp lights, current is present at that point. Test lights should not be used unless specified in the service manual. **Self-powered test lights** can be used to determine if there is continuity in the circuit. They should not be used on solid state circuits.

Jumper Wires

Jumper wires are used to temporarily bypass circuits or circuit components for electrical testing. They consist of a length of wire with an alligator clip or a terminal at each end, **Figure 3-56.** They can be used to test circuit, breakers,

Figure 3-54. A digital multimeter is recommended for electrical checks with diesel engines. (Fluke)

Figure 3-53. Smoke meters measure the amount of particulates in the diesel exhaust to help determine combustion efficiency. (Lucas)

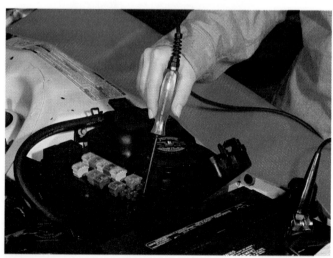

Figure 3-55. A test light is handy to make electrical checks. (Jack Klasey)

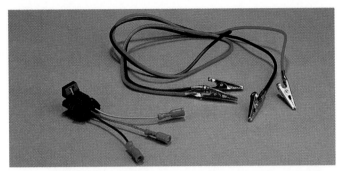

Figure 3-56. Jumper wires are used to feed power or supply a path to ground for a device. Jumper wires can be purchased, created with wire and alligator clips, or terminals cut from the wiring harness of a scrap vehicle.

relays, lights, and other components. To protect circuits from damage, the jumper wire should have an inline fuse or circuit breaker rated for the circuit being tested. Like the test light, jumper wires should only be used when specified in the service manual.

Computer System Diagnostic Tools

The introduction of computer-controlled fuel injection systems on diesel engines has generated the need for tools capable of diagnosing and troubleshooting these systems. The most common diagnostic tool is a hand-held or portable *scan tool,* such as the one illustrated in **Figure 3-57.** The scan tool is designed to communicate with the engine's control computer. It is connected to the engine computer via a data link connector.

When activated, the scan tool will display all active diagnostic trouble codes and all occurrences of inactive faults. Most scan tools also allow the operator to monitor

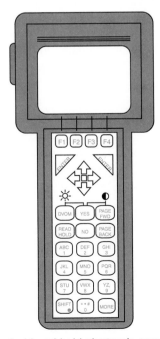

Figure 3-57. Typical hand-held electronic scan tool.

select engine system parameters. In addition to being used to clear stored codes from the computer memory, many scan tools can be used to reprogram the on-board computers with new parameters for such items as high and low idle speeds, road speed, and cruise control speed limits.

Newer software packages allow any PC to be used as an engine diagnostic tool. These powerful and complex PC software programs can quickly perform in-depth engine performance diagnostics.

Fasteners

The selection and proper installation of threaded *fasteners* are an essential part of any diesel engine assembly. A fastener that fails to do its job will certainly cause a problem—often a major one.

Because there are many styles, sizes, and quality grades, it is important that the service technician develop a working familiarity with the fasteners used in engine construction. Remember, each fastener is intended to perform a particular job, and is selected by the engine manufacturer for its suitability. Installing the wrong fastener can cause as much damage as a fastener installed incorrectly. Both under- and overtightening can result in an improperly installed fastener.

Bolt and Screw Terminology

Bolts and *screws* may be identified by type, length, major and minor diameters, pitch, length of thread, material, tensile strength, and wrench size needed. They are measured by the **United States Customary (USC)** or the **metric (SI)** measuring system. However, the points of measure differ.

For example, a bolt's thread pitch in the USC system is determined by the number of threads in one inch of bolt length and is expressed in number of threads per inch. The thread pitch in the metric system is determined by the distance in millimeters between two adjacent threads. To check the thread pitch of a bolt or stud, a **thread pitch gauge** is used, **Figure 3-58A.** Gauges are available in both conventional and metric dimensions.

 Caution: Metric bolts, nuts, and screws are not interchangeable with United States Customary fasteners. The use of incorrect fasteners will cause damage to components.

There are some identifying measurements that are the same for both USC and SI bolts. In both, the bolt length is the distance measured from the bottom of the head to the tip of the bolt. The bolt's major diameter is measured from the top or crest of the threads on one side to the crest on the other. The minor diameter is determined by measuring from the bottom of the threads on one side to the bottom of the threads on the other. If you were to remove the threads, the diameter of the portion left would be the minor diameter.

The metric diameter and length diameter are in millimeters, while USC are in fractions of inches, **Figure 3-58B.**

The bolt's grade, or **tensile strength,** is the amount of stress or stretch it is able to withstand. The type of bolt material and the diameter of the bolt determines its tensile strength. In the U.S. customary system, a bolt's tensile strength is identified by the number of radial lines (grade marks) on the bolt head. More lines mean higher tensile strength, **Figure 3-59.** In the metric system, a number on the bolt head indicates the tensile strength, **Figure 3-60.** The higher the number, the greater the tensile strength.

 Caution: When replacing a fastener, make sure to use one with the same measurements and the same or greater strength than the old one. Incorrect or mismatched fasteners can result in damage to the vehicle and possible personal injury.

Bolts and nuts come in right- and left-hand threads. With common right-hand threads, the fastener must be turned clockwise to tighten. With the less common left-hand threads, turn the fastener in a counterclockwise direction to tighten. The letter "L" may be stamped on fasteners with left-hand threads.

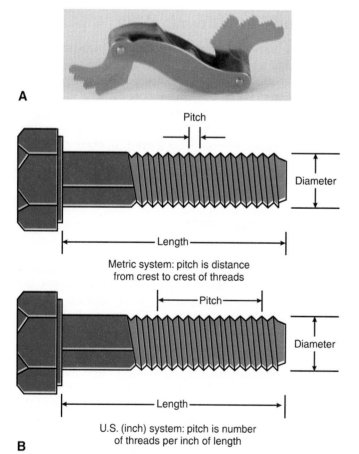

A

B

Metric system: pitch is distance from crest to crest of threads

U.S. (inch) system: pitch is number of threads per inch of length

Figure 3-58. A—Thread pitch gauge is used to determine the type and size of a bolt when it is unknown. This is important if the head of a bolt has broken off. B—USCS and metric bolt thread pitch. (Central Tools, Inc.)

Grade 2 (GM 260-M)   Grade 5 (GM 280-M)   Grade 7 (GM 290-M)   Grade 8 (GM 300-M)

Figure 3-59. USCS bolt grade markings.

4.6   4.8   5.8   8.8   9.8   10.9

Figure 3-60. Metric bolt grade markings.

U.S. and Metric Nuts

Nuts are manufactured in a variety of sizes and styles. They are generally hexagonal shaped (six-sided) and are used on bolts and studs. Obviously, they must be of the correct diameter and thread pitch. To join a nut and bolt or stud, the diameter of the bolt and hole in the nut must be the same. It is just as important that the threads on both be properly matched.

Nuts are also graded to match their respective bolts, **Figure 3-61.** For example, a grade 8 nut must go with a grade 8 bolt. If a grade 5 nut was used instead, a grade 5 connection would result. The grade 5 nut cannot carry the loads expected of the grade 8 bolt. Look for the nut markings which are usually located on one side. Grade 8 and critical applications also require the use of fully hardened flat washers.

Thread Repairs

Thread repairs are sometimes needed when threads have been stripped (broken out of the hole or smashed) or when fasteners have broken inside a hole. The easy way to handle this problem is to use a thread insert or bushing. A **tap and die set** makes thread repair easy, **Figure 3-62.** A typical procedure for installing an insert is as follows:

1. Locate the base and place a drill over the damaged bolt, **Figure 3-63A.**

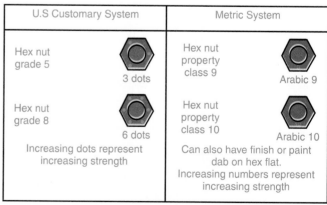

U.S Customary System		Metric System	
Hex nut grade 5	3 dots	Hex nut property class 9	Arabic 9
Hex nut grade 8	6 dots	Hex nut property class 10	Arabic 10
Increasing dots represent increasing strength		Can also have finish or paint dab on hex flat. Increasing numbers represent increasing strength	

Figure 3-61. U.S. and metric nut grading systems.

Figure 3-62. A tap and die set is an important part of any technician's tool chest. It can be used for cleaning and ensuring thread integrity as well as repair.

2. Ream out the old bolt with a reamer, **Figure 3-63B.**

3. Tap the new bolt hole, **Figure 3-63C.**

4. Install the thread repair insert or bushing, **Figure 3-63D.**

5. Cut the top off flush with the surface.

Applying Torque to a Fastener

Properly torquing a bolt requires that it be clean and free from rust, scale, locking compound, or damage. The degree of torque needed depends on whether the threads have been lubricated or not. Read the torque tables carefully to see whether dry or lubricated threads are specified.

Most bolts used in general applications are zinc plate and clear chromate coated. Some platings act as lubricants and care must be taken not to overtorque these fasteners. Most bolts used internally in diesel engines, transmissions, rear axle carriers, transfer cases, power take-offs, and so on, are unplated. When an anti-galling compound is used between steel and aluminum parts, be sure the bolt threads

are cleaned properly. These compounds can act as extreme lubricants, if torque values are misread. The result can be thread stripping or overstretching the fastener.

Torque values are calculated with a 25% safety factor below the yield point. There are some fasteners, however, that are intentionally torqued just barely into a yield condition. This type of fastener, known as a ***torque-to-yield (TTY) bolt,*** will produce 100% yield when torqued to normal values. These fasteners should not be reused unless otherwise specified. Some aftermarket gasket manufacturers will include new TTY bolts. When in doubt, consult the manufacturer's specifications concerning bolt reuse. Used TTY bolts will have a coke bottle shape, referred to as a necked out bolt

The appendix in this text gives standard bolt and nut torque specifications. Follow the manufacturer's specifications precisely. If a TTY bolt is replaced with a new bolt of identical grade, but torqued to a value found in a regular torque table, the clamping force produced will be at least 25% less. This is the reason many manufacturers recommend replacing original bolts with the next higher grade. For example, a grade 8 bolt torqued properly to its 25% safety factor will produce as much or more clamping force than a grade 5 bolt at 100% with no safety factor.

> **Note: If a bolt is stretched to its yield point, it takes a permanent set and will never return to its normal strength. For this reason, torque values are calculated with a margin of 25% below the yield point.**

Most printed torque values are for dry, plated bolts. Lubricants are beneficial when working with engines that have a great deal of oil around and in them. They provide smoother surfaces, allow more consistent and evenly loaded connections, and help reduce thread galling.

A connection that appears to be dry may not actually be dry, because it may have been accidentally lubricated

A   **B**   **C**   **D**

Figure 3-63. Procedure for installing thread insert. A—Drill out the damaged hole. B—Remove the old bolt. C— Tap the hole. D— Install a thread insert. (Kent-Moore Division)

by dirty fingers. If dry torque is applied in this case, the bolt may be torqued into yield or have its threads stripped. This is due to the fact that the bolt receives 75% friction and 25% energy, when only 10% energy may actually be needed to tension the bolt. Dirty or nicked threads can also increase friction, resulting in increased bolt tension and a joint that is not as tight as required, **Figure 3-64.**

meter can be used to check engine blowby and restrictions in both the intake and exhaust systems.

The injector puller provides a positive grip on the injector body where clearance between the valve springs is limited. An injector extractor is a slide-hammer puller designed to remove press-fitted diesel injectors. The selection and proper installation of threaded fasteners is an essential part of any diesel engine assembly or rebuild job. A bolt's tensile strength, or grade, is the amount of stress or stretch it is able to withstand.

Summary

Many tools are available today for testing and servicing diesel engines, but they all have one purpose—to help the technician do the job better and faster. Diesel technicians should have a variety of hand and power tools at their disposal. Torque wrenches allow the diesel engine technician to duplicate the conditions of tightness and stress that meet the manufacturer's specifications.

Measuring tools are precise and delicate instruments. The more precise they are, the more delicate they are. When using any measuring device, always place it so it cannot fall or strike other tools. Precision measuring instruments, especially micrometers, are extremely sensitive to rough handling. Micrometers may be calibrated in either inch or metric graduations. Remember that the micrometer is a delicate instrument and that even slight excessive pressure will result in an incorrect reading.

An inside micrometer is used to measure the distance between two parallel surfaces. The vernier caliper can take inside, outside, depth, and step measurements. Telescoping or snap gauges are used for measuring bore diameters and other clearances. The small hole gauge functions in the same manner as the telescoping gauge. The feeler gauge can be used to measure piston ring side clearance, piston ring end gap, connecting rod side clearance, and crankshaft end play.

An infrared thermometer allows non-contact temperature measurement. A cylinder compression tester checks the compression level in each engine cylinder. A mano-

Important Terms

Hand tools	Small hole gauges
Toolbox	Feeler gauges
Chisels	Dial indicator
Punches	Cylinder bore dial gauge
Saws	Digital tachometer
Knives	Infrared thermometer
Files	Compression gauge
Screwdriver	Manometer
Hammers	Injector puller
Wrenches	Injector extractor
Socket	Diesel injector service set
Socket drivers	Diesel injection pump tester
Allen wrenches	
Pliers	Nozzle pop tester
Cutters	Tip flow gauge
Vise	Glow plug resistance tester
Vise caps	Diesel injection timer
C-clamps	Opacity meter
Torque wrenches	Multimeters
Power tools	Non-powered test light
Air gun	Self-powered test lights
Drill press	Jumper wires
Grinders	Scan tool
Air compressor	Fasteners
Hydraulic press	Bolts
Vehicle lift	Screws
Oxyacetylene welding	United States Customary (USC)
Shielded metal arc welding	Metric (SI)
Arc welding	Thread pitch gauge
Outside micrometers	Tensile strength
Inside micrometers	Nuts
Depth gauge	Thread repairs
Vernier calipers	Tap and die set
Telescoping gauges	Torque-to-yield bolt (TTY)

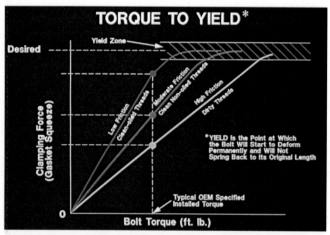

Figure 3-64. This chart indicates how the condition and cleanliness of a threaded hole will affect the amount of clamping force exerted by a torque-to-yield bolt. (Fel-Pro)

Review Questions—Chapter 3

Do not write in this text. Place your answers on a separate sheet of paper.

1. Whenever possible, pull a wrench _____ you.

2. Larger diameter screwdrivers and punches _____ used as prying tools.
   (A) when the need arises, should be
   (B) should always be
   (C) should never be
   (D) None of the above.

3. A _____ is a good tool for loosening tight nuts and bolts.

4. The sleeve of a micrometer contains sleeve marks that represent _____ in _____ increments.

5. The cylinder bore dial gauge is used to determine three measurements which provide basic information about the condition of the cylinder block. Name these three measurements.

6. The most reliable method of measuring cylinder load is to measure the _____ of each cylinder.

7. The computer-controlled electronic systems of today's diesels require the technician to use portable _____ for troubleshooting problems.

8. The scan tool will display all active and inactive _____ registered by the on-board computer.

9. In the U.S. system of fastener designation, slash marks on the bolt heads are used to indicate:
   (A) thread pitch.
   (B) bolt diameter.
   (C) bolt grade.
   (D) bolt length.

10. Torque values are calculated with a _____ safety factor below the yield point.
    (A) 25%
    (B) 33%
    (C) 5%
    (D) 50%

ASE-Type Questions

1. Technician A says hand tools are the simplest tools to handle and operate. Technician B says that more accidents are caused by hand tools that power tools. Who is right?
   (A) A only.
   (B) B only.
   (C) Both A & B.
   (D) Neither A nor B.

2. Technician A says that when the head of a chisel becomes mushroomed, the chisel must be discarded. Technician B says the mushroomed chisel head can be reground to a bevel finish and left in service. Who is right?
   (A) A only.
   (B) B only.
   (C) Both A & B.
   (D) Neither A nor B.

3. Technician A says you can use a screwdriver to center punch a hole. Technician B says a screwdriver should be used for removing and installing screws only. Who is right?
   (A) A only.
   (B) B only.
   (C) Both A & B.
   (D) Neither A nor B.

4. All of the following are soft face hammers, EXCEPT:
   (A) brass.
   (B) copper.
   (C) ball peen.
   (D) plastic.

5. Technician A says that pliers can be used for turning objects into shape. Technician B says that wire cutters can be rocked back and forth to help the cutting process. Who is right?
   (A) A only.
   (B) B only.
   (C) Both A & B.
   (D) Neither A nor B.

6. Technician A says the preferred method of torquing a nut and bolt is to hold the bolt stationary and apply the torque to the nut. Technician B says the nut should be stationary with the torquing force applied to the bolt. Who is right?
   (A) A only.
   (B) B only.
   (C) Both A & B.
   (D) Neither A nor B.

7. Technician A says you should use a micrometer to measure a machined part that is still warm from the machining process. Technician B says the part must be at room temperature for the measurement to be accurate. Who is right?
   (A) A only.
   (B) B only.
   (C) Both A & B.
   (D) Neither A nor B.

8. Technician A says the most bolts used internally in diesel engines, transmissions, rear axle carriers, transfer cases, power take-offs, and so on, are unplated. Technician B says most internally used bolts are plated to provide a small amount of lubricating qualities. Who is right?
(A) A only.
(B) B only.
(C) Both A & B.
(D) Neither A nor B.

9. Technician A says a torque-to-yield bolt will produce 100% yield when torqued to normal values. Technician B says a torque-to-yield bolt will produce 100% yield when torqued to 25% of its safety factor. Who is right?
(A) A only.
(B) B only.
(C) Both A & B.
(D) Neither A nor B.

10. Improper torquing can cause all of the following, EXCEPT:
(A) distorted parts.
(B) leaks.
(C) even gasket compression.
(D) engine failure.

Large vehicles, such as this dump truck, require diesel power to propel the vehicle as well as the heavy loads it must carry. (Caterpillar)

Chapter 4

Principles of Operation

After studying this chapter, you will be able to:
- ❑ Explain the operating principles and interaction of the major diesel engine components, such as the cylinder block, cylinder head, valve train, and accessory items.
- ❑ Explain the difference between two-stroke cycle and four-stroke cycle diesel engines and describe what happens during each piston stroke in both types.
- ❑ Explain the basics of both two-cycle and four-cycle valve timing.
- ❑ Define the most popular diesel engine configurations.
- ❑ Name the two types of combustion chamber designs.
- ❑ Explain the difference between direct and indirect fuel injection.
- ❑ List the functions of the fuel injection system and name the five types of fuel injection systems used in modern diesel engines.
- ❑ Explain the basic engine performance terms and formulas, such as bore, stroke, compression ratio, volumetric efficiency, horsepower, and torque.

To begin to explain the service and repair of diesel engines, it is necessary to understand the function of each part or system and how all systems work together in a diesel engine. This chapter discusses the major components and assemblies of both two-stroke and four-stroke diesel engines.

Major Engine Components

It is often easier to think of an engine as a number of subassemblies or part groups rather than as a single assembly of many components. **Figure 4-1** illustrates the major subassemblies of a typical diesel engine. The largest

engine subassembly is the cylinder block and its related parts, such as the cylinder liners (if used), pistons, connecting rods, crankshaft, vibration damper, flywheel, camshaft, bearings, and camshaft followers.

In order to work on the internal components of the cylinder block, the top and bottom of the engine must be removed. All of the parts above the cylinder block are normally referred to as the **head group.** This group includes the cylinder head, rocker arm assemblies, rocker arm covers, engine brake, air intake manifold, and exhaust manifold.

When the crankshaft must be removed from the cylinder block, the gear cover, oil pan, and flywheel housing must also be removed. The cylinder head, oil pan, gear cover, and flywheel housing are all bolted to the cylinder block. Most joining surfaces must be sealed with gaskets that are designed to withstand the temperatures and pressures generated inside the engine. Accessory items mounted to the outside of the engine include the water pump, fuel pump, oil cooler, turbocharger, supercharger, aftercooler, and fan hub.

A number of other systems are needed for engine operation. These include the cooling, exhaust, fuel metering, fuel injection, and governing systems. All of these systems will be covered in greater detail in the chapters related to their service and repair.

The parts of a diesel engine are so closely related that it is almost impossible for any single part to wear without affecting other components. A skilled diesel technician must be able to visualize all of the engine parts in relation to each other.

Cylinder Block and Cylinder Liners

The **cylinder block** is the foundation of the diesel engine, **Figure 4-2.** It supports all of the other engine components. The block contains openings for each cylinder, internal drilled passages for coolant and lubricating oil, bores for the crankshaft and camshaft, and openings for the push tubes or push rods and cam followers.

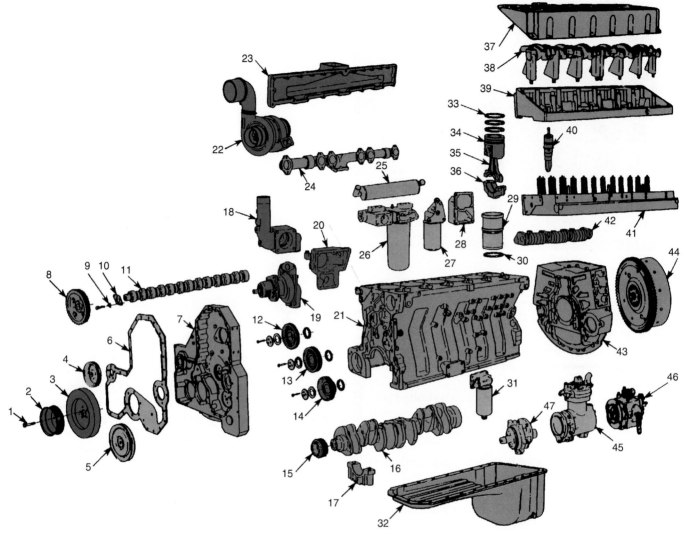

Figure 4-1. *The major subassemblies of a diesel engine. The number and types of components will vary by engine design and manufacturer. (Cummins Engine Co., Inc.)*

1. Capscrew
2. Crankshaft pulley
3. Vibration damper
4. Fan idler pulley
5. Pulley, accessory drive
6. Gear cover
7. Gear cover
8. Camshaft gear
9. Capscrew and lockplate
10. Camshaft thrust support
11. Camshaft
12. Idler gear assembly
13. Idler gear assembly
14. Idler gear assembly
15. Crankshaft gear
16. Crankshaft
17. Cap, main bearing
18. Thermostat housing
19. Water pump
20. Thermostat housing
21. Cylinder block
22. Turbocharger
23. Aftercooler housing
24. Exhaust manifold
25. Lubricating oil cooler
26. Lubricating oil filter assembly
27. Coolant filter
28. Coolant filter head assembly
29. Cylinder liner
30. O-ring, liner
31. Fuel filter
32. Oil pan
33. Piston rings
34. Piston
35. Connecting rod
36. Connecting rod
37. Rocker housing cover
38. Rocker arm assemblies
39. Rocker arm housing
40. Fuel injector
41. Cylinder head
42. Cam follower assembly
43. Flywheel housing
44. Flywheel
45. Air compressor
46. Fuel pump (PT)
47. Fuel pump (transfer)

Cylinder openings are cast in the block. In most cases, **cylinder liners** (sleeves) are placed in these openings to form the walls of the combustion chamber. The top of each liner is sealed by the cylinder head, while the bottom is sealed by the piston. As the engine operates, the moving pistons contact the liners, not the cylinder block itself. When the liner becomes worn, it can be removed from the block and replaced. This is one of the keys to long diesel engine life.

Dry and Wet Liners

Cylinder liners are classified as either dry or wet liners. *Dry liners* fit into the cylinder counterbore in the block and do not make direct contact with the circulating coolant in the water jackets. **Water jackets** are passages in the cylinder block that allow coolant to move around the outside of the cylinder and carry off the excess heat generated by combustion, **Figure 4-3.** In an engine with dry

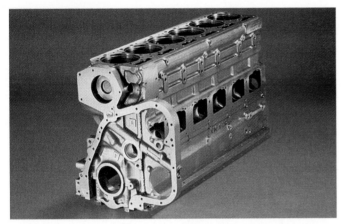

Figure 4-2. Typical inline six-cylinder cylinder block. Note the various openings and passages. (Cummins Engine Co., Inc.)

liners, the heat of combustion must pass through the liner and the cylinder block wall before it reaches the coolant in the water jacket, **Figure 4-4A.**

In contrast, *wet liners* are inserted into the cylinder block and make direct contact with the coolant. This allows for the quickest possible dissipation of excess cylinder heat. The top of the liner is designed to form a tight seal with both the cylinder block and the cylinder head. The bottom of the liner is also sealed with a ring that prevents the coolant in the water jacket from leaking into the oil pan, **Figure 4-4B.**

Pistons

As discussed in **Chapter 1,** *pistons* move up and down in the cylinder liners. When the fuel ignites in the combustion chamber, the hot expanding gases force the piston down. The downward motion of the piston is transferred through a connecting rod to the engine's crankshaft. See **Figure 4-5.**

The piston acts as a movable seal that forms the bottom of the combustion chamber. Grooves cut into the

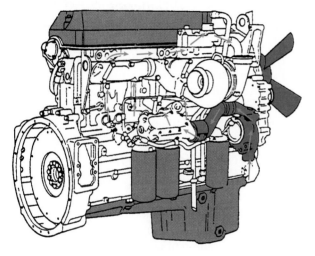

Figure 4-3. The arrows indicate coolant circulation around the outside of the cylinder bores. Coolant also travels through the cylinder head to keep the valves cool. (Cummins Engine Co., Inc.)

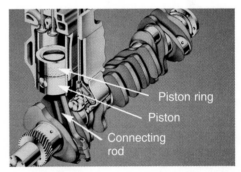

Figure 4-5. The piston, connecting rod, and crankshaft work together to change the energy of combustion into usable motion. (Cummins Engine Co., Inc.)

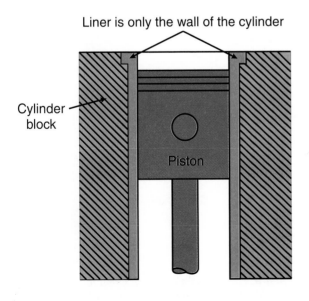

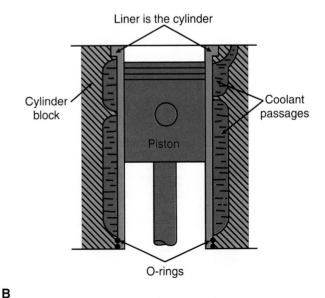

A   **B**

Figure 4-4. The two types of liners used in diesel engines. A—Dry liners are thin and do not make contact with the coolant. B—Wet liners are much thicker since they do not have the walls of the block for support. They come into direct contact with the engine coolant.

side surface of the piston are fitted with **piston rings** that seal in the compression and combustion pressure, and also control the lubricating oil film on the wall of the cylinder liner, **Figure 4-6.**

The top piston ring is a *compression ring,* which forms a seal with the cylinder wall. The middle or *intermediate piston rings* control about 10% of the compression and combustion pressures, but their main job is to reduce oil consumption inside the cylinder. The bottom piston ring is an *oil control ring,* which distributes the lubricating oil over the wall of the cylinder and prevents the oil from entering the combustion chamber.

Crankshaft

The **crankshaft** converts the reciprocal (up and down) motion of the pistons into rotary (turning) motion. This rotary motion can then be used to turn shafts in a transmission or driveline. Since the crankshaft transmits all of the power produced by the engine, it is the engine's strongest part, **Figure 4-7.**

The crankshaft protrudes through the cylinder block at the front and rear of the engine. **Oil seals** are used at these points to prevent the oil inside the block from escaping. The crankshaft has a number of key parts and features, including journals, oil galleries, and timing gears.

Journals

Journals are the sections of the crankshaft that actually ride in the cylinder block saddles. These journals have been ground or polished to a very smooth finish. This allows the journals to rotate easily in their bearings.

A crankshaft has both main bearing journals and connecting rod journals. **Main bearing journals** support the crankshaft as it turns in the cylinder block. Main bearing journals are larger than connecting rod journals. They are also located on a centerline with the ends of the crankshaft. The distance from the crankshaft centerline to the centerline of a connecting rod journal is half the total travel (stroke) of the piston. **Connecting rod journals** are located at the ends of the crankshaft throws. The motion of the pistons and connecting rods is transferred to the crankshaft through the connecting rod journals.

Oil Galleries

The crankshaft also contains a network of drilled passageways along its full length. These passages are called **oil galleries.** Oil galleries feed lubricating oil from the main journals to the connecting rod journals and their bearings. The pressurized oil may also be forced up through galleries in the connecting rods, where it lubricates the piston pins and cools the bottom of the piston heads.

Timing Gears

One of the engine's **timing gears** is mounted on the crankshaft. The other is mounted on the camshaft(s). Additional drive gears for various items, such as the fuel injection pump, oil pump, and air compressor, may be driven off of the crankshaft gear, camshaft gear, or an idler or auxiliary gear meshed with one or the other, **Figure 4-8.**

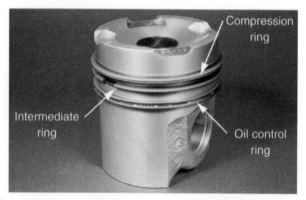

Figure 4-6. *The diesel engine uses three types of piston rings. The top ring is the compression ring and is designed to contain combustion chamber pressures. The middle compression/scraper ring also holds in compression and helps to scrape excess oil from the cylinder walls. The lower ring is called the oil ring or oil control ring. Its job is to keep excess oil from reaching the combustion chamber.*

Figure 4-7. *Typical diesel engine crankshaft. Note the various names of the crankshaft parts. (Cummins Engine Co., Inc.)*

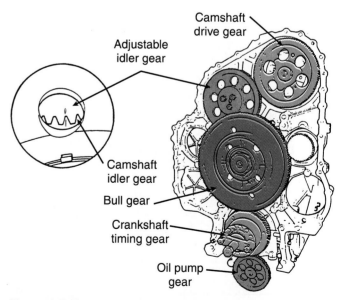

Figure 4-8. *The crankshaft gear is timed to the camshaft gear. Drive gears for the oil pump, fuel injection pump, and other accessories are driven by the crankshaft or camshaft gears. (Detroit Diesel Corp.)*

Connecting Rods

The **connecting rods** form a link between the pistons and the crankshaft, **Figure 4-9.** The connecting rod is attached to the piston with a **piston pin.** The opening or bore in the small end of the rod is fitted with a replaceable bushing that rides on the piston pin. The large end of the connecting rod is attached to the crankshaft with a **rod cap** that encircles the crankshaft's connecting rod journal, **Figure 4-10.**

It is important to remember that rod caps are not interchangeable between connecting rods. Replaceable half-circle bearings are installed between the rod assembly and the crankshaft connecting rod journal. On some connecting rods, there is an oil hole in the big end of the rod opposite the rod cap. This hole is the entrance to an oil gallery that runs up the center of the rod to lubricate the piston pin and to spray oil to cool the piston's underside. Incorrect installation of the rod bearing will cover this hole, blocking the flow of oil, and causing engine failure.

Oil Pan

The **oil pan** forms the bottom of the engine. It is bolted to the bottom of the cylinder block with bolts and sealed with a soft gasket. The area formed by the oil pan and the bottom half of the cylinder block is called the **crankcase,** since it encloses the crankshaft.

One of the oil pan's major jobs is to provide a reservoir for the engine's lubricating oil. The lowest part of the oil pan is called the **sump.** The **oil pickup tube** extends into the sump and is covered by a coarse wire screen that prevents dirt and bits of metal from entering the lubrication system and damaging the engine. There are many different oil pan designs, such as front, center, and rear sump configurations, **Figure 4-11.** By changing the shape of the oil pan, the same engine can be made to run upright, on its side, or even upside down.

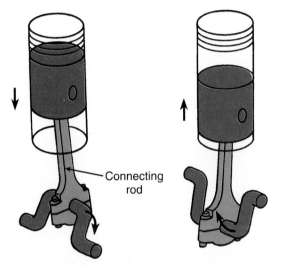

Figure 4-9. *The purpose of the connecting rod is to change the piston's reciprocating motion into the rotational motion of the crankshaft.*

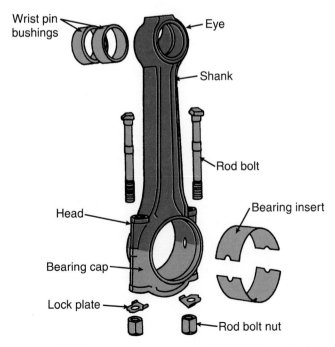

Figure 4-10. *Nomenclature of a crankshaft connecting rod.*

Flywheel

The pistons do not supply a smooth continuous flow of power to the crankshaft. Instead, they provide individual power pulses that result in a rough, jerky turning motion. To compensate for this motion, a **flywheel** is used to store some of the kinetic (or motion) energy of the power strokes.

Flywheels are large, flat disks and are usually made of heavy cast iron or steel. The flywheel uses its momentum to push the pistons during the exhaust and compression strokes. **Top dead center** is the position in the stroke when the piston is at the very top of the cylinder. Heavy flywheels can store more energy than light flywheels.

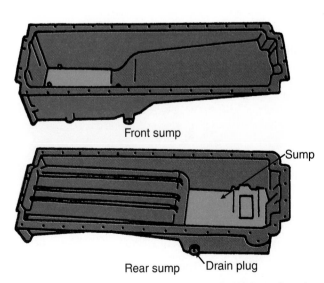

Figure 4-11. *Two common oil pan configurations found on mobile engines. Other designs are used on both mobile and stationary diesel engines.*

The rear surface of the flywheel may be ground smooth and used as one of the contact surfaces for the clutch disks. The ring gear for the starter motor is an integral part of the flywheel assembly, **Figure 4-12.** The *flywheel housing* is mounted to the cylinder block, supports the starter motor, and protects the clutch or torque converter.

Some diesel engines are rigidly coupled to heavy loads that serve the same purpose as a flywheel. These installations include generator set armatures, torque converters, and large fans and propellers.

Vibration Dampers

The *vibration damper* is mounted on the front of the crankshaft next to the accessory pulleys. Its job is to help absorb the *torsional harmonics* (lengthwise twisting) in the crankshaft. This twisting is a result of the multiple power strokes or pulses received from the connecting rods. When subjected to these forces, the crankshaft can begin vibrating like a giant tuning fork along its entire length. The longer the crankshaft, the more difficult it is to control torsional vibration.

The vibration damper may use a ring of viscous fluid or heavy rubber and steel rings to absorb and dissipate crankshaft vibrations. Some larger diesel engines have a complex system of weights and springs in their vibration dampers. Smaller dampers usually consist of a heavy cast iron rim with a rubber ring located between the rim and the damper's hub.

Internal Balancers

An *internal engine balancer* is actually an extra shaft that is geared to the crankshaft. It turns in the opposite direction of crankshaft rotation at a speed that is the same or directly proportional to the crankshaft's rotational speed. The counterweights on the internal balancer create a frequency of vibration that is the exact opposite of the vibrations occurring in the crankshaft. The vibrations produced by the balancer and the crankshaft cancel each other out. Internal balancers are also used to eliminate camshaft vibration.

Camshaft

The *camshaft* coordinates the operation of the intake and exhaust valves, and sometimes the fuel injectors with the action of the pistons. The action of opening and closing the valves (and the injectors) is determined by the cam lobe size and shape. The lobes are precision machined to produce the desired valve action.

The cam lobes are not round like the rest of the camshaft. Instead, they have an eccentric shape. As a result, some parts of the lobe's surface are farther from the centerline of the camshaft than others. The lobe's eccentric shape causes the cam follower, that rides on the lobe, to move up and down as the camshaft rotates, **Figure 4-13.** For each cylinder in the engine, there are several lobes on the camshaft. One lobe operates the intake valve (if intake valves are used), and the second lobe operates the exhaust valve(s). In some engines, a large lobe located between the intake and exhaust lobes controls the fuel injector for the cylinder. See **Figure 4-14.**

To synchronize the action of the valves and injectors with the movement of the pistons, the camshaft timing gear is engaged and timed with the crankshaft timing gear.

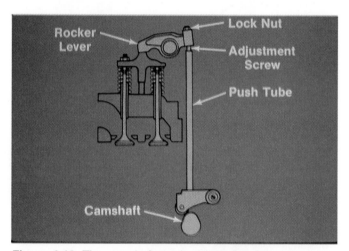

Figure 4-13. *The camshaft, push rod, and rocker arm assemblies coordinate the action of the valves and, in low-pressure injection systems, the fuel injectors. (Caterpillar)*

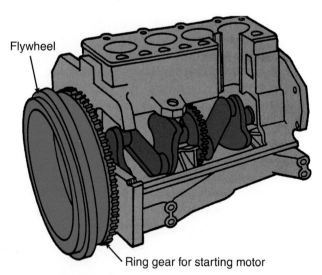

Figure 4-12. *The starter ring gear is part of the flywheel assembly. This gear can be replaced separately from the flywheel in most cases.*

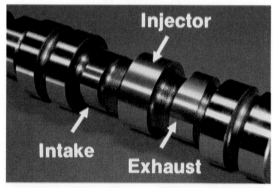

Figure 4-14. *Camshaft lobes are eccentric in shape. Note the individual lobes for the intake valve, exhaust valve, and fuel injector. (Caterpillar)*

During each complete four-stroke cycle, the crankshaft will make two revolutions for every camshaft revolution. Therefore, the camshaft gear in four-stroke engines is twice as large as the crankshaft gear. On a two-cycle engine, the crankshaft and camshaft are the same size because both shafts must turn at the same speed.

Like the crankshaft, the camshaft runs the length of the engine and is supported within the cylinder block. Camshaft journals are fitted with one- or two-piece replaceable bushings. Lubrication to the camshaft journals is provided through holes in the bushings.

Valve Train Components

The **valve train** is comprised of components that perform the opening and closing of intake and exhaust valves, **Figure 4-15.**

Cam followers, also called lifters or tappets, are like connecting rods in that they are motion converters. However, cam followers convert the rotational motion of the cam lobe into a reciprocating motion. This motion is then transmitted through the push tubes (or push rods) and rocker arms to the valves, and in some engines, the fuel injectors. There is one cam follower for each lobe on the camshaft.

The **push tubes** or **push rods** connect the cam followers to the rocker arms (levers). While some engines use solid push rods, many diesels use hollow push tubes made of high-strength alloys. The **rocker arms** (levers) are mechanical fingers that transfer motion from the push tubes to the valve stems.

Cylinder Head

The **cylinder head, Figure 4-16,** is a large casting that bolts to the cylinder block and forms the top of the combustion chamber. The cylinder head contains an elaborate system of passageways that permit coolant to circulate, **intake ports** (four-cycle engines) that allow air to enter the

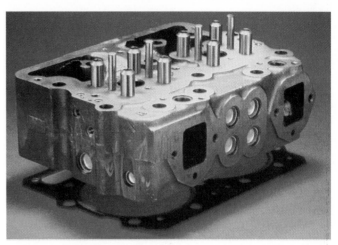

Figure 4-16. Typical cylinder head. Note the numerous passages cast into the head. (Cummins Engine Co., Inc.)

cylinder, and **exhaust ports** which channel the burned combustion gases from the cylinder. The cylinder head also has openings for each cylinder's fuel injector. See **Figure 4-17.**

The cylinder head supports the valve seats, valve guides, valve springs, fuel injectors, push rods, and rocker arm assemblies. Since coolant enters the engine from the bottom and travels toward the top, the cylinder head is the last major engine component to receive coolant. By the time the coolant reaches the head, it is too hot to cool adequately. Nevertheless, a constant, high volume of coolant flowing through the cylinder head is enough to keep it from warping or distorting. This is why the cylinder head, head gasket, valves, and related parts are often the first components damaged when an engine overheats.

When the cylinder head is bolted to the cylinder block, it must form an airtight and watertight seal. A **head gasket** is installed between the head and block to maintain this seal.

Valves

Valves are used to seal the air passages in the cylinder head. In a four-cycle engine, the **intake valves** open to admit fresh air into the cylinders and then close to hold the air in during the compression, power, and exhaust strokes. The

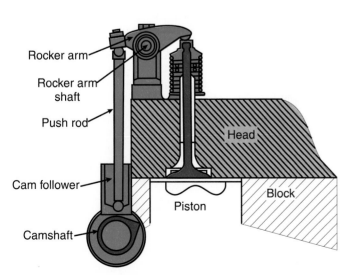

Figure 4-15. Valve train components help time and perform the opening and closing of intake and exhaust valves.

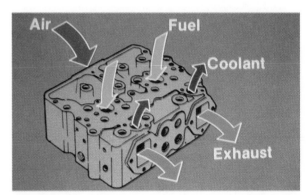

Figure 4-17. The cylinder head passages are used for routing combustion air and coolant, plus mounting openings for the fuel injectors. (Cummins Engine Co., Inc.)

exhaust valves are closed during the intake, compression, and power strokes, but are opened during the exhaust stroke to allow the exhaust gases to exit the cylinder, **Figure 4-18.**

Two-, Three-, and Four-Valve Heads

When an engine is equipped with only two valves for each cylinder, the cylinder head is referred to as a *two-valve head.* In order to provide greater airflow through the cylinder, many engines are designed with three or four valves per cylinder. The cylinder heads for these engines are commonly called *three-* and *four-valve heads.*

Valves operate at very high speeds, opening and closing many times per second. They are also subjected to operating temperatures that often exceed 1400°F (760°C). Valves must be made of extremely high-grade metals in order to withstand these high temperatures. In most cases, the valve face is hardened by the addition of a hard metal, such as Stellite. Some valves are filled with sodium, which helps to transfer additional heat to the top of the valve. Since exhaust valves are subjected to the extreme heat of the exhaust gases, they are often made of higher quality material than intake valves.

Valve Seats and Valve Guides

When closed, the valves must form a very tight seal to contain the high pressure of compression and combustion. This means the face of the valve must precisely match the profile of the opening in the cylinder head. In some engines, this seating surface is machined directly into the cylinder head, **Figure 4-19.** In most modern, high-horsepower diesels, however, a separate **valve seat** is inserted into the head.

During engine operation, the valve stem moves up and down inside a **valve guide.** The valve guide is constructed of soft, pliable metal, such as bronze, and is press fit into the cylinder head. The guide acts as a centering device to keep the valve from tilting and the valve face perfectly aligned with its seat. A coil spring fits around the top of the valve guide and is usually attached to the valve stem by a retainer and a pair of half collets, **Figure 4-20.**

Pressure applied by the rocker arm forces the valve down, opening the valve port in the cylinder head. When the

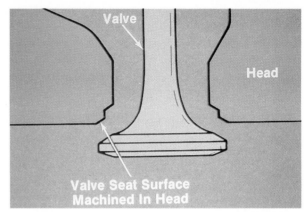

Figure 4-19. *Details of a integral valve seat. Some valve seats are installed as a separate piece in the cylinder head.*

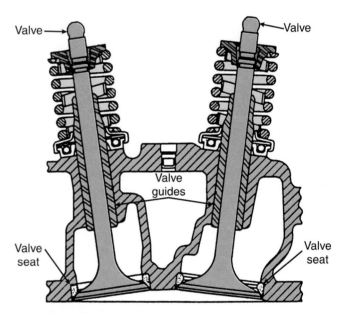

Figure 4-20. *Typical intake and exhaust/related valve setup and components. Note the valve guides which limit valve side-to-side movement in the head.*

rocker arm pressure is released, energy stored in the compressed valve spring forces the valve back up, closing the valve port. On some engines, **valve rotators** are mounted on the valve to turn it with each stroke. The valve rotator tends to help the valve wear more evenly, extending its life.

When two valves are used for the same function in a cylinder, a **crosshead** is added to form a bridge between the valves. This allows the two valves to be opened by a single rocker arm. When a crosshead is used, a guide is installed between the valves to keep the crosshead in the proper position, **Figure 4-21.**

Intake and Exhaust Manifolds

Combustion air and exhaust gases are generally channeled to and from the combustion chamber through the intake and exhaust manifolds that are attached to the outside of the cylinder head. Exceptions to this arrangement will be covered in later chapters.

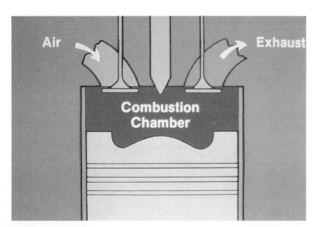

Figure 4-18. *Action of intake and exhaust valves in a four-stroke engine.*

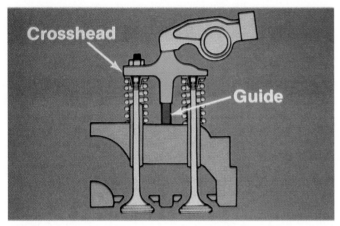

Figure 4-21. *A crosshead or bridge enables a single rocker arm to control two valves. (Cummins Engine Co., Inc.)*

Intake Manifold

The **intake manifold** (air manifold), **Figure 4-22A,** distributes clean, filtered air to the intake ports of each cylinder on a four-cycle engine or to the air box on a two-cycle engine. The intake manifold is made of iron or aluminum and is usually cast in a single piece. The intake manifold uses a soft gasket to maintain a tight seal with the cylinder head.

Exhaust Manifold

The **exhaust manifold** collects burned gases from each exhaust port and channels them to the exhaust pipe on naturally aspirated (air intake) engines. On turbocharged engines, the exhaust gases are routed from the exhaust manifold, through the turbocharger, and into the exhaust pipe, **Figure 4-22B.**

Because they are subjected to rapid and extreme variations in temperature, exhaust manifolds are usually made of cast iron. Marine and large stationary diesel engines may have a water-jacketed exhaust manifold to lower the temperature in the engine compartment. Truck engine exhaust manifolds usually are not liquid-cooled. Therefore, heat-sensitive components must be shielded from the exhaust manifold or located away from it.

Superchargers

A **supercharger** or *blower* is an air pump that is mechanically driven by the engine. See **Figure 4-23.** Its job is to force more air into the combustion chamber. More air permits a larger fuel charge to be burned in the cylinder, creating additional power from the same basic engine. Because the blower is driven by the engine, it robs the engine of power. However, this loss is more than offset by the power gained.

Currently, almost all four-cycle engines use turbochargers, which are more efficient than superchargers. Nevertheless, superchargers are still used on most two-cycle engines, where they scavenge or blow exhaust gases out of the combustion chamber during the period in the exhaust/intake cycle when both the air inlet ports and the exhaust valves are open.

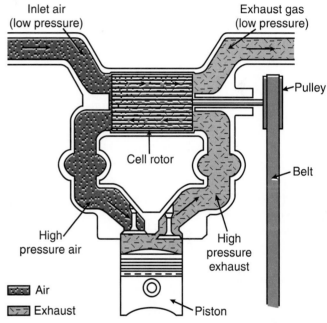

Figure 4-23. *Superchargers increase the amount of combustion air entering the cylinder. Superchargers are either belt-or chain-driven by the engine.*

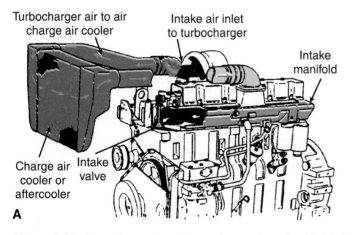

A

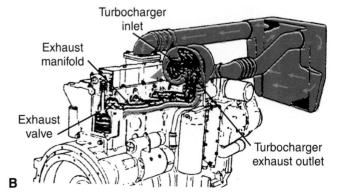

B

Figure 4-22. *Air systems of mobile engine system: A—Air intake manifold brings fresh air to the engine's combustion chambers. B—The exhaust manifold routes the burned gases to the exhaust system. (Cummins Engine Co., Inc.)*

Turbochargers

A *turbocharger* is also an air pump designed to put more air in the combustion chamber. However, it differs from a supercharger in that it is driven by energy that normally would be wasted—the exhaust gases rushing out of the engine, **Figure 4-24.**

One advantage of a turbocharger is that it operates hardest when it is needed the most. For example, when the engine is at idle or coasting downhill, it produces very little exhaust gas. This is fine because extra air and fuel are not needed under these conditions. When the engine is under heavy load, it produces more exhaust, the turbocharger spins faster, and more air is forced into the cylinder.

Turbochargers allow a small diesel engine to produce up to 40% more power than the same size engine could generate if it were naturally aspirated. Turbochargers also help an engine run better at high altitudes, where air is thin. At high altitudes, normally aspirated diesels smoke unless they are equipped with smaller injectors that limit the engine's power. Many two-cycle diesel engines are fitted with a turbocharger and a gear-driven supercharger.

Aftercoolers

An *aftercooler* (intercooler) is simply a radiator that is used to cool intake air. See **Figure 4-25.** A diesel engine will perform better if the intake air is cooled before it enters the combustion chamber. This is because cool air is denser than warm air, and dense air contains more oxygen than thin air.

Remember that a turbocharger will pressurize or compress air. When air is compressed, it heats up and loses some of its density. This air will hold less oxygen and burn

Figure 4-25. *The aftercooler mobile unit is usually located in front of the radiator.*

less fuel. Cooling the air after it passes through the turbocharger increases its density and oxygen concentration. Aftercoolers use either air or coolant as a medium.

Diesel Engine Classifications

Diesel engines can be classified in a number of ways. The broadest classification is by power. Diesel engine power can range from three to four horsepower in the smallest diesels to 40,000 horsepower in giant marine and stationary diesel engines. Speed ratings can also be used to classify engines. Diesel engine speed ratings (rpm) can also vary widely—from 125-250 rpm in large, low-speed diesels to 1800-2200 rpm in high-speed truck engines. Diesel engines can also be classified in the following ways:

- ❑ Engine cycle design.
- ❑ Cylinder arrangement.
- ❑ Combustion chamber design.
- ❑ Fuel injection system design.

Engine Cycle Design

As mentioned in **Chapter 1,** diesel engines can be built to operate using a four-stroke or two-stroke cycle. Four-stroke cycle diesel engines are the most popular.

Four-Stroke Cycle Operation

A *four-stroke cycle engine* (four-cycle engine) requires two complete revolutions of the crankshaft (720° of crankshaft rotation) to complete the four piston strokes that make up one complete combustion cycle. The four piston strokes are the intake, compression, power (combustion), and exhaust strokes. See **Figure 4-26.**

Intake Stroke

During the *intake stroke,* the intake valves are open and the exhaust valves are closed. The piston begins the intake stroke near its top dead center (TDC) position in the

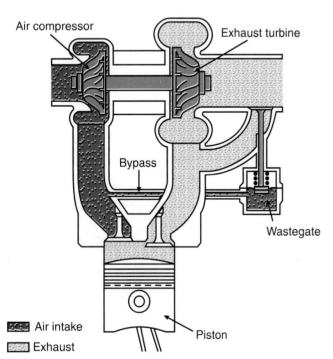

Air compressor

Exhaust turbine

Bypass

Wastegate

▨ Air intake

▧ Exhaust

Piston

Figure 4-24. *Turbochargers are driven by exhaust gases. It is actually more efficient that a supercharger since it does not require engine power for operation.*

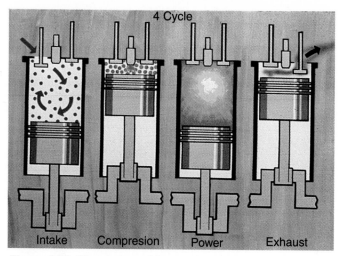

4 Cycle

Intake   Compresion   Power   Exhaust

Figure 4-26. The strokes of four-cycle engine. (Detroit Diesel Corp.)

cylinder. As the piston travels down the length of the cylinder, air is drawn into the combustion chamber.

Compression Stroke

During the **compression stroke,** both the intake and exhaust valves are closed. The piston moves upward in the cylinder, compressing the trapped air, causing it to heat. Common compression levels inside the cylinder between 425 psi-550 psi (2930.4-3792.3 kPa), will elevate air temperatures to 1000-1200°F (537.7-648.8°C). The actual pressure and temperature levels inside the cylinder are affected by many factors, including compression ratio, turbocharger boost pressure, ambient air temperature, valve timing, engine speed, and load.

At the end of the compression stroke, the piston is once again at its top dead center position. The crankshaft has completed another half revolution and is now 360° through its 720° cycle.

Just before the piston reaches the top of the cylinder, the fuel injector is triggered, fuel is sprayed into the cylinder, and combustion begins. This creates a major increase in the pressure and temperature inside the combustion chamber. Fuel is sprayed into the combustion chamber for a specific length of time to maintain this high pressure. The length of fuel injection is based on engine design and the engine's load and speed at the time of injection.

Power Stroke

The combustion chamber is located between the top or crown of the piston and the cylinder head. As combustion gases expand in this confined space, they push on the cylinder head and piston. Since the cylinder head is stationary and airtight, the force directed on the piston pushes it downward in the cylinder during the **power stroke.** Peak cylinder firing pressures in modern, high-speed, heavy-duty truck engines can reach 2300 psi (15 858 kPa), with temperatures ranging between 3000-4000°F (1648-2204°C).

The motion of the piston is transferred through the connecting rod to the crankshaft. The length of the power stroke is controlled by the amount of time the exhaust

valves remain closed. As soon as the exhaust valves open, pressure inside the cylinder is lost, and the piston stops generating pushing power. During the power stroke, the piston moves from the top of the cylinder to the bottom. The crankshaft rotates another half revolution and completes 540° of its 720° cycle.

Exhaust Stroke

As the **exhaust stroke** begins, the exhaust valves are open and the high pressure exhaust gases rush out of the combustion chamber, drawn to the lower atmospheric pressures outside of the cylinder. The upward motion of the piston pushes these gases out of the cylinder. At the end of the exhaust stroke, the piston is in its top dead center position, the crankshaft has completed another half turn for a total of two full revolutions (720° of rotation), the intake valves open and the exhaust valves close. All components have returned to their original positions, and the sequence (cycle) is ready to be repeated.

Four-Cycle Engine Valve Timing

In a four-cycle engine, the camshaft turns at one-half crankshaft speed. Therefore, each valve is opened and closed once during two revolutions (720° rotation) of the crankshaft. The exact time in the cycle the intake and exhaust valves open and close is extremely important to proper engine performance. Engine manufacturers take great care in calculating these times and in manufacturing the camshaft and the valve train parts to exacting specifications.

It may seem logical for the valves to open and close at exactly the top dead center piston positions, but this is not the case. To make certain all exhaust gases are removed or scavenged from a cylinder before the beginning of an intake stroke, the engine is timed to open the intake valves slightly before the upward-moving piston has finished its exhaust stroke. This creates a ram air effect that helps to remove exhaust gases from the cylinder.

Similarly, the exhaust valves are timed to remain open until the piston passes top dead center (TDC) on its exhaust stroke and travels a very short distance downward on its intake stroke. The intake air helps push the remaining exhaust gases out of the cylinder through the exhaust valve. The condition when the intake and exhaust valves are open at the same time during the combustion cycle is called **valve overlap.**

Valve overlap is measured in degrees of crankshaft rotation. For example, if the intake valves open 16° **before top dead center (BTDC)** and the exhaust valves do not close until 16° **after top dead center (ATDC),** the valve overlap is 32°.

Similar timing is designed into the intake valves at the bottom dead center position. **Bottom dead center (BDC)** is the position when the piston is at the very bottom of the cylinder. The intake valves do not close until the piston passes BDC and begins its upward stroke on the compression cycle. Closing the intake valves slightly **after bottom dead center (ABDC)** ensures that the largest possible air charge is drawn into the cylinder.

The true compression stroke can only begin when the intake valves are closed. Fuel is injected earlier in the compression stroke (piston farther from TDC) when the engine is operating under increased speed or light load. Fuel injection is timed to begin later in the compression stroke (piston closer to TDC) when the engine is operating at low speeds or under heavy loads to compensate for the lag time between the start of injection and the resulting increase in expansion pressure.

The power stroke continues until the exhaust valves open. Once again, the exhaust valves are timed to open shortly **before bottom dead center (BBDC).** This gives the exhaust gases extra time to begin moving out through the exhaust ports and into the exhaust manifold. The piston changes direction at BDC and moves up in the cylinder,

displacing all exhaust gases through the exhaust ports. As the piston approaches TDC, the intake valves open and the process repeats itself.

Figure 4-27 illustrates the duration of the various piston strokes, the opening and closing points of intake and exhaust valves, and the duration of the positive valve overlap for a specific engine.

Piston Position

All engines are timed so their pistons will be at different stages in their firing cycles anytime it is running. As mentioned earlier, each firing cycle equals 720° of crankshaft rotation. A six-cylinder, four-cycle engine will have its

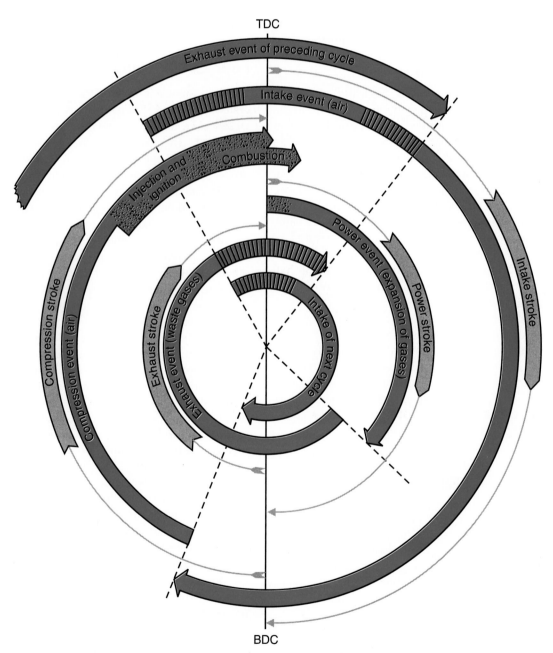

Figure 4-27. Duration of the various piston strokes and valve positions in a typical four-cycle engine. (Detroit Diesel Corp.)

pistons set 120° (720° divided by 6) apart in the firing cycle. **Figure 4-28** illustrates the relative piston firing positions in a six-cylinder, inline, four-stroke engine with a firing order of 1-5-3-6-2-4. As you can see, all six cylinders are in different stages of the combustion cycle.

It is important that the diesel technician understand what one cylinder is doing in relation to the others at any given position of the crankshaft. Knowing the firing order of the engine and the position of each piston at any point in the cycle is important when performing timing and valve adjustments.

Two-Stroke Cycle Engines

As described previously, a four-cycle engine requires two full turns of the crankshaft and four separate piston movements to complete the intake, compression, power, and exhaust strokes of one combustion cycle. A **two-stroke cycle engine** completes this sequence of events in only one complete revolution of the crankshaft (360°) and two separate piston movements. This is done by eliminating the separate intake and exhaust strokes.

Two-cycle engines differ from four-cycle engines in a number of ways. Two-cycle engines do not have intake valves. Instead, each cylinder liner is designed with a series of **intake ports** or holes in its wall. These ports are located about halfway down the length of the cylinder and are open when the piston is at the bottom of its stroke, **Figure 4-29.** Air is free to flow into the cylinder whenever the piston is not blocking the intake ports. In this way, the piston of each cylinder acts as its own intake valve.

The two-cycle engine uses an overhead valve train to open and close the exhaust valves. The valves are timed to open when the intake ports are also open. This creates a strong upward flow through the chamber that pushes out the exhaust gases. Up to four exhaust valves per cylinder are used to maximize scavenging.

Another major difference between two- and four-cycle engines is that two-cycle engines use a positive displacement blower assembly to supply a strong air flow to the engine, **Figure 4-29.** The blower supplies the air flow needed for combustion, exhaust scavenging, crankcase ventilation, and for internal engine cooling. The blower supplies air through an air box built into the engine block. This **air box** runs the length of the block and surrounds all of the cylinder liners.

Operation

The blower supplies a strong flow of air into the cylinder whenever the intake ports are not blocked by the piston. Air flows in one direction only—from the intake ports toward the exhaust valves at the top of the cylinder. Air intake occurs as part of the compression stroke in a two-cycle engine. At the start of the compression stroke,

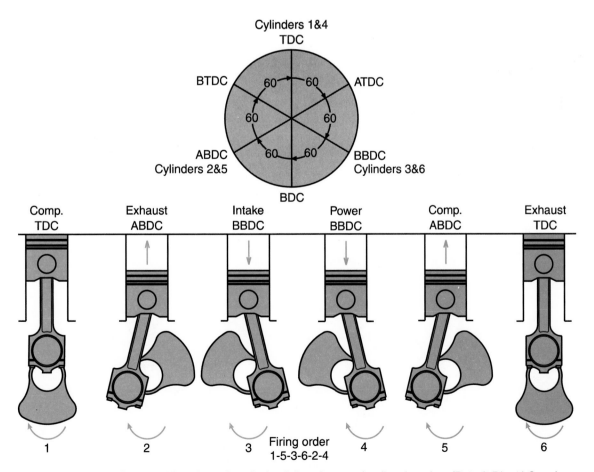

Figure 4-28. Relative piston firing positions in a six-cylinder, inline, four-stroke diesel engine. (Detroit Diesel Corp.).

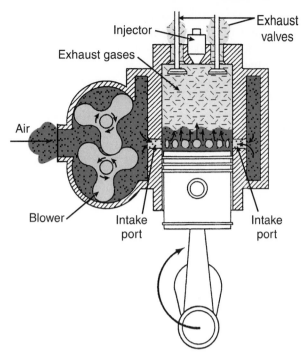

Figure 4-29. *The blower action in a two-stroke engine. The additional compressed air supplies the combustion chamber with fresh air and scavenges any remaining burned gases to the exhaust.*

the piston is at bottom dead center, the intake ports are open, and the exhaust valves are open. Fresh air rushes into the cylinder. The force of the air entering the cylinder

pushes any remaining exhaust gases out of the open exhaust valves, **Figure 4-30A.**

As the piston moves upward, it covers the intake ports. The exhaust valves are timed to close at this point in the cycle. There is now a full charge of clean air in the cylinder and compression begins as the piston continues moving toward top dead center, **Figure 4-30B.** Slightly before TDC, fuel is injected into the cylinder and combustion begins, **Figure 4-30C.** As in a four-cycle engine, the piston passes TDC and moves downward on its power stroke.

The power stroke continues until the piston is about halfway down the cylinder. At this point, the exhaust valves open, and the burned gases escape through the exhaust ports. See **Figure 4-30D.** As the piston continues downward, it uncovers the intake ports, allowing fresh air to rush into the cylinder to scavenge the remaining exhaust gases. The entire combustion cycle is completed in a single (360°) revolution of the crankshaft.

Two-Cycle Engine Valve Timing

A typical valve timing diagram for a two-cycle engine is illustrated in **Figure 4-31.** Remember that all actions occur during a single crankshaft revolution. As in four-cycle engines, timing is very precise and will vary between engine models. The following timing points are examples only.

Fuel injection normally begins about 15° BTDC on the compression stroke and ends about 1.5° ATDC on the power stroke. The power stroke continues until the exhaust valve

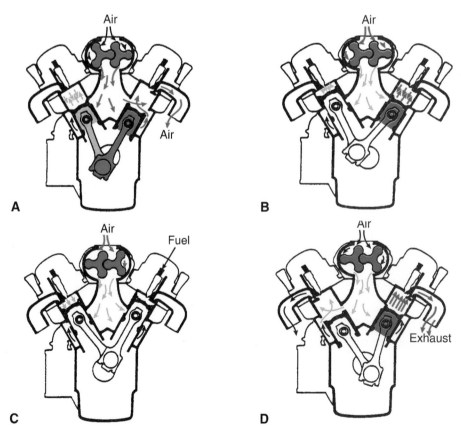

Figure 4-30. *Two-stroke engine cycle. A—Intake. B—Compression. C—Power. D—Exhaust. All four cycles take place in one crankshaft revolution. (Detroit Diesel Corp)*

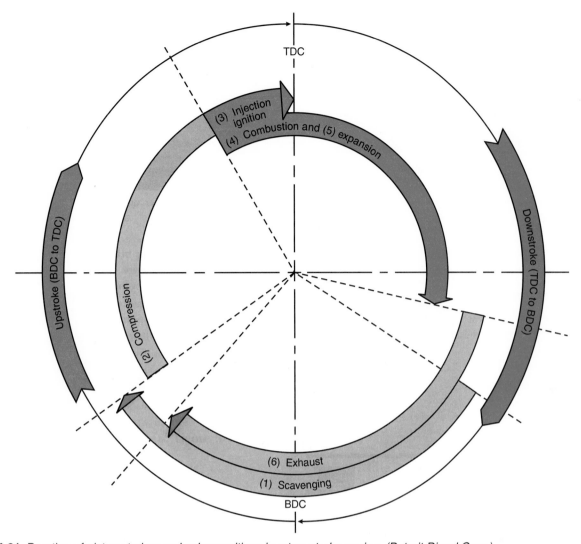

Figure 4-31. *Duration of piston strokes and valve positions in a two-stroke engine. (Detroit Diesel Corp.)*

opens, usually around 90° to 95° ATDC. The exhaust portion of the power stroke continues until the intake ports are opened.

Intake ports normally open around 60° BBDC on the power stroke and remain open around 60° ABDC on the compression stroke. This means the intake ports are open for about 120° of crankshaft rotation during the entire combustion cycle. The exhaust valves are timed to close at about 65° ABDC on the compression stroke. Exhaust valves are usually open for about 155° to 160° of the entire combustion cycle.

Figure 4-32 illustrates some basic differences between typical four-cycle and two-cycle engines. During 720° of crankshaft rotation, a two-stroke engine completes two cycles, while a four-stroke engine completes just one. Two-cycle power strokes last approximately 90° of crankshaft rotation. In contrast, a single power stroke in a four-cycle engine lasts approximately 140° of rotation.

Cylinder Number and Configuration

Diesel engines can be classified by the number of cylinders in the engine. Single cylinder engines are often used for portable power plants or for running irrigation systems. For commercial on- and off-highway vehicles, four-, six-, and eight-cylinder engines are popular. For industrial uses such as power generation, locomotives, and marine propulsion, twelve-, sixteen-, twenty-, and twenty-four cylinder engines are used. The arrangement or configuration of cylinders is also used to distinguish diesel engines from one another.

Inline Engines

The most popular engine configuration is the vertical inline design, with six cylinders being the standard for most trucking and mobile equipment applications, **Figure 4-33.** *Inline engines* are less expensive to overhaul than V-type engines, and their components and accessory items tend to be more accessible and easy to service. Inline engines are long and narrow, which makes them relatively easy to remove and install. However, adequate engine compartment space may be a problem with this design.

In addition to conventional vertical mounting, an inline engine can also be mounted on its side, such as in a bus where the engine is located beneath the passenger floorboard area. Horizontal, **Figure 4-34,** or flat inline

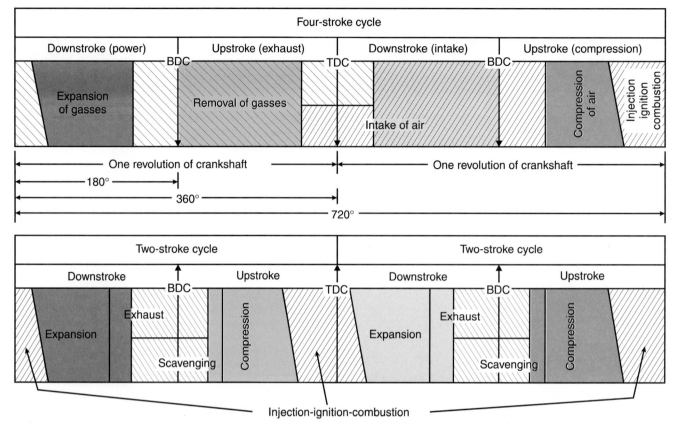

Figure 4-32. *Comparison of the two- and four-stroke cycles. Note the number of degrees it takes for each cycle.*

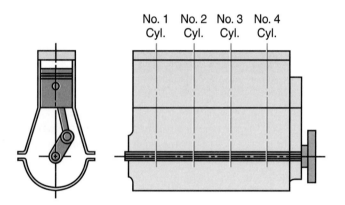

Figure 4-33. *When the cylinders are placed in a single line, the engine is called an inline engine. Most modern inline diesel engines are six-cylinder.*

engines may have all of the cylinders on one side or an equal number of cylinders on each side. When there are an equal number of cylinders on each side of the engine, it is known as a *horizontally-opposed engine.*

V-Type Engines

Another popular engine configuration is the familiar *V-type engine* design. In this arrangement, the cylinders are set at an angle to the crankshaft. Common V-block angles include 45°, 50°, 55°, 60°, and 90°. The angle depends on the number of cylinders and the design of the crankshaft, **Figure 4-35.**

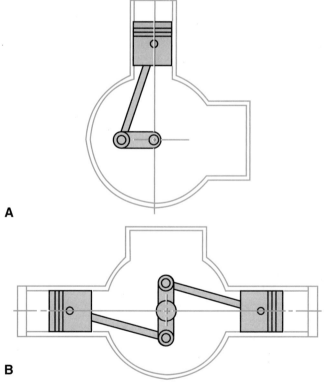

Figure 4-34. *A—When the motion of the piston is in a horizontal position, the engine is known as a horizontal engine. B—When cylinders are placed on each side of the crankshaft, the engine is horizontally-opposed. Such engines may have the cylinders placed either vertically or horizontally.*

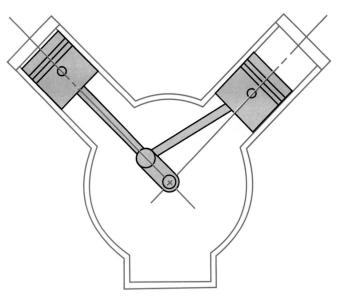

Figure 4-35. A V-type engine has the cylinder placed in two banks and at an angle to each other. The connecting rods operate on a single crankshaft.

V-type engines are commonly available in six-, eight-, twelve-, and sixteen-cylinder designs, but other cylinder configurations can be produced. V-engines are shorter in length and lower in height than comparable inline engines. This makes them desirable when installation space is limited. Also, a V-8 diesel engine can use a smaller bore and stroke than a comparable six-cylinder, inline engine. This means the engine can run faster, increasing power output.

W-Type and Delta Engine Configurations

To form a W-cylinder arrangement, two V-type engines are set side-by-side and operate a single crankshaft, **Figure 4-36.** W-type diesel engines are used primarily in marine applications. Delta configurations have their cylinders formed in a triangle. Such configurations are used in both marine and rail road applications.

Radial Engines

In a radial engine, the cylinders are arranged in a circle around a common crankshaft, **Figure 4-37.** In this type of engine, the connecting rods work on a single crankpin that rotates around the center of a circle. One type of radial engine places four banks of cylinders one above the other and uses a single crankshaft to form a sixteen-cylinder engine.

Double-Acting Piston

The ***double-acting piston engine*** is a modern version of an engine that was originally built by Etienne Lenoir in 1860. In this design, a single piston uses both ends of a cylinder. Explosive pressure is alternately applied to the two faces of the piston to produce power. Pressure is applied on both the up and down strokes. This process requires a very complex intake and exhaust system. The double-acting piston configuration is used on very large, low-speed engines.

Combustion Chamber Designs

Diesel engines use one of several combustion chamber designs, including:
- ❑ Direct injection chambers.
- ❑ Precombustion chambers.
- ❑ Swirl chambers.
- ❑ Energy cells.

Direct Injection Chamber

Most heavy-duty, high-speed diesel engines manufactured today use the ***direct injection combustion chamber*** design. In this design, the fuel is injected directly into an

Figure 4-36. When four banks of cylinders are placed at angles to each other and operate on a single crankshaft, it is a W engine.

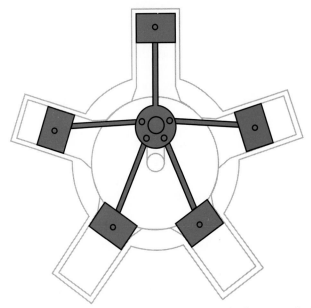

Figure 4-37. Cylinders placed in a circle, operating on a single crankshaft, form a radial engine.

open combustion chamber formed by the crown or top of the piston and the underside of the cylinder head fire deck, **Figure 4-38A.** The fuel injection nozzle is located in the cylinder head and extends directly into the cylinder. The two main advantages of the direct injection design are its high fuel efficiency and its simplicity.

Precombustion Chamber

In the **precombustion chamber** design, a separate precombustion chamber is located in the cylinder head or wall, **Figure 4-38B.** This chamber usually takes up from 25%-40% of the combustion chamber's top dead center volume and is connected to the main chamber above the piston by one or more passages. As air is compressed during the compression stroke, it produces a large amount of turbulence as it passes through the narrow passages into the precombustion chamber.

When fuel is injected into the precombustion chamber, it partially burns, building up pressure. This pressure forces the mixture back through the passageway and into the cylinder (main combustion chamber), where complete combustion is assured. Combustion is very smooth and quiet in these systems. Consequently, indirect injection systems are used on many passenger car diesel engines. However, fuel consumption is 10 to 15% greater in indirect systems when compared to similar open chamber, direct injection systems. This is the main reason that direct injection is almost universally accepted in industrial diesel engine design.

Swirl Chamber

Swirl chamber systems use an auxiliary combustion chamber known as a swirl chamber. This ball- or disk-shaped chamber has a throat area that opens at an angle to the main combustion chamber, **Figure 4-38C.** Unlike a standard precombustion chamber, a swirl chamber contains 50%-70% of top dead center cylinder volume and is connected at a right angle to the main combustion chamber.

A strong air vortex (mass of swirling air) is generated during the compression stroke. The injector nozzle is positioned so that the injected fuel penetrates the swirling air perpendicular to its axis and strikes the opposing chamber side in a hot wall zone.

As combustion begins, the air-fuel mixture is forced into the main combustion chamber through the throat area and mixed with the residual combustion air. Compared to the precombustion chamber design, the flow losses between the main chamber and the auxiliary chamber are lower for the swirl chamber process because the flow cross section is greater. This results in better internal efficiency and lower fuel consumption. The design of the swirl chamber, the arrangement and form of the nozzle jet, and the position of the glow plug (if used) must be carefully matched to ensure proper mixture at all speeds and load conditions.

Energy (Lanova) Cells

In this design, fuel is not injected directly into the chamber. Instead, it is injected across the main combustion

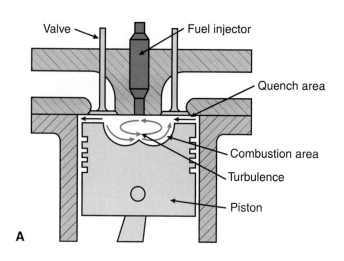

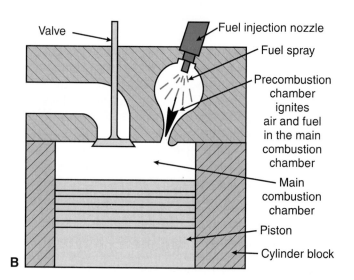

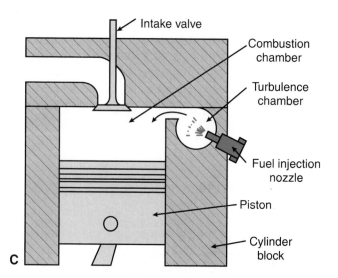

Figure 4-38. Diesel engine combustion chamber designs. A—Direct injection is beginning to see more widespread use. B—The precombustion chamber design has been used for many years and is a popular design. C—Swirl chambers look like the precombustion chamber design, but contains more TDC volume than the precombustion chamber.

chamber and into the energy cell. The **energy cell** is designed to induce a high-energy swirl in the air-fuel mixture in the chamber. It is generally used with pintle-type injection nozzles.

The energy cell contains approximately 15%-20% of the chamber volume when the piston is at top dead center. As shown in **Figure 4-39,** the system contains a main combustion chamber, flat-top piston, small energy cell, and a passageway connecting the main combustion chamber to the energy cell.

The energy cell is actually two separate chambers with a small passageway between them. The entire energy cell is shaped somewhat like a figure eight. It is located directly across the combustion chamber from the injection nozzle. As injection occurs, a portion of the main fuel stream crosses the main combustion chamber and enters the energy cell. The atomized portion of the fuel stream begins to burn in the main combustion chamber.

As the main fuel stream begins to burn in the energy cell chambers, the size and shape of the chambers force the air-fuel mixture back into the main combustion chamber at a very high speed. There is also a great deal of turbulence created in the cell. Nevertheless, the movement of fuel and air from the energy cell to the main combustion chamber occurs at a steady rate, so combustion and engine operation are smooth and fuel economy is excellent.

Diesel Fuel Injection Systems

The **fuel injection system** is the heart of the diesel engine. As you learned in **Chapter 1,** Rudolf Diesel struggled with the problem of fuel injection, finally resorting to a clumsy compressed air system to blow fuel into the combustion chamber. The development of the first reliable

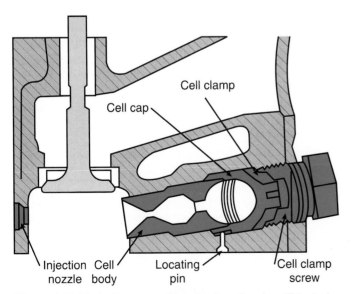

Figure 4-39. Typical energy cell combustion chamber. This design is actually two separate chambers and creates great turbulence which helps the air-fuel mixture to better mix for good combustion.

Cell clamp

Cell cap

Injection Cell　　　Locating　　　Cell clamp
nozzle body　　　　pin　　　　　screw

injectors and pumps by Robert Bosch in the 1920s led to the production of modern high-speed diesel engines.

Injection System Functions

All diesel fuel injection systems must be designed to provide exactly the same amount of fuel to each cylinder. This ensures that the power pulses generated in each cylinder will be equal and that the engine will operate smoothly. The system must also be capable of generating sufficient fuel pressure at the injector nozzles. In some cases, the system's fuel pump or injector unit develops pressures up to 5000 psi (34 475 kPa) to force fuel into the combustion chamber.

Fuel injection must be timed precisely. If fuel is injected at the wrong time, the engine will lose power. If fuel injection occurs early, the combustion chamber temperature will be lower than ideal and ignition will be delayed, resulting in incomplete combustion. If the ignition occurs late (after the piston has passed top dead center), the burning fuel will have more room to expand and some of the potential power in the expanding gases will be lost.

The injection of the fuel must begin abruptly, continue for a specific length of time, and end abruptly. If the fuel is injected too fast, it will have the same effect as fuel that is injected too early. If fuel injection occurs slowly, it will have the same effect as fuel that is injected too late.

The fuel must be atomized when it is injected into the combustion chamber. Atomization distributes the fuel throughout the air charge, increasing the surface area of the fuel exposed to the oxygen in the air. This ensures a complete burning of the air and fuel, which maximizes power output. Atomization occurs as fuel is forced out of the openings in the fuel injector tip. The design of the combustion chamber and the piston head also helps to mix the air and fuel together.

A **governor** is needed to regulate the amount of fuel fed to the cylinders. The governor ensures that there is sufficient fuel delivered at idle to prevent the engine from stalling. It also cuts the fuel supply when the engine reaches its maximum rated speed. Without a governor, a diesel engine could quickly reach speeds that would destroy it. The governor is often included in the design of the fuel injection pump.

Types of Fuel Injection Systems

Diesel engines are equipped with one of five distinct types of fuel injection systems:

- ❏ Individual pump systems.
- ❏ Multiple-plunger, inline pump systems.
- ❏ Unit injector systems.
- ❏ Pressure-time injection systems.
- ❏ Distributor pump systems

Individual Pump Systems

In an **individual pump system,** a small pump, which is contained in its own housing, supplies fuel to one cylinder, **Figure 4-40.** Consequently, there is a pump for each

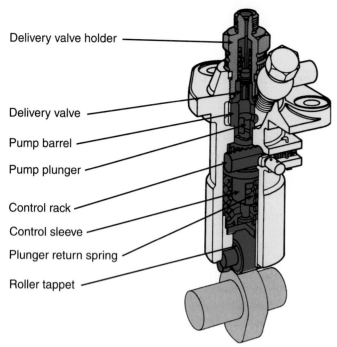

Delivery valve holder

Delivery valve

Pump barrel

Pump plunger

Control rack

Control sleeve

Plunger return spring

Roller tappet

Figure 4-40. *Individual plunger and barrel pump that is driven off of the engine camshaft. These are found on only a few small and very large diesel engines.*

cylinder. This type of system is found on large-bore, slow-speed, industrial or marine diesel engines and on small, air-cooled diesels. Individual pump systems are not used on high-speed diesel engines.

Multiple-Plunger, Inline Pump Systems

The **multiple-plunger, inline pump system** is the most popular fuel injection system in use today. It uses individual pumps that are contained within a single injection pump housing. See **Figure 4-41.** The number of pump plungers in the housing is equal to the number of engine cylinders. The pump plungers in inline pump injection pumps are operated from a pump camshaft.

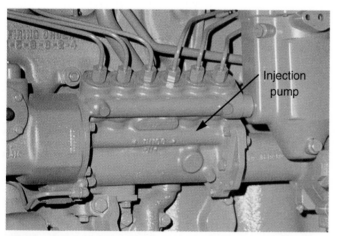

Injection pump

Figure 4-41. *Multiple-plunger, inline fuel injection pump. This pump is used on many mobile applications and is very popular with many engine manufacturers.*

In a multiple-plunger, inline pump system, fuel is drawn from the fuel tank by a fuel supply or transfer pump, passes through primary and secondary filters, and is delivered to the injection pump housing at a pressure between 10 and 35 psi. All individual pumps in the housing are subject to this fuel. Fuel at each pump is timed, metered, pressurized, and then delivered through a high-pressure fuel line to each injector nozzle in firing order sequence.

Unit Injector Systems

In a **unit injector fuel system,** timing, atomization, metering, and fuel pressure generation all take place inside the injector body that serves a particular cylinder. This compact system delivers higher fuel pressures than any other injection system in use today.

Fuel is drawn from the tank by a transfer pump and is filtered. The pressurized fuel (50-70 psi or 344.75-482.65 kPa) then enters a fuel inlet manifold cast within the engine's cylinder head. This inlet manifold feeds all of the unit injectors through a fuel inlet or jumper line. The fuel enters the individual unit injectors, where it is pressurized, metered, and timed for proper injection into the combustion chamber. Unit injector systems use a pushrod-actuated rocker arm assembly or a straight camshaft-operated rocker arm assembly to operate the injector plungers.

Pressure-Time Injection Systems

The **pressure-time injection system (PT system)** is frequently designed by manufacturers for use on their line of diesel engines. Although pressure-time systems are similar to unit injector systems, not all fuel system functions are performed within the injectors.

The PT system derives its name from two of the primary factors that affect the amount of fuel injected per combustion cycle. "P" refers to the pressure of the fuel at the inlet of the injectors. This pressure is controlled by the system's fuel pump. "T" refers to the time available for the fuel to flow through the injector's inlet orifice and into the injector cup. The amount of time that the orifice remains open is controlled by the engine's speed.

This system uses a camshaft-actuated plunger, which changes the rotary motion of the camshaft into the reciprocating motion of the injector. The plunger movement opens and closes the metering orifice in the injector barrel. When the orifice is open, fuel flows into the injector cup. Metering time is inversely proportional to engine speed. The faster the engine is turning, the less time there is for fuel to enter the injector cup.

A third factor affecting fuel delivery amounts in the PT system is the size of the orifice opening. The size of the orifice opening is set through careful calibration of the entire set of injection nozzles.

Distributor Pump Systems

Distributor pump systems, Figure 4-42, are used on small to medium diesel engines. These systems do not have

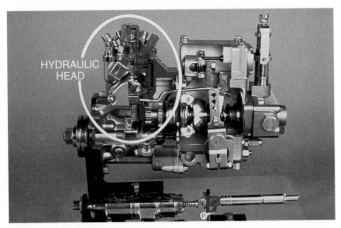

Figure 4-42. *In a distributor fuel injection pump (cutaway shown), a single pumping element dispenses fuel to all injectors. (AMBAC International)*

the capability to deliver sufficient fuel volume or fuel pressures for heavy-duty, large-displacement, high-speed diesel engines, such as those used in trucks. Distributor pump fuel injection systems are sometimes called rotary pump systems. Their operating concept resembles an ignition distributor used on gasoline engines. A rotating part, which is called a rotor, is located inside the pump and distributes fuel at a high pressure to the individual injectors in the proper firing order.

Electronically Controlled Fuel Injection Systems

The electronic engine control revolution that transformed the passenger car and light truck industries during the 1980s has done the same for the diesel engine industry in the 1990s. As in the auto industry, diesel engine electronic control is being driven by increasingly stringent exhaust emission standards.

To meet these emission requirements, diesel engine manufacturers have equipped their engines with *electronic controls* that handle timing, fuel metering, engine speed governing, vehicle speed governing, cruise control, and engine torque shaping and power settings at fixed, full-load speeds.

Multiple-plunger, inline pump fuel injection systems, unit injection systems, and distributor-pump injection systems are now manufactured with electronic controls. Although the actuation of the inline pump plunger or unit injector plunger is still done mechanically to enable the system to attain the high injection pressures needed, metering and timing of the fuel is controlled through the use of electronic, solenoid-operated valves and governors.

Engine Performance Terms and Formulas

The following sections define many of the key terms and formulas used to describe diesel engine operation and calculate engine performance characteristics.

Bore and Stroke

Bore is the diameter of the cylinder. *Stroke* is the distance the piston travels between its top dead center and bottom dead center positions. See **Figure 4-43.** The distance between the crankshaft's main journals and the rod journals determine the length of the stroke.

Displacement

Cylinder displacement is the total volume that is displaced in a cylinder as the piston moves from the bottom of its stroke to the top. Cylinder displacement is measured in cubic inches (cu. in.) or liters (l) and is calculated as follows:

cylinder displacement $= \pi \times r^2 \times S$, where

$\pi = 3.14$
r = bore radius (1/2 bore diameter)
S = length of stroke

Engine displacement is the volume that is displaced by the upward strokes of all the pistons in an engine. To calculate engine displacement, multiply the displacement of a single cylinder by the number of cylinders in the engine.

Compression Ratio

The amount of compression developed in a cylinder is expressed by the *compression ratio.* This ratio compares the cylinder volume when the piston is in the bottom dead center position to the cylinder volume when the piston is in its top dead center position, **Figure 4-44.** The volume of air inside the cylinder that is compressed is known as the *clearance volume.* The clearance volume and cylinder dis-

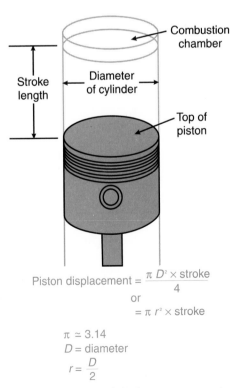

$$\text{Piston displacement} = \frac{\pi\, D^2 \times \text{stroke}}{4}$$
$$\text{or}$$
$$= \pi\, r^2 \times \text{stroke}$$

$\pi \simeq 3.14$
D = diameter
$r = \dfrac{D}{2}$

Figure 4-43. *Engine bore and stroke measurements.*

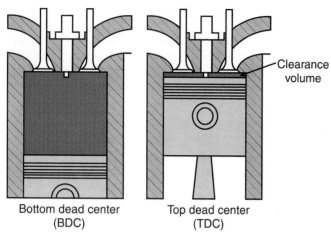

Figure 4-44. *Compression ratio is the relationship between the cylinder volume when piston is at bottom dead center and the cylinder volume when the piston is at top dead center.*

placement can be used to calculate the compression ratio as follows:

$$\text{Compression ratio} = \frac{\text{Cylinder displacement} + \text{clearance volume}}{\text{Clearance volume}}$$

Pressure

Pressure (P) is defined as force per unit area. Gas and liquid pressures are normally stated in pounds per square inch (psi). The metric unit of pressure is the Pascal (Pa). One psi equals 6,895 Pascal's or 6.894 kilopascals (kPa).

Atmospheric Pressure

The weight of the atmosphere, which is called **atmospheric pressure** plays a key role in the operation of naturally aspirated diesel engines. It is constantly exerting force on all objects. At sea level, this force creates a pressure equal to 14.7 psi. At higher elevations, the pressure is less and may cause engine performance problems.

Volumetric Efficiency

The amount of air that is actually pulled into a cylinder compared to the total volume that could theoretically fill the cylinder is known as the **volumetric efficiency** (VE). Volumetric efficiency can be calculated using the following formula:

$$VE = \frac{\text{Volume of air taken into cylinder}}{\text{Maximum possible cylinder volume}}$$

Most naturally aspirated engines have a VE between 80% and 90%. This means the amount of air drawn into the cylinder is slightly less than the cylinder's full capacity.

In a turbocharged or supercharged engine, air is forced into the cylinder at pressures greater than atmospheric pressure. Consequently, the amount of air forced into the cylinder is not directly related to the piston displacement, and volumetric efficiency does not apply. In the case of a supercharged engine, the term scavenge efficiency is used. **Scavenge efficiency** is an indication of how

efficiently the burned gases are removed from the cylinder and replaced with a charge of fresh air.

Specific Gravity

Specific gravity compares the weight of a material with the weight of an equal volume of water. Specific gravity ratings are used when grading fuels and lubricating oils.

Mean Effective Pressure

Mean effective pressure (MEP) is the difference between the compression pressure and the expansion pressure. On naturally aspirated engines, the MEP is between 95-140 psi (667-965 kPa). On a turbocharged engine, MEP can range from 150-300 psi (1035-2068 kPa).

Horsepower and Torque

Engines are rated according to horsepower and torque. The speed and fuel setting of an engine will determine its actual rated horsepower setting. **Horsepower** is the ability to maintain a load at a desirable or fixed speed. It is a measure of how fast work can be done. Horsepower is produced as a direct result of combustion in the cylinder. Combustion pressures force the piston down the cylinder, producing work. The amount of work produced on the power stroke can be determined using the following formula:

Work = Force × Distance
where:
Force = Pressure acting on the piston in psi
Distance = Length of piston stroke

The more fuel released into the cylinder during a given period of time, the greater the amount of horsepower produced.

Horsepower

The term horsepower was first used by the inventor James Watt. He tried to compare the power produced by a steam engine with the power of a horse. Watt determined that a strong horse could pull a 200 lb. load of ore 165 ft. straight up a mine shaft in one minute. This equals 33,000 ft/lb. of work every minute—the modern equivalent of one horsepower.

$$1\ HP = \frac{33,000\ ft/lb.}{Minute}$$

The metric equivalent for horsepower is the watt (W). This unit of measure is more familiar as a unit of electric power. Horsepower can also be measured in kilogram-meters per second. One horsepower equals 746 watts or 76 kg-m/sec.

Not all horsepower measurements are the same. The term **rated horsepower** refers to the horsepower that the engine will produce at its maximum full-load, governed speed in a particular application. **Indicated horsepower (ihp)** is the power transmitted to the pistons from the expanding combustion gases. Indicated horsepower is not measured with a test instrument, it is mathematically calculated. Indicated horsepower does not take into account frictional power losses of moving parts or the power losses caused by accessory items, such as water pumps, blowers, oil pumps, fuel injection pumps, and cooling fans.

The formula for calculating engine indicated horsepower is:

$$ihp = \frac{P \times L \times A \times N}{33,000}$$

where:
P = mean indicated pressure in psi
L = stroke of the piston in feet
A = net piston areas in square inches
N = number of power strokes per cylinder per minute

Brake horsepower (bhp) is the actual usable horsepower produced by the engine. The first step in determining brake horsepower is to connect a dynamometer to the engine's output shaft and measure the torque generated. The following formula is then used to calculate the brake horsepower.

$$\text{Brake horsepower} = \frac{\text{torque} \times \text{rpm}}{5252}$$

Because brake horsepower accounts for power losses due to operating friction and accessory items, it is normally between 10% and 20% less than indicated horsepower for the same engine.

Mechanical efficiency is the difference between indicated horsepower and brake horsepower. Typical mechanical efficiency in a four-cycle engine is between 80% and 90%. This is slightly lower than the mechanical efficiency of a comparable two-cycle engine.

Torque

Torque is the rotational or twisting force around a fixed point. Torque is calculated by multiplying the applied force by its distance from the centerline of rotation. A simple example of this principle is demonstrated whenever a torque wrench is used to tighten a bolt. If a force of 20 lb. is applied perpendicular to the centerline of the bolt's rotation at a distance of 1 ft. from the centerline, 20 ft.-lb. of torque is generated. The torque wrench acts as a lever to apply this force.

If the lever is fastened to a rotating shaft, it is called a crank—hence the term crankshaft. In a diesel engine, the length of the lever or crank is the distance between the centerline of the connecting rod journal on the crankshaft and the centerline of the main bearing journal on the crankshaft. The longer the crank, the greater the amount of torque generated. The amount of time that force is applied to the crank also affects torque generation. Numerous short-duration power strokes will not generate as much torque as fewer longer-lasting power strokes. This is why

torque levels decrease as engine speed levels (rpm) and horsepower increase.

Torque is measured in foot-pounds (ft.-lb.) or inch-pounds (in.-lb.). The metric unit for torque is the Newton-meter (N•m). The formula for determining torque in a diesel engine is:

$$\text{Torque} = \frac{5252 \times \text{horsepower}}{\text{rpm}}$$

To determine the metric equivalent in Newton-meters, multiply your answer by a conversion factor of 1.3558.

Summary

A diesel engine consists of a number of systems and subassemblies that work together to produce power. The cylinder block is the foundation of the engine. It supports all major engine components and has openings for cylinders and passages for coolant and lubricant. It houses the powertrain components, such as the crankshaft, pistons, and connecting rods.

The camshaft is timed to the crankshaft using timing gears. It coordinates the operation of the intake valves, the exhaust valves, and in some systems, the fuel injectors with the action of the pistons. Lobes on the camshaft transfer the rotational motion of the shaft into an up-and-down motion that is transferred to the valves and, if applicable, the injectors.

The cylinder head bolts to the top of the cylinder block and forms the top of the combustion chamber. It has passages for coolant circulation, intake air, and exhaust. It also contains openings for the fuel injectors.

Valves are used to seal the air passages in the cylinder head. Cylinders can be equipped with either two, three, or four valves. Manifolds are used to deliver air to the engine and carry away the exhaust gases produced by combustion.

Turbochargers and superchargers are air pumps that help deliver more combustion air to the engine. Aftercoolers cool the intake air so that it is denser and contains more oxygen.

Four-stroke diesel engines operate using four separate piston strokes to complete one combustion cycle. Two-stroke diesel engines operate using only two piston strokes per combustion cycle. Ports in the cylinder wall take the place of intake valves on these engines and a blower is used to supply air to the cylinder and help clear exhaust gases.

Diesel engines are commonly built in inline and V-type configurations. Double-acting piston engines are also produced. Most high-speed diesel engines use direct fuel injection. Combustion chamber designs used in diesel engines include the precombustion chamber, swirl chamber, and energy cell configurations.

The fuel injection system must deliver the proper amount of fuel for combustion and provide the same amount of fuel to each cylinder to ensure smooth engine operation. Fuel injection systems in use today include individual pump systems; multiple-plunger, inline pump systems; unit injector systems; pressure-time systems; and distributor injection pump systems.

Engines are often rated according to horsepower and torque. Horsepower is the ability to maintain load at a desirable or fixed speed. It is a measure of how fast work can be done. Torque is the measurement of rotational or twisting force around a fixed point.

Important Terms

Head group
Cylinder block
Cylinder liners
Water jackets
Pistons
Piston rings
Crankshaft
Oil seals
Journals
Main bearing journals
Connecting rod journals
Oil galleries
Timing gears
Connecting rod
Piston pin
Rod cap
Oil pan
Crankcase
Sump
Oil pickup
Flywheel
Top dead center (TDC)
Flywheel housing
Vibration damper
Torsional harmonics
Internal engine balancer
Camshaft
Valve train
Cam followers
Push tubes
Push rods
Rocker arms
Cylinder head
Intake ports
Exhaust ports
Head gasket
Valves

Intake valves
Exhaust valves
Valve seat
Valve guide
Valve rotators
Crosshead
Intake manifold
Exhaust manifold
Supercharger
Turbocharger
Aftercooler
Four-stroke cycle engine
Intake stroke
Compression stroke
Power stroke
Exhaust stroke
Valve overlap
Before top dead center (BTDC)
After top dead center (ATDC)
Bottom dead center (BDC)
After bottom dead center (ABDC)
Before bottom dead center (BBDC)
Two-stroke cycle engine
Intake ports
Air box
Inline engines
Horizontally-opposed engine
V-type engine
Double-acting piston engine
Direct injection combustion chamber
Precombustion chamber

Swirl chamber systems
Energy cell
Fuel injection system
Governor
Individual pump system
Multiple plunger inline pump system
Unit injector fuel system
Pressure-time injection system (PT system)
Distributor pump systems
Electronic controls
Bore
Stroke
Cylinder displacement
Engine displacement

Compression ratio
Clearance volume
Pressure
Atmospheric pressure
Volumetric efficiency
Scavenge efficiency
Specific gravity
Mean effective pressure (MEP)
Horsepower
Rated horsepower
Indicated horsepower (ihp)
Brake horsepower (bhp)
Mechanical efficiency
Torque

Review Questions–Chapter 4

Do not write in this text. Place your answers on a separate sheet of paper.

1. The _____ is a large casting that bolts to the top of the cylinder block and forms the top of the combustion chamber.
   (A) intake manifold
   (B) air box
   (C) cylinder head
   (D) piston head

2. List in order the four piston strokes of a four-cycle engine.

3. In a diesel engine, when do valves open and close?

4. Two-cycle engines use a series of _____ to control air flow into the cylinder.

5. The power stroke of a two-cycle engine is _____ than that of a four-cycle engine.

6. List three advantages of the vertical inline cylinder arrangement.

7. List three advantages of V-type cylinder configurations.

8. List the five types of diesel injection systems described in this chapter.

9. Name two ways vibrations can be decreased in a diesel engine.

10. Which engine component converts the rotation motion of a camshaft lobe into an up-and-down motion?
    (A) Rocker arms.
    (B) Push rods.
    (C) Cam followers, tappets, or lifters.
    (D) Crosshead or bridge.

ASE-Type Questions

1. Technician A says the distance from the centerline of the crankshaft to the centerline of a connecting rod journal is equal to half the total travel (stroke) of the piston. Technician B says that this distance equals the total stroke of the piston. Who is right?
   (A) A only.
   (B) B only.
   (C) Both A & B.
   (D) Neither A nor B.

2. The _____ stores some of the kinetic or motion energy of the power strokes and uses its momentum to push the pistons back past top dead center during the exhaust and compression stroke(s).
   (A) vibration damper
   (B) flywheel
   (C) crankshaft
   (D) internal balancer

3. Technician A says that the crankshaft in a two-cycle engine will make two revolutions during each complete combustion cycle, while the camshaft will make only one. Technician B says this is true of four-cycle engines, not two-cycle engines and that in a two-cycle engine both the camshaft and crankshaft turn at the same speed. Who is right?
   (A) A only.
   (B) B only.
   (C) Both A & B.
   (D) Neither A nor B.

4. The camshaft coordinates all of the following, EXCEPT:
   (A) intake valve opening.
   (B) injector opening.
   (C) injector pump operation.
   (D) exhaust valve opening.

5. Technician A says that a supercharger is driven off of the engine while a turbocharger is driven by exhaust gases. Technician B says that turbochargers are more efficient than superchargers because they work hardest when engine speeds are higher and do not draw power from the engine to operate. Who is right?
   (A) A only.
   (B) B only.
   (C) Both A & B.
   (D) Neither A nor B.

6. Technician A says most heavy-duty, high-speed diesel engines now being manufactured use open chamber, direct injection combustion chambers. Technician B says that indirect injection using a precombustion chamber is the preferred method for most diesel engine manufacturers. Who is right?
   (A) A only.
   (B) B only.
   (C) Both A & B.
   (D) Neither A nor B.

7. Technician A says that horsepower is the ability to maintain a load at a desirable or fixed speed and is a measure of how fast work can be done. Technician B says that torque is a rotational or twisting force around a fixed point, and that the longer the crank and the slower the rotational speed, the greater the amount of torque generated. Who is right?
   (A) A only.
   (B) B only.
   (C) Both A & B.
   (D) Neither A nor B.

8. A _____ is used to regulate the amount of fuel fed to the cylinders
   (A) rotor
   (B) plunger
   (C) rack
   (D) governor

9. All of the following takes place inside a unit injector, EXCEPT:
   (A) pressure to draw fuel from the tank.
   (B) injection timing.
   (C) fuel metering.
   (D) fuel atomization.

10. Technician A says that the exhaust manifold of diesel engines used in highway applications are usually cooled with engine coolant circulation. Technician B says that liquid cooled manifolds on marine applications are quite common. Who is right?
    (A) A only.
    (B) B only.
    (C) Both A & B.
    (D) Neither A nor B.

In addition to propelling the vehicle, diesel engines power accessory systems, such as conveyor belts. (Caterpillar)

Chapter 5

Engine Blocks

After studying this chapter, you will be able to:
- ❑ Explain how to gain access to the components in a stationary engine.
- ❑ Explain how to remove a mobile engine from its installation.
- ❑ Describe how to remove the basic parts from a mobile diesel engine block.
- ❑ Name two popular methods used to clean engine blocks.
- ❑ Conduct a cylinder block inspection.
- ❑ Inspect a cylinder and liner.
- ❑ Name the two types of cylinder liners.
- ❑ Describe how to install a cylinder liner.

The cylinder block is the foundation upon which the entire engine is constructed. For an engine to produce maximum power with the best possible efficiency and little friction, the block must be in good shape. This chapter discusses the various engine blocks and components used in diesel engines. Other engine components will be discussed in the chapters to follow.

Cylinder Blocks

The **cylinder block** is the largest single component of the modern diesel engine. It is essentially the framework of the engine. Nearly all engine parts are connected to the cylinder block in some manner.

Blocks for most large stationary diesel engines are of **welded construction.** In this type of construction, **Figure 5-1,** the block is usually fabricated of forgings and/or steel plates that are welded together. Deck plates are generally fashioned to house and hold the cylinder liners. Uprights and other members are welded to the deck plates to form a rigid unit. The cylinder blocks employed in mobile engines are generally of **one-piece construction, Figure 5-2.** This type of block is formed by pouring iron or aluminum into an appropriate sand mold.

All blocks contain openings for the cylinders and bores for the camshaft and crankshaft. They contain passages that allow coolant to circulate around the cylinder liners. Many blocks also have drilled oil passages, that are referred to as **galleries.** Most two-cycle engines also have air passages in the block.

Accessing Diesel Engine Components

When servicing a diesel engine, it is necessary to access various internal parts. The procedure for reaching these parts depends on whether the engine is a stationary unit or a mobile unit.

Figure 5-1. This cylinder block consists of welded steel plates and forgings. Note panels can be removed for engine service. (Stork-Werkspoon)

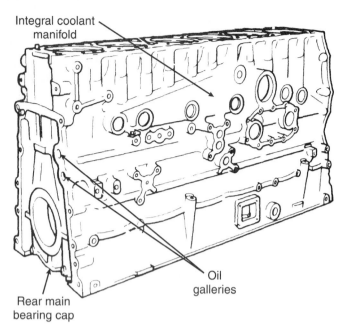

Integral coolant
manifold

Oil
galleries

Rear main
bearing cap

Figure 5-2. The block for most mobile engines is of one-piece construction. (Great Lakes Energy Systems, Inc.)

Accessing Components in a Stationary Engine

The designs of modern stationary engine frames (blocks) differ somewhat from earlier designs. Early frames were often referred to as A-frames, crankcase frames, trestle frames, and staybolt or tie rod frames. Each frame is named according to its shape or the manner in which the parts were fastened together. Many of the features common to early engine frames have been incorporated into newer block designs.

The portion of the block that serves as a housing for the crankshaft is commonly called the *crankcase.* In most engines, the crankcase consists of the lower portion of the cylinder block and an oil pan, sump, or base. The *cylinder assembly* completes the structural framework of an engine. As one of the main stationary parts of an engine, the cylinder assembly, along with various related working parts, serves to confine and release the combustion gases.

In most cases, *stationary engines* are serviced on location and are not removed from their installation. Many stationary engines have openings in the block to permit access to the cylinder liners, main and rod bearings, injector control shafts, and other internal parts. To keep dirt and foreign materials from entering the engine, *access covers* are fitted with gaskets and placed over these openings.

On some very large engines, the access covers, which are sometimes called *explosion doors*, are constructed to serve as safety devices. These covers are equipped with a spring-loaded pressure plate. The spring maintains pressure that keeps the cover sealed under normal operating conditions. In the event of a crankcase explosion or extreme pressure build-up within the crankcase, the excess pressure overcomes the spring tension and the cover's seal is broken. When this occurs, the access opening acts as an escape vent for the pressure, preventing damage to the engine.

The design of stationary engine components will vary considerably from one engine to another. Therefore, it is difficult to give a definite disassembly procedure. Consult the manufacturer's engine service manual for the recommended procedures.

Accessing Components in a Mobile Engine

A *mobile engine* must be removed and disassembled to service many of the components. The exact procedure for removing an engine from a mobile installation will vary. Although you should check the appropriate service manual for specific instructions, the following steps are typical for most installations:

1. Set the vehicle's hand brake, block the wheels, and observe the safety precautions given in **Chapter 2.** Disconnect the battery cables at the battery.

2. Remove the radiator cap. Recover the engine coolant by using an approved coolant recovery unit. If one is not available, open the drain cocks and drain the coolant from the radiator and engine into an appropriate container.

3. Shut off the air supply to the shutterstat (if equipped). The *shutterstat* controls the radiator shutters. Remove the shutterstat before removing the air line. Disconnect the related air line piping around the radiator as well as the shutter cylinder and clamps at the side of the radiator.

4. Disconnect the chassis mounts on the aftercooler. Remove the aftercooler as described in the manufacturer's service manual.

5. Disconnect the upper and lower radiator hoses. Inspect the hoses and replace them if they are worn or defective. Remove the fan shroud, brackets, and hardware. Then, remove the fan.

6. Drain the engine oil.

7. Mark and disconnect all electrical leads, fuel lines, linkages, oil lines, and coolant hoses from the engine. Cap the fuel lines to keep dust out and to prevent spills.

8. Remove the air cleaner-to-engine hose. Disconnect the air control linkage (if so equipped). On some vehicles, it may be necessary to remove the air cleaner as an assembly. Install protective caps to keep foreign objects out of the air induction system.

9. Position jackstands under the transmission. Loosen and remove the bolts that hold the transmission bell housing to the engine flywheel housing, **Figure 5-3.** In most vehicles, it will be necessary to remove or back away the transmission before removing the engine.

10. Remove the radiator. On a vehicle with air conditioning, if necessary, remove the bumper and support or remove the hood. Loosen the bolts that hold the air conditioning condenser in place. Support the condenser so that it is not hanging by its hoses.

*Figure 5-3. Removing the bolts holding the transmission bell housing to the engine flywheel housing.
(Mack Trucks, Inc.)*

Once these steps have been performed, the radiator can be removed without disconnecting the condenser and losing refrigerant.

🔧 **Caution: Do not remove or loosen air conditioning hoses or fittings that will allow refrigerant to escape. If it is necessary to loosen or remove an air conditioning component, be sure to recover the refrigerant charge before opening the system.**

---

11. To lift the engine out of its compartment, attach a spreader bar to the lift brackets or other comparable lifting points. This lifting device should be adjusted so the lifting hooks are directly above the engine lift brackets to prevent bending. Check to be sure everything is disconnected from the engine. Remove the engine mounting bolts and separate the engine from the transmission.

⚠️ **Warning: When removing an engine, serious personal injury or death can occur if an improper lifting method is used. Always follow the engine manufacturer's instructions. Never lift an engine by the rocker arm shafts or by using a strap under the oil pan. To ensure proper weight distribution, all lift brackets provided must be used when lifting the engine. Be sure the spreader bar is of adequate length to prevent the lifter brackets from contacting the rocker cover. If any part of the lifting apparatus fails and the engine drops, allow it to fall.**

---

12. Raise the engine slowly with a mobile crane or a chain hoist. See **Figure 5-4.** Carefully lift the engine out of the engine compartment, making sure that it does not bind or damage related components.

13. Lower the engine onto an appropriate engine repair stand. Whenever mounting an engine to a repair stand, refer to the instructions included with the stand

Figure 5-4. Proper method of lifting a diesel engine. Never stand under an engine while it is being lifted. If the lifting apparatus fails and the engine falls, allow it to fall to the floor. (Mack Trucks, Inc.)

for recommendations on its safe use. Use only grade 8 or equivalent strength metric bolts when mounting an engine to a repair stand.

Engine Disassembly

Once the engine is mounted on the repair stand, **Figure 5-5,** the following general procedure should be observed when disassembling the engine. Refer to the engine service manual for more specific guidelines.

Figure 5-5. A repair stand provides a convenient support when servicing an engine. Note that the engine can be rotated on this stand to facilitate disassembly. (OTC Tools)

1. Remove accessory components, such as the turbocharger (if so equipped), lifting brackets, pulleys, drive belt assemblies, etc., that may hinder access to the block, heads, and rocker boxes (if equipped). Store all parts in a convenient place so they will not be lost. Methods of disassembling these components will be discussed in later chapters.

2. Remove rocker arm covers. Loosen all rocker arm locknuts and the adjusting screws 3-4 turns. Mark the rocker housings according to their location and then remove them. See **Figure 5-6.** Remove the push tubes (push rods) and mark them according to their location. Inspect all push tube sockets and balls for looseness.

3. Carefully remove the injectors, **Figure 5-7,** and place them in an appropriate rack. Cap or cover the injector openings, **Figure 5-8.** The nozzle fuel inlet tube must generally be removed before removing the nozzle holder. If equipped, mark the crossheads as to their locations, loosen the locknuts, and remove the crossheads. Remove the water manifold and thermostat housing and detach the fuel crossover manifolds.

4. Remove the cylinder head bolts with a socket wrench, **Figure 5-9.** Remove the head(s) and gaskets. Always

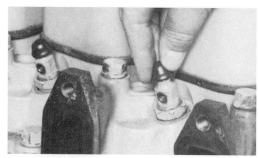

Figure 5-8. *The injector opening should be covered to prevent dirt from entering the fuel system.*

A

B

Figure 5-9. *A—When removing the cylinder head bolts, work from the ends of the head to the center. This prevents any chance of the cylinder head warping during the removal process. B—Then remove the cylinder head and gasket from the block. (Cummins Engine Co., Inc.)*

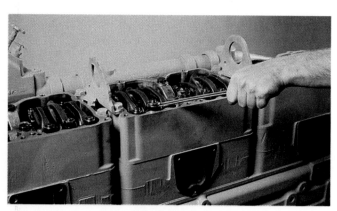

Figure 5-6. *Removing rocker arm housing. Be sure to maintain cleanliness at all times. (Cummins Engine Co., Inc.)*

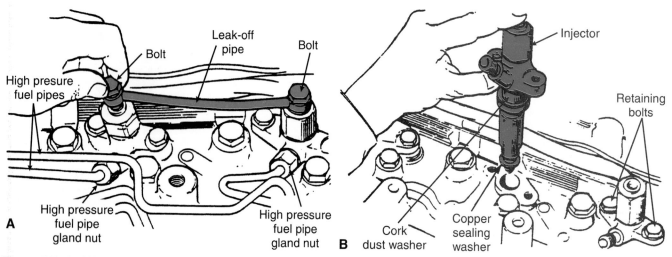

Figure 5-7. *A—When preparing for fuel injector removal, be sure any pressure has been relieved from the system. B—Work carefully as not to damage the injector during removal.*

store cylinder heads on their sides—never on the gasket face. Also, remove the cam follower housings (if necessary), **Figure 5-10.** Mark each housing according to its location. If needed, measure and record the head and cam follower gasket thickness for use as a reference point during reassembly.

5. To detach the flywheel from the crankshaft, lock the crankshaft and remove the flywheel bolts, **Figure 5-11A.** Then remove the flywheel housing bolts and tap the housing loose with a soft-faced hammer, **Figure 5-11B.**

6. Make sure that all of the oil has been drained from the engine before removing the oil pan. Remove all of the oil pan mounting bolts and tap the pan loose with a soft-faced hammer, **Figure 5-12.** Never pry against the finished surfaces of the pan or the block.

7. Extract the rear main seal plate and press the used seal out, **Figure 5-13.**

8. Carefully remove the camshaft bearing supports, keeping the shims together, **Figure 5-14A.** Next, take off the camshaft cover, **Figure 5-14B.**

Figure 5-12. A soft-faced hammer can be used to aid in removing the oil pan. If the oil pan is made of thin material, be sure not to damage the pan. (Cummins Engine Co., Inc.)

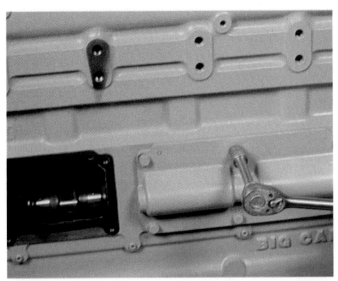

Figure 5-10. Removing the cam follower housing. Keep all parts together so you do not accidentally reinstall them in the wrong location during assembly. (Cummins Engine Co., Inc.)

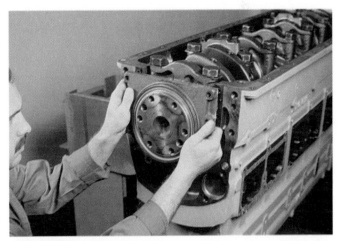

Figure 5-13. Extracting the rear main seal plate. Not all engines use this type of plate. (Cummins Engine Co., Inc.)

A   **B**

Figure 5-11. When removing the flywheel housing, use caution as it is somewhat heavy and cumbersome to handle. A—Removing bolts from the flywheel. B—Use a soft-faced hammer to remove the housing. (Cummins Engine Co., Inc.)

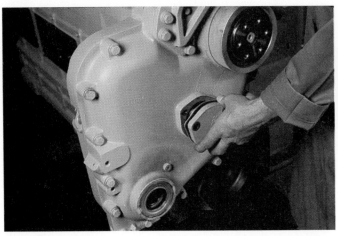

A   B

Figure 5-14. A—Keep all parts together when removing the camshaft cover. If you lose any shims, the cover will not install properly. B—After removing the cover, remove all of the bearing supports. These can usually be accessed through the cam follower housing. (Cummins Engine Co., Inc.)

9. Remove the camshaft bushing and thrust plate from the cylinder block. Remove the camshaft carefully to avoid damaging the cam bushings. During removal, rotate the camshaft while pulling outward with a steady pressure. See **Figure 5-15.** Take care not to drop the thrust washer.

10. Scrape all carbon and gasket material from the top of the cylinder and polish it with a fine emery cloth, **Figure 5-16.** This will prevent damage to the piston lands and rings when removing the piston.

Piston Assembly and Crankshaft Removal

Remove the piston, connecting rod assemblies, and crankshaft as follows:

1. Position the crankshaft throw for the rod being removed at the bottom of its stroke.

2. Mark each rod cap according to the cylinder it is removed from, **Figure 5-17A.** Remove the connecting rod nuts and cap. If necessary, tap the cap lightly with a soft hammer to aid in removal, **Figure 5-17B.**

Figure 5-15. Care should be taken when removing the camshaft bearings from the engine block. (Cummins Engine Co., Inc.)

Figure 5-16. Cleaning the inside wall of each cylinder liner allows easy removal of the piston and rod assemblies. (Cummins Engine Co., Inc.)

3. After the cap has been removed, cover the rod bolts with journal protectors to prevent damage to the crankshaft. If journal protectors are not available, two short pieces of vacuum hose will also do the job.

4. Carefully push the piston and rod assembly out of the cylinder with a wooden hammer handle, **Figure 5-17C.** Support the piston by hand as it comes out of the cylinder. Be sure that the connecting rod does not damage the cylinder during removal.

5. Replace the caps on their respective rods as the assemblies are removed from the cylinders. See **Figure 5-17D.** Check the stampings on each assembly to ensure proper cap positions. Store the rod assemblies in a safe place to prevent damage. Mark rod bearing shells for later inspection.

6. Remove the timing gear from the camshaft. Turn the engine upside down and mark the main bearing caps according to their location. Detach the main bearing caps, lower shells, and all thrust rings, **Figure 5-18.** Mark the components according to their position.

A   B   C   D

Figure 5-17. Typical sequence for removing the piston and connecting rod assemblies. A—Marking a rod cap to aid in identification. B—Striking the cap to aid in removal. C—Pushing piston and rod assembly from a cylinder. D—Replacing cap on rod assembly. (Cummins Engine Co., Inc.)

Figure 5-18. In some cases, a slide hammer is needed to detach the main bearing caps from the engine. (Cummins Engine Co., Inc.)

Figure 5-19. Since most diesel crankshaft are very heavy, they should be lifted with a suitable hoist. (Cummins Engine Co., Inc.)

7. Lift the crankshaft from the block with the appropriate hoist, being careful not to damage the bearing journals, **Figure 5-19.** Support the crankshaft in a V-block to prevent sag, **Figure 5-20.** Remove the main bearings from the block and main bearing caps. Remove the rear main oil seal from the block and the main bearing cap. Examine the bearing inserts for signs of abnormal wear such as embedded particles, lack of lubricant, antifreeze contamination, oil dilution, uneven wear, and incorrect or undersize bearings.

Engine Block Cleaning

The block cannot be thoroughly cleaned unless all oil gallery plugs and freeze plugs are removed. If they are cup-type plugs, drive one end in and use a pair of locking pliers or mechanical fingers to remove them from the block. See **Figure 5-21.**

In some engines, cup-type plugs can be removed easily by pulling them out with a slide hammer or by driving them out from the back side with a long rod. A flat-type plug can be removed by drilling a hole near the center of the plug and inserting a slide hammer to pull it out. Removing threaded

Figure 5-20. The crankshaft should be placed in a V-block so that all crankshaft parts can be inspected carefully. (Cummins Engine Co., Inc.)

oil gallery plugs can be difficult. A drill and screw extractor can be used to help remove these plugs.

After all components have been stripped from the cylinder block, it should be thoroughly cleaned. The cleaning method depends on the material, type, and size of the block. It also depends on the type of cleaning equipment

A

B

Figure 5-21. *Follow these steps in removing freeze plugs. A—Drive the plug in with a punch. B—Remove the plug with mechanical fingers. (Cummins Engine Co., Inc.)*

The open steam cleaning method is no longer used for cleaning blocks due to environmental regulations that stipulate the system must be operated in a closed loop, **Figure 5-22.** The runoff must be contained within the cleaning system and not allowed to enter a public sewer system or soak into the ground. This is because the runoff contains heavy metals and other hazardous material.

Thermal Cleaning

Thermal cleaning ovens have become popular in diesel engine rebuilding shops, **Figure 5-23.** The main advantage of the thermal cleaning system is that it reduces the oil and grease on and in engine blocks, cylinder heads, and other parts. The high temperature inside the oven, generally 650°-800°F (343°C-427°C), oxidizes the grease and oil, leaving a dry, powdery ash on the parts. The ash must be removed by **airless shot blasting** or by washing. Vapors

Figure 5-23. *This block is being placed in a direct-flame, pyrolytic thermal ovens. (Am/Pro Machinery, Inc.)*

available. There are two basic block cleaning procedures: chemical cleaning and thermal cleaning.

Chemical Cleaning

Hot alkaline tank cleaning is a **chemical cleaning** method often used on cast iron blocks. This technique should not be used on aluminum parts because the caustic soda in the tank will dissolve the metal. Aluminum parts must be cleaned in an **emulsion-type cleaning solution.**

Figure 5-22. *Most shops are now required to use closed loop equipment for water-washing of diesel engine block. (AXE Equipment, Inc.)*

and smoke produced by the baking process are consumed by an afterburner in the oven stack.

Only after all components have been thoroughly cleaned can an effective inspection be made or proper machining be done. Therefore, all grease, dirt, oil, scale, and rust must be removed before servicing the engine block.

Cylinder Block Inspection

Once the block is cleaned, it should be visually inspected for cracks. The block can also be checked for cracks using dye or powder tests, x-rays, pressure checking, or magnetic crack detection. These tests are described in **Chapter 8.** If cracks are discovered, replace the block or consult with a rebuilder/machinist about the feasibility of repair.

Study the block for chips, burrs, nicks, scratches, and scores, **Figure 5-24.** The most common locations for defects include:

- ❑ Areas adjacent to freeze plugs.
- ❑ Areas between the cylinders and water jackets.
- ❑ Areas adjacent to the main bearing saddles.
- ❑ Threaded bosses used to attach auxiliary components.
- ❑ The extreme bottom of the cylinders.

Inspect all machined surfaces and threaded holes in the block. Carefully remove any nicks or burrs from the machined surfaces with a file. Clean out tapped holes and repair damaged threads. Check all dowel pins, pipe plugs, expansion plugs, and studs for looseness, wear, or damage. Replace these items as necessary. Coat the parts with *joint sealing compound* before reinstalling them.

Block Deck Warpage and Height Checks

Check for cylinder block *deck warpage* using a straightedge and feeler gauge across both diagonals, along the longitudinal centerline, and across the width, **Figure 5-25.** If warpage exceeds maximum specifications, the block deck must be resurfaced. This can be done by grinding, milling, belt surfacing, or broaching. If warpage exceeds the

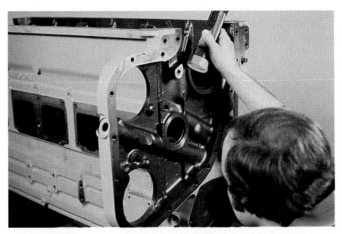

Figure 5-24. *A complete visual inspection should be performed after cleaning the block. (Cummins Engine Co., Inc.)*

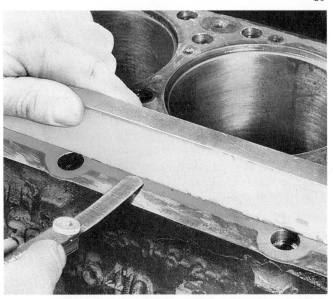

Figure 5-25. *A straightedge and feeler gauge are commonly used to check the block for deck surface warpage and imperfections. (Mack Trucks, Inc.)*

manufacturer's maximum tolerance for material removal, the cylinder block must be replaced.

The *deck height* is the distance from the crankshaft centerline to the block deck. To measure deck height, invert the engine and install the crankshaft, securing it with the center main cap. Measure the distance from the bottom of the crankshaft main journal to the block deck parallel to the cylinder centerline. Measure the main journal end diameter parallel to the cylinder centerline. Divide the diameter in half and subtract the quotient from the initial measurement.

Perform deck height measurements at both the front and rear of the engine block. The front and rear measurements should be identical. If the difference between the measurements exceeds .005" (.127 mm), the deck height should be corrected. Engine block reconditioning requires the skills of a rebuilder or machinist.

After inspection, spray the machined surfaces of the cylinder block with engine oil. If the block is to be stored for a long period, spray it with or dip it in a *rust-preventive solution.* Cast iron will quickly rust when exposed to the atmosphere unless it is greased, oiled, or coated with an appropriate solution.

Cylinder Bores and Liners

The cylinder walls in which a piston moves back and forth may be an integral part of the cylinder block or may be formed by a separate sleeve or liner. The *integral cylinder bore,* often called an "enbloc" or "parent bore", is not replaceable. When excessive wear occurs in a cylinder of this type, then it must be rebored and honed. If wear is too excessive, the block may be sleeved or replaced.

Another disadvantage of integral bores is the inconvenience of removing the entire block from the installation in order to recondition the cylinders. For these reasons, most diesel engines are constructed with replaceable cylinder liners, known as sleeves.

Cylinder Liners

The function of the *cylinder liner,* **Figure 5-26,** is to seal combustion pressure in conjunction with the piston rings and cylinder head gasket, to provide a surface for the transfer of combustion heat to the engine coolant, and to guide the piston during engine operation. Correct extension of the liner's flange above the top deck and a uniform cylinder block counterbore are essential to providing satisfactory liner life and head gasket seal.

Operating conditions of cylinder liners are severe. Not only must they withstand the high pressures of diesel operation and the heavy thrust of the connecting rods, but they must endure extremely high temperatures. Operating temperatures on a two-cycle industrial engine may range from 1100°F (593°C) at the top of the liner, where the cooling conditions are poor, to 400°F (204°C) a short distance below the top. At the bottom end of the liner, the temperature is even lower.

As shown in **Figure 5-27,** two-stroke diesel engines have cylinder liners that contain *air inlet ports.* Two *sealing rings,* which are recessed in the cylinder block, are used to prevent leaks between the liner and the block. The upper and lower portions of the liner are directly cooled by coolant in the water jacket surrounding the liner. At the air inlet ports, the liner is cooled by air introduced into the cylinder through equally spaced ports around the liner.

The liner's air inlet ports are machined at an angle to create a uniform swirling motion as the air enters the cylinder. This motion occurs throughout the compression stroke and facilitates scavenging and combustion.

The success or failure of any replaceable cylinder liner depends on its ability to fit properly in the engine

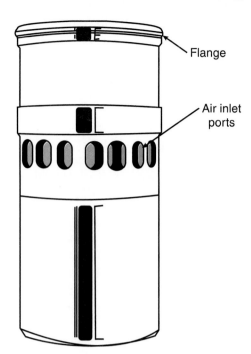

Figure 5-27. *Some cylinder liners have air inlet ports. This type of liner is used on two-stroke engines. (Detroit Diesel Corporation)*

block. The liner provides a cylinder wall on which the piston and rings will operate. The block receiving the replacement liner must be in good condition, and the two parts must be matched. If the cylinder block is worn, a new liner will not function properly. Poor heat transfer, scuffing, scoring, liner movement, fatigue, fracture, and premature failure will result if the block is not in good condition.

The material used in replaceable cylinder liners must withstand the extreme heat and pressure developed within the cylinders and, at the same time, permit the pistons and rings to move with a minimum of friction. The material most commonly used for liner construction is *close-grained cast iron,* however, other steel alloys are occasionally used. Liners are generally plated on the wearing surface with *porous chromium,* which has excellent wear qualities. Also, the pores in the plating tend to hold the lubricating oil. Therefore, they aid in maintaining the lubrication film necessary to minimize friction and promote ring sealing.

In some cases, liners are cast to promote oil retention. A critical factor in the ability of a liner to retain a film of oil is the microstructure of the graphite on the liner's wall.

Inspection of Cylinder Bores and Liners

Except for liner removal and installation, the service procedures for integral bores and replaceable cylinder liners are almost identical. Therefore, the information in this section will apply to both types of cylinders.

Inspect the cylinder walls (bore or liner) for roughness and other signs of wear. Ring and cylinder wall wear can be accelerated if the engine is operated in an abrasive environment. Abrasive particles that get between moving parts grind away at the adjoining surfaces, removing material from the parts. These particles can include metallic debris

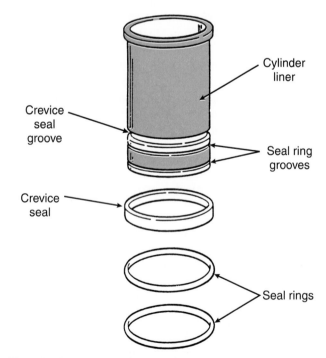

Figure 5-26. *A diesel engine cylinder wet liner uses seal rings to prevent coolant from leaking into the crankcase. Note the position of each ring. (Detroit Diesel Corporation.)*

in the engine as a result of wear, and nonmetallic dirt particles that enter the engine during operation or maintenance. All sources of contamination should be located and corrected to avoid a recurrence of the problem.

Piston ring, piston, and cylinder wall damage can also be caused by scuffing and scoring. Grooves cut in these parts act as passages that allow oil to bypass the rings and enter the combustion chamber. Scuffing and scoring occur when the oil film on the cylinder wall breaks down, allowing metal-to-metal contact of the piston and rings on the cylinder wall. The heat generated by the metal-to-metal contact causes momentary welding or **point welding** of the contacting parts. The reciprocating movement of the piston breaks these welds, cutting grooves in the ring face, piston skirt, and cylinder wall. Cooling system hot spots, oil contamination, and fuel wash are significant causes of scuffing and scoring.

It is important to note that cylinder wear is not uniform from top to bottom. Maximum wear occurs at the top of the ring travel area because pressure on the top ring is at a peak and lubrication is at a minimum when the piston is at the top of its stroke, **Figure 5-28**. Shallow depressions form on the thrust surfaces at the top of the cylinder. As the cylinder wears, it develops an oval shape at the top. **Piston slap** is one of the major problems caused by excess wear at the top of the cylinder. A ridge of unworn material will remain above the upper limit of ring travel. Below the ring travel area, wear is negligible because only the piston skirt contacts the cylinder wall.

Checking for Cylinder Taper

To check a cylinder for wear or taper, either in the case of an integral cylinder bore or a liner, use a cylinder dial bore gauge or an inside micrometer and proceed as follows:

1. Measure the cylinder diameter parallel to the crankshaft at the top of the ring travel zone.

2. Measure the diameter in the same position at the bottom of the ring travel zone.
3. Measure the cylinder diameter at a right angle to the crankshaft at the top of the ring travel zone.
4. Measure the diameter in the same position at the bottom of the ring travel zone, **Figure 5-29**.

Once these measurements are made, compare the measurements obtained in Steps 1 and 3 (subtract the smallest dimension from the largest dimension) to find the out-of-round wear at the top end of the cylinder. Compare the measurements obtained in Steps 2 and 4 to find the out-of-round wear at the bottom end of the cylinder. Subtract the dimension obtained in Step 4 from the dimension obtained in Step 3 to determine the degree of taper present in the cylinder.

The degree of cylinder taper that can be tolerated without reboring or relining the engine depends on the design and condition of the engine and the type of service for which the engine is used. There is no general rule for all diesel engines. Whenever cylinder taper or out-of-roundness exceeds .005" (.127 mm), the engine should be rebored or relined. Follow the engine manufacturer's recommendations for the exact wear limits.

Cylinder Ridge

When inspecting the cylinder, check for a **ridge** near the top of the piston ring travel zone. After extensive engine operation, the cylinder surface will wear in the area where the rings make contact. If the cylinder's initial diameter is the same throughout its length, the increase in diameter in the ring contact area results in the formation of a ridge at the upper limit of piston ring travel.

Remove any ridges from the cylinder using a **ridge reamer**, **Figure 5-30**. To use a ridge reamer, install it so that

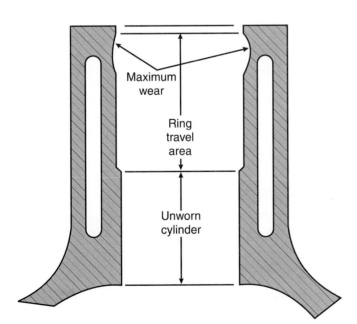

Figure 5-28. Wear pattern of a typical cylinder. This wear pattern is simply due to the normal reciprocal motion of the piston in the cylinder.

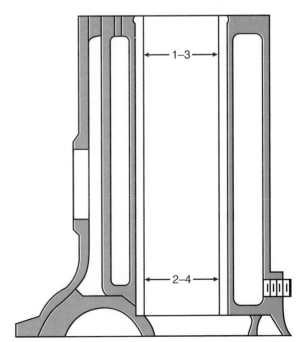

Figure 5-29. When checking for cylinder bore taper, be sure to measure at these points. The difference between these points is the total cylinder taper.

Figure 5-30. *Removing a ridge from a cylinder using a ridge reamer. (Cummins Engine Co., Inc.)*

the cutting guide is below the ridge and the lower part of the cutting tool is tight against the cylinder sleeve. Hold the cylinder sleeve with a clamp to prevent it from turning. Release the spring force; this will force the cutting blades outward and produce a clean cut.

Do not cut down into the ring travel zone when removing the cylinder ridge. It is possible to cut so deeply into the cylinder wall or so far down into the ring travel zone that reboring or replacement of the engine block or liner becomes necessary. Blend the cut made with the ridge reamer so that the area where the worn surface meets the reamed surface is as smooth as possible.

Counterbore

Most modern diesel engine liners have a counterbored area that extends down to the top point of the ring travel zone. The **counterbore** in the block is used to accept the liner's flange. The liner's diameter in the counterbored area is slightly larger than the cylinder's diameter in the piston ring travel area. The counterbore prevents the formation of a ridge or lip on the liner surface at the upper point of ring travel. In most flanged liner applications, the flange should protrude slightly above the block deck when positioned properly in its counterbore, **Figure 5-31.**

Before installing a liner, the block's counterbore should be checked for flatness by measuring its depth in four places, **Figure 5-32.** The variation in depth should not exceed .001″ (.0254 mm). The counterbore must not tilt down toward the center of the bore, but should be square or tilted slightly upward (approximately .25°), **Figure 5-33.**

If the counterbores are reconditioned, be sure that the small chamfer on the inside diameter of the counterbore seat is reestablished. Failure to rechamfer could result in a sharp corner interfering with the radius under the liner's flange. The most common causes of flange breakage are improperly machined or worn counterbores that do not provide a uniform, square contact with the bottom surface of the liner flange. A tool used to recondition counterbores is shown in **Figure 5-34.**

When a cylinder liner or bore is worn beyond the wear limit, the piston generally is also worn beyond its limit. Therefore, replace the piston when the cylinder is reconditioned or the liner is replaced.

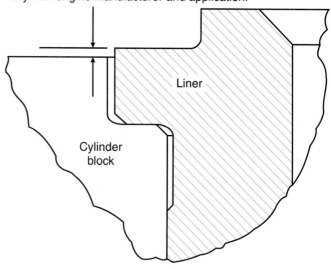

Figure 5-31. *Flange protrusion should protrude slightly, but not an excessive amount.*

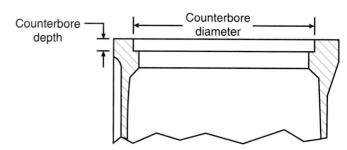

Figure 5-32. *Cylinder liner counterbore dimensions. Be sure the counterbore is clean and free of carbon. This will be important for liner installation.*

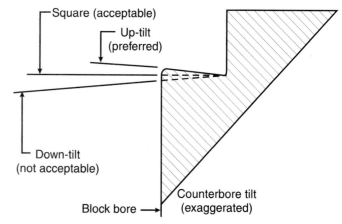

Figure 5-33. *The counterbore should be flat or tilted upward. This tilt is exaggerated for demonstration purposes.*

Types of Cylinder Liners

Cylinder liners can be divided into two general types: dry liners and wet liners. The **dry liner** is not exposed to the engine coolant. Instead, it fits snugly against the wall of the water jacket in the cylinder block bore, **Figure 5-35.** The **wet liner** comes in contact with the engine coolant in the

Figure 5-34. *A counterbore cutting tool is used to resurface the block counterbore. (Mack Trucks, Inc.)*

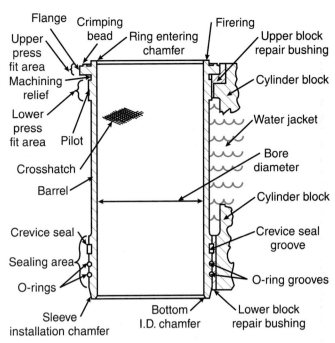

Figure 5-36. *Nomenclature of a wet liner. Study the position of the seal rings at the bottom of the liner.*

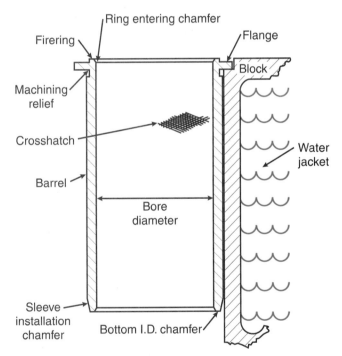

Figure 5-35. *Nomenclature of a dry liner. Note how the liner does not contact the engine coolant.*

water jacket, **Figure 5-36.** Wet liners may have circumference sealing devices or contain integral cooling passages. Liners with integral cooling passages are sometimes referred to as *water-jacketed liners.*

Dry Liners

Dry liners are generally found in smaller diesel engines. Since these liners are supported on their outside diameter by the cylinder block casting, they are usually much thinner than wet liners. Dry liners are installed in some engines with a press fit and in other engines with a loose fit.

The dry liner is usually retained in the block by a small lip or flange that drops into a mating counterbore in the top of the block's deck. This flange prevents the liner from dropping down into the crankcase.

When bolted to the block, the cylinder head and the head gasket hold the liner firmly in position. A dry liner must fit snugly, approximately .001" (.025 mm) tight to .001" (.025 mm) loose in order to obtain effective heat transfer from the liner, through the block, and into the coolant-filled water jacket. If a dry liner does not fit the cylinder casting properly, considerable resistance to heat flow will result.

Wet Liners

In wet cylinder liners, high combustion temperatures are dissipated through the liner walls directly into the water jacket. These liners usually have a close fit between the outside flange diameter and the upper inside diameter of the cylinder block counterbore.

Engine manufacturers have begun to design liners that can be installed with a press fit. Liners that have a press fit just below the block's counterbore are called *lower press fit liners.* These liners are more stable in the block than other designs. Remember that the cylinder wall must withstand the side thrust of the piston caused by the connecting rod acting at an angle. This thrust (piston slap) against the liner may cause vibrations that promote *liner cavitation* (pitting on the outside diameter) and *flange fretting* or breakage.

Wet liners that do not have integral cooling passages have two crucial sealing devices that enclose the water jacket: the top flange and the bottom sealing rings. The top flange is designed to provide a tight seal between the block and the cylinder head. When the cylinder head is torqued, a bead on the liner's lip is embedded in the cylinder head gasket. The press fit in the top counterbore prevents the liner from rocking in the block, reducing operational stress.

Rubber or neoprene **sealing rings** around the bottom (crankshaft end) of the wet cylinder liner prevents coolant from leaking into the oil pan. Most cylinder liners have three sealing rings. The thick top ring is called the *crevice seal*. The crevice seal is designed to withstand high coolant temperatures and prevent foreign particles from damaging the center ring. The *center ring* separates the coolant from the oil. The *bottom ring* helps prevent engine oil from contaminating the upper rings.

Wet liners that do not have integral cooling jackets are constructed to permit lengthwise expansion and contraction. The liner walls are strong enough to withstand the full working pressure of the combustion gases.

Another type of wet cylinder liner, which was mentioned previously, is formed with an integral cooling passage or jacket. This passage may be cast in, shrunk on, or sealed on the liner. Coolant enters through the lower section of the liner's passage and leaves through the top, **Figure 5-37.** This design is used extensively in two-cycle engines as it provides the most effective means of establishing a watertight seal around the ports. It is also used on large, industrial four-stroke diesel engines.

The major disadvantage of any wet liner is the possibility of coolant leaks if the liner is not properly sealed. Sealing procedures for wet liners vary with the type, shape, location, and number of seals used.

Various methods are used to hold the seals in place. In some engines, the liners are grooved to accept the seals. In other designs, the blocks are grooved. Some applications have machined steps or a combination of grooves and steps on the liner and the block. The location of the seals may also vary. Note the seal mounting design and the number of seals used when removing a wet liner. Compare the new liner with the old liner and read any special instruction sheets included with the new liner.

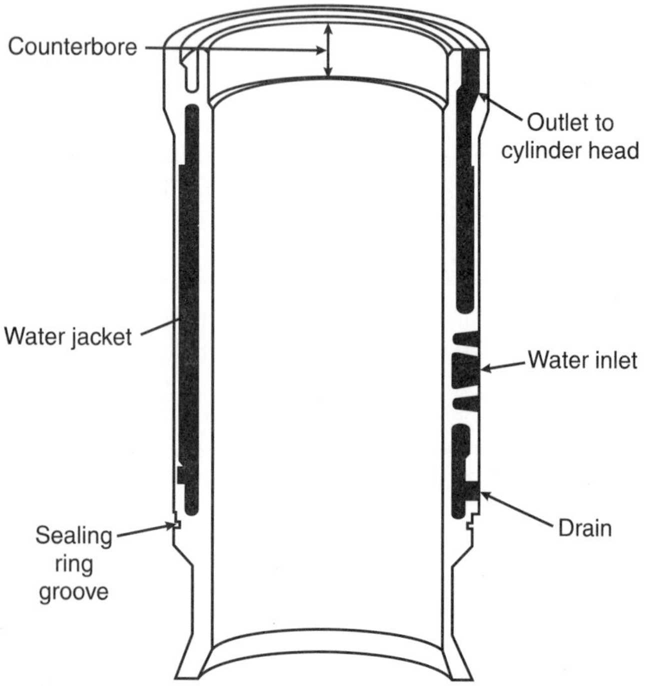

Removing Cylinder Liners

If a liner exceeds out-of-round or taper limits, is worn excessively, or if the **flange height** above the deck is not within specifications, the liner should be removed. Although the specific details of cylinder liner removal will vary with the size and type of engine being serviced, the following basic procedure can be followed when removing a liner:

1. Use a **cylinder liner puller**, **Figure 5-38,** following the tool manufacturer's instructions. Position the puller above the liner and guide the puller's shaft through the liner. To prevent damage to the block, make sure the puller shoes are properly positioned in the bottom of the liner.

⚠ **Warning: Exercise extreme caution when pulling any wet liner. Some hydraulic-powered pullers used to remove cylinder liners can exert a force up to 50 tons. If the puller device is not set up properly, it could slip and fly out with extreme force.**

2. In some cases, the puller will not be able to free the liner. When this occurs, other aids must be used. The

Figure 5-37. *Typical jacket liner has passages for coolant to circulate.*

Counterbore

Outlet to cylinder head

Water jacket

Water inlet

Sealing ring groove

Drain

Figure 5-38. *A cylinder liner puller must be used to remove most liners. Some liners require the use of an pneumatic puller. (Mack Trucks, Inc.)*

use of **inhibited acid compounds** to remove scale from the water jacket may help loosen the liner. Solvents and penetrating oil may also help. In some cases, heat may be used to free the liner. Adding steam or hot water to the water jackets and filling the liner with cold water may help loosen the liner.

3. After the liner has been removed, the block bore should be lightly cleaned to permit accurate measurements and to ensure good heat transfer after the new liner is installed.

4. Check the cylinder block bore with a bore gauge or an inside micrometer. As previously described, take readings at 180° apart (parallel and perpendicular to the crankshaft) at the top and bottom of the bore.

5. If the bore exceeds out-of-round or taper specifications or is excessively scored, rebore the cylinder block to accept the next applicable oversize liner. Liners are generally available in various sizes over standard.

Cylinder Boring

Because there are many types of cylinder boring equipment and each has slightly different operating instructions, it is important to follow the manufacturer's recommendations. **Cylinder boring** should be done only by a machinist. If cylinder boring is done improperly, it will produce a rough surface that may not hone properly. This will lead to a noisy engine (at least during break-in) and accelerated ring wear. Cylinder boring procedures are essentially the same for both integral bores and cylinder liners.

Cylinder Honing

After boring, cylinders are honed to produce a desirable finish. **Honing** is vitally important in obtaining a proper cylinder wall surface to promote ring seating. Adequate stock must be removed from the cylinders to eliminate the peaks and valleys created by the boring equipment. If allowed to remain, these peaks and valleys on the cylinder surface can retain enough oil to retard seating (wear-in) of the piston rings, resulting in poor oil control and blowby, **Figure 5-39.**

Honing also must leave the cylinder with the proper surface so that it will distribute oil, serve as an oil reservoir, and provide a place for worn metal and abrasive particles. At the same time, it must have sufficient flat areas to act as a bearing surface on which an oil film can form.

A conventional rigid hone usually consists of two stones mounted opposite each other on a holder, and is generally used for heavy honing. Flexible hones are used for light work and sometimes to finish the honing job. Either hone rotates at a select speed, usually between 300-400 rpm. When either boring or honing, be sure to keep a steady stream of lubricant on the work, **Figure 5-40.** The hone should be moved over the entire surface of the cylinder.

Honing is a progressive process. A coarse honing stone (150-220 grit) is used first to hone the liner to the total desired depth, **Figure 5-41A.** A fine stone (280-400 grit) is then used to remove the peaks and valleys, leaving

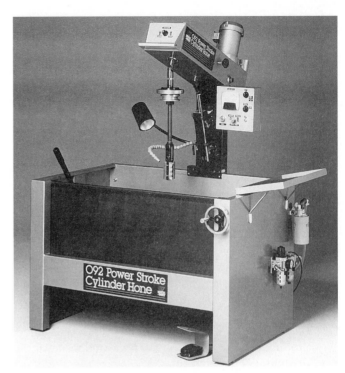

Figure 5-39. *Typical cylinder honing machine. Use care when working with any power tool or machine.* (Kwik-Way Manufacturing Company)

Figure 5-40. *When honing, use a good flow of honing oil to keep the work cool and flush away abrasive particles.*

Figure 5-41. *Plateau honing process. A—Surface scratched by a rough hone. B—Finish smoothed with a fine hone.*

flatter plateaus that cause less ring wear during break-in, **Figure 5-41B.** This removal process is often referred to as **plateau honing.** A basic honing procedure is as follows:

1. If recommended by the hone manufacturer, place oil on the stones. Do not use cutting agents with a dry hone.

2. Insert the hone in the bore or liner and adjust the stones to fit snugly in the narrowest section of the cylinder. When correctly adjusted, the hone will not shake in the cylinder, but will drag freely up and down when it is not running.

3. Start the hone and move it up-and-down in the cylinder with short 1″ (25.4 mm), overlapping strokes. On the first cut, concentrate on the high spots, which will cause an increased drag on the stones. As the high spots are removed, the drag on the hone will become lighter and smoother. Move the hone lightly to avoid too much increase in the cylinder diameter. Some stones cut rapidly, even under low pressure.

4. Clean the stones frequently with a wire brush to prevent stone loading (stone filling with debris). Do not remain in one spot too long or the cylinder will become irregular. Where and how much to hone can be judged by feel. A heavy cut in a distorted cylinder produces a steady drag on the hone and makes it difficult to feel the high spots. Therefore, use a light cut with frequent stone adjustments. Simply moving the hone from the top of the cylinder to the bottom will not correct an out-of-round condition.

5. When the cylinder is fairly clean, remove the hone, inspect the stones, and measure the cylinder diameter to determine if additional honing is necessary. If the cylinder diameter is OK, finish the cylinder by performing a plateau hone.

6. After honing, wash the cylinder with a laundry detergent solution and a stiff bristle brush to remove as much of the honing debris as possible. Rinse the cylinder with hot water and dry it with compressed air immediately.

Caution: Never use gasoline, kerosene, or commercial cleaners to wash the cylinders after honing. These solvents will not remove abrasives from the walls.

7. Coat the cylinder walls with clean engine oil.

8. Wipe the oil from the cylinder walls with a lint free cloth or paper towel. Gray or dark stains on the cloth or towel indicate that honing debris is still present in the cylinder. Repeat the oil application and wiping procedure until no evidence of debris appears on the cloth or towel. The cylinder must be thoroughly cleaned to prevent possible premature ring, cylinder wall, bearing wear, and engine failure.

Deglazing Cylinder Walls

On smaller diesel engines, it is often possible to deglaze the cylinders rather than rebore them. *Deglazing* provides a fine finish by removing scuffs or scratch marks, but does not enlarge the cylinder diameter. Therefore, standard-size pistons may still be used after deglazing.

Combustion heat, engine oil, and piston movement combine to form a thin residue on the cylinder walls that is commonly called *glaze*. In addition to removing glaze, deglazing produces a **crosshatch pattern** on the cylinder walls to provide cavities for holding oil during piston ring break-in. The crosshatch pattern is angled to prevent the rings from catching in the grooves. Various grit deglazing stones are available to produce the desired finish on the cylinder walls. If the cylinder walls are wavy, scuffed, or scratched, deglaze them as follows:

1. Remove the pistons and rods. If the crankshaft cannot be removed, cover the crankshaft with clean rags or damp paper to prevent the abrasives and dirt caused by the deglazing process from falling on the crankshaft. If a cylinder liner is used, it may be removed and placed in a fixture or scrap cylinder block if desired.

Caution: It is highly recommended that the crankshaft be removed from the block during deglazing. If this is not possible, be sure to clean all debris out of the engine.

2. Swab the cylinder walls with a clean cloth that has been dipped in clean engine oil.

3. Mount a recommended deglazing tool in a slow-speed drill (300-500 rpm). A brush-type tool with coated bristle tips often does the best job of deglazing, **Figure 5-42.**

4. Move the tool up and down in the cylinder rapidly enough to obtain the crosshatch pattern shown in **Figure 5-43.** Surface deglaze each cylinder for 10 to 12 complete strokes.

5. After deglazing, thoroughly clean the cylinders. Residue and metal fragments adhering to the cylinder walls are abrasive and will quickly damage the rings, pistons, and cylinders if not removed. After cleaning, coat the cylinder walls with clean engine oil to prevent rust.

Installing Cylinder Liners

The method of installing a cylinder liner in the block depends on the type of liner used. The following sections will

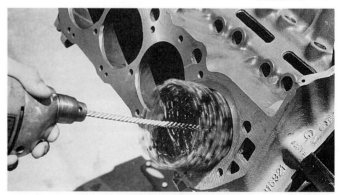

Figure 5-42. Small mobile engines can be deglazed using a bristle hone. Move the hone over the entire bore surface.

Figure 5-43. *The ideal cylinder honing pattern has a cross-hatched pattern. If your cylinders do not have this pattern, more honing is required. (Mack Trucks, Inc.)*

provide general information on installing wet and dry liners. Refer to the engine service manual for specific instructions.

Press-Fit Dry Sleeves

Before installing press-fit dry sleeves, inspect and measure the cylinder diameters, **Figure 5-44.** If the cylinders are distorted, the block must be rebored to accept oversized liners. Otherwise, the sleeves may conform to the distorted cylinders. In some cases, air pockets will form between the sleeves and the block, causing localized hot spots that often result in a breakdown of the contact surfaces. Check the cylinder sleeves's outside diameter with an outside micrometer, **Figure 5-45.**

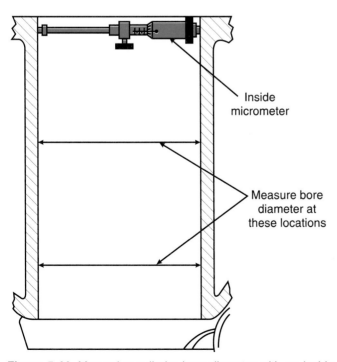

Inside micrometer

Measure bore diameter at these locations

Figure 5-44. *Measuring cylinder bore diameter with an inside micrometer.*

Figure 5-45. *Checking a cylinder sleeves with an outside micrometer.*

Before installing the sleeve, heat the block and pack the sleeve with dry ice. After removing the dry ice from the sleeve, coat it with an appropriate lubricant, **Figure 5-46,** and press or drive it into the cylinder until it touches the lip at the bottom of the cylinder. To drive the sleeve down into the cylinder, use a hydraulic driver or a hammer and a sleeve driver.

 Note: This method differs considerably from dry sleeve installation for certain two-stroke diesels.

When installed properly, the sleeve will extend slightly above the deck of the block. A ***boring bar*** with a ***face tool*** installed in the cutter head is used to machine the top of the sleeve flush with the deck. After installation, the sleeve should be bored to the desired size, just like any other cylinder. The sleeved cylinder is then chamfered and washed as described earlier.

Fitting Dry Liners

The important factors to consider when installing a dry liner with a flange include:

❑ If fitted correctly, a flanged liner should enter the cylinder block one-half to three-quarters of its length and require only a light tapping to gently set it in place against the base of the counterbore.

Figure 5-46. *The sleeve should be lubricated before installation. (Cummins Engine Co. Inc.)*

❑ A liner that is loose will not make good contact with the cylinder block and will result in poor heat dissipation. A loose fit can be spotted by an accumulation of carbon deposits on the liner's outside diameter and pronounced dark areas caused by the restricted heat. This occurs because the deposits will act as a heat dam, which restricts the heat from being dissipated into the block and coolant.

❑ A liner that is tight can cause serious damage—even more damage than one that is too loose. Block distortion, liner collapse, poor heat transfer, hot spots, and scuffing are common results of excessive tightness.

Wet Cylinder Liners

Before installing a wet cylinder liner, clean all deposits from under the liner's flange and the mating counterbore in the cylinder block. The liner must rest flatly in the counterbore to prevent distortion. If the flange surface is uneven, remachine it.

Clean the lower sealing surfaces in the block and on the liner to prevent coolant leaks when the liner is installed. Place new seals on the liner, **Figure 5-47.** Lubricate both the seals and mating surfaces in the block. Check to be sure the seals are not twisted or crimped. A roll or twist increases the density of the seal in the area of the twist. The dense area produced by the twist creates a hard spot that attracts heat. A twisted seal can also distort the liner, reduce the piston operating clearance, and promote failure.

To remove twists from seals, insert the shaft of a small screwdriver under the seal at a right angle to the seal groove and rotate the screwdriver around the liner three or four times. On O-rings, a **parting line** is usually visible in the center of the outside diameter of the seal. This line should be parallel to the groove when the twist is eliminated.

Insert the liner into the cylinder until it contacts the crevice seal, **Figures 5-48** and **5-49.** Place the palms of your hands on the upper end of the liner (180° apart) and push it downward with quick, even pressure.

To continue the installation, tap the liner near its inside diameter with a large, soft-faced hammer. Tap alternately

Figure 5-47. A rubber seal is installed on the liner. Note the parting line on the seal. (Mack Trucks, Inc.)

Figure 5-48. This technician is carefully inserting the liner into the cylinder block. If the liner binds at any point during the installation process, remove it and find out why. (Mack Trucks, Inc.)

Figure 5-49. If all of your checks and work were done correctly, the liner should be able to be partially installed in block like this. (Mack Trucks, Inc.)

from one side to the other, gradually working around the entire circumference of the liner. When the liner is 1.5" (38.1 mm) above the deck, blow compressed air into the counterbore to remove any material that may have accumulated, **Figure 5-50.** Deposits in the counterbore can cause liner distortion. When the counterbore is clean, drive the liner fully into the block, **Figure 5-51.**

After the liner is seated properly, clean the liner flange with a brass wire brush to ensure an accurate measurement of the flange height above the block deck. Check the flange height above the block deck using a sled dial arrangement, such as the one shown in **Figure 5-52.**

Figure 5-50. *Before final liner installation, the counterbore should be blown clean with compressed air. (Mack Trucks, Inc.)*

Figure 5-51. *A soft-faced hammer can be used to drive the liner fully into the block. (Mack Trucks, Inc.)*

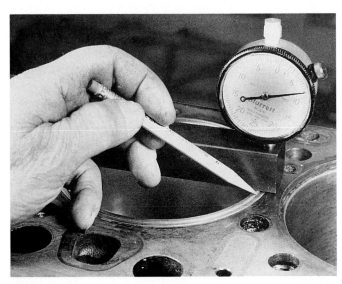

Figure 5-52. *Use a sled dial indicater to measure liner flange height. (Mack Trucks, Inc.)*

If a cylinder block deck is resurfaced, the cylinder block **counterbore depth** must be recut to specifications. If the cylinder block deck has not been resurfaced, but there is excessive pitting or erosion of the cylinder block counterbore, recut the counterbore as required.

Shims are also available to re-establish the correct flange height. Shims are generally manufactured in various thicknesses. Only use the minimum number of shims necessary to obtain the correct liner height. Check the shim's thickness with a micrometer. See **Figure 5-53.** If more than one shim is required, position the thickest shim on the bottom of the counterbore. Position the shim(s) into the cylinder liner counterbore as shown in **Figure 5-54.**

When installation is complete, carefully clean the cylinder liner as previously described. Check the liner diameter using a dial bore gauge. Take readings at 180° apart at two levels as previously described.

After installing a new wet liner, check the top of the block and the cylinder heads for flatness. Check the counterbore depths for waviness and variations. Be sure counterbores are clean and free of dirt and carbon. Also, measure the liner flange height. Follow the engine manufacturer's specifications.

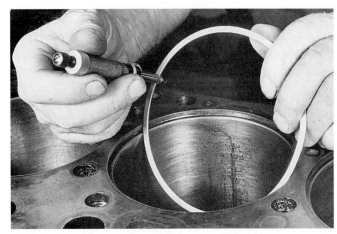

Figure 5-53. *If a shim is needed, check the thickness with a micrometer. (Mack Trucks, Inc.)*

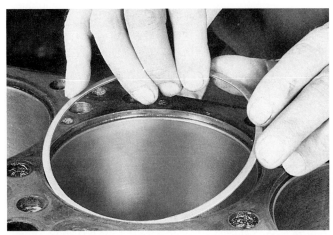

Figure 5-54. *When installing, the shim should be carefully inserted in the counterbore. (Mack Trucks, Inc.)*

Cylinder Liner Safety Features

Two safety features incorporated into many cylinder liners are the firewall and the crimping beads. Do not defeat these safety features by installing parts in engines without following the proper procedures.

Firering

A liner flange may contain a firewall (fire ring) to protect the head gasket from the direct impact of combustion. The *firering*, which is simply a ring that is located next to the liner bore on the top of the flange, greatly reduces the likelihood that the head gasket will be burned by the heat of combustion, **Figure 5-55**.

Crimping Beads

Crimping beads (coining beads) are often machined into liner flanges to promote sealing between the liner and the head gasket. Crimping beads are narrow rings located on the top side of the flange. When the head gasket and cylinder head are bolted down, the crimping beads bite into the gasket, creating a positive seal. See **Figure 5-56**.

Summary

The engine block is the largest single component of the modern diesel engine. Most engine parts are connected to the block in some manner. The block contains openings for the cylinders and bores for the camshaft and crankshaft. It generally contains passages to allow the circulation of coolant around the cylinder liners. Many blocks also have drilled oil passages, which are referred to as galleries. Most two-cycle engines have air passages in the block.

Engine disassembly procedures will vary depending on whether the engine application is stationary or mobile. For mobile applications, the engine is removed from the engine compartment using a hoist and then mounted on an engine stand for further disassembly.

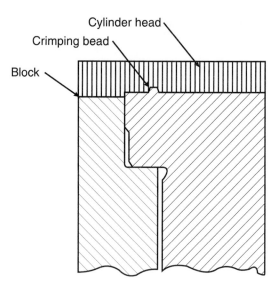

Figure 5-56. *The crimping bead is a common liner safety feature to increase sealing between the liner and head gasket.*

After engine disassembly, the cylinder block and its parts should be thoroughly cleaned and then inspected for cracks or other damage. There are two basic block cleaning procedures: chemical cleaning and thermal cleaning. Thermal cleaning ovens have become increasingly popular. The main advantage to thermal cleaning is that is reduces the total quantity of oil and grease on and in blocks, heads, and other parts.

The block can be checked for cracks by performing dye or powder tests, X-ray, pressure tests, or magnetic crack detection. The block should also be thoroughly inspected for chips, burrs, nicks, scratches, scoring, and warpage.

The cylinder in which a piston moves back and forth may be an integral part of the engine block or it may be a separate, replaceable liner. Precision measurements are needed to determine if the bore or liner exceeds recommended specifications and requires service and/or replacement.

Cylinder liners are divided into two types: dry liners and wet liners. The dry liner does not contact the engine coolant. It fits closely against the wall of the cooling jacket in the cylinder block. With a wet liner, the coolant comes in direct contact with the liner.

Cylinder liners are generally removed from the block with a special puller. If the block bore exceeds out-of-round or taper specifications, is badly worn, or is excessively scored, it can be rebored to accept an oversize cylinder liner or an oversize piston.

Cylinder boring should be done only by a machinist. If boring is done improperly, it will produce a rough cylinder surface that may not clean up when honed. The honing operation, which is performed after boring the cylinders, is vitally important in promoting proper ring and cylinder seating.

It is frequently possible to deglaze the cylinders rather than rebore them. Deglazing provides a fine cylinder wall finish by removing any scuffs or scratch marks, but does not enlarge the cylinder diameter. Therefore, standard-size pistons may still be used after deglazing.

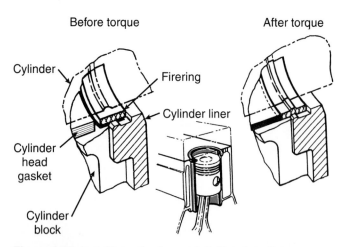

Figure 5-55. *Installing a firering gasket. Note how the compression of the liner fits the firering in the space between the liner and the block.*

The proper installation of cylinder liners is extremely important. Liners that are too loose or too tight in the cylinder block will cause operating problems, such as poor heat transfer, scuffing, block distortion, and liner collapse. Follow the manufacturer's installation instructions closely.

Important Terms

Cylinder block	Piston slap
Welded construction	Ridge
One-piece construction	Ridge reamer
Galleries	Counterbore
Crankcase	Dry liner
Cylinder assembly	Wet liner
Stationary engines	Water-jacketed liners
Access covers	Lower press fit liners
Explosion doors	Liner cavitation
Mobile engine	Flange fretting
Shutterstat	Sealing rings
Chemical cleaning	Flange height
Emulsion-type cleaning solution	Cylinder liner puller
	Inhibited acid compounds
Thermal cleaning ovens	Cylinder boring
Airless shot blasting	Honing
Joint sealing compound	Plateau honing
Deck warpage	Deglazing
Deck height	Crosshatch pattern
Rust-preventive solution	Boring bar
Integral cylinder bore	Face tool
Cylinder liner	Parting line
Air inlet ports	Counterbore depth
Sealing rings	Shims
Close-grained cast iron	Firering
Porous chromium	Crimping beads
Point welding	

Review Questions–Chapter 5

Do not write in this text. Place your answers on a separate sheet of paper.

1. The blocks of most stationary engines are made of _____ .

2. Which of the following should be done first when removing a diesel engine?
   (A) Disconnect the battery.
   (B) Drain the coolant.
   (C) Drain the oil.
   (D) Remove the shutterstat.

3. Which grade of bolts should be used when mounting an engine to a repair stand?
   (A) Grade 4.
   (B) Grade 6.
   (C) Grade 8.
   (D) Grade 5.

4. If any part of the lifting apparatus fails and the engine falls, what should you do?

5. The block cannot be cleaned thoroughly until all _____ and _____ are removed.

6. _____ cleaning solution is often used on aluminum cylinder blocks.

7. What is the main advantage to using thermal cleaning techniques on blocks and other components?

8. List four of the most common areas where cracking occurs in a cylinder block.

9. What is the purpose of the counterbored area in the cylinder that extends from the top point of ring travel to the top of the bore or liner?

10. Explain the difference in design between dry and wet cylinder liners.

ASE-Type Questions

1. When removing a diesel engine from a truck, Technician A says it is OK to disconnect an air conditioning hose, but the refrigerant should be vented to the outside of the shop. Technician B says do not loosen any air conditioning hoses until the refrigerant can be recovered. Who is right?
   (A) A only.
   (B) B only.
   (C) Both A & B.
   (D) Neither A nor B.

2. When lifting an engine out of a vehicle, Technician A says to lift the engine by looping straps around the rocker arm shafts. Technician B says to lift the engine by using a strap under the oil pan. Who is right?
   (A) A only.
   (B) B only.
   (C) Both A & B.
   (D) Neither A nor B.

3. Which of the following methods can be used to check a cylinder block for cracks?
   (A) Dyes.
   (B) Pressure checks.
   (C) Magnetic crack detection.
   (D) All of the above.

4. Which of the following is used to check the cylinder block deck for warpage?
   (A) Straightedge and feeler gauge.
   (B) Inside micrometer.
   (C) Vernier caliper.
   (D) None of the above.

5. The _____ is the distance from the crankshaft center-line to the block deck.
   (A) crankshaft throw
   (B) stroke
   (C) deck height
   (D) deck depth

6. Technician A says that maximum cylinder bore or liner wear occurs at the top of the ring travel area. Technician B says that bore or liner wear is normally uniform from the top to the bottom of the cylinder. Who is right?
   (A) A only.
   (B) B only.
   (C) Both A & B.
   (D) Neither A nor B.

7. Generally, whenever cylinder taper or out-of-roundness exceeds _____, the engine should be rebored or relined.
   (A) .0025" (.0635 mm)
   (B) .005" (.127 mm)
   (C) .010" (.254 mm)
   (D) .050" (1.27 mm)

8. Which of the following should be used to clean a cylinder after boring and honing?
   (A) Household laundry detergent.
   (B) Diesel fuel.
   (C) Kerosene.
   (D) Commercial solvent.

9. Technician A says that deglazing cylinder walls slightly enlarges the cylinder so that oversize pistons must be used. Technician B says that deglazing does not enlarge the cylinder and that standard size replacement pistons can be installed. Who is right?
   (A) A only.
   (B) B only.
   (C) Both A & B.
   (D) Neither A nor B.

10. After installing new cylinder liners, which of the following should be checked before the engine is put back in service?
    (A) Check top of block for flatness.
    (B) Check counterbore depths for waviness and variation.
    (C) Be sure counterbores are clean and free of dirt or carbon accumulations, especially in the lower outside corner.
    (D) All of the above.

Crankshafts

After studying this chapter, you will be able to:
❑ Explain the purpose of a crankshaft.
❑ Recognize common crankshaft terminology.
❑ Describe the various crankshaft classifications.
❑ Inspect a crankshaft for defects.
❑ Describe methods used to recondition a crankshaft.
❑ Describe the purpose of the flywheel and vibration damper.

As the largest and most important moving part in a diesel engine, the crankshaft converts reciprocal (up and down) motion of the pistons into rotary motion. Rotary motion is required to drive wheels, gears, pumps, shafts, etc. This chapter discusses crankshaft service after removal from the engine.

Diesel Crankshafts

As its name implies, the **crankshaft** consists of a series of cranks, or throws, that are formed as offsets on the shaft. The crankshaft is subjected to all the forces developed in an engine, consequently, the shaft must be extremely strong, **Figure 6-1.**

Crankshafts are machined from forged alloy or high-carbon steel. The shafts of most mobile diesel engines are made of steel. Forged crankshafts are nitrided or heat-treated to increase strength and minimize wear. **Built-up crankshafts** (forged in separate sections and flanged together) are used in a few large engines. The crankshafts in most modern engines are of one-piece construction.

Crankshaft Terminology

A variety of terms are used to identify the parts of a crankshaft, however, those shown in **Figure 6-2** are the ones most commonly found in manufacturer's service manuals. The main parts of the crankshaft include journals, throws, and counterweights.

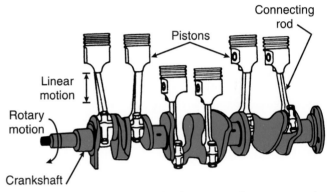

Figure 6-1. Reciprocating motion is converted to rotary motion by the crankshaft.

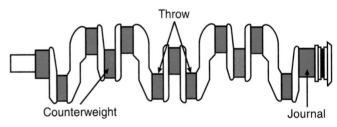

Figure 6-2. Typical diesel engine crankshaft. Note the part names. (Federal Mogul)

Journals

The **main bearing journals** serve as the crankshaft's center of rotation and primary means of support. The journals (both rod and main) are surface hardened, so that a longer-wearing, more durable bearing metal can be used without causing excessive wear of the shaft's bearing surfaces. The crankshaft has a main journal at each end. Usually, there are intermediate main journals between the throws.

Throws

Each crankshaft **throw** consists of three parts—two webs and a connecting rod journal. The **webs** are sometimes called cheeks or arms, and the **connecting rod journals** are often called crank pins. The throws provide points of attachment for the connecting links between main journals. In some crankshafts, particularly those in large engines, the connecting rod journals and main journals are hollow. Hollow journals not only reduce weight, but also provide a passage for the flow of lubricating oil.

Counterweights

On most crankshafts, part of the web extends beyond the main journal to form or to support **counterweights.** The counterweights may be integral parts of the web or may be separate units that are attached to the web by studs and nuts. Counterweights balance the off-center weight of the individual throws, keeping the centrifugal force generated by each rotating crank in equilibrium. Without such balance, severe vibrations would be created by crank action, particularly at high speeds. Since excessive vibration can cause rapid wear and lead to bearing and metal failure, the shaft will be damaged if vibrations are not controlled.

The forces in a diesel engine are produced and transmitted to the crankshaft in a pulsating manner. These pulsations create **torsional vibrations** that are capable of severely damaging an engine if they are not diminished, or "damped", by opposing forces. Counterweights utilize inertia to reduce some of this pulsation effect. However, to ensure satisfactory operation, many engines require an extra damping effect, which is usually provided by torsional vibration dampers mounted on the front end of the crankshaft.

Crankshaft Lubrication

Whether a crankshaft is solid or hollow, the journals, pins, and webs contain passages for lubricating oil. In the crankshaft lubrication system illustrated in **Figure 6-3A,** each **oil passage** runs from a main bearing journal to a connecting rod journal. The oil passages are in pairs that crisscross each other in such a way that two oil holes are on opposite sides of the journal. These holes are in axial alignment with the bearing oil grooves. Since the oil groove in a bearing extends at least halfway around the bearing, a part of the groove will always be aligned with at least one hole.

Pressurized lubricating oil enters the main bearing grooves. The oil lubricates the main bearings and flows through the drilled oil passages to the connecting rod journal bearings. From the journal bearings, the oil may be forced through a drilled passage in the connecting rod to lubricate the piston pin bearing and may then be forced onto the interior surface of the piston crown to help cool the piston.

In the oil passage arrangement shown in **Figure 6-3B,** a passage is drilled straight through each main and connecting rod journal. A single diagonal passage is drilled from the outside of a web to the center of the next main journal. This diagonal passage connects the oil passages in the two adjoining crankshaft and main journals. The outer end of the diagonal passage is plugged.

Lubricating oil under pressure enters the main bearings and is forced through the diagonal passages in the crankshaft to lubricate the connecting rod bearings. From there, it flows through the drilled connecting rods to lubricate the piston pins and cool the pistons.

Engines that use crankshafts with oil passage arrangements like those just discussed have drilled connecting rods that carry lubricating oil to the pistons and piston pins. However, drilled connecting rods are not used in all lubricating systems. In some V-type engines, drilled passages supply oil to the main and connecting rod bearings, but separate lines supply oil for lubricating and cooling the piston assembly. Additional information on engine lubricating systems variations is discussed in **Chapter 10.**

Crankshaft Classifications

The smooth and steady operation of an engine and its power production depends largely on the arrangement of

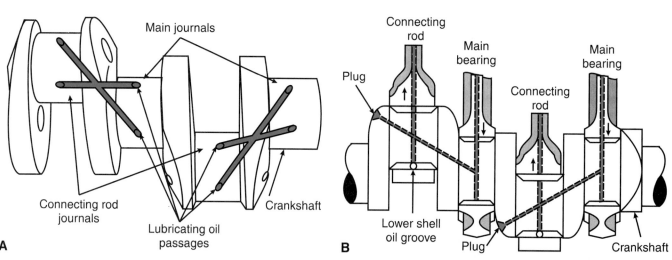

Figure 6-3. Two examples of crankshaft oil passage arrangements. A—Oil passages run from the main bearing journals to the connection rod journals. B—Passages run through the center of each journal and diagonal passages are drilled from the outside webs to the journals.

crankshaft throws and the cylinder firing order. If the crankshaft in a multicylinder engine is to obtain uniform rotation, the power impulses must be:

❑ Equally spaced with respect to the angle of crankshaft rotation.

❑ Timed so that successive combustion does not occur in adjacent cylinders. This is not always possible, especially in two- and four-cylinder engines.

Crankshafts are generally classified by the number of main bearings, which is determined by the design of the engine. See **Figure 6-4**. Since V-block engines use a shorter crankshaft, they generally have fewer main bearings than inline engines with the same number of cylinders. Crankshafts can also be classified according to the number of throws—one-throw, two-throw, etc. The number of throws is determined by the engine design. An inline six-cylinder engine will have six throws; a V-8 engine will have only four throws because each rod journal will be connected to two connecting rods, one from each side of the V.

The arrangement of the crankshaft throws depends on the cylinder arrangement (inline, V-type, flat), the number of cylinders, and the engine operating cycle. The arrangement of throws is generally expressed in degrees. In an inline engine, the number of degrees between throws indicates the number of degrees the crankshaft must rotate to

bring the pistons to TDC in the proper order. This is not true for engines in which each throw serves more than one cylinder.

The way the cylinders are numbered and the firing order affect the crankshaft's shape. The number assigned to each cylinder is also used to identify the corresponding connecting rod journal. However, there are no universal standards for cylinder and connecting rod journal numbering or engine firing order. Therefore, it is important to check the engine service manual for such details.

Crankshaft Service

To inspect and service a crankshaft, it is usually necessary to remove the unit from the engine block as described in **Chapter 5**. While a lift of some type is usually recommended, lightweight crankshafts can be lifted by hand, **Figure 6-5**. If rigging is used, however, be sure to protect machined surfaces of the crankshaft. Avoid laying down a crankshaft because it can warp out of shape or become bent.

When storing a crankshaft, make certain it has proper support along the shaft's entire length to prevent warpage. If the crankshaft is small enough, it may be possible to simply stand it on the flywheel end. Also, do not store a crankshaft or bearing part on any metal surface. When a

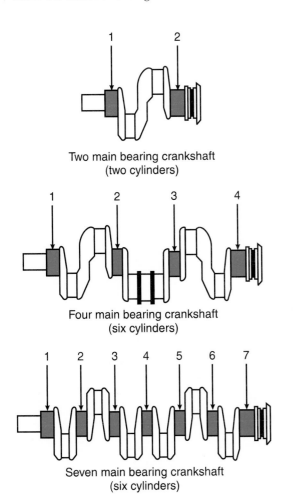

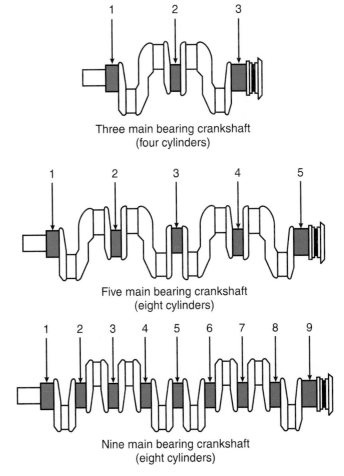

Figure 6-4. Classifications for crankshafts used in various engines. The shaded areas are the main bearing journals.

Figure 6-5. *This technician is removing a crankshaft by hand. In some cases, a hoist should be used. (Cummins Engine Co., Inc.)*

shaft is removed from an engine, it should be placed on a wooden plank with all journal surfaces protected. If the shaft is to be exposed for some time, it is best to protect each journal surface with a coating of heavy grease.

After the crankshaft's auxiliary components have been removed, the shaft cleaned, and placed on a suitable stand, use compressed air to clean out the oil passages. See **Figure 6-6.** Make a visual inspection of the crankshaft, making sure to check for the following:

- ❑ Heat checking or discoloration of the journals caused by high operating temperatures.
- ❑ Worn, ridged, or scored sealing surfaces.
- ❑ Surface cracks or signs of hardness distress.
- ❑ Eroded, fretted, or damaged vibration damper and flywheel mounting surfaces.
- ❑ Damage from previous engine failures.

If any of these conditions are present, note the area(s) of concern on the repair order and mark the components for further evaluation.

Crack Detection

Carefully check the crankshaft for cracks, which often start at an oil hole and follow the journal surface at an angle of 45° to the journal's axis. There are several methods of identifying minute cracks that are not visible to the eye.

Figure 6-6. *Compressed air should be used to clean out oil passages. (Mack Trucks, Inc.)*

Magnetic Particle Test

In a **magnetic particle test,** the crankshaft is magnetized and then covered with a fine magnetic powder or solution. Any flaws, such as cracks, will cause the magnetic particles in the powder or solution to gather, effectively marking the cracks. The crankshaft must be demagnetized after the test. See **Figure 6-7.**

Fluorescent Magnetic Particle Test

The **fluorescent magnetic particle test** is similar to the magnetic particle method, but is more sensitive since it employs magnetic particles that are fluorescent and glow under a "black light." See **Figure 6-8.** Very fine cracks that

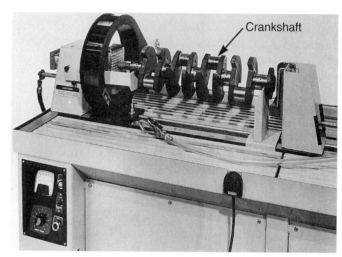

Figure 6-7. *This crankshaft is mounted in a magnetic particle testing equipment. (Magnaflux Corp.)*

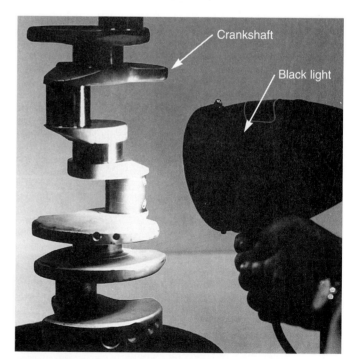

Figure 6-8. *Performing a fluorescent magnetic particle test. Do not use magnetic tests on nonferrous parts. (Atlas Engineering and Manufacturing Co.)*

may have been missed by using the first method will be disclosed under the black light, especially on discolored or dark surfaces.

Fluorescent Penetrant Test

The **fluorescent penetrant** method can be used on both ferrous and nonferrous materials, **Figure 6-9.** A highly fluorescent liquid penetrant is applied to the part. After the excess penetrant is removed from the surface, the part is dried. A developing powder, which helps to draw the penetrant out of the flaws by capillary action is applied to the part. This inspection may also be carried out under a "black light."

The majority of defects revealed by the above testing methods are normal and harmless. Only in a small percentage of cases is the reliability of the part impaired when defects are found. Since inspection reveals harmless defects with the same intensity as the harmful ones, detection is only the first step in the inspection procedure. Interpretation of the defect indications is the most important step.

Crankshaft failures are rare. When a crankshaft cracks or breaks completely, it should be thoroughly inspected for contributory factors. Unless abnormal conditions are discovered and corrected, there will likely be a repetition of the failure.

Crankshaft Loads

There are two types of loads imposed on a crankshaft when it is in service—a bending force and a twisting force. The design of the shaft is such that these forces produce practically no stress over most of the surface. Certain small areas, which are designated as critical areas, sustain most of the load.

Bending Fatigue

Crankshaft failures can result from **bending fatigue**. The crankshaft is supported between each of the cylinders by a main bearing. The load imposed by the combustion pressure on the top of the piston is divided between the adjacent bearings. An abnormal bending stress in the crankshaft, particularly in the crank fillet area, may be the result of main bearing bore misalignment, improperly fitted bearings, bearing failures, a loose or broken bearing cap, or unbalanced drive pulleys. Drive belts that are too tight will also impose a bending load on the crankshaft.

Crankshaft failures resulting from bending generally start at the pin fillet and progress throughout the web, sometimes extending into the main journal fillet. If the main bearings are replaced due to one or more badly damaged bearings, a careful inspection must be made to determine if any cracks have started on the crankshaft. These cracks are likely to occur on either side of the damaged bearing.

Torsional Failure

Crankshaft failures can be caused by torsional vibration, which takes place at high frequencies. A combination of abnormal speed and load conditions may cause the twisting forces to set up a vibration, which is referred to as a torsional vibration. This vibration imposes high stresses at the locations shown in **Figure 6-10.**

Torsional stresses may produce a fracture in either the connecting rod journal or the rear of the crank cheek. **Torsional failures** may also occur at the crankshaft timing gear key slot. Connecting rod journal failures usually occur at the fillet or oil hole and develop at a 45° angle to the axis of the shaft. A loose, damaged, or defective vibration damper; a loose flywheel; or the introduction of improper or additional pulleys or couplings are common causes of this type of failure. Engine overspeeding or overriding the

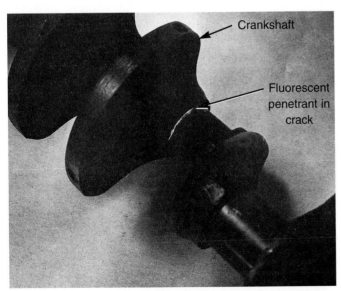

Figure 6-9. Results of a fluorescent penetrant test. Note the crack in the journal area. This test should be used on nonferrous parts, such as aluminum. (Magnaflux Corp.)

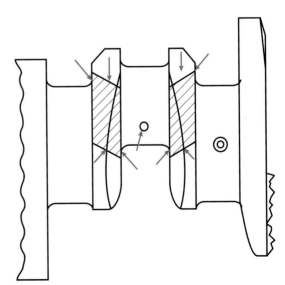

Figure 6-10. Critical crankshaft loading zones should be inspected carefully. These are the areas subject to the greatest amounts of stress and wear.

electronic control system to allow engine overspeeding may also be contributory factors.

As previously mentioned, some of the indications found during crankshaft inspection are harmless. The two types of indications to look for are circumferential fillet cracks in the critical areas and cracks 45° to the axis of the shaft that start from the fillets or the connecting rod journal holes, **Figure 6-11.**

Inspecting Journals

To measure the journals for out-of-roundness and taper, use an outside micrometer, **Figure 6-12.** Taper is measured from one side of the journals to the other. Start by measuring the diameter of each main bearing journal. To determine if the journal is tapered, measurements should be taken at each end of the journal.

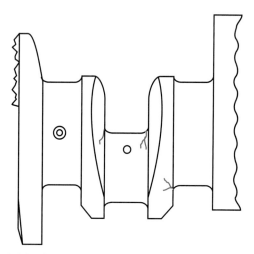

Figure 6-11. *If a crankshaft has fatigue cracks in the areas shown, it must be replaced.*

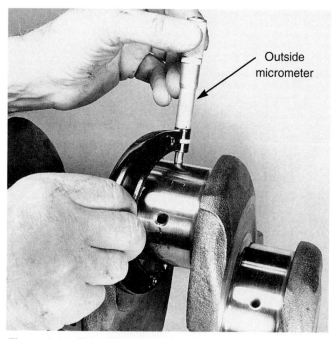

Figure 6-12. *This technician is measuring a crankshaft journal with an outside micrometer. (Mack Trucks, Inc.)*

Measurements should also be taken around the journal in several places to determine if the journal is out-of-round. If any journal measures .001″ (.254 mm) less than the manufacturer's specified diameter, has more than a .001″ (.254 mm) taper, or is more than .001″ (.254 mm) out-of-roundness, the crankshaft should be reground. These tolerances will vary from engine to engine, so consult the manufacturer's service manual for the proper specifications.

The rod journals are measured in the same manner as the main bearing journals. The same general tolerances apply and the same corrective action should be taken if necessary. The crankshaft **thrust surfaces** should be checked for roughness and scoring. Also check the thrust width with an inside micrometer. If the surfaces are damaged or the widths are not within specifications, the crankshaft should be reground.

Compare these measurements with service manual specifications and establish **wear limits** to determine the undersize to which the crankshaft can be reground. If the journals measure within specifications, but visual pits and gouges exist, polish the worst journal with **crocus cloth** (emery) to determine whether grinding is necessary. If the crankshaft does not have to be reground, use fine crocus cloth to polish the machined surfaces of the crankshaft, **Figure 6-13.**

Low ridges on a journal can be removed by working crocus cloth around it. Rotate the cloth (wet with diesel fuel) around the journal. Rotate the crankshaft frequently to prevent an out-of-round condition. If the ridges are greater than .0005″ (.012 mm), use 120-grit emery cloth to clean up the ridge, then use 240-grit emery cloth for finishing. Finally, use wet crocus cloth for polishing. If the ridges are greater than .001″ (.254 mm), the crankshaft may have to be reground.

Crankshaft Straightness

If the crankshaft does not have to be reground, it should be checked for straightness before returning it to service.

Figure 6-13. *If the journal is not worn severely, use a fine crocus cloth to polish the machined surface of the crankshaft. (Mack Trucks, Inc.)*

Check the alignment or straightness of the crankshaft with the shaft supported on its front and rear main journals in V-blocks or in a lathe. Use a dial indicator to check the alignment at the center or intermediate journals, **Figure 6-14.** To prevent scratches while turning, lay a strip of paper in the V-blocks before inserting the crankshaft.

The center or intermediate journals should not vary more than the average oil clearance allowable in the total indicator reading. Refer to the engine specifications for allowable oil clearances. To test either the front or rear main journals, move one of the V-blocks to an intermediate journal.

Ideally, a bent crankshaft should be replaced. However, if it is necessary to straighten a damaged crankshaft, is should be done very carefully. Since straightening requires deformation of the crankshaft metal, the risk of breakage is always present. Improper straightening techniques can lead to disastrous results. For this reason, crankshaft straightening should be performed by a machinist.

Crankshaft Balance

Any rotating mass must be balanced to minimize vibration. The crankshaft must be balanced with great care due to the forces acting on the crankshaft and the speed at which it revolves. The shaft is first balanced statically, then dynamically. **Static balance** is obtained when the weight is distributed equally in all directions from the center of the crankshaft while it is at rest.

Dynamic balance is determined while the crankshaft is turning. It is attained when the centrifugal forces of rotation are equal in all directions at any point on the shaft. The dynamic balancing operation requires special machinery and involves the removal of metal at heavy points or addition of metal at light points.

In addition to balancing the crankshaft itself, the engine rotating assembly must be balanced dynamically. This assembly includes the fan pulley, vibration damper, timing gears, flywheel, and clutch or converter parts. In addition, the connecting rod assemblies, including piston pins, piston, bearings, etc., are also carefully balanced to minimize vibration. Balancing crankshafts and the other rotating components is a precision operation. These com-

ponents can be balanced to a fraction of a gram with **precision balancing equipment, Figure 6-15.**

Reconditioning Crankshaft Journals

There are two methods normally used to recondition the main and rod journals:

- ❑ Grinding the journals to remove material from the surface.
- ❑ Rebuilding the journals by adding material to the surface.

Crankshaft Grinding

When grinding a crankshaft, remove only the amount of material needed to produce a good journal surface. Although journals may be ground to undersizes of .030″ (.762 mm) or more, the more accepted practice is to grind both main and connecting rod journals to .010-.020″ (.254-.51 mm) undersize. It is possible to only regrind the connecting rod journals. It is also possible to regrind just one journal on a crankshaft and leave all the others at the standard size, but this practice is generally discouraged. Crankshafts are normally ground on large machines like the one shown in **Figure 6-16.**

Crankshaft Rebuilding

Crankshaft rebuilding basically involves adding material to the surface of the journals. This is a specialty job and should not be attempted unless the equipment is available and the job is recommended by the engine manufacturer. Several methods of rebuilding are described in the following paragraphs.

Chromium Plating and Metal Spraying

In this process, pure hard chromium is electrolytically plated onto the main journal or connecting rod surfaces.

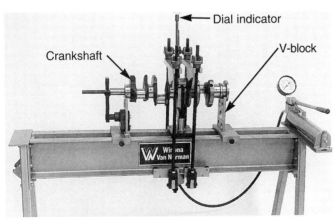

Figure 6-14. Setup for checking crankshaft for straightness. Note that the shaft is mounted in a crankshaft straightening machine. (Winona Van Norman, Inc.)

Figure 6-15. Crankshaft balance is as important as straightness. This is normally done by an electronic crankshaft balancing machine. (Stewart-Warner, Inc.)

Figure 6-16. *A worn crankshaft can be reground on a crankshaft grinding machine. (Rogers Machine Company)*

This nonporous, well-bonded coating is then ground and polished to complete the reconditioning job. With the metal spraying method, molten steel is sprayed on the bearing surfaces as the crankshaft is rotated.

Electro-Welding

In this process, a bead of alloy or high-carbon steel wire material is deposited by an electric arc welding process onto the journal surface, **Figure 6-17.** This continuous weld allows even build-up with a minimum amount of material. Sometimes it is possible to use either submerged arc or inert gas methods when rebuilding the crankshaft. Electro-welding and gas welding are the most common techniques used today.

Bearing Inspection

The crankshaft does not turn directly on the bearings. Instead, it turns on a thin film of oil trapped between the bearing surface and the journal surface. See **Figure 6-18.** Lubricating oil enters the bearings through oil inlet holes and grooves in the shells. The bearing clearance provides space for the film of lubricating oil that is forced between the journals and the bearing surfaces.

Figure 6-17. *An electro-welding machine can be used to add material to worn crankshaft journals. (Storm Vulcan, Inc.)*

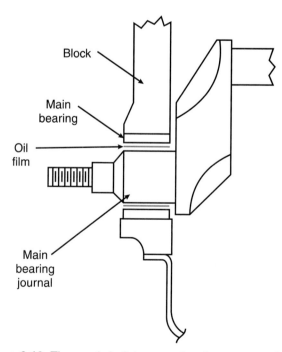

Figure 6-18. *The crankshaft turns on bearings protected by a thin film of oil.*

Under normal operating conditions, the film of oil surrounds the journals at all engine load pressures. If the crankshaft journals become out-of-round, tapered, or scored, the oil film will not form properly and the journal may contact the bearing surface. This will result in early bearing and/or crankshaft failure, **Figure 6-19.** The main and rod bearings are made of softer material than that used to form the crankshaft. By using the soft material, the bearings will wear before the crankshaft journals. Early diagnosis of bearing failure most often will spare the crankshaft and only the bearings will have to be replaced.

Bearings must have the following qualities:

❑ *Seizure resistance* (surface action)—the ability to withstand metal-to-metal contact without being destroyed. It is difficult to avoid some metal-to-metal contact during starting or when the oil film becomes thin during operation.

❑ *Corrosion resistance*—the ability of the bearing material to resist chemical corrosion.

❑ *Temperature strength*—the bearing material's ability to carry its load at high operating temperatures.

❑ *Conformability*—the ability to creep or flow slightly so that the shaft and bearing will conform to each other.

❑ *Embeddability*—the ability to let small dirt and metal particles embed themselves to avoid scratching the shaft.

Other qualities, such as wear rate, cost, thermal conductivity, and the ability to form a good bond with the backing material, must also be considered. Consideration must also be given to the fact that main bearings are subjected to a fluctuating load, as are the connecting rod and piston pin bearings. However, the manner in which main journal bearings are loaded depends on the type of engine in which they are used.

In a two-stroke cycle engine, a load is always placed on the lower half of the main and piston pin bearings in the connecting rod. A load is also placed on the upper connecting rod bearings at the crankshaft end of the rod. This is because the combustion forces are greater than the inertia forces created by the moving parts.

In a four-stroke cycle engine, inertia tends to lift the crankshaft in its bearings during the intake and exhaust strokes. In other words, the load is applied first on one bearing shell and then on the other.

Bearing Material

Most engine bearings are multilayered and use several different types of materials. A simple bearing uses two layers as shown in **Figure 6-20A.** One layer is a steel backing and the other is a liner of bearing material. This type of bearing is called a **bimetal bearing.** Some designs use a third layer, which is an overlay on the bearing material liner. The overlay is bearing material also, but is different than the material used to form the liner. This type of bearing is known as a **trimetal bearing, Figure 6-20B.**

The four most common engine bearing materials include:

❑ Babbitt

❑ Sintered copper-lead

❑ Cast copper-lead

❑ Aluminum

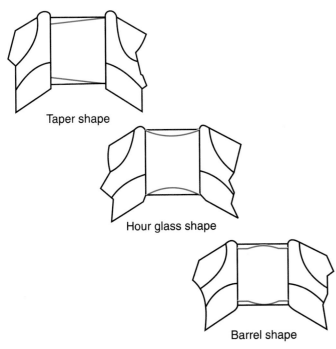

Taper shape

Hour glass shape

Barrel shape

Figure 6-19. Out-of-shape crankshaft journals will not provide an even oil clearance.

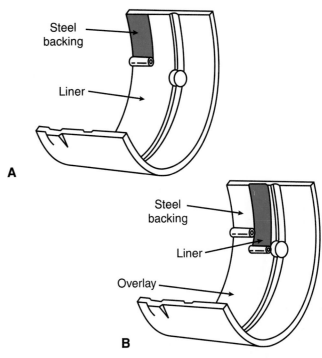

Steel backing

Liner

A

Steel backing

Liner

Overlay

B

Figure 6-20. Typical friction bearing inserts used in engines. A—Bimetal bearings have two layers. B—Trimetal bearings have three layers.

Babbitt is a tin- or lead-based material that can be divided into two categories—conventional babbitt and micro (thin) babbitt. Conventional babbitt bearings differ from micro babbitt bearings in the amount of babbitt laminated on the steel backing. *Sintered copper-lead* bearings are made by a process known as sintering (heating without melting). During this process, lead and copper metal powders are fused to a steel strip. It is available with or without an overlay. *Cast copper-lead* bearings are made by casting copper-lead alloy on a steel backing. It is available also with or without an overlay. *Aluminum* is a widely available, corrosion-resistant material. It is available in solid, bimetal, and trimetal constructions.

Types of Diesel Engine Bearings

In early engines, bearing surfaces were created by casting *babbitt alloy* directly on the block's supporting surfaces. This rough casting was then machined to form the finished bearing. Bearing replacement was a very tedious and time-consuming task.

Virtually all modern bearings are the insert type. This name is derived from the fact that the bearing is made as a self-contained part and then inserted into the bearing housing. This type of bearing provides many advantages, such as relative ease of replacement, greater variety of bearing materials, controlled lining thickness, and improved structure.

There are two types of insert bearings—precision insert bearings and resizable insert bearings. *Precision insert bearings* are manufactured to close tolerances so that no further sizing is required at time of engine assembly. This is the type of bearing most often used in high speed diesel engines. *Resizable insert bearings* are manufactured with an extra-thick lining of bearing material on the inside diameter. This extra material permits the bearing to be machined to the desired size at the time of assembly.

There are two basic designs of insert bearings—full-round bearings and split bearings. See **Figure 6-21.** *Full-round bearings* (one piece) are used when it is possible to slide the journal into place in the bearing. Camshaft bearings are usually full-round types. *Split bearings* (two halves) are used when the bearing must be assembled

around the journal. The bearing housing consists of two parts and includes a cap that holds the assembly together. All connecting rod journal bearings are of the split type.

The *flanged thrust bearing* is a variation of the common split bearing found in most diesel engines. The flanged thrust bearing provides the same support as the split bearing, but it also controls any horizontal (or axial) crankshaft movement. The connecting rod and most main bearings utilize the standard split bearing design. The flanged thrust bearing is used to limit the fore-and-aft movement of the crankshaft. Most thrust main bearings are double-flanged. However, there are some applications where separate thrust members, called *thrust washers,* are used in place of thrust bearings, particularly in heavy duty diesel engines, **Figure 6-22.** They may be supplied as full-round washers or split washers, depending on the engine's design.

Bearing Crush and Spread

Bearing inserts must fit tightly in the main or connecting rod bearing bores. To accomplish this, the distance between the outside parting edges of the bearings is wider than the diameter of the bore in which they are to be used. This condition is called *bearing spread.* See **Figure 6-23.** As the bearing caps are drawn tight, the inserts are compressed, assuring a positive contact between the bearing

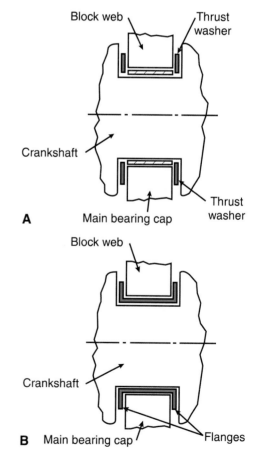

Figure 6-22. Types of thrust bearings configurations.
A—Separate thrust washers. B—Integral thrust flanges.

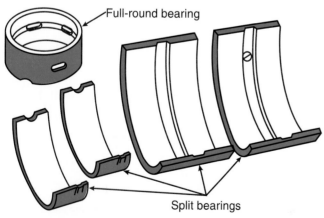

Figure 6-21. Full-round bearings and split bearings are used in diesel and other engines.

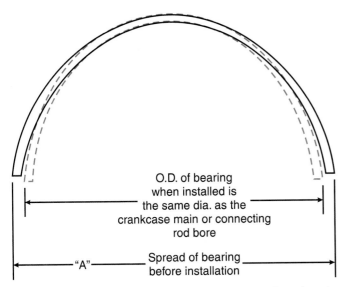

Figure 6-23. *Bearing insert spread. Spread requires that the bearing be snapped into place.*

and the bore. This is necessary to dissipate heat and to give the bearing insert firm support during engine operation.

Main and connecting rod inserts are designed so that the bearing ends protrude beyond the bore, **Figure 6-24.** This is called **bearing crush,** which helps create a tight fit between the insert and the bore. A crush of .025″ (.635 mm) is the minimum on most diesels, but specifications will vary from engine to engine. Some bearing kits will have instructions for proper installation.

Adjusting Bearing Spread

The bearing spread of most inserts can be adjusted. To decrease the spread of thick-walled bearings, such as main bearings, place one end of the bearing insert on a wood block and lightly strike the other end with a soft mallet, **Figure 6-25A.** To increase the spread of thick-walled bearings, place the bearing insert ends on a wood block and lightly strike the back of the insert with a soft mallet, **Figure 6-25B.**

Bearing spread on the thin-walled inserts, such as connecting rod bearings, can be adjusted by hand—either spreading with the thumb and forefinger of both hands to increase spread or by squeezing the insert with the palm of

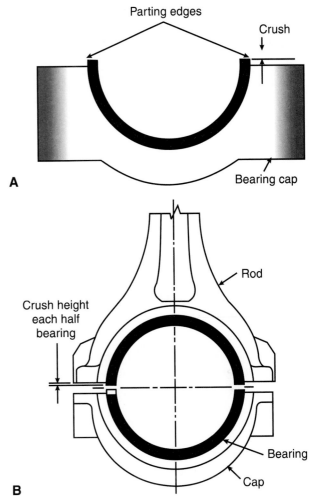

Figure 6-24. *Bearing crush helps produce a tight fit between the bearing bore and the insert.*

the hand to decrease the spread. Check the spread distance often during the adjustment procedure.

Bearing Locks

Engine bearings require some method of preventing them from rotating or shifting sideways in their housings. The most commonly used design is the *locating lug* or *tang.* As shown in **Figure 6-26A,** this design consists of a

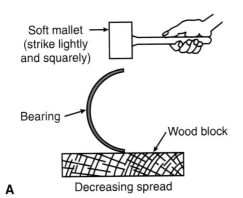

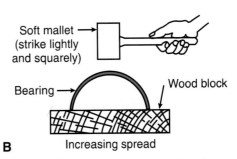

Figure 6-25. *Be careful when altering bearing spread with a hammer. A soft-faced hammer must be used in all cases. A—Tap on the end of the bearing to decrease bearing spread. B—Tapping on the bearing to increase bearing spread.*

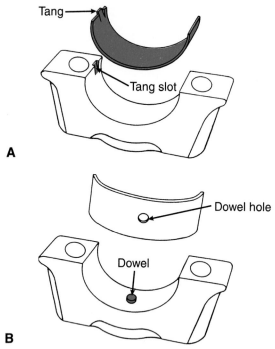

Figure 6-26. *Types of bearing locks. The bearing must snap into the bearing slot. Be sure the oil holes line up or bearing failure will result. A—Bearing with tang. B—Bearing with dowel hole.*

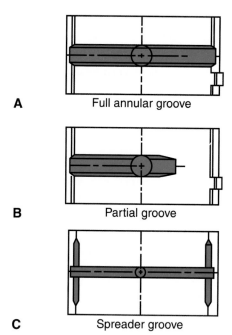

Figure 6-27. *Types of bearing oil grooves. A—Full annular groove. B—Partial groove. C—Spreader groove.*

protruding lug at the bearing parting face, which nests into a slot machined in the bearing housing.

Another design uses a *locating dowel,* **Figure 6-26B.** The dowel may be located on the housing or on the back of the bearing. The dowel is designed to fit into a hole located in the mating part. In other cases, there is a dowel hole in both parts, and the dowel is inserted as a separate piece.

Oil Grooves and Holes

As mentioned earlier, providing an adequate oil supply to all parts of the bearing surface, particularly in the load area, is vitally important. In many cases, some type of *oil groove* must be added to the bearing to ensure adequate lubrication. Some oil grooves are used to provide an adequate supply of oil to adjacent engine parts by means of oil throw-off. Although oil grooves vary in size and shape, a single groove normally works better than multiple grooves. A few typical oil grooves are illustrated in **Figure 6-27.**

Oil holes are designed to permit oil flowing through the engine block galleries to enter the bearing spaces. Oil holes are also used to meter the amount of oil supplied to other parts of the engine. The size and location of oil holes is very important. Be sure that the location and diameter of the oil holes match the original equipment specifications when replacement bearings are installed.

Oil Clearance

A specific *oil clearance* is critical to proper bearing operation. When wear on the internal engine components is minimal, the proper oil clearance specifications can be restored by installing standard size replacement bearing set. However, if the crankshaft is excessively worn, a bearing set

with a greater wall thickness must be used to compensate for the excessive wear. Although these bearings are thicker, they are known as *undersize bearings.* This is because the worn crankshaft journals are smaller in diameter. In other words, they are under the standard size.

Undersize bearings are manufactured in .001″ or .002″ (.254 or 0.51 mm) sizes for shafts that are uniformly worn by that amount. Undersize bearings are also available in thicker sizes, from .010–.030″ (.254–.762 mm), for use with refinished crankshafts.

Using undersize bearings with a refinished crankshaft will return the oil clearance to the original specifications. In cases where a standard undersize bearing is not desirable, a reborable undersize bearing is a good choice. These bearings have a .060″ (1.524 mm) undersize lining, which allows them to be bored to exact crankshaft undersize requirements.

Most bearings are designed to provide an eccentric flow of oil, **Figure 6-28.** To accomplish this, the bearing wall thickness usually becomes thinner in a taper from the crown to the parting faces. *Eccentric bearings* are produced for two reasons. First, the slight eccentricity built into the bearing wall increases the wedge effect needed to build the oil film under the loaded area. Second, eccentric wall bearings allow a reduction of vertical oil clearance without reducing the effectiveness of the oil flow through the clearance. There is no loss of the oil's cooling ability.

Bearing Distress

Bearings can be a continual source of trouble if the recommended engine operation and maintenance procedures are not closely followed. **Figure 6-29** lists the major causes of bearing distress.

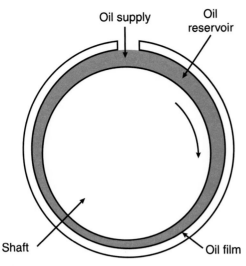

Figure 6-28. *The flow of lubricating oil around eccentric bearings provides a barrier on which the crankshaft and connecting rods move.*

Major Causes of Bearing Failure	
Dirt	43%
Lack of lubrication	15%
Poor assembly	13%
Poor alignment	10%
Overloading	9%
Corrosion	5%
Other causes	5%

Figure 6-29. *Major causes of bearing distress. Note that most of these are due to poor maintenance and abuse.*

Severe bearing failure may be evidenced during engine operation by a pounding noise or by the presence of smoke in the vicinity of the crankcase. Bearing failure may sometimes be detected by a rise in the lubricating oil temperature or, more often, by a decline in the lubricating oil pressure.

Evidence of bearing failures may be discovered during periodic maintenance checks or during engine overhauls. Inspect the bearing shells and backs for pits, grooves, scratches, or corrosion. Also look for the presence of metal particles in the engine oil. Laboratory analysis of used engine oil is an excellent method of early detection of bearing problems.

 Note: Add a thin film of clean lubricating oil to the journals and the bearing surfaces before they are installed, *Figure 6-30.*

Checking Bearing Clearance

The use of leads, shim stock, or other such devices is not recommended for determining the clearance of precision bearings. The soft bearing material may be seriously

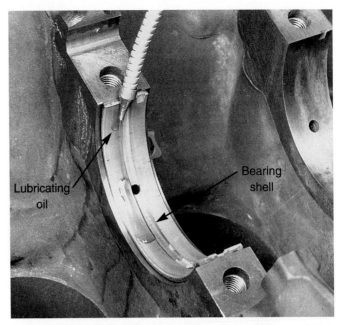

Figure 6-30. *A light coat of lubricating oil should be applied to both shell and journal surface before crankshaft installation. (Mack Trucks, Inc.)*

damaged when using these methods. To measure the thickness of bearing shells, use a micrometer that is fitted with a ball-type seat. The spherical tip must be placed against the inside of the bearing shell to get an accurate reading and to prevent damage to the bearing material. One method of determining oil clearance is as follows:

1. Torque the main bearing cap in its proper position on the block (with bearing removed). Use an inside micrometer or bore indicator to determine the diameter of the bore. This dimension is known as the "bore diameter."
2. Using a micrometer with a ball-type attachment, **Figure 6-31,** measure the thickness of the upper and lower bearing shells. Add these two dimensions together and record the sum as the "combined bearing thickness."

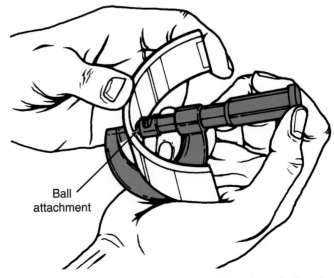

Figure 6-31. *Measuring thickness of bearing shell using micrometer with a ball attachment.*

3. Use the micrometer to measure the outside diameter of the main bearing journal that will fit into the particular bore measured above. Record this figure as the "journal diameter."

4. Subtract the combined bearing thickness from the bore diameter. Then, subtract the journal diameter from the result. This will give you a very close approximation of the bearing clearance.
Example: Bore diameter: 4.000"
Combined bearing thickness: -.024"
Result: 3.976"
Journal diameter: -3.973"
Bearing clearance: .003"

Using Plastigage

An alternate method of measuring clearance is to use *Plastigage.* Plastigage eliminates the possibility of leaving an impression in soft bearing metal because the gauge is softer than the bearing. Clearance can be measured with Plastigage as follows:

 Note: Do not turn the crankshaft during this procedure as this will destroy the Plastigage.

1. Place a strip of Plastigage of the appropriate gauge across the bearing.
2. Assemble the bearing cap and torque it in place.
3. Remove the bearing cap. The flattened Plastigage will adhere either to the bearing insert or the crankshaft.
4. Compare the width of the flattened Plastigage at its widest point with the graduations on the Plastigage envelope. See **Figure 6-32.** Match the number with the graduation to indicate the total clearance.

Approximate taper can be seen when one end of the flattened Plastigage is wider than the other. Measure each end of the flattened Plastigage. The approximate taper is the difference between the two readings. Take measurements whenever the engine is overhauled to establish the amount of bearing wear. In addition, take a number of crankshaft

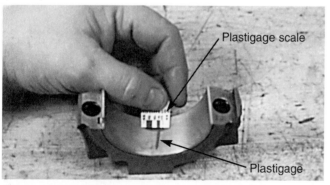

Figure 6-32. Bearing clearance can be determined with Plastigage. The Plastigage must be removed after the clearance has been checked. (Cummins Engine Co., Inc.)

journal diameter measurements at different locations to determine possible out-of-roundness.

Crankshaft Oil Seals

An *oil seal* is used at each end of the crankshaft to retain the lubricating oil in the crankcase. The sealing tips of the oil seals are held firmly, but not tight against the crankshaft sealing surfaces.

The *front oil seal* on many diesel engines is pressed into the crankshaft front cover or into the trunnion assembly. The lip of the seal bears against a removable spacer or vibration damper inner cone on the end of the crankshaft.

A *single-lip seal* is used as the rear main seal of most mobile and industrial diesel engines. A *double-lip seal* is used when there is oil on both sides of the seal. The lips of the two seals face in opposite directions. The *rear oil seal* may be pressed into the flywheel housing.

Many crankshaft oil seals can be serviced without removing the front cover, trunnion, or flywheel housing. This is done by drilling holes directly opposite each other in the seal casing and installing metal screws with flat washers. Remove the seals by prying against the flat washers with suitable pry bars.

Oil leaks often indicate worn or damaged seals. Oil seals may become worn or damaged due to improper installation, excessive main bearing clearances, excessive flywheel housing bore runout, grooved sealing surfaces and out-of-square installation. To prevent a reoccurence of oil seal leaks, these conditions must be corrected.

Inspect the rear end of the crankshaft for wear caused by the rubbing action of the oil seal, dirt buildup, or fretting by the action of the flywheel. The crankshaft surface must be clean and smooth to prevent damaging the seal lip when a new oil seal is installed. The typical installation procedures for both front and rear seals is given in **Chapter 7.** Be sure to check the engine service manual for specific instructions.

Crankshaft Gears

Inspect the *crankshaft gear* for cracks and broken or chipped teeth. If the gear is defective, it must be replaced. To replace the crankshaft gear, proceed as follows:

 Note: Remove crankshaft gear only for replacement purposes.

1. Lightly tap the end of the crank key to remove it from the crankshaft, **Figure 6-33A.** Exercise care while performing this procedure as the crankshaft must not be damaged.
2. Using a standard two-jaw puller, remove the crankshaft external counterweight and the gear.
3. Install the crank key in the crankshaft. Tap the key lightly to set it in place.

4. Heat the new crankshaft gear to approximately 250°F (121°C), using an oven or hot oil immersion. Do not use a torch to heat the gear. Align the keyway and position the large chamfer of the gear toward the rear of the engine with the timing mark facing out.

 Warning: The heated crankshaft gear can cause serious burns. Wear protective clothing and gloves when handling the hot gear.

5. Push the gear onto the crankshaft until it bottoms out. If necessary, gently drive the gear into place. See **Figure 6-33B.**

A

B

Figure 6-33. Installing a crankshaft gear. A—Installing the crank key into the shaft. B—Drive the gear into its proper position on the shaft. (Mack Trucks, Inc.).

Flywheel

The crankshaft rotation speed increases each time the shaft receives a power impulse from one of the pistons. The speed then gradually decreases until another power impulse is received. These fluctuations in speed (their number depends on the number of cylinders firing in one crankshaft revolution) will create vibrations that will have an undesirable effect on the driven mechanism as well as on the engine. Therefore, shaft rotation must be stabilized.

In some engines, this is accomplished by installing a flywheel on the crankshaft. A *flywheel* stores energy during the power stroke and releases it during the remaining events of the operating cycle. In other engines, the motion of the driven units, such as a clutch or generator, serve the purpose. Stationary diesel engines that are rigidly coupled to heavy loads may not require a flywheel. The need for a flywheel decreases as the number of cylinders firing in one crankshaft revolution and the mass of the moving parts attached to the crankshaft increases.

When the speed of the shaft increases, the flywheel absorbs energy. When the speed decreases, the flywheel releases energy in an effort to keep shaft rotation uniform. In doing this, a flywheel:

❑ Keeps variations in speed within desired limits at all loads.

❑ Limits the increase or decrease in speed during sudden changes of load.

❑ Forces the piston through the compression stroke when an engine is running at low or idle speeds.

❑ Brings the engine up to speed when it is being cranked.

❑ In most diesel engines, the flywheel provides a drive from the starting motor via the ring gear and serves as a facing for the engine clutch.

Flywheels are generally made of cast iron, cast steel, or rolled steel. The strength of the flywheel material is very important because of the stresses created in the metal of the flywheel when the engine is operating at maximum designed speed. It is attached to the crankshaft with bolts, **Figure 6-34.** The flywheel may be indexed to the crankshaft and must be installed in its proper position. A *scuff plate* may be used to keep the bolt heads from marring the flywheel surface.

In some engines, a flywheel can have attachments, such as a starting ring gear, turning ring gear, or an overspeed safety mechanism. In some cases, the flywheel rim is marked in degrees and a stationary pointer is attached to the engine. The degree markings can be used to determine the position of the crankshaft when the engine is being timed. In some applications, a *dual mass flywheel* may be used. This type of flywheel can more efficiently absorb the pulsations generated by the engine.

Flywheel Inspection

With the flywheel mounted in its normal position, check the face for runout. Secure a dial indicator to the

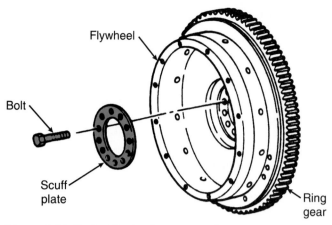

Figure 6-34. *Typical flywheel assembly. The scuff plate protects the flywheel surface from damage by the bolts. (Detroit Diesel Corp.)*

Figure 6-36. *Lock the crankshaft in place before removing the bolts and washers from the flywheel. (Cummins Engine Co., Inc.).*

engine or flywheel housing, making sure the crankshaft does not have excessive end play. If it does, accurate readings cannot be obtained. Place the dial indicator finger against the flywheel. Force the crankshaft back and forth with a pry bar. If end play does not fall within the manufacturer's accepted range, usually .004-.011" (.10-.279 mm), it must be corrected. Keep pushing the crankshaft back and forth during the runout check.

To make the runout test itself, secure the dial indicator to the engine flywheel housing with the gauge finger against the flywheel face outer edge. See **Figure 6-35.** Turn the flywheel one full revolution to obtain a reading. Total allowable indicated runout is generally about .001" per inch of flywheel radius. If runout is excessive, remove the flywheel and check for mounting problems.

The method of removing a flywheel and its housing for inspection generally depends on its weight and size. Remove the mounting bolts from the flywheel, **Figure 6-36.** In many cases, the flywheel can be lifted out by hand. In other cases, it is best to rig the flywheel to an overhead

hoist or crane, **Figure 6-37.** If using a hoist, carefully guide the flywheel away from the flywheel housing. If the housing is to be removed, remove the bolts, **Figure 6-38,** and lift it from the engine.

Once removed from the engine, inspect the flywheel's mounting surface and check the condition of the alignment dowel pins and bolt holes. Use a straightedge to make sure that the friction surface face is flat. Also check the face for burning, scoring, or heat cracks, which would render the flywheel unsafe.

Inspect the flywheel housing's machined surfaces, bolts, holes and dowel locations for cracks or wear. Replace the housing if cracks are evident. Before installing the flywheel housing, examine the crankshaft flange for cracks, surface damage, or the presence of foreign parti-

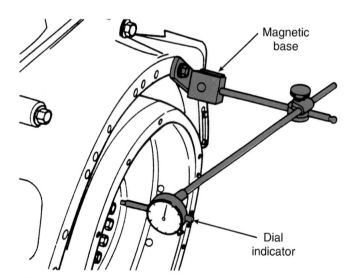

Figure 6-35. *To measure flywheel runout, attach the dial indicator to the flywheel housing. (Detroit Diesel Corp.).*

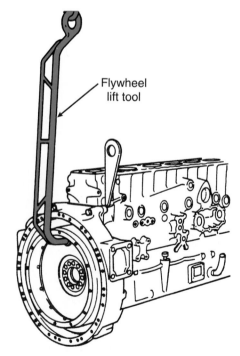

Figure 6-37. *Removing a heavy flywheel using a crane and a flywheel lifting tool. (Detroit Diesel Corp.).*

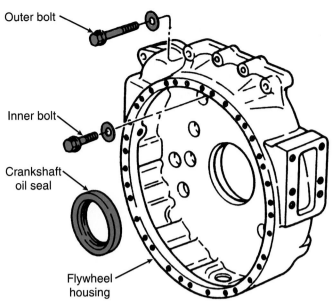

Figure 6-38. *Removing the bolts and oil seal from the flywheel housing. (Cummins Engine Co., Inc.)*

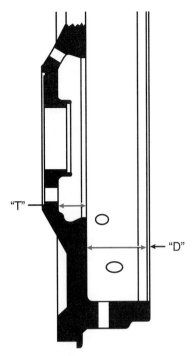

Figure 6-39. *The dimensions illustrated above are critical when resurfacing the flywheel friction surface. (Mack Trucks, Inc.).*

cles. This type of damage could ruin the new seal and lead to oil leakage.

If it is necessary to remove or install a flywheel housing, be sure to follow service manual instructions. As is the case when torquing bolts, be sure they are tightened to the manufacturer's recommended torque and in the proper sequence.

Flywheel Resurfacing

When a flywheel friction face must be resurfaced due to distortion, scoring, heat checks, etc., two critical dimensions must be observed when resurfacing or grinding.

1. Measure the distance between the flywheel friction surface and the clutch cover mounting surface (dimension D), **Figure 6-39.** This dimension must be reestablished after the friction face is resurfaced to ensure trouble-free operation.
2. Measure the distance between the flywheel friction surface and the flywheel mounting bolt surface (dimension T), **Figure 6-39.** If the distance is less than the minimum allowed, do not resurface the flywheel. The maximum amount of material that may be removed from the flywheel friction surface is usually about .070″ (1.78 mm). Check the service manual for the exact amount that can be removed. A typical flywheel grinder is illustrated in **Figure 6-40.**

The *flywheel dowels* (round and diamond-shaped) maintain the alignment of the flywheel on the engine, **Figure 6-41.** This ensures that the transmission is centered with respect to the flywheel. If the dowel holes are not in alignment and the dowels cannot be assembled, drill and ream for oversize dowels. Install the dowel pins in the holes, making the end of the pins flush with the counterbore.

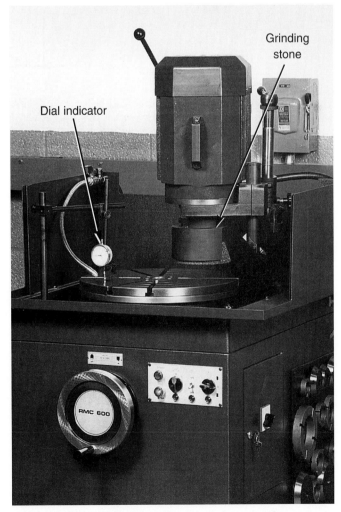

Figure 6-40. *A flywheel grinder is used to resurface a worn flywheel friction surface. (Rogers Machine Company)*

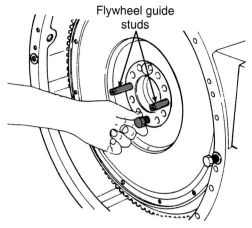

Figure 6-41. *Check alignment of locating guide studs and holes in block and housing. (Detroit Diesel Corp.).*

Starter Gear Inspection

Inspect the **starter gear teeth** for wear. The **starter ring** on most mobile diesel equipment is a replaceable gear that fits onto the outside of the flywheel. To support the flywheel while inspecting the ring gear, use wooden blocks and place the flywheel face or clutch side up. The blocks used to support the flywheel must be smaller than the inside diameter of the ring gear to permit the gear to pass over them.

If inspection shows that it is necessary to replace the starter gear, heat it to 300-350°F (148°-177°C). Be sure to distribute the heat evenly and drive the gear off the flywheel using a brass drift pin, **Figure 6-42.**

The new ring gear must be heated to ease installation. There are two methods of doing this. The preferred method is to heat the ring gear to about 260°F (127°C) in an electric oven. As an alternate method, place the ring gear on a metal plate and apply torch heat to the plate. Heat transfer from the plate to the ring gear provides even distribution. Use a **tempstick** or a piece of solder at three locations. Stop heating the plate when the stick changes color or solder melts. Avoid any localized overheating of the ring gear. *Do not* heat the new ring gear directly with the torch.

Once heated, grasp the ring gear with insulated pliers and position it on the flywheel so that the bevel on the teeth

is toward the crankshaft side of the flywheel, **Figure 6-43.** The gear tooth chamfer should be positioned away from the ring gear retaining shoulder. Tap the ring gear into place with a brass drift or blunt nose chisel and a hammer.

Once the ring gear has been replaced, the housing and flywheel can be reinstalled (assuming they are in good condition). After the flywheel housing is fastened in place, position the flywheel on the guide pins protruding from the crankshaft's rear flange. Secure the flywheel in position with new retaining bolts and washers. Coat the fastener threads with an adhesive/sealant. Install the fasteners with washers and torque them to specifications. On some installations, a special tool is required to lock the crankshaft assembly during the torquing procedure.

Vibration Dampers

A **vibration damper,** or **harmonic balancer**, is necessary to help dampen the normal torsional vibration of the engine crankshaft. It is mounted on the front of the crankshaft, **Figure 6-44.** When the cylinders fire, the pressure "twists" the crankshaft, producing torsional vibrations in the shaft. As mentioned earlier, the crankshaft throws and the flywheel are weighted and arranged to reduce these forces. However, some engines operate under extra stress, such as sudden loads or varying speeds.

The longer the crankshaft, the more difficult it is to control torsional vibration. Most dampers used to prevent

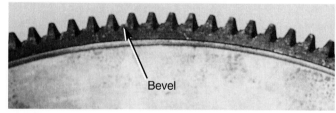

Figure 6-43. *The ring gear must be installed so the bevel on the teeth is toward the crankshaft side of the flywheel. (Cummins Engine Co., Inc.)*

Figure 6-42. *A brass drift pin should always be used to drive the ring gear from the flywheel. (Cummins Engine Co., Inc.).*

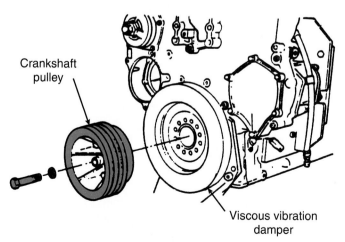

Figure 6-44. *Viscous vibration damper mounting. Use care when handling any vibration damper. (Detroit Diesel Corp.).*

torsional vibration are mounted on the front or free end of the crankshaft. While there are several types of dampers in use, the two most common types are the viscous damper and the two-piece damper.

Viscous Damper

Often called a fluid-type damper, a **viscous damper** contains a chamber of liquid or gel enclosed in a heavy ring, **Figure 6-45.** The liquid in the chamber provides resistance to the crankshaft twisting vibration that results during full combustion. Viscous dampers are easily damaged by striking or dropping them. Always handle viscous dampers carefully to avoid damaging them.

Two-Piece Damper

The **two-piece damper** resembles a miniature flywheel, **Figure 6-46.** A friction facing is mounted between the two hub faces. The two damper surfaces are mounted on the hub face with bolts that go through rubber cones in the flywheel. These cones permit the damper to move slightly on the end of the crankshaft. This reduces the effects of crankshaft torsional vibration.

 Caution: Always use the type of tool specified in the service manual to remove a two-piece vibration damper. The use of a two- or three-jaw puller will damage this type of damper.

Vibration Damper Inspection

To check a viscous damper for proper operation, operate the engine for at least 30 minutes and then shut the

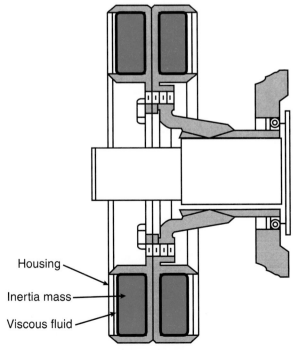

Figure 6-45. Cutaway of a viscous or liquid-type vibration damper. Note the two sections of viscous fluid.

Housing
Inertia mass
Viscous fluid

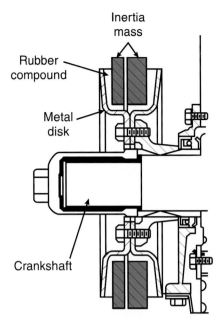

Inertia mass
Rubber compound
Metal disk
Crankshaft

Figure 6-46. Cutaway of a two-piece rubber or solid vibration damper.

engine off. The surface temperature of the damper should be greater than the engine compartment temperature or ambient air temperature. The surface of a functioning damper is approximately 180°F (82°C). When inspecting a viscous-type damper, look for a bulged damper housing.

Check the damper's axial thickness, measuring across from the front face to the rear face at four places spaced 90° apart. Before checking, paint should be removed from the damper with emery cloth at the four places of measurement. Total variation in thickness should not exceed .010″ (.254 mm). If the variation is greater than .010″ (.254 mm), the damper must be replaced. With a two-piece damper, check the friction facing and the inertia ring for wear. If worn, replace the damper.

When inspecting any vibration damper, determine the condition of the mounting holes for elongated wear. Study the vibration damper hub for scoring and check the condition of the flange, threaded holes, and keyway. Hubs showing wear must be replaced or repaired with service sleeves placed around the hub to build them up to their original thickness. Also inspect the damper housing for dents or nicks. Any type of external damage is a sign that the damper must be replaced. An engine that "throws" drive belts from the crankshaft pulley may have a failed damper. Remember that an ineffective damper can also become a source of vibrations.

While checking the damper, inspect the **fan drive pulley** for high polish, indicating belt misalignment or bent sheaves. Inspect the mounting surface and holes. When reinstalling the fan drive pulley, place it over the hub washer and align it with the mounting holes. Then position the vibration damper over the fan drive pulley and hub. Make sure all the mounting holes are in alignment, **Figure 6-47.** Then insert the bolts and washers and torque the fasteners to specifications.

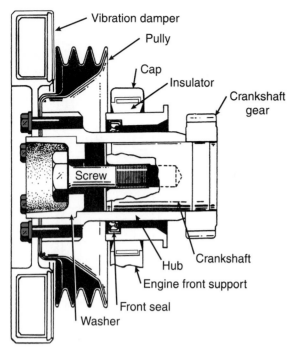

Figure 6-47. *When installing a vibration damper, make sure the mounting holes are aligned properly. (Mack Trucks, Inc.).*

Summary

The crankshaft converts the reciprocating (up and down) motion of the piston into rotary motion. This rotary motion can be used to drive wheels, gears, pumps, shafts, etc.

The crankshaft consists of a series of throws, which are formed as offsets on a shaft. The crankshaft is subjected to all the forces developed in an engine and, therefore, must be extremely strong. It is usually machined from forged alloy or high carbon steel.

The main parts of the crankshaft are the main and rod journals, the crank throws, and the counterweights. The crankshaft journals are lubricated by engine oil that flows through a series of internally drilled passages in the crankshaft.

Crankshaft service includes inspecting for journal heat checking and worn, ridged, cracked, or scored surfaces. Small cracks can be detected using magnetic or fluorescent magnetic particle testing. A micrometer should be used to check journals for taper and out-of-round conditions and to measure crankshaft thrust width.

Crankshaft journals can be reground or rebuilt to restore them to acceptable specifications. The crankshaft must also be checked for proper saddle alignment, straightness, clearance, and end play.

Bearings are used to carry the loads created by the rotating crankshaft. Main bearings support the crankshaft journals that rest in the cylinder block. Connecting rod bearings are located between the crankshaft and the connecting rods.

Most modern bearings are insert-type bearings that are inserted into the bearing housing. Advantages of insert-type bearings include ease of replacement, a greater variety of bearing materials, controlled lining thickness, and improved structure. There are two types of insert bearings: precision insert bearings and resizable insert bearings.

The crankshaft does not turn directly on the bearings. Instead, it rides on a film of oil trapped between the bearing surface and the journal surface. In a two-stroke cycle engine, a load is always placed on the lower half of the main bearings and the lower half of the piston pin bearings in the connecting rod. The load is placed on the upper half of the connecting rod bearings at the crankshaft end of the rod.

In a four-stroke cycle engine, the load is applied first on one bearing shell and then on the other. The reversal of pressure is the result of the large forces of inertia imposed during the intake and exhaust strokes.

An oil seal is used at each end of the crankshaft to retain the lubricating oil in the crankcase. The sealing tips of the oil seals are held firmly against the crankshaft sealing surfaces.

When the speed of the crankshaft increases, the flywheel absorbs energy. When the speed decreases, the flywheel releases energy to the shaft in an effort to keep the shaft rotating at a smooth, even speed. In most installations, the flywheel provides a drive from the starting motor via the ring gear and serves as a facing for the engine clutch. Flywheels are generally made of cast iron, cast steel, or rolled steel.

A vibration damper or harmonic balancer is necessary to dampen the normal torsional vibration of the crankshaft. The longer the crankshaft, the more difficult it is to control torsional vibration. Most dampers used to prevent torsional vibration are mounted on the front or free end of the crankshaft. The two most common dampers are the viscous dampers and the two-piece dampers.

Important Terms

Crankshaft
Built-up crankshafts
Main bearing journals
Throw
Webs
Connecting rod journals
Counterweights
Torsional vibrations
Oil passage
Magnetic particle test
Fluorescent magnetic particle test
Fluorescent penetrant
Bending fatigue
Torsional failures
Thrust surfaces
Wear limits
Crocus cloth

Static balance
Dynamic balance
Precision balancing equipment
Seizure resistance
Corrosion resistance
Temperature strength
Conformability
Embeddability
Bimetal bearing
Trimetal bearing
Babbitt
Sintered copper-lead
Cast copper-lead
Aluminum
Precision insert bearings
Resizable insert bearings
Full-round bearings

Split bearings

Flanged thrust bearing

Thrust washer

Bearing spread

Bearing crush

Tang

Locating dowel

Oil groove

Oil holes

Oil clearance

Undersize bearings

Eccentric bearings

Plastigage

Oil seal

Front oil seal

Single-lip seal

Double-lip seal

Rear oil seal

Crankshaft gear

Flywheel

Scuff plate

Dual mass flywheel

Flywheel dowels

Starter gear teeth

Starter ring

Tempstick

Vibration damper

Harmonic balancer

Viscous damper

Two-piece damper

Fan drive pulley

Review Questions–Chapter 6

Do not write in this text. Place your answers on a separate sheet of paper.

1. Briefly define the following crankshaft parts: journal, throw, and counterweight.

2. What are the two methods normally used to recondition the main and connecting rod journals?

3. List at least three advantages of insert type bearings.

4. Name the two types of insert bearings used in modern diesel engines.

5. What is the most common cause of bearing distress?
   (A) Dirt.
   (B) Lack of lubrication.
   (C) Misassembly.
   (D) Misalignment.

6. Many _____ can be serviced without removing the front cover, trunnion, or flywheel housing.

7. A cracked crankshaft timing gear must be _____.

8. To keep bolt heads from marring the flywheel surface, a _____ is sometimes used.

9. The _____ maintain the alignment of the flywheel housing on the engine.

10. A _____ or _____ is used to dampen the torsional vibration of the crankshaft.

ASE-Type Questions–Chapter 6

1. Technician A says that a six-cylinder inline engine will have six connecting rod journals. Technician B says that a V-8 engine will have four connecting rod journals. Who is right?
   (A) A only.
   (B) B only.
   (C) Both A & B.
   (D) Neither A nor B.

2. Technician A says that a magnetic particle test is the best method of detecting small cracks in a crankshaft. Technician B says that a fluorescent magnetic particle test is a better method for detecting very small cracks. Who is right?
   (A) A only.
   (B) B only.
   (C) Both A & B.
   (D) Neither A nor B.

3. Which of the following is a typical out-of-round specification for main and connecting rod journals?
   (A) .0001″ (.00254 mm).
   (B) .005″ (.127 mm).
   (C) .001″ (.025 mm).
   (D) .010″ (.254 mm).

4. Which type of diesel engine design is more prone to bearing wear on the lower half of the main bearings and lower half of the piston pin bearings?
   (A) Two-stroke cycle engine.
   (B) Four-stroke cycle engine.
   (C) V-block engines.
   (D) None of the above; wear is usually equal on all surfaces.

5. Babbitt, sintered copper-lead, cast copper-lead, and aluminum are all materials used in the construction of _____.
   (A) crankshafts
   (B) bearings
   (C) connecting rods
   (D) flywheels

6. What types of bearings and other components are used to withstand the fore-and-aft movement of the crankshaft in the engine block?
   (A) Straight shell bearings.
   (B) Flanged thrust bearings.
   (C) Separate thrust washers.
   (D) Both B & C.

7. Technician A says you should determine bearing clearance by using a micrometer to measure combined bearing thickness, bore diameter, and journal diameter and then performing a series of simple calculations. Technician B says you should determine bearing oil clearance using Plastigage. Who is right?
   (A) A only.
   (B) B only.
   (C) Both A & B.
   (D) Neither A nor B.

8. Which of the following jobs may be performed by the flywheel?
   (A) Absorbs and gives up energy to keep shaft rotation uniform.
   (B) Helps bring the engine up to speed when it is being cranked.
   (C) Provides a drive from the starting motor via the ring gear, and serves as a facing for the engine clutch.
   (D) All of the above.

9. Technician A says that you should use a jaw puller to remove a two-piece vibration damper. Technician B says that you should use the tools recommended by the service manual to remove a two-piece vibration damper. Who is right?
   (A) A only.
   (B) B only.
   (C) Both A & B.
   (D) Neither A nor B.

10. An engine "throws" drive belts. This indicates _____ .
    (A) belt misalignment
    (B) bent sheaves
    (C) a possible failed damper
    (D) Both A & B.

Diesel powered locomotives transport much of the goods consumed everyday. Railroads are the cheapest, and in some cases, the only means of transporting large quantities of goods.

Chapter 7

Pistons, Rings, and Connecting Rods

Like the crankshaft, the pistons and connecting rods are very important to engine operation. If these components are worn, do not fit properly, or are assembled incorrectly, engine performance will suffer and failure is likely. This chapter will discuss how the pistons, piston rings, and connecting rods should be serviced.

Diesel Pistons

The primary components of the piston assembly include the piston, the piston rings, and the connecting rod, **Figure 7-1.** It forms a movable seal that makes up the bottom of the combustion chamber and transfers the forces of combustion to the crankshaft.

The **pistons** in a diesel engine perform the following functions:

- Receives and transmits combustion forces through the connecting rod to the crankshaft.
- Draws air into the combustion chamber, compresses the air just before the power stroke, and pushes the exhaust gases out.
- Carries the piston rings that seal and wipe the cylinder.

- Supports the side thrust of the connecting rods.
- Absorbs and dissipates heat from the combustion gases.

To accomplish these tasks, the pistons must fit snugly in the cylinder, yet slide freely up and down. They must be able to withstand the force of combustion and the quick stop-and-start action at the end of each stroke. Pistons must be carefully weighed and balanced to prevent vibrations, overcome inertia, and achieve momentum at high speeds.

To meet the demand for improved emission control and to increase output, diesel engine manufacturers have improved combustion efficiency. As a result, higher engine pressures and temperatures, as well as fuel and lubricant variations, have put demands on the piston, piston rings,

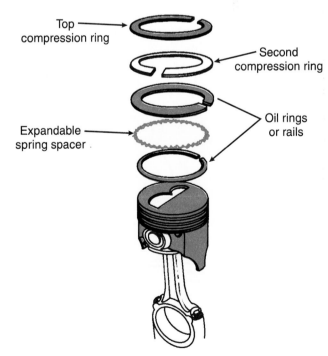

Figure 7-1. *Typical piston assembly. The number of rings and their location will vary according to engine type and piston design.*

and cylinder liner. To produce strong, lightweight pistons for industrial applications, manufacturers often use cast, forged, or malleable iron. Pistons for mobile and marine installations are often constructed of aluminum alloys. All pistons are cylindrical, hollow, and closed on the top. The bottom of the piston is open and is fastened to the connecting rod with a piston pin.

While metal-based piston materials will continue to be used, manufacturers are constantly testing new materials and fabrication techniques and have put them in use in some cases. **Squeeze-cast pistons,** which are manufactured with a ceramic fiber/aluminum alloy composite in the crown area, are now used in some diesel engines. The increased use of **surface treatments** has improved the durability of aluminum pistons. Hard anodizing resists thermal cracking and graphite coatings provide resistance to scuffing. Composite materials, such as ceramic fiber-reinforced aluminum, may also be available in the near future.

To withstand the high mechanical and thermal stresses present in some modern diesel engines, design engineers have added such features as internal oil galleries for cooling and bushing lubrication in the pin bore. Regardless of the type of material used, the piston must be strong enough to withstand the heavy force of expanding combustion gases and light enough to keep bearing loads to a minimum.

Piston Designs

The key parts of a piston are the crown, the ring lands, the ring grooves, the skirt, the piston pin bore, the pin boss, and the combustion cup. See **Figure 7-2.**

Crown

The top section of the piston is known as the **crown** or *head.* The combustion gases exert pressure on the top surface of the crown to force the piston down in the cylinder.

The crown must be thick enough to handle gas pressure without distorting or cracking. The area that is machined or cast out in the crown is called the **combustion cup**. The combustion cup shape causes the fuel to move in a turbulent pattern as it enters the combustion chamber, allowing a more thorough mixture for efficient combustion. Two common combustion cup or bowl designs are the *sombrero cup,* often referred to as a *Mexican hat cup,* and the *turbulence cup,* **Figure 7-3.** Both types are designed to promote air-fuel mixing.

Frequently, the piston crown has **valve reliefs** cast or machined into it. These reliefs provide clearance for the valves when the piston is at top dead center, **Figure 7-4,** allowing the piston to move closer to the cylinder head,

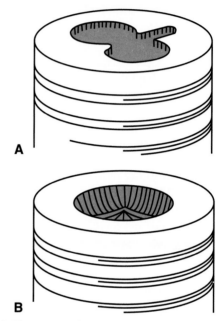

A

B

Figure 7-3. *Two types of crowns used in diesel engine pistons. A–Turbulence cup. B–Sombrero or Mexican hat cup.*

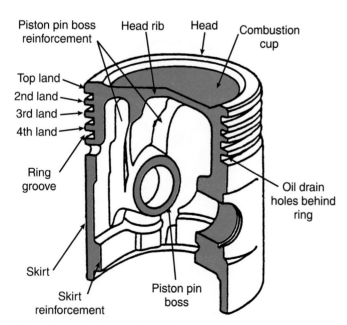

Piston pin boss reinforcement   Head rib   Head   Combustion cup

Top land
2nd land
3rd land
4th land

Ring groove

Oil drain holes behind ring

Skirt

Skirt reinforcement   Piston pin boss

Figure 7-2. *Key parts of a typical piston.*

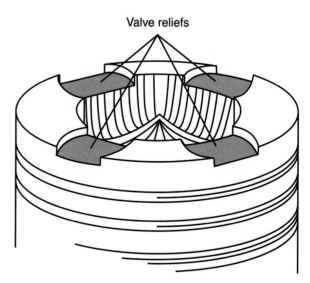

Valve reliefs

Figure 7-4. *Crown valve reliefs permit the piston to move closer to the cylinder head.*

increasing the compression ratio. A piston crown that has valve reliefs is called a **castle-top piston** since the crown area resembles the battlements of a castle. Some loop-scavenged engines have a ridge across the top of the piston to deflect incoming air and to help swirl and discharge the exhaust gases. The thickness also allows heat to flow from the upper surface of the crown to the rings. Ribs inside the piston head add strength and help carry heat from the crown to the piston rings.

Ring Grooves

Ring grooves are cut around the top portion of the piston. The piston rings are housed in these grooves. The ring grooves are shaped to the ring configuration, which controls oil and blowby. The piston contains tiny holes that are located inside and below the bottom ring groove. The oil pulled off the cylinder liner by the rings is forced through these holes and flows back into the crankcase. As the oil flows through the holes, it helps cool the piston. To help the top piston ring form a tight seal, the crown is designed to force the ring against the cylinder liner using combustion pressure.

Piston Lands

The **piston lands** are the areas between the ring grooves. It is important to remember that the piston head is closer to the heat source and, therefore, is hotter than the bottom of the piston (skirt). Since the crown expands more than the skirt, the diameter of the piston lands must be smaller than the skirt diameter. Piston lands are smallest at the top of the piston and increase in diameter as they approach the skirt. They can be either tapered or stepped from the top down.

Skirt

The **skirt** is the piston area that extends from the lowest ring groove down to the bottom of the piston. Its surface rides along the cylinder wall during engine operation. This distributes side forces evenly along the cylinder wall and keeps the piston in alignment. The piston skirt must be strong enough to resist the forces imposed on it by the connecting rod. The rotation of the connecting rod makes the piston skirt rock or slap the sides of the cylinder. Some piston skirts have a sled-runner design to help reduce slap. This design allows the piston to ride closer to the cylinder wall during operation, keeping it straight in the cylinder and ensuring that the rings maintain an adequate seal.

An oil film between the skirt and the cylinder wall normally protects the skirt from sliding friction. To ensure that the oil film is consistent, a threadlike finish is machined into the skirt. This finish carries oil in its shallow grooves to improve skirt lubrication. The oil-cooled webbing inside the piston is needed for both mechanical strength and for heat transmission. Piston skirts are also slightly tapered to allow for thermal expansion. Because the top of the skirt expands more than the bottom, the top is slightly smaller in diameter (about .0015″ or .038 mm).

Types of Pistons

The two basic types of pistons found in diesel engines are the trunk piston and the crosshead piston. Each design has unique features, which will be discussed in the following paragraphs.

Trunk Piston

The **trunk piston** is a one-piece unit commonly used in conventional diesel engines. It is treated with a protective coating that reduces scuffing, prolongs piston life, and permits a close fit between the piston and the cylinder wall. Each piston is fitted with compression and oil control rings. The location of the rings on the piston depends on the engine design and model.

With two-stroke cycle trunk pistons, the piston pin is normally secured by a **cup-shaped retainer** on each side of the piston. The main purpose of the retainer is to prevent the oil, which lubricates and cools the pin bushing and the underside of the piston from reaching the cylinder walls. On two-stroke cycle pistons, the oil control rings are located below the piston pin. Therefore, the oil must not leave the piston pin bores. Upon installation, each retainer should be checked for proper sealing with a vacuum leak detector.

The piston pin used in a four-stroke cycle trunk piston is usually held by **retaining rings** that are locked in grooves in the piston's pin bore. Oil leaving the piston pin bore is not a problem in four-cycle engines, because the oil control rings are located above the piston pins. Trunk pistons in most turbocharged engines employ a piston pin with a highly polished external finish. The pin also has an oil hole that provides oil under pressure to the pin bushings. A groove on each side of the oil hole distributes stress.

Crosshead Pistons

A **crosshead piston** is a two-piece piston consisting of a crown and a skirt. See **Figure 7-5.** This design was originally developed for large stationary engines, which required a sophisticated linkage system to reduce the wear imposed on the trunk piston by the connecting rod. Crosshead pistons are found in many diesel engines, particularly turbocharged two-cycle engines.

A metal oil seal ring is used between the piston's crown and skirt. The two parts are generally held together by the piston pin and bolted from the connecting rod, through the skirt assembly, and into a floating nut in the piston pin. The crosshead piston is designed to absorb greater pressure loads during engine operation and to distribute vertical and horizontal pressures more evenly, reducing distortion. Once the pin is in place in a crosshead piston, retainer cups are installed in the same way as in the trunk piston.

Ring grooves are machined in the crown of the crosshead piston to hold the compression rings. The piston skirt is designed to house the oil control rings, which control oil flow around the piston. It is also plated to prevent scuffing and to allow a close cylinder-to-wall fit.

The internal components of a crosshead piston are lubricated and cooled by engine oil. The oil is pressure-fed

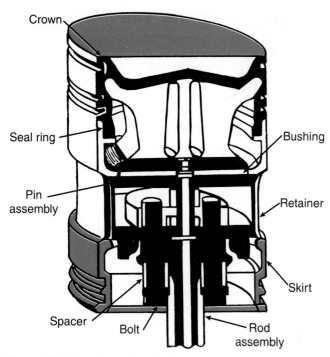

Figure 7-5. *Typical crosshead piston is a two-piece design. The seal ring is located just below the last piston ring.*

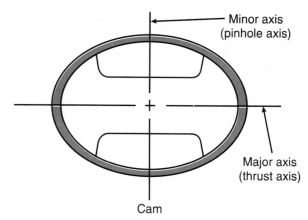

Figure 7-6. *Cam ground pistons are elliptical in shape. However, all pistons do have a major and minor diameter.*

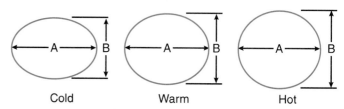

Figure 7-7. *Pistons are often elliptical in shape to compensate for thermal expansion. This illustration is exaggerated for emphasis.*

through a drilled passage in the connecting rod, through an oil tube in the piston pin, and finally through the center hole in the bushing to the underside of the piston crown. In addition, some of the oil is forced into the bushing grooves to lubricate the piston pin. The crosshead piston is designed so that combustion loads pushing down on the head are taken directly by the piston pin and bushing. Because the piston skirt is a separate component, it is free from vertical load distortion. Thermal distortion is also reduced as the piston crown expands. As the connecting rod swings to one side during the piston's downward travel, most of the side load is handled by the piston skirt.

Cam Ground Pistons

A piston has a major axis and a minor axis. The *major axis* is at right angles to the piston pin, and the *minor axis* is parallel to the piston pin. Both trunk and crosshead pistons are frequently cam ground, so that they are slightly elliptical rather than perfectly round. A *cam ground piston* is constructed so that its diameter parallel to the minor axis is less than that along the major axis. See **Figure 7-6.** This is done to compensate for the different rates of thermal expansion that occur in the piston's minor and major axis diameters, **Figure 7-7.** Since the pin bore boss causes the minor axis to be thicker than the major axis, it expands more as it heats up. Therefore, the diameter parallel to this axis is smaller to allow for greater expansion.

Piston Cleaning

After the piston assemblies are removed from the engine and disassembled, they should be cleaned with a

detergent solution or an appropriate solvent, and blown dry with compressed air, **Figure 7-8.** To remove stubborn carbon deposits, use a chemical solvent that does not harm the piston's surface coating. Never scrape the piston with a groove cleaner or a broken piston ring. Hard scraping will scratch the fine surfaces of the piston. The two best methods of cleaning pistons are chemical soaking and bead cleaning.

 Caution: Do not blast clean or wire brush tin-coated pistons. The use of an abrasive cleaning method will remove the tin coating and cause premature failure.

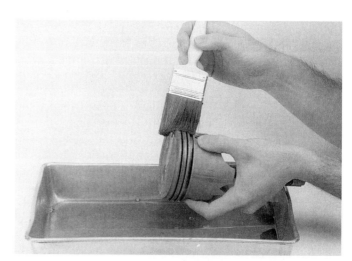

Figure 7-8. *Pistons can be cleaned with an appropriate solvent and a brush.*

Chemical Soaking

Chemical soaking in a recommended piston cleaner will soften the carbon on the piston. The carbon can then be easily removed using a pressure rinse that will minimize any chance of damage. Be sure to follow the solution manufacturer's instructions for proper cleaning.

Bead Cleaning

In **bead cleaning,** the pistons are blasted with glass beads, crushed walnut shells, or other medium using a machine such as the one shown in **Figure 7-9.** As with chemical soaking, be sure to follow the manufacturer's instructions. After the pistons have been cleaned, lightly engrave the proper bore number on the bottom of piston. See **Figure 7-10.** This will ensure that the piston is reinstalled in the proper location.

Inspecting Pistons

After cleaning, carefully inspect the piston for scoring, cracks, worn or damaged ring grooves, plugged oil drain holes, or heat damage on the piston coatings. Pistons with slight scoring can generally be reused, but a more seriously damaged piston must be discarded.

 Caution: Do not attempt to repair a cracked piston.

Closely inspect the inside of the piston for cracks across the struts. Discard the piston if cracks are found. Inspect the top of the piston head for signs of overheating, such as burned spots or carbon formations, which are normally found on the underside of the crown. These signs may indicate a lack of oil spray on the underside of the piston.

Figure 7-9. Typical glass bead blaster. (Kansas Instruments)

Figure 7-10. Marking a piston with an engraver will help ensure proper installation. (Mack Trucks, Inc.)

Checking Piston Wear

To check for **piston wear,** measure the piston diameter at a right angle to the piston pin bore. Take a reading at both the top and bottom of the skirt. Compare these measurements with the standard dimensions given in the service manual. The difference between the skirt measurements and the specifications represents piston wear.

Checking Piston Clearance

Piston clearance is the area between the piston and the cylinder walls. Proper clearance is necessary to permit piston expansion and to allow the formation of an oil film on the cylinder walls. Since most pistons wear with use, be sure to check the piston clearance during every engine overhaul. Pistons should be replaced if their clearance exceeds the manufacturer's specifications. To measure piston clearance, proceed as follows:

1. Using a cylinder dial gauge, an inside micrometer, or a telescoping gauge with an outside micrometer, measure the cylinder diameter at right angles to the crankshaft in the lower or least worn area of the cylinder.

2. Measure the piston diameter across the thrust faces with an outside micrometer, **Figure 7-11.** The difference between these two measurements is the piston clearance.

An alternate method is to remove the rings from the piston and use feeler gauges to carefully measure the clearance between the piston and the cylinder wall. When doing this, be certain not to scratch the cylinder wall.

Piston Rings

Piston rings are precision made seals designed to seal tightly against the cylinder bores while fitting snugly into

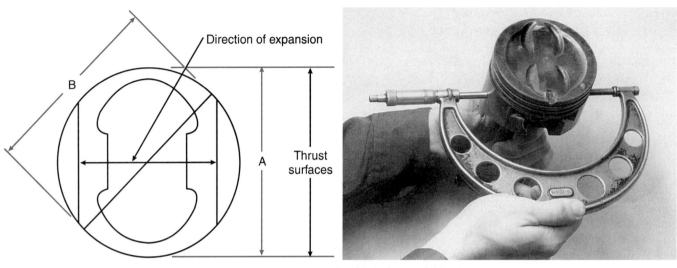

Figure 7-11. An outside micrometer is used to measure across the thrust faces of this piston.

the piston ring grooves. A properly formed piston ring will exert an even pressure and maintain contact with the cylinder wall throughout its entire circumference. The piston ring has three main tasks:

- ❏ To form a tight seal between the piston and cylinder.
- ❏ To transfer heat in order to cool the piston.
- ❏ To help maintain proper lubrication between the piston and the cylinder wall.

Although there are many different piston designs, there are only two main classifications: compression rings and oil rings.

Compression Rings

The *compression rings* are designed to prevent gases from leaking by the piston during the compression and power strokes, **Figure 7-12.** These rings form a seal on two surfaces—on their face against the cylinder wall and on their bottom surface against the ring groove. They provide

a sliding seal to keep the highly pressurized gases in the combustion chamber from leaking into the crankcase. This leakage, which is called *blowby,* can decrease engine performance and contaminate engine oil, **Figure 7-13.**

In addition to sealing combustion pressure, compression rings transfer heat from the top of the piston to the liquid-cooled cylinder walls. About one-third of all heat absorbed by the piston passes directly through the rings and into the cylinder wall. For this reason, piston rings must be made of heat-conducting materials, such as ductile iron or alloyed cast iron. In a new engine, the ring face and cylinder wall surface will wear in or seat after several hundred miles. Special chrome, molybdenum, or plasma sprays can be used on the ring face to protect against abrasion and scuffing once the ring face has seated.

The *top piston ring* (often called the *fire ring*) is a compression ring. The top ring usually has a barrel face for more positive and uniform seating against the cylinder wall. This ring may also have a keystone shape, which is very common in diesel applications. The keystone shape helps keep the ring from sticking under extreme temperatures and high pressure. The keystone-shaped grooves are designed to promote self-cleaning action and to prevent carbon build-up. In addition, the top ring groove is positioned to keep its temperature below the carbonization

Figure 7-12. Compression rings are available in many shapes and configurations. Engine design and application determines the type of ring used.

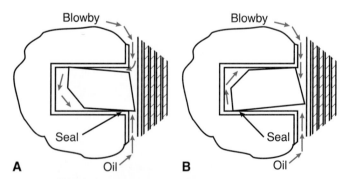

Figure 7-13. A–A positive twist ring is used as the top combustion ring because it controls blowby effectively. B–Reverse twist ring is used as a second ring to control oil flow to the top ring.

temperature of engine oil. Although raising the location of the top ring can improve engine emissions, it also increases the temperature of the ring, which can decrease its useful life.

The wear that high loads place on the top piston ring and ring groove can be reduced by using top ring groove inserts. **NI-resist inserts** made of high-nickel alloy iron are normally cast into the top and second ring grooves to extend their life. The long wearing grooves, in turn, extend the piston life. The expansion qualities of NI-resist inserts are similar to those of aluminum.

Intermediate Ring

The piston's *intermediate rings* are also compression rings. They control about 10% of the compression and combustion pressure, while helping to reduce oil consumption. In a sense, the intermediate rings act as a backup for the primary compression and oil control rings. Intermediate rings vary in design. The type of ring used depends on the type of engine.

All compression rings provide outward spring tension against the cylinder walls. During the compression, exhaust, and intake strokes, this tension prohibits leakage past the ring faces. However, during the power stroke, expanding combustion gases force the top ring down against the bottom of the piston ring groove. The resulting clearance at the top side allows compression gases to travel behind the ring, exerting additional outward pressure, forcing the ring firmly against the cylinder wall. This action helps to create a good side and face seal.

Compression rings are split to allow easy installation on the piston. The split ends do not form a perfect seal, so more than one ring must be used. Compression ring gaps must be wide enough that the ends do not touch when the ring heats up and expands during normal engine operation. The three most common ring joints are the step joint, the straight joint, and the angled joint. See **Figure 7-14.** The step joint seals best, but the straight joint or the angled joint is often used when joint leakage is not a major problem. The end gaps of each ring must be arranged so they are not aligned vertically. Aligning all ring gaps vertically will cause a compression leak through the end gaps. Follow the manufacturer's directions or space the gaps 120° apart on the piston. Typical piston ring spacing is shown in **Figure 7-15.**

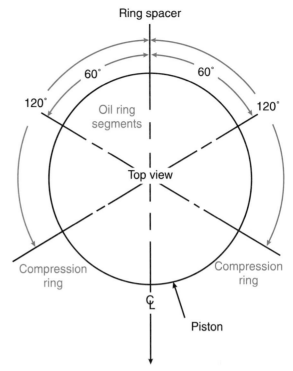

Figure 7-15. *Positioning ring gaps. This shows the location of the ring gaps with respect to the piston's centerline.*

Oil Control Rings

The piston ring found in the lowest ring groove is the **oil control ring.** This ring is designed to push most of the oil to the bottom of the cylinder through drain holes in the oil ring groove, **Figure 7-16.** Although some oil is needed on the cylinder wall to properly lubricate the compression rings, excess oil in the combustion chamber can ignite during the power stroke and cause damage. The oil control ring operates in such a way that it brings oil to the cylinder liner on the upstroke for proper lubrication and pushes excess oil to the cylinder bottom on the downstroke, **Figure 7-17.**

Oil control rings are usually made of cast iron and are chrome-plated. They have two narrow faces to obtain maximum wall pressure. Excess oil flows through ventilating slots located between the two faces. From the slots, the oil travels through holes in the oil ring groove and back into the crankcase, **Figure 7-18.** A coil or serpentine **expander** is located behind the ventilating slots, **Figure 7-18.** This expander exerts outward pressure against the ring, helping it to wipe oil from the cylinder wall. The amount of force the ring face exerts against the wall determines the amount of oil that remains. The amount of tension the spring expander/oil ring system must produce depends on the individual engine and operating conditions.

The piston cooling system in some diesel engines, including most heavy-duty types, consists of a specially designed, high-rpm **oil nozzle** assembly for each cylinder. See **Figure 7-19.** An oil pressure relief valve is usually mounted on the cylinder block. This valve controls the amount of oil entering the piston cooling passages. Cooled oil leaves the passages and is sprayed through a nozzle into the underside of the piston. The oil then drains back to the sump.

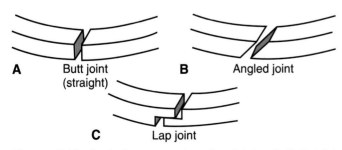

Figure 7-14. *Typical compression ring joints. A–Butt joint, B—Angled joint. C—Lap joint.*

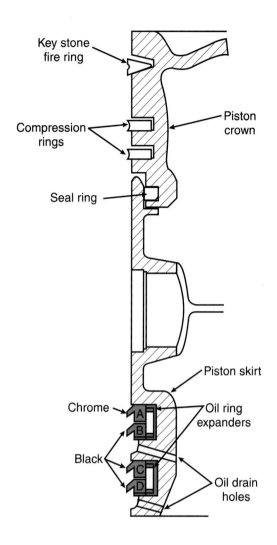

Figure 7-16. *Location of oil holes in the piston head. Note the holes in the lower ring groove. (Mack Trucks, Inc.)*

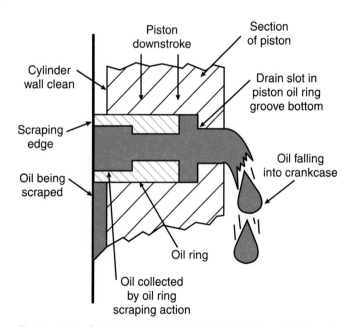

Figure 7-17. *Oil control rings scrape oil from the cylinder walls and channel it back into the crankcase.*

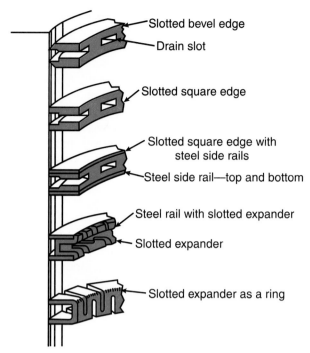

Figure 7-18. *Typical coil or serpentine expanders are located between the oil control rings.*

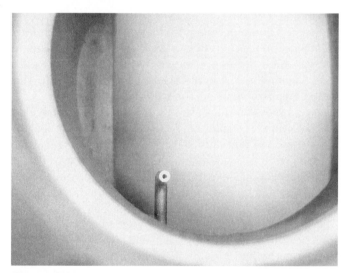

Figure 7-19. *View of spray nozzle from the top side of the engine block. (Mack Trucks, Inc.)*

The location of the oil spray on the underside of the piston is important to ensure adequate heat dissipation. It is important to make certain that the crankshaft counterweights or the connecting rods do not strike the oil nozzle. If the nozzle is bent during assembly, it will be necessary to retarget the spray. If the nozzle is badly deformed, it must be replaced.

Checking Ring Grooves for Wear

When the ring grooves show signs of excessive wear, the pistons should be replaced or the grooves reconditioned. Normally, excessive wear occurs when side clearance is more than .008″ (.203 mm). However, different limits apply to many two-cycle and large-bore diesels. Refer to the manufacturer's recommendations for each application.

Although the top piston groove normally receives the most wear, grooves can wear unevenly. Therefore, it is best to check the grooves at several locations. To check for piston groove wear, install a new ring in the groove. Then insert a feeler gauge between the upper surface of the new ring and the land to check the clearance, **Figure 7-20.** Many ring manufacturers provide special gauges for checking ring groove wear. These gauges are available for checking both rectangular and keystone grooves.

If the engine manufacturer recommends regrooving, the piston can be remachined with a ring groove cutting tool, **Figure 7-21.** After machining, heat-treated steel spacers must be installed above the ring to compensate for the metal removed by regrooving. The spacers are flat and supply a bearing surface along the full depth of the groove. This ensures a good seal and proper heat transfer between ring and groove. Top groove spacers can be easily installed without special tools.

Piston Ring Installation

Before installing piston rings, the piston should be cleaned and the ring grooves checked for carbon or dirt deposits. See **Figure 7-22.** Rings must be installed with the top side up to provide proper oil control, **Figure 7-23.** Refer to the engine service manual for instructions on installing different ring types properly. Generally, each new ring set contains an instruction sheet.

Install the compression rings on the piston and stagger the ring gaps. To avoid overstressing the rings during installation, do not spread the rings more than needed to slip them on the piston. Using a *ring expander* will make this task much easier, **Figure 7-24.**

The use of a ring expander prevents the possibility of overspreading the rings. Select the correct ring for the ring groove. Use the ring expander tool to expand the piston

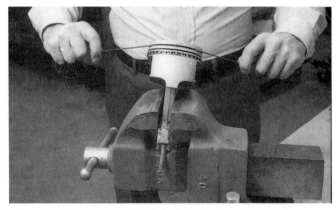

Figure 7-22. *This technician is removing carbon from ring grooves. (Goodson Shop Supplies.)*

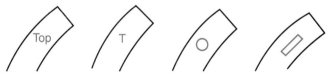

Figure 7-23. *Identification marking on rings should face the top of the piston. (Mack Trucks, Inc.)*

Figure 7-20. *A feeler gauge can be used to check ring clearance. (Cummins Engine Co., Inc.)*

Figure 7-21. *Piston ring grooves can be remachined with a special tool. (Mack Trucks, Inc.)*

Figure 7-24. *A ring expander tool can be used to aid in ring installation. (Mack Trucks, Inc.)*

ring, then, place the ring and tool over the piston. Set the ring in the groove and release the tool. Double-check the installation to make sure that the ring is in the right groove and is not upside down.

Generally the oil control rings are installed with the scraping edge down. Make sure that the ring ends do not overlap. Install the bottom ring with the gap positioned 45° from the top oil ring gap. Remember to check that the ring ends do not overlap. A common method for checking the ring gap end clearance is shown in **Figure 7-25**.

Piston Pins (Wrist Pins)

The **piston pin** or *wrist pin* is the link between the connecting rod and the piston. It can be fastened in one of three

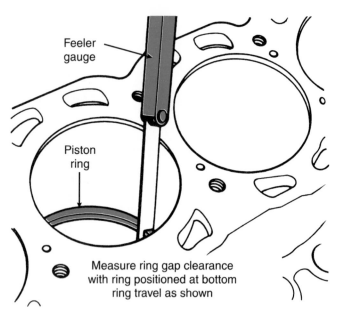

Figure 7-25. *Piston ring end gap is measured by placing the ring squarely in the cylinder bore and measuring the gap with a feeler gauge. (Mack Trucks, Inc.)*

ways. A **fixed pin** is fastened to the piston and moves in the connecting rod. A **semi-floating pin** is fixed to the connecting rod and moves in the piston. **Full-floating pins** move in both the connecting rod and piston. See **Figure 7-26**.

Most diesel engine piston retainer pins are the full-floating type, with bearings in both the piston and the connecting rod. The pin is usually secured in the piston by spring retainers that fit into grooves in the piston pin hole. Apart from this restraint, the piston pin is free to move in both the piston bearing and rod bushing. Several other methods of attaching the connecting rod to the piston using retaining pins is shown in **Figure 7-27**.

The **pin boss** is the reinforced area around the piston pin hole. High load on the top of the piston requires the use of tapered pin bosses, which are designed to fit with the tapered small ends of the connecting rods. This tapered design allows for additional bearing surfaces in the heavily loaded areas of the piston and connecting rod.

Inspecting Piston Pins

Piston pins should be checked for looseness, scoring, etching, and excessive damage or wear. To prevent piston damage, always determine the type of pin and rod assembly by referring to the engine service manual prior to disassembly. Replace any piston pins that are damaged.

Full-floating pins should be checked for clearance in both the piston and connecting rod. The fit in the piston is checked by inspecting for any noticeable rocking movement between the pin and the piston, however, the piston should move freely. The pin fit in the connecting rod is checked by inserting the pin into the rod. Try to rock the pin up and down. Any noticeable movement indicates excessive wear in the rod bushing. Wear will generally occur in the rod area, since there is less bearing surface area than in the piston. The wear is corrected by fitting the small end of the connecting rod with a new bushing and honing to size. The maximum allowable pin-to-piston clearance is normally .001″ (.254 mm).

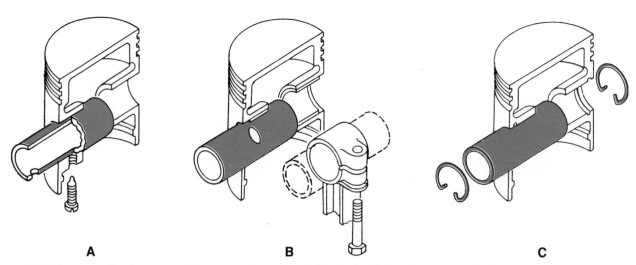

Figure 7-26. *Various methods are used to secure piston pins. A–Fixed pin. B–Semifloating pin. C–Full-floating pin. (Deere & Company)*

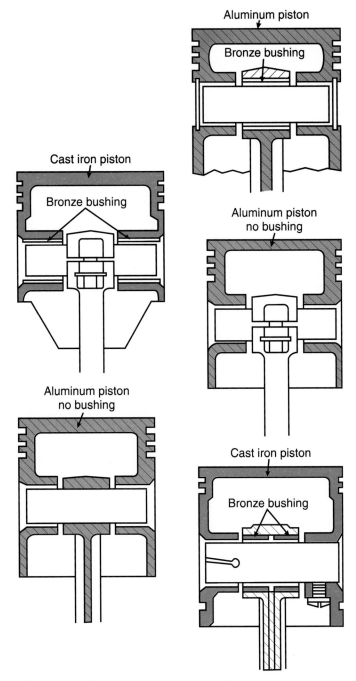

Figure 7-27. *Various methods to attach diesel pistons to the connecting rods. Some methods do not require the use of a bushing.*

The piston pin retainers on some two-cycle engines serve a second purpose, to prevent air from entering the combustion chamber from the air box. In such cases, the service manual will suggest that the technician use a leak detector gauge to check for air leaks after the retainer pin is installed, **Figure 7-28.**

Connecting Rods

The **connecting rod** connects the piston to the crankshaft. This rod must be very strong for two reasons—to transmit the force of the piston to the crankshaft and to

Figure 7-28. *Checking a piston pin retainer for air leaks by using a hand-held vacuum pump.*

withstand the directional changes of the piston itself. See **Figure 7-29.** Connecting rods normally have an I-beam design, which provides the highest strength at the lowest weight. The lower end of the rod (cap) bolts to the upper end (yoke), securing it to the crankshaft, **Figure 7-30.** There is a bore in the small end of the rod that holds the piston pin. The bore has a bushing, called a rod or *pin bushing,* in which the piston pin moves.

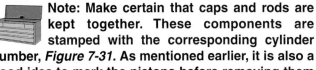 **Note: Make certain that caps and rods are kept together. These components are stamped with the corresponding cylinder number,** *Figure 7-31.* **As mentioned earlier, it is also a good idea to mark the pistons before removing them from the cylinders.**

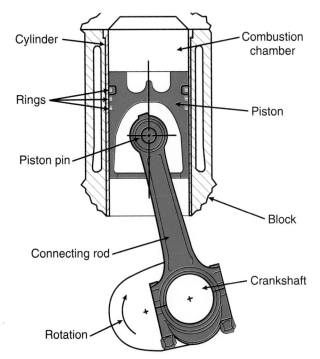

Figure 7-29. *Operation of the piston and connecting rod assembly in the cylinder block.*

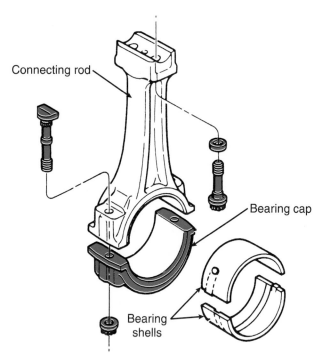

Figure 7-30. Parts of a typical crosshead connecting rod. (Detroit Diesel Corp.)

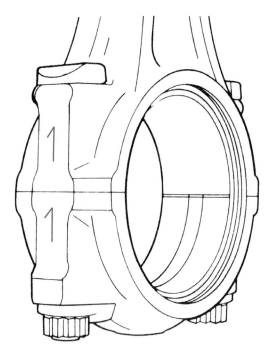

Figure 7-31. The connecting rod bearing caps are usually numbered according to the cylinder position. Matching numbers are stamped on the connecting rod yoke.

Most connecting rods are forged in one piece. Then the cap is cut off and fitted to the yoke. The parting line between the cap and the yoke is normally angled to allow clearance for rod removal through the cylinder. The upper end of the rod (yoke) is threaded in this design. Yokes and caps have a precision-machined tongue and groove to provide perfect alignment of upper and lower bearing bore halves.

As mentioned previously, many connecting rods are drilled through their entire length to provide piston pin lubrication. This oil passage can also be designed to cool the piston. In this design, the top of the rod has a spray nozzle that is directed at the underside of the piston. See **Figure 7-32.**

Due to the forces of inertia that act on the piston and connecting rod, the weight of these parts must be kept to a minimum. It is also important to keep the weight of the piston and connecting rod in balance to avoid excessive vibration. In order to control the piston weight, a ***balancing pad*** is added to the bottom of the pin bosses. The balancing pads are normally machined so all the pistons are the same weight.

Connecting Rod Inspection

Because the connecting rod must handle a great deal of stress, it must be thoroughly inspected for signs of wear. The rod should be carefully checked using the magnetic particle testing methods described in **Chapter 6. Figure 7-33** gives the magnetic particle inspection limits for connecting rods.

The large-end bore can be checked for out-of-roundness with a dial bore gauge. See **Figure 7-34.** A telescoping gauge or an inside micrometer can also be used. Another instrument used to check for out-of-round conditions is the special bore gauge shown in **Figure 7-35.** The bearing bores must also be examined for taper.

Generally, connecting rods do not bend or twist during normal operation. These conditions usually result from either poor machining or mishandling during engine reconditioning. A twisted connecting rod not only places uneven loads on the connecting rod bearings, but also places loads on the piston that can lead to scuffing. The method of checking for bends or twists is shown in **Figure 7-36.**

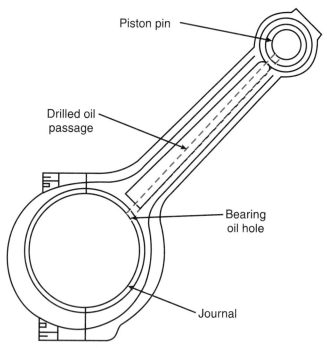

Figure 7-32. Typical connecting rod lubricating system. Oil flows from the bearings to the piston pin.

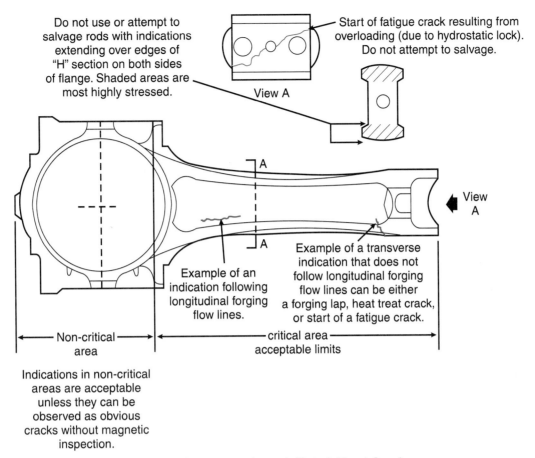

Do not use or attempt to salvage rods with indications extending over edges of "H" section on both sides of flange. Shaded areas are most highly stressed.

View A

Start of fatigue crack resulting from overloading (due to hydrostatic lock). Do not attempt to salvage.

View A

Example of an indication following longitudinal forging flow lines.

Example of a transverse indication that does not follow longitudinal forging flow lines can be either a forging lap, heat treat crack, or start of a fatigue crack.

Non-critical area

critical area acceptable limits

Indications in non-critical areas are acceptable unless they can be observed as obvious cracks without magnetic inspection.

Figure 7-33. Magnetic particle inspection limits for a connecting rod. (Detroit Diesel Corp.)

Figure 7-34. The connecting rod bore should be checked for an out-of-round condition with a bore-type gauge. (Sunnen Products Company)

The connecting rod must be properly aligned within specified limits. The crankshaft bearing bore and the piston pin bushing bore should be parallel to each other (within .001" in 6" or .254 mm in 152.4 mm). Connecting rods are normally reused after they are inspected. However, the old bushing is generally pressed out and a new one is inserted, **Figure 7-37.** Before installing a new bushing, chamfer the end of the pin to remove the sharp edge that receives the bushing. Otherwise, the sharp edge will remove stock from the bushing, loosening the pin in the bore. Check the engine service manual for the recommended method of securing the piston pin in the piston and connecting rod.

Connecting the Piston to the Rod

After all pistons and rods have been cleaned, checked, and reconditioned or replaced as required, install a piston pin retaining ring on one side of the piston. Locate the piston with the arrow or marking on the crown in the same position as the forged *Front* marking on the connecting rod. Align the connecting rod piston pin bore with the piston. Lubricate the piston pin and hand press it through the piston pin bores, **Figure 7-38A.** Sometimes, heat or mechanical pressure is needed to drive the pin into place. After the pin is fully inserted, install the remaining retaining ring, making certain it is properly seated in the groove and the sharp edge faces outward, **Figure 7-38B.**

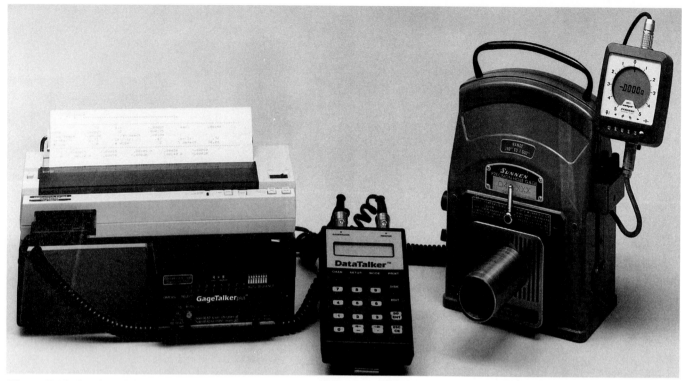

Figure 7-35. *An electronic bore and clearance gauge can be used to check for out-of-round conditions. (Sunnen Products Company)*

Figure 7-36. *Checking a connecting rod for twist and bend with a honing machine. This type of machine can also be used to bore the rod.*

Assembling the Main Bearings

As mentioned in **Chapter 6,** the crankshaft bearing shells are precision made and generally do not require machining. They consist of an upper bearing shell that is seated in each cylinder block, main bearing support, and a lower bearing shell that is seated in each main bearing cap. The upper and lower bearing shells are located in the respective block and bearing cap by a tang at the parting line on one end of each shell. The tangs are offset from the center to aid installation. Bearing shell sets are supplied as a matched assembly and should not be mixed.

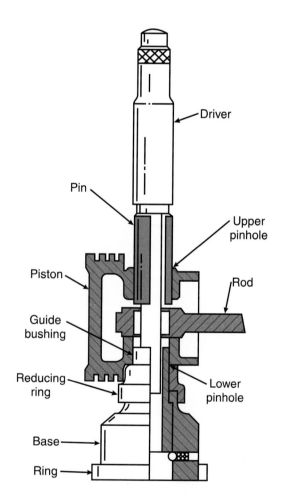

Figure 7-37. *Installing a new bushing in a connecting rod. (Mack Trucks, Inc.)*

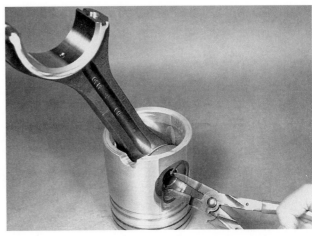

A   **B**

Figure 7-38. Steps usually needed to connect the piston to the connecting rod. A–Inserting the piston pin. B–Installing a retaining ring on the pin. (Mack Trucks, Inc.)

A hole in each upper main bearing shell aligns with a vertical oil passage in the cylinder block. Lubricating oil passes from the cylinder block oil gallery, through the bearing shells, and into the crankshaft's drilled passage. It then travels to the connecting rods and connecting rod bearings. In addition to the oil hole, the upper bearing shell is grooved. The lower main bearing shells have no oil holes or grooves. Therefore, the upper and lower main bearing shells must not be interchanged.

To assemble a main bearing, select the appropriate bearing shells and confirm their size with a micrometer. Then proceed to install the crankshaft main bearings as follows:

1. Thoroughly clean each bore and the back of each bearing. Position the block upside down and lay the upper main bearing shells in the block, **Figure 7-39.**
2. Thoroughly lubricate the upper main bearing shell faces with clean engine oil. Then use a suitable sling

or hook to lower the crankshaft carefully into position on the bearings.

3. Lubricate the thrust washers and install the upper front and rear thrust washers at the proper main bearing locations, **Figure 7-40.**
4. Clean the bore in the bearing caps and the back of the lower bearings. Insert the lower bearings in the cap bores, making sure that the location lugs fit into the notches provided. Apply a light coat of lubricating oil to the lower bearing shell surfaces and install the bearing caps to their proper position, **Figure 7-41.**

 Caution: Do not mix bearing shells or caps. Bearing caps are numbered to match the corresponding numbers stamped adjacent to their location on the cylinder block.

---

5. To aid in the installation of the main bearing caps, use a plastic faced hammer or equivalent to tap the caps in place. This will ensure proper seating on the cylinder block saddles, **Figure 7-42.** Keep in mind that

Figure 7-39. Procedure for installing a main bearing. Make sure the oil holes line up and the bearing tang fits in the block. (Mack Trucks, Inc.)

Figure 7-40. Installing the thrust washers. Make sure you place these in the correct location. (Cummins)

Figure 7-41. *Install the main caps after lubricating the bearing inserts. (Mack Trucks, Inc.)*

Figure 7-43. *Apply a light coat of oil to the main bolts. Do not use an excessive amount. (Mack Trucks, Inc.)*

tapping the main bearing caps may loosen the bearing shell. Make sure the bearing shell is in place before installing the bolts.

6. Lightly lubricate the main bearing bolt threads and lock plates with engine oil to allow good lubrication, **Figure 7-43.**

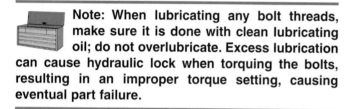

Note: When lubricating any bolt threads, make sure it is done with clean lubricating oil; do not overlubricate. Excess lubrication can cause hydraulic lock when torquing the bolts, resulting in an improper torque setting, causing eventual part failure.

7. Align the main bearing cap thrust faces. Install and tighten the bolts alternately with a short-handled ratchet wrench. Then torque all bolts to the manual's specified torque. See **Figure 7-44.**

Checking Crankshaft Clearances

To check for *crankshaft end-play,* install a dial indicator against the side of the crankshaft counterweights. Pry the crank toward the front of the engine and "zero" the indicator. Then pry the crankshaft toward the rear of the engine and read the indicator, **Figure 7-45.** The indicator reading must be within specifications. If end-play is not correct, loosen the bearing bolts and shift the crank toward the front of the block and back again. Retorque and recheck the end-play, **Figure 7-45.**

The *main bearing clearance* should be checked by using Plastigage. If checking clearance with the engine in

Figure 7-42. *In some cases, it may be necessary to tap the bearing cap with a hammer. Tap the cap lightly; do not strike it with a heavy blow as this can distort the bearing insert. (Mack Trucks, Inc.)*

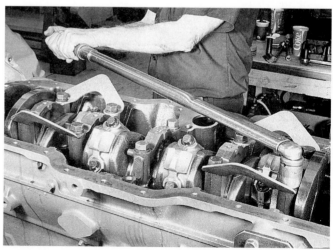

Figure 7-44. *Torque each main cap. Follow the order given in the service manual. (Mack Trucks, Inc.)*

Figure 7-45. This technician is checking the crankshaft for end-play. (Mack Trucks, Inc.)

an upright position (such as in the vehicle), the weight of the crankshaft must be removed from the lower half of the bearing being checked. This can be accomplished by jacking up the adjacent areas of the crankshaft or by shimming one or two adjacent bearings.

After shimming, snug the caps on the adjacent bearings until the journal in the main bearing being checked seats against the upper bearing shell. Do not fully torque the screws on the adjacent caps. The bolts on the bearings being checked should be torqued to specifications after the Plastigage strip has been positioned on the bearing shell. See **Figure 7-46.** Support the crankshaft at bearings 3 and 5 for checking 1, 4, and 7 bearings.

After the proper main bearing clearance is obtained, remove the Plastigage, lubricate the bearings, and reinstall the caps. Lubricate the bolts (including buttress bolt if used) and torque all fasteners to the manufacturer's specifications.

 Note: The crankshaft must rotate freely when installation is completed.

Connecting the Piston Assembly to the Crankshaft

The piston assembly is installed in the cylinder block as follows:

1. Make sure that the cylinder liner is clean. Then, apply a coat of clean lubricating oil to the piston rings, the inside diameter (ID) of the cylinder liner, and the ID of the ring compressor tool.

2. Slip a *ring compressor* over the piston rings as shown in **Figure 7-47.**

 Caution: If using a strap-type ring compressor, such as the type shown in *Figure 7-48*, make sure the inside end of the strap does not hook on a ring gap and break the ring.

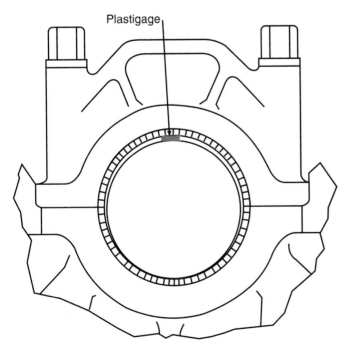

Plastigage

Figure 7-46. Placement of Plastigage to check for main bearing clearance.

Figure 7-47. Ring compressors come in many sizes. When installing the compressor, make sure the rings are properly positioned. This step is critical, so do not hurry. (Mack Trucks, Inc.)

Figure 7-48. *Strap-type ring compressors are used by many technicians. Take care when installing this type of compressor.*

3. Turn the engine to the vertical or side position and rotate the crankshaft so any two crank throws are at bottom center, **Figure 7-49.** On some two-cycle engines, the piston and rod assembly are loaded into the liner *before* installation. The liner and piston assembly is then installed as a unit into the block.

4. Position the rod so the numbered side is toward the camshaft side of the engine. Check the service manual for exact positions. Push the piston through the ring compressor until all the rings are inside the liner, **Figure 7-50.** Use a firm, steady pressure. If binding occurs, remove the assembly and make the necessary corrections. Binding generally indicates an incorrectly aligned ring. Never force a piston into a liner. Finally, working from the bottom of the block, pull the rod to within 1″ (25.4 mm) of the rod journal.

Caution: Use extreme care when installing the piston and rod assembly to avoid bending the piston cooling tube (if so equipped) or damaging the connecting rod bearing and/or crank journal.

Figure 7-49. *Turn the crankshaft to align the rod journals for piston installation. The crankshaft should turn with no binding. (Cummins)*

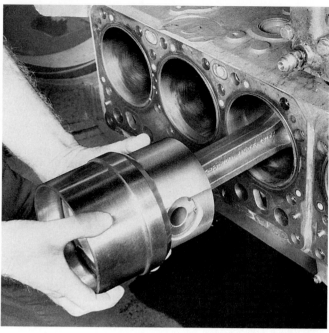

Figure 7-50. *Carefully install the piston in the engine. It should slide in with little force needed. If the piston binds in the cylinder, remove it. (Mack Trucks, Inc.)*

5. Coat the bearing shell face with clean engine oil. Do not lubricate the steel back or ends of the shell. Install the bearing shell into the rod bore with the locking tang seated properly and the oil hole in the shell indexed with the oil gallery in the connecting rod, **Figure 7-51.** Pull the rod down against the journal carefully so that the bearing shell is not dislodged or mislocated. See **Figure 7-52.** Inspect the bearing location and realign the bearing if needed.

6. Coat the lower bearing shell face with clean engine oil and seat it in the rod cap, **Figure 7-53.** Make sure that the stamped numbers on the rod and cap are identical and that they are facing the correct side of the engine.

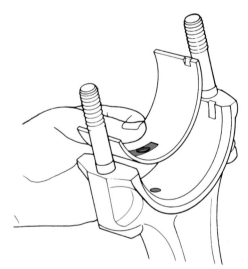

Figure 7-51. *When installing connecting rod bearings, make sure the tang and oil hole match. Many of the same rules for main bearing installation also apply to rod bearings.*

Figure 7-52. *Connecting the piston assembly to the crankshaft. Work carefully to avoid scratching the crankshaft journals. (Cummins)*

Figure 7-54. *Torque each rod cap, then rotate the engine by hand. It should turn with no binding. (Cummins)*

Figure 7-53. *Installing connecting rod caps. Make sure the bearings are properly lubricated before installation. (Cummins)*

Figure 7-55. *Checking bearing side clearance with a feeler gauge. (Mack Trucks, Inc.)*

7. Lubricate the bolt threads with clean engine oil and insert the bolts into the rod. Push the cap onto the dowels and then tighten the bolts alternately and evenly. Finally, torque the bolts to the proper specifications, **Figure 7-54.** Never drive a cap onto the dowels since, in most cases, this will dislocate the bearings.

8. Using only hand pressure, make sure that the rod is free to move sideways on the crank journal. If it is not free, remove the cap and check for improper bearing size, dirt, etc. Then, check side clearance with a feeler gauge, **Figure 7-55,** or a dial indicator and compare the clearance to specifications.

Diesel engine bearings will not function properly if they are not installed correctly. In many cases, incorrect assembly will result in the premature failure of a bearing. **Figure 7-56** shows typical assembly errors that often occur during engine bearing installation. After securing and checking all piston and rod assemblies, turn the engine to the horizontal position. Make sure the freeze plugs and oil gallery pipe plugs are installed in the block.

Always replace both front and rear *oil seals,* **Figures 7-57** and **7-58.** They keep oil from leaking at the crankshaft.

Since there are a variety of types and designs, it is best to purchase a manufacturer's seal kit, which is made specifically for the engine being worked on. Carefully follow the instructions that come with the kit. Regardless of the kit used, the seal lip and the sealing surface on the crankshaft must be clean and free from all residue to prevent leaks.

Summary

The pistons receive the combustion forces and transmit them through the connecting rods to the crankshaft. Piston

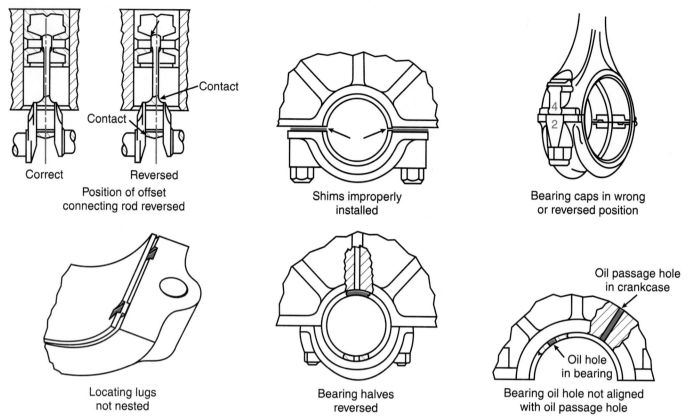

Figure 7-56. *These are common mistakes made by technicians during bearing installation. In most cases, mistakes are made through carelessness or working too quickly.*

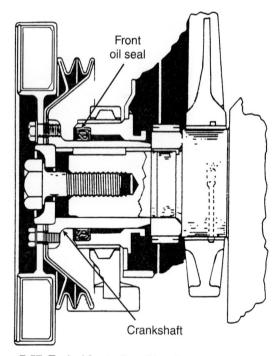

Figure 7-57. *Typical front oil seal location.*

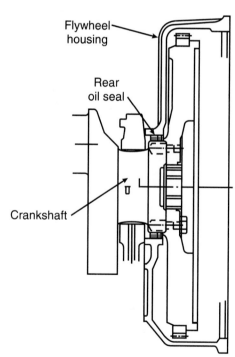

Figure 7-58. *Typical rear oil seal location.*

movement also draws air into the combustion chamber, pushes exhaust gases out, and compresses the air prior to the power stroke.

The piston carries the piston rings, which seal combustion pressure and wipe oil from the cylinder walls. The

pistons absorb heat from the combustion gases and dissipate this heat through the piston rings to the cylinder walls and the oil.

The key parts of a piston are the crown or head, the lands, the grooves, the skirt, the pin bore, the pin boss, and the combustion cup. There are two basic types of pistons found in diesel engines: trunk pistons and crosshead pistons. Trunk pistons are one-piece units commonly found in conventional diesel engines. Crosshead pistons are two-piece pistons, consisting of a crown and a skirt. Crosshead pistons are found in many diesel engines, particularly turbocharged two-cycle engines.

Both trunk and crosshead pistons are frequently cam ground so that they are slightly elliptical. A cam ground piston is constructed so that its diameter along the minor axis is less than that of the major axis. This is done to compensate for the different rates of thermal expansion that occur in the piston's minor and major axis diameters.

After the pistons are removed from the engine and disassembled, they should be cleaned with a detergent or solvent and blown dry with compressed air. To remove stubborn carbon deposits, use a chemical solvent that does not harm the surface coating of the piston or perform bead cleaning with the proper equipment.

After cleaning, carefully inspect the piston for scoring, cracks, worn or damaged ring grooves, plugged oil drain holes, or heat damage on the piston coatings. Pistons with only slight scoring can be reused. To check for piston wear, measure the diameter of the piston at right angles to the piston pin bore. Take a reading at both the top and bottom of the skirt.

Piston rings are precision-made seals, designed to seal tightly against the side of the piston grooves while fitting into the cylinder bores. The piston ring forms a gas-tight seal between the piston and cylinder, transfers heat in order to cool the piston, and maintains proper lubrication between the piston and the cylinder wall.

There are two main classifications of piston rings: compression rings and oil control rings. Top and middle piston rings are compression rings, while the bottom ring is an oil control ring design. Ring grooves in the piston must be carefully cleaned prior to installing new piston rings. Special gauges are used to check the grooves for excessive wear. In some cases, worn grooves can be recut or reconditioned, but often the piston must be replaced. Piston rings are installed using a ring expander tool.

The piston pin (wrist pin) is the link between the connecting rod and the piston. Most diesel piston pins are the full-floating type, with bearings in both the piston and the connecting rod. The piston pin is usually fastened in the piston by retaining clips that fit into grooves in the piston pin hole.

The connecting rod connects the piston to the crankshaft. Connecting rods normally have an I-beam design, which gives great strength with low weight. The lower end of the rod (cap) bolts to the upper end of the rod (yoke) to secure the rod assembly to the crankshaft. There is a bore in the small end of the rod that holds the piston pin or pin bearing.

The underside of the piston is cooled with oil that is supplied through passages in the connecting rods. Some piston cooling systems consist of a specially designed, high-rpm oil pump, oil passages in the cylinder block, and an oil spray nozzle assembly for each cylinder. The nozzles direct cooling oil onto the underside of the pistons.

Because the connecting rods handle a great deal of stress, they must be carefully inspected for signs of wear. The bearing bores must also be examined for taper and out-of-round conditions. Connecting rods are normally reused after they are inspected, but the old bushings are pressed out and new bushings are installed. Replace all oil seals.

Important Terms

Pistons	Piston clearance
Squeeze-cast pistons	Piston rings
Surface treatments	Compression rings
Crown	Blowby
Combustion cup	NI-resist inserts
Valve reliefs	Oil control ring
Castle-top piston	Expander
Ring grooves	Oil nozzle
Piston lands	Ring expander
Skirt	Piston pin
Trunk piston	Fixed pin
Cup-shaped retainer	Semi-floating pin
Retaining rings	Full-floating pins
Crosshead piston	Pin boss
Major axis	Connecting rod
Minor axis	Balancing pad
Cam ground piston	Crankshaft end-play
Chemical soaking	Main bearing clearance
Bead cleaning	Ring compressor
Piston wear	Oil seals

Review Questions–Chapter 7

Do not write in this text. Place your answers on a separate sheet of paper.

1. What are the two basic piston types used in diesel engines?

2. The two best methods of cleaning pistons is _____ and _____.

3. Describe the three major jobs performed by piston rings.

4. The bottom piston ring or fire ring is an _____.

5. The top piston ring is also called a _____.

6. The ring found in the middle of the piston is a _____.

7. The _____ is the link between the connecting rod and the piston.

8. The piston pin on two-cycle engines prevent _____ from the _____ from entering the combustion chamber.

9. _____ and _____ are matched sets and should never be interchanged.

10. Connecting rods are generally reused provided the _____ is replaced.

ASE-Type Questions

1. The Mexican hat and the turbulence cup are two types of _____.
   - (A) piston crown designs
   - (B) ring grooves
   - (C) bearings
   - (D) camshaft lobes

2. Which type of piston is a two-piece design?
   - (A) Trunk.
   - (B) Crosshead.
   - (C) Mexican hat.
   - (D) Bi-metal.

3. Technician A says that both trunk and crosshead pistons are manufactured to be perfectly round. Technician B says that these pistons are cam ground to be slightly elliptical in shape. Who is right?
   - (A) A only.
   - (B) B only.
   - (C) Both A & B.
   - (D) Neither A nor B.

4. The portion of the piston that extends from the bottom of the piston body up toward the lowest piston ring groove is known as the _____.
   - (A) piston land
   - (B) piston crown
   - (C) piston ring
   - (D) piston skirt

5. Technician A says you should install an oil control ring with the scraping edge of the ring facing down. Technician B says you should install the oil control ring with the scraping edge facing up. Who is right?
   - (A) A only.
   - (B) B only.
   - (C) Both A & B.
   - (D) Neither A nor B.

6. Always install the bottom ring with the gap positioned _____ from the top oil ring gap.
   - (A) 15
   - (B) 20
   - (C) 45
   - (D) 5

7. Fixed, semi-floating, and floating are three types of _____ designs.
   - (A) piston crown
   - (B) piston pin
   - (C) connecting rod
   - (D) piston ring

8. Technician A says that full-floating piston pin designs use bearings in both the piston and connecting rod. Technician B says that this type of piston pin uses no bearings. Who is right?
   - (A) A only.
   - (B) B only.
   - (C) Both A & B.
   - (D) Neither A nor B.

9. Technician A says that you should use a dial bore gauge to check a connecting rod large bore for out-of-roundness. Technician B says that you should use an inside micrometer to check a connecting rod large bore for out-of-roundness. Who is right?
   - (A) A only.
   - (B) B only.
   - (C) Both A & B.
   - (D) Neither A nor B.

10. A hole in each upper main bearing aligns with a _____ in the cylinder block.
    - (A) vertical oil passage
    - (B) slot
    - (C) groove
    - (D) horizontal oil passage

Chapter 8

Cylinder Heads and Related Componenets

After studying this chapter, you will be able to:
- ❏ List the parts that are attached to the cylinder head.
- ❏ Explain cylinder head disassembly procedures.
- ❏ Explain how to check for cracks and leaks in a cylinder head.
- ❏ Name five problems common to cylinder heads.
- ❏ Describe the service of valves and related components.

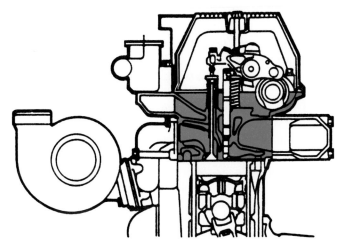

Figure 8-1. Typical cylinder head assembly for a mobile diesel engine. (Detroit Diesel Corp.)

An important part of the engine and combustion chamber is the cylinder head. The cylinder head must be in good shape to withstand the rapid changes of temperature and pressure taking place in the combustion space, as well as the stress resulting from being bolted securely to the cylinder block. This chapter discusses the cylinder head and the components installed on and in it. **Chapter 9** will cover the valve train in more detail.

Cylinder Head

The typical diesel *cylinder head* contains intake and exhaust valve ports and fuel injector bores. It also contains coolant passages or water jackets, which help cool the valve ports, injector bores, and the combustion chamber areas, **Figure 8-1.** The cylinder head is made of a heat-resistant cast iron alloy or aluminum. Engine parts that are essential to the diesel engine's operation may be found in, or attached to the cylinder head, **Figure 8-2.**

The cylinder head houses intake and exhaust valves, valve guides, and valve seats. Many modern mobile engines use two intake and two exhaust valves per cylinder, **Figure 8-3.** This arrangement permits a crossflow pattern, with minimal restriction and maximum exposure to coolant flow. Rocker arm assemblies are frequently attached to the

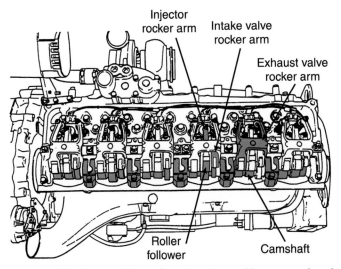

Figure 8-2. In an inline diesel engine with an overhead camshaft, the head assembly includes the camshaft bearings, camshaft, and rocker arm assemblies, which transmit cam motion directly to the valves and injectors. (Detroit Diesel Corp.)

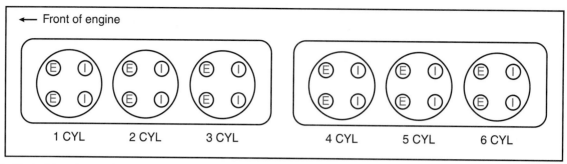

Figure 8-3. Typical heads used on an inline engine. Note that there are four valves used per cylinder: two intake valves and two exhaust valves. (Mack Trucks, Inc.)

cylinder head. The fuel injection components are almost always located in the cylinder heads of a diesel engine. The cylinder heads of large diesel engines may also be fitted with air starting valves, indicator cocks, and safety valves. These parts are covered in detail later in the book.

The number of cylinder heads found on diesel engines varies considerably. Small inline engines generally have one head for all cylinders. V-type engines typically have one cylinder head for each bank. Large diesel engines generally have one cylinder head for each cylinder. However, some large engines use a single head to cover two or three cylinders.

The cylinder head passages are designed to provide a sufficient flow of coolant to prevent warping or distortion. The coolant enters the head from the block or the cylinder liners. The connections between the cooling passages in the block or liners and the head will vary. These connections may consist of ferrules or similar connections. In some cases, the connections are made with separate lines. See **Chapter 11** for additional information on diesel engine cooling systems.

Cylinder Head Studs and Gaskets

The seal between the cylinder head and the block must be air and watertight. In most engines, this seal depends principally on the cylinder head studs, hold-down nuts, and gaskets. The studs and nuts secure the cylinder head to the block. A gasket between the head and the block is compressed to form a seal when the nuts are properly tightened. In some cases, bolts are used instead of studs and nuts.

Studs

Alloy steel rods with threads cut on both ends are generally used for *cylinder head studs.* The threads that screw into the block generally fit much tighter than those that receive the *hold-down nuts.* The tight fit in the block helps prevent the studs from unscrewing when the stud nuts are removed. Usually, studs that are in good condition should not be removed from a cylinder block.

All stud nuts should be tightened to the manufacturer's *torque specifications.* Overtightening is as damaging as undertightening. Sometimes, studs that are relatively inaccessible are neglected during periodic tight-

ness checks. Such an oversight may result in the studs loosening and failing.

When installing stud nuts, carefully clean the threads of the studs and the nuts with a wire brush and an approved solvent. Cleaning will minimize thread wear and distortion caused by dirt. It will also increase the accuracy of the torque wrench readings. A higher torque reading will be necessary to reach the required tension when the threads are dirty than when they are clean and well lubricated.

When installing stud bolts or nuts, tighten them finger tight. Then, use a torque wrench to tighten each bolt or nut about one-half turn at a time. Start at the center of the head and alternately work out to each end until the specified torque is reached. If the engine manufacturer's bolt tightening sequence is not available, use the sequence shown in **Figure 8-4.** This sequence distributes the tightening force evenly to provide a proper seal.

Removing Broken Studs or Bolts

Broken studs or bolts can be removed from the block or the head with a *stud extractor* (unthreaded) or *locking pliers* (threaded). Penetrating oil will often aid in loosening

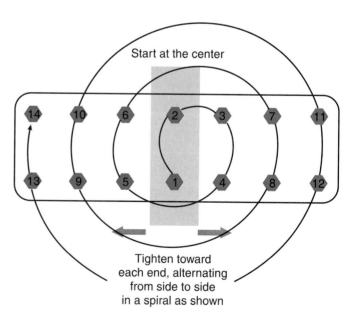

Figure 8-4. This sequence can be followed to properly tighten hold-down nuts or cylinder head bolts. (Deere & Company, Inc.)

frozen threads. In cases where the broken fastener is either flush with or below the surface, proceed as follows:

1. Drill a hole approximately 1/2 diameter of the broken stud or bolt. The hole depth should be no more than the length of the flutes on the screw extractor.

2. Select a screw extractor of the proper size. Using a hammer, lightly tap the extractor into the stud or bolt until it reaches the bottom of the drilled hole.

3. Turn the extractor counterclockwise to remove the broken stud or bolt.

4. If the screw will not extract, drill through the stub shell and tap out the hole. Repair any damaged threads with Helicoil®.

Gaskets

Although the mating surfaces of a cylinder block and a cylinder head appear to be quite smooth, irregularities can be seen in the surfaces when they are highly magnified. Unless steps are taken to fill these slight irregularities, combustion gases, oil, and/or coolant will leak from between the mating surfaces. Therefore, compressible **gaskets** are used to fill the irregularities and prevent leakage.

There are many materials used in the manufacture of gaskets. Gaskets can be made from copper and other relatively soft metals, laminated steel sheets, fiber, cork, rubber, synthetic rubber, or a combination of these and other materials. Combinations of gaskets, seal rings, grommets, or similar devices may be used to prevent leakage between the cylinder block and cylinder head. Two popular gaskets for mobile diesel engines are shown in **Figure 8-5.** Some diesel engine gaskets are made of solid steel plates. See **Figure 8-6.**

Finding Gasket Leaks

Leakage is the main problem encountered with gaskets. Leakage becomes apparent when the compression pressure becomes low, resulting in starting difficulties, or

Figure 8-6. *Solid steel gaskets are used in some diesel engines. (Victor/Dana)*

when fluid escapes between the head and block. Sometimes, there is no external leakage from a defective gasket. Instead, leakage occurs between internal passages and chambers. Although a compression check that reveals abnormally low compression on two adjacent cylinders may indicate a leaking head gasket, the only way to verify **internal leakage** is to remove the cylinder head.

Leaking gaskets may be the result of permanently compressed gasket material, improper tightening of cylinder head stud nuts, careless gasket installation, or the installation of a damaged or defective gasket.

If leakage occurs, the leak can occasionally be stopped by torquing the hold-down nuts. However, tightening will not stop the leakage if the gasket is in very poor condition. If a gasket is torn or burned across areas that must be sealed, it should be replaced. Gaskets should be removed and replaced according to the engine manufacturer's instructions.

Gaskets may be damaged by tightening the hold-down nuts unevenly or insufficiently. The portion of the

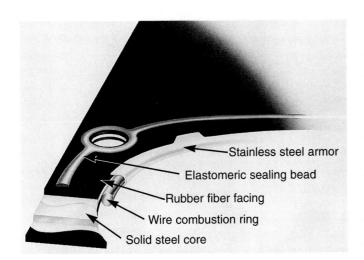

A

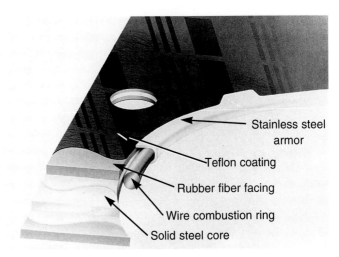

B

Figure 8-5. *Construction design of two popular diesel engine gaskets. Both are covered with stainless steel armor. A—Gasket with elastomeric sealing bead. B—Gasket with Teflon coating. (Fel-Pro Inc.)*

gasket below the tightest nuts may be pinched or cut, particularly if the mating surfaces of the head and block are not perfectly parallel.

If the stud nuts are not tightened to the specified torque, the gasket may not be compressed adequately to conform to the irregularities in the mating surfaces. This will prevent proper sealing and allow blowby of combustion gases and burning of the gasket material, resulting in gasket failure.

When installing a gasket, be careful not to bend, tear, or break the gasket material on the head studs. In some installations, the gasket does not surround the studs, but is placed in recesses in the cylinder block or the cylinder head. The gasket must be positioned properly in the recesses to avoid cutting or pinching it when the head is tightened down.

It is extremely important to use the correct gasket. A gasket of the wrong type or size may prevent proper sealing between the cylinder head and the cylinder block, resulting in stud breakage and burned surfaces. Also, vital coolant, water, or oil passages may be blocked if an improper gasket is used. When installing a cylinder head, a new gasket should always be used. Never reuse a head gasket.

Disassembling a Cylinder Head

To start cylinder head disassembly, remove all components attached to the head, such as the intake and exhaust manifolds and the turbocharger. Then, remove the valve (rocker) cover or covers. The **valve covers** completely enclose the valve and injector operating mechanisms on the top of the cylinder head. Depending on the installation, the valve cover may be a one-, two-, or three-piece design. Regardless of the design, clean the exterior of the cover before removal to keep dirt and debris out of the valve gear chamber. A **crankcase vent breather element** is generally located in the valve cover. Remove the wire mesh element from the cover, **Figure 8-7.** Wash the breather element in clean fuel oil and blow dry with com-

pressed air. Clean the breather recess in the valve cover and reinstall the element.

To continue the disassembly procedure, it may be necessary to remove the rocker arm components. When removing the **rocker arm assemblies,** remember that each manufacturer may have a different procedure. It is best to check the service manual for the specific procedure. Keep the pushrods and the rocker arm assemblies in exact order if they are to be reused. These parts are mated to each other by wear and should be reassembled in the same position on the camshaft. Use an organizing tray or label the parts with a felt-tipped marker or tags to facilitate reassembly. Check the lifters for either a dished bottom or for scratches, which indicate poor rotation.

Fuel Injector Removal

To prevent damage to the fuel injectors and **fuel injection nozzles,** it is good practice to remove them before removing the head. Clean the area around the nozzles carefully to prevent dirt from entering the engine. Some nozzles are screwed into the head; others are held in place with a spring clamp. Always protect the nozzle tips to prevent damage. Remove the **glow plugs** (if used) by unscrewing them from the combustion chamber. After the nozzles are removed, all fuel lines should be carefully plugged or capped.

Cylinder Head Removal

 Caution: Do not remove a cylinder head that is still warm. Doing so may cause the head to warp. Always allow the head to cool before disassembly.

When removing the cylinder head from the block, loosen the hold-down nuts or head bolts one or two turns

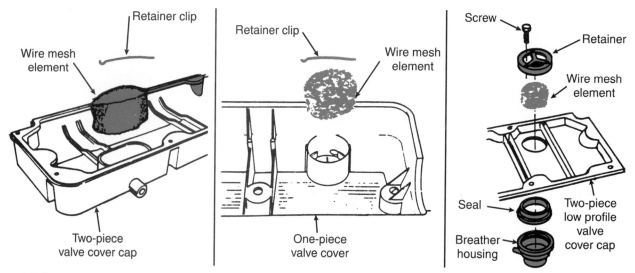

Figure 8-7. *Installation details for various crankcase vent breathers. A— Two-piece valve cover cap. B—One-piece valve cover cap. C—Two-piece low profile valve cover cap. (Detroit Diesel Corp.)*

each, working from the outside of the cylinder head inward. This procedure helps prevent distortion, which can occur if nuts are loosened completely. After loosening each nut slightly, they can be completely removed, again following the specified sequence.

With the nuts or bolts removed, the cylinder head can be lifted from the block. Do not pry on the contact surfaces when removing the head. If necessary, tap the head lightly with a soft mallet to help loosen it. In many cases, it is necessary to install some type of cylinder head **lift bracket** before separating the head from the block. Use a suitable lifting device to lift the cylinder head from the block, **Figure 8-8.** Place the head on a bench or a suitable fixture.

Valve Removal

To remove the valves from the cylinder head, the valve spring retainers and the valve springs must be removed. To aid in spring and retainer removal, a **valve spring compressor** is used to compress the spring, **Figure 8-9.** First, adjust the tool to fit the valve assembly. Then, proceed as follows:

> ⚠ **Warning: Valve springs are under considerable tension and can fly from the cylinder head with considerable force. Wear proper face and eye protection when compressing springs and removing valve keepers.**

1. Depress the handle of the spring compressor tool until the valve spring is compressed, **Figure 8-10.** Remove the valve spring keepers, **Figure 8-11.** You may need to use a magnet or a screwdriver. Carefully unload the valve spring compressor.
2. Remove the upper retainer from the valve stem.
3. Remove the valve spring, **Figure 8-12.**
4. Remove the lower spacer (if used) and remove the valve from the head, **Figure 8-13.**

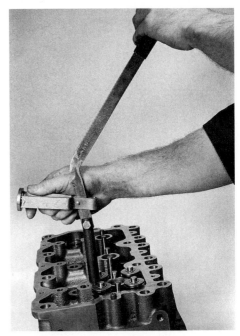

Figure 8-9. Begin the valve removal sequence by installing the valve spring compressor. (Mack Trucks, Inc.)

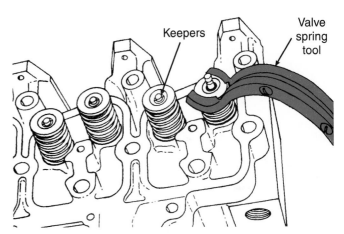

Figure 8-10. Be careful when using the valve spring compressor. Use a magnet to remove the valve keepers. (Cummins Engine Co.)

Figure 8-8. This cylinder head is being removed from the block. Note the lifting fixture, which is bolted to the head.

Figure 8-11. Remove the valve spring retainers. Keep all of the valve spring parts for each valve assembly together. (Fel-Pro)

Figure 8-14 shows an intake valve assembly and an exhaust valve assembly. Removal procedures are the same for both intake and exhaust valves. Keep each value and spring assembly separated.

Figure 8-12. Removing the valve spring. (Mack Trucks, Inc.)

Figure 8-13. Removing the lower spacer. Check this spacer and all other valve spring parts for wear. (Mack Trucks, Inc.)

Figure 8-14. Complete valve assemblies. (Mack Trucks, Inc.)

Cylinder Head Inspection and Service

Cylinder heads should be cleaned before they are inspected and serviced. Cylinder heads can be cleaned in the same manner as described in **Chapter 5.** Valves can be cleaned in a solvent tank or with a wire wheel mounted on a bench grinder.

 Warning: A face shield must be worn when using a powered wire wheel.

Checking for Cylinder Head Damage

Conditions requiring repair of a cylinder head are, in many ways, similar to those encountered when servicing cylinder liners. These conditions can be loosely grouped as cracks, warpage, fouling, and burning.

Cracks

Cylinder head **cracks** generally occur in narrow metal sections, such as the areas between valves and injectors. The cracks may be caused by the addition of cold water to a hot engine, restricted cooling passages, obstructions in the combustion space, or improper stud tightening.

The symptoms of a cracked cylinder head are the same as those of a cracked liner. The four methods commonly used to check a head for cracks or leaks are the pressure method, magnetic crack detector method, fluorescent penetrant method, and spot-check dye method.

In the **pressure method,** the head openings are sealed with plates or plugs, and the head is connected to an air hose with the appropriate fittings. Then the head is immersed in hot water (180-200°F or 82-93°C) for 15 minutes. Leaks are revealed by air bubbles that appear in the water. This testing technique is commonly used to check for leaks in areas that cannot be seen, such as internal passages and bores. See **Figure 8-15.**

A variation of the pressure method is the so-called universal system, which combines the principles of both air and water testing. The universal system is usually a bench model that comes with a tank and all necessary controls, **Figure 8-16.** Some units are available with a rollover fixture for V-type diesel engines. Once the pressure plate designed for the head being tested is bolted on the head, air is supplied to the assembly. The head and plate assembly is submerged in a tank of water. Once the regulated air supply is turned on, watch for a stream of bubbles to point directly to the crack regardless of the location.

When using the **magnetic crack detector method,** a horseshoe-shaped detector is placed over the suspected area, setting up a magnetic field in the head. Fine white or yellow metallic powder is then sprinkled over the area and the detector is rotated 90°, **Figure 8-17.** After the excess

Figure 8-15. *Universal pressure test bench combines both air and water testing to check engine parts for cracks and other defects. (Axe Equipment Co.)*

Figure 8-16. *Typical water pressure tester with circulator tank and pump. (Irontite Products Co., Inc.)*

Figure 8-17. *Magnetic powder is applied to the surface of a cylinder head during a magnetic particle test. (Magnaflux Corp.)*

powder is blown from the head, cracks are clearly displayed in white or yellow, depending on the color of the powder used.

The **fluorescent penetrant testing method** can be used on nonferrous materials, such as aluminum and stainless steel. In this test, a highly fluorescent liquid dye penetrant is applied to the head. Then, the excess penetrant is wiped off and the part is dried. After the part is dry, a developing powder is applied to help draw the penetrant out of the flaws by capillary action. Inspection is carried out under an ultraviolet light.

In the **spot check dye method,** a dye is sprayed onto the suspected area. After the dye is wiped off, the area is sprayed with a developer. Cracks will show up brightly. While this method can be used on both magnetic and nonmagnetic surfaces, spot check dyes will only indicate surface cracks. Therefore, structural cracks below the surface may escape detection.

Although techniques for repairing a cracked cylinder head are available, most manufacturers recommend replacement if cracks are found.

Warpage

Excessive heat caused by engine overheating or compression leaks can warp and distort the cylinder head. **Warpage** may occur if the head is still hot when it is removed from the engine block. Cylinder head distortion may also be caused by the use of improper techniques to repair cracks or by the improper tightening of cylinder bolts or hold-down nuts. Occasionally, new heads are warped by improper casting or machining processes. Severe cylinder head warpage or distortion is apparent when the mating surfaces of the head and block fail to conform. If distortion is extreme, the head will not fit over the studs.

Most warpage is slight and not to such an extreme as to be apparent to the naked eye. The machined surface of the head should be checked for warpage as follows:

1. Use a gasket scraper or a wire brush to clean the machined surfaces of the head.
2. Place a heavy, accurate straightedge across the machined surface. Using feeler gauges, determine the clearance at the center of the straightedge.
3. Check for warpage across the diagonals, along the longitudinal centerline, and across the width of the cylinder head at several points. Be sure to check the flatness of the intake and exhaust manifold mounting surfaces on the head.
4. In addition to checking for warpage, look for dents, scratches, and corrosion around the water passages. This is especially important on aluminum heads.

If warpage exceeds the maximum specifications given in the engine service manual, you must decide whether to reface the head or to discard it. Refacing can be done by surface grinding, milling, belt surfacing, or broaching. See **Figure 8-18.** The procedure used will be determined by the amount of metal that can be removed. Consult the engine service manual for refacing limits. After refacing, the sur-

Figure 8-18. Surface grinder in operation. Refacing a block or head may not work if the casting is severely warped. (Winona Van Norman Machine Co.)

face must be smooth enough to form a proper seal and prevent leakage, but rough enough to provide a good seal. Ideally, a severely warped head should be replaced.

Fouling

Combustion chamber **fouling** will decrease the combustion efficiency. Combustion chambers are designed to create the desired turbulence for mixing the air and fuel. Any accumulation of carbon deposits in the combustion chamber will impair both turbulence and combustion by altering the shape and decreasing the chamber volume. Many diesel engines are equipped with a removable **precombustion chamber** or *swirl chamber,* located in the cylinder head. The fuel injector and the glow plug assemblies are normally installed so that their operating ends are exposed in this chamber. The chamber capacity is normally equal to 75% of the total volume of air in the cylinder at the end of the compression stroke.

As discussed in **Chapter 4,** the purpose of the swirl chamber is to produce a strong turbulence of air, which progressively increases as the piston approaches top dead center (TDC) on the compression stroke. When fuel is injected, it is swirled by the air in the chamber. As soon as ignition begins, the pressure in the chamber increases, forcing the air, the burned gases, and the unburned fuel towards the cylinder. This causes the piston to descend. The descending piston reverses the direction of the swirl, causing more intense turbulence. The intense turbulence aids combustion, lowers fuel consumption, and minimizes emissions.

Symptoms of fouling in the combustion spaces include smoky exhaust, loss of power, and high compression. Such symptoms may indicate extensive carbon formation or clogged passages. In some engines, these symptoms indicate that the shutoff valves for the auxiliary precombustion chamber are stuck. Faulty injection equipment, improper assembly procedures, or excessive lubricating oil getting past the piston rings can also foul combustion chambers.

Cleaning fouled combustion spaces generally involves removing the carbon accumulation. The most effective cleaning method is to soak the dirty parts in an approved solvent and then scrape or scrub off all traces of carbon. Use a scraper or a wire brush to remove the carbon, but be extremely careful not to damage the surfaces. If lubricating oil is causing carbon formation, check ring, bearing, piston, and liner wear. Replace or recondition excessively worn parts. Carbon formation resulting from improperly assembled parts can be avoided by following the procedures described in the engine service manual.

Burning and Corrosion

Burning and **corrosion** of the cylinder head and the block mating surfaces may be caused by a defective gasket. Although regular maintenance will help prevent burning and corrosion, this type of trouble may take place under certain conditions. When corrosion and burning occur, there may be a loss of power as a result of combustion gas or coolant leakage into the combustion space. Other symptoms of leakage include hissing or sizzling sounds in the area of the head where gases are leaking, bubbles in the expansion tank sight glass (marine applications), or overflow of the expansion tank.

Gaskets and grommets that seal combustion spaces and coolant passages must be in good condition. Otherwise, fluid leakage will cause corrosion or burning of the areas adjacent to the leaks. Improper coolant treatment is another factor that may accelerate the corrosion rate. Check for deposits in the water passages. If necessary, use a recommended solution and dip the head to clean out scale and lime.

Valves

A diesel engine must take in air and expel exhaust gases at precise intervals. The valves control this operation by opening and closing the intake and exhaust ports to the cylinder. See **Figure 8-19.** The valves must seal the combustion chamber when they are closed.

The valves used in internal combustion engines are called **poppet valves.** The word "poppet" comes from the popping action of the valve as it opens and closes. A poppet valve has a round head, a tapered face, and a cylindrical stem. A slot machined in the top of the stem houses the valve spring retainers. Intake valve heads are generally larger than exhaust valve heads, **Figure 8-20.** The size or diameter of the valves is determined by the engine design.

 Note: Some manufacturers identify the valves by forging an "I" (intake valve) or an "E" (exhaust valve) on the combustion face.

The **valve head** seals the intake or exhaust port. This seal is made as the valve face contacts a valve seat in the

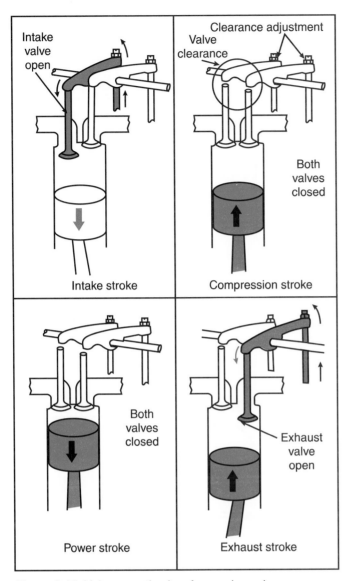

Figure 8-19. *Valve operation in a four-cycle engine. (Deere & Company, Inc.)*

port. The **valve face** is the tapered area machined on the valve head. The angle of the taper depends on the engine model. Valve face taper may also vary between intake and exhaust valves on the same engine.

The area between the valve face and the valve head is called the **margin.** The margin allows for some machining of the valve face, which is often needed to restore its finish. The margin also provides the valve with an extra capacity to hold heat.

Valve Stem

The **valve stem** guides the valve during its up-and-down movement and provides a place to connect the valve spring to the valve. Valve spring retainers and keepers are used to secure the spring to the stem. The stem rides in a guide that is either machined into the cylinder head or pressed into the head as a replaceable part. Virtually all diesel engines use replaceable valve guides, **Figure 8-21.**

Valve Seat

The area of the cylinder head contacted by the valve face is known as the valve seat. Although the seat may be machined in the head, it is generally a separate component that is pressed into the head to allow for easy replacement.

Figure 8-22 shows the dimensions that must be known to recondition valves, seats, and guides properly. Refer to the engine service manual for the correct dimensions for specific engines.

Valve Construction

Exhaust valves are usually made of siliconchromium steel or steel alloys. A high content of nickel and chromium is usually included in the steel or the alloy to resist the

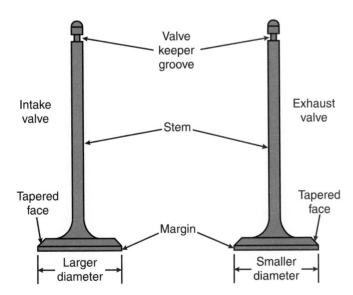

Figure 8-20. *It is important to be familiar with the terminology used to identify the various parts of a common poppet valve.*

Figure 8-21. *Several valve guide designs are used in diesel engines. (Goodson Shop Supplies)*

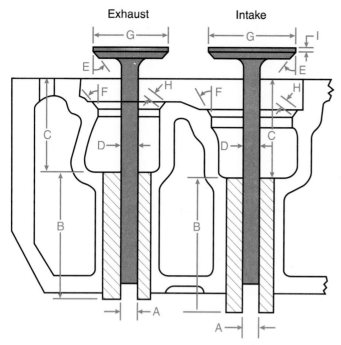

Figure 8-22. Several dimensions must be checked when servicing valves. A—Inside diameter of valve guide. B—Length of valve guide. C—Distance from guide to head. D—Outside diameter of valve stem. E—Angle of valve face. F—Angle of valve seat. G—Width of valve head. H—Width of valve seat. I—Margin. (Deere & Company, Inc.)

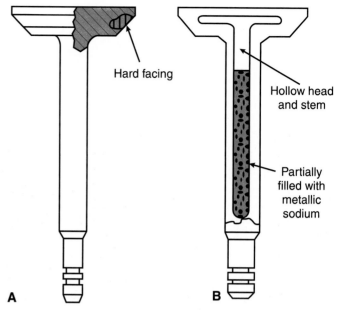

Figure 8-23. Two common poppet valves. A—Hard-face valve. B—Sodium-filled valve.

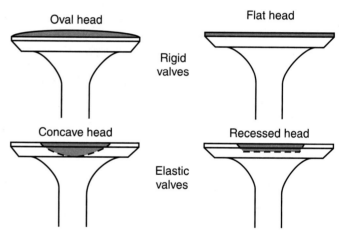

Figure 8-24. Valves used in diesel engines may vary from one engine to another.

corrosion caused by the hot combustion gases. A hard alloy, such as Stellite, is often fused to the seating area of the valve head and to the tip of the valve stem. The alloy increases the wearing qualities of these surfaces.

There are two ways for an exhaust valve to dissipate heat. First, as the valve face contacts its seat, the heat from the valve will be transferred to the cylinder head. The second method of heat dissipation is through the valve stem, to the valve guide, and on to the cylinder head. To aid in heat transfer, some exhaust valve stems are filled with metallic sodium, **Figure 8-23.** The sodium absorbs and transfers heat much better than steel.

> ⚠ **Warning: Sodium-filled valves cannot be resurfaced and must be disposed of properly to avoid a fire hazard. Sodium reacts violently with water on contact and is considered an explosive hazard. Never dispose of sodium-filled valves in a haphazard manner.**

Because they are not subjected to the corrosive action of the hot exhaust gases, intake valves are generally made of low-alloy steels. Several valve designs are used in modern diesel engines. The most common designs are shown in **Figure 8-24.** Although valves are available in several shapes that are designed to meet various specifications, they can be grouped in two basic types: rigid valves and elastic valves. The rigid valve is fairly strong, holds its shape well, conducts heat quickly, and wears slowly. Unfortunately, it is more likely to leak and burn than an elastic valve. Therefore, elastic valves are used for exhaust

and intake valves when high seating pressures are needed. High seating pressures help crush or wipe off deposits to prevent valve leakage.

Valve Inspection

Inspect each valve face for burns, pits, cracks, grooves, or scores, **Figure 8-25.** The tip of the valve stem should show an even wear pattern. The keeper grooves should show no signs of wear or damage. Replace a valve if the face or head is cracked, pitted, worn, or badly burned. Valves with worn or bent stems or damaged keeper grooves should also be replaced. If the valves are to be reused, soak them in solvent to soften the carbon deposits. Then, remove the deposits with a wire buffing wheel, taking care not to nick the valves when using the buffing wheel.

Once the valve is clean, it can be resurfaced. There are two surfaces on the valve that are reconditioned—the valve face and valve tip. The face is resurfaced before the

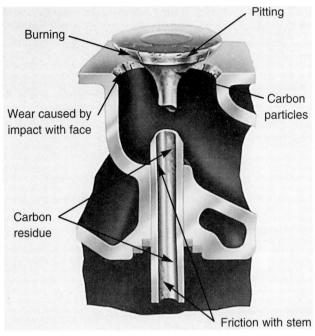

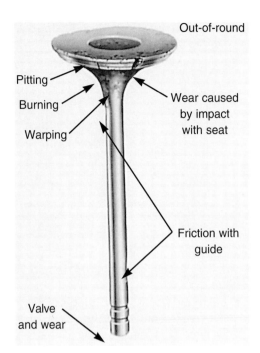

Figure 8-25. *Common valve problems that should be identified when servicing cylinder heads. (Sioux Tools, Inc.)*

valve tip. Valves can be resurfaced by using either *valve grinding machines* or *valve cutting machines.* See **Figure 8-26.** Although both machines can be used to resurface valves and smooth and chamfer valve stem tips, grinding is the most popular reconditioning technique. Large valve grinding machines are also capable of doing other valve train reconditioning tasks, such as resurfacing certain types of rocker arms. Computerized valve grinding machines are becoming popular in large shops.

Grinding Valves

When using a valve grinding machine, make certain the grinding wheel is properly dressed as directed by the manufacturer. When installing the valve in the grinding machine, chuck the valve as close to the head as possible to eliminate stem flexing from grinding wheel pressure, **Figure 8-27.** Take light cuts, using the full wheel width. Most grinding machines pump coolant onto the valve and the grinding wheel. Make sure that the coolant is striking the contact point between the valve face and the grinding wheel. When grinding, remove only enough metal to clean the valve face. A sharp edge between the face and the head is not desirable. It will burn easily and may cause preignition, **Figure 8-28.** As a general rule, it is not advisable to grind a valve face to a point where the margin is reduced by more than 25% or to where it is less than .045" (1.14 mm) on the exhaust valve and .030" (.771 mm) on the intake valve.

Figure 8-26. *Valves are commonly resurfaced on a valve grinding machine. (Sioux Tools, Inc.)*

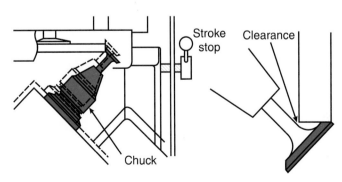

Figure 8-27. *Once a valve grinding machine is set up, the valve is secured in the chuck and the proper stroke is made. Make sure that there is clearance between the grinding stone edge and the valve so that the stem does not contact the stone. (Sioux Tools, Inc.)*

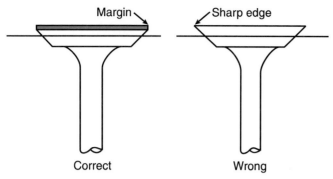

Figure 8-28. When a valve face is resurfaced, the correct margin must be maintained. (Sioux Tools, Inc.)

Once grinding is complete, check the valve face runout. Total indicated runout (TIR) should not exceed .002″ (.0508 mm). The valve face should have no chatter marks or unground areas. After grinding, re-examine the valve face for cracks. Sometimes, fine cracks in the face are visible only after grinding. In some cases, cracks occur during the grinding process due to inadequate coolant flow or excessive wheel pressure.

 Warning: Always wear eye protection when operating any type of grinding equipment.

On large diameter valves and hard-faced valves, it might be necessary to make a finish grind. Once the valve face is ground, decide whether the valve tip requires grinding. To determine this, place the valve in the cylinder head and check the stem height. Using the manufacturer's specifications, determine whether, and to what extent, the tip must be ground.

The valve tip should be ground so that it is perfectly square with the stem. Because valve tips have hardened surfaces that are up to .030″ (.762 mm) deep, only .010″ (.254 mm) should be removed during grinding. If more than .010″ (.254 mm) is removed from the tip, the valve must be replaced. If too much material is removed from the valve tip, there may be interference between the rocker arm and the spring retainer or the valve rotator. Follow the manufacturer's specifications for the allowable grinding limits.

Resurfacing procedures for a valve grinding machine and a valve cutting machine are quite similar. The guidelines given here for grinding a valve face can be followed when using most valve cutting equipment.

Valve Guides

Valve guides support the valves in the cylinder head. They are machined to a diameter that is a few thousandths of an inch larger than the diameter of the valve stems, providing a very small clearance between the stem and the guide. This clearance is important for a number of reasons. It prevents the lubricating oil from being drawn past the valve stem and into the combustion chamber during the

intake stroke, prevents burnt gases from entering the crankcase area past the valve stems during the exhaust stroke, and helps keep the valve cool, **Figure 8-29.** The small valve guide clearance also keeps the valve face in perfect alignment with the valve seat.

Although valve guides may be cast integrally with the head, removable valve guide inserts are used in most cases. Removable guide inserts are press-fit into the cylinder head. See **Figure 8-30.**

Checking Stem-to-Guide Clearance

Before checking *valve stem-to-guide clearance,* clean the valve stem with solvent to remove all gum and varnish. Clean the valve guides with solvent and/or a wire-type expanding valve guide cleaning tool. Then, insert the proper valve into its guide and hold the head against the valve seat tightly. Mount a dial indicator on the valve

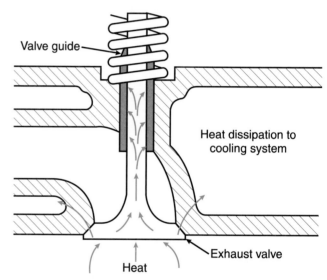

Figure 8-29. A valve guide helps the valve dissipate heat. Note that heat is also dissipated through the valve head. (Deere & Company, Inc.)

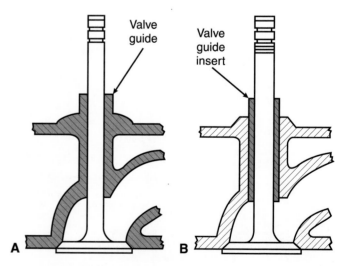

Figure 8-30. Two types of valve guides used in diesel engines. A—Integral guide. B—Insert guide.

spring side of the cylinder head so that the indicator's foot rests against the valve stem at a 90° angle. See **Figure 8-31.** Move the valve slightly off its seat and measure the valve guide-to-stem clearance by moving the stem back and forth to actuate the dial indicator. If the dial indicator reading is not within the manufacturer's specifications, measure the valve stem with a micrometer. Compare the valve stem diameter to the manufacturer's specifications to determine if the stem or the guide is responsible for the excessive clearance.

In an alternate method of checking the valve stem-to-guide clearance, the dial indicator is mounted on the combustion side of the cylinder head. Before mounting the indicator, move the valve head away from its seat a predetermined distance, either by a special collar tool, which is placed on the valve stem between the head of the valve and the guide, or by simply measuring the distance between the valve head and the seat with a scale. After positioning the valve and mounting the dial indicator, move the valve head back and forth to actuate the indicator. If the indicator shows more clearance than specified, measure the valve stem with a micrometer. Compare the stem's diameter to specifications to determine whether the problem is caused by a worn valve guide or a worn valve stem.

Other measuring techniques are available to the engine technician. A bore gauge, a micrometer, **Figure 8-32,** or an inside caliper-type small hole gauge can be used to determine valve guide wear. Remember, guides do not wear uniformly. Therefore, plugs, valve stems, or pilots should never be used to measure valve guides. Measure the inside diameter of the guides at several different points. Careful measurement and inspection of the guides will help detect bell mouthing or elliptical wear. These conditions normally occur at the ends of the guides, **Figure 8-33.**

A

B

Figure 8-32. Checking the condition of a valve guide. A—Telescoping gauge. B—Outside micrometer.

Figure 8-31. A small bore gauge can be used to obtain accurate bore measurements. (Mack Trucks, Inc.)

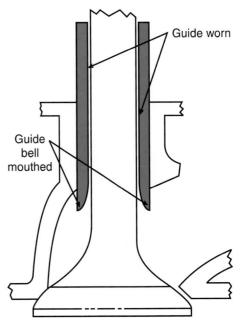

Figure 8-33. Worn and bell-mouthed guides will cause improper valve seating.

Valve Guide Reconditioning

Valve guide reconditioning is an important part of head reconditioning. There are five techniques that can be used to restore worn valve guides, including knurling, reaming, the use of thin wall liners or threaded bronze inserts, and replacement of the entire guide.

Knurling

Knurling is one of the fastest methods of restoring the inside diameter of a worn valve guide. It is the method usually chosen to recondition integral guides. Knurling raises the surface of the guide's inside diameter by cutting tiny furrows through the metal. As the knurling tool cuts into the guide, metal is raised or pushed up on either side of the cuts. This decreases the diameter of the guide hole. A burnisher or reamer is then used to press the ridges flat and shave off the peaks. The final result produces the proper-size hole and restores the correct guide-to-stem clearance, **Figure 8-34.** Knurling can be done with either a tap-type knurling tool or a wheel-type (roller) knurling tool. Either tool can be hand or power driven. When power knurling, a drill speed reducer, such as the one shown in **Figure 8-35,** is usually required.

Reaming

Reaming increases the diameter of the guide hole so it can be fitted with a oversize valve (valve with a larger stem). Reaming can also be used to restore the guide to its original diameter after installing inserts or knurling. The advantage of reaming for an oversized valve is that the finished product is totally new, **Figure 8-36.** The guide is straight, the valve is new, and the clearance is accurate. Installing an oversized valve is generally considered to be superior to knurling. The process is also relatively quick and easy. The only tool required is a reamer.

Thin Wall Inserts

The use of **thin-wall guide liners** offers a number of important advantages over knurling or reaming. The liners provide the benefits of a bronze guide surface, including

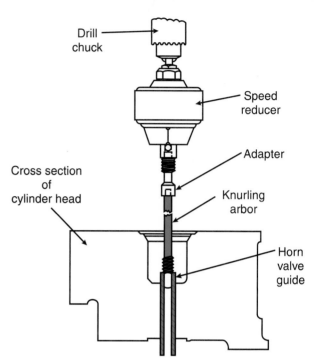

Figure 8-35. Using a drill and a speed reduction arrangement to knurl a valve guide. (Mack Trucks, Inc.)

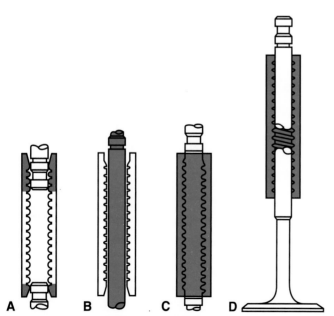

Figure 8-34. Valve guide knurling procedure. A—Initial knurl. B—Initial ream. C—Final knurl. D—Finished guide. (Sunnen Products Co.)

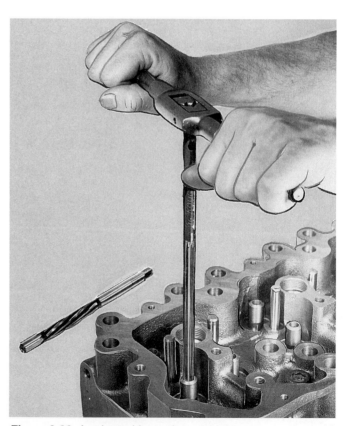

Figure 8-36. A valve guide can be reamed to accept valves with oversized stems. Reaming is also performed after a new guide is installed. (Mack Trucks, Inc.)

better lubrication, reduced wear, and tighter clearances. Thin-wall liners can be used with either integral or replaceable guides (cast iron or bronze). Thin-wall liners are faster, easier, and less expensive to install than new guides. Installing thin-wall guide liners also maintains guide centering with respect to the valve seats.

Threaded Bronze Inserts

The use of **threaded bronze inserts** is another method used to restore worn guides. In this method, the worn guides are tapped and the threaded bronze inserts are installed in the tapped holes. They provide better lubrication, excellent wear qualities, and tight clearance, but they are more expensive and difficult to install.

Valve Guide Replacement

The fifth method of repairing worn valve guides involves replacing the entire guide. As already mentioned, there are two basic types of valve guides: integral guides and insert guides. Because aluminum is relatively soft, aluminum cylinder heads require a valve guide insert that is made of a hard material.

To replace an integral valve guide, bore out the old guide and drive a replacement guide into the hole. Portable or stationary seat-and-guide machining equipment can be used for this process. Freeze the replacement guides and drive in cold with a press fit of approximately .002" (.0508 mm). Use an assembly lubricant to prevent galling. Keep the centerline of the guide concentric with the valve seat so that the rocker arm-to-valve stem contact area is not disturbed. If the new guide is not concentric with the valve seat, install a new seat or regrind the valve face to correct the problem.

To replace an old valve guide insert, position a proper-sized driver so that its end fits snugly into the guide. The driver's shoulder should be slightly smaller than the guide's outside diameter so that it will go through the cylinder head. Use a heavy ball peen hammer or a press to drive or press the guide out of the cylinder head deck. See **Figure 8-37.**

When removing and installing a valve guide, take care not to break the guide or damage the guide hole in the head. Cast-iron guides have a tendency to gall aluminum heads, and once the guide hole is damaged, it must be bored to accommodate an oversized guide. Some engines use a **stepped exhaust guide,** which must be driven out from the combustion chamber side to prevent cracking the head. Many heads have **flanged guides** that must be driven out from a particular side. Be sure to check the manufacturer's manual for specific instructions.

As a rule of thumb, straight guides (no taper or flange) should be driven out toward the combustion chamber. Before installing new guides, chill them in a freezer or with dry ice to create the tightest possible interference fit between the guide and the head. Then, install the guide using the following general procedure:

1. Insert the valve guide into the valve guide installation tool.

Figure 8-37. A valve guide insert can be pressed from the cylinder head. (Mack Trucks, Inc.)

2. Oil the outside surface of the new guide before installation. This will ease installation and help prevent galling, **Figure 8-38A.**
3. Press the new valve guide into the top of the cylinder head.
4. Using a depth gauge, check the extension of the valve guide from the valve spring to the top end of the guide, **Figure 8-38B.** Then ream the new valve guide to the dimension given in the service manual.

Keep in mind that these are general instructions. Always use the procedure recommended in the engine manual to prevent damage to the head.

Valve Seats

The **valve seat** is the machined surface on which the valve rests when it is closed. The valve seat may be made of the same material as the block or head or it may be made of a hardened material. The seat is ground to form a tight seal with the valve face.

The most critical sealing area in the valve train assembly is the surface between the valve face and its seat when the valve is closed. Leakage between these surfaces reduces engine compression and power. It can also lead to valve burning. The valve seat must be of the correct width to form a proper seal with the valve, be properly located on the valve face, and to be concentric with the guide.

Most diesel engine manufacturers specify the amount of contact between the valve seat and the valve face. This

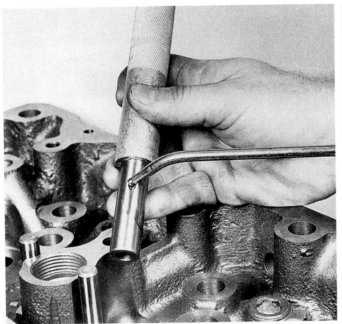

A

B

Figure 8-38. *Procedure for installing a valve guide. A—Apply lubricant to the new guide and press the guide into the head. B—Checking the valve guide extension. (Mack Trucks, Inc.)*

is known as the ***valve seat width.*** When specifications are not available, a general rule is that the contact area should be .0625″ (1.59 mm) for intake valves and .094″ (2.38 mm) for exhaust valves.

A proper overhang is also necessary, **Figure 8-39.** The ***overhang*** is the area of the valve face between the contact area and the margin. It should be .017″ (.40 mm) for both intake and exhaust valves. When the engine reaches its normal operating temperature, the valves expand slightly more than the

seats. This causes the contact area to move down on the valve face. Seats that contact the valve face too low may lose partial contact at normal operating temperatures.

Types of Valve Seats

As with valve guides, there are two types of valve seats—integral seats and insert seats. ***Integral seats*** are part of the head casting, while ***insert seats*** are added to the cylinder head after casting. See **Figure 8-40.** In some cases,

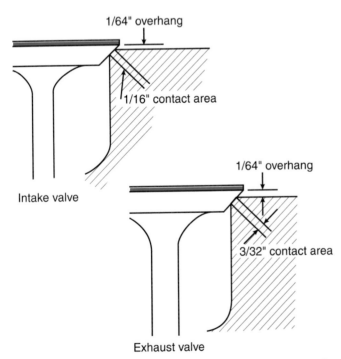

Figure 8-39. *Valve overhang is necessary for proper valve operation. (Sunnen Products Co.)*

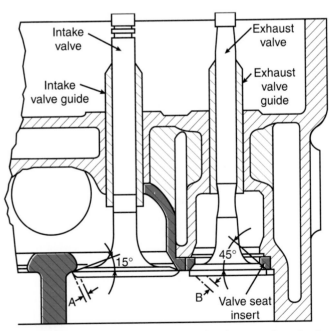

Figure 8-40. *Cross section of valves showing an exhaust valve seat insert and an integral intake valve seat. The seat angles are labeled "A" and "B."*

insert seats are installed as replacements for worn integral seats. The insert seat is a ring of special high-grade alloy metal that is placed in the block or head. It reduces wear, prevents leakage, and reduces valve grinding frequency. It is also relatively easy to replace.

Valve Seat Angles

The valve seat for a typical mobile diesel engine is approximately .060" (1.52 mm) wide. The seat begins .030" (.76 mm) from the valve margin. A properly ground seat has three angles: the top angle is 15°; the seat angle is 45° or 30°; and the throat angle is 60°. Typically, the 45° seat angle forms a better seal than the 30° seat angle, therefore, it is used most often. Using three seat angles maintains the correct seat width and sealing position on the valve face, **Figure 8-41.** Sealing pressure and heat transfer from the valve to the seat are also affected by the seat angles.

In most modern diesel engines, the valve face and the seat are machined to slightly different angles. Generally, there is a variance of .5° to 1.5° between the face and the seat, **Figure 8-42.** This variance forms an *interference angle* and creates a narrow line of contact between the valve face and the seat. The narrow line of contact provides for better sealing on the combustion chamber side of the seat. This helps prevent combustion pressures from escaping between the valve and the seat when the valve is closed. When no interference angle is provided, the possibility of a *negative interference angle* (open on the combustion chamber side of the seat) exists. A negative interference angle allows combustion pressures to unseat the valve. In some diesel engines, an interference angle is not used. Instead, these engines have a relatively narrow seat. See **Figure 8-43.**

Valve seat width should be measured when the seat location is checked. It should be noted that wide seats improve heat transfer, while narrower seats help crush valve deposits. An interference angle is best for engines with high deposits and for use in some seat distortion conditions because it provides a positive seat on the combustion side of the valve and seat. However, an interference angle will cause the valve to run hotter.

A valve seat runout gauge is similar to a dial indicator. It features a gauge face that is divided into .001" (.03 mm) increments, an arbor that centers the instrument in the valve guide bore, and an indicator bar that can be adjusted so it rests on the valve seat. The tool is slowly rotated around the circumference of the valve seat to check the seat's concentricity (runout). Compare the runout gauge reading to the manufacturer's specifications to determine if valve seat service is necessary. See **Figure 8-44.**

To check valve seat inserts for tightness, position a drift (or similar blunt-nose tool) on the top of the insert and lightly tap the end of the drift with a hammer. See **Figure 8-45.** If a ringing sound is heard, the insert is tight. If a dull, "thud-like" sound is heard, the insert is loose.

Replacement of Valve Seats

Valve seat replacement is needed when a valve seat is cracked, burned, pitted, loose, or recessed in the cylinder

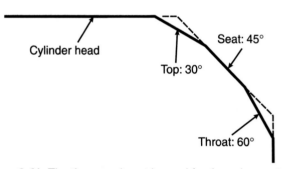

Figure 8-41. The three-angle cut is used for the valve seats on most diesel engines.

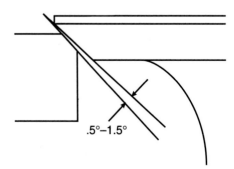

Figure 8-42. Grinding the valve to an angle of .5°-1.5° less than the seat produces an interference angle. This provides a high-pressure seal on the combustion chamber side of the valve and the seat. (Sioux Tools, Inc.)

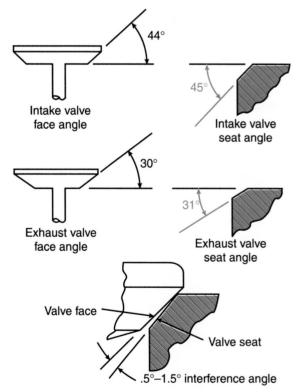

Figure 8-43. The difference between the valve seat angle and the valve face angle is known as the interference angle. This angle results in a line contact between the valve and the seat. (Sunnen Products Co.)

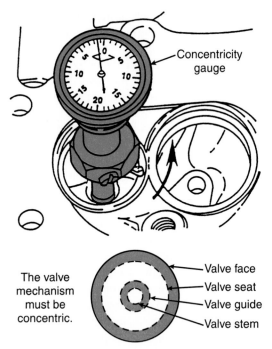

Concentricity gauge

The valve mechanism must be concentric.

Valve face
Valve seat
Valve guide
Valve stem

Figure 8-44. Checking valve seat concentricity with a special gauge. The gauge is turned in the direction of the arrow. As indicated, the valve face, seat, guide, and stem must all be concentric. (TRW, Inc.)

Figure 8-45. A hammer and a small drift punch can be used to check valve seat inserts for looseness. (Mack Trucks, Inc.)

head. If the valve seat inserts are defective, they must be removed from the cylinder head with a special tool. To aid in the removal of a valve seat insert, a groove must be ground around the inside diameter of the insert. Remove valve seat inserts as follows:

1. Measure the distance from the surface of the cylinder head to the point on the insert where the groove should be ground (depth of the groove). See **Figure 8-46.** The amount to be ground can be found in the service manual.
2. Install a grinding wheel into the grinder and set the depth of the grinder base to the dimension determined in the proceeding step. Then, attach the base to the grinder.

3. Grind a groove around the inside diameter of the valve seat insert. This groove should be ground to the depth determined in step 1.
4. Attach the collet to the T-handle and shaft assembly of the seat removal tool. Locate the collet in the valve seat insert so that the collet ridge is inside the groove. Turn the T-handle to fully expand the collet.
5. Position the lifting bridge under the crank handle and turn the handle to lift the insert from the head. Finally, turn the T-handle slightly to release the insert from the collet. See **Figure 8-47.** After the valve seat insert has been removed, clean the surface of the counterbore with a wire brush. Then, check the counterbore depth and diameter, **Figure 8-48.**

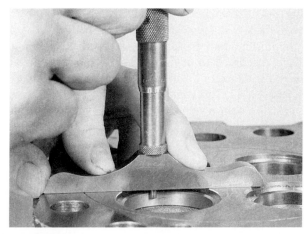

Figure 8-46. Determining the valve seat insert groove depth.

Figure 8-47. Removing a valve seat insert. Secure the collet in the insert, remove the insert from the counterbore, and unscrew the T-handle to remove the insert from the collet. (Mack Trucks, Inc.)

To install a new valve seat insert, position the insert over the counterbore in the block. Using a driver handle, drive the insert into the head. Grind the insert to the angle recommended in the engine service manual, **Figure 8-49.** When grinding the valve seat insert, keep the following precautions in mind:

❑ Do not grind the insert excessively. It only takes a few seconds to recondition valve seats. Avoid grinding off too much material.

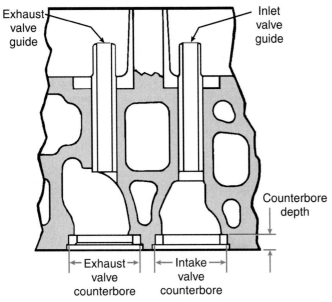

Figure 8-48. *Check counterbore depth by using a depth micrometer. (Mack Trucks, Inc.)*

Figure 8-49. *After positioning the seat and driver in the counterbore, grind the insert until good concentricity is achieved. (Mack Trucks, Inc.)*

❑ Do not use too much pressure. While grinding, support the weight of the grinder to avoid excess pressure on the stone.

❑ Keep the work area clean. Precision workmanship cannot exist in a dirty work area. Always keep the tools and the work area clean.

After the grinding has been completed, clean the valve seat insert thoroughly. Determine the *concentricity* of each valve seat insert relative to the valve guide. Valve seat runout must be within the specifications given in the service manual.

To determine the valve seat's concentricity, apply a dab of Prussian blue at four points on the valve face (90° apart), **Figure 8-50A.** Lower the stem of the valve in the valve guide and rotate the valve 90° in its seat, **Figure 8-50B.** This procedure will indicate the area of contact between the valve head and the valve seat insert. If the valve seat is ground properly, Prussian blue should be visible around the entire insert diameter.

After the valve seat inserts have been ground and checked, thoroughly clean the cylinder head before reinstalling the valves.

Valve Stem Seals

Valve stem seals are designed to prevent oil from leaking past the valve stems and into the combustion chambers. If leakage occurs past the valve stems, oil consumption will rise. Therefore, all valve stem seals should be replaced when a head is rebuilt. Seals tend to wear, crack, and become brittle due to age and operation at elevated temperatures. There are dozens of valve stem seal designs, **Figure 8-51,** but they all fall into the following general categories: deflector seals, positive seals, and O-ring seals.

❑ Deflector seals deflect oil away from the valve stem. They act like an umbrella, moving with the valve stem to shield the valve guide from excess oil. This type of valve stem seal is commonly called an *umbrella seal.*

❑ Positive seals are attached to the valve guide boss. They work like a squeegee, wiping oil from the valve stem. If positive stem seals are installed after machining the guide ends, follow the instructions covering the use of original equipment seals and the positive seal.

❑ O-ring seals, like the other two types, prevent excessive oil consumption. They are also frequently used in conjunction with umbrella seals or positive seals.

Valve stem seals are generally made of synthetic rubber. This material is inexpensive, easy to install, and accommodates oversize valve stems. The most common types of synthetic rubber used include polyacrylate, Viton, and nitrile.

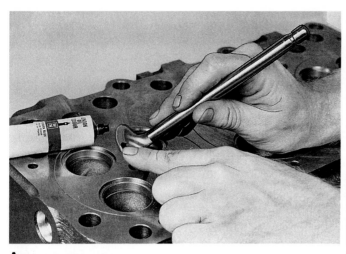

A   **B**

Figure 8-50. Checking valve seat concentricity using Prussian blue. A—Apply Prussian blue to the valve face. B—Insert the valve into the guide. (Mack Trucks, Inc.)

Figure 8-51. Common valve stem seal designs. (Goodson Shop Supplies, Inc.)

> **Note: Always use the type of valve stem seal specified by the engine manufacturer. Valve stem seals are significantly different in lubrication properties and can fail if installed incorrectly or used in the wrong application.**

Valve Springs

The main function of the *valve spring* is to close the valves and to keep them closed and seated until they are opened by the camshaft. Cylindrical valve springs are used by most engine manufacturers. Some engines use one spring per valve, while others use two or three springs. The additional springs reduce vibration and valve flutter. In a three-spring arrangement, two coil springs are usually separated by a flat spring.

Common valve spring designs are illustrated in **Figure 8-52.** Some springs have *dampers* or *damper coils* to assist in reducing valve spring surge at high speeds.

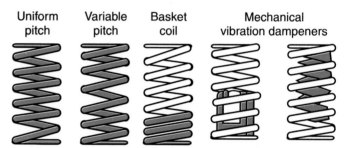

| Uniform pitch | Variable pitch | Basket coil | Mechanical vibration dampeners | |

Figure 8-52. Typical valve spring designs. (Sealed Power Corp.)

Special spring dampers are used to reduce vibration and humming. Damper coils on the springs should be installed toward the cylinder head. However, before installing any spring, the valve seal must be placed over the valve stem.

The valve spring itself is held in place by the *valve spring retainer,* which is locked onto the valve stem with two wedged-shaped parts called valve keepers or collets, **Figure 8-53.**

Corrosion and metal fatigue are common causes of valve spring failure. Broken valve springs cause excessive valve noise and may cause erratic exhaust gas temperatures. When a valve spring breaks, serious damage can occur. When a spring breaks, it may collapse enough to allow the valve to drop into the cylinder, where it can cause damage. Similarly, the valve stem keepers may release the valve and allow it to drop into the cylinder, causing severe damage to the piston, cylinder head, cylinder, and adjacent parts.

A number of precautionary measures can be taken to prevent the causes of valve spring breakage. Reasonable care must be taken during assembly and disassembly. Before a valve assembly is reassembled, the valve spring must be thoroughly cleaned and inspected. A suitable solvent should be used for cleaning valve springs. Never use an alkaline solution, which will remove the spring's protective coating. The best indication of impending valve spring failure is the condition of the spring's surfaces.

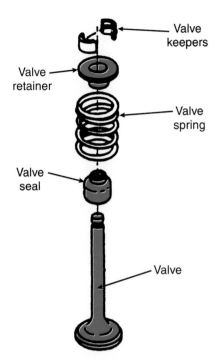

Figure 8-53. *Components of a typical valve assembly. Note that this assembly is equipped with an umbrella-type valve seal.*

Figure 8-54. *A dial spring pressure tester can be used to check a valve spring open and close pressure. (Sealed Power Corp.)*

Magnetic crack detection should also be used to find any cracks. Springs with cracks or surface corrosion must not be reinstalled in the engine.

The *free length* of a spring should be within the limits specified by the manufacturer. If such information is not available, compare a new spring to a used spring. If the used spring is more than 3% shorter than the new spring, the used spring must be replaced. It should be noted, however, that the loss of spring pressure will not always be reflected in a proportional loss of overall length. Springs of the proper length may have lost enough tension to warrant replacement.

Checking Valve Spring Tension

To check a spring's open and close pressures (tensions), use either a torque tester or dial spring tester, **Figure 8-54.** *Close pressure* guarantees a tight seal; *open pressure* overcomes valve train inertia and closes the valve at the appropriate time. All spring pressures must be within 10% of the service manual's specifications. If the protective coating on the spring is nicked, a new coating should be applied. Corrosive conditions can be minimized by using clean lube oil, eliminating water leaks, and keeping vents open and clean.

Valve Rotators

Many heavy-duty diesel engine applications use some type of mechanism to rotate the exhaust valves. Rotating the valves prevents carbon from building up between the valve face and the seat. Carbon buildup can hold a valve partially open, causing it to burn. The most common types of *valve rotators* include release-type rotators and positive-type rotators.

Checking Release-Type Rotators

With the *release-type rotator,* **Figure 8-55,** spring tension is momentarily released, allowing the valve to rotate. During each valve cycle, the tappet or rocker arm forces the keepers away from the shoulder on the valve stem. This action releases the spring load from the valve. Engine vibration and moving gases cause the valve to rotate freely when the spring load is released.

Check the release-type rotator for clearance built into the *valve tip cup.* This built-in clearance should be main-

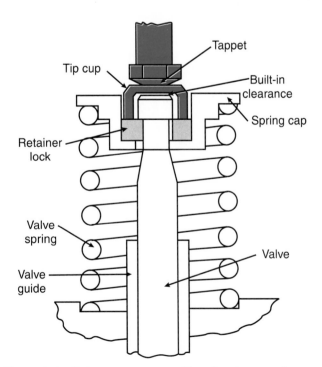

Figure 8-55. *Release type rotator. Note the clearance between the tip cup and the valve stem. (Deere & Co., Inc.)*

tained. Wear is uniform up to the wear limit. After the wear limit is reached, wear rapidly increases and may cause excessive valve clearance. To check tip cup clearance, use a special micrometer, such as the one shown in **Figure 8-56** and proceed as follows:

1. Remove the tip cup from the valve end.

2. Place the micrometer on the end of valve stem and set it at the "0" mark. Hold the plunger pin firmly on top of the valve stem tip and tighten the clamp screw.

3. Place the tip cup against the plunger pin and check reading on the micrometer spindle barrel. Readings to the right of "0" indicate too much clearance, and readings to the left of "0" indicate too little clearance.

Plastigage can also be used to check valve tip cup clearance (see **Chapter 6**). When using this method, a Plastigage strip is inserted between the tip cup and the valve stem tip. With the keepers held firmly against the shoulder on the valve, the cup is pressed in place. The width of the Plastigage is then checked. To reduce clearance, grind the open end of the tip cup. To increase clearance, grind the valve stem tip.

Positive-Type Rotators

The **positive-type rotator** is a self-contained unit that is installed in the same manner as an ordinary valve spring retainer. Rotation occurs when the valve moves and is dependent on the difference in spring loads between the valve's open and closed positions. See **Figure 8-57.** When the valve is closed, the spring washer load is located at points 1 and 2. As the valve starts to open, the extra spring load causes the washer to flatten, transferring the load from point 2 to point 3. This forces the balls to roll down the inclined

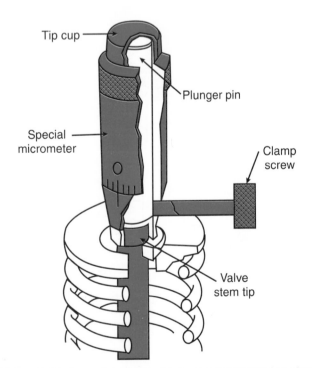

Figure 8-56. A special micrometer is commonly used to check the clearance of release-type rotators. (Deere & Co., Inc.)

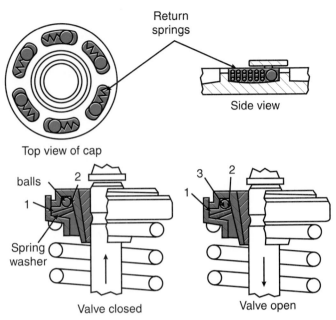

Figure 8-57. A positive-type rotator causes the valve to rotate with changes in valve spring pressure. (Deere & Co., Inc.)

races, rotating the whole assembly and transmitting the movement to the valve. As the valve closes again, the spring washer is released from the balls. The balls then return to their original position by the action of the return springs.

There is no maintenance required with positive-type rotators. However, when valves are replaced or reground, the rotators should also be replaced. Check for proper rotator opening by looking for carbon build-up on the valve or by inspecting the seat. (Rotation can also be checked by watching the valve with the engine idling.) On valves that use rotators, inspect the valve stem, valve stem cap, and the keepers for wear. Wear at these points results in false tappet settings and rapid increases in tappet clearance.

Valve Keepers

Valve keepers (locks) fit in the **keeper grooves** that are machined in the end of the valve stem. The keepers attach the entire spring assembly to the valve. Inspect the keeper's outside diameter and the beads for wear. If necessary, replace worn parts. It is advisable to replace all the keepers if several are worn. If an unworn keeper is inadvertently mated with a worn keeper, **Figure 8-58,** the retaining cap may shift. This may cause the valve tip to break.

Excessive keeper groove wear is a common cause of valve failure. To check the keeper groove for wear, simply run your finger around the groove. A sharp edge indicates the groove is worn.

Worn valve keepers and retainers may result if **valve stem caps** (used in some engines) are improperly fitted. Such caps are provided to protect and increase the service life of the valve stems. Trouble occurs when the cap does not bear directly on the end of the stem, but on the valve keepers. In some cases, spring retaining force is transmitted from the cap to the keepers or the retaining washer before reaching

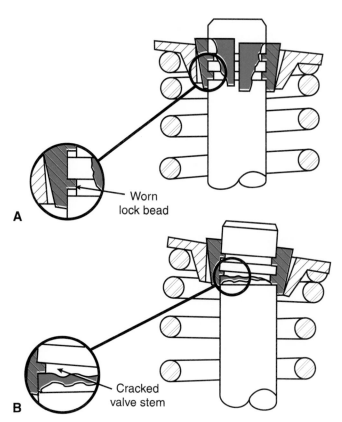

Figure 8-58. *Worn valve keepers may cause valve assembly problems. A—Worn lock bead. B—Cracked valve stem. (Sealed Power Corp.)*

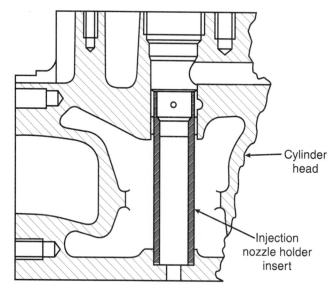

Figure 8-59. *Typical injection nozzle holder insert. Note that the water jacket surrounds the holder.*

the stem. This force causes the stem grooves and the valve keepers to wear extensively. As a result, the retaining washers may loosen and the valve stems may be broken.

The improper fit of a valve stem cap may result from the omission of spacer shims or the use of improper parts. *Steel spacer shims,* which are required in some caps to provide proper clearance, are placed between the ends of the valve stems and the caps. Omitting the shims will result in the cap's shoulder spring contacting the keepers. When disassembling a valve assembly, determine whether shims are used. If so, record their location and exact thickness. Check that the valve caps are of the proper size. The wrong caps will cause problems similar to those resulting from shim omission.

Injection Nozzle Holder Inserts

When inspecting the cylinder head, it is wise to check the *fuel injection nozzle inserts and bores.* These bores are usually machined in the cylinder head to provide a press fit for the injection nozzle, **Figure 8-59.** The holder inserts are checked in basically the same manner as valve guides. If they do not meet the manufacturer's specifications, they can generally be removed as follows:

1. Tap the inside diameter of the nozzle insert with an appropriate tap.
2. Install a nozzle sleeve puller to a slide hammer.
3. Thread the puller into the threaded end of nozzle holder insert.

4. Using a slide hammer, remove the insert from the bore.

After the insert has been removed, thoroughly clean the nozzle bore area and use a small magnet to pick up any metal chips, **Figure 8-60.** When installing the replacement injection holder insert, proceed as follows:

1. Install the insert on a nozzle sleeve installer tool.
2. Apply an adhesive/sealant to both the upper and lower ends of the holder insert. See **Figure 8-61A.**
3. Using an appropriate driving tool, drive the insert and the installer tool into the cylinder until it bottoms out on the lower counterbore face, **Figure 8-61B.**

Figure 8-60. *After tapping the holder insert, thread the puller in and remove the holder insert. Then remove all metal fragments with a magnet. (Mack Trucks, Inc.)*

A

B

Figure 8-61. Procedure for installing a nozzle holder insert. A—Place the insert on an installer tool and apply sealant to the insert. B—Driving the insert into the bore. (Mack Trucks, Inc.)

Summary

The cylinder head is mounted on top of the cylinder block. It houses a number of important engine parts, including intake and/or exhaust valves, valve guides, and valve seats. Rocker arm assemblies are frequently attached to the cylinder head. The fuel injection components are almost always located in the cylinder head (or heads) of a diesel engine.

The cylinder head must withstand the rapid changes of temperature and pressure taking place in the combustion space, as well as the stresses resulting from being bolted securely to the block. Passages in the cylinder head allow coolant to circulate around critical components.

Small inline engines commonly use one cylinder head for all cylinders. V-type engines typically have one cylinder head for each bank. Large diesel engines often have one cylinder head for each cylinder. However, some engines use one head to cover two or three cylinders.

The seal between the cylinder head and the block is made using studs, hold-down nuts, and gaskets. In some cases, bolts are used instead of studs and nuts.

When removed from the engine, the cylinder head must be carefully inspected for cracks, damage, warpage, burned or fouled combustion chambers, and corroded coolant passages.

A diesel engine must take in air and expel exhaust gases at precise intervals. The valves control this operation by opening and closing intake and exhaust ports to the cylinder. When the valves are closed, they must seal the combustion chamber.

A stem guides the valve during its up-and-down movement. The stem rides in a guide that is either machined into the head or pressed into the head as a replaceable part. Virtually all diesel engines use replaceable valve guides.

The valve seat is the area of the cylinder head that is contacted by the valve face. The seat may be machined in the head, but is most often pressed in as a replaceable unit.

There are two basic types of valve designs: rigid valves and elastic valves. Rigid valves are fairly strong, hold their shape well, conduct heat quickly, and wear slowly. Elastic valves are frequently used as exhaust valves. They are also used as intake valves in engines where high seating pressures are needed.

Valve faces and valve stem tips can be resurfaced on either grinding or cutting machines. The five techniques that can be used to restore worn valve guides are knurling, reaming, installing thin-wall liners, installing threaded bronze inserts, and replacing the entire valve guide.

Valve springs are used to close the valves and keep them closed until they are opened by the camshaft. Cylindrical springs are used by most engine manufacturers. In some cases, two springs are used for each valve to reduce spring vibration and valve flutter. Special spring dampers may also be used to reduce vibration and humming.

Rotators are frequently used to make the exhaust valves rotate. This keeps carbon from building up between the valve face and the seat. Carbon buildup can hold the valves partially open, causing them to burn.

Valve keepers fit in the keeper groove section of the valve stem and hold the spring assembly to the valve. Injection nozzle holder inserts should be inspected when servicing a cylinder head. Procedures for replacing the holders are similar to those for replacing valve guides.

Important Terms

Cylinder head

Cylinder head studs

Hold-down nuts

Torque specifications

Stud extractor

Locking pliers

Gaskets

Internal leakage

Valve covers

Crankcase vent breather
element

Rocker arm assemblies

Fuel injection nozzles

Glow plugs

Lift bracket

Valve spring compressor

Cracks

Pressure method

Magnetic crack detector
method

Fluorescent penetrant
testing method

Spot check dye method

Warpage

Fouling

Precombustion chamber

Burning

Corrosion

Poppet valves

Valve head

Valve face

Margin

Valve stem

Valve grinding machines

Valve cutting machines

Valve guides

Valve stem-to-guide
clearance

Knurling

Reaming

Thin-wall guide liners

Threaded bronze inserts

Stepped exhaust guide

Flanged guides

Valve seat

Valve seat width

Overhang

Integral seats

Insert seats

Interference angle

Negative interference
angle

Concentricity

Valve stem seals

Valve spring

Dampers

Damper coils

Valve spring retainer

Free length

Close pressure

Open pressure

Valve rotators

Release-type rotator

Valve tip cup

Positive-type rotator

Valve keepers

Keeper grooves

Valve stem caps

Steel spacer shims

Fuel injection nozzle
holder bores

Review Questions—Chapter 8

Do not write in this text. Place your answers on a separate sheet of paper.

1. To prevent damage, it is good practice to remove the _____ before removing the head.

2. List three possible causes of a cracked cylinder head.

3. Name two causes of cylinder head warpage.

4. Elastic valves are less prone to _____ than rigid valves.

5. It is not advisable to grind a valve face to a point where the margin is reduced by more than _____ .

6. When grinding a valve, it is important to:
   (A) chuck the valve close to the valve head to prevent stem flexing.
   (B) take light grinding cuts, using the full wheel width.
   (C) make certain coolant is striking the cutting area.
   (D) All of the above.

7. Most diesel engines use _____ valve guides.

8. Name at least three methods used to restore the inside diameter of worn valve guides.

9. All _____ should be replaced when a cylinder head is rebuilt.

10. A compression check that shows abnormally low compression on two adjacent cylinders may indicate a _____ .

ASE-Type Questions

1. Technician A says the best time to remove a cylinder head is when it is still warm, as the heat will help loosen frozen head bolts. Technician B says the head must always be cool before removal to prevent warpage. Who is right?
   (A) A only.
   (B) B only.
   (C) Both A & B.
   (D) Neither A nor B.

2. Technician A says that intake valves are normally larger than exhaust valves. Technician B says that exhaust valves are made of siliconchromium steel, which is capable of resisting corrosion caused by high temperatures. Who is right?
   (A) A only.
   (B) B only.
   (C) Both A & B.
   (D) Neither A nor B.

3. Hollow exhaust valve stems are normally filled with _____ to help the valve transfer and dissipate heat.
   (A) engine coolant
   (B) engine oil
   (C) metallic sodium
   (D) ceramic

4. The _____ are the parts that support the valves in the head.
   (A) valve seats
   (B) valve guides
   (C) rocker arms
   (D) locking tabs

5. Technician A says that you should tighten cylinder head stud nuts in the sequence illustrated in the engine service manual. Technician B says that you should perform at least two or three rounds of tightening before bringing all the nuts up to the specified torque. Who is right?
   (A) A only.
   (B) B only.
   (C) Both A & B.
   (D) Neither A nor B.

6. Technician A says when disassembling a diesel engine, keep the pushrods and rocker arm assemblies in exact order so they can be reassembled in the same position on the camshaft. Technician B says that these are not mated assemblies and that this step is not needed. Who is right?
   (A) A only.
   (B) B only.
   (C) Both A & B.
   (D) Neither A nor B.

7. Technician A says that a sharp edge between the valve face and the head is desirable. Technician B says that a sharp edge will not burn easy and will prevent preignition. Who is right?
   (A) A only.
   (B) B only.
   (C) Both A & B.
   (D) Neither A nor B.

8. Technician A says that many heads have flanged guides that must be driven out from a particular side. Technician B says that it does not matter which side you drive out a flanged guide. Who is right?
   (A) A only.
   (B) B only.
   (C) Both A & B.
   (D) Neither A nor B.

9. A properly ground valve seat has which of the following three angles for the top angle, seat angle, and throat angle, respectively?
   (A) 10°, 20 or 25°, 30°.
   (B) 15°, 30 or 45°, 60°.
   (C) 15°, 20 or 30°, 60°.
   (D) 10°, 30 or 45°, 60°.

10. Smoky exhaust, loss of power, and high compression are all symptoms of what cylinder head related problem?
    (A) Cracked head.
    (B) Leaking gasket.
    (C) Fouling in the combustion chamber.
    (D) None of the above.

Chapter 9

Camshaft and Valve Train Components

After studying this chapter, you will be able to:
- ❑ Remove and replace a diesel engine camshaft.
- ❑ Explain the operation of the valve train components.
- ❑ Determine the causes of valve guide wear.
- ❑ Describe the operation of a gear train.
- ❑ Identify and explain the purpose of various components of the valve timing drive assembly.
- ❑ Explain how to assemble valve train components.

The camshaft is responsible for the proper operation of the valves, and the injector pump in some cases. If there is any problem with the camshaft or in the valve train, engine performance will be noticeably affected. This chapter discusses camshaft and valve train service and repair.

Valve Train Operating Mechanisms

A diesel engine's intake and exhaust valves are opened and closed by movement of the valve train. Typical valve train components include the crankshaft, camshaft, lifters, push rods, rocker arms, bridges, and valve springs. In most cases, a gear on the camshaft is timed to and driven by a gear on the front of the crankshaft. Crankshaft rotation causes the camshaft to rotate. As the camshaft turns, the rotating lobes force the cam followers to move up and down. This movement is transferred to the push rods, which move the rocker arms. Rocker arm movement opens and closes the intake and exhaust valves at specific intervals. In some diesel engines, the valve train also operates the fuel injection system.

Camshafts

The *camshaft* is often called the "brain" of the diesel engine. See **Figure 9-1.** It acts as a mechanical computer,

synchronizing the operation of the valves and fuel injectors with the action of the pistons. A camshaft's operation is often considered the opposite of a crankshaft's operation. As mentioned in **Chapter 6,** the crankshaft converts the reciprocating motion of the pistons into rotary motion. A camshaft, on the other hand, converts rotary motion into the reciprocating motion of the rocker arms and/or valve stems. Another major difference between the two is that a crankshaft controls motion in two directions (pulling and pushing), and a camshaft controls motion in only one direction (pushing).

The camshaft is "programmed" by the precise machining of each lobe. The signals for opening and closing the valves and for triggering the injectors are determined by the cam lobe design. Instead of being round like the rest of the shaft, the lobes have an egg or *eccentric shape.* As a result, some parts of the lobe's surface are farther from the camshaft centerline than others. The eccentric shape of the lobes causes the followers, which ride on the lobes, to move up and down as the camshaft rotates.

Cam Lobes

There are several lobes on the camshaft for each cylinder in the engine. In a three-lobe arrangement, for example, one lobe operates the intake valve, one operates the exhaust valve, and a relatively large center lobe operates the fuel injector.

Originally, *cam lobes* were manufactured as separate pieces and were fastened to the camshaft. In most modern

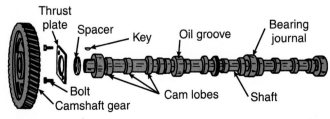

Figure 9-1. *Typical camshaft. Note the names of the various parts. (Detroit Diesel Corp.)*

diesel engines, however, the lobes are forged or cast as an integral part of the camshaft. The exception to this is found in some large engines. In these engines, the camshaft may be built up in sections or individual cam lobes that are assembled on the shaft. The camshaft extends along the entire length of a mobile or stationary diesel engine and is supported by bearing assemblies.

The bearings on mobile diesel engines are precision inserts that are split at their centerlines, **Figure 9-2.** The lower bearing shell is positioned in a saddle that is an integral part of the cylinder head. The upper bearing shell is held in place by a machined camshaft cap. Upper and lower crankshaft bearing shells have locating tangs and oil holes. The upper bearing shell oil holes are indexed with

oil supply holes in the camshaft caps. The camshaft caps are precision line bored after assembly to the cylinder head. Consequently, the caps are not interchangeable. Caps are often numbered and must not be interchanged with other caps of the same part number from stock or from a different cylinder head.

To reduce wear, camshafts are made of low carbon alloy steel. The cam and journal surfaces are often given a special coating before the final grinding is done. Accurately ground cams ensure efficient, quiet cam follower operation. After grinding, the lobes and journals are heat treated to provide a hard wear surface.

The camshaft thrust plate is sealed to the gear case with an O-ring that fits into a groove machined into the thrust plate outer diameter. A rubber seal fits into another groove in the plate face and seals the thrust plate to the cylinder head and the number one camshaft cap.

Camshaft Gear

In a four-stroke engine, the camshaft gear is twice as large as the crankshaft gear. This is necessary because the crankshaft must make two revolutions during each complete cycle, while the camshaft makes only one revolution during each cycle. On the other hand, the camshaft in a two-stroke engine turns at the same speed as the crankshaft. Therefore, the camshaft gear in a two-cycle engine is the same size as the crankshaft gear.

In most mobile diesel engines, the camshaft *drive gear* is keyed and pressed onto the camshaft. See **Figure 9-3.** The camshaft drive gear is fastened to the end of the camshaft by a retaining bolt. The camshaft keyway is indexed to the gear by a dowel. The camshaft drive gear rides on a thrust plate, which is retained by two bolts.

Valve Operation

The cam lobe shape determines when the valve will begin to open, how fast it will open, how much it will open (valve lift), and how fast it will close. The maximum lift

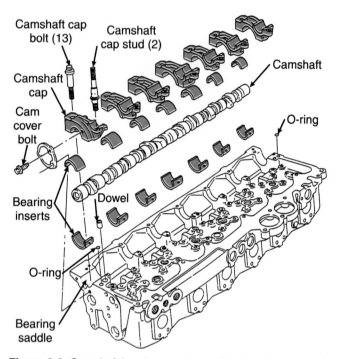

Figure 9-2. Camshaft bearings and caps. The bearings are split for ease of installation. Camshaft bearings installed in the block are one piece.

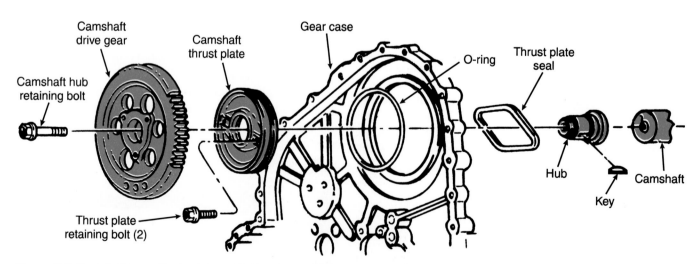

Figure 9-3. Camshaft gear, thrust plate, and related parts. The gear's size will vary depending on the type of engine.

occurs at the apex (nose or toe) of the cam lobe. The distance from the apex to the base circle is the **cam lift**. See **Figure 9-4.** When the valves are opened directly by the camshaft, the cam lift will be the **valve lift.** When rocker arms are used to open the valves, the valve lift can be varied by changing the ratio of the lengths of the valve rocker. The ratio is known as **rocker arm ratio.**

The cam lobe **duration angle** determines the amount of time the valve is open. It is measured from the point the valve first begins to open to the point where it closes on the valve seat. The duration angle is usually measured in degrees of crankshaft rotation. Note that crankshaft rotation is used instead of camshaft rotation to describe valve opening (duration) in relation to piston travel. See **Chapter 4** for more information on valve timing.

Camshaft Location

The camshaft can be found in either the cylinder block or cylinder head. If the camshaft is positioned in the block, it is positioned above the crankshaft and the valves are opened through lifters, pushrods, and rocker arms. As each cam lobe rotates, it pushes a lifter, which, in turn, lifts up on a pushrod. The pushrod moves one end of a rocker arm up and forces the other end downward until the valve opens. As the cam lobe continues to rotate, the valve spring forces the valve closed and maintains contact between the valve and the rocker arm. The spring also keeps the other components of the valve operating mechanism in contact with each other. Engines with camshafts in the engine block and the valves in the cylinder head are referred to as **overhead valve (OHV)** engines.

Overhead Camshaft

Overhead camshaft (OHC) engines have the camshaft mounted in or on the cylinder head. Most OHC engines do not use pushrods. As the camshaft rotates in an engine of this type, the lobes ride directly above the valves. The lobes open the valves either by directly depressing the valve or by depressing the valve through the use of a cam follower. The closing of the valves is still the job of the valve springs.

An advantage of the overhead camshaft engine is that valve train inertia is relatively low and there is little deflection in the system. Inertia is the tendency of an object to resist changes in momentum. The heavier the object and the greater its speed, the higher the inertia will be. Inertia is reduced on some OHC engines by eliminating the pushrods and the followers. With lower inertia, the chance of valve float at higher engine speeds is reduced.

Valve float occurs when the valve is momentarily thrown free from the direct influence of the cam lobe. This can occur at high engine speeds when the inertia of the lifters and pushrods momentarily overcomes the force of the valve springs, allowing gaps to form in the valve train. When the spring pressure finally resists additional movement, the parts come together forcefully and quickly. Valve float may affect the performance of the engine and may cause damage. Because of the reduced inertia in OHC engines, OHC camshafts can be machined to provide more rapid valve opening and closing than standard OHV camshafts. This results in increased volumetric efficiency, which improves engine efficiency.

Camshaft Drive Systems

The **camshaft drive system,** which is often referred to as the valve timing drive assembly, is the group of components that takes power from the crankshaft and transmits it to the camshaft. The drive mechanism does not change the type of motion in an engine, but it may change the direction of the motion. A gear, chain, or belt drive mechanism may be used to transfer power, **Figure 9-5.** Today, most mobile diesel engines are equipped with a gear drive mechanism, but some stationary engines are equipped with chain assemblies. A combination of gears and chains

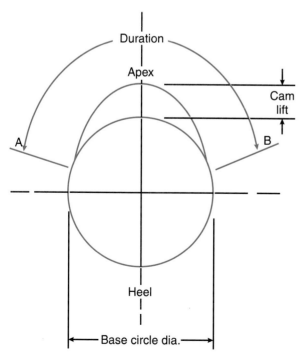

Figure 9-4. Cam lobe nomenclature. The duration determines the amount of time the valve is open. The longer the duration, the longer the valve stays open. (Detroit Diesel Corp.)

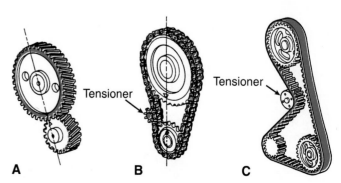

Figure 9-5. Engine timing drives. A—Direct drive gear drive. B—Chain and gear drive. C—Gear and belt drive.

is sometimes used as the drive mechanism on larger industrial and marine diesel engines.

Some engines have a single drive mechanism that transmits power for the operation of the cam and other engine parts and accessories. In other engines, there may be two or more separate drive mechanisms. When separate drive mechanisms are used, the one that transmits power for the operation of the accessories is called the *accessory drive,* or *auxiliary drive.* Some engines have more than one accessory drive.

Gear Train

When the drive mechanism consists only of gears, it is commonly referred to as a **gear train.** A typical gear train for a four-stroke diesel engine with various accessories is shown in **Figure 9-6.** As previously mentioned, the camshaft gear is twice the size of the crankshaft gear in the four-stroke engine. On many two-stroke engines, there is a camshaft gear and a balance shaft gear. These gears mesh with each other and rotate at the same speed as the crankshaft. Depending on the direction of engine rotation, either of the gears may be driven from the crankshaft timing gear through an idler gear. The function of an idler gear is to convey driving force. The idler gear mates with the left hand gear in engines that rotate clockwise and with the right hand gear in engines that rotate counterclockwise.

Some engines employ one or more idler gears between the crankshaft and camshaft timing gears, as well as between other gears in the gear train. In some engines, accessories such as the fuel injection pump, governor, air compressor, fan drive, blower, and tachometer drive are driven by an auxiliary shaft, which is part of the gear train.

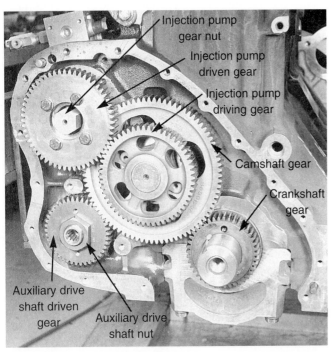

Figure 9-6. *Four-stroke diesel engine gear train. The camshaft gear is larger than the crankshaft gear. (Mack Trucks, Inc.)*

Injection pump gear nut

Injection pump driven gear

Injection pump driving gear

Camshaft gear

Crankshaft gear

Auxiliary drive shaft driven gear

Auxiliary drive shaft nut

Auxiliary Drives

On some industrial and marine two-stroke diesels, the accessory drives are located at the rear of the engine to accommodate both gear-driven and belt-driven accessories. For example, the drive for a direct gear-driven accessory, such as a raw water pump, is made up of a spacer (when used), an accessory drive plate, and a drive coupling. The drive plate and spacer are bolted to the camshaft or balance shaft gear. The accessory is bolted to the flywheel housing and driven by the drive coupling, which is splined to both the accessory shaft and drive plate.

The gear train of most mobile diesel engines is equipped with idler gears to drive the fuel-injection pump, the compressor, and/or the blower. The idler gears rotate on bushings or bearings on a dummy shaft or hub. The bushings or bearings are pressure lubricated. The idler gears have an odd number of teeth to prevent meshing of identical teeth with each revolution. As a result, tooth life is increased and gear noise reduced.

Some gear trains have a **bull gear.** This large gear is usually driven directly by the crankshaft timing gear. The bull gear directly or indirectly drives all the gear-driven engine accessories. In two-cycle diesel engines, a small camshaft idler gear is often located behind the bull gear and turns at the same speed as the bull gear. Of course, not all gear train assemblies have an idler or bull gear.

The difference between the number of teeth on a driving gear and the number of teeth on a driven gear determines the number of times the driven gear turns for each revolution of the driving gear. This difference in gear revolutions is called the **gear ratio.** As a rule, the gear ratio for each gear is given in relationship to the crankshaft timing gear, **Figure 9-7.**

Gear trains are generally lubricated by overflow oil from the camshaft and balance shaft pockets spilling into the gear train compartment. In most two-stroke engines, a certain amount of oil also spills into the gear train compartment from the idler gear bearing and the camshaft and balance shaft end bearings. The blower drive gear bearing is often lubricated through an external pipe leading from the main cylinder block oil gallery to the gear hub bearing support. The idler gear bearings are usually pressure lubricated by oil passages in the idler gear hub. These passages are connected to the oil gallery in the cylinder block.

Chain Drive

In **chain drive systems,** the camshaft and crankshaft sprockets are linked by a continuous chain. In most industrial applications, both the camshaft and crankshaft sprockets are made of steel.

Engine Timing

The relationship between the crankshaft and camshaft must be maintained to properly control the opening and closing of the intake and exhaust valves, operate the fuel injection system, and maintain engine balance. The crankshaft timing gear on mobile equipment engines is generally mounted in only one position on the shaft because one attaching bolt hole in the shaft is offset. Similarly, the camshaft gear can be mounted in only one position due to the location of the gear key. When the engine is properly timed, the *timing marks* on the various gears will match, **Figure 9-8.**

Figure 9-9 shows typical valve timing diagrams for both four- and two-cycle diesel engines. For a particular engine, the valve timing depends on several factors, including cam

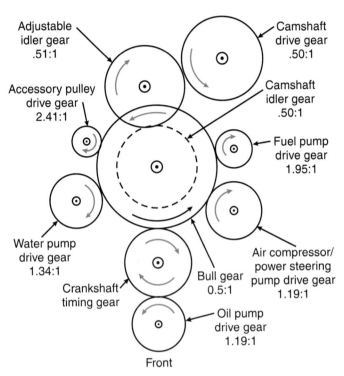

Figure 9-7. *Two-stroke diesel engine gear ratio and rotation featuring a bull gear. (Detroit Diesel Corp.)*

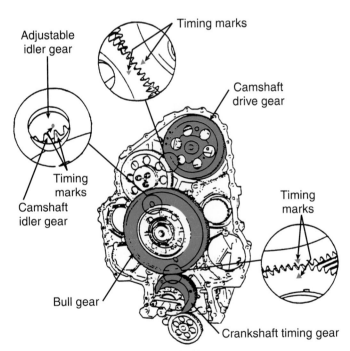

Figure 9-8. *Matching the timing marks on the various gears is critical for engine timing. Remember there are no standards for timing marks, therefore check the service manual for the exact timing for the engine being worked on. (Mack Trucks, Inc.)*

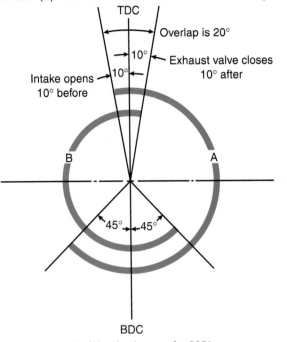

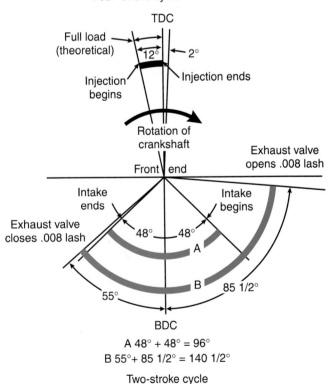

Figure 9-9. *Typical crankshaft-valve timing diagram of four-cycle and two-cycle engine.*

profile, valve lift, and engine speed. In the four-cycle timing diagram illustrated, the intake valve opens at 10° *before top dead center (BTDC)*. It stays open until 45° *after bottom dead center (ABDC)*. During this time, the engine takes in its full charge of air and the piston begins its compression stroke. After the air is sufficiently compressed, the fuel is injected into the cylinder. The high temperature of the compressed air ignites the fuel, and the power stroke begins.

When the piston reaches a point 45° *before bottom dead center (BBDC)*, the exhaust valve opens. The valve does not close until the piston has traveled to 10° *after top dead center (ATDC)*. Note that an increase in valve clearance will delay, or retard, the opening of a valve and will speed up, or advance its closing. Therefore, an increase in clearance will shorten the time the valve is open. A decrease in clearance will have the opposite effect. An excessive decrease in clearance, however, may prevent the valve from seating firmly, causing loss of power, valve seat burning, and other problems.

The cam profile consists of the face, which is the opening ramp on the cam lobe; the nose, which holds the valve open; and the backside, which allows the valve to close. As shown in **Figure 9-10,** the slope of the cam lobe determines how fast the valve opens and closes.

❑ Cam number 1 has curved face, nose, and back surfaces that cause the valve to open faster at the start and to hold the valve open wide until the closing surface is under the valve tappet.

❑ Cam number 2 provides a rapid opening and closing with a long period at wide open.

❑ Cam number 3 is sometimes employed in high speed engines to hold the valve open as long as possible.

The amount of time that a valve should remain open for any given diesel engine is determined by manufacturers' tests. Remember that an engine that is out of time may result in poor idle and a loss of power.

Cam Follower Assemblies

There are two basic types of cam follower assemblies used in diesel engines: lifter followers (tappets) and finger followers (bucket followers). On overhead camshaft (OHC) engines, the follower movement will move the rocker arms. When the camshaft is located in the block (OHV engines), the movement of the followers must be transferred through pushrods to move the rocker arms.

Although lifter cam followers can be used with most engine configurations described in **Chapter 4,** finger or inverted bucket followers are used in most overhead camshaft (OHC) diesel engines. With either type, the cam followers are essentially motion converters.

Operation of Lifter Cam Followers

As the end of the *lifter cam follower* in an OHV engine moves over the camshaft lobe surface, it translates the lobe's eccentric shape into a reciprocating motion that travels

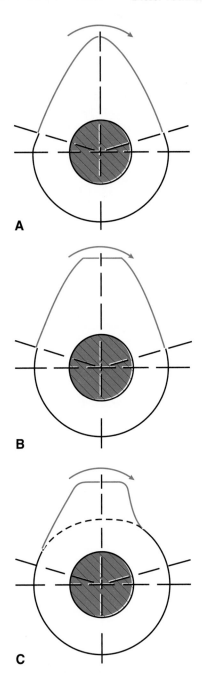

A

B

C

End view of camshafts

Figure 9-10. *Cam profiles determine how fast the valve opens. A—The curved face will cause the valve to open fast and remain open until the closing surface is under the tappet. B—This profile provides great wide open duration as well as rapid opening and closing. C—This profile will hold the valve open for a wide duration. This is necessary on high speed diesel engines.*

through the push rod and rocker arm, **Figure 9-11.** Actually, the rocker arm pivots on the rocker arm shaft to transmit motion from the push rod to the valve stem. On many modern mobile diesel engines, an additional push rod and rocker arm is used to activate the injector, **Figure 9-12.**

Valve lifters in most inline diesel engines are of the solid type. This means they provide a rigid connection between the camshaft and the valves. Solid lifters require a clearance between the parts in the valve train. This clearance allows for expansion of the valve train components as

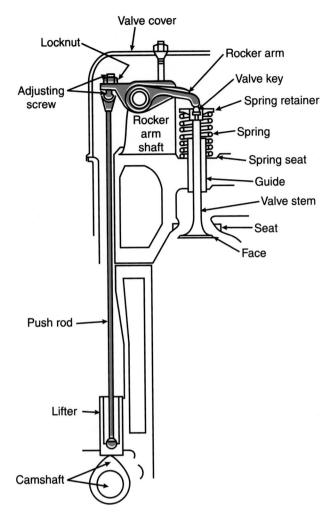

Figure 9-11. *Valve train components. The parts shown here are used in overhead valve engines.*

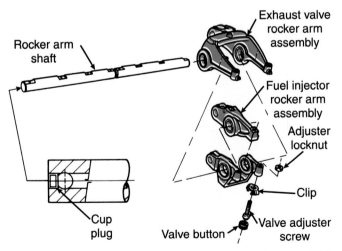

Figure 9-12. *The rocker arm and its related parts. This assembly uses a separate rocker arm for the fuel injection system. (Detroit Diesel Corp.)*

the engine gets hot. Periodic adjustment of this clearance must be made, and excessive clearances may cause a clicking noise. This clicking noise may also indicate the hammering of valve train parts against one another, which will result in reduced camshaft and lifter life. A socket is

usually machined to accept the end fitting of the push rod in the top of the lifter. In many lifters, this socket is fitted with a replaceable insert of hardened metal.

Operation of Finger Cam Followers

As previously mentioned, some OHC engines have **finger cam followers.** With this type of follower, the camshaft acts directly on the follower assembly. In such an arrangement, **push rods** or lifters are not needed, because the rocker arms usually incorporate the follower rollers, **Figure 9-13.** The followers ride directly on the camshaft lobes and transmit cam motion directly to the valves and the injectors.

Rocker Arms

Rocker arms can be used in both overhead cam (OHC) and overhead valve (OHV) engines. They are usually made of forged cast iron or laminated steel. The arms have drilled (or cast) sockets or ball sockets screwed into them to allow for injector adjustment, **Figure 9-14.** The end of the rocker arm that contacts the valve stem is hardened or has an insert of high strength steel. This area is ground so that a minimum of side thrust is exerted on the valve stem as the rocker arm pivots. A bushing is pressed into the rocker arm bore to reduce friction and increase service life. The fulcrum point of the rocker arm is generally closer to the push rod end of the arm to multiply the cam shaft lift to a ratio that ranges from 1:1 to 1:1.75. See **Figure 9-15.**

Generally, there is a set of three rocker arms for each cylinder. The intake rocker arm opens the intake valves, the injector rocker arm works the fuel injector (some engine designs), and the exhaust rocker arm opens the exhaust valves. Some large engines have two exhaust valves and are equipped with two exhaust rocker arms. The injector rocker arm is generally heavier than the intake and exhaust rocker arms because greater force is required to operate the fuel injectors than to operate the valves.

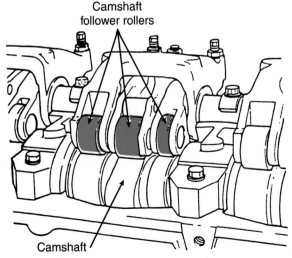

Figure 9-13. *Typical view of an overhead camshaft cam follower arrangement. (Detroit Diesel Corp.)*

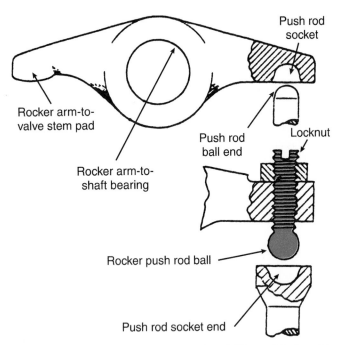

Figure 9-14. *Rocker arm socket and ball. The rod ball end is slotted and held by a locknut to allow for valve adjustment.*

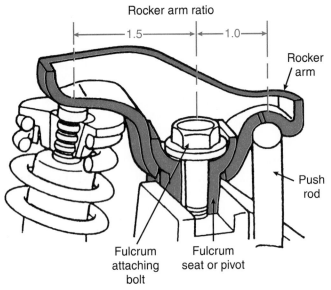

Figure 9-15. *Rocker arm ratio is determined by the distance from the center of the rocker pivot to the contact points of the push rod and valve stem tip. The greater the distance from the bolt to the end that touches the valve, the greater the ratio.*

Rocker Arm Shafts

The **rocker arm shaft** may be solid or hollow and is made of hardened steel. It is supported either in a separate rocker arm housing or by two or more rocker arm shaft pedestals. Holes are drilled into the rocker arm shaft at precise locations to help lubricate each rocker arm bushing. On OHV engines, the push rods or tubes form a critical link between the lifters and the rocker arms.

When two valves are used for the same function, a **crosshead** is often added to form a bridge between the valves so they can be opened by a single rocker arm. When a crosshead is used, a guide is installed between the valves to keep the crosshead properly positioned. The rocker arm contacts the crosshead instead of the valve stem tip. See **Figure 9-16.** The crosshead then exerts force on both valves, causing them to open simultaneously.

Valve Train Lubrication

The valve and injector operating mechanism is usually lubricated from a longitudinal oil passage on the camshaft side of the cylinder head, which connects with the main oil gallery in the cylinder block. Oil from this passage usually flows through drilled passages in the rocker shaft bracket bolts, through passages in the rocker arm shaft, and to the rocker arms, **Figure 9-17.** Some rocker arms are cross-drilled to supply lubrication to the push rods, cam followers, and valves.

Overflow oil from the rocker arms lubricates the exhaust valves, valve crossheads, and cam followers. The oil then drains from the top deck of the cylinder head and flows into the camshaft pockets in the cylinder block. From the camshaft pockets, the oil flows back to the oil pan. The cam follower rollers are lubricated with oil from the followers, camshaft lobes, and oil pressurized from milled slots in the camshaft bearings.

In some diesel engines, the cam follower assemblies are mounted in separate housings and are installed through an opening in the side of the engine block. Many engines use individual rocker arm assemblies, **Figure 9-18.** In addition to forming an oil-proof seal, the follower housing gasket also acts as a shim or spacer. Changing the gasket thickness advances or retards the valve and fuel injector timing by altering the point of contact between the cam followers and the camshaft.

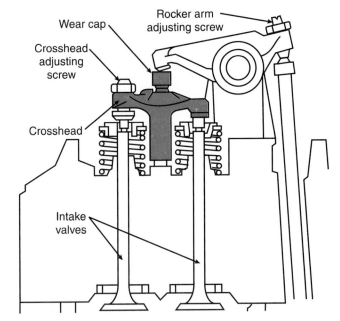

Figure 9-16. *Rocker arm contacts on a crosshead bridge. This setup has two adjustments. (Deere & Company, Inc.)*

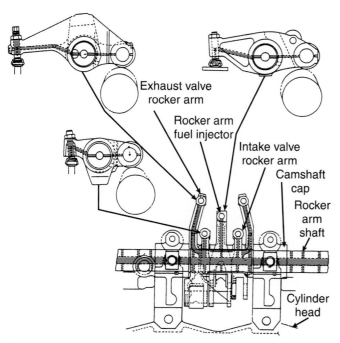

Figure 9-17. Cylinder head lubrication schematic. (Detroit Diesel Corp.)

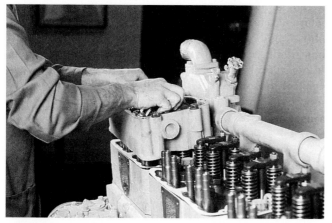

Figure 9-18. Installing a rocker assembly on an engine block. Many engines use individual rocker arm assemblies. (Cummins Engine Co., Inc.)

Rocker Housing Cover

In most engines, a separate cam follower housing is not used. Instead, the valve train is located under a *rocker housing cover.* This cover is fitted with a gasket and is secured to the cylinder head with bolts. Rocker housing covers are usually designed to keep dirt out of the engine. Most covers rest on a gasket and are held in place by bolts that must be tightened in the proper sequence to ensure equal seating pressure and to prevent gasket damage.

To equalize the pressure between the atmosphere and the crankcase, a filtered air vent or breather is often used on top of the rocker housing covers, **Figure 9-19A.** On some engines, the filter is eliminated and crankcase ventilation is provided by the breather body and a vent tube, **Figure 9-19B.** Many rocker housing covers also have a filler tube through which oil can be added to the engine.

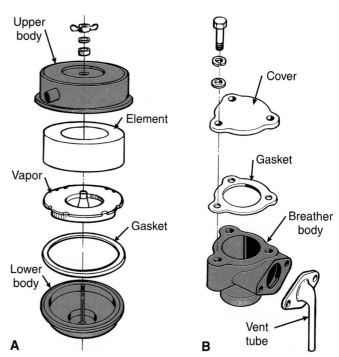

Figure 9-19. Two types of rocker housing vents. A—Round type. B—Breather body type.

The filler tube is plugged by an oil filler cap. Usually, a sealing cap is used. See **Figure 9-20.** Rotating the lever on top of the cap expands a rubber grommet to form a seal between the filler tube and the filler cap.

Inspecting and Servicing the Valve Train

Once the rocker housing is removed, the valve train components can be disassembled for inspection and service. Always remove the components as directed in the

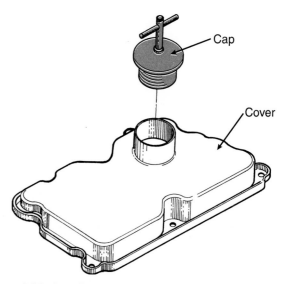

Figure 9-20. A sealing cap is used on many diesel engines. The lever on the cap expands a seal that prevents dirt and other foreign substances from entering the crankcase.

service manual. As shown in **Figure 9-21,** special tools are sometimes required during the removal procedure. It is important to tag or label all cam follower parts as they are removed from the engine with the appropriate cylinder number as well as their relative position to the other valve train parts, **Figure 9-22.** Many of the cam follower parts are interchangeable. However, it is best to reinstall them in the position from which they were removed to maintain established wear patterns.

The valves and fuel injectors operate only as well as their actuating mechanisms. Therefore, each valve train component must be checked and replaced or repaired if necessary. Most of these components are replaced rather than repaired. The exception is the machining of some types of rocker arms.

Camshaft Service

Remove the camshaft as shown in **Figure 9-23,** and place it on a V-type holder frame. Thoroughly clean the shaft with a solvent and a lint-free cloth. After removing the

Figure 9-23. Removing the camshaft. (Perfect Circle/Dana)

camshaft, thoroughly check the camshaft bearings. See **Figure 9-24.** To do this, proceed as follows:

1. Check the ID of the bearings with a telescoping gauge or inside micrometer, **Figure 9-25A.**

2. If the ID exceeds the tolerance given in the service manual or if there is evidence of scratching or scoring, all bearings should be replaced. When removing a bearing, use the tool recommended by the engine manufacturer.

3. Drive out the front bearing first, then remove the remaining bearings in sequence, **Figure 9-25B.**

4. If there is any indication the bearing has turned in the block, measure the bearing bore diameter with a telescoping gauge or an inside micrometer as in step 1 and check the finish in the block.

5. To complete the job, use clean solvent and a brush to clean the main bearing-to-cam bore oil drilling.

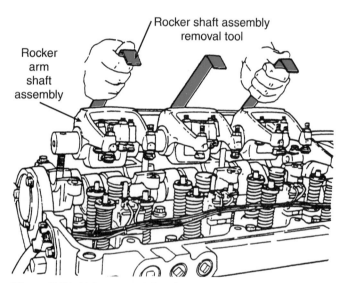

Figure 9-21. Using a special rocker arm shaft assembly removal tool to remove OHC cam followers. (Detroit Diesel Corp.)

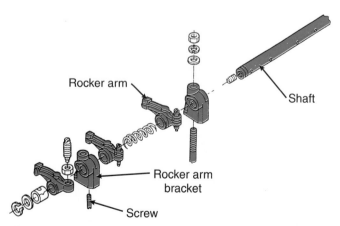

Figure 9-22. Typical cam follower/valve group. Label each part as you remove them from the engine.

Figure 9-24. Measuring the camshaft bearing with a snap gauge. (Mack Trucks, Inc.)

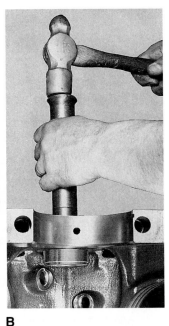

A   B

Figure 9-25. Checking and servicing the camshaft bearing. A—Measuring bearing wear with a dial bore gauge. Drive any worn bearings out using the proper tools. B—Carefully drive the bearing out. Remove each bearing in the engine from front to back. (Mack Trucks, Inc.)

After cleaning the camshaft, check each lobe for scoring, scuffing, fracturing, or other signs of wear. See **Figure 9-26.** Premature wear is usually caused by metal-to-metal contact between the cam lobe and lifter bottom. When this occurs, the nose will be worn from the cam lobes, and the lifter bottoms will be worn to a concave shape. In some cases, the lifter bottoms are worn completely away. This type of failure usually begins within the first few minutes of engine operation. It is the result of insufficient lubrication or use of an oil that does not meet the engine manufacturer's requirements for viscosity and service grade. See **Chapter 10.**

Checking Camshafts for Wear

The simplest way to measure cam lobes for wear is to use an outside micrometer. Position the micrometer to measure from the heel to the nose of the lobe. Measure the lobe at 90° from the original measurement. Record the dimensions measured for each intake and exhaust lobe. Any variation in lobe height indicates wear. Check all measurements against the manufacturer's recommended lobe height. A cam lobe lift loss of .005″ (.13 mm) or less is generally acceptable. If desired, the cam lobes can also be checked with a dial indicator, **Figure 9-27.**

Measure the camshaft journals in several places with an outside micrometer to determine if they are badly worn. If any journal diameter is .001″ (.025 mm) or more below the manufacturer's specifications, the camshaft should be reground to a standard undersize or replaced. See **Figure 9-28.**

Check the camshaft for straightness with a dial indicator. Place the shaft on V-blocks and position the dial indicator on the center bearing journal. Slowly rotate the camshaft. If the dial indicator shows a runout variation of more than .002″ (.050 mm), the camshaft is not straight. If the shaft is not excessively bent, it can usually be straightened by an experienced engine rebuilder.

 Note: A bent camshaft should be replaced, if possible.

Figure 9-26. Inspecting the camshaft for abnormal wear. Note the worn lobe. (Mack Trucks, Inc.)

Figure 9-27. Checking camshaft lobes with a dial indicator. The same arrangement can be used to check a camshaft for straightness. (Goodson Shop Supplies, Inc.)

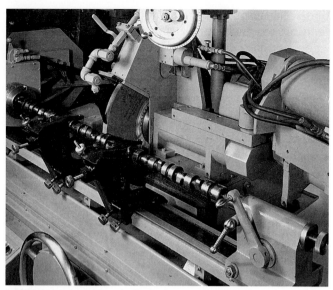

Figure 9-28. A worn camshaft can be reground to an acceptable undersize. Ideally, any worn or bent camshaft should be replaced. (Storm Vulcan, Inc.)

Inspect the camshaft drive gear and the injection pump drive gear for pitting, scoring, broken teeth, or unusual wear. Examine these components carefully. Look for cracks at the root of their teeth. Inspect all the timing gears in the same manner. If it is necessary to take a closer look, a gear can be removed from its shaft by removing the bolt(s) that hold it in place, **Figure 9-29,** and pulling the gear from the shaft, **Figure 9-30,** or, by removing the shaft and pressing the gear off.

If there is evidence of wear, the lobes must be reground or the camshaft replaced. In certain cases, the lobes can be rebuilt by welding a layer of metal onto the lobe surfaces and regrinding the built-up lobe. After grinding, the camshaft lobe surfaces are covered with a hard surface overlay. The overlay is harder and more slippery than the base metal of the camshaft and prevents rapid lobe wear. Grinding or rebuilding a camshaft is only done in a specialized rebuilding shop.

Figure 9-30. Removing a gear from an engine using a puller. (Cummins Engine Co., Inc.)

Cam Follower Service

When removing cam followers for inspection, mark each one so it can be returned to the bore from which it was removed, **Figure 9-31.** After the followers have been removed, clean all parts thoroughly with solvent and dry them with compressed air. Look on the face of each follower for signs of unusual wear, such as galling, pitting, scoring, or cracking. If the follower is scuffed or "dished," it must be replaced. Additionally, the camshaft may also need to be replaced when these conditions are present in the followers.

Use clean solvent and a brush to clean the main oil rifle or hole to the overhead oil drilling, **Figure 9-32.** Measure the lifter bores (if so equipped) with a dial bore gauge to make sure they conform to the manufacturer's specifications, **Figure 9-33.** Also, check the bores for score marks. If necessary, remove any marks with crocus cloth.

Normal follower operation will cause some surface disintegration, usually as a result of wear. Nicks and dents on rollers will also start disintegration. Check periodically for defective roller surfaces. Also look for nicks, scratches, or dents

Figure 9-29. Removing the bolt from a gear. It may be necessary to use a prybar to hold one of the other gears to prevent the engine from turning. (Cummins Engine Co., Inc.)

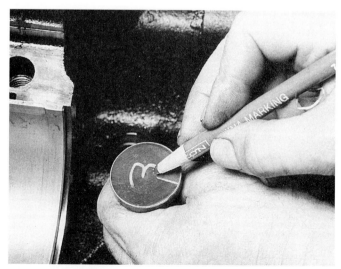

Figure 9-31. Tag the lifter so that it can be returned to its original location. Also, check the valve lifter for unusual wear patterns. (Mack Trucks, Inc.)

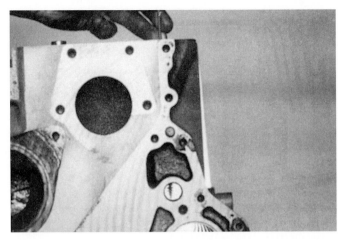

Figure 9-32. Cleaning the main oil rifle to overhead oil drilling with a small brush. (Cummins Engine Co., Inc.)

Figure 9-33. Checking a valve lifter bore with a dial bore gauge. (Mack Trucks, Inc.)

in the camshaft. When a defective cam follower is discovered, it should be replaced. Roller cam followers must be replaced if the follower body, guide or needle bearings (if used) are worn. The rollers must turn smoothly and freely on their pins, and they must be free of flat spots or scuff marks. If the rollers do not turn freely or have been scored or worn flat, examine the cam lobes on which the rollers operate. When a camshaft is replaced, the cam followers should also be replaced.

Lifter cam followers that have a flat surface instead of a roller should be checked for excessive wear and pitting.

Rocker Arm Service

After removing the rocker arms, wash them in clean solvent and dry them with compressed air. Use a wire brush to clean the oil passages in the rocker arms (if used).

Inspect the rocker arm shaft for scratches, scores, burns, or excessive wear at points of the arm contact. Examine each rocker arm for cracks.

Check the rocker arm surface where it contacts the valve stem or the bridge. Look for cupping or concave wear on this surface. See **Figure 9-34**. Any concave wear will make it difficult to adjust the followers properly. A worn rocker arm may also create side thrust on the valve stem, causing wear. Replace or recondition the rocker arms when any noticeable grooving or wear occurs.

Examine the rocker arm adjusting screw and nut (if used) for damage. Use a small hole gauge to check the ball ends on the adjusting screws. If the adjusting screw threads or ball ends are damaged or flattened, the screw must be replaced. Also examine the springs between the shaft and the rocker arms. Make sure the springs are strong enough to exert a positive pressure on the arms.

Measure the rocker arm bushings for wear. The repair procedure for worn rocker arm bushings is the same as that for piston pin bushings. Excessively worn bushings must be replaced. Installation of a new bushing generally requires that the bore be reamed and an oversize bushing installed.

Push Rod Service

Check the push rod ends for wear or damage. Some rods extend through the cylinder head. All push rods are subject to tip wear and any badly worn or plugged push rods should be replaced. Check for bends by placing the rod on a flat metal plate or holding fixture. Rotate the rod and measure the deflection with a dial indicator, **Figure 9-35**. If the rod is bent, it should be replaced.

If a failed valve has been replaced, also carefully inspect its mating rocker arm and push rod. Although not immediately noticeable, the rocker arm and push rod (tube) are often bent or otherwise damaged when the valve fails.

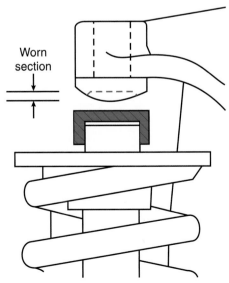

Figure 9-34. Worn rocker arm tip. Note the curved (concave) wear on the rocker arm surface. (Deere & Company, Inc.)

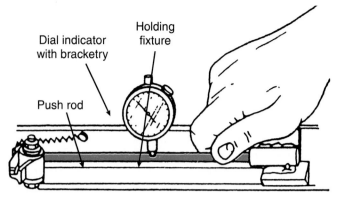

Figure 9-35. *Checking for push rod runout and bends. (TRW, Inc.)*

Note: Occasionally, a push rod will develop a small crack and become filled with oil. The extra weight of the oil in the push tubes causes increased wear of the cam follower and the camshaft. A push rod suspected of containing oil can be tested by lightly striking one end of the tube against a hard surface. If the tube makes a ringing sound, it is good. However, if it makes a dull sound, the tube is probably full of oil and should be replaced.

In some diesel engines, the exhaust valve push rods are fitted with collars that are engaged by a compression release shaft. The purpose of the compression release is to help the engine start when it is cold by holding the exhaust valves open, reducing compression in the cylinders. On some engines, guide plates are used to limit side movement on the push rods. The *guide plates* hold the rods in alignment with the rocker arms. When the rods pass through holes in the cylinder head or intake manifold, guide plates are not needed.

Assembling the Valve Train Components

After all valve train parts have been cleaned, inspected, and reconditioned or replaced, they can be reassembled on the cylinder head. All cylinder head components and valve train parts should be reinstalled in the reverse order of removal. Apply the cylinder head core sealant to the freeze plug bores as shown in **Figure 9-36A** and install the freeze plugs in the cylinder head, **Figure 9-36B**.

Camshaft Installation

Before installing the camshaft, the bearings, thrust washer, and timing gear must be replaced if they were removed. See **Figure 9-37**. To install the camshaft bearings, use the installation tool recommended by the engine manufacturer. Install the camshaft bearings from the back of the

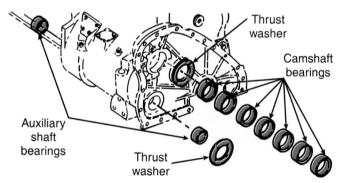

Figure 9-37. *Install new camshaft and auxiliary shaft bearings and thrust washers in the block. Make sure the oil passages line up. Carefully drive in the bearings using the proper driver tool. (Mack Trucks, Inc.)*

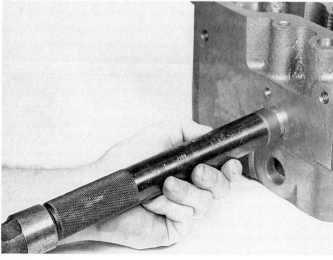

A   **B**

Figure 9-36. *Installing new freeze plugs. A—Apply a small amount of sealant to the plug. Do not use an excessive amount. B—Install the plug until it is just flush with the engine part. (Mack Trucks, Inc.)*

engine to the front (reverse order of removal). The oil supply holes in the block must be aligned with the oil holes in the bearings.

If the camshaft bearing journals were reground, remember that the new bearings must be thicker than the original bearings. Most diesel engines use camshaft bearings that have the same outside diameter. However, a few engines use bearings with different size outside diameters. It is a good idea to measure the inside diameter of each bearing bore with a telescoping gauge, **Figure 9-38.** Compare the measurements with the outside diameters of the bearings.

After installing the bearings, install the camshaft thrust washer or plate, which limits the end play of the camshaft at the front of the block in the counterbore. When required, secure the washer with the proper retaining pin. Drive the pin into the block using an appropriate punch, making certain the retaining pin does not protrude above the surface of the thrust washer. Do not use extreme force when installing the pin.

If the camshaft timing gear was removed during engine disassembly, the gear should be replaced as follows:

1. Lightly tap the key into the keyway on the camshaft.
2. Install the thrust washer on the camshaft.
3. Heat the timing gear to approximately 425°F (218.3°C) in an oven or on a hot plate, **Figure 9-39.**
4. Once the timing gear has reached the proper temperature, align the keyway and position the gear on the shaft with its timing marks facing out. Then push downward with a quick, even pressure until the gear bottoms on the thrust washer, **Figure 9-40.**

 Warning: Wear protective gloves when handling the heated timing gear.

A few camshaft timing gears are press-fit on the camshaft. When replacing this type of gear, it must be pressed with a bearing press before shaft installation.

Before installing the camshaft, use clean engine oil to lubricate the cam journals and lobes. Also, lubricate the camshaft bearings and the thrust washer face in the cylinder

block. Carefully install the camshaft through the cam bearings to prevent damage to the inside diameter of the bearing, **Figure 9-41.** Once the camshaft is completely installed in

Figure 9-39. Heat the camshaft gear to expand the material. (Mack Trucks, Inc.)

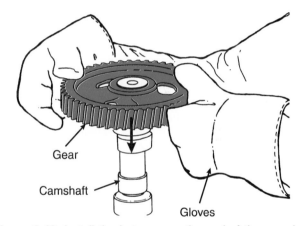

Gear

Camshaft

Gloves

Figure 9-40. Install the hot gear on the end of the camshaft. Use gloves to avoid a painful burn.

Figure 9-38. Measure the camshaft bearing bore before installing new bearings and the camshaft. (Mack Trucks, Inc.)

Figure 9-41. Take your time while installing the camshaft. It is very easy to scratch one of the camshaft bearings. If a bearing is damaged, it will have to be replaced.

the block, you should be able to turn it by hand. If the camshaft does not turn, binding might be the cause. Binding is the result of a damaged bearing, a nick on a cam journal, or a slight misalignment of the block journals.

The camshaft should be tested for both end play and runout. End play is checked by making certain that clearance between the camshaft boss (or gear) and the thrust washer is within the manufacturer's specifications. Make this check with a feeler gauge.

Before assembling the serviced valve train operating mechanisms, there are two critical measurements that must be carefully checked: the installed valve stem height and the installed valve spring height. To check OHC camshaft runout at various bearing journals, use a dial indicator with a magnetic base, **Figure 9-42.** If the camshaft runout exceeds specifications, the camshaft must be replaced.

Assembling the Camshaft Gears

When installing camshaft gears, follow the sequence given in the service manual. Here are the instructions for camshaft gear, hub, and thrust bearing installation as given in a typical service manual.

1. Lubricate the contact surfaces of the hub and camshaft thrust plate with clean oil and install the hub to the camshaft thrust plate.
2. Install the key into the keyway on the hub.
3. Apply a thin film of lubricant to the bore of the camshaft gear.
4. Support the hub from the engine side. Align the keyway in the camshaft drive gear with the key in the hub and press the gear onto the hub until it is firmly seated against the shoulder. Make sure all the timing marks on each gear are aligned, **Figure 9-43.**
5. While supporting the camshaft thrust plate, position the camshaft drive gear assembly with the engine side facing down. Assemble a dial indicator and magnetic base so that the indicator stem rests on face of the camshaft drive gear just inboard of the drive gear teeth. Zero the dial to the indicator.

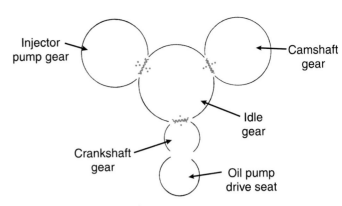

Figure 9-43. *When installing a typical gear drive train, be sure to line up the timing marks.*

6. Rotate the drive gear two full rotations. As the gear is rotated, the dial indicator needle may register both to the left and to the right of zero. The total amount that the dial indicator needle moves to the left and right of zero, added together, gives the ***total indicated runout (TIR).*** Maximum allowable TIR is given in the manual.
7. Lubricate the camshaft thrust plate O-ring with clean engine oil and install it into its groove in the camshaft thrust plate.
8. Coat one side of the square cut diamond seal with a spray adhesive and install it into its groove on the engine side of the thrust plate.
9. Be sure the sealing surface of the gear case is clean and free of burrs, **Figure 9-44.** Install the hub and camshaft thrust plate to the opening in the gear case. The depression in the rear face of the camshaft thrust plate must be positioned up at the 12 o'clock position to align the bolts in the thrust plate with those in the head and number 1 cam cap.
10. Tap the camshaft gear at 90° intervals, toward the engine with a fiber mallet or plastic hammer, until the thrust plate bolts can be started in the cylinder head and number 1 camshaft cap.

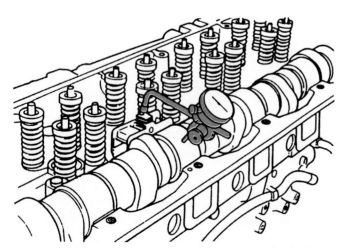

Figure 9-42. *Checking camshaft runout with a dial indicator. (Detroit Diesel Corp.)*

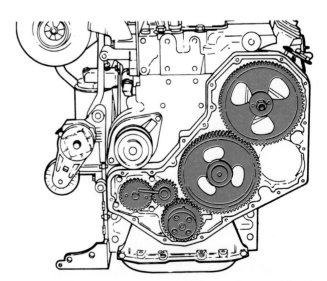

Figure 9-44 *The timing gear cover sealing surface must be clean before sealant or a gasket can be applied for cover installation.*

11. Using a socket and ratchet, tighten the thrust plate bolts alternately and evenly to draw the thrust plate straight in the gear case. Tighten the bolts to manufacturer's specified torque.

When assembling any crankshaft assembly, always be sure to check the manual. Always check the gear backlash of the mating gear teeth. Backlash is the amount of "play" between two gears in mesh.

When installing the gears, follow the sequence given in the manual. Always check the gear backlash of the mating gear teeth, **Figure 9-45**. Backlash is the amount of "play" between two gears in mesh. The gear train should be assembled in the following manner:

1. With either a block- or head-mounted camshaft, the camshaft drive gear must be installed so that the camshaft and crankshaft are in time with each other. To accomplish this, index the timing mark on the camshaft gear with the mark on the crankshaft gear. Follow the procedures for camshaft gear installation described earlier in this chapter.

2. To check the backlash, insert a dial indicator between the crankshaft gear and the camshaft driving gear or camshaft/injection pump driving gear.

3. Check the backlash between the camshaft gear and the auxiliary drive gear shaft, if used.

4. Check the backlash between all the gears in the gear train. Check the engine service manual for the allowable backlash between the gears.

After assembling the gear train and checking the gears for backlash, apply a silastic sealant on the block surface. Then, position the timing gear cover on the engine, **Figure 9-46.** Install the timing gear cover bolts and torque them to specifications.

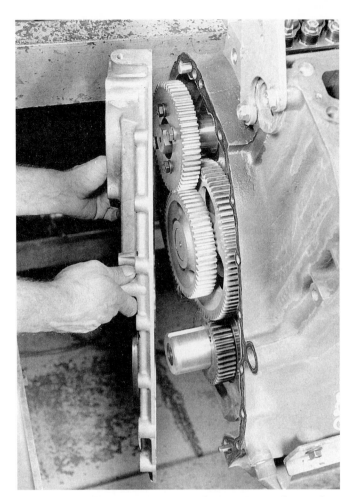

Figure 9-46. Place a gasket or sealant on the sealing surface and install the timing cover. (Mack Trucks, Inc.)

Installing the Cylinder Head

Before installing the cylinder head on the block, check the head depth of each valve, if needed. This can be done with a dial indicator or a depth gauge, **Figure 9-47.** If the depths are within specifications, proceed with the installation of the cylinder head as follows:

1. Install a new gasket on the cylinder block, locating it over the guide pins or dowels, **Figure 9-48A.** Some gaskets will have instructions such as "Top, Front," or "This Side Up" stamped on one surface. Remember to follow these instructions when installing the gasket. A reversed gasket can easily cause low oil pressure, overheating, and engine failure. Many of today's gaskets are pre-coated and do not require any type of sealing compound. Remember that a head gasket should never be reused.

2. Sort out the cylinder head bolts. Many engines use head bolts of different lengths. The service manual will usually identify where the long and short bolts should be installed. If the service manual specifies that the old head bolts should not be reused, be sure to obtain new ones.

3. Position the cylinder head(s) on the cylinder block, locating the guide holes over the pins or dowels in the block. Lubricate the cylinder head bolt threads, the underside of the bolt head, and washers. See **Figure 9-48B.** Install the head bolts in the appropriate positions.

Figure 9-45 Checking timing gear runout. The total amount that the dial indicator moves is the total indicated runout.

Figure 9-47. *Checking valve head depth with a depth gauge.* (Mack Trucks, Inc.)

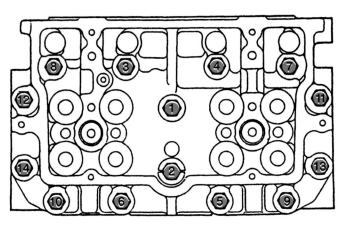

Figure 9-49. *Cylinder head tightening sequence for one cylinder head.*

ened to 35 ft.-lbs. (47.43 N•m) in the first stage, 70 ft.-lbs. (94.86 N•m) in the second stage, and 100 ft.-lbs. (135.58 N•m) in the final stage. Some manufacturers recommend that the head bolts be retorqued after the engine has been tested and is hot.

Start the assembly of the remaining valve train components by installing the push rods in their passages, **Figure 9-50.** Before inserting the push rods, coat them with clean

4. Using a straight edge on the exhaust mounting surface, align both heads to manual specifications if noted in the service manual.

5. After the heads are aligned properly, torque the cylinder head bolts to the manufacturer's specifications, **Figure 9-48C.** Cylinder head bolts must be tightened in the correct order and to the proper torque. A tightening sequence chart is usually provided in the service manual, **Figure 9-49.** The chart shows the sequence in which the bolts should be tightened.

Most cylinder heads are tightened in a sequence that starts in the middle and moves out to the ends. The bolts are generally tightened in two or three stages. For example, if the final torque specified for cylinder head bolts is 100 ft.-lbs. (135.58 N•m), the bolts may be tight-

Figure 9-50. *Installing a push rod. (Mack Trucks, Inc.)*

A

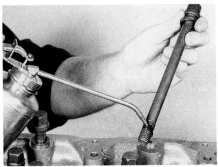

B

C

Figure 9-48. *Steps in assembling a cylinder head to the cylinder block. A—Install a new head gasket. B—Locate the proper installation point of each head bolt and coat the threads of each head bolt with clean engine oil. C—Torque all cylinder head bolts following the specifications outlined in the service manual. (Mack Trucks, Inc., Cummins)*

engine oil. When inserting the push rods, make certain that they are seated properly in the valve lifters.

Next, assemble the components of the valve rocker arm shaft assembly, including the fuel injector arm if so equipped. (The fuel injector installation procedure is given in **Chapter 18.**) Install the shaft assembly components in the reverse order of removal. Install the rocker arm shaft assembly on the head, **Figure 9-51.** After all the bolts have been installed in the rocker arm shaft brackets, torque them to the manufacturer's specifications.

Adjusting Valve Clearance

After checking the valve stems and springs as directed in the service manual, the next step is to adjust the valve clearance. When the valves are properly adjusted, there will be a small clearance between the valve stem and the end of the rocker arm. This clearance is sometimes referred to as *valve lash,* tappet clearance, or valve clearance. The valve clearance allows for heat expansion of the parts. Without clearance, expansion of the heated parts would cause the valves to remain partly open during operation.

The valve clearance may vary, depending on the engine model and whether the engine is hot or cold during adjustment. Always follow the engine manufacturer's procedure for correct valve clearance adjustment. The typical adjustment procedure for an OHV engine is performed as follows:

1. Turn the crankshaft to bring the number 1 piston to top dead center (TDC). Most engines have timing marks on the flywheel or fan drive pulley to mark TDC or other timing points. A good way to determine when a piston is at TDC is to remove the injection nozzle and hold your finger over the nozzle opening. On the compression stroke, air will be forced out against your finger until the piston reaches the TDC position. All timing marks must line up for the engine to function properly.

2. With the number 1 piston at TDC, check the valve clearance using a feeler gauge, **Figure 9-52.** If neces-

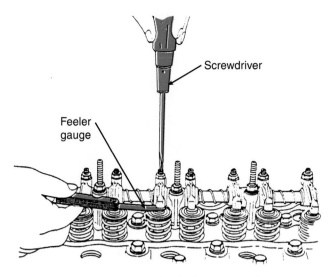

Figure 9-52. *Adjusting the valve lash on the engine.*

sary, adjust clearance by turning valve adjusting screw up or down until it is within specifications. Be sure to differentiate between the intake and the exhaust valves, because the clearances are usually different for the two.

> **Caution: On two-cycle engines do not attempt to adjust the crossheads while installed on the cylinder head. They must be removed and adjusted in a fixture to prevent damage to the alignment dowel.**

3. Rotate the engine crankshaft in its firing order and adjust the clearances for the appropriate valves as each piston reaches TDC of its compression stroke. Often, two or three sets of valves can be set at a time with one rotation of the crankshaft. Refer to the service manual for the recommended adjustment procedures for specific engines. Recheck the valve clearance after allowing the engine to "run-in" and retighten the cylinder head(s), if needed.

> **Note: On engines with two intake and two exhaust valves per cylinder, there are generally adjusting screws on both the crosshead and the rocker arms, *Figure 9-53*. The crosshead adjustment is to ensure both intake or both exhaust valves are actuated at the same time. The rocker arm adjustment is used to set the valve clearance.**

The rocker arm adjustment locations for a typical OHC diesel engine are shown in **Figure 9-54.** To adjust valve clearance, follow the procedure given in the service manual.

Valve Cover

The rocker or valve housing cover completely encloses the valve and injector (if used) operating mechanism on the top of the cylinder head. Although the cover

Figure 9-51. *Installing the valve rocker arm shaft assembly on the cylinder head. (Mack Trucks, Inc.)*

A

B

Figure 9-53. A—Lubricate the cross guides. Place the crossheads with the adjusting screws away from the camshaft side of the engine in their proper location. B—Adjust the crossheads and check the clearance between crosshead and the valve spring retainer. (Cummins Engine Co., Inc.)

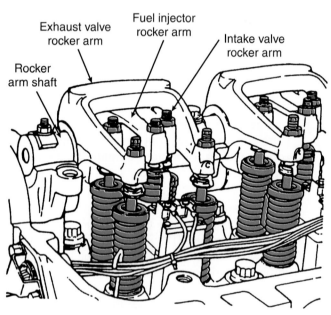

Figure 9-54. Location of adjusting screws in a typical OHC diesel engine. (Detroit Diesel Corp.)

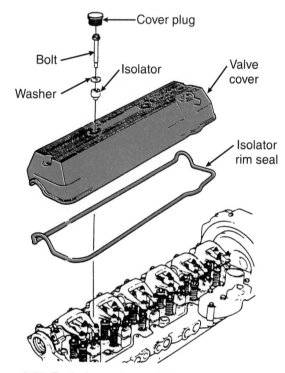

Figure 9-55. Typical one-piece cylinder head cover. (Detroit Diesel Corp.)

can be of two- or three-piece design, the one-piece type shown in **Figure 9-55** is used on most mobile diesel engines.

Most rocker housing covers require a rim seal gasket, or isolator. Install the rim seal gasket in the rocker housing cover groove Check the gasket contact surfaces to be sure that they are dry and free from dust. Then, install the gasket and rocker housing cover(s) on the engine. Lubricate the cylinder head bolts and torque them to specifications.

Summary

The camshaft rotates on bearings located in the cylinder block or the cylinder head. The crankshaft drives the camshaft, which may drive other engine components,

such as the fuel injection pump, governor, air compressor, fan drive, blower, tachometer drive, etc.

The rotation on the camshaft is timed to crankshaft rotation to control the operation of the valves and, in some cases, the fuel injectors. The correct relationship between the crankshaft and the camshaft must be maintained during service and reassembly. The diesel crankshaft and camshaft timing gears are made of steel and are of a helical design.

A camshaft converts rotary motion into the reciprocating motion of the rocker arms and/or the valve stems. The camshaft is "programmed" by the precise machining of each lobe on the shaft. The signals for opening and closing the valves and for injecting the fuel are permanently "engraved" in the cam lobe design.

There are several lobes on the camshaft for each cylinder in the engine. In a three lobe arrangement, two of the lobes operate the valves, and the relatively large center lobe operates the fuel injector.

The cam lobe shape determines when the valve will begin to open, how fast it will open, how much it will open (valve lift), and how fast it will close. Service to the camshaft includes inspecting it for damage, measuring the cam lobes for wear, checking the shaft for straightness, checking for camshaft end play, and replacing the camshaft bearings as required.

The rocker arms are commonly activated by the upward lift of push rods (or push tubes), which is controlled by the rotating camshaft. The rocker arms and the push tubes are held in contact with each other by an adjustment screw that fits through the end of the rocker arm.

When two valves are used for the same function in a cylinder, a crosshead is often added to form a bridge between the valves. The bridge allows both valves to be opened by a single rocker arm. Most valve train components are replaced rather than repaired. The exception is the machining of some types of rocker arms.

Some engines have valve clearances specified for a cold engine, others have clearances specified for a hot engine. Generally, clearances are set with the piston for the related valves at TDC. The clearance between the top of the valve stem and tappet bearing surface is measured with a feeler gauge. If adjustment is needed, the nut on the tappet screw is loosened and the screw is turned until the feeler gauge will just slide freely in the gap. The nut is then tightened. Since tightening the nut tends to increase the clearance, the gap must be rechecked after the nut is tightened. All valve train parts should be reinstalled in the reverse order of removal.

Important Terms

Camshaft	Chain drive systems
Cam lobes	Timing mark
Drive gear	Lifter cam follower
Cam lift	Finger cam followers
Valve lift	Pushrods
Rocker arm ratio	Rocker arm
Duration angle	Rocker arm shaft
Overhead valve (OHV)	Crosshead
Overhead camshaft (OHC)	Rocker housing cover
Valve float	Guide plates
Camshaft drive system	Total indicated runout
Gear train	(TIR)
Bull gear	Valve lash
Gear ratio	

Review Questions—Chapter 9

Do not write in this text. Place your answers on a separate sheet of paper.

1. What is the purpose of a diesel engine camshaft?
2. Valve lifters in most inline diesel engines are of the _____ .
3. On _____ the push rods form a critical link between the lifters and rocker arms.
4. _____ occurs when the valve is momentarily thrown free from the cam's influence.
5. Install the camshaft bearings from _____ to _____ .
6. Worn or plugged push rods should be _____ .
7. Which components should always be inspected when a valve fails and must be replaced?
8. The camshaft should be tested for _____ and _____ .
9. Torquing a cylinder head normally occurs in _____ or _____ stages.
10. Generally, tappet or lifter clearances are set with the piston for the related valves at _____ .

ASE-Type Questions

1. During a combustion cycle, what determines when the valve will begin to open, how fast it will open, how much it will open (valve lift), and how fast it will close?
   (A) The speed of the engine
   (B) The shape of the cam lobe
   (C) The stroke of the piston
   (D) All of the above.

2. Which of the following is the most common camshaft drive mechanism on diesel engines?
   (A) Gear.
   (B) Chain.
   (C) Belt.
   (D) None of the above.

3. Which method of measuring cam lobe wear requires that the camshaft be removed from the engine?
   (A) Dial indicator.
   (B) Outside micrometer.
   (C) Both A & B.
   (D) Neither A nor B.

4. When checking a camshaft for straightness, a runout deviation of _____ or more indicates a bent shaft.
   (A) .001″ (.025 mm)
   (B) .002″ (.051 mm)
   (C) .003″ (.076 mm)
   (D) .005″ (.127 mm)

5. Technician A says you should mark each lifter as it is removed so it can be reinstalled in the bore from which it was removed. Technician B says that this procedure is not necessary. Who is right?
   (A) A only.
   (B) B only.
   (C) Both A & B.
   (D) Neither A nor B.

6. The shop foreman taps the end of a push tube on a concrete floor. It produces a dull thumping sound. Technician A says this is normal. Technician B says that the tube probably has a hole in it and is full of engine oil. Who is right?
   (A) A only.
   (B) B only.
   (C) Both A & B.
   (D) Neither A nor B.

7. Technician A says that engine manuals specify tappet or lifter clearances for a cold engine. Technician B says that this is not always the case and that some manuals specify the clearances for a hot engine. Who is right?
   (A) A only.
   (B) B only.
   (C) Both A & B.
   (D) Neither A nor B.

8. Before starting to assemble the valve train operating components, the two critical measurements that must be carefully checked are the _____ .
   (A) installed valve stem height and the installed valve spring height
   (B) valve lash and camshaft end play
   (C) rocker arm clearance and tappet clearance
   (D) timing gear backlash and camshaft end play

9. Which tool can be used to check OHC camshaft runout at various bearing journals?
   (A) Depth micrometer.
   (B) Vernier caliper.
   (C) Telescoping gauge.
   (D) Dial indicator with magnetic base.

10. All of the following can occur if the head gasket is installed incorrectly, EXCEPT:
    (A) overheating.
    (B) low oil pressure.
    (C) engine failure.
    (D) poor timing.

Since they produce more power for the amount of fuel consumed, one of the first uses of diesel power was for marine applications, such as this fire boat.

Chapter 10

Lubrication Systems

After studying this chapter, you will be able to:
- ❏ Explain how oil flows through the lubrication system.
- ❏ Name the four primary tasks of lubricating oil.
- ❏ Describe the two types of lubrication systems employed in diesel engines.
- ❏ Explain the operation of the various types of oil pumps used in diesel engines.
- ❏ Explain the purpose of an oil cooler.
- ❏ Name the different classifications of oil filters and explain how they operate.
- ❏ Describe how different types of oil filtration systems function.
- ❏ Explain the operation of various oil pressure indication systems.
- ❏ Explain the purpose of oil additives.

Any part that moves requires some sort of lubricant to reduce friction and remove heat. Without a lubrication system, moving parts would overheat and eventually stop. The lubrication system is vital to a diesel engine's operation and extended service life. This chapter covers the operating principles and components of the diesel engine lubrication system.

Lubricating Systems

The proper operation of the diesel engine's *lubrication system* is extremely important. If an engine does not receive a proper supply of lubricating oil, it may be damaged beyond repair. The lubrication system assists the engine in the following ways:
- ❏ Holds an adequate supply of oil for the engine.
- ❏ Delivers oil to all necessary engine components.
- ❏ Reduces friction between moving parts.
- ❏ Absorbs and dissipates heat.
- ❏ Cleans and flushes contaminants from moving parts.
- ❏ Removes contaminants from the oil.
- ❏ Seals the piston rings and cylinder walls.
- ❏ Helps to reduce engine noise.

Lubricating oil cools, cleans, seals, and lubricates the engine. Without lubricating oil, the engine could not function.

Lubrication System Components

A typical lubrication system in a mobile diesel engine includes the following components:
- ❏ Oil pan or sump.
- ❏ Oil level dipstick.
- ❏ Oil pump.
- ❏ Oil cooler.
- ❏ Oil filter.
- ❏ Lubricating pressure regulating valves.
- ❏ Oil pressure indicators.
- ❏ Oil lines to and from the turbocharger (if used).
- ❏ Oil passages in the cylinder block.

During normal operation, oil from the oil pan or the sump is drawn to the oil pump through the suction pickup tube, **Figure 10-1.** The pump sends oil to the oil cooler and then to the oil filter. From the filter, oil travels to the galleries in the cylinder block. From the galleries, the oil takes several paths. In turbocharged engines, some of the oil travels through a supply line to the turbocharger. Oil from the turbocharger flows back through a return line into the oil pan or sump. Oil is also sent through drilled passages in the block to the main and camshaft bearings. The oil travels from the main bearing journals to the connecting rod bearings through drilled holes in the crankshaft. A small amount of oil is sent through the jet tubes to cool the pistons. Oil then travels to grooves in the camshaft

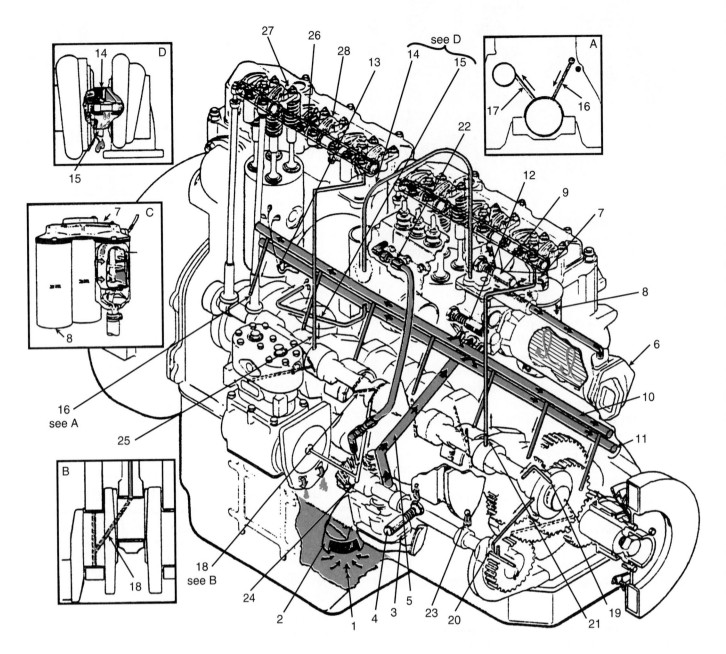

Figure 10-1. *Typical diesel engine lubrication system. Note the flow of oil through the engine. (Mack Trucks, Inc.)*

1. Oil sump
2. Oil pump inlet tube
3. Oil pump
4. Ol pressure relief valve (located in pump)
5. Oil pump discharge line to oil cooler
6. Lube oil cooler
7. Oil filter mounting adapter
8. Oil filter(s) (see inset "C")
9. Oil filter by-pass valve
10. Main oil gallery.
11. Piston oil cooling gallery
12. Oil pressure relief valve piston cooling gallery)

13. Piston cooling oil spray nozzle
14. Lube feed to turbocharger (external)
15. Turbocharger lube oil return (external)
16. Oil supply from main oil gallery to crankshaft main bearings (see inset "A")
17. Oil flow from main bearings to camshaft bearings (see inset "A")
18. Main bearing and connecting rod bearning lube oil passages (see inset "B")

19. Line to injection pump drive assembly from camshaft to bushing number 1
20. Line to auxiliary shaft from camshaft bushing number 1
21. Oil supply from camshaft bushing number 2 to rocker arm shaft
22. Line to injection pump (external) and governor from camshaft bushing number 4

23. Fuel injection pump lube oil drains
24. Line to auxiliary shaft rear bushing and compressor from camshaf bushing number 4
25. Oil supply from camshaft bushing number 5 to rocker arm shaft
26. Rocker arm shaft
27. Valve rocker arm
28. Oil supply from rocker arm shaft to rocker arm tip

bearings and into oil passages that lead to the valve lifter bores. These passages supply oil under pressure to lubricate the valve lifters.

Oil also travels through a passage to supply pressurized lubrication to the rocker arm shaft(s). Oil then flows through holes in the rocker arm shaft(s) to lubricate the valve system components in the cylinder head. Oil travels to other engine components, such as the air compressor, idler gear, fuel injection pump and governor, and automatic timing advance unit, from various passages in the block and related components.

A *pressure regulator valve* controls the oil pressure as it flows through the pump. If too much oil is pumped into the system, the pressure goes up and the valve will open, allowing excess oil to return to the inlet oil passage. Once the lubricating oil has traveled through the entire system, it returns to the engine oil pan or sump. Some large industrial engines have a separate reservoir instead of a sump.

An engine's oil level can be checked using a calibrated dipstick. It is best to check the oil level when the engine is cold, because oil expands when the engine is hot.

There must always be enough oil in the engine to supply the oil pump under all operating conditions.

Types of Lubrication Systems

There are two types of diesel engine lubrication systems currently in use—the internal force-feed system and the circulating splash system. Although most modern diesel engines use the internal *force-feed lubrication system,* **Figure 10-2,** the circulating splash system is found in some older industrial applications.

In the *circulating splash lubrication system,* an oil pump supplies oil to a splash pan located under the crankshaft, **Figure 10-3.** As the connecting rods revolve, scoops located on the rod caps dip into the splash pan troughs. The oil splashes, lubricating the moving parts nearby. Additional oil is splashed into collecting troughs and is gravity fed through channels or lines to lubricate moving parts that are farther away. Finally, as the connecting rods revolve, an oil mist forms that lubricates the upper parts of the cylinders, pistons, and pins.

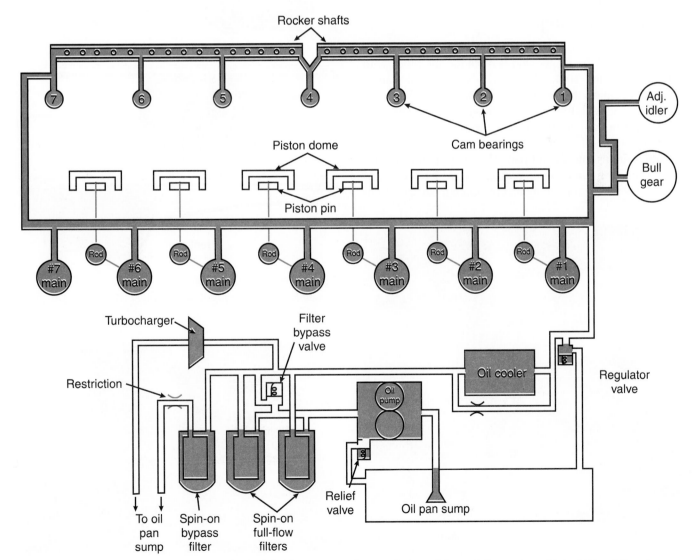

Figure 10-2. Schematic of a typical force-feed lubrication system. A pump forces oil through the system. (Detroit Diesel Corp.)

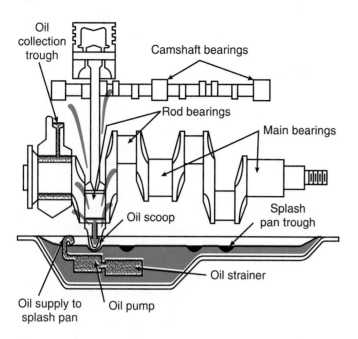

Figure 10-3. *The circulating splash lubrication system can be found on older diesel engine applications. (Deere & Company, Inc.)*

Oil Pans and Sumps

On most mobile diesel engines, the *oil pan* is located at the bottom of the engine and is bolted to the block with bolts. See **Figure 10-4.** The area where the oil pan is attached is called the *crankcase,* because it encloses the crankshaft and seals the bottom of the engine. In addition to sealing the bottom of the engine, the oil pan provides a reservoir for the oil. After the oil has circulated through the engine, it drains into the oil pan. The lower part of the oil pan is called the *sump.* An *oil pickup line,* which is covered by a coarse wire screen that prevents foreign particles from

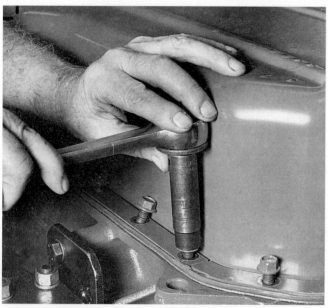

Figure 10-4. *The oil pan is bolted to the bottom of the engine block. Remove all of the bolts before prying on the pan. (Mack Trucks, Inc.)*

damaging the engine, extends into the sump. Industrial diesel engine sumps are discussed later in this chapter. There are many different types of oil pans. Depending on the engine, the oil pan can have a front, center, or rear sump, **Figure 10-5.** The capacity of the oil pan can also vary with the size of the engine.

Oil Pan Construction

Most mobile engine pans are made of stamped steel, cast iron, or aluminum. Either a one-piece oil pan, **Figure 10-6,** or a two-piece pan (upper and lower pan sections bolted together) may be used. Certain large V-type engines are equipped with an oil pan that consists of an upper section and two lower sections. Depending upon the model application, oil pans may be designed to permit an engine inclination of up to 45°. One or more *drain plugs* in the bottom of the oil pan are used to drain the oil at required intervals.

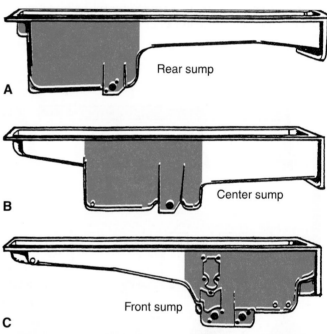

Figure 10-5. *Common oil pans designs. A—Rear sump. B—Center sump. C—Front sump. (Cummins Engine Co., Inc.)*

Figure 10-6. *Typical one-piece oil pan assembly. (Mack Trucks, Inc.)*

Caution: The stamped metal oil pans used on many marine engines have a thin protective coating to shield the metal against the corrosive action of salt water. Therefore, do not rest, slide, or rock the engine on its oil pan. If the surface of the oil pan is scratched, the protective coating will be damaged and the pan will corrode. Always perform engine repairs carefully to avoid scratching the outer surfaces of the pan.

Sealing the Oil Pan

A soft gasket is used to form a seal between the engine block and oil pan. Either a one-piece gasket or a sectional gasket consisting of two side sections and two end sections may be used. Before installing the oil pan gasket, use a straightedge to check the flanges for warpage, **Figure 10-7A.** If a straightedge is not available, lay the pan on a flat surface (flange side down) and place a flashlight underneath. Light will be visible between the flanges and the surface if there is warpage. Carefully check areas around bolt holes. Minor damage can be corrected with a hammer and a block of wood, **Figure 10-7B.** If the flanges are bent excessively, the pan must be replaced.

Large stationary and propulsion systems have tanks that collect oil that has been used for lubricating and cooling so it can be recirculated through the engine. Some systems also have a sump tank or drain tank located under the engine. These tanks collect the oil as it drains from the engine crankcase. Storage and sump tanks are not commonly used in mobile or auxiliary engines. Instead, the oil is stored directly in the oil pan.

Dipstick

A calibrated **dipstick** is used to check oil levels on most mobile diesel engines. The dipstick is usually located in the side of the cylinder block or the oil pan, **Figure 10-8.** Always maintain the oil between the full and low marks on the dipstick. Operation below the low mark may expose

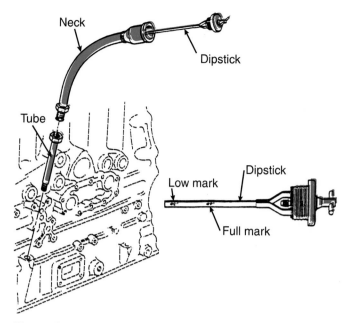

Figure 10-8. An oil dipstick is used to check the oil level in the crankcase. (Detroit Diesel Corp.)

the pump pickup, causing aeration and/or loss of oil pressure. Overfilling occurs when the oil rises above the full mark. In this case, the excess oil will be churned by crankshaft throws, causing foaming or aeration.

Before checking the oil level, the engine should be off for a minimum of twenty minutes to allow oil from various engine components to drain back into the oil pan. Dipsticks are normally designed to be used only when the vehicle or equipment is on a level surface. An improper reading will be taken if the vehicle or equipment is on a grade. If necessary, fill the crankcase with oil as follows:

1. Fill the oil pan to the full mark on the dipstick.
2. Start the engine and allow it to run for about ten minutes.
3. Stop the engine. Wait at least twenty minutes and recheck the oil level on the dipstick.
4. If necessary, add enough oil to reach the full mark on the dipstick.

A

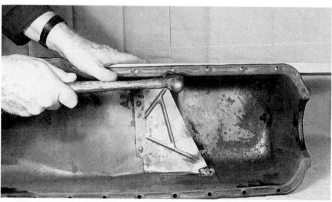

B

Figure 10-7. A—Checking the oil pan flanges for flatness. B—Removing small dents from the flanges with a ball peen hammer and a piece of wood. (JF Industries, Inc.)

Dipsticks in marine engines are designed to measure the proper oil level at any angle applicable to a specific boat. In a properly filled crankcase, the oil level must be below the crankshaft rear oil seal when the boat is at rest.

Oil Pumps

The force-feed lubrication system has one prime requirement, to develop sufficient oil pressure to lubricate the engine's moving parts. To maintain this pressure, an *oil pump* is used. See **Figure 10-9.** Lubricating oil is pumped through many passages in modern pressurized systems. The required oil pressure for most diesel engines ranges from 25-40 psi (170-275 kPa), although the requirements for some industrial diesel engines may be as high as 65 psi (450 kPa). All oil pumps used in mobile diesel engines have a positive displacement design.

Positive Displacement Pumps

The *positive displacement pump* utilizes the force of a mechanical part (a gear) that is confined within a chamber to force fluid from an inlet port to an outlet port. In other words, rotational movement allows the chamber to expand and take in fluid. Once the fluid is taken in, the chamber contracts, expelling the fluid to the outlet port. The rate of discharge depends on the speed of rotation. The pump has a pressure relief valve that prevents too much pressure from building up in the system. Excess oil pressure could cause part failure.

A positive displacement pump is normally very efficient. However, if the pump clearances are too tight or the parts are not properly lubricated, the resulting increase in friction can lower efficiency. Also, if there is too much internal clearance, there will be leakage from the high-pressure side of the pump to the low-pressure side, reducing efficiency.

External Gear Pumps

Of all positive displacement pumps, the **external gear pump** is the most commonly used in diesel lubricating systems. Normally, an oil pump is mechanically driven by the diesel engine. The external gear pump is usually driven from the crankshaft. On larger engines, however, an elec-

tric motor mounted on the engine may be used to drive an auxiliary oil pump.

The external gear oil pump consists of an idler (internal) gear, which is connected to the driveshaft and a drive gear, **Figure 10-10.** The idler gear turns the driven gear. Both gears trap oil between their teeth and the pump body wall, moving it from the inlet to the outlet. Output volume for each gear revolution depends on gear length and tooth depth. The machined surfaces of the pump body seal the gears. As oil enters on the inlet side of the external gear pump, it flows to the edge of the gears in the spaces between the gear teeth. As the gear teeth mesh together, the oil is forced out the discharge side of the pump. As the teeth come out of mesh on the inlet side, additional oil is drawn into the spaces between the teeth and the cycle repeats itself.

Internal Gear Pumps

Another type of oil pump used in diesel engines is the **internal gear pump.** The internal gear pump most commonly used is of the gerotor design. See **Figure 10-11.** It consists of a four-lobe inner rotor, which is usually driven

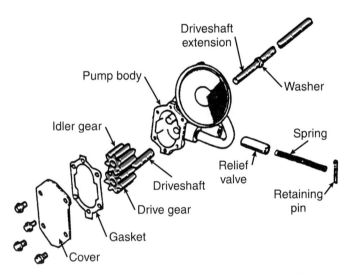

Figure 10-10. Exploded view of a typical external gear oil pump. (TRW, Inc.)

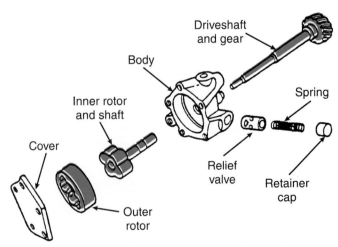

Figure 10-11. Typical rotor-type internal gear pump. (TRW, Inc.)

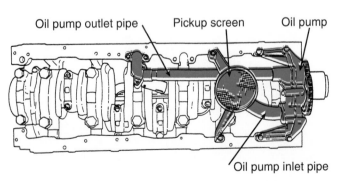

Figure 10-9. Typical oil pump location. Note the pickup screen and outlet pipe. (Detroit Diesel Corp.)

by the camshaft, and a five-lobe outer rotor, which is driven by the inner rotor. As the turning rotor lobes unmesh, oil is drawn in from the pan; trapped between the lobes, cover plate, and top of the pump cavity; and moved to the outlet, where the meshing lobes force the oil out. Output for each revolution depends on the rotor diameter and thickness. Another type of gear pump is the **crescent pump,** or **trochoidal pump,** which has an idler gear with internal teeth that spins around a drive gear, **Figure 10-12.** As the idler gear turns, it walks around the drive gear to move the oil in the space between them.

Herringbone Gear Pump

Still another style of gear pump is the **herringbone gear pump, Figure 10-13.** This pump is driven by external gears and a drive shaft coupled to it. It is normally found only on large stationary diesel engines. The sliding vane type shown in **Figure 10-14** is sometimes found in smaller mobile equipment diesels. It consists of a rotor housed eccentrically in a bored hole machined in the pump body. Four sliding vanes are located in grooves, machined in the rotor periphery. The vanes are held against the pump body bore by centrifugal force while the pump is operating. The inlet port is connected to the sump and the outlet port is connected to the oil galleries in the engine. As the rotor revolves, the vanes pass over the inlet port and oil is drawn into the space between the rotor and the pump housing. Oil is carried between the vanes to the outlet port where it is forced out into the engine oil galleries, since the space between the rotor and pump bore decreases.

A suction pickup tube is mounted on the oil pump inlet and runs to the low part of the sump. The tube, which is equipped with a screen to keep coarse dirt and debris out of the pump, is attached to the pump by a flange.

Servicing Oil Pumps

When disassembling an oil pump, follow the steps given in the engine service manual. When handling an oil

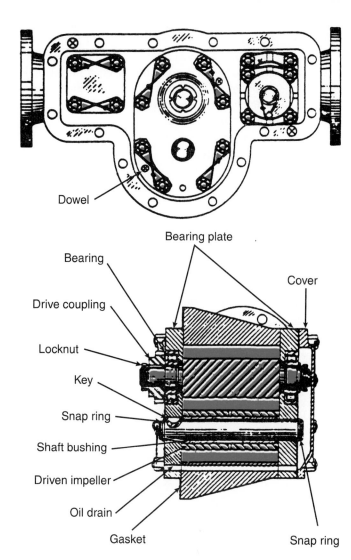

Figure 10-13. *Herringbone gear oil pump. (Colt Industries, Inc.)*

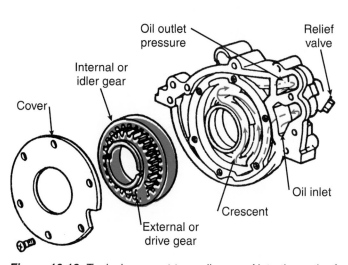

Figure 10-12. *Typical crescent-type oil pump. Note the path of oil flow in the pump. (TRW, Inc.)*

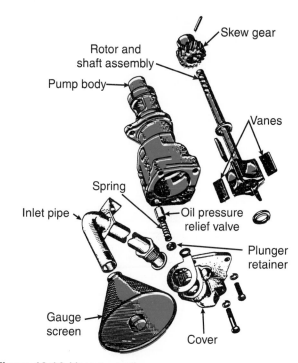

Figure 10-14. *Vane-type oil pump.*

pump, be careful not to damage the mounting surfaces of either the housing or the cover. These machined surfaces are normally the only seal, as most pumps do not use a gasket.

To inspect and service an external gear pump, it is usually necessary to remove the pump from the engine. The typical procedure for disassembling and servicing an external gear oil pump is as follows:

1. Place the oil pump assembly in a soft-jaw vice and unscrew the oil relief cap, **Figure 10-15.** The relief valve has one function, which is to turn on piston cooling oil at higher engine speeds, while maintaining adequate oil pressure at idle. When the engine oil pressure reaches approximately 20 psi (137.9 kPa), the plunger is moved off its seat. As the pressure increases, the plunger uncovers a hole in the sleeve and oil is directed into the piston cooling gallery. The oil then flows through the piston cooling oil spray nozzle. A key holds the drive gear on the oil pump shaft, **Figure 10-16.**

2. After the oil relief valve cap is loose, remove the washer, spring, and plunger. The operation of this oil pressure relief valve is discussed later in this chapter.

3. Before removing the oil pump driving gear, the retaining ring must be removed. This can be accomplished by prying the ring away from the shaft with a pointed tool, **Figure 10-17A.** After removing the ring, place the oil pump in a table press, supporting the gear with two strips of angle iron. See **Figure 10-17B.** Place the bar between the press and the end of the shaft, and remove the gear.

 Note: Make certain the oil pump drive gear key is removed from the shaft. Then remove the oil pump drive shaft and the gear.

Figure 10-15. Removing an oil pressure valve from an external gear oil pump. (Mack Trucks, Inc.)

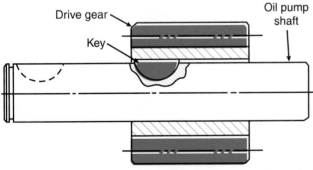

Figure 10-16. Some oil pump driveshafts use a gear key to hold the drive gear in place.

A

B

Figure 10-17. A—Removing the driven gear using a table press. B—Removing the gear retainer from the driver shaft. (Mack Trucks, Inc.)

4. With the oil pump housing placed in a soft-jaw vise, loosen and remove the oil pump housing screws. After the screws have been removed, take off the cover to service the gears, **Figure 10-18.**

Checking Oil Pump Components

After disassembling the pump, wash the components in oil and dry them with compressed air. If the housing is damaged, it must be replaced. The greatest amount of wear in an external gear pump is generally found on the idler and driven gears. This wear can be minimized by keeping the lubricating oil clean and acid-free. If dirt and sludge are allowed to accumulate, it may increase wear on the gears. Proper oil filter service will also help extend gear life.

The idler gear can be checked for free play by spinning it on its shaft. If it is binding, check the housing bore for burrs and nicks, and study the gear teeth edges for burrs. If either the housing or gear is damaged, it must be replaced. Next, insert the driven gear in the housing, **Figure 10-19,** and check for free play by spinning the gear. If it binds, check the housing for burrs or nicks. Replace the housing if it is damaged. If the gear is damaged, replace the gear and shaft.

With the gears mounted in the housing, check the end clearance by placing a straightedge across the face of the gears and using a feeler gauge to measure the clearance between the gears and the straightedge, **Figure 10-20A.** If this measurement does not meet the specifications given in the engine service manual, replace the gears.

To check side clearance, insert a proper size feeler gauge between the side of the gears and the housing, **Figure 10-20B.** If the gauge drags or the measurement exceeds service manual specifications, replace the gears. Check backlash by inserting a feeler gauge between the gear teeth, **Figure 10-20C.** If the gauge binds or if clearance exceeds specifications, replace the gears.

Finally, check the face of the oil pump housing cover for nicks and burrs. If it is damaged and cannot be

Figure 10-18. Removing the oil pump cover. (Mack Trucks, Inc.)

Figure 10-19. Installing a driven gear in an oil pump housing. (Mack Trucks, Inc.)

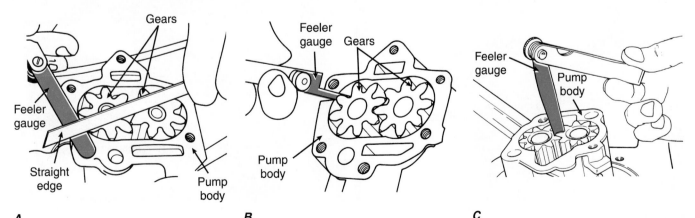

Figure 10-20. Various checks should be made before putting an external oil pump back in service. A—Checking end clearance. B—Measuring the side clearance. C—Checking backlash.

repaired, it must be replaced or rebuilt. Position the cover over the housing. Apply a sealant to the threads of the cover bolts and tighten the bolts. Kits are available from the vehicle manufacturer that contain all the parts and necessary instructions to rebuild a pump, **Figure 10-21.**

Assembling Gear Pumps

If the timing drive gear or the external idler gear was removed from the oil pump shaft, reassemble it as described in **Chapter 9.** After the gear has been heated, align the keyway in the gear with the key in the shaft and press the gear on the shaft. Position the gear so that there is an appropriate clearance between the gear and the oil pump housing, **Figure 10-22A.** Install the oil pump drive and idler gears (if used) in the gear train system. Then check the oil pump driving gear for backlash. See **Figure 10-22B.**

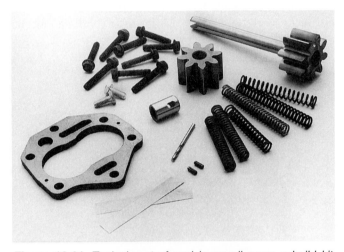

Figure 10-21. *Typical parts found in an oil pump rebuild kit. (Mack Trucks, Inc.)*

Prevent movement of the adjoining gears when checking the backlash, or the reading will be incorrect.

Assembling Rotor Pumps

In the rotor-type pump, wear will occur between the lobes on the inner rotor and the lobes on the outer rotor. Using a micrometer, measure the widths of these parts and measure the pump drive shaft. Check all measurements against pump specifications.

Before making any clearance checks, it is a good idea to clean all parts in solvent and use compressed air to dry them. Inspect the pump housing and rotor drive shaft for damage and excessive wear. Next, make the following clearance checks, comparing all measurements to the engine service manual specifications.

- ❏ Measure the tip clearance between inner and outer planetary rotors.
- ❏ Measure the clearance of the outer planetary rotor to the cover plate.
- ❏ Measure the clearance of the inner rotor planetary to the drive body.

If measurements are not within specifications listed in the service manual, replace the rotors.

Most oil pumps have a screen covering the inlet to strain out foreign material. If possible, remove the screen from the inlet pipe and clean both parts with solvent. Then use compressed air to dry the parts. Finally, inspect all bushings in their housings and replace those that show excessive wear.

Oil Cooler

The lubricating systems of most diesel engines must have a *heat exchanger,* or an **oil cooler,** to keep the oil temperature in the most efficient range. As oil passes

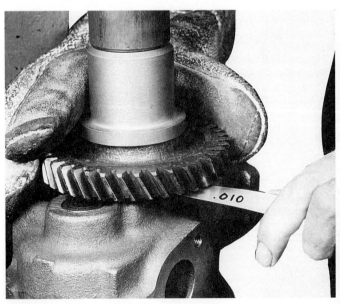

A

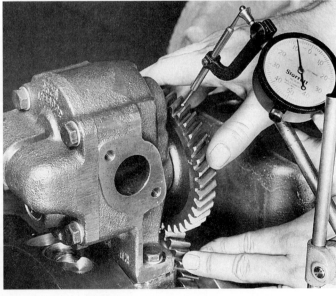

B

Figure 10-22. *Reassembling the oil pump driving gear to the timing system. A—Checking clearance between the driving gear and the pump housing. B—Checking the driving gear for backlash. (Mack Trucks, Inc.)*

through the engine, it absorbs heat from the metal parts it contacts along the way. Because the oil is continuously recirculated, it is constantly absorbing additional heat. At extremely high temperatures, the oil will oxidize quickly and form carbon deposits. These high temperatures will also increase the rate of oil consumption. Unless some method is provided for removing this heat, the oil temperature will become excessive. An oil cooler is designed to help retain the lubricating qualities of a diesel engine oil by removing excess heat. See **Figure 10-23.**

There are two types of oil coolers—the tube-type cooler and the plate-type cooler. The size of an oil cooler depends mainly on the quantity and temperature of the oil to be cooled. However, other important factors are the location of the cooler, the surrounding temperature, the oil viscosity, and the temperature of the coolant.

Tube Oil Cooler

The **tube oil cooler, Figure 10-24,** is usually mounted externally on the side of the engine block. Both coolant and lubricating oil are pumped through the cooler. In a typical system, the coolant flows from the engine's water pump through a passage in the cooler's front cover. It passes through the tubes in each section of the oil cooler and then travels back through the outlet passage in the front cover to the oil galleries in the cylinder block.

A bypass valve is generally provided to permit the oil to flow directly through the oil cooler if the cooling element becomes clogged. The valve can sense low oil pressure as well as full-flow filter restrictions. The valve may be

connected to a signal light on the vehicle's instrument panel to warn the operator of either condition. Bypass valves are covered in detail later in this chapter.

Plate Oil Cooler

The **plate oil cooler** is usually mounted in the engine or in the crankcase. See **Figure 10-25.** Oil from the oil pump flows through a passage in the oil housing to the oil filter, passes through the oil cooler core, and then travels through the outlet passage in the cooler housing to the cylinder block oil galleries. The engine coolant is pumped through the oil cooler to completely surround the oil cooler core.

To maintain oil flow if the filter becomes restricted, a bypass valve is built into the cooler housing. If the pressure drop across the filter exceeds 20 psi (138 kPa), the bypass valve opens to allow the oil to flow through the cooler.

Maintenance of Oil Coolers

Normal maintenance of the cooling system (see **Chapter 11**) should keep the coolant passages in the oil cooler clean. When cleaning the system, remove the cooler and clean the oil passages with solvent. The following sections outline maintenance procedures for both tube-type and plate-type oil coolers. Be sure to check the appropriate engine service manual for specific instructions.

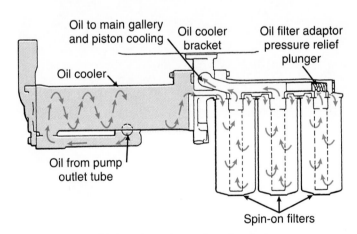

Figure 10-23. Oil flow through an oil cooler system. (Mack Trucks, Inc.)

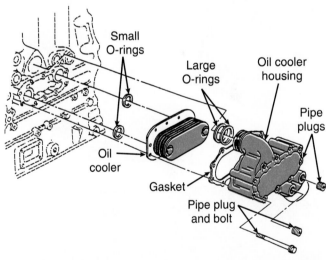

Figure 10-25. Parts of a typical plate oil cooler. This cooler is mounted in the engine or the crankcase. (Detroit Diesel Corp.)

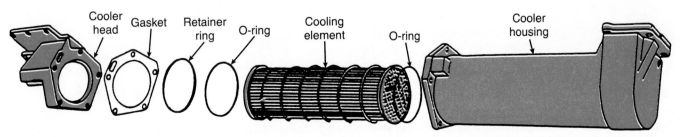

Figure 10-24. Exploded view of a typical tube oil cooler. It is normally mounted outside of the engine. (Mack Trucks, Inc.)

Tube Oil Cooler Maintenance

Inspect the oil cooler to be sure the gasket, O-ring, and core are not damaged. Replace any parts that show signs of wear. To clean the coolant section of the oil cooler, do a periodic reverse flush of the cooling system.

The presence of engine oil in the coolant may indicate loose, worn, or damaged parts within the oil cooler. Oil in the coolant can also indicate internal engine problems. If it is necessary to disassemble a tube-type oil cooler for service, do so in the following manner:

1. Remove the cooler head, **Figure 10-26A.**
2. Remove the cooling element from the cooler housing, **Figure 10-26B.**
3. The cooling element should be checked for leaks by submerging it in water and applying air pressure at about 80 psi (551.6 kPa), **Figure 10-26C.** Special pressure plates (available as a manufacturer's special service tool) must be fastened at each lock to keep the pressure within the element. If leaks develop, the element must be replaced.
4. Remove all oil cooler gaskets and scrape the gasket residue from the housing, **Figure 10-26D.**

To reassemble the oil cooler after it has been cleaned and inspected, proceed as follows:

1. Lubricate the O-ring and install it on the water inlet end of the housing. See **Figure 10-27A.** Spread gasket sealer on the top of the O-ring.
2. Spread gasket sealer on the cooler head (inlet water) flange before installing the gasket. Once the gasket is placed, coat it with gasket sealer, **Figure 10-27B.**
3. Install the cooling element, **Figure 10-27C.** Tap the last .5″ (12.7 mm) of the element into the housing with a soft-faced hammer. See **Figure 10-27D.**
4. Place a gasket on the tube bundle end and spread gasket sealer on the top of the gasket before installing the end cap.

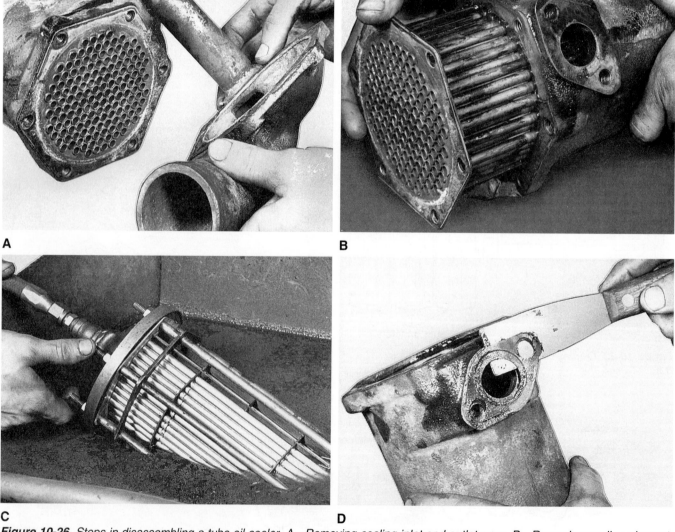

A

B

C

D

Figure 10-26. Steps in disassembling a tube oil cooler. A—Removing cooling inlet and outlet caps. B—Removing cooling element. C—Checking element for leaks by applying air pressure. D—Remove gaskets with a gasket scraper. (Mack Trucks, Inc.)

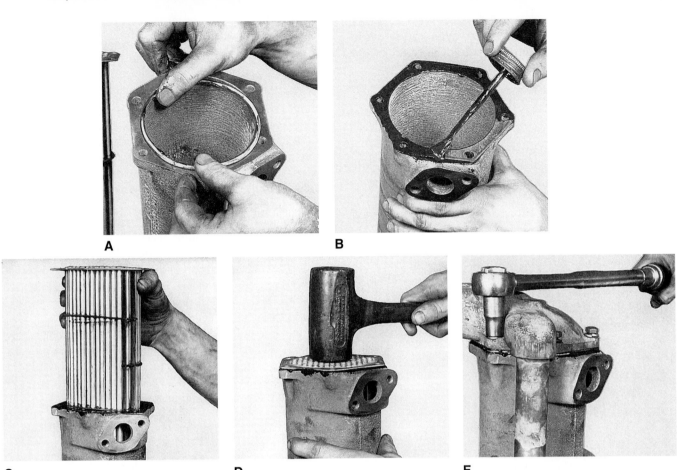

Figure 10-27. *Reassembly of a tube oil cooler. A—Installing the O-ring on the inlet. B—Coating the gasket with sealant. C—Installing the cooling element. D—Using a hammer to tap the element into the housing. E—Tightening the inlet bolts. (Mack Trucks, Inc.)*

5. Install the water inlet cap. Then torque the inlet bolts to the specifications given in the engine service manual, **Figure 10-27E.**

6. To complete the assembly of the oil cooler, give the unit an *air pressure test.* Install a steel plate with a rubber gasket and an air line fitting on the oil cooler head opening. Connect an air line and an air pressure regulator to the fitting. See **Figure 10-28.** To start the test, immerse the oil cooler in water and apply air pressure at 2-5 psi (13.8-34.5 kPa) maximum. Slowly increase the pressure while checking for leaks until it reaches 80 psi (551.6 kPa). If the cooler passes this air pressure test, reinstall it on the engine. If it fails the test, replace it.

⚠️ **Warning: When conducting the air pressure test, be sure all persons in the area are adequately protected against any stream of pressurized air that could come from a rupture in the fitting, hose, or oil cooler core. Always use an air pressure regulator.**

Plate Oil Cooler Maintenance

Once a plate oil cooler is removed from the cylinder block, the cooler element can be cleaned as follows after the coolant has been drained:

1. Clean the oil side of the coil by circulating a solution recommended in the service manual through the core passages with a hand force pump or compressed air to remove carbon and sludge. Clean core before the sludge hardens. If the oil passages are badly clogged, circulate an appropriate cleaning solution through the core and flush thoroughly with clean, hot water.

Figure 10-28. *Connecting an air line to the fitting to conduct an air pressure test. (Mack Trucks, Inc.)*

Warning: Protect your eyes and avoid breathing the fumes or direct contact of the solution with your skin.

After the oil cooler has been cleaned, check for leaks as follows:

1. Purchase from the manufacturer a special tool or make a suitable plate and attach it to the flanged side of the cooler core. Use a gasket made from rubber to assure a tight seal. Drill and tap the plate prior to mounting to permit an air hose fitting to be attached at the inlet side of the core.

2. Attach an air hose to the air hose fitting. Regulate air pressure to 75-100 psi (517-689.5 kPa) and submerge the oil cooler and plate assembly in a container of water heated to 180°F (82°C), **Figure 10-29.** Any leaks will be indicated by air bubbles in the water. If leaks are indicated, replace the core.

3. After the check is completed, remove the plate and air hose from the cooler core, then dry the core with compressed air. If the core shows a leak, it should be replaced.

Reinstall the oil cooler as follows:

1. Reconnect the oil cooler gasket, element, and cooler cover to the cylinder block. If a new element is used, be sure to remove the shipping plugs. Torque the bolts to the specifications given in the engine service manual.

2. Reconnect the turbocharger oil supply line (if used).

3. Refill the coolant system as described in **Chapter 11.**

Oil Pressure Relief Valve

The *oil pressure relief valve* mentioned earlier in this chapter stabilizes lubricating oil pressure at all speeds, regardless of oil temperature. The valve is usually installed at the end of the vertical oil gallery near the front of the

cylinder block opposite the oil cooler. See **Figure 10-30.** The oil pressure regulator consists of a valve body, hollow piston-type valve, spring, spring seat, and pin to retain the valve assembly within the valve body.

The oil pressure relief valve is held on its seat by the spring, which is compressed by the pin in back of the spring seat. The entire assembly is bolted to the lower flange of the cylinder block. When the oil pressure at the valve exceeds approximately 50 psi (345 kPa), the valve is forced from its seat and oil from the engine gallery is bypassed to the engine oil pan. In this way, the lubricating oil pressure is stabilized.

On most engines, pressure regulating valve adjustments are made using shims and washers. To raise oil pressure, add shims behind the spring. This increases spring tension, which in turn increases the amount of pressure required to open the valve. To lower oil pressure, remove the shims to decrease spring tension. This will result in the valve opening at a lower pressure.

Some regulating valves have an adjustment screw. Turn the screw in to raise spring tension and the pressure setting. Turn the screw out to lower spring tension and the pressure setting. **Figure 10-31** lists the possible causes of excessively low or high engine oil pressure.

Oil Filters

Oil contamination reduces diesel engine efficiency more than any other factor. To maintain engine efficiency, *oil filters* are used in all modern diesel engine lubrication systems. There are two basic filter designs: the cartridge design and the spin-on design.

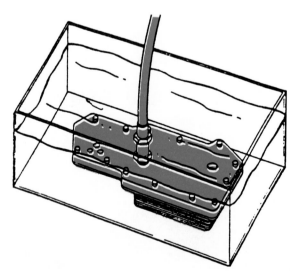

Figure 10-29. Pressure testing a plate oil cooler. (Cummins Engine Co., Inc.)

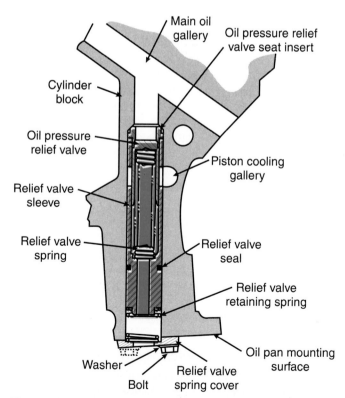

Figure 10-30. Typical oil pressure relief valve. (Mack Trucks, Inc.)

Possible Causes of Low Oil Pressure	Possible Causes of High Oil Pressure
Oil in crankcase is too thin	Regulator valve misadjusted
Low oil level in crankcase	Regulator valve sticking
Worn oil pump	Improper regulator valve spring tension
Regulator valve spring failure	
Regulator valve misadjusted	Defective pressure gauge or sender
Oil filter leak	
Oil pump leak	Oil in crankcase too heavy
Clogged oil filter	Restricted oil gallery

Figure 10-31. Possible causes of oil pressure problems.

Cartridge Design Oil Filters

Figure 10-32 shows the various components in a typical **cartridge filter,** or **bolt-on filter.** The foundation of the cartridge filter element is the center tube, which is made from strong perforated metal. The center tube has a collapse pressure of 100 psi (689.5 kPa), although it can be higher for certain applications.

A pleated paper element surrounds the center tube. The tube supports the paper element. The pleated paper normally consists of cellulose fibers that are bound together with a phenolic resin. However, for special applications, the filter paper may be a blend of cellulose and synthetic fibers. The paper's porosity and thickness will vary from filter to filter, depending on the filter's application. Filter paper is designed to handle the normal operating temperatures and contaminants found in the lubricating system. Folding or "pleating" the paper may increase the amount of exposed surface area. However, too many or too few pleats can have a detrimental effect on filter performance. A perforated metal body surrounds the pleated element to protect it during shipment and installation. The metal adds strength to the overall filter element assembly.

While cartridge oil filters vary in filter element design and in size, they all basically function in the same manner. As shown in **Figure 10-33**, the cartridge type filter may be a free-standing unit outside the engine in some industrial systems.

Spin-on Design

The most popular type of filter used today on mobile type diesel engines is the **spin-on filter, Figure 10-34.** The term "spin-on" refers to the way that the unit is installed—

Figure 10-33. A free-standing industrial cartridge filter (Waukesha Engine Division)

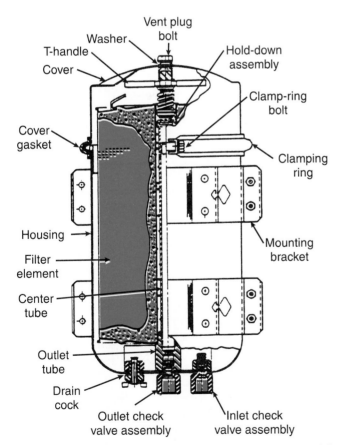

Figure 10-32. Typical cartridge filter assembly found in mobile diesel engines. (Freightliner Corporation)

Figure 10-34. Spin-on oil filters. (Mack Trucks, Inc.)

it spins onto a threaded bushing and base on the engine, and is then tightened to proper specifications, **Figure 10-35.** Spin-on filters for mobile diesel engine use come in a variety of sizes. There are three types of spin-on filter elements: surface type, depth type, and combination type.

Surface Filter Elements

The basic material used in *surface filter elements* is pleated paper. As oil from the crankcase enters the filter, it flows over the surface of the paper. Microscopic pores in the paper allow the oil to flow through, while trapping contaminants that are too small to be seen by the naked eye. See **Figure 10-36.**

The typical surface-type spin-on filter housing is a seamless high strength steel cylinder that is open at one end. A coiled steel spring holds the inner element in position and is sealed to the threaded plate regardless of the

pressure conditions. The inner element consists of the top and bottom metal endcaps, the pleated paper filter media, and the center tube. The pleated paper is sealed to the endcaps with an adhesive. An anti drain-back valve automatically closes when the engine is stopped to help prevent oil from draining out of the filter assembly. The threaded plate is made of thick steel to prevent damage from minor pressure surges. A gasket retainer and sealing gasket are also part of this type of filter assembly.

Depth Filter Elements

The construction of a *depth filter element,* **Figure 10-37,** is similar to that of the surface element. However, the

Figure 10-36. Cutaway view of a surface-type filter. (Wix Corporation)

A

B

Figure 10-35. A—Threaded bushing used for mounting spin-on filters. B—Spin-on filters should be hand tightened. (Mack Trucks, Inc.)

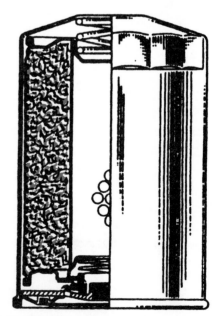

Figure 10-37. Cutaway view of a typical depth-type oil filter. (Wix Corporation)

depth filter element normally consists of a blend of cotton thread and resilient supporting fibers. The depth media is contained in a perforated metal can with metal endcaps. A media migration barrier is wrapped around the center tube of the filter.

A depth filter is designed so that oil entering the filter penetrates the filtering media. As the oil flows through the filter, the contaminants are trapped in the media. Every drop of oil must flow through this uniform mass of filtration media, which has a high capacity for dirt retention. While the efficiency of a depth filter is ideal for some applications, it has a limited flow rate when compared with surface filters of the same size. A depth filter cannot handle the high flow rates that many engines must have to maintain proper cooling and lubrication.

In some cases, combination filter elements are used. These filter elements combine both the surface element and depth element properties into one filter.

Degree of Filtration

In addition to selecting the proper type of filter, determining the degree of filtration needed is also important. This is done by measuring the filter element to determine the smallest particle it can remove from the oil. The unit of measure most commonly used for this purpose is the micron. One micron is approximately .00004″ (400 millionths of an inch). Because the smallest particle that can normally be seen with the naked eye is 40 microns in diameter, much of the dirt filtered by the lubrication system is invisible.

It is difficult to determine the number of particles a filter will collect. The openings in a filter element will gradually decrease as particles build up on its surface. As particles collect in each hole of a filter element, the holes become smaller. Eventually, the holes will plug, and the filter must be replaced. The rate at which particles build up on the filter element determines the useful life of a filter. Some engines have only one lubricating oil filter; others may have two or more. See **Figure 10-38.** There are two major types of filtration systems employed in modern diesel engines—bypass systems and full-flow systems.

Bypass Oil Filtration Systems

Figure 10-39 is a schematic of the *bypass oil filtration system.* It illustrates the various components in the flow path of the oil. The oil flows through the pickup screen, which acts as a strainer to catch large particles. It then flows to the oil pump, which forces the oil to the gallery or main oil channel. Pressure in the system is controlled by the pressure regulating valve, which is located in the oil pump. This valve is normally set to maintain pressures of 40-70 psi (276-483 kPa) in most diesel engines.

Approximately 90% of the oil is pumped directly to the engine bearings and other moving parts before it drains back to the oil pan. The remaining oil is diverted to the filter housing. The oil is cleaned as it passes through the filter cartridge and then flows out through the restrictor orifice to the sump. The *restrictor orifice* provides a con-

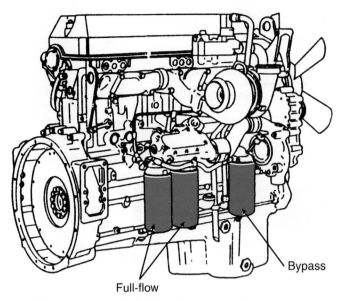

Figure 10-38. Many lubrication systems contain multiple filters. (Detroit Diesel Corp.)

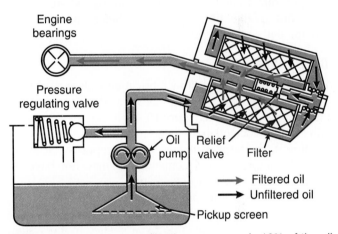

Figure 10-39. In a bypass filtration system, only 10% of the oil is filtered during each circulation cycle. (Deere & Company, Inc.)

trolled restriction of the oil flow to maintain normal pressure in the gallery.

Since the filter is designed to remove contaminants from the engine oil, every effort should be made to ensure that the filter does not become clogged. When a filter becomes clogged, unfiltered oil is pumped directly to the engine. The contamination level increases as the unfiltered oil continues to circulate through the system. For this reason, the filter cartridge should be changed periodically to ensure a constant supply of clean oil to the engine's moving parts. The filter is normally changed every time the oil is changed. While the bypass system helps control the level of contaminants, it has one shortcoming: it filters only 10% of the oil during each complete circulation cycle.

Full-flow Oil Filtration Systems

As illustrated in **Figure 10-40,** the full-flow system is very different from the bypass system. In the *full-flow filtration system,* the oil is drawn through the pickup

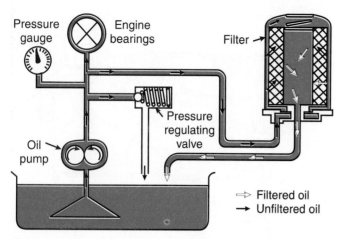

Figure 10-40. Full-flow filtration systems filter all the oil before it reaches the bearings. (Deere & Company, Inc.)

screen and into the oil pump. The oil pump forces the oil under pressure to the main oil channel or gallery. Pressure is controlled by a regulating valve. All of the oil is then pumped directly into the oil filter housing, where it must pass through the filtering element before going to the moving parts of the engine.

Bypass Valves

Every filter in a full-flow lubrication system must have a **bypass valve.** This valve may be located in the filter or in the filter mounting pad, **Figure 10-41.** In larger engines, the valve can be located in the engine itself. When replacing a full-flow filter, be sure to install the proper replacement filter. An improper replacement filter may not have an identical bypass valve arrangement.

There is very little pressure drop through a new filter. As a filter is used, it gradually clogs up, increasing pressure on its inlet side. If there is no bypass valve in the system and the filter becomes completely clogged, the pressure regulating valve will open completely. The oil will then return directly to the crankcase instead of going to the engine, resulting in serious engine damage.

The filter bypass valve is actually a safety device. The valve is normally set to open before the filter becomes completely clogged to ensure that oil will reach the engine bearings. Although bypass valves allow unfiltered oil to be delivered to the engine, unfiltered oil is preferable to oil starvation. As previously mentioned, bypass valves can also be used with oil coolers. If the oil cooler becomes obstructed or clogged, oil can flow through the valve and back into the lubrication system.

As the cooler becomes clogged during use, pressure builds on the head of the valve. When enough pressure builds up, the valve opens. This allows oil to bypass the cooling element and flow to the oil filters. If the oil filters are also obstructed, the pressure causes another bypass valve to open and the unfiltered oil flows directly to the oil gallery for lubrication. If oil pressure in the oil gallery is excessive, a pressure regulating valve spool opens, allowing the oil to flow back to the oil pan.

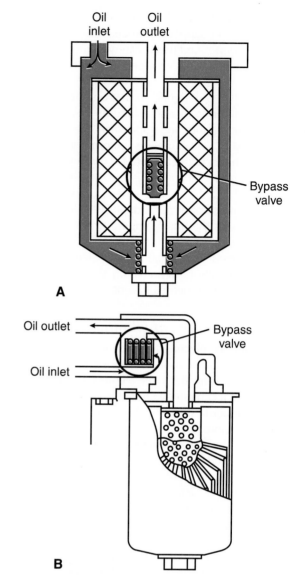

Figure 10-41. Filter bypass valve locations. A—Bypass valve in the oil filter. B—Bypass valve in the mounting pad. (Deere & Company, Inc.)

Other Oil Filtration Systems

While a little more complex, both full-flow and bypass filtering systems may be used in large stationary diesel applications, **Figure 10-42.** In addition, stationary engines may also use shunt and sump filtering systems.

In a **shunt-type filtering system,** oil is taken from the sump by the pressure pump and flows through a strainer, a filter, and a cooler before being delivered to the engine. The pump delivers a constant amount of oil per revolution, but the resistance in the strainer and the filter varies, depending on the condition of these units and the temperature of the oil. To ensure that an adequate flow of oil is delivered to the engine at all engine speeds, the filter and the strainer are each fitted with a spring-loaded bypass valve through which a percentage of the oil flows.

If the filter becomes clogged or if the oil is cold, a relatively large percentage of the lubricating oil is shunted through the bypass. Strainers and filters in a shunt-type fil-

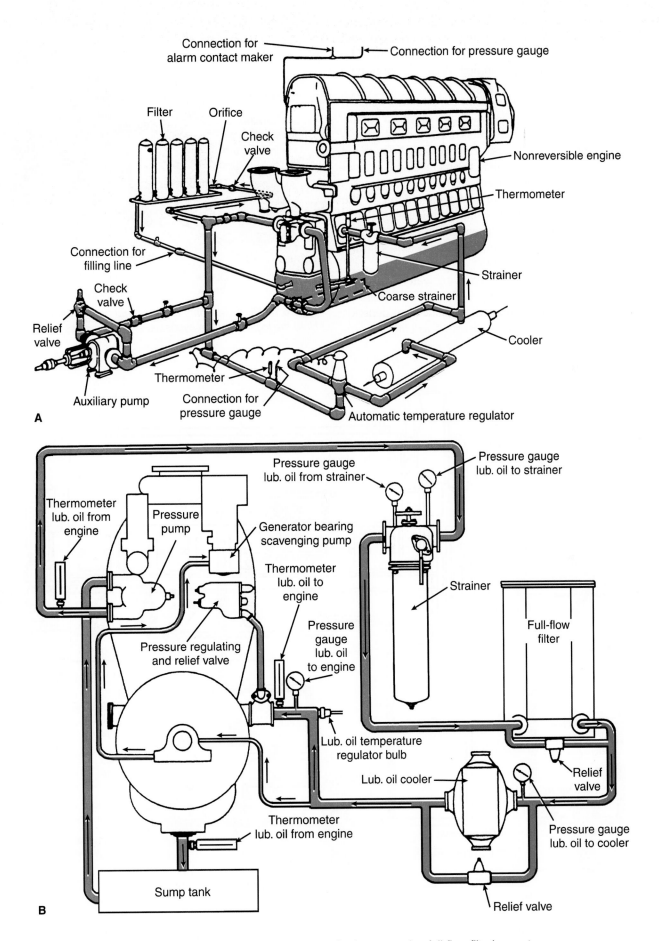

Figure 10-42. *A—Stationary engine bypass filtering system. B—Stationary engine full-flow filtering system.*

tering system may also be manually bypassed. Three-way valves are provided for bypassing each unit, so that strainers may be cleaned or filter elements may be replaced while the engine is running. A typical shunt-type filtering system is illustrated in **Figure 10-43**.

In a *sump-type filtering system,* the oil filter is placed in a separate recirculating system that includes a motor-driven pump. In this system, the oil is drawn from the sump by an electrically operated motor. After passing through the oil cover, one line passes the oil to the metal edged strainer and onto the engine. Another line from the cooler passes oil through the filter and returns the filtered oil to the engine sump. The oil can be circulated while the engine is stopped since the pump is independent of engine operation. The path of the oil through a sump-type filtering system is shown in **Figure 10-44**.

Purifying lubricating oil by removing water and solids is extremely important, especially in large diesel installations. The amount of wear on internal engine parts depends on the purity of the oil used. Solids, such as carbon parti-

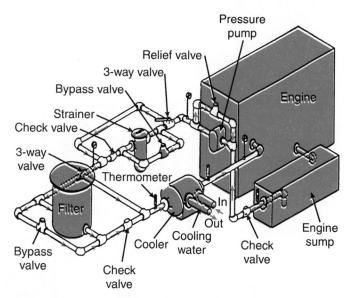

Figure 10-43. *A stationary engine shunt-type oil filtering system. Note the location of the bypasses and the bypass valves.*

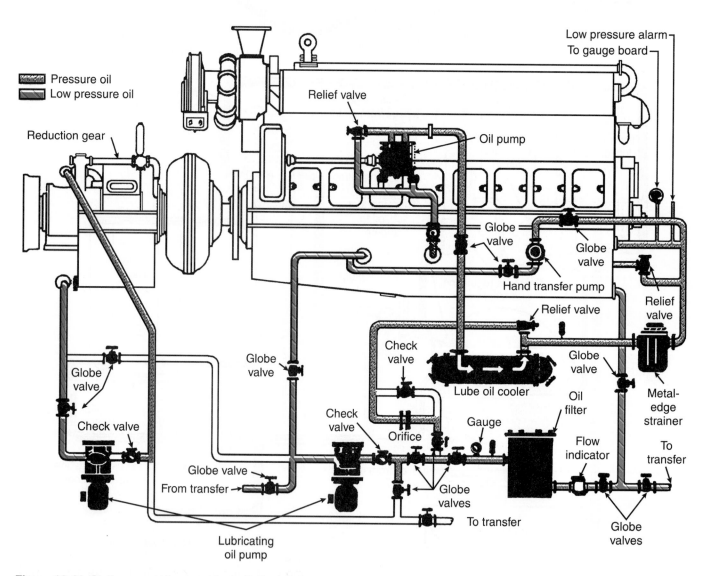

Figure 10-44. *Stationary engine sump-type oil filtering system.*

cles and bits of metal, may pass through the bearings, scoring them and wearing their surface area. Water and sludge can interfere with the ability of an oil to provide a proper lubricating film between bearing surfaces. There are two methods of purifying a lubricating oil: the batch method and the continuous purification method. In the **batch purification method,** oil is placed in a tank and heated, causing the water and solids to settle out. In the **continuous purification method,** a centrifugal purifier is used. The purifier uses centrifugal force to remove water and solids from the oil under normal operating conditions.

Oil Filter Maintenance

To ensure maximum protection and ensure protection from a filter, it is essential that proper service and installation procedures are followed. Be sure to adhere to all service instructions included with the filter. The following service procedures for mobile diesel engines are for the two basic types of filters: spin-on filters and cartridge filters.

Spin-on Filter Service Procedures

To remove and install a spin-on type filter, proceed as follows:

1. Drain the oil only when it is hot and the contaminants are in suspension. Operate the engine until the coolant temperature reaches approximately 140°F (60°C). Shut the engine off and place a container under the oil drain plug to catch oil. Then remove the oil drain plug. See **Figure 10-45A.**

 Warning: Hot oil can cause personal injury if it contacts skin.

2. Clean the area around the oil filter head and remove the filter. Clean the gasket surface of the filter head, **Figure 10-45B.** Make sure the O-ring has been removed from the filter head, as it has a tendency to cling to the top. Also, be sure to clean and dry the relief valve if it is removable.
3. Fill the filter with clean lubricating oil, **Figure 10-46C.** Apply a light film of lubricating oil to the gasket sealing surface before installing the filter(s). Install a new gasket and seal rings.
4. Install the filter as specified by the filter manufacturer. Keep in mind that overtightening may distort the threads or damage the filter element seal. See **Figures 10-45D** and **10-45E.**
5. Check and clean the oil drain plug threads and the sealing surface. Install the drain plug and tighten it to specifications.
6. Fill the engine the proper level with clean oil. Check the oil level on the dipstick.
7. Idle the engine and check for leaks at the filter(s) and the drain plug.

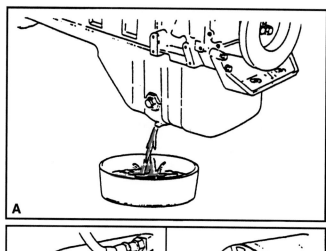

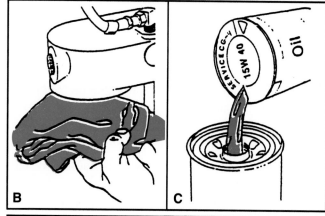

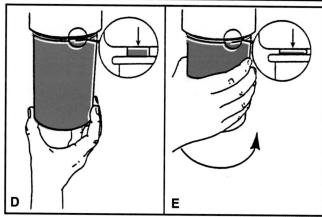

Figure 10-45. *Typical spin-on filter service procedure. A—Draining the used oil. B—Cleaning the filter head gasket surface. C—Filling the new filter with clean oil. D—Install the new filter after lubricating the seal with a light coat of fresh oil. E—Filter should be tightened until the gasket is compressed as shown. Overtightening can damage the gasket. (Cummins Engine Co., Inc.)*

Cartridge Filter Service Procedure

To remove and install a cartridge element filter, proceed as follows:

1. With the engine oil warm enough (not hot) to flow normally, place a container under the oil pan. Remove the drain plug by turning it counterclockwise with the proper size wrench, **Figure 10-46A.** Allow all the oil to drain from the engine and replace the plug.
2. Place the drain pan under the filter housing and remove the filter drain plug. After the oil has drained

from the housing, replace the plug and loosen the center bolt. Remove the shell, the used element, and the gasket from the mounting base.

3. Thoroughly clean the shell assembly and the mounting base with solvent and dry the base with a clean cloth, **Figure 10-46B.**

4. Install the sealing gasket in the base along with any other seals or gaskets provided for the shell assembly.

Note: Never use oil or grease on gaskets unless otherwise instructed by the service manual.

5. Place the new element in the shell and position the shell/element onto the mounting base. Tighten the assembly to the recommended torque. Refill the crankcase with the proper amount of oil.

6. Start the engine, allow the oil to warm up, and check for leaks around the filter and the drain plug.

Cartridge Filters with Removable Covers

To service element filters with removable covers, **Figure 10-47,** proceed as follows:

1. With the engine oil warm (not hot), place a container under the oil pan. Remove the drain plug by turning it counterclockwise with the proper size wrench. Allow all the oil to drain from the engine and replace the plug.

2. Remove the clamp-ring bolts. Remove the clamp ring and lift the cover off of the filter housing.

3. Unscrew the hold-down assembly T-handle and remove the assembly from the housing. Check the O-ring and spring for damage. Replace if needed.

4. Remove and discard the oil filter element.

5. Clean the inside of the housing, and the hold-down assembly, with solvent.

6. Install the drain plug.

7. Install a new filter element.

8. Install the hold-down assembly and securely tighten the T-handle.

9. Check the cover gasket for damage, and replace it if needed. Position the gasket on the housing.

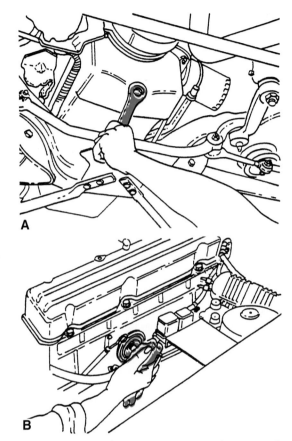

Figure 10-46. Typical cartridge filter service procedure for mobile diesel engines. A—Removing the drain plug. B—Cleaning the filter mounting base. (Wix Corporation)

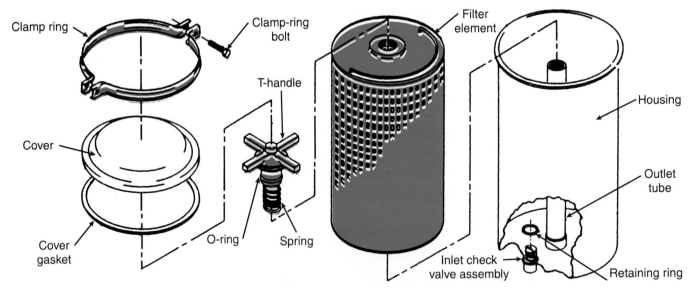

Figure 10-47. Parts of a cartridge filter with a removable cover. (Freightliner Corporation)

10. Position the cover on the housing. Install the clamp ring and bolts. Alternately tighten the bolts to ensure a uniform seal between the cover and gasket.
11. Start the engine and let it idle until normal operating oil pressure is reached.
12. Add oil as needed to bring the oil level up to the full mark on the dipstick.

On many industrial engines, a replaceable cartridge filter element is housed in a stand-alone permanent base, and is connected to the engine by piping. When servicing this type of system, use a solvent to clean both the housing and the mounting base, and dry them thoroughly with compressed air. Edge-type industrial metal strainer filters should be cleaned with solvent and a soft brush.

Oil Pressure Indication Systems

There are several systems used to warn the operator of low oil situations. The most common are a warning light or gauges. The gauges may be mechanical, electrical, or electronic.

Oil Pressure Warning Light

The purpose of the *oil pressure warning light* is to alert the operator to low oil pressure conditions. Occasionally, the light will not go out after changing the oil and the filter. This problem is normally caused by an air lock between the pump and filter. To eliminate the air lock, simply loosen the filter until there is a small space between the gasket and mounting base. Next, disable the fuel pump and turn over the engine until oil flows past the filter seal. Finally, retighten the filter, start the engine, warm it up, and check for leaks. If this procedure does not solve the problem, it may be necessary to check the lubrication system further. An oil pressure warning light may be either mechanically or electrically actuated.

Mechanical Indicating System

The major component of a *mechanical oil pressure indicating system* is a bourdon tube gauge. This gauge is attached by tubing to the pressure source, which is usually the main oil passage in the engine block.

The *bourdon tube gauge* consists of a tube made of spring bronze or spring steel and a pinion-and-sector mechanism, **Figure 10-48**. One end of the tube is attached to the pressure inlet of the gauge, and the other end is attached to the pinion-and-sector mechanism. Oil pressure inside the tube causes it to straighten out. This, in turn, causes the pinion-and-sector mechanism to move, rotating the pointer on the face of the gauge. Oil pressure can then be determined by reading the gauge.

Electrical Indicating System

The *electrical indicating system* has two major components: a sending unit at the pressure source and an indi-

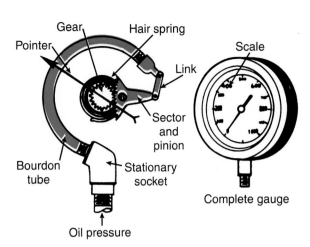

Figure 10-48. *Bourdon tube gauge used to indicate oil pressure. As the oil pressure straightens the tube, the sector-and-pinion mechanism moves, rotating the pointer on the gauge. (Deere & Company, Inc.)*

cating device on the control or instrument panel. The four common types of electrical indicating systems include the electromagnetic coil type, the heating coil type, the pressure switch type, and the electronic type.

Electromagnetic Coil Indicating System

The sending unit in an *electromagnetic coil indicating system* consists of a resistor, a sliding contact called a wiper, an actuating lever, and a spring-loaded diaphragm enclosed in a housing, **Figure 10-49**. Oil pressure works against the diaphragm, causing the lever to move. The lever then changes the position of the wiper, which moves along the resistor to vary resistance. The indicating device or gauge consists of two coils and an armature with a pointer. A magnetic field is created in the coils as current flows from the battery to coil #1, then through a parallel circuit to coil #2, and to the resistor in the sending unit. As the pressure against the diaphragm changes, the resistance in the sending unit also changes. This change in resistance causes current at coil #2 to change. As the current changes in coil #2, its magnetic field becomes stronger or weaker. This change in the magnetic field controls armature rotation. The armature then rotates the pointer on the face of the gauge.

Heating Coil Indicating System

The sending unit in a *heating coil indicating system* contains a heating coil that is wound around a bimetal strip. See **Figure 10-50A**. Oil pressure deflects a diaphragm in the sending unit, causing the contacts to close. When the contacts close, current flowing through the coil creates heat, which deflects the bimetal strip until the contacts open. As the bimetal strip cools, it returns to its original position, closing the circuit. This opening and closing cycle repeats continuously.

The indicating gauge contains a similar heating coil wound around a bimetal strip. This coil is connected in series to the coil in the sending unit. As the sending unit coil heats up, the gauge coil heats up. The bimetal strips in

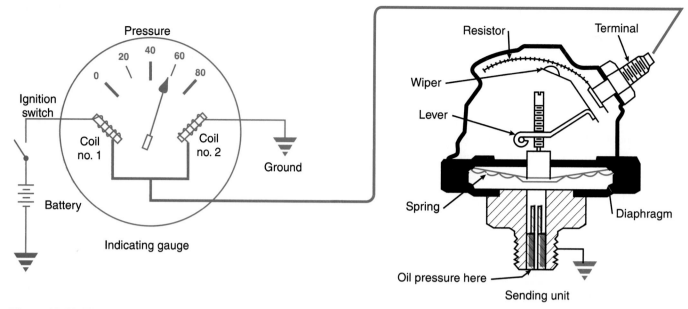

Figure 10-49. Electromagnetic coil oil gauge system. As the wiper moves, the resistance of the unit changes. (Deere & Company, Inc.)

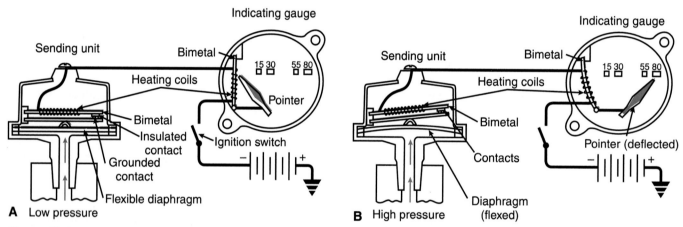

Figure 10-50. Heating coil system for indicating oil pressure. A—Oil pressure causes the contacts in the sending unit to close. B—Contacts in the gauge close, which causes the needle to deflect and rise or fall. (Deere & Company, Inc.)

both coils deflect simultaneously. As the bimetal strip in the gauge deflects, it causes the pointer to move on the face of the gauge. As oil pressure increases, the sending unit diaphragm is deflected more, **Figure 10-50B.** More current is needed to heat the bimetal strip and open the contact. The increase in current causes greater deflection of the bimetal strip in the gauge, which moves the pointer to indicate the increase in oil pressure.

Pressure Switch Indicating System

The **pressure switch indicating system** is different than the two systems just mentioned. Although it has a sending unit, a light bulb is used as the indicator instead of a gauge, **Figure 10-51.** This type of system can indicate whether oil pressure is low, but it does not measure the actual pressure in the lubrication system.

The sending unit consists of a set of contacts and a diaphragm. Before the engine is started, the contacts are closed and the indicating bulb is lit. When the engine is

started, oil pressure builds and pushes the diaphragm up. This separates the contacts and breaks the circuit, causing the bulb to go out. As oil pressure drops, the diaphragm falls, closing the contacts and completing the circuit. When the circuit is complete, the bulb lights, indicating low oil pressure.

Electronic Indicating System

The **electronic indicating system** operates in the same manner as the pressure switch system. However, a relay assembly is used in place of the diaphragm, and a digital gauge is used in place of the warning light. At present, the digital type gauge is becoming more popular for all types off applications.

Ventilation Systems

Most diesel engines are equipped with a **ventilation system** to vent the internal spaces that are associated with the

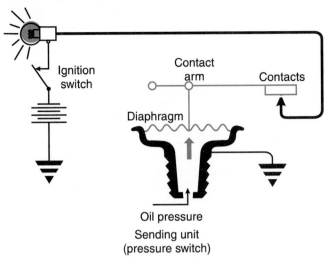

Figure 10-51. *Pressure switch system for indicating oil pressure. (Deere & Company, Inc.)*

Figure 10-52. *A breather pipe can be used to vent the crankcase. (Detroit Diesel Corp.)*

lubrication system. The system may be vented directly into the atmosphere or through the engine intake-air system. Marine installations normally use a system that vents through the intake-air system. Venting the heated, fume-laden air directly to the atmosphere in a compartment will seriously contaminate the compartment air and may create a fire hazard. If the lubricating oil system is not vented in some way, combustible gases may accumulate in the crankcase and oil pan. Under certain conditions, these gases could explode.

Harmful vapors formed in the engine crankcase, gear train, and valve compartments must be removed by a *continuous pressurized ventilation system.* In this type of system, a slight pressure is maintained in the engine crankcase. This pressure is created by the seepage of a small amount of air from the air input system past the piston rings. This air sweeps up through the flywheel housing and enters the valve compartment through cavities in the lifter brackets and vent casings. Ventilating air in the valve compartment is then drawn off by the governor control housing through a breather pipe attached to either the valve rocker cover or the side of the cylinder block.

In mobile equipment installations, there are several methods of providing ventilation to the crankcase. In some cases, the vapors are routed out of the crankcase and into the air induction system. A special oil cap with a small built-in air cleaner is another method of circulating air through the crankcase. Some systems use the main air cleaner as the inlet.

A breather or **vent tube** is sometimes used on mobile engines as the outlet for a ventilation system. See **Figure 10-52.** As the machine moves, the air moving past the tube opening causes the pressure at the opening to be lower than the pressure at the breather. Because air in the engine naturally flows to this low pressure area, the tube becomes an outlet for the system. Some vent tubes use a wire mesh screen to filter particles out of the vapors.

The intake manifold can also serve the dual function of circulating the air and acting as the vent outlet. In this type of ventilation system, a tube connects the crankcase to the intake manifold. The intake vacuum draws the air through the engine, into the manifold, into the cylinders, and out through the exhaust system.

In many large diesel engines, an impeller pump circulates air to ventilate the crankcase. The air is taken from the main air cleaner and pumped through the engine. The outlet can be either a vent tube or the intake manifold. The crankcase breather on some diesel locomotives is mounted on the top of the oil pan. Minute particles of lubricating oil in the moving vapors are trapped in an oil separator located in the breather. The particles are eventually returned to the crankcase, and the vapors are released to the atmosphere through the breather tube.

Engine Oils

The final consideration in the diesel lubrication system is the choice of the correct **lubricating oil.** Diesel lubricating oils are carefully formulated to reduce the adverse effects of diesel engine operation. A quality lubricating oil should be able to do the following:

- ❑ Flow properly under various environmental conditions.
- ❑ Provide cooling for diesel engine components.
- ❑ Prevent formation of sludge and varnish by resisting the absorption of oxygen.
- ❑ Reduce friction and wear.
- ❑ Prevent deposits on internal engine components.
- ❑ Protect all metal surfaces from corrosion.
- ❑ Provide lubrication without foaming.
- ❑ Neutralize harmful acids.

To make the correct lubricating oil selection, there are three basic lubricating oil properties that must be understood: API service classification, viscosity, and total base number (TBN). Always refer to the engine manufacturer's service manual for guidance in making the proper oil selection.

API Service Classification

The American Petroleum Institute (API) governs oil classifications, **Figure 10-53.** Designations such as CH-4 is an *API service classification,* which tell the user what type of performance tests the oil have passed.

> **Note: The standard or *S* service classification is for gasoline engines that use spark ignition, while the *C* service classification refers to compression ignition used by diesel engines.**

Engine oils used in low-emissions diesel engines must meet the requirements of API CG-4. This certification ensures the oil provides deposit control and guards against soot-related problems such as valve train wear, viscosity increase, and filter plugging. Lubricants used in four-stroke diesel engines operating on high-sulfur fuels must meet the requirements of API CF-4. Lubricants used in modern two-stroke diesel engines must meet the requirements of API CF-2.

Viscosity

Viscosity is a measure of an oil's thickness or resistance to flow. It is an indication of the ability of the oil to maintain proper lubricating quality under various speed, temperature, and pressure conditions. Viscosity is important because it is directly related to how well the oil protects against metal-to-metal contact. Because oils thin as temperature increases and thicken as temperature drops, ambient air temperature must be taken into consideration when choosing a proper viscosity.

The Society of Automotive Engineers (SAE) has established an oil viscosity classification system that is accepted throughout the industry. In this system, the heavier weight (thicker) oils receive higher numbers or ratings. For example, an oil with an SAE rating of 50 is heavier and flows slower than an oil with an SAE rating of 10. Heavyweight oils are normally used in warm climates, while lightweight oils are used in low-temperature areas. **Figure 10-53** outlines the API classifications for diesel engine oil. Both SAE and API viscosity ratings can be found on the oil container.

Multi-Viscosity Oil

Multi-viscosity oils are classified with two viscosity numbers, such as 10W-30, 15W-40, etc. These oils have been developed to cover a wider range of operational temperatures than single viscosity oils, **Figure 10-54**. For example, a 10W-30 oil has the same viscosity characteristics as a 10-weight oil at 0°F (-18°C), but it will duplicate flow the properties of a SAE 30-weight oil at 212°F (100°C). This oil will provide easy starting in cold weather and good protection at all operating temperatures.

Total Base Number

Most diesel fuel contains some degree of sulfur (see **Chapter 14**). A byproduct of sulfur, sulfuric acid, acts as a corrosive to engine components. As mentioned earlier, one

Diesel Engine Oil Classifications	
Oil Classification	Recommended Usage
CA-CE	No longer in use
CF-4–1990 Diesel Engine	Designated for high-speed, four-stroke cycle diesel engines. Should be used in place of API CE oils. Particularly suited for on-highway, heavy-duty truck applications.
CF–Indirect-Injected Diesel Engine	Designated for indirect-injected diesel engines and other diesel engines that use a broad range of fuel types, especially fuel with high sulfur content (over 0.5% wt). May be used when API Service Category CD is recommended.
CF-2–Severe-Duty Two-Stroke Cycle Diesel Engine	Designated for two-stroke cycle diesel engines requiring highly effective control over cylinder and ring-face scuffing and deposits. May also be used when API Engine Service Category CD-2 is recommended.
CG-4–1994 Severe-Duty Diesel Engine	Used in high-speed four-stroke cycle diesel engines. Primarily used in heavy-duty on-highway (0.05% wt sulfur fuel) and off-highway (less than 0.5% wt sulfur fuel) applications. May also be used in engines requiring API Service Categories CD, CE, and CF-4.
CH-4–1998 Severe-Duty Diesel Engine	Suitable for high-speed, four-stroke diesel engines designed to meet 1998 exhaust emission standards. Compounded for use with diesel fuels with sulfur content up to 0.5% weight.

Figure 10-53. *Diesel engine oil classifications.*

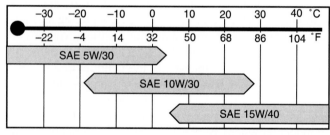

Figure 10-54. *Temperature/viscosity range of different grades of multi-weight oil.*

of the functions of lubricating oil is to neutralize this acid using alkalinity additives. An oil's *total base number (TBN)* is a measure of its reserve alkalinity. As a general rule, a CG or CH oil with a TBN that is twenty times greater than the fuel sulfur level should be used in diesel engines. The higher the oil's TBN, the greater its neutralizing ability. When an oil's TBN drops below a certain level, it can no longer adequately protect the engine and it should be changed.

Total Acid Number

The **total acid number (TAN)** is a companion to the TBN. It measures the acidity of an oil. The higher an oil's TAN number, the more acidic it is. The effects of high TAN values in oil include corrosion, oil thickening, formation of deposits, and accelerated wear. In some applications, a high TAN can be caused by acid contamination from the environment. Using the wrong lubricant or burning-high sulfur fuel may also result in high TAN values.

Resistance to oxidation is an important characteristic governing the service life of lubricating oil. **Oxidation** occurs when the hydrocarbons in the oil chemically combine with oxygen. Although **antioxidants** are added to most oils, they are gradually used up during service. Heat, pressure, and an abundance of air speed up the oxidation process. Other common causes of excessive oxidation include overextended oil drain intervals, the wrong oil for the application, excessive combustion gas blowby, and high-sulfur fuels. As oil oxidizes, corrosive acids can form and deposits may accumulate on critical engine parts, inhibiting operation and accelerating wear.

Synthetic Oil

The introduction of **synthetic oil** dates back to World War II. Synthetic oil is manufactured in a laboratory, rather than being refined from natural crude oil. It offers a variety of advantages over natural oils, including stability over a wide range of temperatures and operating conditions, better fuel economy, and longevity. However, synthetic oil is not recommended by the Automotive Engine Rebuilder's Association (AERA) for the break-in period. This oil's ability to reduce wear by virtually eliminating friction between moving components, while normally desirable, is not ideal for a break-in oil. Certain predictable amounts of friction are required for the proper break-in of pistons and piston rings. Consequently, AERA does not recommend the use of synthetic motor oils for the first 5000 miles (8000 km) of service after a rebuild.

Oil Additives

To help a lubricating oil achieve maximum performance, additives are generally recommended by lubricating oil manufacturers. **Additives** are blended with the oil for many reasons—all of which are based on the type of service expected from the oil. Some of the more common additives used with lubricating oils include:

❑ Oxidation inhibitors to keep oil from oxidizing, even at very high temperatures. Because they prevent the oil from absorbing oxygen, they also help to prevent varnish and sludge formation.

❑ Anti-rust agents prevent rusting of metal parts when the engine is not in use. They form a protective coating that repels water droplets and neutralizes harmful acids.

❑ Detergent additives help keep metal surfaces clean and prevent deposits. These additives help suspend particles of carbon and oxidized oil in the oil. The suspended contaminants are removed from the system when the oil is drained.

❑ Anti-corrosion agents protect metal surfaces from corrosion. These agents are combined with oxidation inhibitors.

❑ TBN additives, which are acid molecule neutralizers, are blended into the oil to reduce the harmful effects of the acid compounds produced as a byproduct of the combustion process.

Unfortunately, additives gradually lose their effectiveness with use. Always drain the oil before the additives are totally depleted. Also remember to service filters at regular intervals. Engine oils are blended to a very delicate chemical balance. The addition of aftermarket additives can upset this balance and adversely affect engine performance. Be sure to follow manufacturer's recommendations when using any additive with a lubricating oil.

Oil Analysis

The heavy metals used in some additives can cause reactions that are damaging to emission control equipment, such as catalytic converters and/or particulate traps. Oil analysis is becoming a very important factor in determining the right time to drain the system's oil. Testing by a chemical lab is the best way to check oil contamination. However, for a quick shop test, a simple tester such as shown in **Figure 10-55** can be used.

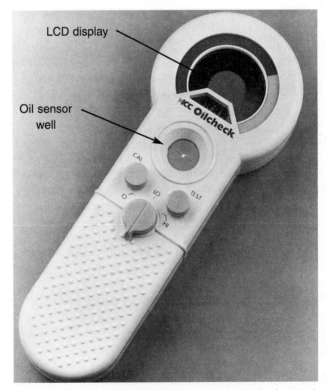

Figure 10-55. *Various instruments are available for testing engine oil for quality and contamination. (ICC Federated)*

When a clean oil sample is placed in the sensor well and the test button is pressed, the tester will zero on the clean sample. After wiping the sensor well with a clean, dry rag and replacing the clean oil with a contaminated sample, press the test button again. The LCD display will indicate the actual level of contamination, relative to the tricolor band (green/yellow/red) on the outer edge of the display. This is a suggested threshold of acceptability only. A different threshold may be selected, based upon user experience. In these cases, the new threshold can be marked on the tester outer band.

Summary

The lubricating system assists the engine by absorbing and dissipating heat, reducing friction between moving parts. It also cleans and flushes moving parts, removing contaminants from the engine, sealing piston rings and cylinder walls, and reducing engine noise. A typical lubrication system consists of an oil pan or sump, oil level dipstick, oil pump, oil cooler, oil filter, lubricating valves, oil pressure indicators, oil lines to and from the turbocharger (if used), and oil passages in the cylinder block.

Oil drawn from the oil pan or sump travels through the suction pickup tube to the oil pump. The oil pump sends warm oil to the oil cooler and then to the oil filter. From the oil filter, oil travels to the oil galleries in the cylinder block, where it proceeds to the oil supply line and then to the turbocharger or through passages to the main bearings and rod bearings. Oil from the turbocharger flows back through the oil return line to the oil pan or sump.

There are two types of diesel engine lubricating systems currently in use—the internal force-feed system and the circulating splash system. Most modern diesel engines use the internal force-feed system. The circulating splash system is found in some older industrial applications.

On most mobile diesel engines, the oil pan is located at the bottom of the engine and is bolted to the block. The area where the oil pan is attached is called the crankcase, because it encloses the crankshaft.

The oil pan provides a reservoir for the lubricating oil. The lower part of the oil pan is called the sump. The oil pickup line extends into the sump.

Large stationary and propulsion-type systems have tanks that collect oil that has been used for lubricating and cooling so it can be recirculated through the engine. Some systems also have a sump or drain tank located under the engine. These tanks collect the oil as it drains from the engine crankcase.

The oil level must be maintained between the full and low marks on the dipstick. Operating at low oil levels exposes the pump pickup, causing aeration and/or a loss of pressure. Operating with excess oil causes foaming or aeration.

The required oil pressure for most diesel engines ranges from 25-40 psi (170-275 kPa), although some industrial diesel engines may go as high as 65 psi (450 kPa). All oil pumps have a positive displacement design.

Normally, an oil pump is mechanically driven by the diesel engine. The external gear pump is usually driven from the crankshaft, but on larger engines, an electric motor attached to the engine may be used to drive an auxiliary pump for prelubrication.

Disassemble an oil pump following the steps given in the engine service manual. When handling an oil pump, be careful not to damage the mounting surfaces of either the housing or the cover.

The lubricating systems of most diesel engines have a cooler (or heat exchanger) to keep the oil temperature in the most efficient range. There are two types of oil coolers: the tube type and the plate type.

The oil pressure regulator valve stabilizes lubricating oil pressure at all speeds, regardless of the oil temperature. The valve is usually installed at the end of a vertical oil gallery located near the front of the cylinder block on the side opposite the oil cooler. On most engines, pressure regulating valve adjustments are made using shims and washers.

There are also two basic filter configurations: the cartridge filter and the spin-on filter. The three types of filtering elements are the surface type, the depth type, and the combination type.

There are two major types of filtration systems employed in modern diesel engines—bypass systems and full-flow systems. The bypass system filters only a percentage of the oil during each complete circulation cycle. In a full-flow system, all of the oil is pumped directly into the oil filter housing, where it must pass through the filtering element before traveling to the moving parts of the engine.

Every filter in a full-flow lubrication system must have a bypass valve. Bypass valves can also be used in oil coolers. Some stationary engines use shunt and sump filtering systems. In order to ensure maximum life and protection from a filter, it is essential that proper service and installation procedures be followed.

Most diesel engines have some method of ventilating the internal spaces that are associated with the lubricating oil system. The system may be vented directly to the atmosphere or through the engine intake-air system.

To make the correct lubricating oil selection, there are three properties that must be understood: API service classification, viscosity, and total base number (TBN). Common oil additives include oxidation inhibitors, anti-rust agents, detergent additives, anti-corrosion agents, and TBN additives.

Important Terms

Lubrication system	Oil pan
Lubricating oil	Crankcase
Pressure regulator valve	Sump
Force-feed lubrication system	Oil pickup line
	Drain plugs
Circulating splash lubrication system	Dipstick

Oil pump

Positive displacement pump

External gear pump

Internal gear pump

Crescent pump

Trochoidal pump

Herringbone gear pump

Oil cooler

Tube oil cooler

Plate oil cooler

Oil pressure relief valve

Oil filters

Cartridge filter

Bolt-on filter

Spin-on filter

Surface filter elements

Depth filter element

Bypass oil filtration system

Restrictor orifice

Full-flow filtration system

Bypass valve

Shunt-type filtering system

Sump-type filtering system

Batch purification method

Continuous purification method

Oil pressure warning light

Mechanical oil pressure indicating system

Bourdon tube gauge

Electrical indicating system

Electromagnetic coil indicating system

Heating coil indicating system

Pressure switch indicating system

Electronic indicating system

Ventilation system

Vent tube

Lubricating oil

API service classifications

Viscosity

Multi-viscosity oils

Total base number (TBN)

Total acid number (TAN)

Oxidation

Antioxidants

Synthetic oils

Additives

Review Questions—Chapter 10

Do not write in this text. Place your answers on a separate sheet of paper.

1. List at least five jobs performed by a diesel engine's lubricating system.

2. *True or False?* Most oil pumps are mounted to the engine block using a gasket.

3. In a rotor pump wear will occur between the _____ on the inner and outer rotors.

4. Name the two types of oil coolers.

5. A(n) _____ stabilizes oil pressure at all engine speeds.

6. List the three types of spin-on filter elements.

7. When replacing a spin-on filter, you should apply a light film of _____ to the gasket sealing surface before installing the filter.

8. An oil with an SAE viscosity rating of 50 is _____ than an oil with an SAE viscosity rating of 10.

9. The Automotive Engine Rebuilder's Association (AERA) does not recommend _____ during the break-in period of a diesel engine.

10. *True or False?* Most diesel engines have some method of ventilating the internal spaces that are associated with the lubricating system.

ASE-Type Questions

1. During a combustion cycle, what determines when
1. Technician A says that most modern diesel engines use the internal force-feed system. Technician B says that the circulating splash system is found in some older industrial applications. Who is right?
   (A) A only.
   (B) B only.
   (C) Both A & B.
   (D) Neither A nor B.

2. Depending on the application, oil pans may be designed to permit an engine inclination of up to _____ .
   (A) 15°
   (B) 30°
   (C) 45°
   (D) 90°

3. The _____ gear pump is the most common type of oil pump used in a diesel lubricating system.
   (A) internal
   (B) external
   (C) rotor
   (D) crescent

4. Which type of oil cooler is usually mounted internally in the engine or the crankcase?
   (A) Tube type.
   (B) Plate type.
   (C) Gear type.
   (D) All of the above.

5. Which of the following is not normally a cause of low engine oil pressure?
   (A) Worn oil pump.
   (B) Low oil level in crankcase.
   (C) Oil in crankcase too heavy of grade.
   (D) Leaking oil filter or pump.

6. Which of the following oil filter designs is used in modern diesel engines?
   (A) Surface type.
   (B) Depth type.
   (C) Combination type.
   (D) All of the above.

7. Technician A says that every filter in a full-flow lubrication system must have a bypass valve. Technician B says that no filter in a full-flow lubrication system must have a bypass valve. Who is right?
(A) A only.
(B) B only.
(C) Both A & B.
(D) Neither A nor B.

8. Technician A says you should allow the engine oil to cool before draining it for oil and filter maintenance. Technician B says you should drain the oil with the engine warm or hot. Who is right?
(A) A only.
(B) B only.
(C) Both A & B.
(D) Neither A nor B.

9. Technician A says that an oil's total base number (TBN) is an indication of how well the oil will flow at various temperatures. Technician B says that TBN gives an indication of how well the oil can neutralize sulfuric acid that enters the oil during engine operation. Who is right?
(A) A only.
(B) B only.
(C) Both A & B.
(D) Neither A nor B.

10. After the oil and filter are changed on a diesel engine, the oil warning light remains on. Technician A says that the most likely cause is an air lock between the oil pump and the filter. Technician B says the most likely cause is a bad oil gauge light. Who is right?
(A) A only.
(B) B only.
(C) Both A & B.
(D) Neither A nor B.

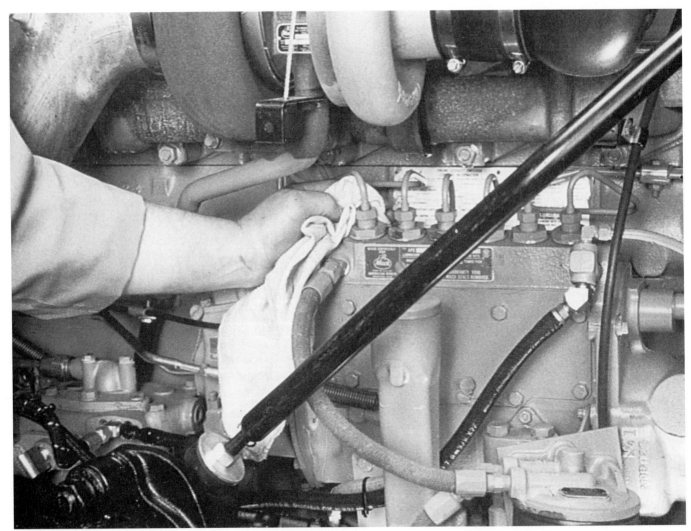

Dirt is the greatest contributor to engine wear and failure. Make sure the outside of the engine is clean before beginning work, no matter how minor the job. (Mack Trucks, Inc.)

Chapter 11

Cooling Systems

After studying this chapter, you will be able to:

❑ Explain the purpose of a diesel engine cooling system.

❑ Describe the cooling system's basic components and explain their operation.

❑ Describe the difference between air and liquid cooling systems.

❑ Describe the major components of a diesel liquid cooling system.

❑ Explain how a thermostat regulates engine temperature.

❑ Explain why cooling system filters and conditioners are used.

❑ Describe coolant pump operation.

❑ Explain how an industrial diesel engine's heat exchanger works.

❑ Name the four types of cooling systems used with marine diesel engine applications.

A great deal of heat is generated within a diesel engine during operation. Although combustion produces the majority of this heat, compression within the cylinders and friction between the moving parts add to the total amount of heat developed in an engine. This chapter discusses the methods of heat removal used in diesel engines.

Cooling Systems

Of the total heat created by burning fuel in the cylinder of an engine, only about one-third is transformed into useful work. An equal amount is removed in the exhaust gases. This leaves approximately 30-35% of the combustion heat in the engine. Since the temperature of combustion itself is approximately twice that at which iron melts, without some means of dissipating heat an engine would operate for a very

limited time. This excess heat is transferred from the engine through coolant, lubricating oil, air, and fuel.

The function of a diesel engine's cooling system is to carry away excess heat in order to prevent the engine's metal from melting and the lubricating oil from breaking down. Heat dissipation would be relatively simple if it were not essential that the cooling system maintain a relatively uniform temperature at any engine speed, load, and varying ambient temperatures. At high engine torque output and ambient temperature, the system operates at its maximum capacity to maintain a temperature of approximately 180°F (82°C). It must also maintain this temperature while the engine is idling in subzero temperatures.

In a diesel engine, fuel ignites when it is injected into the hot compressed air in a cylinder. Correct engine temperature is critical for efficient diesel operation. Without a properly maintained cooling system, a diesel engine can easily run too cool in the winter. A defective cooling system may appear operative, when in fact, it is allowing the temperature to fall to 150°F (65°C) or less. Such low temperatures can cause a number of problems, including:

❑ Engine oil turning to sludge.

❑ Deposit build-up in the cooling system, reducing coolant flow to the cylinder and accelerating wear.

❑ Condensation in the lubrication system, contaminating the oil and inducing premature wear.

❑ Poor combustion, causing horsepower and torque loss.

❑ Reduced fuel economy, raising operating costs.

❑ Reduced oil viscosity, limiting ring and bearing lubrication.

❑ Lower combustion temperatures, which leaves unburned fuel in the cylinder, causing carbon build-up, engine oil dilution, excessive exhaust smoke, and excessive exhaust emissions.

To avoid these problems, which can lead to unscheduled downtime, reduced engine durability, or possible mechanical failure, diesel engine technicians must be familiar with cooling system operation and maintenance. Additional cooling system monitoring should be done by the engine operators. Operators should monitor the engine temperature gauge on a regular basis. The data should be collected over several months and compared. Wide variances in readings may indicate a cooling system problem.

Types of Cooling Systems

There are two types of cooling systems used in modern diesel engines: air-cooled systems and liquid-cooled systems. While these systems perform the same function, they are very different in components and operation.

Air-Cooled Systems

Air-cooled systems are designed to quickly transfer heat from the combustion chamber to the outside of the engine. Air, which is supplied by forward vehicle movement and/or by a fan or blower, cools the engine by transferring the heat to the atmosphere, **Figure 11-1.**

When cool air passes over a hot metallic surface, the heat near the part's surface reduces the velocity of the cool air. Engines with an air-cooled system are equipped with **cooling fins.** These fins are needed to maintain the correct temperature and air velocity around the cylinder(s). The size and spacing of the cooling fins depends on the amount of heat to be dissipated, the diameter of the cylinder, the type of metal used for the fins, and the velocity and temperature of the cooling air, **Figure 11-2.**

Many short fins are used rather than a few large fins. Short fins that are close together cause the cooling air to become warmer, but fewer fins are needed for cooling. The cylinder structure must be rigid enough to prevent fin distortion. An engine-driven cooling fan is used to provide the

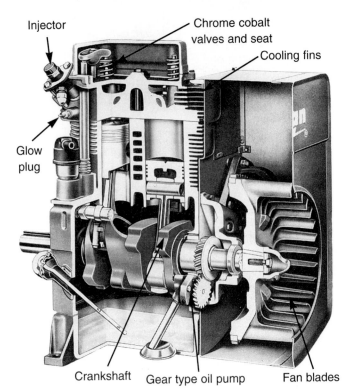

Figure 11-2. *Sectional view of a two-cylinder air-cooled engine. Note the size of the cooling fins and the fan blades on the flywheel. (Petters, Ltd.)*

proper air volume. It is important that an air-cooled engine has all shields and shrouds properly installed and sealed with gaskets or sealant as needed.

> **Note: Because very few modern diesel engines use air-cooled systems, only liquid cooling systems will be detailed in this chapter.**

Liquid Cooling Systems

The **liquid cooling system** uses a liquid medium—generally a mixture of ethylene glycol (antifreeze) coolant and water—to cool the engine. In cold weather, the antifreeze solution mixed with the water prevents it from freezing. Coolant circulates through the engine in jackets (passages) around the cylinders and other high friction areas. The liquid coolant absorbs heat from these areas and then flows to the radiator. See **Figure 11-3.** Airflow through the radiator cools the liquid coolant and the heat is dissipated into the air. The liquid coolant then recirculates through the engine to absorb more heat and the cooling cycle is repeated.

Types of Liquid Cooling Systems

The two major types of liquid cooling systems are the open system and the closed system. The **open system** is used primarily in stationary diesel installations. In this system, water travels through the water jackets in the block

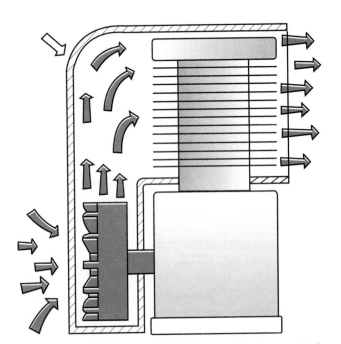

Figure 11-1. *Air cooling system operation. (Wix Corporation)*

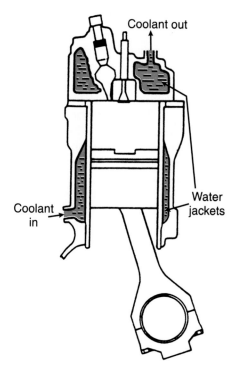

Coolant out

Coolant in

Water jackets

Figure 11-3. *Liquid coolant passes through water jackets (passages) cast in the cylinder head and block to carry away excess heat.*

and then flows to an outside source, such as a city main, a cooling tower, or spray pond, where it is cooled by the air and returned to the engine's cooling system, **Figure 11-4.** In open systems, it is important that any water lost through

evaporation be replaced and that it is operated following the proper safety regulations.

> **Note: Many diesel engine manuals, as well as many of the technicians in the trade, still use the word "water" instead of coolant when referring to cooling systems components. For example, the coolant pump is frequently called the water pump. This is a carryover from the days before modern coolants were invented.**

In the *closed system*, which is normally used in transportation and construction vehicle applications, the diesel engine has coolant (water) jackets or passages in the engine block to permit coolant to pass over hot combustion chamber surfaces. This coolant, which is circulated by a water pump, acts as a liquid conveyor to carry the excess heat from the combustion chamber surfaces to the radiator. This heat is removed by cool outside air flowing through the radiator. An engine thermostat controls the amount of coolant entering the radiator. When the engine is cold, the coolant circulates within the engine. The coolant bypasses the thermostat and radiator through a passage that prevents excess heat from building up in the engine.

When the coolant temperature reaches the temperature rating of the thermostat, the thermostat opens and allows the coolant to flow through the radiator. The hotter the coolant gets, the more the thermostat opens, allowing additional coolant to flow through the radiator.

Once coolant has passed through the radiator and has been "cooled" (or has given up its excess heat), it re-enters

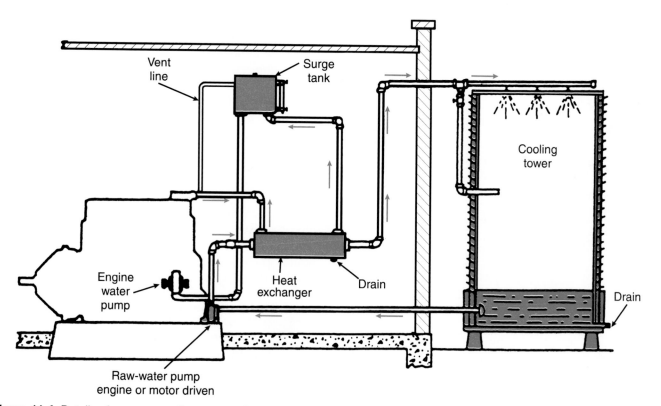

Vent line

Surge tank

Cooling tower

Engine water pump

Heat exchanger

Drain

Drain

Raw-water pump engine or motor driven

Figure 11-4. *Details of an open cooling system. (Waukesha Motor Company)*

the water (coolant) pump and the cooling cycle begins again. The water pump pushes the coolant up through the water jackets and passages surrounding the combustion chambers to pick up the excess heat, **Figure 11-5.**

Liquid Cooling System Operation

Figure 11-6 shows the major components typically found in the closed liquid cooling systems used in the transportation field, as well as those used with portable and semi-portable equipment.

Beginning at the water (coolant) pump, the coolant is sent into the cylinder block. The cylinder block routes the coolant to three different places. The majority of coolant flows into the cylinder head. Some of the coolant flows to

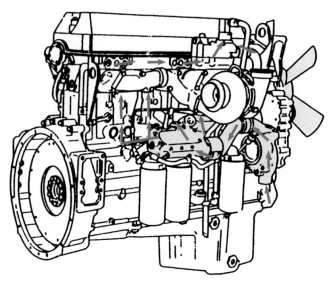

Figure 11-5. Coolant flow through a mobile diesel engine. (Detroit Diesel Corp.)

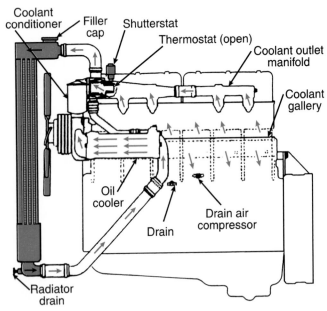

Figure 11-6. Components of a liquid cooling system used on mobile applications. (Mack Trucks, Inc.)

the air compressor and some flows to the coolant conditioner. From the air compressor, the coolant flows back to the water pump. The cylinder head directs its coolant into the water manifold. The water manifold contains a temperature sending unit (shutterstat), which assists the cooling process by controlling the flow of air through the shutters to the radiator. The water manifold and the coolant conditioner direct coolant to the thermostat area. Some coolant in the water manifold flows through a hose into the heater core. A hose from the heater core channels this coolant back to the water pump.

Meanwhile, the thermostat directs coolant flow to the bypass tube during warm-up and to the radiator once the thermostat reaches the proper temperature, generally about 195°F (91°C). The bypass tube and radiator feed the oil cooler. The oil cooler routes the coolant back to the water pump, making the cycle complete.

Some diesel engines have additional components, such as a coolant recovery system, coolant filters, fan shrouds, a second coolant pump, coolant heaters, and a torque converter oil cooler.

Coolants

Coolant, as previously stated, is the liquid medium that circulates through the cooling system to absorb excess heat. The proper coolant is a very important part of any cooling system. Plain water can cause rust and corrosion in the cooling system. Hard water leaves scale deposits on the cylinder liner and water jacket surfaces that can interfere with heat transfer. Distilled water does not leave scale deposits, but can still cause corrosion. For this reason, most diesel engine manufacturers recommend a coolant mixture of an **ethylene glycol** or **propylene glycol** based antifreeze, clean water, and a reliable corrosion inhibitor.

Antifreeze is normally mixed with an equal amount of water. The mixing should always be done before adding the solution to the cooling system. In extremely cold temperatures, the mix can contain up to 70% antifreeze. However, too much antifreeze actually reduces cold weather protection. Also remember that keeping antifreeze in a diesel engine is important at any time of the year. Antifreeze not only protects against the winter cold, but acts as a summer coolant in extremely high temperatures. While water boils at 212°F (100°C), ethylene glycol antifreeze has a boiling point of 223°F (106°C). In addition, a pressurized cooling system will raise the boiling point an additional 3°F (1.6°C), for every 1 PSI (6.895 kPa) of pressure in the system.

 Note: Although ethylene glycol-based antifreeze is sometimes labeled "permanent," this does not mean it should never be replaced. Ethylene glycol contains inhibitor additives that do not last. Antifreeze should be periodically replaced.

Corrosion Inhibitors

A **corrosion inhibitor** is a water-soluble chemical compound that protects metal surfaces of the cooling system against corrosion. The importance of adding a good corrosion inhibitor to coolant cannot be overstressed. Some of the common corrosion inhibitors include chromates, borates, nitrates, and nitrites. (Soluble oil is not recommended as a corrosion inhibitor.) Inhibitors are gradually depleted in normal use, therefore they should be periodically replaced.

There are numerous chemical reactions that affect engine coolants, particularly when these coolants have been used for an extended period of time. Initially, engine coolant is a mixture of ethylene or propylene glycol, water, and anticorrosion and lubrication additives. Both propylene glycol and ethylene glycol are derivatives of petrochemicals. The effects of the two glycols are similar, but their physical properties are different.

When mixed with water, the major difference is that an ethylene glycol solution has a slightly lower freezing point and a propylene glycol solution has a slightly lower surface tension. Because the heat of a working engine accelerates chemical reaction with metal components, some metals dissolve into the coolant solution. Oxygen in the system becomes a damaging factor, by promoting a chemical reaction that results in rusting or pitting. While coolant additives inhibit the process, they do eventually break down. Therefore, they need to be replaced periodically. As a result of normal breakdown, the dissolved solids build up and increase the conductivity of the coolant to damaging levels.

Environmental Concerns

The environmental concern over the disposal of coolant comes primarily from the metal content of the substance. The EPA and state authorities enforce stringent regulations governing the disposal of used engine coolant. According to the Environmental Protection Agency, any used coolant with a lead content in excess of 5 PPM is considered a hazardous waste. While safe disposal is usually possible, there is an increasing trend toward recycling used coolant.

State regulations vary from state to state. Many states currently follow EPA guidelines when defining substances as hazardous wastes. In compliance with the Clean Water Act, no used coolant can be disposed of in septic systems or streams. Similarly, coolant should never be dumped on the ground. Used coolant can poison any animal that drinks it. Any antifreeze headed for landfills must be solidified. In some cases, area sewage treatment plants must be notified prior to the dump so they are prepared to handle the substance.

Filters used in recycling or regenerating coolant may also end up as hazardous waste, depending on the type and amount of substance they have collected. These filters must also be disposed of in compliance with applicable laws.

Recycling Coolant

The recycling of coolant can be done by an off-site facility or by an in-shop unit. Once recycled, new inhibitors are added to the reclaimed coolant. The end product is a high quality, highly concentrated solution of glycol and additives. To check the purity of recycled coolant, it can be sent to a test laboratory. Test kits are also available for this purpose. When using a test kit, follow the manufacturer's instructions closely.

Radiator Systems

The radiator has two main purposes in the cooling system.

❑ It allows the coolant's heat to be released into the atmosphere, while keeping the coolant under pressure.

❑ It provides enough liquid storage to ensure that the cooling system works properly.

The radiator assembly, **Figure 11-7,** includes the surge or recovery tank, a radiator, the radiator cap, and often, a metal or plastic shroud.

Surge Tank

The **surge tank,** or *recovery tank,* found in many mobile diesel cooling systems provides space for the coolant and expansion space for heated coolant. The cooling system must be designed so that a reasonable loss of coolant from the normal level will not result in aeration. The capability of the system to handle the coolant loss is determined by a draw-down test. The minimum draw-down for systems having a total capacity between 40-100 quarts (38-95 liters) is 10% of the capacity. If minimum draw-down capacity cannot be obtained and space is not available to increase the volume of the integral top/surge tank or the remote surge tank, the additional capacity may be obtained with a coolant recovery tank.

The coolant recovery tank should be mounted as high as the installation permits, and it must be sized to provide cooling expansion volume above the added draw-down capacity (6% of total volume, including

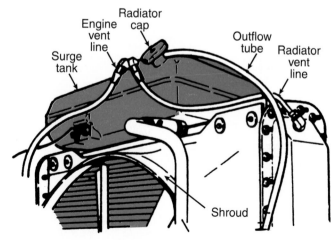

Figure 11-7. *Typical diesel engine radiator assembly with surge tank and fan shroud. (Caterpillar Inc.)*

heater and lines). The pressure cap used with this system provides a vacuum-tight seal on the top of the top/surge tank filler neck.

As the engine temperature increases, the coolant and air in the system expands and builds pressure. The valve in the radiator pressure cap unseats and allows the coolant to flow into the coolant recovery tank, **Figure 11-8A.** When the engine starts to cool down, the air and coolant contract, creating a vacuum in the system. The vacuum unseats another valve in the radiator pressure cap, allowing the coolant to be drawn back into the expansion tank or radiator, **Figure 11-8B.** The coolant level in the recovery tank should be checked daily.

Caution: Any leak in the cooling system (radiator, top tank, pressure cap, hose connection, and so on) will prevent the coolant recovery system from functioning, and visual inspection of the recovery tank will give a false indication of coolant level. Therefore, a periodic check of the top tank and/or surge tank coolant level under ambient conditions is recommended. Under normal operation, a small amount of coolant will be lost due to evaporation. If an abnormal amount of coolant is being used, check the cooling system for leaks.

Radiators

There are two basic types of **radiators** found in mobile diesel equipment: high-flow (traditional) radiators and low-flow radiators. High-flow and low-flow radiators are either single-pass, vertical-flow radiators (with top and bottom tanks) or single-pass, cross-flow radiators (with side tanks).

In *single-pass, cross-flow radiators,* **Figure 11-9,** coolant enters the inlet at the top of the left end tank, flows across the core to the right end tank, and exits the radiator at the bottom of the right end tank. In *single-pass, vertical-flow radiators,* coolant enters the inlet on the top tank, flows down through the core, and exits the radiator at the bottom tank, **Figure 11-10.**

On *low-flow radiators,* either the top tank or right end tank has a check ball to limit coolant flow through the vent line. Low-flow radiator cores have lubricated tubes with internal baffles that constantly mix the coolant, providing more efficient heat transfer.

Coolant flowing from a low-flow radiator is pumped more slowly and is cooler than that from a high-flow radiator. This coolant is routed directly to the aftercooler, where it lowers the charge-air temperature more than the coolant from a high-flow system would.

The radiator core consists of a series of tubes and fins, **Figure 11-11.** The core exposes hot coolant to as much

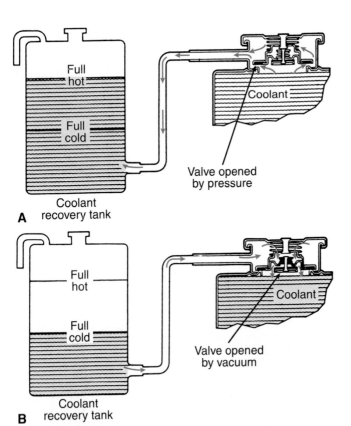

Figure 11-8. Operation of coolant recovery tank. A—As coolant temperature increases, coolant flows into the recovery tank. B—After the engine is shut off, the coolant is pulled in by vacuum as the engine cools. (Detroit Diesel Corp.)

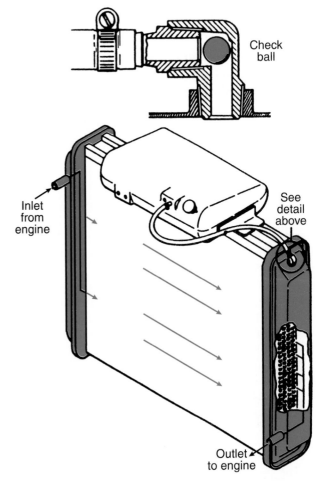

Figure 11-9. Typical cross-flow radiator. (Caterpillar Inc.)

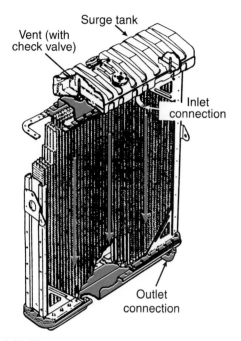

Figure 11-10. *Typical vertical-flow radiator. (Caterpillar Tractor Company)*

surface area as possible, maximizing the amount of heat that is released to the passing air. Radiators are also classified by the number of tube layers and the number of fins per inch. The more tube layers and fins per inch, the more effectively the radiator will dissipate heat. For example, an 11-fin core has 11-fins per inch of tubing and will cool more quickly than a 9-fin core. When heat dissipation is a major concern, the number of fins per square inch can be increased. However, adding more fins increases air resistance, thus requiring an increase in the amount of horsepower needed to drive the radiator fan.

When air enters a liquid cooling system, it can become one of the major causes of corrosion. Air in the system also reduces the coolant's ability to dissipate heat and can cause the coolant pump to lose its prime. To reduce aeration in the cooling system, some radiators are equipped with a solid baffle that divides the top tank into upper and lower sections. Coolant enters the lower section from the cylinder head and the upper section from a radiator vent connected to the recovery tank. The upper section of the tank also serves as a reserve space that compensates for minor coolant losses and provides space for thermal expansion.

Most radiators are equipped with at least one drain valve, or petcock located at the bottom of the radiator. There is also generally one inlet connection in the top of the tank, overflow tube, and filler opening, which is sealed by the radiator pressure cap.

Radiator Fan Shroud

To increase the cooling efficiency of the radiator, a *fan shroud* is placed around the fan. There should be an airtight seal between the fan shroud and the radiator to prevent recirculation of the hot air drawn through the radiator. Hot air permitted to pass around the sides or bottom of the radiator and then drawn through the unit again will cause engine overheating.

Radiator Pressure Caps

At one time, a *radiator cap* (or purge tank cap) was simply designed to keep liquid coolant from splashing out of the radiator. Today's radiator caps are equipped with pressure springs and atmospheric vents. This pressure control valve/cap is normally a closed valve. The cap is designed to permit a pressure in the cooling system equal to the rate stamped on the top of the cap. A cap with a "9" on the top allows the cooling system to develop 9 psi (62 kPa pressure) before the valve opens, **Figure 11-12A.** This system pressure raises the boiling point of the coolant and reduces coolant loss. The maximum allowable coolant temperature, regardless of the pressure cap used, is 210°F (99°C). When pressure in the system exceeds the speci-

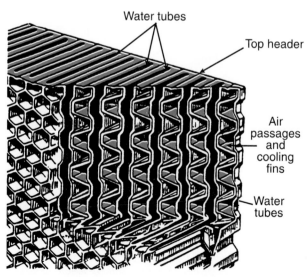

A

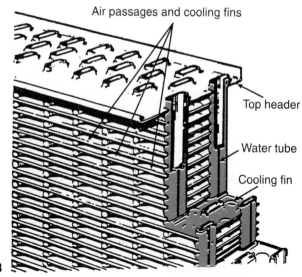

B

Figure 11-11. *Radiator core construction details. A—Honeycomb or cellular type. B—Tube and fin. (Modine Mfg. Company)*

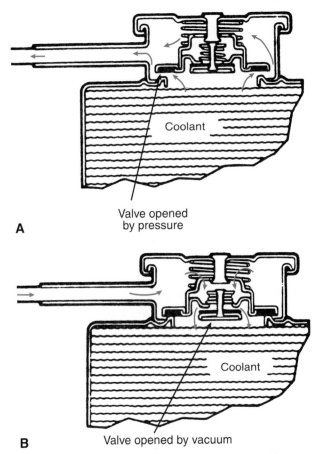

A Valve opened by pressure

B Valve opened by vacuum

Figure 11-12. Operation of a typical pressure radiator cap. A—Pressure valve open. B—Vacuum valve open. (Detroit Diesel Corp.)

fied limits, the seal between the cap and the radiator filler neck opens, allowing the coolant to vent into the recovery (surge) tank.

An atmospheric vent or valve in the cap ensures that atmospheric pressure will not crush the radiator as the coolant cools and contracts, **Figure 11-12B**. In addition, if the vehicle is equipped with a coolant recovery system, the vent will also allow the liquid from the overflow tank to return to the radiator. This refills the system and normalizes the pressure.

All radiator caps are designed to meet SAE standards for safety. These standards require that all caps have a detent position, which allows pressure to escape from the cooling system without allowing the hot coolant to escape.

 Warning: Escaping steam and water can cause scalding and extreme burns.

A lever or button may be found on the cap to release the pressure prior to removing the cap. Only after all pressure has been released should any attempt be made to remove the cap from the filler neck.

Radiators and caps may be tested with a relatively simple, inexpensive tester. This tester consists of an air

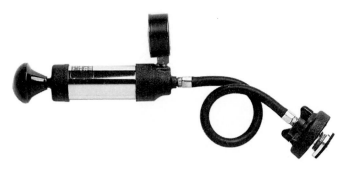

Figure 11-13. Radiator pressure tester. (Stant Manufacturing, Inc.)

pump, a pressure gauge, and a filler neck adapter, **Figure 11-13**. The cap is placed on the tester's filler neck and air pressure is applied with the pump. The gauge will indicate the "pop-off" pressure of the cap. This reading should correspond with the pressure specifications shown on the face of the cap. **Figure 11-14** shows how the pressure tester is used to check the cooling system for leaks.

Radiator Shutters

Many heavy-duty vehicles employ radiator shutters to regulate the flow of cool air through the radiator, maintaining engine temperature within a predetermined range. The radiator shutters can be operated by one of the following methods: air, vacuum, or thermostat.

Air Shutter System

The **air shutter system** contains a shutter assembly, which consists of a series of vanes and springs, **Figure 11-15**. The springs open the vanes, then an air cylinder closes them with the aid of a **shutterstat valve, Figure 11-16**. Other system components include a lightweight fan, an on-off fan clutch, and a control that synchronizes the fan and the shutters.

Vacuum Shutter System

The **vacuum shutter system** is normally found on vehicles with vacuum-hydraulic brakes. It consists of a

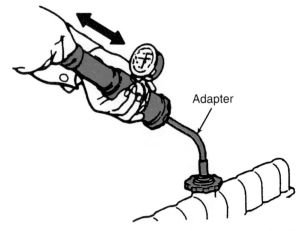

Figure 11-14. Applying pressure with a tester to check the cooling system for leaks. (Stant Manufacturing, Inc.)

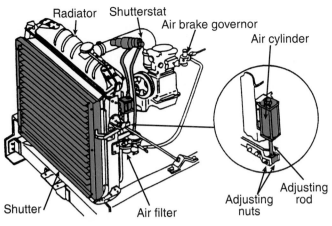

Figure 11-15. Air-operated, shutter-type radiator.

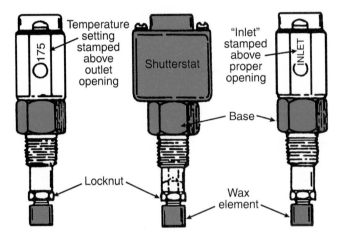

Figure 11-16. Typical shutterstat.

shutter assembly, a thermostatically operated shutterstat, and a vacuum-powered cylinder that controls the shutter assembly, **Figure 11-17.** Vacuum shutter system operation is almost identical to the air shutter system operation except that the power source is vacuum instead of air.

Thermostatic Shutter System

The **thermostatic shutter system** consists of a shutter control assembly, control pull rod, and shutter assembly, **Figure 11-18.** The shutter control assembly has a power element that works like a thermostat and controls the shutter movement according to engine coolant temperature. When it is necessary to service any type of radiator shutter system, follow the instructions given in the manufacturer's service manual.

Hoses

Obviously, the primary function of a cooling system **hose** is to carry coolant between the different system components. Most systems have an upper radiator hose, a lower radiator hose, and two heater hoses. In addition, some systems may have one or more bypass hose(s). Hoses are usually made of butyl or neoprene rubber, **Figure 11-19A.** Some hoses that connect metal tubing between the engine and the radiator are reinforced with wire, **Figure 11-19B.**

Nearly all OE-type radiator hoses have a molded, curved design although some are wire-reinforced, flex-types, **Figure 11-19C.** The flex-type hose allows the same

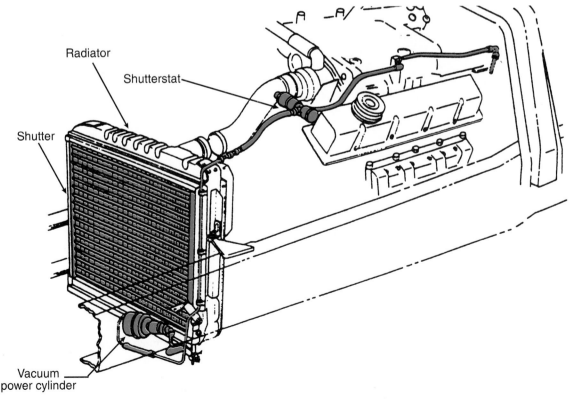

Figure 11-17. Vacuum radiator shutter system. Note the thermostatically operated shutterstat.

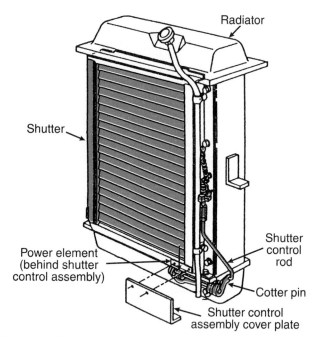

Figure 11-18. *Thermostatic radiator shutter system.*

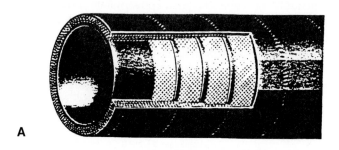

A

B

C

Figure 11-19. *Three common types of coolant hoses. A—Straight hose. B—Wire reinforced. C—Flex hose. (Gates Rubber Co.)*

part to be used on a greater number of vehicles, but it cannot be used on a vehicle requiring extreme bending or shaping. Lower radiator hoses are wire reinforced to prevent collapse during water pump suction.

The upper radiator hose is subjected to the roughest service life of any hose in the cooling system. First, it must absorb more vibration than most other hoses. In addition, it is exposed to the coolant at its hottest stage and is insulated by the hood or engine covers. These three conditions make the upper radiator hose much more likely to fail than the other hoses.

Normally, a hose will deteriorate from the inside. Pieces of deteriorated hose will then circulate through the system. These pieces usually end up clogging the radiator core. Any hose showing signs of external bulging or cracking should be replaced. A soft and spongy or hard and brittle hose is likely to fail and should also be replaced. Be sure to inspect all hoses periodically for signs of failure.

Radiator Outlets

The connection between the upper radiator hose and the engine is usually made through either a gooseneck or straight-pipe **coolant outlet.** Made of cast iron or aluminum, this outlet is designed to house the thermostat on the engine—although some outlets also direct coolant through a bypass. The outlet is connected to the top of the engine by bolts and is easily located. This unit can become damaged due to flange breakage during installation of a new thermostat. On some aluminum units, erosion may be a problem as well.

Cleaning and Replacing Cooling System Components

A dirty, obstructed, or leaking cooling system, a malfunctioning thermostat, or an inoperative water pump will cause the engine to overheat. When cooling system problems develop, they must be repaired immediately to prevent serious damage caused by engine overheating. The external cleanliness of the radiator should be checked if the engine overheats and there are no other visible problems.

Hose Replacement

When replacing a coolant hose, drain the cooling system below the level of the hose. Loosen or remove the old clamp and slide the old hose off the fitting. If the hose is stuck, do not pry it off. This might damage the inlet-outlet nipple or the attached component. Use a knife to cut off any pieces of the hose, then clean the remaining particles with a wire brush, **Figure 11-20.** The fitting should be clean and smooth when you install a new hose. Burrs or sharp edges could cut into the hose and lead to premature failure. Once the fitting is clean, locate the raised bead on the end of the nipple. Slide the new hose over the fitting.

A   B

Figure 11-20. Removing a defective radiator hose. A—Cut the old hose off, do not pull it off. B—Use a wire brush to remove any corrosion build-up on the radiator and engine inlets. (Gates Rubber Co.)

The clamp must be positioned over the end of the hose and the bead, **Figure 11-21**. A clamp tightened on top of the bead eventually will cut into the hose.

 Note: Heat and water can damage other parts in the cooling system, such as the rubber seals in water pumps and the O-rings that seal the lower end of wet cylinder sleeves. To avoid damage from coolant leakage, be sure to replace these parts according to the manufacturer's specifications

Cleaning a Radiator

The radiator should be cleaned whenever foreign deposits hinder the flow of air around the cooling fins or the transfer of heat to the air. In a hot, dusty area, periodic cleaning of the radiator will prevent loss of efficiency and extend engine life.

When cleaning a radiator, the shroud and grill should be removed, if possible, to facilitate cleaning of the radiator core. An air hose with a pressure nozzle can be used to remove loose dirt and debris from the radiator core. If oil is present on the core, an appropriate solvent should be used to loosen the dirt. The use of kerosene or fuel oil is not recommended as a solvent. A spray gun can be used to apply

the solvent to the radiator core. After applying the solvent, use compressed air to remove the remaining dirt. Repeat this process as many times as necessary, then rinse the radiator with clean water and dry it with compressed air.

 Warning: Provide adequate ventilation in the working area to avoid possible toxic effects of the solvent spray.

Scale deposits inside the radiator are normally the result of using hard, high-mineral content water in the cooling system. Heat affects the minerals in the water, causing scale to form on metal surfaces within the radiator. The scale reduces the transfer of heat from the coolant in the radiator to the surrounding air. Instead of forming scale, some hard water will produce a silt-like deposit, which restricts the flow of water. This type of deposit must be flushed out at least twice a year.

Inspecting and Testing the Cooling System

Clean all radiator parts thoroughly, removing dirt, scale, and other deposits. Since most diesel engine cooling systems are pressurized, it is important to be sure all components are tightly connected and free of damage. If the system pressure is not maintained, overheating and loss of coolant will result. To test the cooling system, install a pressure tester on the radiator according to the manufacturer's specifications. After pressurizing the system, carefully inspect the radiator, water pump, hoses, drain cocks, and cylinder block for leakage.

Inspect the radiator for bent fins and straighten any that are bent. Make sure that there are no cracked or corroded tubes. Study the radiator for kinks, dents, and fractured seams. If leaks cannot be found visually, test the radiator for leaks as follows:

1. Remove the radiator from the vehicle and install the radiator cap. (Instructions for radiator removal can be

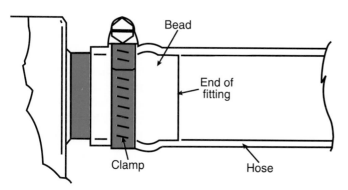

Bead

End of fitting

Clamp   Hose

Figure 11-21. Installed hose on radiator.

found in the service manual.) Plug the overflow tube and the outlet pipe. Attach the air hose to the inlet connection.

2. Fill the radiator with 7-10 psi of compressed air and submerge it in a tank of water.

 Caution: Do not overpressurize the radiator. Excessive pressure can damage the unit.

3. Look for any bubbles rising up through the water. The bubbles will help pinpoint the source of leaks.

Repair leaks by soldering the holes after thoroughly cleaning the surface area. If necessary, the repairs can be performed by a radiator repair shop. Inspect the rubber cushion washers and replace them as necessary.

Coolant Filters

Many modern mobile diesel engines use a **coolant filter** in the cooling system. Chemical inhibitors alone cannot fully protect the cooling system. Abrasive contaminants are always present in the cooling system and must be removed by a filter. A properly installed and maintained coolant filter and conditioner provides a cleaner engine cooling system, and contributes to longer life through greater heat conductivity.

The filter provides mechanical filtration by means of a closely packed element through which the coolant passes. Any impurities such as sand and rust particles suspended in the coolant will be removed by the straining action of the element. The removal of these impurities will contribute to longer water pump life and proper thermostat operation. The filter also serves to condition the coolant by softening the water to minimize scale deposits, maintaining an acid-free condition, and also acts as a rust preventative.

Like oil filters, coolant filters are of two basic types: the spin-on type and the cartridge type. Several types of **spin-on coolant filters** are available—with and without chemical additives, **Figure 11-22**. Filters without additives can be used for long periods of time. Liquid cooling system treatments must be used in these systems. Spin-on coolant filters with built-in chemicals, often called "coolant conditioners," must be changed periodically so a fresh "charge" of chemicals can be introduced into the cooling system. Before using certain filters with chemicals, the cooling system must be "precharged" with the liquid coolant treatment. Always check instructions included with the filter for specific usage recommendations.

Most **cartridge filters** contain a pleated paper media, which removes solid contaminants, and chemical additives, which protect the cooling system. After the filter is installed on the engine, the chemical mixture dissolves into the coolant solution. Radiator sealant chemicals, often referred to as "stop leak," should never be used in a cooling system that contains a filter. The stop leak will plug the filter, rendering it useless.

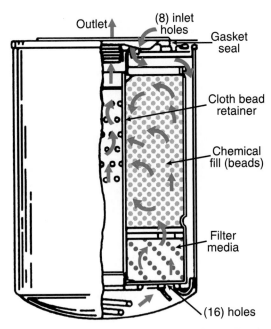

Figure 11-22. *Typical spin-on coolant filter. (Mack Trucks, Inc.)*

Coolant filters are generally used only in bypass-type cooling systems. Normally, coolant filters are connected to the heater system, however, there are other points in the cooling system where the filters can be connected. Shut-off valves are normally provided at the inlet and outlet connections of the filter assembly, allowing it to be replaced without draining the system.

Coolant Filter Service

To replace a spin-on coolant filter, proceed as follows:

1. Relieve the cooling system pressure by pushing the release button or lifting the lever on the top of the radiator cap, then close the inlet and outlet valves at the filter assembly (if so equipped).

2. Remove the used filter and thoroughly clean the mounting base. Dispose of the used filter properly.

3. Rub a thin film of oil (not grease) on the filter sealing gasket. Install the filter and tighten it to specifications.

4. Open the inlet and outlet valves.

5. Start the engine and allow the coolant to warm up. If the filter does not feel warm, check the lines between the filter and the cooling system. Also check the restrictor orifices to be sure they are clean and open.

⚠️ **Warning: Make sure the engine is off when changing any type of coolant filter. To avoid injury when removing the radiator cap, turn it counterclockwise to the first stop, but do not depress it. Be careful of hot steam and coolant which can cause scalding and burns. After the pressure has dissipated, press the cap downward and continue to turn it until it is free.**

Cartridge-type coolant filters may be a bolt-on type, which is fastened in the cooling system similar to the oil filter cartridge-type or the spin-on type with a special base, **Figure 11-23.**

To install this filter:

1. Relieve the coolant system pressure by removing the radiator cap, pushing the pressure release button, or lifting the lever on the top of the cap. Then close the inlet and outlet valves at the filter assembly. If the system is not equipped with valves, squeeze or clamp the hoses to stop coolant flow.
2. Remove the bottom drain plug from the filter housing and catch the coolant in a container. Some systems have a shut-off valve before and after the filter.
3. Remove the filter cover bolts and remove the cover.
4. Remove the used filter and the spring. Dispose of the filter properly.
5. Thoroughly clean the interior of the filter housing and spring.
6. Replace the drain plug in the filter housing.
7. Install the spring and new filter cartridge. Be sure to remove all protective wrapping from the cartridge.
8. Install the new cover gasket, place the cover on the housing and tighten the cover bolts evenly.
9. Open the inlet and outlet valves.
10. Start the engine and allow it to warm up. Check the filter for leaks and coolant flow. If the housing does not get warm, check the lines for proper hookup and the restrictor orifice for foreign debris.

The first two steps for removing a spin-on cartridge filter are the same as for bolt-on types. Once the pressure is relieved, the filter can be spun off the base and the old cartridge removed. Once the new filter is in place, the unit can be spun back on its base.

Before converting a cooling system from one type of filter to another, or changing the type of chemical inhibitor being used, always check with the engine or filter manufacturer. If this is not possible, it is a good idea to thoroughly flush the cooling system and add new coolant solution before switching filters. Follow the instructions provided with the chemical inhibitors and filters. Failure to do this could result in incompatibility between chemicals, leading to cooling system problems.

Thermostats

The coolant temperature in a diesel engine cooling system is controlled by a **thermostat** that is located somewhere between the cylinder head and the radiator top tank. Its exact location varies, depending on the type and configuration of the engine. Normally, the thermostat is located inside a housing in either end of the cooling manifold or cylinder block, **Figure 11-24.** Some larger engines

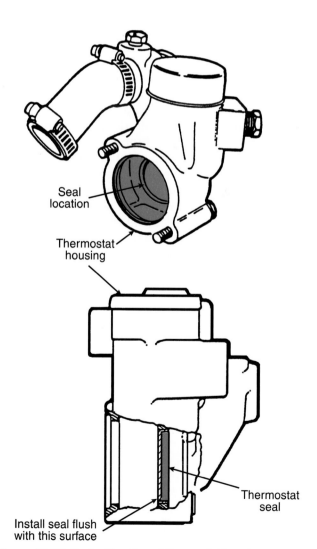

Figure 11-24. *Typical location of thermostat housing. (Detroit Diesel Corp.)*

Figure 11-23. *Installing a spin-on cartridge-type coolant filter. (Wix Corporation)*

have dual thermostats, **Figure 11-25,** while a few engines may have four thermostats. In these systems, all thermostats operate as a single unit.

At coolant temperatures below approximately 170-195°F (76.6-90.5°C), most thermostat valves remain closed, blocking the flow of coolant to the radiator, **Figure 11-26A.** During this period, all of the coolant in the system is circulated through the engine and is directed back to the suction side of the water pump via the de-aeration bypass tube. During warm-up, enough coolant to vent the system is bypassed to the radiator top tank by means of a separate external de-aeration line and then flows back to the water pump without going through the radiator core. As the coolant temperature rises above 170-195°F (76.6-90.5°C), the thermostat valve starts to open, restricting the bypass system and permitting a portion of the coolant to circulate through the radiator, **Figure 11-26B.** When the coolant temperature reaches approximately 185°F (85°C), the thermostat valve opens fully, the bypass system is completely blocked off, and all of the coolant is directed through the radiator.

A defective thermostat that remains closed or only partially opens will restrict the flow of coolant and cause the engine to overheat. A thermostat that is stuck open may not permit the engine to reach its normal operating temperature. The incomplete combustion of fuel that is caused by cold engine operation will result in excessive carbon deposits on the pistons, rings, and valves.

Note: While most diesel engines are equipped with high-temperature thermostats that operate in the range of 170-195°F (76.6-90.5°C), always check the exact service range needed. If the engine operating temperature is not maintained in the normal range, the efficiency of the engine may be reduced. A defective thermostat should always be replaced with one of the proper temperature rating.

Inspecting and Testing a Thermostat

Always make sure the engine's thermostat is in good working condition. When the thermostat is not working properly, the engine can run too hot or too cold. Overheating can damage the thermostat, causing improper operation of its valve. Rust can also interfere with thermostat operation.

Never operate a diesel engine without a thermostat. Discard any thermostat that is damaged. Be sure that the thermostat is of the proper temperature range and type listed in the service manual. The number stamped on a thermostat indicates the approximate temperature it must reach before it starts to open.

To remove the thermostat from its housing, follow the instructions given in the engine service manual. The following removal procedure is for a typical thermostat:

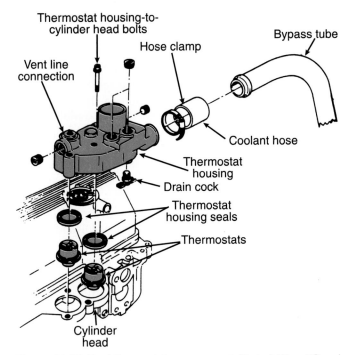

Figure 11-25. Dual thermostat arrangement. (Detroit Diesel Corp.)

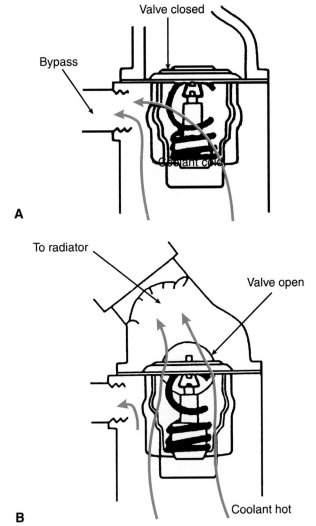

Figure 11-26. Typical thermostat operation. Note the coolant flow in both stages. A—Thermostat closed. B—Thermostat open.

1. Unbolt the housing and remove the thermostat. Clean the thermostat housing and the mating cylinder head gasket surfaces with a scraper.
2. Remove all gasket material.
3. Wash the housing and dry it with compressed air.

After removing the thermostat from the housing, check these components as follows:

1. Visually inspect the thermostat seal for lip damage. If necessary, replace the seal as directed in the service manual.
2. Visually inspect the thermostat for debris or a loose fit at the seat, the wrong seal location, rough or uneven wear of the seal seat, or an uneven seat. These conditions may cause improper closure.

To check a thermostat for operation (for example, a thermostat with a 180°F (82.2°C) start-to-open rating and a 200°F (93.3°C) full-open rating), proceed as follows:

1. Fill a metal or glass container with water and place a heat source under the container. Tie a line to the valve end of the thermostat.
2. Manually open the thermostat enough to insert a nylon ribbon under the thermostat seat. Suspend the thermostat by the ribbon and lower it into the container. Make sure the thermostat does not touch the bottom of the container. Hold the suspension line loosely, as shown in **Figure 11-27A.**
3. Heat the water in the container. Observe the thermometer and record the temperature as soon as the thermostat drops from the nylon ribbon. This reading is the start-to-open temperature. In this case, the start-to-open temperature should be approximately 180°F (82.2°C).
4. Continue to heat the water to the full-open temperature, in this case, 200°F (93.3°C). Observe the thermostat sleeve movement when full-open temperature is achieved, **Figure 11-27B.** Minimum sleeve travel at full-open temperature is generally given in the engine service manual.

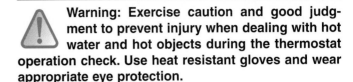

Warning: Exercise caution and good judgment to prevent injury when dealing with hot water and hot objects during the thermostat operation check. Use heat resistant gloves and wear appropriate eye protection.

5. Remove the thermostat from the water. While the sleeve is off the seat, inspect the seat area for pitting and foreign deposits.
6. Replace the thermostat if it does not operate as described earlier or if the seat is pitted.

When installing a thermostat, position it in its housing with the expansion element toward the engine. Some thermostats are marked with an arrow that points to the radiator or the engine block housing, while others are marked "back" or "front." Be sure to install the thermostat so that the frame does not block the water flow.

Clean all gasket surfaces to help prevent leakage. When the outlet casting and gasket have been properly located, tighten the nuts evenly and securely. A new thermostat housing gasket does not normally need to be cemented.

Cooling System Service

The most important consideration in any cooling system is keeping it clean. This involves the proper maintenance of filters and the draining of the system at recommended intervals. The system should be drained as follows:

1. Drain the coolant from the system by opening the drain cocks in the water outlet elbow, the oil cooler housing, the fresh water pump, the radiator and, on certain engines, the water hole cover located on the blower side near the rear of the cylinder block. Cooling system components that do not have a drain cock are generally drained through the oil cooler housing drain cock.

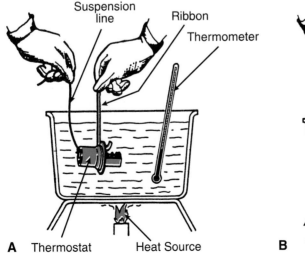
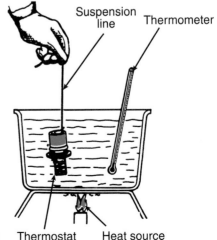

Figure 11-27. Thermostat function test. A—Start to open temperature (ribbon test). B—Full open temperature (minimum sleeve travel test). (Mack Trucks, Inc.)

2. Release any pressure in the system and remove the cooling system filler cap to permit the coolant to drain completely.

3. To ensure that the coolant is drained completely from an engine, all cooling system drains should be opened.

 Note: When freezing weather is expected, units not adequately protected by antifreeze must be drained. If any water that may be trapped in the cylinder block or radiator freezes, it will expand and may cause damage. Leave all drain cocks open until refilling the cooling system.

Filling a Cooling System

The following procedure should be used when filling a typical diesel engine cooling system:

1. Before starting the engine, close all of the drain cocks and fill the cooling system with the proper concentration of water and coolant. Use clean, soft water to reduce the formation of scale. Hard, mineral-laden water should be treated using water softening chemicals before it is poured into the cooling system. These water softeners modify the minerals in the water and greatly reduce or eliminate the formation of scale.

2. When filling the system, air must be vented from the engine coolant passages. Open the engine venting petcock and, in aftercooled engines, the aftercooler petcock.

3. Start the engine and, after the normal operating temperature has been reached, check the coolant level. The coolant level should be within 2" (51 mm) of the top of the filler neck.

4. If a daily loss of coolant is observed and there are no apparent leaks, gases may be leaking past the cylinder head water seal rings and into the cooling system. The presence of air or gases in the cooling system may be detected by connecting a rubber tube from the overflow pipe to a water container. Bubbles in the water during engine operation will indicate combustion gas leakage. Another method for observing trapped combustion gas in the cooling system is to insert a transparent tube in the water outlet line.

Operating a diesel engine with a partially full cooling system can result in aerated coolant or an air-bound water pump. Be sure that the cooling system is full before undertaking any drain and refill operation. Bleed all air from the system, including the heater core and heater hoses. If a heater shutoff valve exists, it must be opened during the refill operation. With the pressure cap and coolant recovery tank cap removed, run the engine at 1500 rpms for 10 minutes after the thermostat operating temperature has been reached to allow coolant system de-aeration. If necessary, add coolant to the radiator and install the pressure cap. Fill the coolant recovery tank to the "hot" level mark and install the recovery tank cap.

The coolant value or condition can be tested by chemical test means or by using the engine coolant tester shown in **Figure 11-28.** To use the tester, place a few drops of coolant from the radiator on the prism, look through the eyepiece and read the scale. The scale reads the coolant's strength in degrees.

Flushing a Cooling System

If a properly maintained coolant filter is used, the cooling system does not normally need to be flushed. When a filter is not used, the cooling system should be flushed each spring and fall. If the cooling system is contaminated, flush the cooling system as follows:

1. Drain the coolant from the engine as previously explained.

2. Refill the system with soft clean water.

 Caution: Allow the engine to cool before filling. Filling a hot engine with cold coolant or water may cause the block or head(s) to crack..

3. Start the engine and operate it for 15 minutes after the thermostats have opened to thoroughly circulate the water.

4. Drain the cooling system completely after engine has cooled.

5. Refill the system with clean water. Start the engine and operate it for 15 minutes after the thermostats have opened.

6. Again, drain the system completely after cooling..

7. Fill the cooling system with an appropriate mixture of antifreeze and water. Also add a precharge of inhibitor.

If the engine overheats and fan belt tension, coolant level, and thermostat operation have been found to be satisfactory, it may be necessary to remove scale formation by using a descaling solvent. Immediately after using the descaling solvent, neutralize the system with a neutralizer, which is usually packaged with the descaling solvent. It is important that the directions printed on the container of the descaler be followed closely. After the solvent and neutralizer have been used, completely drain the engine and radiator. Reverse flush the system as described in the next section before refilling it.

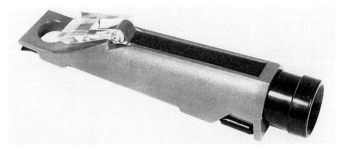

Figure 11-28. *Engine coolant tester. This tool will check the condition as well as the freeze point.*

Reverse Flushing a Cooling System

Before attempting to reverse flush a cooling system, the water pump should be removed from the system. The radiator and the engine should be reverse flushed separately to prevent dirt and scale deposits from clogging the radiator tubes or being forced through the pump. Reverse flushing is accomplished by forcing hot water through the radiator or the block in a direction opposite of the normal flow of coolant. This loosens dirt and scale deposits and forces them out of the components. A radiator is reverse flushed as follows:

1. Remove the radiator inlet and outlet hoses and reinstall the radiator cap.
2. Attach a hose to the inlet of the radiator to lead water away from the radiator.
3. Attach a hose at the bottom of the radiator and insert a flushing gun in the hose.
4. Connect the water hose of the flushing gun to a water source and the air hose to a compressed air outlet.
5. Turn on the water and, when the radiator is full, turn on the air in short blasts, allowing the radiator to fill between blasts.
6. Continue this procedure until only clean water is expelled from the radiator.

The cylinder block and cylinder head water passages are reverse flushed as follows:

1. Remove the thermostats and the water pump from the system.
2. Attach a hose to the water inlet of the oil cooler housing to drain water away from the engine.
3. Attach a hose to the water outlet at the top of the cylinder head (thermostat housing) and insert the flushing gun in the hose.
4. Turn on the water. When the water jackets are full, turn on the air in short blasts. Allow the engine to fill with water between air blasts.
5. Continue the flushing procedure until the water from the engine runs clean.

In addition to flushing the system, the cooling system components should be inspected periodically to keep the engine operating at peak efficiency. The thermostat and the radiator pressure cap should be checked and replaced if they are found defective.

When coolant seals and hoses are installed, be sure the connecting parts are properly aligned and the seal or hose is in its proper position before tightening the clamps. All external leaks should be corrected as soon as they are detected. The fan belt must be adjusted to provide the proper tension, and the fan shroud must fit tightly against the radiator core to prevent recirculation of air, which may lower the cooling efficiency.

Water Manifolds

Some larger diesel engines—both mobile and stationary types—are equipped with external **water manifolds.**

These manifolds are designed to ensure a more even temperature through the entire engine, **Figure 11-29.**

Coolant entering the water manifold leaves the cylinder head through openings directly over each exhaust port. Because the outlets are usually located near the exhaust ports, the hottest coolant is continually removed. As the coolant flows from the rear of the engine toward the radiator or heat exchanger, it becomes warmer. The water manifold collects the coolant from each cylinder or pair of cylinders and routes it to the thermostats and the radiator. This results in cylinder temperatures being more equal.

Water Pumps

The **water pump,** or coolant pump, is the "heart" of the cooling system. It circulates coolant through the engine block, the cylinder heads, the radiator or heat exchanger, and the oil cooler, **Figure 11-30.** It may be either gear- or belt-driven from the crankshaft.

Centrifugal pumps and *gear pumps* are the principle types of water pumps used in diesel engine cooling systems. Some engines are equipped with a rotary-type pump, which uses an impeller that has flexible vanes.

Centrifugal water pumps are more common in engine cooling systems than pumps of other types, particularly in mobile diesel engines, **Figure 11-31.** There are many types of centrifugal pumps. However, all centrifugal pumps are designed so that coolant drawn into the center of the

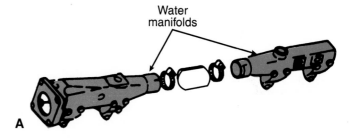

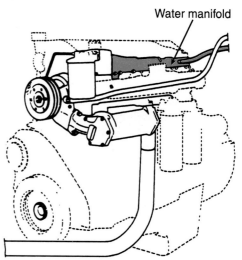

Figure 11-29. *A—Typical water manifold. B—Manifold location on the engine block. (Mack Trucks, Inc.)*

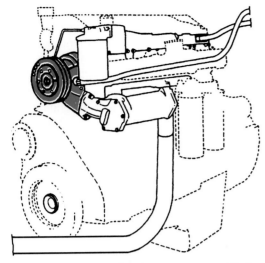

Figure 11-30. *Typical centrifugal water pump and its location in the cooling system. (Mack Trucks, Inc.)*

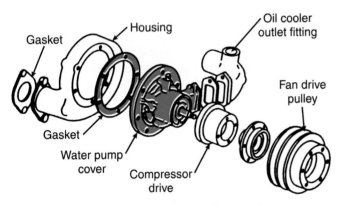

Figure 11-31. *Exploded view of a centrifugal water pump. (Mack Trucks, Inc.)*

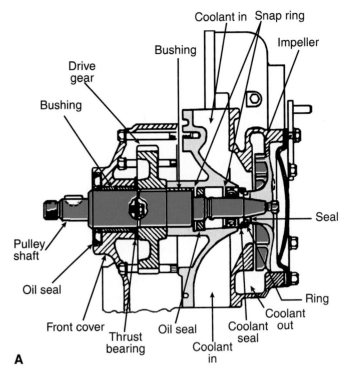

A

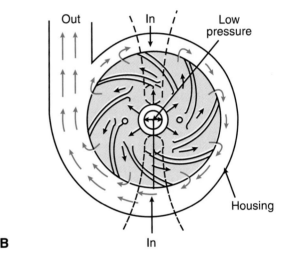

B

Figure 11-32. *A—Sectional view of a coolant pump. B—Note how coolant is thrown off by the impeller blades. (Caterpillar, Inc.)*

impeller is thrown at high velocity into the surrounding casing. As the coolant hits the casing, the velocity decreases and pressure increases.

As shown in **Figure 11-32,** a bearing and shaft assembly support the drive hub on one end and the impeller on the other. A cartridge-type seal prevents coolant from reaching the bearing. This seal generally consists of a spring-loaded unit with a rotating ceramic or sintered metal washer that rides on a hard carbon washer. These washers form a strong seal that is only broken under extremely abrasive conditions.

 Caution: Care must be taken not to touch the ceramic seal (if used). Oil from your skin will contaminate the seal and cause it to leak.

The *impeller* is usually either die cast or machined to promote smooth coolant flow. As the impeller casting erodes, cavitation can occur, reducing pump efficiency. The rate of erosion depends on the pump materials. Aluminum wears much more quickly than cast iron, so while a cast iron pump can be rebuilt, an aluminum pump must normally be replaced.

The majority of coolant pump failures are caused by leaks in the system. When the pump seal fails, coolant can seep out of the weep hole in the casting. The seals may simply wear out due to abrasives in the cooling system. When a pump is operated without coolant, the seals become red hot. If cold water is added to the pump too quickly, the hot seals will crack. This change in temperature is known as thermal shock. Adding cold water to a hot engine can also cause other parts, such as the cylinder heads, to crack. Other failures can be attributed to bearing and shaft problems and, occasionally, a cracked casting. Hubs and impellers can also come loose, but they generally cause only minor pump damage.

In many cases, water pumps are simply replaced. However, some pumps can be rebuilt and reused. To rebuild and reassemble a typical centrifugal pump, follow the directions given in the service manual.

V-Belt Drives

While some industrial applications may drive the coolant pump off the camshaft or the gear train, nearly all mobile diesel systems drive the pump using a V-belt drive.

The V-belt drive system is popular for a number of reasons. It is less expensive and is quieter than chains and gears. Because the V-belts are flexible, they absorb shock loads to cushion the shaft bearings from excessive loads. Also, V-belt drive systems have a long service life and are easy to repair.

Problems occur with V-belts because they are not checked until they either make excessive noise or break. Today's heavy duty fans and water pumps place extra tension on V-belts. Greater output demands from charging systems have also added strain. Additionally, air conditioning units have added the most impact to V-belts, due to the magnetic clutch operation, which engages the unit at any engine rpm. However, the V-belt is still reliable enough to be used in late-model diesel engines.

Heat is the most common cause of belt deterioration. Excessive heat causes rubber belts to harden and crack. Heat is caused by belt slippage, which is often the result of insufficient belt tension or oily conditions. As the belt wears, it rides deeper in the pulley groove and loses tension. When the belt slips, excess heat from the belt travels through the drive pulley and down the shaft to the support bearings. The bearings can be damaged if the slippage problem is not corrected. Periodically adjusting the belt tension will help prevent slippage.

Many drive units consist of two "matched belts" that run over parallel pulleys. Even if the belts are the same length, one belt can wear excessively. Consequently, the belts will run at slightly different speeds. This results in one belt that constantly slips. Matched belts should always be replaced in pairs so that they wear together, reducing slippage problems.

Belt problems are easily discovered either by visually inspecting the belts for cracks and splits or by listening for a screeching noise caused by belt slippage. The fan belts should be adjusted to the correct tension. A belt that is too tight adds an extra load to the fan bearings. This shortens the life of the bearings and the belt.

Checking Fan Belt Tension

One of the major causes of engine overheating is fan belt slippage, which is caused by incorrect belt tension, worn fan belts, worn belt pulley grooves, or the use of fan belts of unequal length in a matched belt system. The belt tension and condition of the belts should be checked periodically as follows:

1. Measure the belt deflection at the longest span of the belt. Normal maximum deflection is 3/8"-1/2". Check the belt tension using a gauge similar to one shown in **Figure 11-33**. The gauge tension should usually be 80-110 pounds. Check the engine service manual for exact specifications.

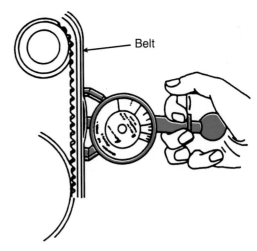

Figure 11-33. *Checking the condition and deflection of fan drive belts. (Cummins Engine Co., Inc.)*

2. To check the condition of the drive belt, tensioner bearing, and fan hub, remove the drive belt and complete the following steps:

   ❑ Inspect the belt for damage.
   ❑ Check the tensioner bearing. The tensioner pulley should spin freely with no rough spots detected under hand pressure.
   ❑ Check the fan hub bearing. The fan should spin freely without wobble or excessive end play.

3. Install the drive belt following the service manual instructions.

Fans and Fan Clutches

The cooling system efficiency is based on the amount of heat that can be transferred from the system to the surrounding air. At highway speeds, the "ram" air traveling directly through the radiator is enough to maintain proper cooling. However, at low speeds and idle, the air movement is low enough to require the use of a fan.

 Warning: Do not stand next to a fan that is running. If a fan blade comes loose, serious injury or death could result. Also keep hair, loose clothing, fingers, etc., away from belts and fan when the engine is running.

The cooling system *fan* is usually located between the radiator and the engine, and is mounted on the front of the engine block. The fan in most mobile applications is driven by a V-belt from the engine crankshaft. Fan designs vary with the system air requirements. Diameter, pitch or angle of the blades, and number of blades may vary, depending on the amount of air flow required.

To increase the cooling efficiency of the radiator, a metal or plastic fan shroud is usually placed around the fan. The fan shroud must fit tightly against the radiator to prevent recirculation of the hot air through the radiator. The fan can either be a suction type or a blower type, **Figure 11-34,** depending upon the design of the system.

Suction Fans

Suction fans pull air through the radiator and push it over the engine. This design permits the use of a smaller fan and radiator than is needed with blower-type fans. One of the more popular suction fans is the thermo-modulated fan assembly. This assembly is designed to regulate the fan speed, maintaining an efficient engine coolant temperature regardless of variations in engine load or outside air temperature.

The thermo-modulated fan assembly is a compact, integral unit that requires no external piping or controls and operates on a simple principle. This principle involves transmitting torque from the input shaft to the fan by the shearing of a silicon fluid film between input and output plates in a sealed, multiplate, fluid-filled clutch housing. The thermostatic control element, which is an integral part of the fan drive, reacts to changes in engine temperature varying the fluid film thickness between the plates, changing the fan speed. Proper selection of the control element setting (as determined by the manufacturer) helps maintain optimum cooling. No adjustment is necessary.

Blower Fans

Blower fans or pusher fans pull air across the engine, then push it through the radiator or heat exchanger. These fans are used in industrial equipment, slow-moving machines, and on machines where harmful materials could be drawn into the radiator with a suction-type fan.

In either case, the fan moves air across the radiator core to cool the liquid in the radiator whenever the engine is running. Reversible fans, which can blow the air in either direction are used in some systems.

As mentioned earlier, fan air is necessary at idle and low-speed operation. At higher speeds, however, the fan is not necessary on mobile equipment. For this reason, most fans are designed to slow down at higher speeds. This reduced fan speed allows more horsepower to travel to the drive wheels if needed. It also cuts down on the noise that the fan would produce at higher speeds.

Fans of the standard stamped steel variety, or flex fans, actually allow the blades to bend. This decreases the pitch and the subsequent horsepower losses and reduces noise levels as engine speed increases. It is very important that none of the blades are bent, nicked, or misaligned. If any of these conditions exist, the fan should be replaced. Never attempt to straighten bent fan blades. As a rule of thumb, no blade tip should be more than 3/32" out of the plane of the other blades. This test can be made by laying the fan on a flat surface and measuring the distance between the blades and the surface.

Fan Clutch

A more popular way of controlling fan noise and horsepower loss is by using a **fan clutch, Figure 11-35.** This unit couples the fan to the drive pulley—usually on the water pump shaft. The basic principle is that the clutch will "slip" at high speeds and, therefore, will not turn at full engine speed.

There are two basic types of fan clutches, temperature-controlled clutches and speed-controlled clutches. The advantage of the temperature-controlled unit is that it can sense when air is needed to cool the system. Both units have the same basic design, **Figure 11-36.**

A silicone fluid couples the fan blade drive plate to a driven disk through a series of annular grooves in both pieces. Fluid fills the grooves to drive the fan until differential torque between the fan and drive disk makes the fluid shear or slip. Both the speed-controlled and temperature-controlled units operate in this way.

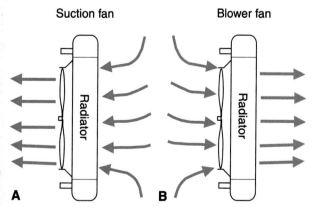

Figure 11-34. *Two types of fan arrangements. A—Suction fan. B—Blower fan. (Deere & Company, Inc.)*

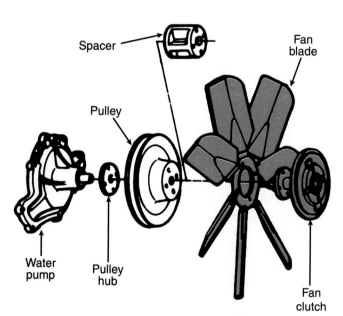

Figure 11-35. *Location of the fan clutch on the fan drive assembly.*

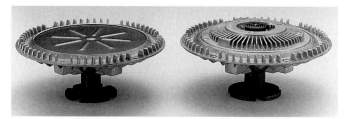

Figure 11-36. A—Non-thermal fan clutch. B—Thermal fan clutch. (Stan Manufacturing, Inc.)

The temperature-controlled unit, however, has an additional bimetal element that senses the air temperature behind the radiator. This element is calibrated to open and close a valve in the clutch that dispenses the silicone into the drive disk at a particular temperature. The silicone in the groove area of the temperature-controlled unit allows less slippage at higher engine speeds than does the speed-controlled unit. When the fluid is returned to its reservoir, the temperature-controlled unit permits the fan to free-wheel until more cooling is needed. Testing has shown that maximum cooling is necessary less than 10% of the time. Therefore, the temperature controlled fan clutch can save fuel and reduce noise 90% of the time.

Coolant Heaters

To keep fuel flowing freely in cold weather, *coolant heaters* are frequently used. There are three common types of coolant heaters used on mobile diesel equipment. These are described in the following paragraphs.

Immersion Block Heater

The *immersion heater,* **Figure 11-37** is installed directly into the engine block in a location normally determined by the engine manufacturer. The heated coolant circulates around the cylinder walls by convection. The most common immersion heaters are powered by 120 or 240 volts ac and are available in wattages up to 2500 watts.

Circulating Tank Heater

The *circulating tank heater,* **Figure 11-38,** sets up circulation through the engine by the way it is installed. The

Figure 11-37. Typical immersion block heater. (Philip Temro, Inc.)

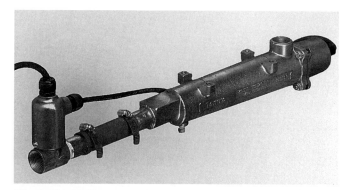

Figure 11-38. Typical circulating tank heater. (Philip Temro, Inc.)

coolant leaves the bottom of the block and travels to the heater. The heated coolant rises and is plumbed back into the top of the block. Generally, tank-type heaters should be of higher wattage than immersion heaters because some heat is lost from the hoses and tank body itself. Circulating tank-type heaters are most commonly available in 120- or 240- volt ac versions and are available in wattages up to 3500 watts. There are also nonelectric models, which are usually propane fired.

Fuel-Fired Heater

This type of coolant heater, **Figure 11-39,** is used to heat the engine and cab/sleeper unit via the coolant by burning diesel fuel. These units burn less fuel than would be consumed by idling the engine all night.

Industrial Engine Cooling Systems

The basic purpose of an industrial cooling system is the same as for the mobile cooling system. The heat absorbed from the engine components is contained in the coolant and must be removed. The coolant removes heat by an increase in temperature or a change of phase (water to steam). A system which removes heat by increasing the coolant temperature is known as a "solid water" system, and a system which uses a change of phase is known as an "ebullient" system.

Radiator

As in mobile cooling systems, the radiator removes heat from the coolant by exchanging it with ambient air which is cooler than the engine coolant. They are typically used when an application has no use for the engine heat and therefore releases heat to ambient air, **Figure 11-40.** The cooling fans on a radiator can be engine driven or powered by electric motors. Electric motors are used when a radiator is located in a position which is difficult to drive from the engine. Engine driven fans may be either a suction type or blower type.

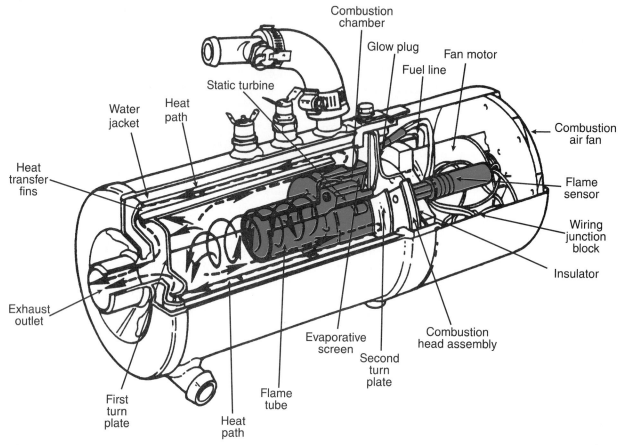

Figure 11-39. *Typical diesel fuel-fired heater.*

Figure 11-40. *The industrial coolant fan and radiator serve the same function as they do in the mobile diesel application, but are much larger. (Waukesha Engine Division)*

Heat Exchanger

A *heat exchanger,* **Figure 11-41A,** removes heat from the engine coolant and transfers the heat to a secondary fluid. Heat exchangers are often used where installation of a radiator is not practical, or where there is a use for the heat. Common heat exchangers are shell and tube, and plate type.

A *shell and tube heat exchanger* consists of a bundle of tubes with coolant flowing through them, **Figure 11-41B,** the shell is around these tubes and contains the other heat transfer fluid. Heat is transferred through the walls of the tubes. The highest heat transfer rate for a single pass shell and tube heat exchanger is generally achieved with a counter flow design as shown in **Figure 11-42.** Counter flow also allows the plant water or raw water outlet to have a higher temperature than the engine inlet temperature which provides more useful, higher temperature plant heat. Some heat exchangers use the parallel flow which permit lower temperature outputs. A *plate heat exchanger* consists of a series of formed metal plates sandwiched together. Engine coolant is on one side of the plate while plant coolant, or raw water is on the other with counter flow in the plates. See **Figure 11-43.**

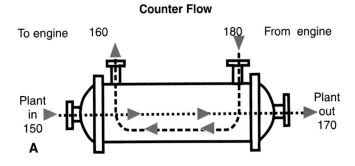

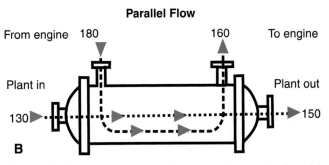

Figure 11-42. *The two types of coolant flow paths found in industrial heat exchangers. A—Counter flow. B—Parallel flow. (Waukesha Engine Division)*

A

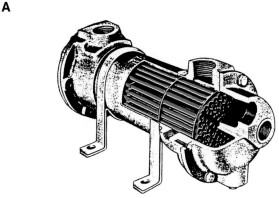

B

Figure 11-41. *A—Typical industrial heat exchanger. B— Cutaway of an industrial heat exchanger. (Young Radiator Company)*

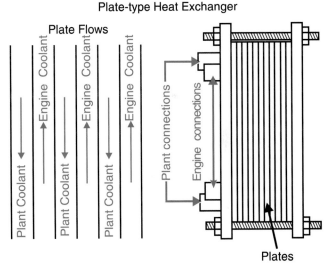

Figure 11-43. *Operation of plate-type heat exchanger. (Waukesha Engine Division)*

Cooling Tower

An open system employing a heat exchanger and a **cooling tower** is shown in **Figure 11-44.** The arrows show the direction of coolant flow. Note that the coolant passes through the heat exchanger and the cooling tower after passing through the engine water jacket. Various forms of cooling towers, or cooling ponds may be used. But, as stated earlier in the chapter, open cooling systems may be required to meet environmental regulations. Because of this and the cost of operations, the cooling towers are seldom used in modern industrial installations.

One or more water pumps force coolant to flow through the other components in solid water cooling sys-

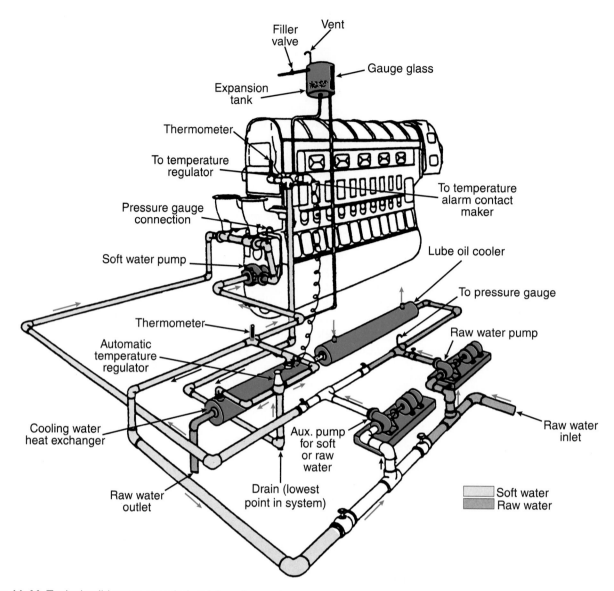

Figure 11-44. Typical solid water open industrial cooling system.

tems, including the engine, piping, surge tanks, thermostats, and other devices. The cooling system piping must be sized to allow coolant flow without excessive restriction. The piping material must be suitable for the temperatures and pressures encountered as well as vibration from the operating engine. Flexible connections (bellows type or rubber hose) are recommended at all connection points to the engine to isolate the engine and piping components from high stresses due to vibration.

Engines mounted on spring isolators or other soft mounting systems must have flexible cooling system connections with sufficient flexibility for the motion normally encountered.

Expansion and Surge Tanks

Expansion and surge tanks perform several functions in a cooling circuit. Some of these functions are:

- ❑ De-aerate coolant.
- ❑ Control cooling system pressure.
- ❑ Allow coolant expansion.
- ❑ Provide coolant reserve.

An expansion tank is a single chamber tank located at the highest point in the cooling system. See **Figure 11-44.** Vent lines are connected to form high points in the cooling system to the expansion tank below the water line. These vent lines allow trapped air to escape to the expansion tank where the air bubbles out of solution thus de-aerating the coolant.

A surge tank performs the same function as an expansion tank but adds a full flow chamber. In this chamber the coolant velocity is reduced, allowing air trapped in the solution to escape to the high point in the chamber. A bleed hole between the full flow chamber and expansion chamber allows air to escape to the expansion chamber.

Sight Glass

A **sight glass** in an expansion tank, or the expansion chamber of a surge tank allows visual monitoring of coolant level. A circuit filled to the top of the expansion area will overflow coolant on start-up due to coolant expansion. After that the coolant level should not overflow. Designing the filler neck to extend into the expansion tank will prevent overfilling the tank. A pinhole in the filler neck will allow the coolant in the filler neck to drain into the tank after filling.

A pressure cap is required to prevent coolant evaporation losses and to prevent boiling in the system. The pressure cap must have a vacuum relief function to prevent a vacuum from forming in the tank during load reduction or cool-down operation. Only a single pressure cap can be used in a cooling system and must be located at the highest point on the expansion tank or on the surge tank's expansion chamber.

Caution: Do not assume a pressure cap will pressurize the tank to the cap's rating. Pressure in the tank can range from atmospheric pressure to the pressure cap's rating. The exact expansion tank pressure can be affected by many factors including: ambient temperature, engine load, and expansion volume. It is difficult to predict the exact tank pressure.

Thermostat

A thermostat controls temperature by directing coolant in different branches of a cooling circuit to maintain the required temperature, **Figure 11-45**. The thermostat in an industrial cooling system operates basically the same as those in a mobile system except that temperature ranges vary according to their use in the cooling system.

Like in a mobile cooling system, debris in the coolant can block cooling passages, erode cylinder liner packing areas, wear out water pump seals, and do several other types of damage. This debris may be from the fabrication cooler or piping. Cleaning the cooling passages during assembly or prior to start up will not remove all of the debris. Bypass coolant filtration shown in **Figure 11-46** can remove debris from the cooling system on any industrial diesel engine.

Heat rejection from an engine can be up to 70% of the total energy input. This energy is lost when an engine is utilized for its crankshaft power only. Heat recovery allows the capture of the majority of this otherwise lost energy and puts it to use in place of boilers and other heating devices in a plant. Solid water heat recovery systems circulate coolant through the engine jacket and exhaust heat recovery equipment to pick up the heat energy from the engine. This engine is then piped to areas in the plant where it can be used.

Ebullient Cooling Systems

Ebullient cooling systems cool the diesel engine by absorbing heat during a phase change from liquid to vapor (steam). This process absorbs heat more efficiently than having coolant absorb heat through a temperature increase. Because of its efficiency, an ebullient cooled engine can operate at a higher temperature while sufficiently removing heat, **Figure 11-47**.

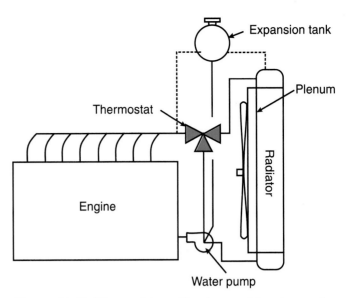

Figure 11-45. Thermostat position in a solid water cooling system. (Waukesha Engine Division)

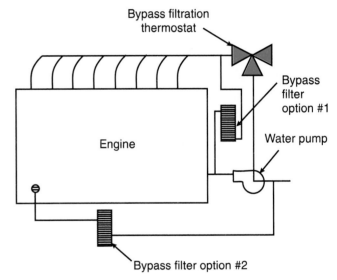

Figure 11-46. Typical bypass coolant filtration system. (Waukesha Engine Division)

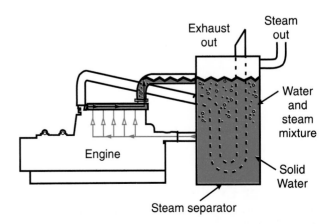

Figure 11-47. The typical ebullient cooling system in operation. (Waukesha Engine Division)

The effectiveness of this system is better understood when the following is considered. It takes 180 BTUs (190 000 J) to raise the temperature of 1 pound (.45 kg) of water from 32°F (0°C) to 212°F (100°C). It will take another 970 BTUs (1 023 000 J) to convert that same pound of water at 212°F (100°C) to steam at 212°F (100°C). Note that the temperature of the coolant did not rise, only the state of the coolant changed from liquid to steam.

Ebullient circulation is developed by coolant vaporizing at the hot surfaces in the engine. This vapor is much lighter than the surrounding coolant and raises to the highest point in the system. Coolant will be carried along with the vapor bubbles causing some solid water circulation as well as steam formation. The steam bubbles and coolant travels out of the engine and into the steam separator.

Steam Separator

In the **steam separator,** the flow velocity becomes very slow allowing the steam bubbles to separate from the surrounding coolant and collect at the top of the separator chamber. The steam is put to use in other areas. The remaining liquid will circulate to the bottom of the separator chamber and return to the engine inlet. A condensate or feed water pump adds water to the steam separator to maintain the liquid level and replace coolant lost to the process. Steam at the top of the separator is removed from the chamber to be used for various needs. The steam exits through a modulating backpressure control valve. This valve is very important for preventing boil-over of the coolant.

The backpressure control valve maintains pressure in the steam separator at the steam pressure required, usually 15 psi at 250°F (103.4 KPa at 121.1°C) at the engine outlet flange, regardless of downstream demand. If the steam pressure in the steam separator were allowed to fluctuate, a rapid decrease in pressure would cause instant boiling in the engine and leave the engine coolant level very low, resulting in overheating damage to exhaust manifolds and cylinder heads. The damage is almost instantaneous, so even a separator coolant level switch cannot prevent the damage. The loss of coolant is similar to the removal of a radiator cap from a hot engine.

Marine Engine Cooling System

Marine diesel engine cooling systems may use any of the following configurations:
- ❏ Raw-water cooling systems.
- ❏ Heat exchanger cooling systems.
- ❏ Keel cooling systems.
- ❏ Radiator cooling systems.

Raw-Water Cooling System

In the basic **raw-water cooling system,** fresh water or salt water is used as a coolant. The liquid is directly circu-

lated through the engine to cool it and is then dumped back overboard. The major advantage of this cooling system is its simplicity—it has relatively few components. The raw-water inlet, **Figure 11-48,** serves the following functions:
- ❏ Provides a low restriction connection for the water inlet plumbing.
- ❏ Provides a connecting point for the water shutoff valve installed between the water inlet and the water inlet plumbing.
- ❏ Provides a way to separate air from the water required for cooling. "Sea chests" must have connections to allow air, forced under the hull during maneuvering, to be purged before it is able to reach the seawater pump.

The closed cooling system of a marine engine actually consists of two separate circuits (sometimes called systems)—an internal or freshwater (coolant) circuit and an external or salt water circuit. The fresh water circuit is a self-contained system, similar to the closed cooling system in a mobile installation. One of the primary differences between a marine installation and a mobile installation is the use of a cooler instead of a radiator. The cooler is mounted in the fresh water system. Heat is carried away from the cooler by raw-water in the salt water system.

There are three common types of raw-water pumps presently in use. These are described in the following paragraphs.

Rubber impeller seawater pumps are characterized by excellent priming characteristics, though they often suffer relatively short life in abrasive waters. *Water ring seawater pumps* utilize a novel pumping principle, originally used for vacuum pumps. Their priming characteristics are outstanding, easily permitting water to be lifted 15' (4.56 m). They are made entirely of corrosion resistant metals, with no elastomeric components.

Centrifugal seawater pumps are the most commonly used, but they must be installed with their inlet below the boat's light waterline. Air allowed to enter centrifugal seawater pumps will likely result in a loss of prime and probable engine damage due to loss of cooling. Do not start an engine equipped with a centrifugal pump unless the pump and priming chamber are full of water.

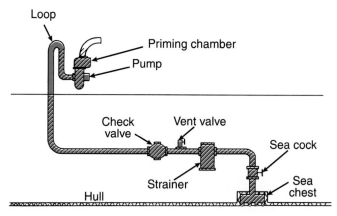

Figure 11-48. *A basic raw-water marine cooling system. (Caterpillar, Inc.)*

Materials Used in Marine Heat Exchangers

While copper nickel piping is considered the best against corrosion, it is the most expensive. Black iron pipe is often used in raw-water service, however, replacement should be planned every two or three years. If it is necessary to use pipe or other cooling system components of more than one material, avoid letting the dissimilar metals touch, even by mutual contact, with an electrically conductive third material.

Strainers protect the raw-water pump, heat exchanger and other cooling system components from foreign material in the seawater. The foreign material can plug or coat heat transfer surfaces, causing overheating of the engine. If abrasive, foreign material will erode pump impellers and soft metal parts, reducing their effectiveness. Strainers should be installed below the boat's waterline and as close to the sea water intake or sea chest as possible (adjacent to the sea cock). The strainer must be installed so it can be easily cleaned, even in the worst weather conditions.

Salt Water Applications

In salt water installation, **Figure 11-49,** scale and corrosive salts tend to build up in the hottest parts of the cooling system, especially around the cylinder walls. This tends to reduce cooling efficiency. The rate of scale formation is proportional to coolant temperature and accelerates when temperatures are above 160°F (71.1°C). As a general rule, raw-water cooled engines are kept at lower temperatures (about 140°F or 60°C) than engines with closed cooling systems. However, these lower operating temperatures cause water to condense in the cylinders and tend to reduce the overall thermal efficiency of the engine. Most marine diesel engines are best operated at a temperature of about 185°F (85°C).

Another major problem in the raw salt water system is **electrolysis** (chemical change between two dissimilar materials). To reduce the effects of electrolysis, an arrangement of sacrificial **zinc anodes** must be installed in the system.

While fresh water does not contain corrosive salt, it may be muddy, containing dirt, silt, and dissolved minerals that can coat the inside of the engine when the water heats up. Keep in mind that in any installation using raw-water, it is more difficult to regulate the engine temperature because the temperature of the raw-water can range from the freezing point in northern climates to 90°F (32.2°C) or more, in tropical climates.

Heat Exchanger Cooling Systems

The heat exchanger cooling system combines the raw-water and conventional diesel engine cooling systems into a single system. The major components of the heat exchanger system include a water-cooled exhaust, an engine coolant pump, a raw-water pump, a heat exchanger and an oil cooler, **Figure 11-50.**

A circulating raw-water system is used to control operating temperature in a heat exchanger system. Raw water circulates over the outside of a heat exchanger, while the engine coolant circulates through the inside of the heat exchanger. Consequently, the raw water removes heat from the coolant. Coolant normally enters the engine at a low point and flows up around the cylinders. The heated raw

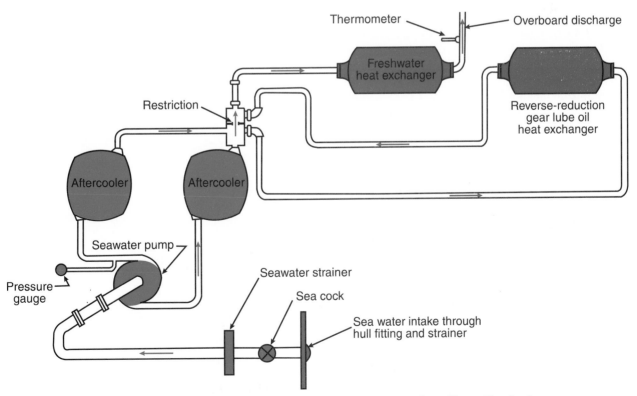

Figure 11-49. Closed marine cooling system that uses sea water as the cooling medium. (Caterpillar, Inc.)

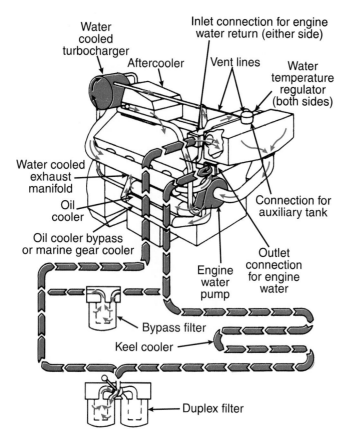

Figure 11-50. *Typical keel heat exchanger cooling system.*

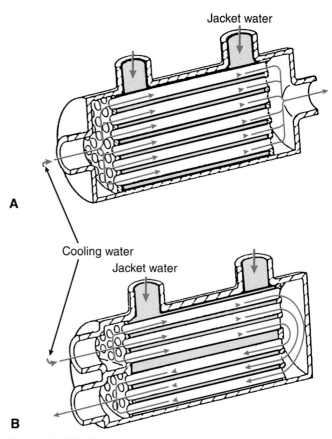

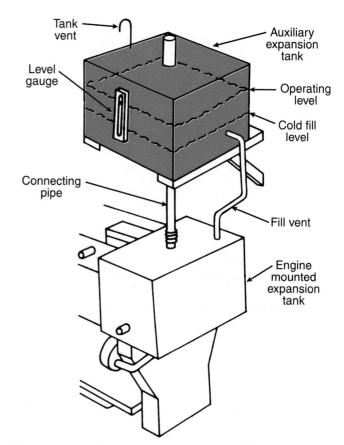

Figure 11-51. *Marine heat exchangers. A—Single-pass heat exchanger. B—Two-pass heat exchanger. (Caterpillar, Inc.)*

Figure 11-52. *Expansion tank location in a marine heat exchanger cooling system. (Caterpillar, Inc.)*

water is normally ejected from the system through an exhaust pipe. This system is similar to conventional mobile equipment with the notable exception that it replaces the air-to-water cooling process with a water-to-water process.

Most inboard heat exchangers are shell and tube type. They require a water pump to circulate sea water through the heat exchanger tubes. It is good design practice to always flow the raw-water through the tubes. The tubes can be cleaned by pushing a metal rod through them; the shell side requires chemical cleaning which is only available at shore-side facilities.

Most shell and tube heat exchangers are of either the single-pass or the two-pass type, **Figure 11-51.** This designation refers to the flow in the cold water circuit of the exchanger. In the two-pass type, the cold water flows twice through the compartment where it is circulated in the jacket water. When using a single-pass exchanger, the cold water should flow through the exchanger in a direction opposite to the flow of jacket coolant to provide maximum heat transfer. This results in improved heat exchanger performance. In a two-pass exchanger, cooling will be equally effective using either of the jacket water connection points for the input and the other for return.

Heat exchangers should always be located at a lower level (elevation) than the coolant level in the expansion tank. Actually, the expansion tank is located at the highest point in the system and is connected to the system by piping, **Figure 11-52.** This permits air to leave the coolant and collect on the tank. If air remains in the cooling water, it will promote rusting of metal parts. The air in the expansion tank allows the water to expand when it is heated. Additional water or chemicals needed in the system are added at the expansion tank.

Keel Cooling Systems

The **keel cooling system** is a closed system that consists of the water-cooled exhaust manifold, a high-capacity coolant pump, an expansion tank, and the keel cooling coil. Actually, a keel cooler is an outboard heat exchanger which is either attached to or built as part of a ship's hull below the light waterline, **Figures 11-53** and **11-54.** The heated water from the engine(s) circuit(s) is circulated through the cooler by the engine-driven water pump(s).

Coolant is drawn to the coolant pump from the bottom of the expansion tank. It then travels through the cylinder block and cylinder head into the manifold. Part of the coolant is bypassed through the rear of the cooling manifold, travels into the rear of the exhaust manifold jacket, and finally flows into the expansion tank.

When the thermostat is open, most of the coolant in the water manifold passes through the thermostat housing and goes to the keel cooling coils. When the thermostat is closed, however, the coolant travels from the thermostat housing to the engine pump inlet, where it remixes with coolant from the exhaust manifold jacket. The coolant then circulates through the cylinder block and cylinder head. A closed thermostat means a quicker warm up because coolant bypasses the keel cooling coils.

The keel cooler is located in a well protected area of the hull. Frequently, two or more keel coolers are employed to achieve the greatest possible heat transfer. A separate keel cooler for the aftercooler may be located low on the hull, forward of the keel cooler. Heated water from the aftercooler should enter the keel cooler at the rear end and be discharged from the cooler for return to the engine

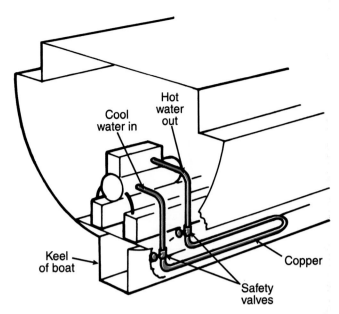

Figure 11-54. *This keel-mounted heat exchanger runs the coolant tube lengthwise through the keel. This type of system must be well protected. (Caterpillar, Inc.)*

at the cooler's forward end. This arrangement ensures maximum heat transfer with the vessel either dead in the water or moving ahead.

While the area immediately forward of the propeller(s) is a region of high water velocity and high enough on the hull to be protected from grounding damage, one must consider the effects on the keel cooler from sandblasting (from the propeller(s) during backing maneuvers).

The cooler and its through-hull connections are located so that the length of water piping will be kept to a minimum and so that the cooler will be well vented. Water piping should extend downward from the engine to the keel cooler, without high points.

The major advantage of the keel cooling system is that there is no raw water to contend with, reducing the possibility of electrolysis or water freezing. A disadvantage of this type of system is the amount of external piping needed and the possibility of damage to the keel cooling coils on the bottom of the vessel.

Radiator Cooling

With radiator cooling, the hot water from the engine jacket flows to the radiator core where it is cooled by air being pushed or pulled over the core fins by a fan, **Figure 11-55.** The cooled water is then pumped back through the engine; circulation is maintained by a gear- or belt-driven water pump, similar to those used in mobile diesel engine systems. Radiator cooling is used to cool engines that must be located well above the vessel water line or for emergency generator sets that require completely independent support systems.

Care must be taken in marine radiator systems to ensure engine exhaust gases are not drawing into the radiator. Additionally, the radiators must be arranged so the hot air discharge of one radiator does not recirculate to the

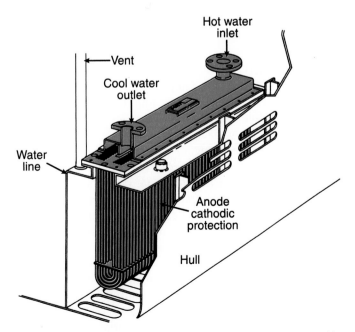

Figure 11-53. *Typical box keel-mounted heat exchanger and its components built into the hull. (Caterpillar, Inc.)*

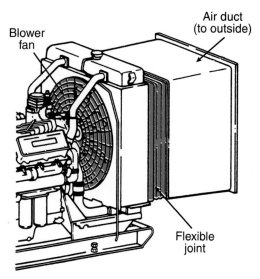

Figure 11-55. Marine radiator cooling system. (Caterpillar, Inc.)

inlet of another radiator. Also, for maximum efficiency, the direction of radiator air flow should not be against the direction of strong prevailing winds.

When an engine-mounted radiator is used and the engine is installed in the center of the room, a blower fan can be used and a duct provided to the outside.

Summary

The function of a diesel engine's cooling system is to carry away excess heat, protecting the metal of the engine from melting and the lubricating oil from breaking down. There are two basic types of cooling systems used in modern diesel engines: air-cooling systems and liquid-cooling systems.

In an air-cooling system, air passes around the engine to dissipate heat. Cooling fins are needed to maintain correct air temperature and velocity. When servicing an air-cooled engine, make sure all shields and shrouds are properly installed and sealed with gaskets or sealant as needed.

The two major types of liquid-cooling systems are the open system and the closed system. The open system is used primarily in stationary diesel installations. In this system, water travels through the water jackets to an outside source, such as a city main, a cooling tower, or spray pond. In the closed liquid-cooling system, which is normally used in transportation and construction services, the diesel engine has coolant (water) jackets inside the engine block to permit coolant to pass over hot combustion chamber surfaces. This coolant, circulated by a water pump, acts as a liquid conveyor to carry the excess heat to the radiator. This heat is removed from the coolant by the cooler outside air flowing through the radiator. An engine thermostat controls the amount of coolant entering the radiator.

Once coolant has passed through the radiator and has been "cooled" (or has given up its excess heat), it re-enters the water pump. The water pump pushes it up and through the water jackets surrounding the combustion chamber to pick up the excess heat and start the cycle once again.

The closed-type cooling system of a marine engine actually consists of two entirely separate circuits—a fresh water circuit and a salt water circuit. Components of a typical closed liquid cooling system used in the transportation field include the radiator, pressure cap, fan, fan belt, coolant (water) pump, engine oil cooler, aftercooler, and connecting hoses and pipes. The cylinder head and cylinder liner, with their cast passages, are also part of the system.

For coolant, most diesel engine manufacturers recommend a mixture of an ethylene glycol-based antifreeze, water, and a reliable corrosion inhibitor. The radiator allows the coolant's heat to be released into the atmosphere while keeping the coolant under pressure. It also provides enough liquid storage to ensure that the system works properly. The radiator assembly includes the surge (recovery) tank, the radiator, and the radiator cap.

There are two types of radiators found in mobile diesel equipment: high-flow (traditional) radiators and low-flow radiators. High-flow and low-flow radiators are either single-pass, vertical-flow radiators (with top and bottom tanks) or single-pass, cross-flow radiators (with side tanks). The radiator or heat exchanger consists of a series of tubes and fins that expose heat from the coolant to as much surface area as possible.

Many heavy duty vehicles employ radiator shutters to regulate the flow of cool air through the radiator, which helps maintain engine temperature within a predetermined range. The radiator shutter can be operated by one of the following methods: air, vacuum, or thermostat.

Cooling system hose carries coolant between system components. Hoses are usually made of butyl or neoprene rubber. Some hoses that connect metal tubing between the engine and the radiator are wire reinforced. Normally, a hose will deteriorate from the inside.

Most modern mobile diesel engines use a coolant filter in the cooling system. Like oil filters, coolant filters are of two basic types: the spin-on type and the cartridge type.

The temperature of the engine coolant is automatically controlled by a thermostat located somewhere between the cylinder head and the top tank of the radiator. Normally, the thermostat can be found in either end of the cooling manifold. Some larger engines have dual thermostats, while a few have four.

Some large diesel engines—both mobile and stationary types—are equipped with external water manifolds (one per cylinder head). These manifolds are designed to ensure a more even temperature control through the entire engine.

The water pump circulates coolant through the engine block, the cylinder heads, the radiator or heat exchanger, and the oil cooler. It may be either gear- or belt-driven from the crankshaft.

One of the major causes of engine overheating is slippage of the fan drive belts, which is caused by incorrect belt tension, worn belts, worn pulley grooves, or the use of fan belts of unequal length when two or more belts are used. The cooling system fan is usually located between the radiator and the engine. The fan in most mobile applications is generally driven by a V-belt from the engine crankshaft.

The most important consideration in any cooling system is keeping it clean. This involves the proper maintenance of filters and the draining and flushing of the system at recommended intervals.

Some marine and industrial diesel engine installations use raw-water cooling, heat exchangers, or keel coolers. In the basic raw-water cooling system, fresh water or salt water is used as a coolant. The liquid is directly circulated through the engine to cool it and is then dumped back overboard.

Important Terms

Air-cooled system
Cooling fins
Liquid cooling system
Open system
Closed system
Coolant
Ethylene glycol
Propylene glycol
Corrosion inhibitor
Surge tank
Radiators
Fan shroud
Radiator cap
Air shutter system
Shutterstat valve
Vacuum shutter system
Thermostatic shutter
 system
Hose
Coolant outlet
Coolant filter
Spin-on coolant filters

Cartridge filters
Thermostat
Water manifolds
Water pump
Impeller
Fan
Suction fans
Blower fans
Fan clutch
Coolant heaters
Immersion heater
Circulating tank heater
Heat exchanger
Cooling tower
Sight glass
Ebullient cooling systems
Steam separator
Raw-water cooling system
Electrolysis
Zinc anodes
Keel cooling system

Review Questions—Chapter 11

Do not write in this text. Place your answers on a separate sheet of paper.

1. An engine that operates at too low a temperature tends to:
 (A) use more fuel.
 (B) have less horsepower.
 (C) turn oil to sludge.
 (D) All of the above.

2. _____ cooling systems are generally used in mobile diesel engine applications.

3. Scale deposits on cylinder liners and water jackets will cause what type of problem for the coolant system?

4. Name the two main purposes the radiator has in the cooling system.

5. Chromates, borates, nitrates, and nitrites are all types of:
 (A) antifreeze/coolants.
 (B) coolant system corrosion inhibitors.
 (C) oil additives.
 (D) solvent.

6. The _____ radiator hose is subjected to the roughest service life of any hose in the cooling system.
 (A) lower
 (B) upper
 (C) vent cap
 (D) Both A & B.

7. The majority of coolant pump failures are caused by _____.

8. _____ is not an acceptable solvent for cleaning oil and dirt build-up on a radiator core.

9. Ebullient circulation is developed by coolant vaporizing at the _____ of the engine.

10. To reduce the effect of electrolysis on marine diesel cooling systems, _____ must be arranged in the system.

ASE-Type Questions

1. Technician A says that in extremely cold conditions, pure antifreeze should be installed in an engine's coolant system. Technician B says that a solution of 70% antifreeze and 30% water is the maximum amount of antifreeze that should be used in any system. Who is right?
   (A) A only.
   (B) B only.
   (C) Both A & B.
   (D) Neither A nor B.

2. A coolant recovery tank should be mounted as _____ as the installation permits and must provide cooling expansion volume.
   (A) high
   (B) low
   (C) close to the radiator
   (D) Both B & C.

3. Coolant flowing from a low-flow radiator has been pumped more _____ and is _____ than that from a high-flow radiator.
   (A) quickly/warmer
   (B) quickly/cooler
   (C) slowly/warmer
   (D) slowly/cooler

4. Radiator shutters _____.
   (A) shield the radiator from mud and road debris in mobile applications
   (B) regulate the flow of coolant through the core
   (C) regulate the flow of cool air through the radiator to maintain engine temperature at predetermined levels
   (D) None of the above.

5. Technician A says you should remove an old radiator hose by cutting it at the connection and folding it back. Technician B says you should not cut the hose, but simply pull it straight out from the connecting point. Who is right?
   (A) A only.
   (B) B only.
   (C) Both A & B.
   (D) Neither A nor B.

6. At coolant temperatures below approximately _____, most typical thermostat valves remain closed and block the flow of coolant to the radiator.
   (A) 120°F/49°C
   (B) 150°F/66°C
   (C) 170°F/77°C
   (D) 190°F/88°C

7. _____ fans pull air across the engine, then push it through the radiator or heat exchanger.
   (A) Suction
   (B) Blower
   (C) Both A & B.
   (D) Neither A nor B

8. As the cooling system is being filled with coolant, Technician A says you should open the engine venting petcock and the aftercooler petcock. Technician B says you should keep these petcocks closed during filling. Who is right?
   (A) A only.
   (B) B only.
   (C) Both A & B.
   (D) Neither A nor B.

9. Technician A says that no special precautions are needed before reverse flushing a cooling system. Technician B says you should remove the water pump and reverse flush the engine and radiator separately. Who is right?
   (A) A only.
   (B) B only.
   (C) Both A & B.
   (D) Neither A nor B.

10. All of the following should be done when servicing marine diesel heat exchangers EXCEPT:
    (A) use piping and cooling system components made from the same type of metal.
    (B) avoid contact between cooling system components made from more than one type of metal.
    (C) use either copper nickel or black iron piping.
    (D) allow cooling system components made from dissimilar metals to touch.

Chapter 12

Air Intake Systems

After studying this chapter, you will be able to:

❑ Explain the purpose and operation of diesel air intakes.
❑ Define scavenging and supercharging.
❑ Name the major air cleaner types and describe their operation.
❑ Explain the purpose of an air silencer.
❑ Describe the two principal types of blowers used in diesel air systems.

All diesel engines require an amount of clean fresh air to operate properly. Each cylinder should contain more air than required to produce combustion under ideal and controlled conditions. Increasing the amount of fuel injected by raising fuel rail pressure or by metering more fuel will not result in more power if there is not enough air to combine with it. An insufficient amount of air may result in power loss, smoke, excessive exhaust temperature, and reduced engine life. This chapter discusses the various air intake systems used on diesel engines

Air Intakes

The ability of a four-cycle diesel engine to fill a cylinder with clean fresh air is based on the engine's *volumetric efficiency.* For example, consider a cylinder that can hold 100 cubic inches of air at atmospheric pressure when the piston is at bottom dead center and the engine is off. If only 80 cubic inches of air enters the cylinder at a certain engine speed, the volumetric efficiency is 80% at that particular speed.

In the two-cycle diesel engine, the amount of fresh air retained, or trapped in the cylinder is more important than how much air it will hold. The amount of air trapped will directly affect the amount of fuel that can be burned and, consequently, the power output. A two-cycle engine's

ability to trap air in its cylinders is commonly known as *scavenging efficiency.*

Factors Affecting Air Intake Efficiency

There are many variables in diesel air induction systems that directly affect how well the engine "breathes." Engine design—smoothness of the internal air passages, degree and number of bends in the air passages, and size of the ports—plays an important part in air induction efficiency.

Valve timing can also have a pronounced effect on breathing efficiency. If the valves in a four-cycle engine close too soon, part of the residual exhaust gas can be pushed out through the intake port during the valve overlap period. This condition may dilute or restrict the flow of clean incoming air. Also, intake valves and ports are often overheated when exhaust gas escapes in this way. The abnormally hot intake surfaces will heat incoming air, making it less dense. This means that less oxygen will be available to support combustion. Premature opening of the intake valves tends to have a similar effect. For these reasons, valve adjustments should be performed as part of a regular routine maintenance. Since changes in rocker arm clearance result in changes in valve timing, periodic valve adjustments are extremely important.

Although the primary function of a diesel engine *air intake system* is to supply the air needed for combustion, the system also cleans the air and reduces the noise created as it enters the engine. The basic air intake system components found in a naturally aspirated four-cycle engine include the air cleaner, intake manifold, and related piping. Some systems also contain an aftercooler, also known as an air-to-air cooler or intercooler. If a turbocharger is employed, its compressor side is considered part of the intake system, **Figure 12-1.**

A two-cycle engine air intake system is similar to the system used in a four-cycle engine. The components of a two-cycle engine intake system include the air cleaner, connecting piping, air box (intake manifold), and blower.

If used, the aftercooler and the compressor side of the turbocharger are considered intake system components. See **Figure 12-2**.

Scavenging and Supercharging

Many diesel engines use the processes called scavenging and supercharging. In some four-cycle diesel engines, air enters a cylinder due to the change in pressure created as the piston moves away from the combustion space during its intake stroke. The air is forced into the cylinder due to greater atmospheric pressure outside of the engine This is referred to as a ***naturally aspirated intake.***

A device called a blower is used to increase airflow into the cylinders on all two-cycle engines and some four-cycle engines. The blower compresses air and forces it into an air box or manifold connected to the engine cylinders. This provides an increased amount of air under constant pressure to be used during engine operation.

Scavenging Principles

The air produced by the blower fills the cylinder with fresh air, clearing it of combustion gases. This process is known as ***scavenging.*** The air forced into the cylinder is called scavenge air and the entrance ports are referred to as scavenge ports.

Scavenging can take up only a small portion of the operating cycle, however, the length of time it takes is different in two-cycle and four-cycle engines. In a two-cycle engine, scavenging takes place during the end of the downstroke (expansion) and the beginning of the upstroke (compression). In a four-cycle engine, scavenging occurs as the piston is passing top dead center during the end of the exhaust upstroke and the beginning of the intake downstroke. Intake and exhaust openings are both open during the scavenging process. This allows the air from the intake to pass through the cylinder and into the exhaust manifold, removing exhaust gases from the cylinder and cooling hot engine parts in the process.

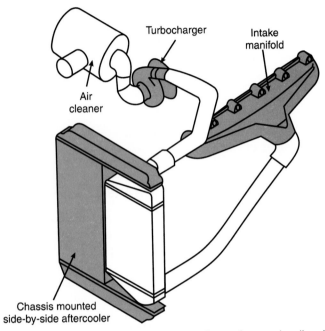

Figure 12-1. *Typical intake system for a four-cycle diesel engine. The turbocharger is considered part of the exhaust system. However, its compressor wheel is considered part of the intake system. (Mack Trucks, Inc.)*

Types of Scavenging

Scavenged air must be directed when entering the cylinder of an engine so that the waste gases are removed from the remote parts of the cylinder. The two principle methods for accomplishing this are sometimes referred to as ***port scavenging*** and ***valve scavenging.*** Port scavenging may be of the direct (cross-flow) or uniflow types. See **Figure 12-3A** and **12-3B.** Valve scavenging is of the uniflow type, **Figure 12-3C.** These scavenging air actions also cool the internal components, such as the piston, liner, and valves, with approximately 30% of engine cooling being provided by this airflow. This leaves the cylinder full of fresh air for combustion purposes when the piston covers the liner ports.

Supercharging Principles

Increased airflow to the cylinders can also be used to increase the power output of an engine. Since engine power develops from burning fuel, increasing the power requires more fuel. The increase in fuel, in turn, requires more air in order for combustion to take place. ***Supercharging*** is a process that supplies more air to the intake system than is normally taken in under atmospheric pressures.

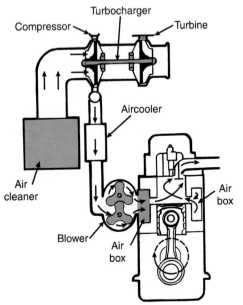

Figure 12-2. *Typical intake system for a two-cycle engine. This engine uses both a turbocharger and a blower. (Detroit Diesel Corp.)*

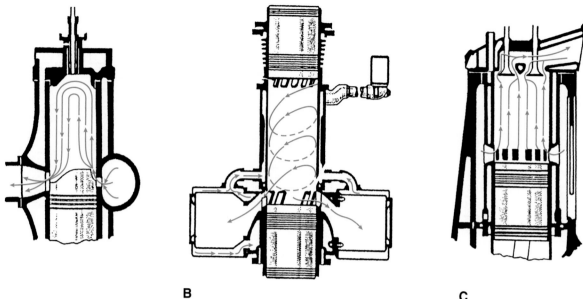

Figure 12-3. *Methods of scavenging used in diesel engines. A—Port direct scavenging. B—Port uniflow scavenging. C—Valve uniflow scavenging.*

In some two-cycle diesel engines, the cylinders are supercharged during the air intake simply by increasing the amount and pressure of the scavenge air. The same blower can be used for both supercharging and scavenging. Scavenging admits air under low pressure into the cylinder while the exhaust ports or valves are open. Supercharging takes place after the exhaust ports close, allowing the blower to force more air into the cylinder and increasing the amount of air available for combustion. The increase in air pressure created by the compression action of the blower usually ranges from 1-5 psi (6.9-34.5 kPa). This increase in pressure results in greater combustion efficiency. Consequently, a supercharged engine can produce more power than an engine of the same size that is not supercharged.

A blower is generally used to supercharge a four-cycle engine. Valve timing in a supercharged four-cycle engine is different from valve timing in a four-cycle engine that is not supercharged. In a supercharged engine, the intake valve opening is advanced and the exhaust valve closing is retarded so that there is a much greater valve overlap. This overlap will increase power. The amount of the increase depends on the pressure the supercharger applies. The increased valve overlap allows the air pressure created by the blower to remove gases from the cylinder during the exhaust stroke.

Study **Figure 12-4** to see how the opening and closing of the intake and exhaust valves affects both scavenging and supercharging. Also, note the differences between these processes in two- and four-cycle engines. In the illustration, the circular pattern represents crankshaft rotation.

 Note: Some of the events occurring in the cycles are shown as degrees of shaft rotation.

When checking the timing of a specific engine, refer to the owner's manual for instructions. Keep in mind that the crankshaft in a four-cycle engine makes two complete revolutions in one operating cycle, while the crankshaft in a two-cycle engine makes only one revolution per cycle. Also, remember that the exhaust and intake strokes in a two-cycle engine do not involve complete piston strokes.

Four-Cycle Supercharging

Figure 12-4A shows the operating process of a four-cycle engine with a centrifugal-type blower, which supplies the cylinders with air under pressure. In a supercharged four-cycle engine, the intake and exhaust valves remain open much longer than the valves in a non-supercharged engine. The compression and power strokes are shorter, allowing more time for scavenging. After the exhaust stroke ends, the blower charges the cylinders (fills them with air under pressure), before the compression stroke begins.

To fully understand the relationship between scavenging and supercharging, start with the cycle of operation with the piston in the top dead center position at the beginning of the power stroke. At this point, the peak compression has been reached, fuel injection is almost complete, and combustion is in progress. Power is delivered during the downstroke of the piston through 125° of crankshaft rotation. At this point in the downstroke (55° before bottom dead center), the power stroke ends and the exhaust valves open.

The exhaust valves will stay open through the rest of the downstroke (55°), through the next upstroke (180°), and through 85° of the following downstroke—a total of 320° of crankshaft rotation. At a point 75° before the piston reaches top dead center, the intake valves open and the supercharger begins forcing fresh air into the cylinder. For 160° of shaft rotation, the air passes through the cylinder and travels out the exhaust valves, clearing the burnt waste gases. The scavenging process continues until the exhaust valves close at 85° past top dead center.

After the exhaust valves close, the intake valves remain open for an additional 140° of crankshaft rotation (45° after bottom dead center). From the time the exhaust valves close until the piston reaches approximately bottom dead center, the cylinder is being filled with air from the blower. During this time, there is little increase in pressure due to increasing volume of cylinder space (the piston is in downstroke). However, when the piston reaches bottom dead center and starts the upstroke, the volume of space decreases as the blower continues to force air into the cylinder. Supercharging is the result, with a pressure increase of 3-5 psi (20.4-34.5 kPa) by the time the intake valves close.

During the rest of the upstroke (after the valves are closed), the supercharged air is compressed. Fuel injection begins several degrees before top dead center and ends shortly after top dead center. The actual length of the injection period in a specific engine depends on the engine speed and load. When the piston reaches top dead center, the cycle is complete, and the engine is ready to begin the cycle again.

Two-Cycle Supercharging

Compare **Figures 12-4A** and **12-4B**. Note that the length of the supercharging and scavenging periods in a two-cycle engine are not the same as those in a four-cycle engine. Also note that the location of the piston differs for each process. In a four-cycle engine, scavenging takes place while the piston is traveling through the end of the upstroke and the beginning of the downstroke. Four-cycle supercharging occurs when the piston is near bottom dead center. In a two-cycle engine, however, both scavenging and supercharging take place while the piston is in the lower part of the cylinder. The piston in the four-cycle engine does much of the intake and exhaust work, while the piston in the two-cycle engine does very little of this work. This is why two-cycle engines normally require a blower to assist in clearing out exhaust gases.

As shown in **Figure 12-4B**, two-cycle engine operation begins with the piston at top dead center. Fuel is injected, ignition has taken place, and combustion has started. The power that results forces the piston through the power stroke until it is 92.5° past top dead center—slightly more than halfway through the downstroke. At this point, the exhaust valves open, exhaust gas escapes through the manifold, and cylinder pressure quickly drops.

When the piston reaches 48° before bottom dead center, the intake ports are exposed by the piston as it moves downward and scavenging begins. The scavenging air swirls upward through the cylinder under blower pressure, clearing the cylinder of exhaust gases. Scavenging continues until the piston is 44.5° after bottom dead center (for a total of 92°). At this point, the exhaust valves close. Note that the exhaust valves are open during 132° of crankshaft rotation, compared to 320° in the four-cycle

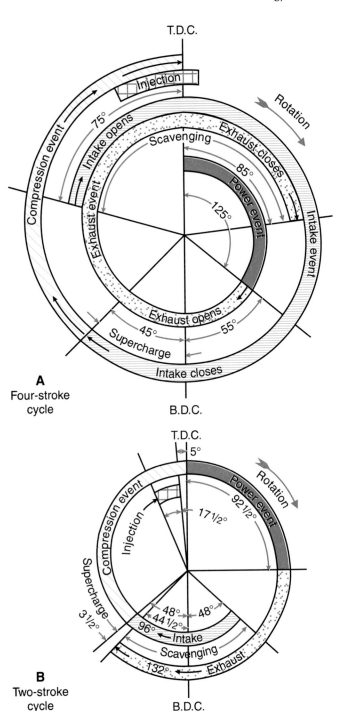

A
Four-stroke cycle

B
Two-stroke cycle

Figure 12-4. *The scavenging periods for four-cycle and two-cycle diesel engines.*

engine example. The scavenge ports remain open for another 3.5° of crankshaft rotation and the blower continues to force air into the cylinder. Even though the ports are open for only a short time after the exhaust valves close, there is enough time for the blower to supercharge the system before the compression stroke begins.

The piston closes the intake ports at 48° past bottom dead center. The compression stroke takes place during the rest of the upstroke, with injection and ignition occurring near top dead center. This completes the cycle, and another cycle is ready to begin. In a two-cycle engine, some type of air pump must always be used to provide the scavenging.

Air Cleaners

All internal combustion engines require clean air for proper operation. Diesel engines in particular need a lot of air—about 8500 gallons of air for every gallon of diesel fuel.

Contaminants in the air may cause excessive engine wear, low engine power, poor fuel consumption, reduced turbocharger life, increased maintenance, excessive downtime, and short engine life. Airborne contaminants, such as dust, dirt, moisture, and soot, must be removed before the air is introduced into the engine. This is the task of the *air cleaner.* The air intake stacks on mobile diesels can be mounted externally on the cowling of long-nose conventional trucks, **Figure 12-5A;** or as a "snorkel" type behind the cab of a cabover vehicle, **Figure 12-5B.** An underhood intake air system keeps the air cleaner canister out of the windstream to improve aerodynamics, **Figure 12-6.**

A **B**

Figure 12-5. Two types of air intake cleaner assemblies. A—Air intake stacks. B—Snorkel air cleaner.

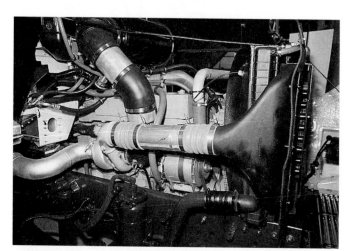

Figure 12-6. An underhood intake air cleaner keeps the canister out of the windstream to improve aerodynamics. (Heavy Duty Trucking/ Andrew Ryder)

The air cleaner must be properly maintained if it is to operate effectively. A plugged or dirty *air filter* can upset the air-fuel ratio, causing incomplete fuel burning. This causes cylinder washdown and contaminates the oil with partially burned fuel. It is very important that the air filter be installed properly.

The air filter must be sealed properly in order to be effective. When checking a filter, be sure that no part of an old gasket has been left on the filter housing. Replace the gaskets and warped lids on the housing to ensure a complete seal. Check the clean side of a filter for dirt patterns, which indicate that the filter has not been properly sealed. Use care to avoid denting or puncturing a filter element during handling. A damaged filter cannot protect the engine. Leaks in the air duct leading from the filter provide a path for dirt to get inside the engine. Make sure all duct joints are tight. Air filters for mobile and some stationary equipment are often classified by the amount of dust they must handle:

❑ Light dust air filters are for over-the-highway trucks, many stationary diesel engines and light industrial vehicles generally operated in low-dust environments.

❑ Medium dust air filters are found on equipment used in medium-dust environments such as agriculture and medium construction applications.

❑ Heavy dust air filters are required for dusty environments on equipment such as haul trucks, heavy crawlers/bulldozers, and other off-highway construction machines.

Multiple air cleaner installations are sometimes used in applications where engines must operate under extremely dusty air conditions or where two small cleaners must be used in place of a larger one.

Precleaners

Precleaners are usually installed at the end of a pipe that is extended upward into the air from the air cleaner inlet. This location keeps them in an area that is relatively free of dust. A precleaner is designed to remove larger particles of dirt and other foreign materials from the air before it enters the main air cleaner. A precleaner reduces the load on the main air cleaner, which, in turn, allows longer intervals between servicing. Most precleaners also have a prescreener that prohibits lint, chaff, and leaves from entering the air intake. See **Figure 12-7.** Centrifugal force in the bowl separates up to 75% of incoming dust before entering the air intake system. Precleaners are generally used on off-round vehicles and equipment under medium to heavy dust environments.

Figure 12-8 shows an air precleaner that is installed on mobile vehicles used in heavy dust and debris environments. The built-in louvers spin the intake air to separate up to 90% of incoming dirt and debris from the air intake system. Working in conjunction with an exhaust ejector scavenge flow system (ejector check valve), it continuously expels contaminants from the exhaust system flow.

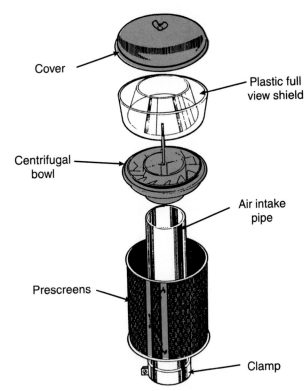

Figure 12-7. *Precleaner and prescreener assembly. (Deere & Company, Inc.)*

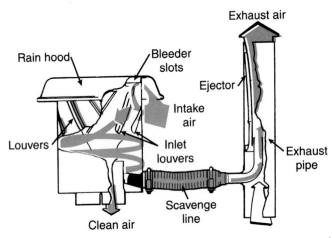

Figure 12-8. *Cutaway of a precleaner installed on vehicles used in severe dust conditions. (Donaldson Company)*

Excessive moisture is also a problem in the air intake system. Combined with dust, it can lower the efficiency of an air cleaner, while also reducing engine performance and life. If moisture is in the vicinity of the air intake, it will be pulled in. Some of the air cleaners, discussed later in this chapter, are designed with some water expelling capabilities. However, these design features have limitations and excessive moisture situations may require additional accessories to avoid saturating the air filter element.

One of the simplest ways to reduce water problems is the *ram inlet hood.* As shown in **Figure 12-9,** the moisture filled air enters the hood and is naturally forced against the rear wall by air pressure, often called the ram effect. Thus moisture on the wall will then drain out. Virtually moisture

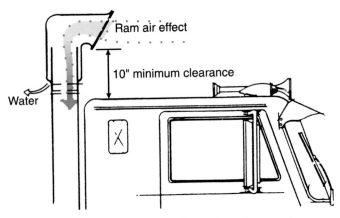

Figure 12-9. *Ram inlet prefilter. (Donaldson Company)*

free air (acceptable for good filter life and engine performance) passes to the air cleaner, **Figure 12-10.**

There are other input water removing devices in addition to the stack mounted ones just described, in areas where water problems exist. One of the simplest is the *inline moisture skimmer* shown in **Figure 12-11.** As the incoming moisture filled air enters the intake system, it hits a 90° tube elbow and its forward inertia causes the suspended moisture to collide with the tube wall. This results in the separation of water from the air. Air also slows down in the turn and allows moisture to fall out of the airflow and collect on the tube following the elbow. Moisture collects on and flows along the intake tubing. This moisture then gets trapped in the skimmers ducting canal and flows into a vacuator valve tube where it is finally released automatically. The virtually moisture free air can pass on to the air cleaner filter.

There are several other inline moisture separators on the market for either vertical or horizontal mounting.

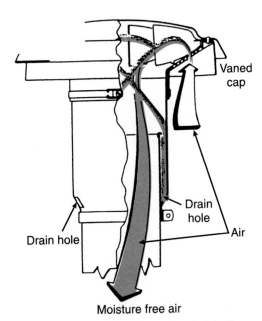

Figure 12-10. *The ram effect forces the air into the wall by air pressure. Virtually moisture free air enters the filter. (Donaldson Company)*

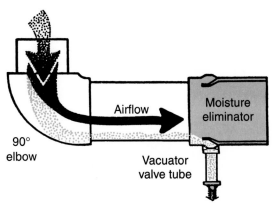

Figure 12-11. The inline skimmer is a simple design that uses the angle of a 90° elbow to separate the water from the incoming air. (Donaldson Company)

Figure 12-12 shows the centrifugal type which operates in the following manner:

- ❏ Moisture filled air enters the separator.
- ❏ Built-in vanes (stationary) cause the air to spin.
- ❏ Moisture is forced to the outside wall where it separates from the air.
- ❏ Moisture collects in the vacuator valve tubing and is automatically released by the vacuator.
- ❏ Virtually moisture free air flows on to the air cleaner.

The other common inline water separators work in basically the same manner as the centrifugal type.

Impingement Separation Method

In the ***impingement separation*** method, water travels along a winding path, where it bumps into many surfaces. It clings to these surfaces, forming water droplets that run down into a collection chamber with drain holes. As water builds up in the chamber, its weight overcomes atmospheric pressure and the water drains out.

Inertial Separation Method

In the ***inertial separation*** method, air velocity is changed to a much lower speed than the air velocity at the inlet. Water-laden air is guided into a plenum. Moisture condition, vehicle design, and the routing of the air intake to the air filter will help determine which method of water separation, if any, is used.

Intake Air Cleaners

As stated earlier, the air cleaner is designed to remove foreign matter from the air, pass the required volume of air for proper combustion, and maintain efficient operation for a reasonable period of time before requiring service. The importance of keeping dust and grit-laden air from entering the cylinders is essential to satisfactory engine operation and long engine life. Should dust in the air supply enter the engine, it would be carried into the cylinders and due to its abrasive properties, cause premature wear of the moving parts. Dirt, which is allowed to build up in the air cleaner passages, will eventually restrict the air supply to the engine and result in heavy carbon deposits on the valves and pistons due to incomplete combustion. The air cleaner must have a capacity large enough to retain the material separated from the air to permit operation for a reasonable length of time before cleaning is required.

There are two basic types of intake air cleaner filters: oil bath type and dry-type.

Oil Bath Air Cleaners

The ***oil bath air cleaner*** basically consists of a metal-wood screen element that is saturated with oil and supported inside a cylindrical housing. Air that is drawn into the cleaner passes through the element, where dust and other foreign matter are removed. The air then travels

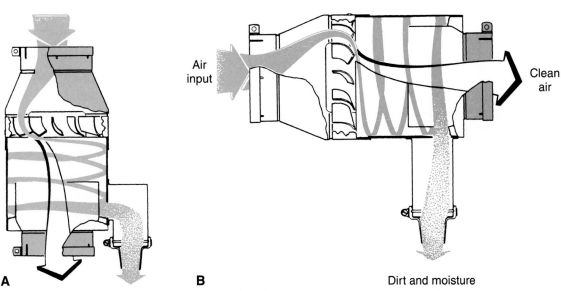

A   **B**   Dirt and moisture

Figure 12-12. A—Vertical moisture separator. B—Horizontal moisture separator. Note the circular flow of moisture as it separates from the air in both designs.

down the central tube in the cleaner and into the blower inlet. This lower portion of the air cleaner serves as a chamber to reduce air intake noise.

Light Duty Oil Bath Air Cleaner

The *light duty oil bath air cleaner* usually consists of a wire screen filter element that is supported inside a cylindrical housing. The oil bath is located in the housing directly below the element, **Figure 12-13.** Air that is drawn through the cleaner passes over the top of the oil bath. The air stream direction reverses when the air impinges on the oil in the sump and is then directed upward by baffle plates. During this change in the direction of airflow, much of the foreign matter is trapped by the oil and is carried to a receptacle or sump. The air travels up through the metal-wool elements, which remove most of the dust and the entrained oil. A second change of air direction at the top of the cleaner directs the air down through the center tube and into the blower inlet housing.

Light duty oil-bath filters, often called viscous type, are used in automotive diesel applications, where there is relatively little dust. The lower portion of a viscous air cleaner usually consists of a hollow chamber that quiets air intake noise.

Heavy Duty Oil Bath Air Cleaners

In all *heavy duty oil bath air cleaners,* air is drawn through the air inlet hood, which acts as a cleaner, and travels down through the center tube, **Figure 12-14.** At the bottom of the tube, the direction of airflow is reversed, and oil is picked up from the oil reservoir cup. The oil-laden air is carried up into the separator screen, where the oil and dirt is separated from the air.

A low pressure area is created near the center of the air cleaner as the air passes a cylindrical opening formed in the outer perimeter of the central tube and the inner diameter of the separator screen, **Figure 12-15.** This low pressure area is caused by the difference in air current velocity across the opening. The low pressure area, the effect of gravity, and the inverted cone shape of the separator screen cause the oil and dirt mixture to drain to the

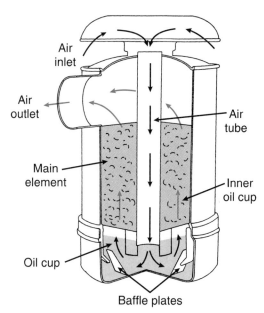

Figure 12-14. *This oil bath air cleaner is used in many heavy duty applications. (Detroit Diesel Corp.).*

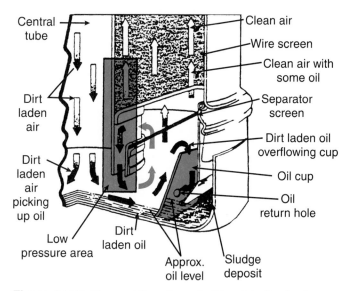

Figure 12-15. *Trace airflow through this oil bath air cleaner. Note the looping cycle in the center of the filter.*

center of the cleaner cup. This oil is again picked up by the incoming air, causing a looping cycle.

Gravity and inertia cause the larger dirt particles to settle at the bottom of the overflow cup as a sludge deposit. As the oil and air mix in the oil cup and cause a vapor, the smaller dirt particles are carried with the oil mist (vapor) upward through the filter element or elements. When the oil mist enters the filter, the heavier particles (oil mixed with dust) cling to the wire mesh and oil mist is reduced continuously until clean, oil-free air leaves the filter.

Depending on operating conditions, an air inlet hood or a prescreener must be used with heavy duty air cleaners. These components normally require more frequent cleanings than the main air cleaner. The usual installation employs an air inlet hood that prevents rain, paper, leaves,

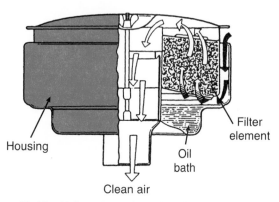

Figure 12-13. *Airflow through typical light duty oil bath air cleaner. (Wix Corp.)*

etc., from entering the air cleaner. The smaller heavy duty cleaners have a spherical-shaped hood. Air enters the lower portion of the hood through a heavy screen. The air is reversed in the hood and pulled down into the air cleaner. The hood is mounted on the air cleaner inlet tube.

Large heavy duty air cleaners use a dome-shaped hood. The hood is mounted on the air inlet tube of the air cleaner and is secured by a screw clamp. A heavy screen inside the hood prevents large pieces of foreign material from entering the cleaner. The openings in the hood should be kept clear to prevent excessive airflow restriction.

A prescreener on the inlet tube of the air cleaner is often used instead of an inlet hood for those applications where the air cleaner elements fill with lint or chaff. The purpose of the prescreener is to remove as much lint or chaff as possible before the air enters the cleaner.

Marine Oil Bath Air Cleaners

Oil bath air cleaners used on marine units are similar to heavy duty units except that the air is drawn into the cleaner through a series of slots around the perimeter of the cleaner body just above the oil cup assembly. The air passes over the oil, is directed upward by baffles, and then passes through a removable screen assembly. During this change in the direction of airflow, the larger particles of foreign matter, such as lint, chaff, and leaves, are removed from the air by the oil and the screen assembly. These particles settle in the oil cup sump. The air continues upward through metal-wool element, which removes the finer particles and the entrained oil. The clean air is then discharged through a side outlet near the top of the cleaner body and flows through the air inlet housing to the intake side of the blower assembly.

To service either a light or heavy duty oil bath air cleaner, check the specific procedures given in the engine service manual. The following are the major service recommendations:

❑ Keep the oil cup filled to the proper level with the correct weight of oil.

❑ Replace the oil when it gets dirty or thickens, reducing its ability to remove dirt particles from the air.

❑ Keep the filter element clean. Depending on the diesel application, the element may require cleaning as often as three times a day.

Dry Air Filters

Dry air filters come in a variety of sizes and shapes and are engineered to meet the performance requirements of individual engines and specific operating conditions. When dry air filters become plugged with contaminants, they must be replaced with a new element. The maximum restriction recommended for dry-type air filters varies with engine type and design.

In recent years, dry air filters have been used in place of the oil bath type in on-highway vehicles. However, the oil bath air cleaner is still used in off-highway, marine, and industrial applications.

Like the oil bath air cleaners, there are two basic types of dry air filters: light duty and heavy duty. *Light duty air filters* are generally used on truck engines. Heavy duty filters are found in most other diesel applications.

Heavy duty air filters are available in a variety of shapes and sizes, **Figure 12-16.** Heavy duty filters have a minimum efficiency rating of 99.9%. Because these filters are much larger than light duty types, a tighter, more efficient paper can be used and still allow a sufficient volume of air to pass through with a minimum of restriction. Plated perforated or expanded metal inner and outer bodies protect the paper and strengthen the filter assembly. A round single-stage air filter is illustrated in **Figure 12-17.**

Figure 12-16. Various types of heavy duty dry air filters are used in diesel engines. (Wix Corp.)

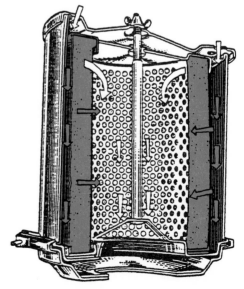

Figure 12-17. Round single-stage air filter. (Wix Corp.)

A round *two-stage air filter* is actually a secondary filter inside the primary filter, **Figure 12-18.** In some cases, the secondary filter has a fin assembly near its top, **Figure 12-19.** When used in the appropriate housing, air is directed to the top of the fin. The air is then forced through the fin causing it to swirl around the filter. The heavier dirt particles continue in a downward path to a collection area. Some of the fine dirt particles will follow the air to the pleated paper where they will be caught. Since the majority of the dirt particles are removed by centrifugal action, the life of the pleated-paper media is substantially increased. Some housings are equipped with a vacuator to automatically eject dirt collected in the bottom pan.

In extremely severe operating environments a *safety element* or inner element may be used inside the primary air filter. **Figure 12-20** shows a typical housing arrangement using an inner safety element. The main purpose of the inner element is to prevent dirt from entering the engine during installation and removal of the primary filter. Under normal conditions, the inner element should be replaced every third primary replacement. Cleaning of inner elements is not recommended.

Another type of filter, **Figure 12-21,** is called the *multi-tube filter.* This type of filter is made of small pleated-paper cylinders that are held together with a molded top and bottom. These filters are either square or rectangular, and the number of tubes in the filter determines airflow capacity. The smallest filter contains two tubes and the largest contains 64 tubes. Each tube is actually an individual filter that traps dirt as the air flows from the inside of the tube to the outside.

Panel air filters, **Figure 12-22,** are used as engine intake air filters on such applications as mobile pressurized cab filters on farm and construction equipment. These filters may be square or rectangular and consist of metal screen and pleated-paper. They usually have a 99.9% effi-

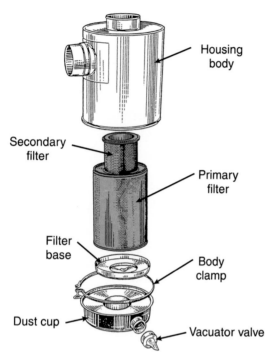

Figure 12-18. Parts of a vertical round two-stage air filter. (Wix Corp.)

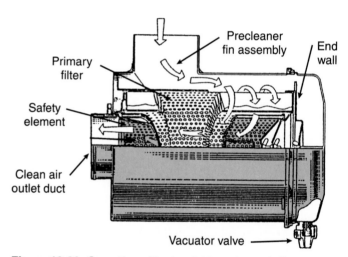

Figure 12-20. Operation of horizontal two-stage air filter. (Wix Corp.)

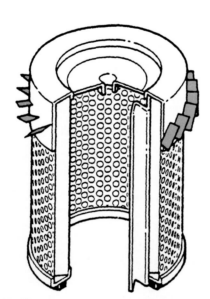

Figure 12-19. Fin assembly on a secondary filter. (Wix Corp.)

Figure 12-21. Typical multi-tube air filters. (Wix Corp.)

Figure 12-22. *Various panel-type air filters. (Wix Corp.)*

ciency rating. Cleaning a panel-type air filter for reuse is not normally recommended. Sealing gaskets are permanently attached where needed.

Another dry type air filter is the *spiral rotor* design shown in **Figure 12-23.** Air first enters the precleaner fin assembly where the angled veins turn the entering air, spinning out the larger contaminants. The centrifuged contaminants are carried along the wall of the cleaner and then ejected at the baffled dust cup. Dust remaining in the precleaned air is removed by the primary filter. This chemically treated filter element is pleated and embossed for maximum surface area. A perforated metal shell protects the element, giving it rigidity and strength. Air flows through both the primary and safety elements. In the case of accidental perforation of the primary filter, a safety element protects the engine. It also guards the engine during the primary element cleaning. The clean air outlet duct has fittings for mounting the restriction indicator. The vacuator valve ejects dust and water continuously, eliminating regular cap servicing.

Filter Housing

The **filter housing** is a box with a tubular outlet. The inlet is an integral part of the sealing cover, **Figure 12-24.** Two types of sealing covers or panels are used depending on whether a moisture eliminator and a precleaner is employed in the intake line. Generally, the moisture eliminator cover panel is used on over-the-road vehicles, while the precleaner cover panel is used on off-highway vehicles.

Servicing Dry Filters

The interval for servicing an air cleaner can be established according to the severity of the dust conditions encountered and the mileage accumulated. A more accurate method of determining the need for service is through the use of a **dial restriction indicator.** The type shown in

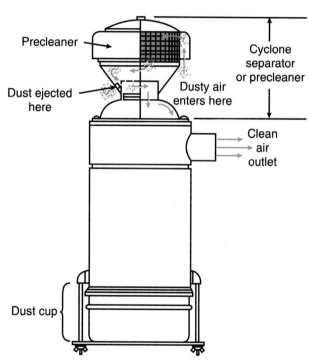

Figure 12-23. *A spiral-rotor designed air cleaner. (Donaldson Company)*

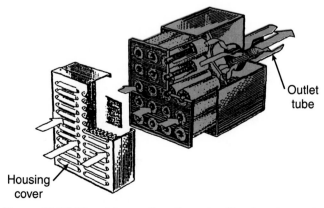

Figure 12-24. *The airflow pattern through a filter housing and a multi-tube filter. (Wix Corp.)*

Figure 12-25A may be mounted directly on the air cleaner or remotely mounted on the instrument panel.

A *restriction indicator* shown in **Figure 12-25B** is mounted on the instrument panel to permit continuous monitoring for a restriction while the engine is in operation. The restriction (service) indicator may be a piston or dial type. The piston is gradually drawn downward into view as the element loads with dirt. It locks into full view only after the restriction caused by a dirty element reaches the rated value of the indicator.

A *water manometer,* **Figure 12-26,** may also be used to check air inlet restriction. On turbocharged engines, the manometer should be connected to the air intake pipe, one to two pipe diameters upstream from the turbocharger inlet, in a straight section of pipe, **Figure 12-27.** Turbocharged engines should be under full load for a time to allow the turbocharger to reach maximum allowable air induction system restriction for efficient operation. These values are shown in a chart in the service manual. If these charts are not available, a good rule of thumb for diesel engines is not to exceed a reading of 20″ of water on the manometer.

> **Note: A damaged element gasket, ruptured element, incorrectly installed element or oil leakage in the engine air intake ducts and piping will cause a low restriction reading and cannot be identified as such. Care should be taken to check for the above when servicing the air cleaner.**

Additional Service Tips

The vacuator valve should be removed and cleaned periodically. Valve lips must be kept free of debris. Mud and chaff can lodge in the lips and hold them open during operation. The valve lips should be open only when the engine is at a slow idle or stopped. The lips must point downward in order to function properly.

If a valve is drawn inward (collapsed), check for a restricted air inlet. Another cause may be that the operator may have used it as an entry point to inject ether into the intake system by putting their fingers into the valve, and not pulling the lips back into position when completed. This also causes road dirt to be drawn into the air cleaner and will plug up the element prematurely. When installing the valve, posi-

Figure 12-26. A water manometer can also be used to check for air restrictions.

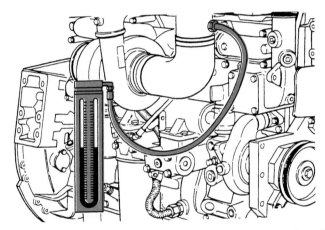

Figure 12-27. Checking a turbocharger air intake for air blockage with a water manometer. (Cummins Engine Co., Inc.)

tion it so that the lips will be facing 90° to the flat spring inside the vacuator valve mounting outlet of the air cleaner.

In cases where the air cleaner manufacturer recommends cleaning or washing the filter elements, the maximum service life is still one year or maximum restriction. Cleaning, washing, and inspection must be done per the manufacturer's recommendations. Inspection and replacement of the cover gaskets must also be done per the manufacturer's recommendations.

The following additional maintenance procedures will ensure efficient air cleaner operation:

- ❑ Keep the air cleaner housing tight on the air intake pipe.
- ❑ Be sure the correct filter replacement is used.
- ❑ Keep the air cleaner properly assembled so the joints are tight.
- ❑ Repair any damage to the air cleaner or related parts immediately.
- ❑ Inspect and clean or replace the elements as operating conditions warrant.

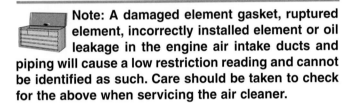

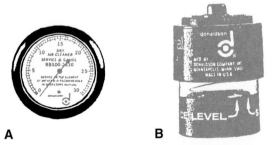

A   **B**

Figure 12-25. Various gauges can be used to monitor air filter restrictions and detect any blockage of the air intake system. A—Dial restriction gauge. B—Restriction service gauge. (Donaldson Company)

❑ Carefully inspect the entire system periodically. Enough dust-laden air will pass through an almost invisible crack or opening to eventually cause damage to an engine.

❑ Check duct joints for airtight connections. See that all clamps, flange joints, and air cleaner mounting bolts are tight. Seal any leaks immediately. All duct joints from air cleaner to engine must be tight.

❑ Check all rubber elbows and hoses for hardness. After a period of time, these parts become hard and lose much of their ability to maintain a good seal. They should always be replaced at engine overhaul.

Changing Air Filter Elements

Service intervals for air filters may be based on the manufacturer's recommended intervals or by monitoring the filter restriction. When replacing a filter, be sure that it is the same model number as the old one. Most filter manufacturers can advise you on the correct replacement. The following recommendations should be kept in mind when changing an air filter element:

1. Remove the old element as gently as possible. Accidentally bumping it can contaminate the clean side of the filter housing.

2. Clean the inside of the housing carefully. Dirt left in the housing may enter the engine. Use a clean, damp cloth to wipe every surface. Check the housing visually to make sure it is clean before putting in a new element.

3. Clean the gasket sealing surfaces of the housing. An improper gasket seal is one of the most common causes of engine contamination. Make sure all dirt is completely removed from all gasket sealing surfaces.

4. Check for uneven dirt patterns on the old element. A pattern on the element's clean side is a sign that the old element was not properly sealed or that a dust leak exists. Make certain the cause of a dust leak is identified and corrected before replacing the element.

5. Examine the gasket(s) on the new filter element for the following:

   ❑ Ridges caused by poor splicing.

   ❑ Uncoated, open-cell sponge material that may allow dust to pass through the gasket.

   ❑ Improper gasket thickness, which may change the quality of the seal.

   ❑ Insufficient adhesion to the end cover, which may allow dust to migrate between the gasket and the end cover of the filter.

   ❑ Improper position on end cover, which may provide for an improper seal.

6. Make sure the gasket is seated evenly, there will be no protection if it is not. Also make sure that the sealing surface of the housing is clean and that the element is the correct model number.

Intake Air Silencers

The air that enters the diesel engine intake system must be as clean as possible. It must also do so relatively quietly. A diesel engine uses great quantities of air, and the rush of this air through the air-cleaning devices often creates an extremely high-pitched whistle. Various silencers are used to reduce the noise of intake air.

In most mobile and industrial diesel applications that employ them, the **intake air silencer** devices are generally incorporated into the design of the air cleaner. When a separate air silencer unit is used, **Figure 12-28,** it should be installed between the air cleaner and the engine, as close to the intake manifold as possible. When installing the silencer, be sure that no air leaks are in the air intake system.

Pipes and Hoses

Pipes and hoses are used to transport the air from the air cleaner to the intake manifold or turbocharger, **Figure 12-29.** They must be airtight, rugged , and flexible. Additionally, they must be designed in such a way that they do not restrict airflow. Therefore, intake piping should be as short and as straight as possible. Bends should have a long sweep to reduce the resistance to airflow.

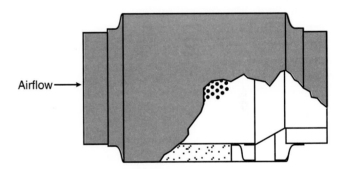

Figure 12-28. An air silencer is sometimes placed near the intake manifold to reduce noise.

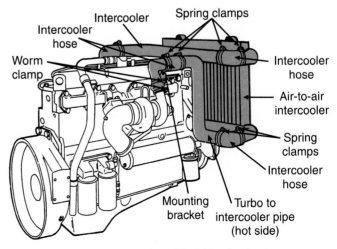

Figure 12-29. Air intake piping. (Mack Trucks, Inc.)

Blowers

As discussed earlier, superchargers or **blowers** are needed on most two-cycle engines to force scavenging air through the cylinders. In addition, a two- or four-cycle supercharged engine must have a blower to fill the cylinder with fresh air at a pressure above atmospheric pressure before the compression starts. The primary function of an engine blower is to deliver a large volume of air at a low pressure (1-5 psi or 6.89-34.5 kPa). There are two basic types of blowers: positive displacement blowers and centrifugal blowers.

Positive Displacement Blowers

A positive displacement blower is usually gear driven directly by the engine, while a centrifugal blower is usually driven by an exhaust-gas turbine. Positive displacement blowers can be divided into two groups: the multiple-lobe type, commonly called the lobe or Roots-type blower, and the axial-flow type. Since there are many variations in blower designs, we will concentrate on the Roots-type blower, which is by far the most popular type used.

Roots Blower

The **Roots blower** is essentially a gear pump similar to the lubricating pump described in **Chapter 10,** with two rotors of either two, three, or four lobes enclosed in a housing, **Figure 12-30.** The rotors do not come in contact with each other or with the housing. One of the rotors is driven directly through gears from the crankshaft and is connected to the other through gearing, **Figure 12-31.**

In operation, air from the blower passes into an air manifold, or air chamber, and enters the cylinder through ports that extend completely around the cylinder, **Figure 12-32.** This provides better air distribution throughout the cylinder and results in scavenging. When the piston is at the end of its downstroke and at the very beginning to the upstroke, both intake and exhaust ports are open. Air

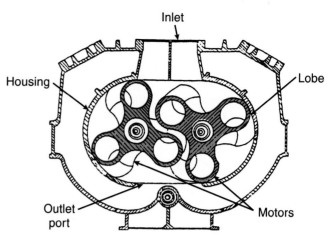

Figure 12-31. *Typical Roots-type blower. Note the location of the rotors.*

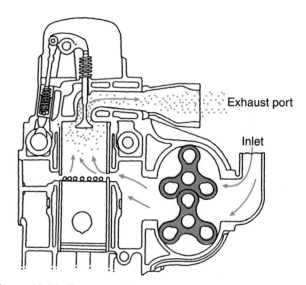

Figure 12-32. *Roots-type blower in operation. (Detroit Diesel Corp.)*

blown into these ports pushes the exhaust gases from the previous cycle out through the exhaust valves. When the piston is about one-quarter of the way up the cylinder, the valves close. The ports are covered by the pistons as they move. Exhaust gases will have been expelled and the cylinder will be full of fresh air. The rest of the stroke is an ordinary compression stroke at the end of which fuel is injected and combustion takes place.

One of the major advantage of the Roots blower is that the air delivery is almost directly proportionate to engine speed, ensuring a proper supply of air to the engine. Normal operating speed of the Roots blower ranges from 2000 to 6000 rpm.

Centrifugal Blowers

Unlike positive displacement blowers, which are driven through a gear train by the engine crankshaft, the **centrifugal blower,** or *turbocharger,* takes its power from the exhaust gases and thereby makes use of energy that would otherwise be wasted, **Figure 12-33.** Turbocharger

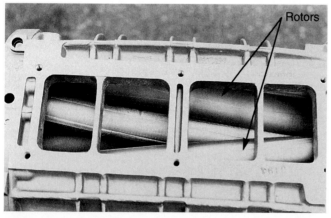

Figure 12-30. *Supercharger rotors in their housing. (Penn Diesel)*

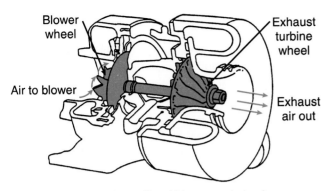

Figure 12-33. Typical centrifugal blower, or turbocharger.

operation is relatively simple. Gases from the exhaust manifold drive a turbine. The turbine drives an impeller (on the same shaft), which supplies air to the cylinder for scavenging and supercharging. Turbocharger operation is covered fully in **Chapter 13.**

Intake Air Passages

Air must pass through a number of passages to reach the combustion spaces within an engine. So far, we have considered components that clean, silence, and compress the intake air. From the blower, however, the air is discharged into a unit or passage to route the air to the intake valves or ports of the cylinders. The design of such a unit, as well as the terminology used to identify it, differs depending on the type of engine.

In two-cycle engines, the passage that conducts intake air to the cylinders is generally called an **air box.** The air box surrounds the cylinders and, in many engines, is built into the block. For example, the air box in two-cycle, V-type diesel engines consists of the space (within the block) between the two banks of the V-construction and the open space between the upper and lower deckplates of each bank.

The scavenging air passages in an opposed-piston engine are referred to as the air receiver. These passages are in the upper part of the block and surround the cylinder liners. In some two-cycle engines, the passage that serves as a reservoir for intake air from the blower is called the air header.

Drains are generally provided in air boxes, receivers, and headers to drain off any liquids that accumulate. Vapors from the air charge may condense and settle in the air box, or a small amount of lubricating oil may be blown into the air box as the piston passes the ports on the downstroke that follows the power event.

On some engines, the drains are vented to the atmosphere. In others, specifically in stationary or marine applications, a special drain tank is provided to collect the liquids from the air box. The purpose of the tank is to prevent drainage of the liquids to the engine room. A small connection from the fuel pump also carries fuel oil that leaks past the pump seals to the drain tank.

Drains are very important when the air cooler is installed between the blower discharge and the air intake

manifold, or receiver. The drains are usually left open during engine operation to prevent condensation or water from leaky coolers from being carried into the engine cylinders with the combustion air.

Ingestion of road salt and water into the air intake and blower has increased as more truck manufacturers adopt frontal air systems or mount air inlets close to the ground. But even systems equipped with air deflectors, in which the stack is located behind the cab, are not immune to the problem. Because of the vacuum generated behind a deflector, the concentration of moisture may be higher there than at the frontal air intake point.

Some diesel engines have an air shutdown valve(s) that will shut off the air supply and stop the engine whenever abnormal operating conditions require an emergency shutdown. The valve is located in the air shutdown housing, which is usually mounted on the side of the blower and may serve as a mounting for the air cleaner or air cleaner duct. In a few engines, the air shutdown valve is located in the intake manifold, **Figure 12-34.**

The intake air passage from the blower to the cylinders in four-cycle diesel engines differs from the intake air passage in two-cycle engines. The air passage may or may not be an integral part of the four-cycle block. In some four-cycle engine designs, a separate cast iron or die cast aluminum intake or inlet manifold is attached to the block for the purpose of conducting intake air to the engine cylinders, **Figure 12-35.** External **intake manifolds** have the advantage of being easier to remove when repairs are necessary. On the other hand, integral intake manifolds also have advantages, including:

❑ The intake passages are surrounded by the engine coolant, which warms the incoming air before it reaches the cylinders.

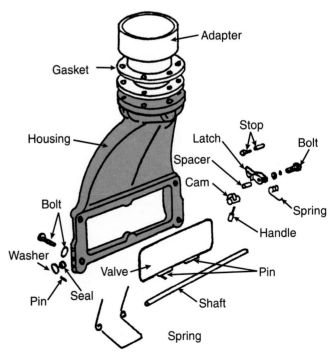

Figure 12-34. Exploded view of a typical air shutdown valve. (Detroit Diesel Corp.)

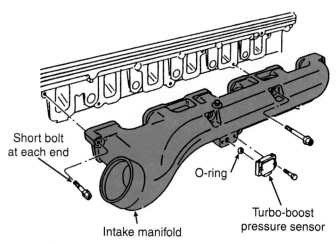

Figure 12-35. *Typical intake manifold.*

❏ There are fewer joints in integral manifolds. Therefore, fewer gaskets are required, and there is less leakage.

Intake Installation

When sealing the intake manifold to the cylinder head, take maximum advantage of the available bolt torque. The load distribution of the bolts, manifold, and head may be uneven, but the gasket can be designed to

overcome this problem by filling much of this unevenness to provide a good seal. Before installing a gasket, check the intake manifold to make sure that the flanges are flat. Also check for cracks or foreign material in the manifold. Then proceed as follows:

1. Install the new intake manifold gasket. Use a small amount of contact adhesive to hold the gasket in place during manifold installation. Align the gasket properly before the adhesive dries, **Figure 12-36A**.

2. After allowing the adhesive to dry completely, position the lockwashers on the bolts and lubricate the threads, **Figure 12-36B**.

3. Position the manifold on the block and insert the bolts. Torque the bolts to the specifications given in the engine service manual, **Figure 12-36C**.

4. Tighten all manifold hose clamps, **Figure 12-36D**.

Summary

All diesel engines require more clean air in the cylinders than is required for combustion. Lack of sufficient combustion air results in power loss, smoke, excessive exhaust temperature, and reduced engine life. The primary function of a diesel engine air intake system is to supply the

A

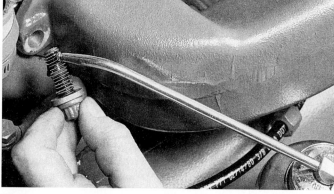

B

C

D

Figure 12-36. *Reinstalling an intake manifold. A—Installing the gasket. B—Oiling the bolts. C—Torquing the bolts to the manufacturer's specifications. D—Connecting the hose clamps. (Mack Trucks, Inc.)*

air needed for combustion. It also cleans the air and reduces the noise created as the air enters the engine.

The basic air system components found in a naturally aspirated four-cycle engine include the air cleaner, intake manifold, and piping. An air charge cooler (also known as an aftercooler or intercooler) can also be part of the intake system. When a turbocharger is employed, its compressor side is considered part of the intake system.

The components of a two-cycle diesel engine's air intake system include the air cleaner, connecting piping, intake manifold or air box, and blower. Aftercoolers and turbochargers can also be used. In some four-cycle diesel engines, air enters the cylinder due to a change in pressure created as the piston moves away from the combustion space during intake stroke. This is referred to as naturally aspirated intake.

A blower is installed in all two-cycle and some four-cycle engines to increase the airflow into the cylinders. The blower compresses the air and forces it into an air box, or manifold, which is connected to the engine cylinders. This provides an increased amount of air under constant pressure to be used during engine operation.

The process of using a blower to fill the cylinder with fresh air to clear it of combustion gases is known as scavenging. In a two-cycle engine, scavenging takes place during the end of the downstroke (expansion) and the beginning of the upstroke (compression). In the four-cycle engine, scavenging occurs when the piston is nearing and passing top dead center during the end of the exhaust stroke and the beginning of the intake stroke.

Supercharging is the process used to supply any air above the normal amount taken into the system under atmospheric pressures when the energy required to supply the additional air is produced mechanically. To supercharge a four-cycle engine, a blower must be added to the system since exhaust and intake are normally performed by piston action alone.

An air cleaner removes fine materials, such as dust, sand, chaff, and lint, from the air. A plugged or dirty air filter can upset the air-fuel ratio so that the fuel does not completely burn. Precleaners are usually installed at the end of a pipe that extends from the air cleaner inlet.

The oil bath air cleaner consists of a metal screen element that is saturated with oil and supported inside a cylindrical housing. Air that is drawn into the cleaner passes through the element, where dust and foreign matter are removed.

Dry air filters come in a variety of sizes and shapes, and are engineered to meet the engine's performance requirement. Dry filters are replaced when dirty. In recent years, dry air filters have replaced oil bath filters in popularity in on-highway vehicles.

Service intervals for air filters may be based on the manufacturer's recommended intervals or by monitoring the filter restriction. The maximum restriction recommended for dry-type air filters varies with engine type and design.

Silencers, which are used to reduce the noise of intake air, may be incorporated into the design of the air cleaner or they may be separate units. Pipe and hoses are used to transport air from the air cleaner to the intake manifold or turbocharger. They must be airtight, rugged, flexible, and designed in such a way that they do not restrict airflow.

There are two basic types of blowers: positive displacement blowers and centrifugal blowers. A positive displacement blower is driven directly by the engine, while a centrifugal blower is usually driven by an exhaust-gas turbine. Positive displacement blowers can be divided into two groups: the multiple-lobe type, commonly called the lobe or Roots blower, and the axial-flow.

In two-cycle engines, the passage that conducts intake air to the cylinders is called the air box. Drains are generally provided in air boxes, receivers, and headers to drain off any liquids that accumulate. Moisture separators are also available to remove moisture from the intake air. In four-cycle diesel engines, the intake air passage from the blower to the cylinders is a separate intake manifold and is attached to the block.

Some diesel engines have an air shutdown valve(s) that will shut off the air supply and stop the engine whenever abnormal operating conditions require an emergency shutdown.

Important Terms

Volumetric efficiency	Dry air filter
Scavenging efficiency	Two-stage air filter
Air intake system	Safety element
Naturally aspirated intake	Multi-tube filter
Scavenging	Panel air filter
Port scavenging	Filter housing
Valve scavenging	Dial restriction indicator
Supercharging	Restriction indicator
Air cleaner	Water manometer
Air filter	Intake air silencer
Precleaners	Blowers
Ram inlet hood	Roots blower
Inline moisture skimmer	Centrifugal blower
Impingement separation	Air box
Inertial separation	Intake manifolds
Oil bath air cleaner	

Review Questions—Chapter 12

Do not write in this text. Place your answers on a separate sheet of paper.

1. Name the three main functions of a diesel engine air intake system.

2. In a two-cycle engine, supercharging takes place _____ the exhaust ports _____ .
   (A) after/close
   (B) before/close
   (C) after/open
   (D) None of the above.

3. In a four-cycle supercharged engine, the intake-valve opening is _____ and the exhaust valve closing is _____ to achieve a much greater overlap.
   (A) retarded/advanced
   (B) advanced/retarded
   (C) advanced/advanced
   (D) retarded/retarded

4. A plugged or dirty air filter can upset the _____ , causing incomplete burning of the fuel.

5. The replacement or service interval for air filters may is based on _____.
   (A) miles traveled
   (B) hours of operation
   (C) filter restriction
   (D) None of the above.

6. The maximum restriction recommended for dry air filters varies with _____ .

7. Centrifugal, impingement, inertial, and skimmer are all of _____ methods.
   (A) moisture separation
   (B) supercharging
   (C) blower pump
   (D) turbocharger

8. Name four additional maintenance procedures that will ensure efficient air cleaner operation.

9. With a Roots blower, the amount of air delivered to the engine is almost directly proportional to the _____ .

10. What is a more common name for a centrifugal blower?

ASE-Type Questions

1. Using an external blower to clear a cylinder of combustion gases is known as _____ .
   (A) turbocharging
   (B) supercharging
   (C) scavenging
   (D) blowby

2. _____ is a process that supplies more air to the intake system than is normally taken in under atmospheric pressures.
   (A) Aspirated
   (B) Supercharging
   (C) Scavenging
   (D) Blowby

3. Technician A says that contaminants in the air will cause excessive engine wear if they are not removed. Technician B says that contaminants in the air will cause poor fuel consumption. Who is right?
   (A) A only.
   (B) B only.
   (C) Both A & B.
   (D) Neither A nor B.

4. Which type of air filter is the most popular in on-highway diesel engine applications?
   (A) Oil bath.
   (B) Dry.
   (C) Magnetic.
   (D) None of the above.

5. All of the following is part of normal service procedures for an oil-bath air cleaner EXCEPT:
   (A) replace the oil in the cup when it gets dirty or thickens.
   (B) keep the oil cup filled to the proper level.
   (C) clean all filter elements no more than once a month.
   (D) keep the filter element clean.

6. Heavy duty diesel air filters have a minimum efficiency rating of _____ .
   (A) 99.9%
   (B) 90%
   (C) 100%
   (D) 80%

7. All of the following is true about air boxes used on diesel engines EXCEPT:
   (A) drains are vented into the engine.
   (B) vapors from the air charge may condense and settle in the air box.
   (C) drains used in marine diesel engines are vented to a tank.
   (D) the drains are left open during engine operation.

8. Technician A says that too many bends and turns in the air system piping can restrict airflow in the system. Technician B says that leaks in the air intake system are a primary cause of dirt entering the engine. Who is right?
   (A) A only.
   (B) B only.
   (C) Both A & B.
   (D) Neither A nor B.

9. Technician A says that a positive displacement blower is usually gear-driven off of the engine. Technician B says this type of blower is normally powered by an exhaust-gas driven turbine. Who is right?
   (A) A only.
   (B) B only.
   (C) Both A & B.
   (D) Neither A nor B.

10. Technician A says that in a naturally aspirated engine, a pressure difference between the outside atmosphere and the inside of the cylinder forces air into the cylinder. Technician B says that this pressure difference is created by movement of the piston. Who is right?
    (A) A only.
    (B) B only.
    (C) Both A & B.
    (D) Neither A nor B.

Chapter 13

Exhaust Systems

After studying this chapter, you will be able to:
- ❑ Explain the purpose of exhaust systems used on diesel engines.
- ❑ List the main components of a typical exhaust system.
- ❑ Define back pressure and describe its effect on the exhaust system.
- ❑ Understand the precautions that must be taken when routing exhaust pipes.
- ❑ Describe the parts and operation of a turbocharger.
- ❑ Explain the purpose of an intercooler or aftercooler.
- ❑ Identify the causes of turbocharger failure.
- ❑ Explain the operation of an exhaust pyrometer.
- ❑ Name some of the more common diesel engine emission controls.

The exhaust system collects combustion gases from the diesel engine and carries them away. Removing spent combustion gases from the cylinders with as little resistance as possible is as important for effective combustion as filling the cylinders with fresh air. In addition to removing combustion gases, the exhaust system is designed to remove engine heat, muffle exhaust noise, quench sparks, remove solid material from exhaust gases, and in some cases, furnish energy to a turbocharger.

Exhaust System

The **exhaust system** of a mobile diesel engine generally consists of the exhaust manifold, exhaust piping, muffler, and outlet pipe (the latter is called the *stack* in heavy duty vehicles or the *tailpipe* on cars and light trucks). Gases must be pushed around bends, elbows, adapters and connectors in the piping system, and of course, through the muffler. Some systems contain a resonator and some type of emission control device, **Figure 13-1.** On many engines, a turbocharger is installed in the exhaust system and driven by exhaust, **Figure 13-2.**

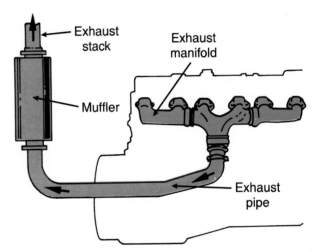

Figure 13-1. *The diesel engine exhaust system shown in this illustration is used for many mobile applications. Emission control devices are not shown in this illustration.*

A stationary diesel exhaust system usually contains an exhaust manifold, exhaust piping, silencer and/or muffler and outlet pipe, **Figure 13-3.** The system is designed to discharge the exhaust in a manner that is not objectionable or dangerous. The system must also conduct and discharge exhaust without creating a fire hazard at the installation site or surrounding structures. The exhaust system must be in compliance with environmental regulations and any applicable building codes, while efficiently removing exhaust gases without creating excessive back pressure.

Back Pressure

Back pressure is the resistance to the free flow of gas. This resistance can be caused by friction between the gases and the system piping, undersized exhaust piping, a sudden constriction in the piping, and pipe elbows and bends.

The higher the back pressure, the harder it is for the hot exhaust gases to get out of the engine. Excessive back pressure leads to high valve temperatures, water jacket overheating, and premature engine wear. Another result of

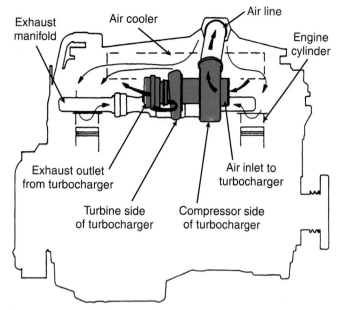

Figure 13-2. *Turbochargers are commonly used on diesel engines. (Caterpillar Tractor Company)*

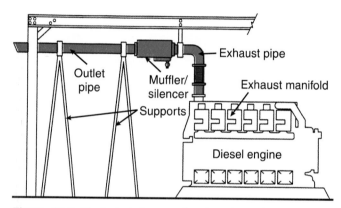

Figure 13-3. *Typical stationary diesel engine exhaust system. Designs will vary with each installation and application. (Waukesha Engine Division)*

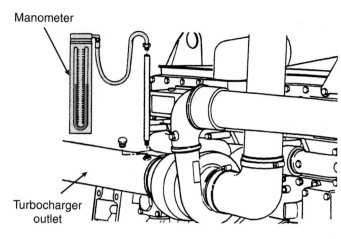

Figure 13-4. *A manometer is used to check diesel exhaust back pressure. (Cummins Engine Co.)*

Exhaust System Components

While the design of exhaust system components varies from one manufacturer to another, the primary objectives of all systems are the same. Exhaust systems are designed to provide optimum noise silencing, minimum back pressure, and long life. They are also designed to be as light as possible without sacrificing durability.

The exhaust system on a heavy duty truck, for example, has to carry out its tasks in a harsh environment. It must survive engine and vehicle twist, internal and external temperature extremes, as well as shock and bounce. Components must be mounted away from exposure to wheel splash and other external hazards and be accessible for repair. Mountings also should allow for maximum trailer swing clearance behind a tractor cab, a wide range of body types for straight trucks, and maximum ground clearance for both trucks and tractors.

 Note: It is important to remember that the intake air system and the exhaust system are very closely related and must work together. The intake system brings fresh air into the engine and the exhaust system gets rid of the burnt gases, *Figure 13-5.* **Therefore, there is a certain overlap of data between this chapter and** *Chapter 12.*

Exhaust Manifolds

The *exhaust manifold* collects engine exhaust gases from the cylinder ports and carries them to an exhaust pipe. The exhaust manifold is essentially a large pipe that has exhaust ports, or openings from each cylinder of the engine feeding into it. The exhaust ports and passages in the exhaust manifold are large enough to allow the escaping gases to flow freely and expand. This permits better scavenging of the engine cylinders. If burned gases are left in the cylinders following the exhaust stroke, the

excessive back pressure is loss of power. The hot combustion gases trapped in the cylinder dilutes the fresh, incoming air. Less oxygen is available to support fuel combustion, and engine power output drops. The harder the pistons have to work against the buildup of the combustion gases, the less power for turning the crankshaft. A 2 psi increase in back pressure can cause an engine to lose approximately 4 horsepower. A low restriction exhaust system minimizes the amount of energy needed to pump combustion gases from the cylinders, leaving more power for operating the engine. This results in better fuel economy at any given speed.

Back pressure is measured in inches of mercury, or water, with a manometer. The back pressure limits set by the engine manufacturer are given in the service manual and will vary with engine use. On medium and heavy duty mobile engines, for example, the acceptable maximum might be 3" of mercury (or 13.9" of water). **Figure 13-4** shows a manometer connected to check the exhaust restriction near the turbo outlet.

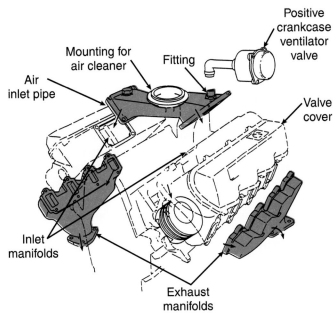

Figure 13-5. *The air intake system and the exhaust system are closely related. (Caterpillar Tractor Company)*

Figure 13-7. *Heat reduction is important in small marine engine applications. (Viking Yachts Inc.)*

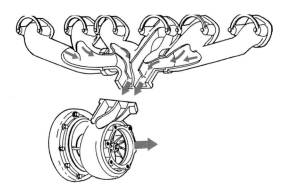

Figure 13-8. *The exhaust manifold serves as passageway for the gases traveling from the combustion spaces to the exhaust inlet of the turbocharger.*

amount of air mixture that can be taken in on the next intake stroke will be limited. This reduces engine power and increases fuel consumption.

While most exhaust manifolds are made of cast iron or nodular iron, some are made of steel plate with welded joints and branch elbows that are made of steel castings. Other manifolds are made of cast aluminum. Most cast iron manifolds are made of two or more pieces with slip joints to allow for heat expansion, **Figure 13-6.** On many marine engines, the exposed surfaces of the exhaust manifold and related parts may be insulated with layers of spun glass held in place by woven insulation covers, **Figure 13-7.** This insulation aids in reducing heat radiation. **Slip joints,** sometimes referred to as expansion joints, are generally provided between the manifold sections and the turbocharger or other outlet connections on most marine and mobile diesel engines.

The exhaust manifold shown in **Figure 13-8** serves as the passage for the gases from the cylinders to the exhaust inlet of the turbocharger. The turbine end of the turbocharger is considered part of the exhaust system because it forms a part of the passageway for the escape of gases to

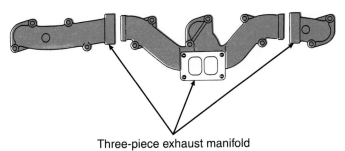

Three-piece exhaust manifold

Figure 13-6. *This three-piece manifold contains slip rings that allow movement during heat expansion. (Detroit Diesel Corp.)*

the exhaust outlet. A similar arrangement is used on other turbocharged diesel engines.

Exhaust Manifold Service

When servicing an exhaust manifold, check it carefully for cracks, burnout, and damaged threads. Be sure the inner passages are free of carbon buildup. Remove any metal chips or loose carbon from inside the manifold. If excessive carbon buildup exists, use a carbon solvent or carefully scrape the area clean. In severe cases, a combination of solvent and scraping must be used.

It is not advisable to remove a hot manifold from an engine, as this can promote warpage. Check for a warped manifold by placing a straightedge across the exhaust ports, **Figure 13-9.** The maximum allowable clearance between the manifold and the straightedge is usually about .004″ (.101 mm). Check the engine service manual for the exact clearance. If warpage is severe enough to prevent a good seal, the manifold must be machined or replaced.

Exhaust Manifold Gasket

An *exhaust manifold gasket* is used to prevent leakage between the mating surfaces of the engine and the manifold. While some engines are designed for no gasket, a gasket may eventually be required to compensate for

Figure 13-9. A straightedge should be used when checking exhaust ports for warpage. (Mack Trucks, Inc.)

warpage. Always clean the mating surfaces before installing a new manifold gasket, **Figure 13-10.** Use proper bolts or studs and torque them in the sequence recommended by the manufacturer. After the initial warm-up and cool-down, it may be necessary to retorque the bolts.

 Caution: Never reuse an exhaust manifold gasket.

Mufflers

The **muffler** is the primary component responsible for minimizing the harsh sound waves created as combustion gases are forced out of the cylinders through the exhaust ports. Its function is to remove acoustic energy from the gases without unnecessarily obstructing their flow to the outlet pipe. In industrial installation, the muffler is often called a silencer.

Types of Mufflers

Mufflers are grouped into two general categories: dry mufflers and wet mufflers. There are two main silencing strategies used in dry mufflers: dispersive and reaction.

Dispersive mufflers

Dispersive mufflers, or straight-through mufflers, **Figure 13-11A,** are designed to force advancing sound waves through small openings of high restriction, reducing their intensity. These mufflers consist of a perforated inner pipe that is enclosed by an outer pipe roughly three times larger in diameter. The space between the pipes is sometimes filled with a heat-resistant, sound-absorbing material.

Reactive mufflers

Reactive mufflers, or reverse-flow mufflers, **Figure 13-11B,** use various assemblies, such as tubes, chambers, and baffles to cause a portion of the advancing sound waves to be reflected. This decreases the intensity of the waves as they travel back-and-forth through the muffler.

Most modern muffler designs actually use some combination of dispersion and reaction in their acoustic packaging. Nevertheless, even a well-designed muffler will produce some back pressure in the system. However, a small amount of back pressure can be used intentionally to allow a slower passage of exhaust gases through the particulate oxidizer catalytic converter (described later in this chapter), which results in a more complete conversion of harmful gases.

An exhaust system that contains many bends can use up as much as half the engine's total allowable back pressure before the exhaust gases even reach the muffler. Such a complex system is hardly ideal from an efficiency standpoint. Nevertheless, exhaust system engineers can prevent excessive back pressure in the system as a whole by using the proper muffler.

The newest muffler design, which is presently being used on some heavy duty trucks, is the electronic type. This unit, **Figure 13-12,** consists of sensors and microphones that

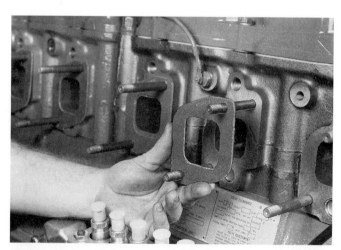

Figure 13-10. After cleaning the mating surface, install the gasket. (Mack Trucks, Inc.)

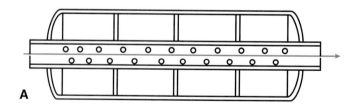

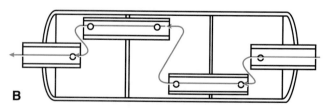

Figure 13-11. Two types of dry muffler designs. A—Straight-through muffler. B—Reverse-flow muffler. (Deere & Company, Inc.)

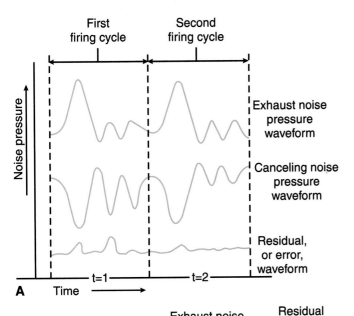

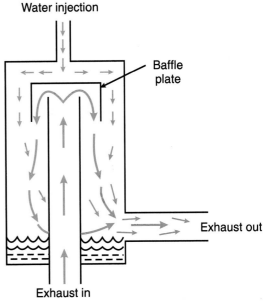

Figure 13-13. *Cutaway view of a wet muffler. (Caterpillar Tractor Inc.)*

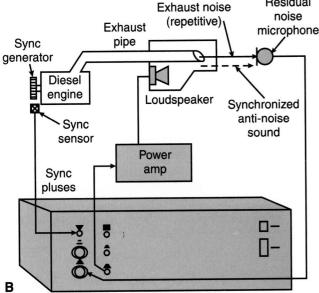

Figure 13-12. *Operation of an electronic muffler. A—The production of a mirror image, out-of-phase waveform will cancel out sound. B—Schematic shows how the sensors, microphones, and speakers in an electronic muffler are combined to cancel noise.*

pick up the pattern of the pressure sound waves emitted by the exhaust pipe and send this information to the vehicle's on-board computer for analysis. A mirror-image pattern of pulses is instantly created by the computer and sent to speakers mounted near the exhaust outlet. The speakers create out-of-phase waveforms that cancel out the noise.

Marine Mufflers

Since raw water is readily available in marine applications, most modern marine engines use a wet muffler. A cutaway view of a wet muffler is shown in **Figure 13-13.** *Wet mufflers* are usually constructed of cast iron or sheet iron and contain a system of internal baffles that break up the exhaust gas pulsation. This produces a silencing effect without back pressure. The water used in the wet muffler also aids in reducing noise. The water cools the exhaust

gases, causing them to contract. This decrease in volume reduces the velocity of the exhaust gases and, therefore, reduces the exhaust noise. The water itself also absorbs some of the sound.

Some marine engines are equipped with *dry mufflers*. In both wet and dry marine mufflers, circulating water reduces the temperature of the exhaust gases. The difference between the two is that the exhaust gases in a dry muffler do not contact the cooling water and the water in a dry muffler does not flow through the muffler compartment. Instead, it flows through a jacket around the muffler. In wet mufflers, the gases are expanded into the muffler and directly contact the water. In passing through the baffles and the water, the gases are cooled, condensed, and effectively silenced.

Water Removal

Water accumulation in the exhaust system may result in rusted pipes and fittings, early engine component failure, and excessive noise caused by internal muffler corrosion. Vertically installed mufflers used on mobile installations are often equipped with drain holes, which prevent moisture from entering the engine. A hinged rain cap is often attached to the top of each stack to prevent water from rain, snow, or washing from entering the system. The cap opens only when the engine is running and closes when the engine is shut off. A simple alternative to the rain cap is a curved outlet tip, which is standard on most new trucks.

Mufflers Used on Turbocharged Engines

The internal design of most mufflers varies depending on engine and exhaust system design. On turbocharged engines, the turbine blades are located between the engine and the muffler. As the blades are turned by the exhaust gas, they tend to chop up the sound waves. For this reason, mufflers for turbocharged diesels are simpler and less restrictive than those for naturally aspirated diesels. Mufflers designed for dual applications may differ in baffling and flow capability from mufflers designed for a specific application.

Muffler as a Spark Arrester

In addition to acting as silencers, most mufflers also function as spark arresters—an important safety factor when operating mobile and stationary diesel equipment in fire-prone areas. In some dry mufflers, a device is incorporated to trap burning carbon particles and soot, **Figure 13-14.** In wet mufflers, the water serves as a spark arrester. As with any exhaust system part, spark arresters must be matched to the displacement and horsepower of the engine.

The relatively low exhaust gas temperatures found in newer engines allow more moisture to remain in the exhaust system. This moisture and the acids that result from fuel sulfur may contribute to internal corrosion. To combat this internal corrosion, many of today's mufflers have internal components made of stainless steel—even when the shell is a more common material, such as aluminized steel. Polished stainless steel shells are available and give the best overall life. Chromed steel shells are not as desirable from the standpoint of appearance and longevity— one drawback being their tendency to turn blue when subjected to excessive heat.

Resonator

On some mobile vehicles, there is an additional muffler, known as a **resonator** or silencer, to further reduce the exhaust decibel level. When used in the exhaust system, it may be located before or after the muffler. The resonator looks like a muffler and is constructed like a straight-through muffler. It is connected to the muffler by an intermediate exhaust pipe.

Industrial Silencers

On naturally aspirated engines, a silencer will be required to reduce exhaust noise and a spark arrester is recommended. By placing the silencer or muffler as close to the engine as possible, interferences resulting from pressure waves are reduced, **Figure 13-15.** While large silencers are usually covered with insulation, **Figure 13-16,** they must be fitted with some means of draining condensation and cleaning trapped soot from the bottom of the unit. Clean silencers operate with less restriction. More importantly, if not drained away, condensed water could react with the products of combustion, forming acids which would quickly corrode system components. To reduce heat transfer within the engine room, both the silencers and piping are insulated. The insulation also can help to reduce noise. Silencers may not be required on some turbocharged engines due to the silencing effect of the turbocharger. The same is true for spark arresters, but local regulations may dictate the use of one of the above in all applications.

To dispose of heat radiated by exhaust pipes, sheet metal ducts are sometimes provided for the exhaust pipe with an inch or two spacing between the duct and pipe. This will provide a chimney effect and help carry away a portion of the heat. Water-cooled exhaust manifolds and pipes are used particularly in the case of marine installations. Water-cooled exhaust systems are used in some

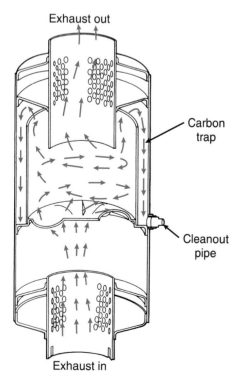

Figure 13-14. Exhaust flow through a spark arrester.

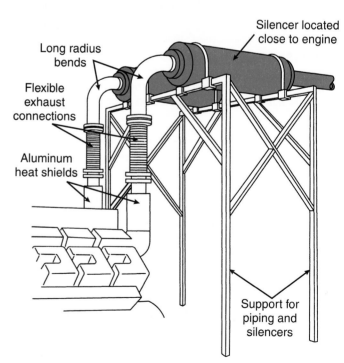

Figure 13-15. Location of silencer in an industrial application. (Waukesha Engine Division)

Figure 13-16. *Silencers and exhaust pipe covered with insulation to reduce heat. (Advanced Thermal Products)*

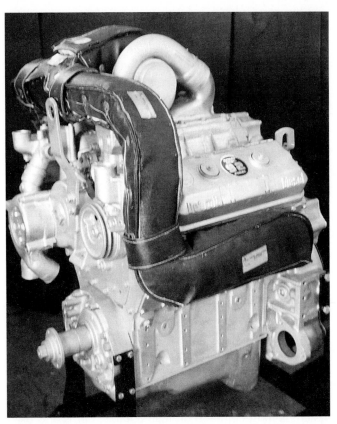

Figure 13-17. *Insulation kit applied to a mobile diesel exhaust system. (Advanced Thermal Products)*

industrial applications. In such cases, coolant in the closed system is recirculated.

Heat can also be a problem in mobile construction equipment. The turbo and exhaust system on a diesel engine operates at extremely high temperatures in excess of 1000°F (538°C). When running hard, pulling grades, or towing a vehicle, these temperatures increase to where the exhaust may actually glow red hot. A thermal insulation blanket kit such as shown in **Figure 13-17,** will contain this heat within the system, forcing it to exit the tailpipe or outlet pipe.

There are some instances when the exhaust system's heat can be put to use. The silencer used in stationary diesels, in addition to operating as a noise reducer, may

also operate a heat recovery unit to provide heat for other plant operations, **Figure 13-18**. After exhaust gas enters the heat recovery silencer, it travels through a perforated tube and two silencing chambers. It then reverses its path and goes through the same silencing chamber for maximum noise reduction. It also transfers substantial heat energy to the inner body. This core section is a critical part of improving the heat transfer capability.

In V-engines used in industrial and marine applications, separate exhaust systems are usually provided for each bank

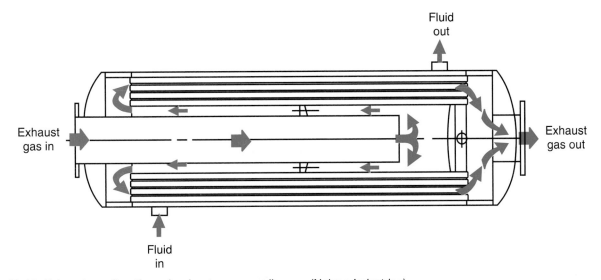

Figure 13-18. *Exhaust gas flow through a heat recovery silencer. (Nelson Industries)*

of cylinders. Two manifold branches should not be brought together in a single pipe with a T-connection, as this leads to pulses of exhaust from one bank interfering with exhaust from the other. Extremely high back pressure may develop.

Exhaust Piping

Total system back pressure is greater than that of the muffler alone. In addition to muffler restriction, every bend and joint in the exhaust system adds resistance to exhaust flow. While truck builders would ideally prefer to install short, straight pipes with a minimum of bends and clamps, most truck exhaust systems must accommodate frame rails, battery boxes, air dryers, fuel tanks, cab mounts, and dozens of other components. The resulting pipe routing is almost always a compromise between what the engine ideally needs and what will fit on the truck. The exhaust systems of mobile diesel applications can be grouped into two main types.

Vertical Muffler

The *vertical muffler* system takes the exhaust gases from the exhaust manifold turbocharger outlet, pipes them to the rear of the cab, then upward through a single muffler and stack into the atmosphere, **Figure 13-19A.** This system is most common on road tractors, but is also used on straight trucks. It directs fumes up and out and allows easy access to the muffler and stack.

Vertical stacks might be pleasing to the eye, but they cost more than underframe systems and are more vulnerable to damage from jackknifing and overhead obstacles, such as tree branches. They can present clearance problems with close-coupled trailers or truck bodies that are snug to the cab. When attached to the cab instead of to a separate vertical support, vertical stacks can transmit noise and vibration to the driver's compartment. As newer diesel engines breathe cleaner to meet federal emissions regulations, the need to get the exhaust outlet high up into the air is becoming less important.

Horizontal Muffler

Horizontal mufflers with short tailpipes (tailspout) that are aimed downward are shorter, simpler, and lighter than vertical mufflers and stacks, **Figure 13-19B.** Though it is not common practice, many heavy duty trucks may be equipped with horizontal systems in the near future. Drawbacks to horizontal systems include higher restriction from mufflers that must be short and compact, and increased noise from sound waves reflected off the pavement. Also, thermal shock from road splash tends to shorten the life of mufflers mounted close to the road. Because the outlet pipe generally bends toward the ground, the strong exhaust flow can kick up dust from dirt or dry gravel surfaces. This can be especially bothersome for vehicles regularly running on such surfaces.

A variation of the horizontal muffler is hung on the frame but the outlet pipe winds upward along the cab's corner or side, **Figure 13-19C.** The muffler's outlet can be

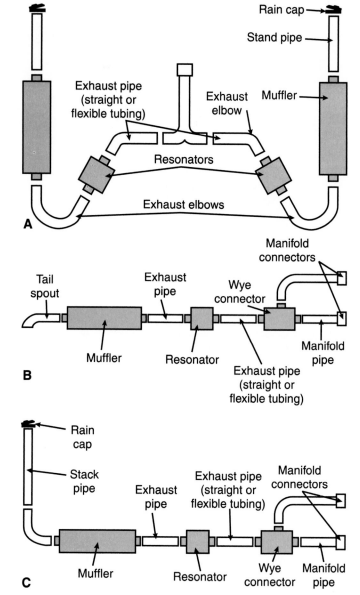

Figure 13-19. *Three popular exhaust system layouts used in mobile diesel engine installations (Nelson Industries)*

at the end opposite the inlet, or on the same end—the so-called "reverse-flow" muffler described in this chapter. The latter is used in extreme cases of cab-body restrictions, as on some vans or dump trucks, **Figure 13-20.** One advantage of this configuration is the reduction of bulk along the cab's corner. A 2" or 3" pipe is less of a problem than a 10" vertical muffler. Hanging the muffler on the frame also does away with the stout brackets needed to support the vertical muffler, reducing some of the vehicle's weight.

Disadvantages include exposing the muffler to road contaminants and the possibility of increased back pressure. Mobile diesel vehicles may have either a single or double exhaust stack, **Figure 13-21.** A dual is more symmetrical and traditional appearance is its aesthetic advantage. Since they are plumbed from a single pipe that extends backward from the turbo, dual mufflers and stacks can be more cosmetic than functional.

Figure 13-20. *Exhaust systems often must be routed around various obstacles.*

A

B

Figure 13-21. *Mobile applications use one of two vertical exhaust applications A—Single vertical stack. B—Twin (double) vertical stack.*

They will also reduce back pressure if set up correctly. If the muffler and piping added from the "Y" split is the same size as the single original, back pressure can be cut by one-fourth which means fuel saving. Dual stacks become more viable with an increase in engine size. The bigger the engine, the higher the flow of gasses. For smaller engines, a more cost effective solution might be to use a single low-restriction exhaust. In the noise-versus-back pressure tradeoff, dual stacks/mufflers give the best silencing for a given level of restriction.

Considerations when Installing an Exhaust System

In some cases, choosing a high outlet for diesel exhaust is a matter of obeying local regulations that require them. Whether vertical stacks are used for appearance, low restriction, or to comply with local law, there is another consideration when designing exhaust system piping: muffler orientation. Where possible, vertical stacks should be given vertical mufflers. Nevertheless, a horizontal muffler with a vertical outlet pipe is usually chosen for packaging reasons on a straight body truck because limited space in the tractor design prevents the use of a straight-through muffler design. When given room to stretch out vertically, mufflers can do their job of silencing with less restriction than horizontal mufflers of similar sound ratings.

Regardless of the exhaust system configuration in mobile installation, the exhaust outlet should always be located so that it prevents the exhaust gases from circulating behind the truck cab. Vertical stacks need to be high or outboard enough to ensure that the gases will be carried away from the truck by the airflow. It is important to keep the air intake from "rebreathing" exhaust soot, which can plug the air filter in a very short period of time. Horizontal tailpipes should expel exhaust gases behind the rear axle or between the axles on tandem rear ends. Exhaust flow should be aimed downward and away from under-vehicle components.

Selecting Exhaust Pipes and Related Parts

In both industrial and mobile applications, care must be taken when selecting the exhaust pipe to be certain that it is the recommended diameter, that there are no sharp bends, and that it is adequately supported. It is also important to provide independent support for the exhaust system. No strain should be imposed on the engine exhaust manifolds. This is especially important on a turbocharged engine. Stress on a turbocharger could distort the housing, leading to turbocharger failure, **Figure 13-22.**

Figure 13-22. Typical turbocharger mounting on a mobile diesel engine. (Penn Diesel)

Cast iron, wrought iron, or steel exhaust pipes are preferred for industrial applications because they can stand up to hot and cold cycling. However, these materials are generally considered too heavy for mobile diesel applications. Stainless, galvanized, and aluminized steels are preferred in mobile applications.

Figure 13-23 shows a typical method of supporting a vertical muffler, vertical outlet configurations on a mobile

vehicle. Hangers, brackets, and clamps must be sturdy enough to withstand expansion forces and severe vibration and shock loads.

In industrial installations, the design of the hangers must allow for thermal expansion of the pipe (10' of pipe will lengthen almost one inch when hot). Heavier exhaust system components, such as silencers, should be independently supported. The design of the installation must allow for expansion on both sides of the silencer. The exhaust system must be supported from below when overhead cranes or booms will be used in the engine room. Flexible (flex) connections capable of absorbing vibration, shock, and expansion forces should be installed when needed in the exhaust system. These connections must be large enough to isolate engine vibration, and to compensate for thermal expansion. If the connector is not large enough, thermal expansion would compress the connection, resulting in damage to the engine manifolds and turbochargers, and in excess exhaust system vibration.

The piping of a marine diesel engine system, like those of a muffler, can be either wet or dry. Wet exhaust systems are characterized by the following:

❑ Generally the exhaust gases are mixed with the water which is discharged from the water side of the engine's heat exchanger.

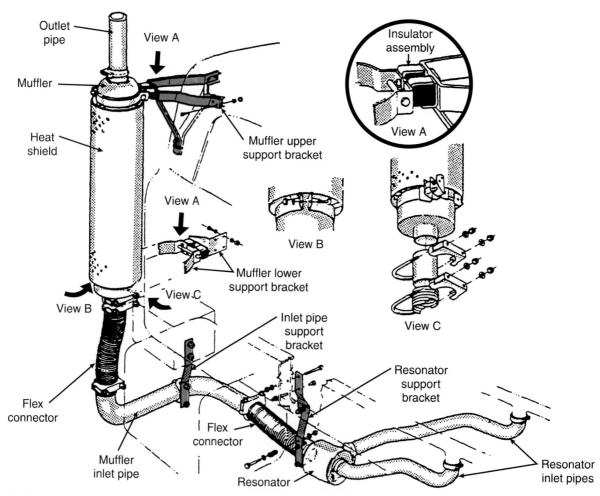

Figure 13-23. Method of supporting a vertical exhaust stack on a mobile application.

❑ Particulate and condensable/soluble gaseous emissions from the exhaust system are effectively scrubbed from the exhaust gases, reducing the possibility of atmospheric pollution.

❑ Exhaust piping which is cool enough to be made of uninsulated fiberglass reinforced plastic (FRP) or rubber.

❑ The moisture of exhaust gases and water is discharged from the boat at or slightly below the vessel's waterline.

❑ With the relatively small elevation difference between the engine's exhaust discharge elbow and the vessel's waterline, it is difficult to design a system which will always prevent water from entering the engine through the exhaust system.

While a number of proprietary exhaust components are available to help avoid this problem, the most common generic methods are exhaust risers and water lift mufflers.

Preventing Water Intrusion in Marine Applications

One way to minimize the possibility of water entering the engine from backflow in the wet exhaust system is to have a steep downward slope on the exhaust piping, downstream of the engine. *Exhaust risers* are pipes which elevate the exhaust gases, allowing a steeper slope in the downstream piping. The risers must be insulated or water-jacketed to protect personnel in the engine compartment from the high temperature exhaust gas. The water is not injected into the exhaust gases until downstream of the top of the riser, so the upward-sloping portion is dangerously hot if not insulated or water-jacketed. Another way to minimize the possibility of water entering the engine from backflow in the wet exhaust system is by using a wet muffler, described earlier in this chapter.

The traditional method of preventing water from entering an idle engine is to locate the engine far enough above the water line that breaking waves do not reach the height of the exhaust elbow, **Figure 13-24**. While the relative elevation of the engine to the water line is fixed and unchangeable, it is possible for the exhaust system to protect the engine from ingesting water. Features of such an exhaust system include the following:

❑ Sufficient elevation difference between the water line and the highest point in the exhaust piping to prevent even small amounts of water from reaching the engine.

❑ Some method of dissipating the kinetic energy of the waves as they enter the exhaust piping. The more effective the method of wave energy dissipation, the lower the elevation difference required.

A valve located where the exhaust piping penetrates the hull can keep waves from entering the exhaust piping when the engine is not running. The valve mechanism should not include any components which rely on sliding contact to maintain flexibility. This type of action has proven troublesome in an environment of salt water and exhaust gas. A flexible strip of one of the chemically inert plastics can provide hinge action.

In dry marine exhaust systems appropriate drain connections, rain caps, or other means must be provided to protect the engine from rainwater or sea spray entering the engine through the dry exhaust piping, **Figure 13-25**. Long runs of exhaust piping require traps to drain moisture. Traps should be installed at the lowest point of the line near the exhaust outlet to prevent rain water from reaching the engine. Slope exhaust lines from the engine and silencer to the trap, so condensation will drain. Traps may be built by inserting a vertical pipe, with a drain petcock, down from a tee section in the line.

The last few feet of the exhaust pipe discharge should be sloped to prevent rain water or spray from entering the pipe. Alternatively, some form of rain cap should be fitted to a vertical exhaust pipe section. As with the wet exhaust system, a dry system's exhaust pipe must be isolated from the engine with flexible connections. The dry mufflers and other large dry exhaust system components would be best mounted outside the engine compartment.

Exhaust System Service

Exhaust system components are subject to both physical and chemical damage. Any physical damage to an exhaust system part that partially restricts or blocks the exhaust system usually results in a loss of power and poor fuel economy. In addition to improper engine operation, a blocked or restricted exhaust system causes increased vehicle noise and air pollution. Leaks in the exhaust system caused by either physical or chemical (corrosion) damage may result in illness, asphyxiation, or even death. Remember that vehicle exhaust fumes are very dangerous.

Most exhaust system components, especially the exhaust pipe, muffler, and outlet pipe, are subject to rusting and cracking. These parts are also susceptible to other types of damage, particularly on trucks. Clamps can break or work loose, permitting parts to separate. Components may fail due to the impact that occurs when the vehicle hits ruts and potholes in the road. Parts may also be damaged by rocks thrown up by the wheels.

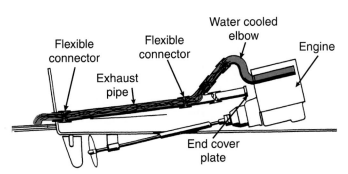

Figure 13-24. *Marine wet exhaust systems use water that actually contacts the exhaust gases. (Caterpillar Trucking Co.)*

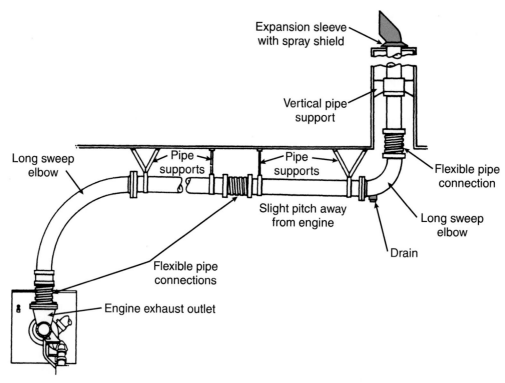

Figure 13-25. Marine dry exhaust systems also use water, but the gases do not make contact. (Caterpillar Trucking Co.)

> ⚠️ **Warning: On all inspection and repair work, be sure to wear appropriate eye protection. Also, remember that exhaust system components get very hot.**

Before making a visual inspection, listen closely for any hissing or rumbling sound that may indicate the beginning of exhaust system failure. With the engine idling, slowly move along the entire system and listen for leaks. It is normally not difficult to locate the source of a leak in this way.

Noise Levels

In the past years, federal, state and local governments have adopted noise control laws that affect on-highway vehicles. Enforcement of these laws take place through certification of newly manufactured trucks, through vehicle registration controls, or by actual roadside noise monitoring stations where citations may be issued.

Federal regulations require total decibel readings not to exceed 80 decibels (dB) at 50'. Engineers try to reduce the engine exhaust's contribution to less than 75 dB (a measure of sound intensity). The turbocharger cuts about 6 dB and the muffler 10-20 dB. The measure of a decibel's effects are progressive. A 3 dB drop equals about one-half the sound power, but is only the threshold of what the average person can perceive as "quieter." It takes a 10 dB reduction for that person to perceive that something is "twice as quiet."

There are five major sources that contribute to the noise level of a highway truck:

❑ Excessive fan speed is one of the biggest noise source in this area. This can be controlled by the use of a temperature controlled or thermostatic fan. Other noise concerns are bent fan blades and broken or missing shrouds.

❑ Both the type of air intake system and the location of the inlet affect the truck noise level. Air inlets that open to the side are generally noisier than those that do not. Intake air silencers, described in **Chapter 12,** are available.

❑ The engine and drive train are a major source of noise. Operating speed, type of engine, and drive train all affect the total noise level. Shielding the exposed areas with acoustic barriers will reduce noise levels. Generally, any part of the engine or drive train that can be seen while standing away from the truck will contribute toward the total vehicle noise level.

❑ Tires are often a major noise source at high speeds. The condition of the tires and the tread pattern affect the noise level. At low speeds, this is a less noticeable noise source.

❑ Worn or inadequate exhaust systems are the greatest and most frequent contributor to high noise levels. A visual inspection and repair of leaking connections or replacement of failed components will noticeably reduce noise levels. If the noise level is still too high, the addition of resonators, or a change in muffler is recommended.

Reducing Exhaust Noise

To control engine exhaust noise without a corresponding decrease in performance or fuel economy is important. Here are several specific ways to reduce exhaust noise levels on mobile vehicles.

Worn or leaking flexible tubing is a common noise source. Replace the part if necessary. The service life of flexible tubing can be improved if it is installed in a relaxed position. Bending, stretching, and compression all reduce life because they limit the tubing's ability to "flex." The entire exhaust system should be well supported. This will reduce the noise generated by the exhaust pipes and the muffler shell. Care must be taken to isolate engine vibration from the exhaust system and to provide for expansion when the system is hot.

The addition of a universal resonator, silenced "wye" or splitter muffler in the system will generally reduce the exhaust noise level from 3 to 6 dB. Both items have a minimal effect on engine back pressure. Change from a single to a dual system. For maximum benefit, when changing from a single to a dual system, special "dual only" mufflers are recommended. Lower system back pressure is an additional benefit of dual systems.

On horizontal systems, the tail spouts should be pointed toward the center of the roadway. This helps reduce the noise reflected off the road surface toward curbside observers. On vertical systems, straight stacks will yield lower sound levels than curved stacks. Straight stacks direct the noise upward, while curved stacks direct more noise toward the observer.

Turbochargers

Turbochargers can be used on both two- and four-cycle diesel engines. They utilize exhaust energy, which is normally wasted, to drive a turbine-powered centrifugal air compressor that converts air velocity into air pressure to increase the flow of air to the engine cylinders. The air drawn into a naturally-aspirated engine cylinder is at less than atmospheric pressure. A turbocharger packs the air into the cylinder at more than atmospheric pressure. A turbocharger improves combustion, resulting in decreased exhaust emissions, smoke, and noise. Increased power output and the ability of the turbocharger to maintain nearly constant power at high altitudes are also benefits of a turbocharged engine.

Fuel economy is another very significant reason for using a turbocharger with a diesel engine. Today, conserving fuel has become nearly as important as controlling air pollution. The extra air provided by the turbocharger allows increased horsepower output without increasing fuel consumption. Lack of air is one factor limiting the engine horsepower of naturally aspirated engines. As engine speed increases, the length of time the intake valves are open decreases, giving the air less time to fill the cylinders and lowering volumetric efficiency.

Turbocharger Components

The turbocharger normally consists of three components, **Figure 13-26.** There is a radial inward flow turbine wheel and shaft, a centrifugal compressor wheel, and a center housing, which supports the rotating assembly, bearings, seals, turbine housing, and compressor housing. The center housing has connections for oil inlet and oil outlet fittings.

The **turbine wheel** (hot wheel) is located in the turbine housing and is mounted on one end of the turbine shaft. The **compressor wheel,** or impeller, (cold wheel) is located in the compressor housing and is mounted on the opposite end of the turbine shaft to form an integral rotating assembly.

Other parts of the rotating assembly include the thrust bearing (or spacer), backplate, and wheel retaining nut. The rotating assembly is supported on two pressure lubricated bearings that are retained in the center housing by snap rings. Internal oil passages are drilled in the center housing to provide lubrication to the turbine shaft bearings, thrust washer, thrust collar, and thrust spacer, **Figure 13-27.**

The **turbine housing** is a heat-resistant alloy casting that encloses the turbine wheel and provides a flanged exhaust gas inlet and an axially-located turbocharger exhaust gas outlet. The compressor housing, which encloses the compressor wheel, provides an ambient air inlet and a compressed air outlet. In a typical installation, the turbocharger is located to one side, usually close to the exhaust manifold. An exhaust pipe runs between the engine exhaust manifold and the turbine housing to carry the exhaust gases to the turbine wheel. Another pipe, called the **crossover tube,** conducts fresh compressed air to the intake manifold, **Figure 13-28.**

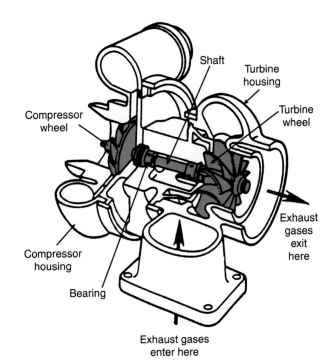

Figure 13-26. The components of a turbocharger. Turbochargers are not normally field-serviced. (Garrett Group)

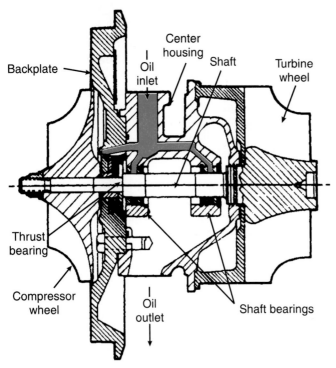

Figure 13-27. *Oil flow through a turbocharger's center bearing section. (Detroit Diesel Corp.)*

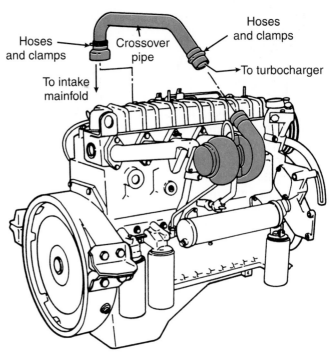

Figure 13-28. *The location of the crossover pipe or tube. (Mack Truck)*

The flow of exhaust gas from each cylinder occurs intermittently as the exhaust valves open. This results in fluctuating gas pressures (pulse energy) at the turbine inlet. With a conventional turbine housing, only a small amount of this pulse energy is available for use.

To better utilize these pulses, some designs have an internal division in the turbine housing and an exhaust manifold that directs the exhaust gases to the turbine wheel, **Figure 13-29.** There is a separate passage for each half of the engine cylinder exhaust. Using a fully divided exhaust system, plus a dual scroll turbine housing, results in a highly effective air output velocity. This produces higher turbine speeds and manifold pressures than can be obtained with an undivided exhaust system turbo.

Turbocharger Operation

After the diesel engine is started, the exhaust gases flowing from the engine and through the turbine housing cause the turbine wheel and shaft to rotate. The gases are discharged into the atmosphere after passing through the

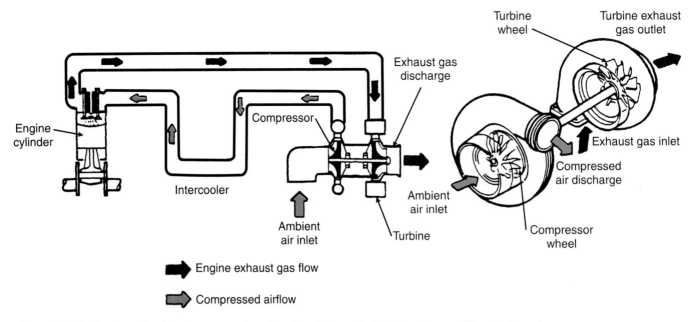

Figure 13-29. *The flow of exhaust gases and combustion air through the turbocharger. (Garrett Group)*

turbine housing. As mentioned, the compressor wheel, which is mounted on the opposite end of the turbine wheel shaft, rotates with the turbine wheel. The compressor wheel draws in fresh air, compresses it, and delivers high pressure air through the engine blower to the engine cylinders.

During operation, the turbocharger responds to the engine load demands by reacting to the flow of the exhaust gases. As the engine power output increases or decreases, the turbocharger delivers the required amount of air under all conditions. Under normal atmospheric conditions, air is drawn into a naturally aspirated engine at a maximum of 14.7 psi at sea level. Turbo boost is the term used to describe the positive pressure increase created by a turbocharger. For example, 10 psi of boost means the air is being fed into the engine at 24.7 psi (14.7 psi of atmospheric pressure plus 10 pounds of boost).

Turbocharger Advantages

The turbocharger offers a distinct advantage to a diesel engine operating at high altitudes. The turbocharger automatically compensates for the normal loss of air density and power as the altitude increases. With a naturally aspirated engine, horsepower drops off 3% per 1000 feet because of the 3% decrease in air density per 1000 feet. If fuel delivery is not reduced, smoke level and fuel dilution will increase with altitude.

With a turbocharged engine, an increase in altitude also increases the pressure drop across the turbine. Inlet turbine pressure remains the same, but the outlet pressure decreases as the altitude increases. Turbine speed also increases as the pressure differential increases. The compressor wheel turns faster, providing approximately the same inlet manifold pressure as at sea level, even though the incoming air is less dense. However, there are limitations to the amount of altitude compensation a turbocharged engine can produce. This is primarily determined by the amount of turbocharger boost and the turbocharger-to-engine match.

Turbocharger Lubrication

Lubricating oil for the turbocharger is supplied under pressure through an external oil line extending from the cylinder block to the top of the center housing, **Figure 13-30.** From the oil inlet in the center housing, the oil flows through drilled oil passages in the housing to the shaft bearings, thrust ring, thrust bearing, and backplate or thrust plate. Gravity forces the oil to the engine oil pan through an external oil line extending from the bottom of the turbocharger's center housing to the cylinder block.

When the turbocharger is operating, the exhaust pressure behind the turbine wheel and the air pressure behind the compressor wheel are greater than the pressure inside the bearing housing. To prevent pressurized exhaust gas and air from entering the bearing housing, sealing rings are used. These rings fit snugly in the bearing housing and do not rotate with the shaft. A heat shield and, in some cases,

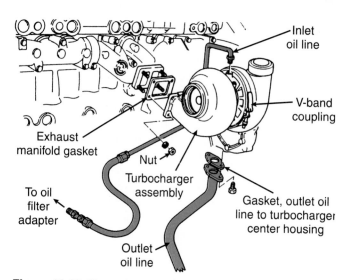

Figure 13-30. *The turbocharger uses engine oil for lubrication, usually supplied by one or more lines that thread into the block. (Garrett Group)*

an insulating pad are used on the turbine end of the bearing housing to prevent the turbine exhaust gas temperature from transferring into the bearing housing.

Turbocharger Controls

While there are several types of turbocharger controls, they fall into two general groups: those that limit turbocharger speed and those that limit compressor outlet pressure, or boost. Controls that limit turbocharger speed keep the turbocharger from destroying itself. Those that limit boost keep the turbocharger from damaging the engine. Since the typical modern turbochargers can produce more pressure than any engine can use, most controls are designed to limit the amount of boost.

One of the most common methods of limiting turbocharger boost and speed is with a **waste gate,** or turbine-bypass valve, **Figure 13-31.** It can be either a butterfly valve or a diaphragm. The valve may be operated manu-

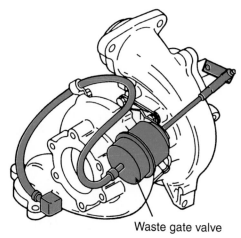

Waste gate valve

Figure 13-31. *The waste gate valve is used to control the speed of the turbocharger. (Garrett Group, Inc.)*

ally, by intake manifold pressure, or by a servo motor. If a servo motor is used, it can be controlled manually or by a device that senses turbocharger speed, gauge pressure, absolute pressure, density, or air flow. Turbocharger speed can also be sensed and controlled electronically.

Regardless of the type of valve used, the method of operation is the same. When the valve is closed, the exhaust gases cannot get through the bypass. All the exhaust gases flow over the exhaust turbine. The turbocharger turns as fast as the exhaust gases will drive it. The turbocharger speed and pressure, however, are limited. When the waste gate is opened, some exhaust gas is routed through the bypass instead of over the turbine. This slows the turbine and lowers the compressor output pressure.

Scavenging Pump and the Turbocharger

A turbocharger may be used in addition to the regular scavenging system in some engines. In these cases, air is drawn into the scavenging pump or blower and is then compressed to scavenging pressure. It then passes the exhaust to the turbocharger, where it is raised to supercharge air pressure. At light loads when there is little energy available to drive the turbocharger, the mechanically-driven blower alone puts the scavenging air into the cylinders. At increased loads, the turbocharger speeds up and takes in so much air that its inlet pressure drops to atmospheric level, causing a check valve in the blower to open. When this occurs, the blower becomes unloaded (saving engine power) and the turbocharger alone provides scavenging and supercharging. **Figure 13-32** shows a two-stroke engine having both a turbocharger and scavenging blower.

Under the most favorable conditions, the engine's starting air contains enough energy to start the tur-

bocharger. However, the turbocharger frequently requires some method for supplying additional scavenging air when the engine is being started and when it is running at low speeds. There are two common methods used to supply this air: the mechanical method and the jet air method.

In the *mechanical starting* method, a blower is required in addition to the turbocharger. Both units are driven from the crankshaft during starting and at slow speeds. When engine speed increases, a drive coupling disconnects and the turbocharger operates on exhaust gas only. The *jet air starting* method is used only during starting. Air is blown through jets into the turbocharger compressor. Air passing through the compressor travels into the engine to assist in scavenging.

Two- or Four-Stage Turbocharging

To increase both the torque range and the mean effective pressure to an even greater degree, some engines may employ either two or four turbochargers and charge-air coolers (one for each exhaust manifold, or use two turbochargers in series and a charge-air cooler), **Figure 13-33**, or use four turbochargers in series and two charge-air coolers. The series of turbochargers operates at lower-than-normal pressure ratios and reduced turbine speeds.

Turbocharger Inspection and Troubleshooting

Figure 13-34 lists common turbocharger problems and probable causes. If a defective turbocharger is suspected, always make sure the turbocharger is really at fault. Repairs are sometimes performed on the turbocharger when the real source of the trouble is a restricted air cleaner, a plugged crankcase breather, or deteriorated oil lines.

Common symptoms that may indicate turbocharger trouble include:

- ❑ A lack of engine power
- ❑ Black smoke
- ❑ Blue smoke
- ❑ Excessive engine oil consumption
- ❑ Noisy turbocharger operation.

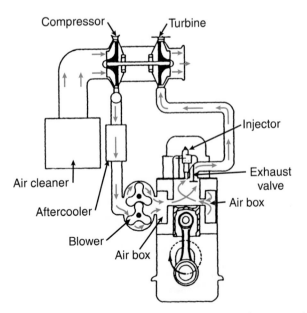

Figure 13-32. This engine contains both a turbocharger and a blower.

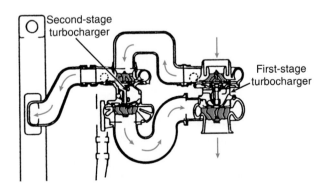

Figure 13-33. Two-stage turbocharging utilizes two turbochargers. (Garrett Group, Inc.)

A lack of engine power and black smoke can both result from insufficient air reaching the engine and can be caused by restrictions in the air intake or air leaks in the exhaust system or the induction system. The first step in troubleshooting any turbocharger is to start the engine and listen to the sound the turbocharging system makes. As a techni-

TURBOCHARGER TROUBLESHOOTING GUIDE

Condition	Possible Causes Code Numbers	Cause Description by Code Numbers
Engine lacks power	1, 4, 5, 6, 7, 8, 9, 10, 11, 18, 20, 21, 22, 25, 26, 27, 28, 29, 30, 37, 38, 39, 40, 41, 42, 43	1. Dirty air cleaner element 2. Plugged crankcase breathers 3. Air cleaner element missing, leaking, not sealing correctly; loose connections to turbocharger 4. Collapsed or restricted air tube before turbocharger
Black smoke	1, 4, 5, 6, 7, 8, 9, 10, 11, 18, 20, 21, 22, 25, 26, 27, 28, 29, 30, 37, 38, 39, 40, 41, 43	5. Restricted damaged crossover pipe, turbocharger to inlet manifold 6. Foreign object between air cleaner and turbocharger 7. Foreign object in exhaust system (from engine, check engine) 8. Turbocharger flanges, clamps, or bolts loose 9. Inlet manifold cracked; gaskets loose or missing; connections loose
Blue smoke	1, 2, 4, 6, 8, 9, 17, 19, 20, 21, 22, 32, 33, 34, 37, 45	10. Exhaust manifold cracked, burned; gaskets loose, blown, or missing 11. Restricted exhaust system
Excessive oil consumption	2, 8, 15, 17, 19, 20, 29 30, 31, 33, 34, 37, 45	12. Oil lag (oil delay to turbocharger at startup) 13. Insufficient lubrication 14. Lubricating oil contaminated with dirt or other material 15. Improper type lubricating oil used
Excessive oil– turbine end	2, 7, 8, 17, 19, 20, 22, 29, 30, 32, 33, 34, 45	16. Restricted oil feed line 17. Restricted oil drain line 18. Turbine housing damaged or restricted 19. Turbocharger seal leakage
Excessive oil– compressor end	1, 2, 4, 5, 6, 8, 19, 20, 21, 29, 30,33, 34, 45	20. Worn journal bearings 21. Excessive dirt buildup in compressor housing 22. Excessive carbon buildup behind turbine wheel
Insufficient lubrication	8, 12, 14, 15, 16, 23, 24, 31, 34, 35, 36, 44, 46	23. Too fast acceleration at initial start (oil lag) 24. Too little warmup time 25. Fuel pump malfunction
Oil in exhaust manifold	2, 7, 17, 18, 19, 20, 22, 29, 30, 33, 34, 45	26. Worn or damaged injectors 27. Valve timing 28. Burned valves
Damaged compressor wheel	3, 4, 6, 8, 12, 15, 16, 20, 21, 23, 24, 31, 34, 35, 36, 44, 46	29. Worn piston rings 30. Burned pistons 31. Leaking oil feed line 32. Excessive engine pre-oil
Damaged turbine wheel	7, 8, 12, 13, 14, 15, 16 18, 20, 22, 23, 24, 25, 28, 30, 31, 34, 35, 36, 44, 46	33. Excessive engine idle 34. Coked or sludged center housing 35. Oil pump malfunction 36. Oil filter plugged
Drag or bind in rotating assembly	3, 6, 7, 8, 12, 13, 14, 15, 16, 18, 20, 21, 22, 23, 24, 31, 34, 35, 36, 44, 46	37. Oil-bath-type air cleaner: • Air inlet screen restricted • Oil pullover • Dirty air cleaner • Oil viscosity low • Oil viscosity high
Worn bearings, journals, bearing bores	6, 7, 8, 12, 13, 14, 15, 16, 23, 24, 31, 35, 36, 44, 46	38. Actuator damaged or defective 39. Waste gate binding 40. Electronic control module or connector(s) defective 41. Waste gate actuator solenoid or connector defective
Noisy	1, 3, 4, 5, 6, 7, 8, 9, 10, 11, 12, 13, 14, 15, 16, 18, 20, 21, 22, 23, 24, 31, 34, 35, 36, 37, 44, 46	42. EGR valve defective 43. Alternator voltage incorrect 44. Engine shut off without adequate cool down time 45. Leaking valve guide seals
Sludged or coked center housing	2, 11, 13, 14, 15, 17, 18, 24, 31, 35, 36, 44, 46	46. Low oil level

Figure 13-34. Turbocharger troubleshooting guide. This guide is very useful in diagnosing many turbocharger performance related problems. (Garrett Group, Inc.)

cian becomes more familiar with this characteristic sound, he or she will be able to identify an air leak between an engine and a turbocharger by a higher pitched sound. If the turbo sound cycles, or changes in intensity, a plugged air cleaner, loose material in the compressor inlet ducts, or dirt buildup on the compressor wheel and housing may be the cause.

After listening to the system, check the air cleaner for a dirty element. If in doubt, measure for restrictions according to the engine manufacturer's recommendations. The next step is to stop the engine and remove the ducting between the air cleaner and the turbocharger in order to look for dirt buildup or damage. Also, check for loose clamps on the compressor outlet connections and inspect the engine intake system for loose bolts, leaking gaskets, etc. Disconnect the exhaust pipe and look for restrictions or loose material. Examine the engine exhaust system for cracks, loose nuts, or blown gaskets.

Checking the Turbocharger

Rotate the turbo shaft assembly. It should rotate freely without signs of rubbing or wheel impact damage. There is normally radial shaft play in the turbo shaft assembly. However, if either of the wheels touch the housing when the shaft is rotated by hand, excessive play is indicated. If none of these symptoms are present, the low power complaint is probably not being caused by the turbocharger. Consult engine manufacturer's troubleshooting procedures in this case.

Checking for Oil Consumption

Blue smoke is an indication of oil consumption and can be caused by either turbocharger seal leakage or internal engine problems. First, check the air cleaner for restrictions per the engine manufacturer's instructions. Excessive air cleaner restriction can cause compressor oil seal leakage. Next, stop the engine, remove the turbo ducts, and check the shaft assembly for free rotation, rubbing, and damage. Inspect the oil drain line for restrictions or damage, which can cause seal flooding and leakage. Also, check for high crankcase pressure. If in doubt, measure crankcase pressure, which must be within engine manufacturer's specifications. Finally, loosen the exhaust manifold duct and check for oil in the engine exhaust. If oil is present, refer to the engine manual for appropriate repair procedures.

If there is excessive oil consumption without smoke, check the air cleaner for restrictions. Inspect the compressor discharge duct for loose connections, check the crankcase pressure, and examine the turbo shaft assembly for free rotation. Also, look for evidence of the wheels rubbing on the housing walls. If no fault is found, consult the engine manual for further troubleshooting procedures.

In the case of a noisy turbocharger, inspect all pressure connections for tightness—compressor discharge ducting, exhaust manifold, etc. Check the turbocharger shaft for looseness, look for evidence of wheel rubbing, and inspect the blade for impact damage caused by foreign material. If evidence of rubbing or impact damage is found, replace the turbocharger.

Turbocharger Removal and Installation

To remove a turbocharger, proceed in the following manner:

1. Loosen the connector clamp and position the flexible connector so that the crossover tube can be removed, **Figure 13-35A.**
2. Disconnect the drain tube from the bottom of the turbocharger, **Figure 13-35B.**
3. Remove the turbocharger drain line from the drain tube in the cylinder block, **Figure 13-35C.** Then remove the oil supply line.
4. Remove the boots, exhaust clamp, turbocharger, and gasket, **Figure 13-35D.**

If a turbocharger must be rebuilt, take it to a shop that specializes in this type of work. To install a new or remanufactured turbocharger, simply reverse the removal procedure. Prime the turbocharger before engine start-up by removing the oil supply inlet fitting and adding approximately .5 pint (.235 L) of clean engine oil to the turbocharger, **Figure 13-36.** Operate the engine at low idle for a few minutes before operating it at higher speeds. Finally, check the system for leaks.

Special Turbocharger Precautions

Even the best engines and turbochargers will not perform optimally if the operator is not trained to use them properly. Turbocharged engines require proper shutdown procedures to prevent bearing damage. If the engine is shut down from high speed, the turbo will continue to rotate after engine oil pressure has dropped to zero. Always idle the engine for several minutes before shutting it down.

Turbocharger damage can also be caused by oil lag. *Oil lag* is a lack of lubrication that occurs when oil pressure is not sufficient to deliver oil to the turbocharger bearings. Before running the engine up to high rpm, operate it at low speeds for at least 30 seconds after initial start-up to allow the oil flow to become established. Additional time should be allowed when the outside temperature is below freezing (32°F or 0°C). After replacing a turbocharger, or after an engine has been unused or stored for a significant period of time, there can be a considerable lag after the engine is started.

Aftercoolers (Intercoolers)

Although the turbocharger improves volumetric efficiency, it also increases the temperature of the intake air. This increase in temperature is due to the fact that the turbocharger compresses the air. The increase in temperature is directly proportional to the amount of air pressure produced. To help counteract this temperature increase, an *aftercooler,* or *intercooler,* is used.

Aftercoolers can reduce the temperature of the compressed air by as much as 90°F (32°C), making the air

Figure 13-35. *Removing a turbocharger. A—Loosening the clamp. B—Disconnecting the drain tube. C—Removing the drain line. D—Removing the fasteners securing the turbocharger. (Cummins Engine Co., Inc.)*

Figure 13-36. *Always prime the turbocharger with oil before engine start-up. Starting an engine without prelubricating the turbocharger will drastically shorten the turbo's life or even destroy it. (Mack Trucks, Inc.)*

denser and allowing more air to be packed into the combustion chambers. This results in:

❑ More power as the additional air allows the fuel to burn more completely, resulting in higher horsepower.

❑ Because the fuel is burned more completely, more power is produced from a given amount of fuel.

❑ Quieter combustion by controlling warm air for air-fuel mixing. A smoother pressure rise in the engine cylinder is produced.

❑ Reduced particulate emissions as a result of more complete fuel burning.

Types of Aftercoolers

There are two types of aftercoolers currently in use with turbocharged engines: coolant aftercoolers and air-to-air aftercoolers. In coolant aftercoolers, engine coolant flows through the aftercooler core tubes, **Figure 13-37.** As the hot compressed air from the turbocharger passes around the coolant tubes, it is cooled since the coolant temperature is lower than the air temperature. Air flows through the aftercooler to the cylinders in the opposite direction of coolant flow. In some aftercoolers, the coolant can actually pass through the core several times. This design usually provides better air cooling for improved combustion. Additional cooling fins can also be used to improve aftercooler efficiency.

The air-to-air aftercooler (also called an air-charged or charge-air system) is attached to the front of the radiator, **Figure 13-38.** This type of aftercooler is similar to a radiator. Outside air that passes through the aftercooler core cools the

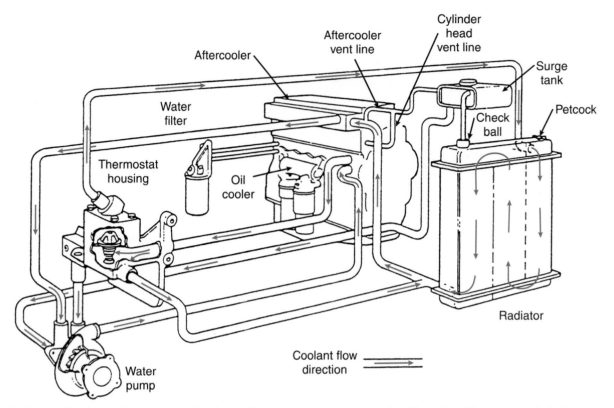

Figure 13-37. Coolant is used as a part of the aftercooler system as can be seen by this coolant flow schematic (turbo not shown). (Cummins Engine Co., Inc.)

engine's intake air charge. The air charge leaving the turbocharger is approximately 275-375°F (129-190°C). Some aftercooler systems can reduce this temperature to about 110°F (43°C) before the air charge enters the intake manifold. Air-to-air systems are more popular than coolant-type aftercoolers in modern mobile diesels. The system shown in **Figure 13-39** uses both air and water to cool the intake air.

Aftercooler Service

The servicing of an aftercooler is basically the same as for radiators. Clean the aftercooler core openings of debris, leaves, and other foreign matter with compressed air or water spray. Always clean the core from the engine side. Never use a wire or a screwdriver. Hose clamps must be kept tight and must be replaced when they deteriorate.

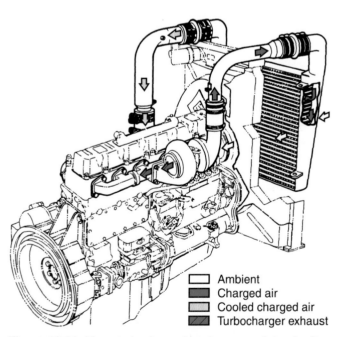

Figure 13-38. Many turbocharged engines use air-to-air aftercooler systems. (Mack Trucks, Inc.)

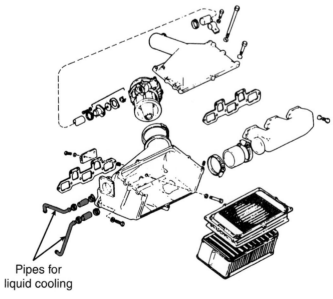

Figure 13-39. This aftercooler uses both air and water to cool the turbocharger intake air.

The aftercooler (as well as the rest of the cooling system) must be kept clear of sludge, scale and rust. These deposits may be loosened by the use of an approved cooling system cleaner. For specific instructions on aftercooler maintenance refer to the engine service manual.

Exhaust Pyrometers

Exhaust pyrometers measure the temperature of the exhaust gases in each cylinder. The pyrometer's gauge is generally located in the dash or instrument panel, **Figure 13-40.** As the pyrometer is switched from one cylinder to another, the exhaust temperature should be about the same for each cylinder. If the temperature does not fluctuate, the engine is operating normally. However, the exhaust temperature will not be uniform when the load is not balanced among the cylinders, the valves and piston rings are not tight, the nozzles are in poor condition, or other defects are present. Therefore, an exhaust pyrometer helps the operator keep the engine in good condition and properly adjusted. Careful observation of the exhaust pyrometer can prevent serious damage to the turbocharger and to the rest of the engine.

Exhaust Brakes

An *exhaust brake* is another item that is found in some exhaust systems. Like other engine brakes and retarding devices, it assists the service brakes on diesel-powered vehicles. Remember that an exhaust engine brake is used to only slow down a vehicle, not to stop it. It is not a substitute for the vehicle's brakes. The brakes must be used to bring the vehicle to a complete stop.

Diesel Engine Emissions

The actual levels of oxides of nitrogen (NO_x) and particulates in diesel engine exhausts have dropped dramatically since the late 1980s. Concern over how these substances affect human health and the environment has led to stricter federal truck emissions standards—standards that are expected to tighten through the 1990s. Between 1988 and 1998, oxides of nitrogen (NO_x) limits will drop by about 60% and particulate limits by 80%.

Emissions Reduction

Today, diesel engines are realigned to meet Clean Air Act standards for *hydrocarbon (HC), carbon monoxide (CO),* and *oxides of nitrogen (NO_x)* emission levels. The problem lies with *particulates.* Particulates are solid particles of carbon that are emitted as black smoke from a diesel engine. While turbochargers and acceleration smoke controls (see **Chapter 19**) can decrease particulates to meet standards, lowering particulates to future standards will require a new generation of emission control equipment.

Particulate generation is complex and depends on many factors, including the NO_x/particulate tradeoff—the fact that if an engine is designed to reduce particulate emissions by prolonging combustion, the higher temperatures that result will increase NO_x emissions. Reducing NO_x by such methods as injection retardation can lower in-cylinder temperature. However, less fuel is burned, resulting in an increase in emissions of HC and particulates.

The tradeoff can also affect other engine design strategies. High pressure fuel injection normally improves combustion by creating smaller fuel droplets that burn more completely, but this also tends to increase NO_x emissions. To minimize the NO_x/particulate tradeoff, engine manufacturers have introduced electronic engine controls. These new electronic systems allow flexibility under various

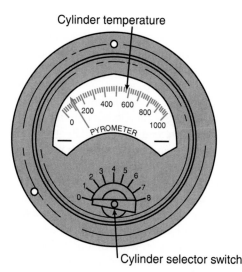

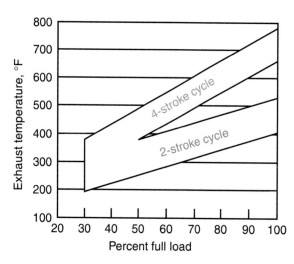

Figure 13-40. *The exhaust pyrometer measures the exhaust gas temperature. This measurement can be used when diagnosing emissions and performance problems. (John Deere)*

operating conditions by balancing key factors such as injection pressure and ignition timing.

Particulate Traps

Particulate traps often contain a separation medium with tiny pores that capture particles. However, as trapped material accumulates, resistance increases. This resistance equates to a pressure drop across the oxidizer bed that increases back pressure and fuel use.

The trapped mass is mostly dry soot or soot with absorbed **soluble oil fraction (SOF).** When a sensor indicates that resistance to flow in the trap is excessive, the filter medium is regenerated by heating it to burn off the soot. Filter efficiency is normally between 60% and 90%. Although particulate traps can have either a ceramic or wire filter medium, the ceramic type is used more often because it withstands higher temperatures and has a higher trapping efficiency.

Particulate Oxidizer Materials

There are several particulate trap oxidizer material presently in use. The system shown in **Figure 13-41** is a single trap configuration that consists of a filter and a regeneration system. The filter is positioned in the exhaust stream to collect particulate emissions while allowing the exhaust gas to pass through the system. The particulate matter is collected and oxidized in the trap filter. A computer monitors engine operation and trap loading status, initiates the regeneration process (6 to 15 minutes) upon reaching a desired point, and controls the regeneration

sequence. The electric heater is positioned close to the front face of the trap. The heater's proximity to the trap is important because its main method of heating the trap is by radiative transfer.

The electric heater is designed to heat the entire face of the trap. This uniform heating ensures complete, simultaneous ignition at the front face of the trap. A small blower is incorporated in the air supply line to support the combustion process during **trap regeneration.** An important feature of the blower is its inlet filter. Oversized for longer system life, the filter removes dust that could otherwise be ingested by the trap, causing a drop in trap capacity.

Electronic Sensors

Trap pressure loss is measured by a **differential pressure sensor,** while the inlet airflow is measured by a **mass flow sensor** located in the engine compartment, downstream of the air cleaner. For vehicles with an existing mass flow sensor, an additional sensor is not needed.

The temperature at the trap face is measured with a **thermocouple.** This reading can be used with the pressure differential and flow rate signals to indicate the condition of the trap independent of engine speed, load, or temperature. The temperature signal is also used to indicate selected trip points in the heating stage of regeneration. Valves are used either to direct exhaust flow through, or to bypass a particular trap. Standard exhaust brake valves have adjustments to minimize closed position leakage, yet avoid possible valve sticking. The valves are located directly upstream of the inlet section of the trap.

Another particulate trap regeneration system features three main parts: a compact ceramic core exhaust filter, which replaces the vehicle's muffler; a regeneration control valve, which is located in the turbocharger's discharge housing or an external valve located next to the turbocharger; and a microprocessor controller.

The microprocessor control unit is the key element in the successful operation of the regeneration control value which is sometimes called a back pressure valve, **Figure 13-42.** It allows particulate trap regeneration without interfering with the engine operation. For example, if the microprocessor control senses that full engine power is needed, the regeneration process is blocked temporarily.

This system offers several other advantages, such as rapid engine warm up, improved engine operating capabilities in extremely cold environments, and an optional part- or full-time exhaust brake. It is a continuous-flow design that uses back pressure engine exhaust heat to periodically ignite the soot particles trapped by the ceramic filter. The control valve and trap are installed downstream of the turbocharger and require only a small amount of space. When trap regeneration is needed, the valve and back pressure creates heat that causes normal exhaust temperatures of 600°F (315.5°C) to rise to 1200°F (649°C) and burn off the trapped soot. The control senses trap pressure differential condition, operator input (throttle position), and various engine parameters to modulate back pressure, maximizing driveability while controlling the regeneration process.

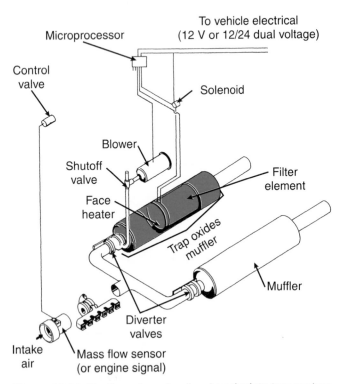

Figure 13-41. Configuration of a diesel particulate trap system. (Donaldson Company, Inc.)

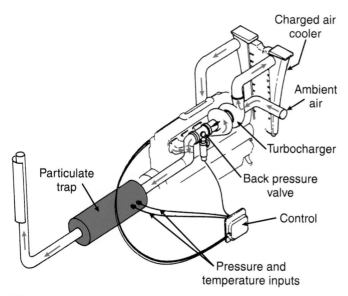

Figure 13-42. Popular particulate trap regeneration system for use on mobile diesel engines. (Garrett Group)

Figure 13-43 shows a self-cleaning diesel soot filter designed to remove visible smoke and soot for use on medium- and heavy-duty diesel powered mining and construction applications. It catalytically burns trapped soot particles. The filter is a ceramic flow-through design housed in a retractable stainless-steel canister. The self-cleaning process revolves around a proprietary precious metal catalyst coating that lowers the temperature at which trapped particles vaporize. Without the catalyst, temperatures upward of 1200°F (649°C) would be required for complete vaporization.

Another system used to reduce soot is the **puff limiter,** which is designed to reduce the characteristic smoke puffs

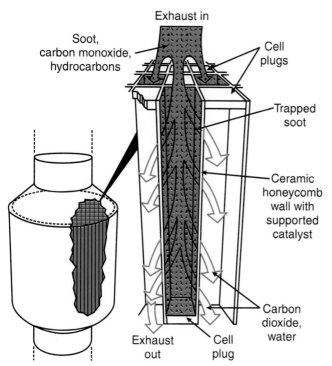

Figure 13-43. Cutaway of a self-cleaning diesel soot filter. (Englehard Corp.)

on turbocharged diesel engines. The system includes a reversing relay valve mounted on the inlet manifold and an outlet pressure line which connects the reversing relay to the air cylinder on the injection pump.

Diesel Oxidation Catalysts and Catalytic Converters

Diesel oxidation catalysts (DOC) help eliminate many of the gases, sulfates, and heavy organic fumes in the exhaust. Depending on the type of engine and the content of its exhaust, DOCs can oxidize 30% of the gaseous hydrocarbons (HC) and 40-90% of the carbon monoxide (CO) present. They do not, however, reduce NO_x emissions. DOCs have been used in diesel fork lift trucks and mining vehicles since 1967 to control HC and CO emissions. They are combined in one housing referred to as a **catalytic converter.**

Although diesel catalytic converters have little effect on dry soot (carbon), engine tests show that they typically remove 30-50% of the total particulate load. This is achieved by oxidizing 50-80% of the sulfates and heavy organic fumes (SOF) present. DOCs are less effective in engines that produce particulates with a very low SOF content.

DOCs are similar to their gasoline engine counterparts because they use a monolithic honeycomb support. Gases flow through the honeycomb with minimum pressure drop and react with the catalyst within the unit. The catalyst typically consists of platinum, palladium, or a mixture of both dispersed on a carrier with a large surface area, that maximizes contact with the gases and liquid organics. Both metal and ceramic supports may be used. Metal supports allow for thinner walls, more cells per unit area, lower back pressure, and greater strength. Ceramic substrates offer stronger catalyst adhesion, less sensitivity to corrosion, and lower cost. Ceramics are often used in automotive converters to help improve their performance.

The ceramic honeycomb support acts as a heat sink, holding the exhaust gas heat effectively. Thus, when an engine is left under idle conditions, the low exhaust gas temperature (a condition that reduces conversion) is compensated for and exhaust purification can still be effective.

The catalyst coating is designed to reduce the SOFs in the exhaust. Although diesel soot is best known for carbon particles, as previously mentioned, SOFs and sulfates that absorb onto the carbon have recently caused a lot of concern. The most common DOC presently in use is the *particulate oxidizer catalytic converter.* As shown in **Figure 13-44,** it uses a monolithic element, located between the exhaust manifold and the turbocharger. Its catalytic coating is very effective. This catalytic converter is simple, but more complex designs are on the way as new ideas are being put into use.

 Note: Whenever a catalytic converter is replaced, be sure to turn it over to a recycler for recovery of the elements in the converter. Do not discard a used converter as scrap.

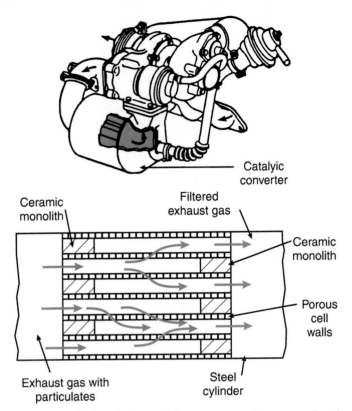

Figure 13-44. *Particulate oxidizers are used in many diesel exhaust systems to control particulate emissions.*

Water as a Performance Enhancer

The use of diesel fuel combined with water to improve power output and decrease fuel consumption can be traced back to the early 1920s when scientists began experimenting with diesel fuel and water emulsions. The injection of water into the engine's intake manifold (fumigation), and the separate injection of water and fuel directly into the combustion chamber were tried.

More recently, research has shown the beneficial environmental effects of introducing water into the cylinder. The diesel/water emulsion system shown in **Figure 13-45** is capable of reducing emissions of NO_x particulates, hydrocarbons, and carbon monoxide up to 90% from stationary and mobile diesel engine applications. **Figure 13-46** shows the unit, whose major components include an electronically controlled mechanical mixing unit (about the size of a shoe box) situated between the fuel filters and the injection pump, which produces a consistent and uniform emulsion of diesel fuel and water. The percentage of water in the emulsion can be pre-set from five to more than 30%. An electronic control module (ECM) maintains this percentage as a function of fuel flow.

The emulsion is created by shear forces between the fuel and the water in a pear-shaped vortex chamber inside the mixing unit. The fuel is pumped to the mixing unit where it enters a ring channel that surrounds the vortex

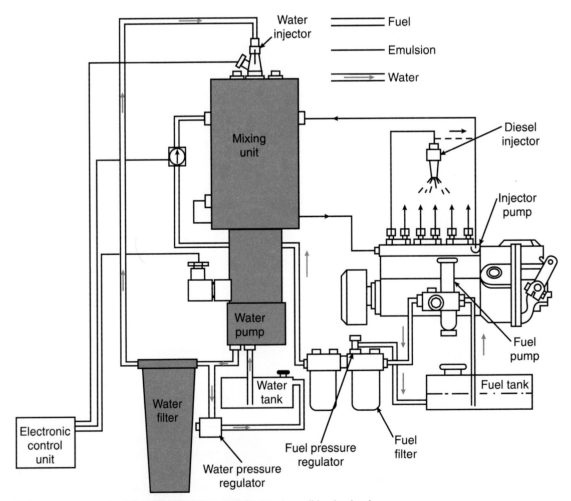

Figure 13-45. *The components of the diesel/water emulsion system. (Harrier Inc.)*

Figure 13-46. The major mixing component of the diesel/water emulsion system. (Harrier Inc.)

chamber. From this ring channel, the fuel flows into the vortex, at a high speed of rotation.

The water is injected into the center of the vortex, parallel to its longitudinal axis. The high rotational speed of the fuel and the shape of the vortex causes the water and fuel to bond well before the end of the vortex. As the emulsion exits the vortex it flows through a circulation pump to the engine's injection pump. Only part of the emulsion enters the combustion chamber. Most is returned to the injection pump where it is continuously recirculated throughout the mixing unit.

At this point, the diesel fuel has formed a skin around the water droplets. When the emulsion is injected into the combustion chamber, the water droplets expand or "spon-taneously evaporate," rupturing the fuel skin and dispersing it into much smaller droplets. This causes a significant reduction in exhaust emissions by improving combustion efficiency and fuel atomization, increasing the fuel's fluid momentum, and cooling the combustion chamber's flame zone.

Catalytic Converter/Silencer

The industrial application of a combination catalytic converter/silencer is shown in **Figure 13-47,** which provides simultaneous reduction of exhaust emissions and exhaust noise of diesel engines. It controls particulates, eliminates odor, reduces carbon monoxide, and changes exhaust to almost clear. The unit is cylindrical in configuration and is larger than a common catalytic converter. The design of some units feature a "catalyst portal" which provides quick access to the catalyst module.

The catalyst substrate is made of stainless steel with chevron-shaped corrugations which form the cell structure. The catalyst module is strategically located where the sealing and mechanical support structure provides a large safety factor against damage by engine backfire. Air-fuel ratio control ensures a continuous high conversion rate and minimizes fuel cost. An automatic air-fuel ratio controller governs system balance and follows load changes.

Alternative Fuels

The use of alternative fuels will become increasingly important as a way of controlling diesel emissions. Compression engines such as diesel types using compressed natural gas (CNG), liquefied petroleum gas (LPG), liquefied

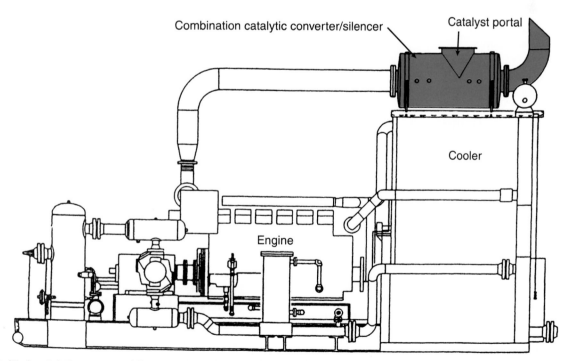

Figure 13-47. A catalytic converter/silencer unit is often used with stationary diesel engines. This unit has a catalyst portal. (Houston Industrial Systems)

natural gas (LNG), methanol, ethanol, biodiesel, or other fuels may offer the best long-term way to meet exhaust restrictions. These alternative fuels produce fewer particulates and can be run lean, therefore having a better chance of meeting future CO, HC, and NO$_x$ standards. More information on alternative fuels can be found in **Chapter 14.**

Summary

The exhaust system of a diesel engine, which is used to rid the engine of exhaust gases, consists of the exhaust manifold, exhaust (header) pipe, muffler, emission control devices, and tailpipe. In some cases, a spark arresting device is installed in the system. On many engines, a turbocharger is installed in the exhaust system.

Horsepower is required to expel exhaust gases from the cylinders. The harder the engine's pistons work against the buildup of these gases, the less power is left for turning the crankshaft. A low-restriction exhaust system minimizes the amount of energy wasted pumping gases.

The exhaust manifold collects engine exhaust gases from cylinder ports and carries them to an exhaust pipe or to some other location away from the engine. The exhaust ports in the engine and the passages in the exhaust manifold are large enough to allow free flow and expansion of the escaping gases.

The muffler is the primary component responsible for minimizing the harsh sound waves created by the combustion gases being forced out of the cylinders through the exhaust ports. Its function is to remove acoustic energy from the gases without unnecessarily obstructing their flow to the outlet pipe. In addition to acting as silencers, most mufflers also function as spark arresters.

Total exhaust system back pressure is greater than that of the muffler alone because every bend and joint adds resistance to exhaust flow.

Most big-bore diesels are equipped with 5- or 6-inch diameter exhaust tubing, which is adequate to evacuate gases from even the largest displacement turbocharged diesel engines.

A turbocharger can be used on both two- and four-cycle diesel engines. It uses exhaust energy to drive a turbine-powered centrifugal air compressor, which is used to convert air velocity into air pressure and increase the flow of air to the engine cylinders. Turbocharging also reduces exhaust emissions and improves fuel economy.

The turbocharger normally consists of three basic systems, a turbine wheel and shaft, a centrifugal compressor wheel, and a center housing.

During operation, the turbocharger responds to the engine load demands by reacting to the flow of the exhaust gases. The turbocharger automatically compensates for the normal loss of air density and power as the altitude increases. A turbocharger can be used in addition to the regular scavenging system.

Common symptoms that may indicate possible turbocharger trouble include a lack of engine power, black smoke, blue smoke, excessive engine oil consumption, and noisy turbocharger operation.

Aftercoolers, or intercoolers, are used to reduce the temperature of the compressed air generated by the turbocharger by as much as 90°F. This makes the air denser, allowing more to be packed into the combustion chambers. This results in more power, greater fuel economy, and quieter operation. There are two types of aftercoolers presently in use with turbocharged engines: coolant-type aftercoolers and air-to-air aftercoolers.

An exhaust engine brake is an optional device that assists the service brakes on diesel vehicles.

Diesel engines must meet 1994 Clean Air Act standards regarding the emission of hydrocarbons, carbon monoxide, and nitrogen oxide levels. Solid particulate emissions must also be controlled. While turbochargers and acceleration smoke controls can control particulates to 1994 standards, lowering particulates to future standards will require a new generation of emission control equipment. The use of alternative fuels will also become increasingly important as a way of controlling diesel emissions.

Important Terms

Exhaust system	Oil lag
Back pressure	Aftercooler
Exhaust manifold	Intercooler
Slip joints	Exhaust pyrometer
Exhaust manifold gasket	Exhaust brake
Muffler	Hydrocarbons (HC)
Dispersive muffler	Carbon monoxide (CO)
Reactive muffler	Oxides of Nitrogen (NO$_x$)
Wet muffler	Particulates
Dry muffler	Particulate traps
Resonator	Soluble oil fraction (SOF)
Vertical muffler	Trap regeneration
Horizontal mufflers	Differential pressure sensor
Exhaust risers	
Turbochargers	Mass flow sensor
Turbine wheel	Thermocouple
Compressor wheel	Puff limiter
Turbine housing	Diesel oxidizing catalysts (DOC)
Crossover tube	
Waste gate	Catalytic converter

Review Questions—Chapter 13

Do not write in this text. Place your answers on a separate sheet of paper.

1. Define the term exhaust back pressure.

2. Even a well-designed muffler will produce some _____ in the exhaust system.

3. In wet mufflers used in marine applications, the _____ contact the cooling water.

4. A turbocharger _____ the engine's exhaust emissions, smoke, and noise.

5. Which of the following is an advantage of a turbocharged engine?
(A) Decreased emissions.
(B) Better fuel economy.
(C) Increased power output.
(D) All of the above.

6. A turbocharger automatically compensates for the normal loss of _____ and _____ at higher altitudes.

7. Aftercoolers, or intercoolers, cool the compressed air from the turbocharger using _____.
(A) engine coolant
(B) cool outside air
(C) Both A & B.
(D) oil from the engine

8. Briefly describe the operation of an exhaust engine brake.

9. At higher engine operating temperatures _____.
(A) fewer solid particulates and more NO_x gases are produced
(B) more solid particulates and fewer NO_x gases are produced
(C) fewer solid particulates and NO_x gases are produced
(D) more solid particulates and NO_x gases are produced

10. A _____ provides simultaneous reduction of diesel engine exhaust emissions and noise.

ASE-Type Questions

1. It has been estimated that each 2 psi of back pressure causes a loss of about ___ engine horsepower.
(A) 1
(B) 2
(C) 4
(D) 6

2. Technician A says that the exhaust ports in the engine and the passages in the exhaust manifold are large to allow free flow and expansion of the escaping gases. Technician B says that ports and passages are rather narrow to promote the fast movement of exhaust gas needed to turn the turbocharger turbine. Who is right?
(A) A only.
(B) B only.
(C) Both A & B.
(D) Neither A nor B.

3. Technician A says it is not advisable to remove a manifold while it is still hot, as this can promote warpage. Technician B says to clean the carbon buildup on the inside of the manifold use solvent and a scraping tool. Who is right?
(A) A only.
(B) B only.
(C) Both A & B.
(D) Neither A nor B.

4. Technician A says that it is permissible to reuse a exhaust manifold gasket if it is less than 1 year old and not cracked. Technician B says it may be necessary to retorque exhaust manifold bolts after initial warm-up and cool-down. Who is right?
(A) A only.
(B) B only.
(C) Both A & B.
(D) Neither A nor B.

5. All of the following are causes of blue smoke from a turbocharged engine, EXCEPT:
(A) air cleaner dirty.
(B) seal leakage.
(C) air cleaner element missing.
(D) excess carbon build-up.

6. Technician A says that the job of the waste gate or turbine bypass valve is to provide a method of limiting turbocharger speed. Technician B says that the waste gate and turbine bypass valve actually divert exhaust gas to warm intake air during cold operating conditions. Who is right?
(A) A only.
(B) B only.
(C) Both A & B.
(D) Neither A nor B.

7. Which of the following problems can create symptoms that appear to be turbocharger related?
(A) Restricted air cleaner.
(B) Plugged crankcase breather.
(C) Deteriorated oil lines.
(D) All of the above.

8. Aftercoolers provide all of the following to turbocharged engines, EXCEPT:
(A) it makes the air less dense.
(B) reduces particulate emissions.
(C) quieter combustion.
(D) provides more power.

9. An exhaust brake is _____.
(A) a substitute for the vehicle's brakes
(B) used to help slow the vehicle
(C) found in all exhaust systems.
(D) All of the above.

10. Particulate traps have _____.
(A) a separation medium with tiny pores that capture particles.
(B) a wire filter medium.
(C) a ceramic medium.
(D) All of the above.

Truck manufactures strategically locate components to reduce drag and improve fuel economy. Note the shields next to the exhaust stacks; these reduce the amount of wind resistance between the cab and the trailer. (W. Scott Gauthier.)

Chapter 14

Diesel Fuels

After studying this chapter, you will be able to:
- ❑ Explain diesel fuel grades.
- ❑ Discuss fuel properties and characteristics.
- ❑ Explain the reason for the use of fuel additives.
- ❑ Describe the proper procedures for handling and storage of diesel fuel.
- ❑ Name the various alternative fuels and their properties.

All engines have a few things in common. One of them is that they all need fuel in order to operate. Most mobile and stationary engines use fuels derived from crude oil. Diesel engines use several variations of diesel fuel, depending on their application. In this chapter, you will learn about the different types of diesel fuel used in diesel engines.

Hydrocarbon Fuels

The liquid fuel used to operate diesel engines is obtained from crude oil. Crude oil consists of a mixture of **hydrocarbons** (hydrogen and carbon) and compounds such as benzene, petane, hexane, heptane, toluene, propane, and butane. These compounds have different **relative volatility** points because they will vaporize, or flash, at different temperatures. Hydrocarbon fuels are separated in a fractionating column.

As crude oil is heated, these hydrocarbons are given off as a vapor, **Figure 14-1.** After the natural gas is vaporized from the crude oil, the applied temperature is raised and the hydrocarbon with the next highest vapor point is obtained. The first hydrocarbon engine fuel obtained is high octane aviation gasoline. If everything is working properly and you continue to raise the temperature, other hydrocarbon fuels such as commercial gasoline, kerosene, diesel fuel, domestic heating fuel, industrial fuel oil,

lubricating oil, and paraffin are obtained. Finally, only coke and asphalt remain.

The type of hydrocarbons obtained will vary depending on the original geographic location of the crude oil. This heating of crude oil to obtain various hydrocarbons is known as **distillation.** It is a highly complicated process and precision control of pressure and temperature is required.

Diesel Fuel Grades

While the American Society of Testing Materials (ASTM) has divided diesel fuels into three classifications, only two recommended grades are considered acceptable for use in high-speed trucks and buses in North America.

These are the number 1D and number 2D classifications. Grade number 4D, the heaviest diesel fuel, is used in large stationary constant-speed engines or in some marine applications. Number 4 and bunker fuels are not used in high-speed mobile diesel engines, which are continually accelerating and changing speed. Grade number 3D was discontinued a number of years ago and is obsolete.

The Canadian government 3-GP-6D diesel fuel has its own fuel specifications recognizing five categories of diesel fuels, with even more restrictive standards set by ASTM.

Each individual refiner and supplier attempts to produce diesel fuels that meet as closely as possible with ASTM and American Petroleum Institute (API) standards, **Figure 14-2.** Depending on the crude oil source, the diesel fuel end product may be on either the high or low end of the prescribed heat energy scale in Btus per gallon. This is why individual diesel fuels grades may vary slightly from one supplier to another.

Grade 1D is generally the most refined and volatile diesel fuel available. It is a premium fuel used in high rpm engines requiring frequent changes in load and speed. Grade 2D is more widely used in truck fleets due to its greater heat value per gallon, particularly in warm to moderate climates. Although Grade 1D fuel has better properties for cold-

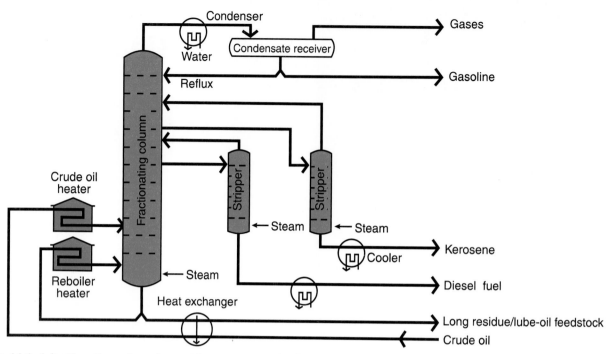

Figure 14-1. *A fractionating column is used to separate crude oil into the various hydrocarbon products. As the vaporized products rise in the tower, they settle onto trays at different levels and are then piped out to other equipment for further refining. (Allis-Chalmers Engine Div.)*

General Fuel Classification	No. 1 ASTM 1D	No. 2 ASTM 2D
BTU/Gal average	137,000	141,800
Gravity (°API)	40–44	33–37
Flash point, minimum [°F(°C)]	100 (38)	125 (52)
Viscosity, kinematic at 100°F (40°C) (cSt)	1.3–2.4	1.9–4.1
Cloud point (°F)	10°F	10°F
Sulfur content, maximum (wt%)	0.5	1.0
Carbon residue on 10% maximum, (wt%)	0.15	0.35
Accelerated stability total insolubles, maximum (mg/100 ml)	1.5	1.5
Ash, maximum (wt%)	0.01	0.02
Cetane number, minimum	45	40
Distillation temperature [°F (°C)]		
Initial boiling point, typical	350 (177)	375 (191)
10%, typical	385 (196)	430 (221)
50%, typical	425 (218)	510 (256)
90%	500 (260) maximum	625 (329) maximum
Endpoint	550 (288) maximum	675 (357) maximum
Water and sediment, maximum (%)	0.05	0.05

Figure 14-2. *General classification chart for diesel fuel. (Detroit Diesel)*

weather operation, many fleets still use Grade 2D in the winter. Other cold weather aids include a fuel heater/water separator for easier starting, as well as fuel additive conditioners that can be added directly to the fuel tank.

Like gasoline, diesel fuels are blended on a seasonal and geographical basis to satisfy anticipated temperature conditions. It is usually best and cheapest to burn the heaviest fuel that will work under given circumstances. Heavier grades of diesel can usually produce more energy than light grades, just so long as the increased viscosity does not make the fuel too thick to flow and inject properly.

It is important to remember that the wrong grade of diesel fuel can affect the operation of the engine. If an engine is not developing the proper horsepower, an improper grade of diesel fuel could be the cause. An easy way to confirm this is to use a diesel fuel quality tester such as shown in **Figure 14-3.**

Diesel Fuel Properties

In a diesel engine fuel system, the fuel itself performs three functions. It supplies chemical energy to be transformed into mechanical energy, lubricates precision parts in the fuel system components, and cools metal surfaces operating in conditions of friction.

The properties or characteristics of diesel fuel must meet these three if the engine is to perform with reliability.

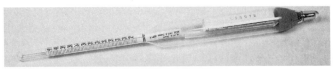

Figure 14-3. A typical diesel fuel quality tester. (Kent-Moore Tools)

Fuel processors, as well as engine manufacturers, run laboratory tests on all fuel used in diesel engines. These measured properties give a good indication of the way the fuel will perform, however, there is no real substitute for an actual engine test. The major diesel fuel properties affecting engine performance are:

❑ Heat value.
❑ Specific gravity.
❑ Flash point.
❑ Volatility.
❑ Cetane number rating.
❑ Pour point.
❑ Cloud point.
❑ Viscosity.
❑ Carbon residue.
❑ Sulfur content.
❑ Fungus and bacterial contaminants.
❑ Oxidation and water.

Heat Value

The **heat value** of fuel is a general indication of how heat energy is supplied to an engine and how well the engine converts heat energy into work. The heat value can be found by testing with a calorimeter. With this test, a premeasured amount of fuel is burned and the amount of heat emitted is carefully measured in Btus per pound of fuel. A **British thermal unit (Btu)** is the amount of heat required to raise the temperature of one pound of water one degree Fahrenheit. The metric equivalent of this unit is known as a **joule.** To convert Btus into joules, multiply by 1054.8.

Specific Gravity

The **specific gravity** of fuel is a ratio of the fuel density to the density of water. It is measured using a hydrometer. Specific gravity affects the fuel's spray penetration as it is injected into the combustion chamber. Because water is the standard, it has a specific gravity of one. Since oil floats on water, a diesel fuel's specific gravity is always less than one. Diesel fuel's specific gravity ranges from 0.8 to 0.94. Specific gravity is also a factor in measuring the heat value of the fuel. In general, heavier fuels usually have a greater heat value per gallon (Btus) than lighter fuels. Thus, specific gravity is a good indicator of the amount of heat (Btus) available in a given amount of fuel.

The American Petroleum Institute (API) employs another scale to determine specific gravity. Water has a specific gravity of 20 on the API gravity scale. Ten is the lowest value on this scale, the reverse of the system just described. Diesel fuels generally range from 20 to 45 on the API gravity scale, with most ranging between 34 to 36 at 60°F (15.5°C).

Flash Point

A fuel's **flash point** is the lowest temperature at which it will give off flammable vapors in sufficient quantity to flash or momentarily ignite when brought into contact with an open flame. The flash point has no effect on engine performance or on its ignition qualities. It is specified simply as an index of fire hazard—a fuel oil with an extremely low flash point is dangerous to store and handle. Diesel fuel flash points are not an indication of how they will ignite in an engine cylinder, however. This depends on the ignition quality of the fuel. For example gasoline, which has a very low flash point, would be a very poor diesel fuel due to its ignition quality.

Volatility

Volatility is a fuel's ability to change to a vapor. It is indicated by the air-vapor ratio that is formed at a specific temperature. Diesel fuel volatility is indicated by a 90% distillation temperature (the temperature at which 90% of the fuel is distilled off). As volatility decreases, carbon deposits and engine wear increase. Depending on such factors as the combustion chamber condition, more smoke will also affect power output, performance, starting, and warm-up.

Cetane Number Rating

The ease of diesel fuel oil ignition and the manner in which it burns determine the ignition quality of the fuel oil. Diesel fuel oil is injected into the combustion chamber in liquid form. The fuel must then be able to vaporize quickly and ignite without a flame or spark. This ability to vaporize and ignite easily is called **ignition quality.**

The ignition quality of a diesel fuel is determined by its **cetane number rating,** or cetane value. The cetane rating or value of a diesel fuel is based on the ability of the fuel to ignite. The cetane rating of a fuel is determined by comparing it with pure cetane, which is a test fuel, and is identified by a cetane number, **Figure 14-4.** This cetane number represents the percentage of pure cetane in a reference fuel which will exactly match the ignition quality of the fuel being tested. The cetane rating scale ranges from 0 to 100, with 100 being the highest ignition quality.

In general, the higher a fuel's cetane rating, the lower the emissions. Currently, a 40 cetane or above rating is standard for all on-highway diesel engines. (In some areas, 50 cetane and higher are current standards.) Newer diesel engines may require higher cetane fuel. The diesel engine service manual will specify what cetane number to use.

Improving Ignition Quality

Fuels with poor ignition qualities can be improved or reformulated by blending them with fuels that have good ignition properties. The cetane number of such blends are an average of the cetane numbers of the individual fuels. To meet the cetane number required by most on-highway vehicles, cetane improvers are added to the blends. The lower cetane compounds are less responsive to these improvers than the higher cetane paraffin fuels.

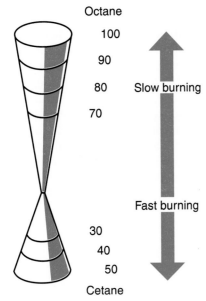

Figure 14-4. Comparison of the relationship between octane and cetane. Note that as the cetane number increases, the burn rate increases.

The improvers promote early, uniform ignition of fuel and prevent high pressure increases in the combustion cycle. Depending on the amount of cetane components in the base fuel, typical alkyl nitrate additive treatment can increase cetane by three to five numbers (1:1000 ratio). With high natural cetane premium base fuels (containing a high percentage of paraffins) and a 1:500 alkyl nitrate treatment ration, cetane may increase as much as seven numbers.

Most improvers contain alkyl nitrates, which break down readily to provide extra oxygen for combustion. They also break down and oxidize fuel in storage. However, they also generate organic particulates, water, and sludge, all of which degrade fuel quality. Recently, several new cetane improvers or reformulaters without alkyl nitrates have been used and, properly blended, will improve oxidation stability while providing a cetane increase of two to five increments.

It is only recently that engineers have learned how important the cetane level is in meeting the emissions equation. The air-fuel mixing time in a direct-injected diesel is about one-tenth the time of a carbureted gasoline engine. Early ignition promotes smooth and complete combustion, leading to reduced emissions. Excessive delay can produce very high peak cylinder pressures which make for rough and noisy engine operations. The higher the cetane number, the shorter the **ignition delay**, which is the time between the start of fuel injection into the cylinder and the actual start of combustion.

Cloud Point

Diesel fuel contains paraffin, and in cold temperatures wax crystals can start to form, accumulate, and clog engine filters. The temperature at which this happens is referred to as the **cloud point** or *wax appearance point (WAP)*. Generally, if the fuel's cloud point is at least 10° below the

ambient temperature, engine performance will be satisfactory. The most commonly available and recommended fuel oil, grade 2D, has a cloud point or WAP of 10°F (-12°C), while the thinner grade 1D has a cloud point of -20°F (-28.8°C). In extremely cold areas, engine clogging can be minimized by using only grade 1D in winter. In most places, mixing a base of grade 2D with grade 1D, or other special additives will work well in colder weather. As a rule, grade 2D's cloud point drops two degrees for every 10% of grade 1D that is added up to a ratio of 50:50.

One of the most effective chemical means for lowering the fuel's cloud point is by blending #2 diesel fuel with kerosene. Fuel suppliers will often do this themselves and market it as "winter fuel." Although this fuel will be more viscous at lower temperatures, its performance will still vary depending on where it was purchased. A 40/60 blend in one area may have a higher cloud point that a 40/60 blend in another. A fuel additive with a cloud point depressant will have limited effect. At most, it will lower the cloud point by 3-4°.

Pour Point

Pour point is another common way of measuring a diesel fuel's performance in cold weather. It is the minimum temperature at which fuel can flow and is expressed as the temperature 5°F (8.8°C) above the level at which the oil becomes solid or refuses to flow. The pour point averages about 10° lower than the cloud point. Fuel treated with special additives called flow improvers or wax modifiers keep the wax crystals from forming clumps and choking fuel lines. They will generally give satisfactory performance at 9° lower than untreated fuel. However, fuel improvers do not have any impact on the fuel's cloud point, since they do not prevent the formation of wax or paraffin crystals themselves.

Viscosity

Viscosity or stiffness is the property of a diesel fuel that resists the force which causes the fluid to flow. It is related closely to specific gravity and pour point. Two common methods of measuring viscosity are the Saybolt test and the kinematic centistokes test. Both tests involve heating the oil to an exact temperature and measuring its flow rate through a standard sized orifice.

The viscosity of diesel fuel is generally specified at 100°F (38°C). Viscosity of fuels for medium-speed and high-speed engines normally ranges from 2.4 to 4.1 centistokes (cSt), or about 39 seconds Saybolt Universal (SSU). In general, any fuel with a viscosity lower that 2.4 cSt or 34 SSU, when measured at 100°F (38°C), will be too thin and could damage injectors as well as other parts of the fuel system.

Diesel fuel also has a low viscosity index, meaning that it is thin when hot but gets thick when cold. A diesel fuel that can go through an injection system easily in warm weather may get too thick to flow properly in cold weather. As mentioned previously, diesel fuel is supplied in several grades. Grade 1D is a winterized diesel fuel. It is thinner or less viscous than grade 2D, which is the fuel used at normal operating temperatures. Diesel fuel viscosity also affects the spray pattern in the combustion chamber, **Figure 14-5.** Low viscosity creates a fine mist, while high viscosity results in coarse or heavy atomization. In other words, viscosity affects:

❑ Lubrication capability at various temperatures.

❑ Atomization, or spray capability.

❑ Ignition and burning characteristics.

Therefore, viscosity is a very important consideration when selecting a diesel fuel.

Carbon Residue

Ash or **carbon residue** is the deposit left in the combustion chamber due to incomplete combustion or the use of fuels made from residual blends. It can be measured in the laboratory by heating a measured fuel sample in a closed container in the absence of air. Carbon residue is the final product that remains in the container after heating. This product is then expressed as a percentage by weight of the original sample.

The amount of carbon residue considered acceptable in diesel fuel oil varies depending on the combustion chamber design, the injector adjustments, and the general condition of the engine. This amount is generally more critical in small high speed engines than in large slow speed industrial engines.

Soot ash, which originates from either additives or the crude oil itself, also causes wear of fuel injection components, the pistons, and the piston rings. Standard requirements allow for a maximum of .001% soot ash content. Due to air quality concerns, future legislation may further limit the amount of soot ash a diesel engine can produce.

Sulfur Content

The wear of pistons, rings, and cylinders in a diesel engine will generally increase when there is an excessive amount of **sulfur** in the fuel. Excess sulfur content will also

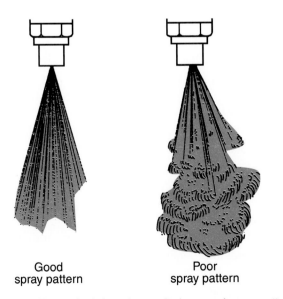

Good
spray pattern

Poor
spray pattern

Figure 14-5. Proper fuel viscosity results in a good spray pattern. However, improper viscosity will give a poor spray pattern.

cause varnish to form on the piston skirt and will create oil sludge in the engine crankcase. Sulfur also combines with water to form corrosives as a result of the combustion process. These corrosives can etch finished surfaces, accelerate engine wear, attack softer metals (such as bearings), and deteriorate engine oil. Similar corrosion damage from sulfur is frequently found in the engine's exhaust system.

Sulfur and aromatic content of diesel fuel are most responsible for harmful exhaust emissions. Reducing sulfur will reduce sulfur dioxide emissions and the sulfate fractions of diesel particulates that also will indirectly reduce ambient levels of sulfate particles. The reduction of aromatics will reduce carbonaceous particle emissions.

In an effort to reduce sulfur and aromatic content of diesel fuel, regulations mandate a sulfur content of no greater than .05% by weight and a maximum aromatic content of 35% in all on-highway diesel fuels used. Only non-highway diesel engines can use higher sulfur content fuels. These include farm tractors, boats, trains, and stationary engines. The non-highway markets represent nearly half of the total diesel fuel consumption.

To ensure nobody tries to use high-sulfur diesel, which is not subject to the same taxes as truck fuel, refiners and marketers are required to add a dye to all non-highway diesel fuel. This dye is either bright blue or red. If the dye shows up in the exhaust smoke or the fuel tanks of a vehicle, the owner can be fined.

Sulfur Removal

Sulfur is removed from diesel fuel in a process called **hydrotreating.** Hydrogen is used with a catalyst at temperatures between 500-800°F (260-427°C) to react with the sulfur compounds present. The reaction forms hydrogen sulfide, which is separated from the hydrocarbon and sent to a sulfur plant to be converted to elemental sulfur. The sulfur content can only be determined by chemical analysis of the fuel.

As already stated, the aromatic content which is a product of the distillation process must also be reduced. Many states require lower aromatic content than EPA maximums. For example, some states require major fuel producers and marketers in the state to reduce the aromatic components in diesel fuels from 30-35% down to 10%. To meet requirements for lower aromatic fuels, fuel marketers have found that reformulating the basic formula by 15% to 19% with increased cetane (55 to 59) will achieve the same results as a 10% aromatic. Although the alternative fuel is cheaper than a 10% aromatic type, it is more expensive than a national low-sulfur diesel fuel.

Although low-sulfur, low aromatic and higher cetane number diesel fuels are more expensive than the minimum standard types, they enhance engine performance because improved engine combustion results in easier starting, smoother running, less noise, and less smoking during start up.

Fungus and Bacterial Contaminants

Fungus and **bacteria** are the most common contaminants found in diesel fuels. Fungus and bacteria live in the water and feed on the hydrocarbons found in diesel fuel. These contaminants are called *hydrocarbon utilizing microorganisms (HUM)*. HUM will spread through a fuel system where moisture or trace amounts of water are present. The resulting bacterial problems can shorten engine filter life. Draining the fuel system will reduce HUM activity, but will not eliminate it. The only way to eliminate HUM growth entirely is to treat the fuel system with a biocide.

Oxidation and Water

Precipitants and particulates are noncombustible materials formed when either two incompatible fuels are mixed or when a fuel oxidizes. Oxidation can occur when an unstable fuel circulates through a diesel engine. Precipitants and particulates will normally settle out in low or slow flow areas of the fuel system or become part of the tank's bottom sludge. Both can get mixed into the fuel system during fuel delivery and can plug fuel filters.

Water causes the greatest concern because it is the most common form of diesel fuel contaminant. Water is found in the fuel system in two forms, free and dissolved. Most diesel fuels contain some dissolved moisture. Diesel fuel has a saturation level of water at any given temperature. As the temperature goes down, the fuel will hold less dissolved moisture. There is currently no common method of removing dissolved water from diesel fuel.

Free or nondissolved water in the fuel tank usually comes from bulk storage tanks, from condensation, or from dissolved water. Free water mixes with the fuel as storage tank bottoms become agitated while dispersing or receiving fuel. When water in vapor form is present in air, it is called humidity. As air replaces fuel in storage tanks or vehicle tanks when the tanks disperse or receive fuel, moisture in the form of humidity finds its way into the fuel system. From there, moisture may condense as hot fuel returning from the injectors flows back into the cooler fuel tank. Free water in liquid form is heavier than diesel fuel and settles on the low flow or slow flow areas of the fuel system. The reverse is true when water freezes. Ice is lighter than diesel fuel and floats in the system to create plugs in fuel separators, filters, even going as far as the fuel pump injectors.

Fuel Additives

Generally, no fuel additives are necessary when a good quality, clean, and properly selected fuel is used. Certain fuel characteristics, however, can be improved by treatment with a fuel additive or conditioner. As already mentioned, additives are used to improve fuel flow properties under cold or winter-like conditions, increase the cetane number of the fuel, and to introduce a biocide to the fuel to prevent fungi and bacteria from growing in the diesel fuel.

There are a number of commercially available fuel additives that will reduce smoke and corrosion of vital

parts; and others that are oxidation inhibitors. In addition, remember that geographical locations, operating conditions, type of fuel storage, handling methods, and maintenance procedures are all factors that can determine whether or not a fuel additive might help a diesel fuel. In most cases, the fuel supplier or marketer will select and add the additive they believe will best improve the quality of their fuels. Fuels containing special additives may be more expensive than those without additives.

Fuel Handling and Storage

The importance of clean fuel in the operation of a diesel engine cannot be overemphasized. Diesel fuel is generally delivered clean and free of impurities. However, every time it is transferred or handled, the risk of contamination increases. While fuel storage and its handling are not normally within the engine technician's job classification, the results of improper storage and handling certainly are. A problem often found by the engine technician when troubleshooting a diesel engine is dirty fuel.

One of the major problems either in storage or in the engine's fuel system is **leakage.** Each connection or fitting in either system is a potential source of leakage. There are two phases to the leakage problem:

- ❑ The visible leakage of fuel from a line during operation and shutdown.
- ❑ The opening in a suction line or fitting which pulls foreign material into the system during operation and may or may not present a visible leak during shut-down.

Visible leakage should be stopped as soon as it is discovered. Loss of flow volume generally means a loss of pressure. The leak which allows fuel contamination by pulling in air, water, or dirt is more difficult to locate and can be more damaging. **Foreign materials,** such as air, sediment (dirt), and water present problems and can enter the fuel system in various ways.

Air

Air drawn into the fuel system may make pick-up of fuel much more difficult or prevent it entirely. In addition, it may show up as:

- ❑ Low power.
- ❑ Gear pump wear.
- ❑ Rough operation.
- ❑ Soft or non-responsive throttle.

Air enters the fuel in any one or more of the following ways:

- ❑ Suction leaks (lines, fittings, filters, or tank).
- ❑ Fuel turbulence, especially when coupled with poor tank venting.
- ❑ Fuel pick-up point near return.
- ❑ Combustion gases entering injector.

Dirt

Dirt is as damaging to moving parts in the fuel system as it is in any other part of the engine. Dirt can:

- ❏ Clog filters.
- ❏ Abrade metal surfaces.
- ❏ Increase combustion chamber deposits.

Dirt enters the fuel when:

- ❏ A suction leak is in an exposed area.
- ❏ Dirt collection in the tank allows it to be picked up with the fuel.
- ❏ Careless filling and handling permits mud, dust, or grime to enter tank or lines.

The technician may unintentionally add dirt to the fuel system by:

- ❏ Using containers and tools exposed to dirt.
- ❏ Using lines and fittings that have accumulated dirt or dust without cleaning them out.
- ❏ Allowing dirt to enter a line while it is removed for repair. This is especially critical between filter and pump.

Water

Water in the fuel may show up as low power, also, but it is even more damaging as already stated when it simply contributes to corrosion on fuel system components. Water enters the fuel system through:

- ❏ A suction leak in an exposed location.
- ❏ As condensation due to warm fuel.
- ❏ Careless storage and handling.

Careless storage and handling of diesel fuels can be a major concern in the engine's operation, **Figure 14-6.** It is important to periodically perform inspection, cleaning and maintenance checks on all fuel handling and filtering equipment. Many state and local regulations include record keeping and periodic inspections.

Storage Tanks

Diesel fuel tanks may be **underground storage tanks (UST),** or **above-ground storage tanks (AGST)** systems. Fuel stored in AGSTs will deteriorate more quickly than the fuel stored in USTs due to sunlight exposure and the greater temperatures found above ground.

Tanks are made of steel, fiberglass, or steel clad with fiberglass. Galvanized copper tanks are not used to store diesel fuel. There is a chemical reaction between the fuel and copper that produces a residue which will quickly clog fuel filters. The term "tank systems" includes the piping connected to the tank(s).

In some large industrial operations, diesel fuel is fed from the main supply tank located (above or in the ground) outside of the building to a one day tank in the engine room, **Figure 14-7.** This tank has a sufficient capacity to provide fuel for one day of operation. The main storage normally stores only a one week supply of fuel. Fuel stored for longer periods of time can form gum which, along with corrosive compounds, causes deterioration. Many refineries are also using additives that help stabilize the fuel in order to reduce gum formation. Moisture condensation inside a fuel storage tank increases as the tank size increases. Condensation in a large tank can be reduced by using a series of smaller tanks and connecting them into the line as needed. The size of tanks is generally controlled by the Bureau of Fire Underwriters and local regulations.

Alternative Fuels

The Clean Air Act of 1990 requires that all centrally fueled on-highway transportation operations of 10 or more vehicles in certain high-emission areas are to phase in **alternative fuel vehicles** beginning in 1998. The act does not mandate any specific alternative fuels such as gasoline

A   B

Figure 14-6. Care must be taken whenever filling a storage tank. A—Below ground tanks can present as much danger as above-ground tanks. B—Always take care when filling any vehicle's tank. (Heavy Duty Trucking and Andrew Ryder)

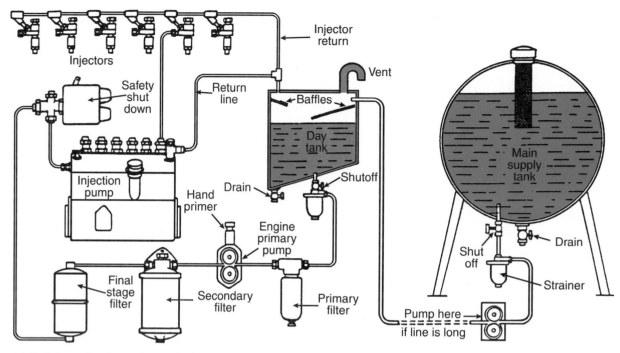

Figure 14-7. Schematic of a stationary fuel system with day tank and storage tank. (Waukesha Engine Div.)

or diesel. The alternative fuels now being considered for use with compressed (diesel) type engines are **compressed natural gas (CNG), liquefied petroleum gas (LPG), liquefied natural gas (LNG),** and alcohol based products such as methanol and ethanol.

The Clean Air Act regulations are closely tied to clean fuel vehicle programs in each state's air quality implementation plan. Since the supply of certain fuels varies across the country, not all states favor the same fuel. Southern states are a big supporter of natural gas, while some western states favor liquid petroleum gas, and a number of midwestern states actively tout ethanol.

Some engine companies are devoting research and testing to natural gas. Many of their initial products are being offered for urban buses, which are on a stricter emissions schedule than medium and heavy duty trucks. Engine makers face the challenge of designing new engines with efficiency as close as possible to that of a diesel without making major component changes, **Figure 14-8.**

Compressed and Liquefied Natural Gas

Natural gas has been used for 60 years to power industrial internal combustion engines. Natural gas is one of the lightest fractions (parts) of crude oil. Chemically, it is very similar to gasoline. Unlike gasoline, natural gas contains several light gases, heavy gases, and other impurities. For natural gas to be used as an alternative fuel for on-highway vehicles, it must be refined into either compressed natural gas or liquefied.

Compressed natural gas (CNG) is composed primarily of methane and hydrocarbons that have a high carbon-to-hydrogen ratio. Hydrogen is an ideal fuel that burns well while producing little pollution. It is the main reason CNG is a very desirable fuel. It produces good power, economy, and low exhaust pollution levels.

On-board fuel tanks for the storage of CNG while the vehicle is on the road is a problem that needs to be addressed. For example, experimental taxi cab operations, using two or three tanks placed in the trunk, had a range of only 120-160 miles (192-256 km).

Liquefied petroleum gas (LPG) is mainly propane and butane, along with small amounts of other gases. It has combustion qualities equal to or better than high grade diesel. LPG is a vapor or gas at normal room temperature and atmospheric pressure. This presents some design problems. In fact, since LPG is not a liquid, the entire fuel system must be redesigned to handle it, **Figure 14-9.**

Figure 14-8. Natural gas engines are used in place of basic diesel engines in some mobile and stationary applications. (Diesel & Gas Turbine Publications)

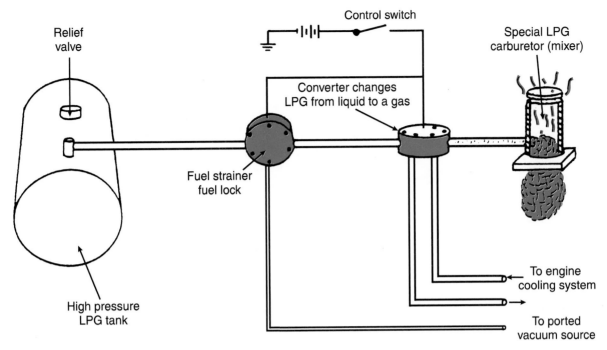

Figure 14-9. *Simple schematic for a liquefied petroleum gas (LPG) fuel system.*

LPG has operating characteristics almost identical to those of gasoline and diesel. However, because it is already a gas, the problem of breaking up liquid fuel is eliminated. Since the fuel enters the intake manifold and combustion chamber as a vapor, combustion is much more efficient. For this reason, LPG is generally an excellent alternative fuel. Liquefied natural gas (LNG) has the same characteristics as LPG.

A heavy-duty truck would need too many fuel storage tanks to make it a feasible concept at this time. Conversions on city buses, however, are presently being done in several big cities, since fuel tank storage is not as great a problem, **Figure 14-10.** In most cases there is adequate clearance underneath the bus, **Figure 14-11,** to allow for several tanks. In addition, most city buses do not accumulate as many miles in a day as would a long-haul tractor/trailer. Buses can also be fueled up locally at their

Figure 14-10. *Buses in some major cities are now able to operate on compressed natural gas (CNG). (Pierce Transit, Inc.)*

depot by a fast-charge facility, which takes about the same length of time as it would to fuel up a diesel-powered bus.

Alcohol-Based Fuels

Alcohol-based fuels are being studied as an alternative fuel for the transportation industry. The two types used to power internal combustion engines are ***ethyl alcohol*** (ethanol) and ***methyl alcohol*** (methanol). Alcohols are especially desirable as an automotive fuel because they can be manufactured from sources other than crude oil. Alcohol intended as an automotive fuel must be almost pure. Quite often, several refining steps are needed to approach this purity.

Denatured ethanol or "grain alcohol" is colorless, harsh tasting, and highly flammable. It can be made from numerous farm crops such as wheat, corn, sugar cane, potatoes, fruit, oats, soy beans, or any material rich in carbohydrates. Crop wastes are also a source. One of the major drawbacks of using ethanol, clean burning as it is, is that it does emit carbon dioxide (CO_2).

Methanol or wood alcohol can be made from wood chips, coal, oil shale, tar sands, cornstalks, garbage, or even manure. Like ethanol, methyl alcohol is a colorless, odorous, flammable liquid. Methanol is currently being used in several urban diesel powered bus fleets with some success. It is one of the major contenders as a feasible alternative fuel for heavy-duty high-speed diesel engines.

Biodiesel Fuels

Biodiesel is a comparative latecomer to the alternative fuels scene, but as on-highway, off-highway, stationary and even marine engine manufacturers continue to investigate any and all alternative sources of fuel, biodiesel is quickly

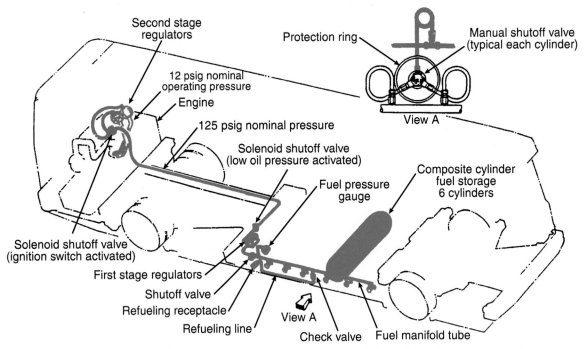

Figure 14-11. *Schematic showing typical CNG tank arrangement under a bus. (Diesel & Gas Turbine Publications)*

gaining support in the transportation industry. At present, the most popular is a biodiesel blend of number 1 diesel fuel and soybean oils. There is also especially strong interest in biodiesel from the agricultural community because used cooking oils in addition to soybean and processed animal fats can be used in the making of biodiesel fuels.

Industry interest is high in biodiesel since it is not a stand-alone fuel like the other alternative fuels, but rather a blend with existing diesel fuels. This means little or no alterations to the fuel injection system are necessary, nor is any special training of service personnel needed. The two most popular blends today are 20-30% soybean oil to 80-70% diesel oil. In physical properties, biodiesel has almost the same btu/gallon ratio as number 1 diesel. Further, it has slightly better lubricating qualities than number 1 diesel, can handle winterizers and has a gel point slightly

higher than number 2. It has very low toxicity if ingested and is actually biodegradable by some standards.

The main drawback to biodiesel is its relative high cost in comparison to regular diesel fuel. Biodiesel is now being tested in some city buses and is seeing increased popularity for use in marine diesel engines. In the near future, biodiesel fuels may play a major role in the future in reducing engine exhaust emissions, **Figure 14-12.**

More money and time must be spent on alternative fuel engine development, on-board fuel storage, and fueling facilities before they can become a reality. At the present, diesel fueled engines are still cheaper and more efficient than any of the proposed alternatives. With the federal mandate of low-sulfur diesel fuel, as well as reformulation procedures, diesel engines will be performing well into the 21st century.

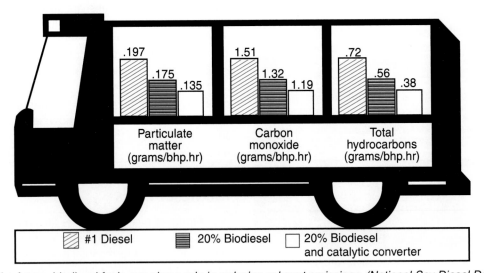

Figure 14-12. *In the future, biodiesel fuels can play a role in reducing exhaust emissions. (National Soy Diesel Development Board)*

Summary

Only two recommended grades of diesel fuel are considered acceptable for use in high-speed trucks and buses in North America. These are the number 1D and number 2D classifications. Grade 1D is generally the most refined and volatile diesel fuel available. It is a premium quality fuel used in high rpm engines requiring frequent changes in load and speed. Grade 2D is more widely used in truck fleets due to its greater heat value per gallon, particularly in warm to moderate climates.

Diesel fuels are blended on a seasonal and geographical basis to satisfy anticipated temperature conditions. In a diesel engine fuel system, the fuel-supplied chemical energy is transformed into mechanical energy. Diesel fuel also cools and lubricates precision parts in the fuel system.

The heat value of fuel is a general indication of how heat energy is supplied to an engine, and thus, how well the engine converts heat energy into work. Specific gravity is a ratio of the density of fuel to the density of water.

The cetane rating or value of a diesel fuel is based on the ability of the fuel to ignite. There is currently no common method of removing dissolved water from diesel fuel. Free or nondissolved water in the fuel tank usually comes from bulk storage tanks, from condensation, or from dissolved water. Free water gets into the fuel as storage tank bottoms become agitated while dispersing or receiving fuel.

Fuel additives are commonly used to improve diesel fuel flow properties under cold or winter-like conditions. Additives can also help eliminate fuel contaminants.

The importance of clean fuel in the operation of a diesel engine cannot be overemphasized. Foreign materials such as air, sediment (dirt), and water presents various problems and can enter the fuel system in various ways such as suction leaks, careless filling and handling, and dirty fittings, and lines. It is important to periodically perform inspection, cleaning, and maintenance checks on all fuel handling and filtering equipment.

The method of storing the fuel depends on the size of storage tanks or drums, the frequency and method of fuel delivery, climatic conditions, and local regulations. Fuel stored above ground will deteriorate more quickly than fuel stored underground due to the greater temperature ranges found above ground. Underground storage tanks as well as above ground storage tank systems must meet rigid safety and environmental regulations.

Storage tanks should be made of either steel, fiberglass, or steel clad with fiberglass. Galvanized and copper tanks should not be used to store diesel fuel. The most promising alternative fuels are compressed natural gas, liquefied petroleum gas, liquefied natural gas, methanol and ethanol.

Important Terms

Hydrocarbons	Fungus
Relative volatility	Bacteria
Distillation	Leakage
Heat valve	Foreign materials
British thermal unit (BTU)	Air
Joule	Dirt
Specific gravity	Water
Flash point	Underground storage tank (UST)
Volatility	
Ignition quality	Above-ground storage tank (AGST)
Cetane number rating	
Ignition delay	Alternative fuel vehicles
Cloud point	Compressed natural gas (CNG)
Pour point	
Viscosity	Liquefied petroleum gas (LPG)
Carbon residue	
Soot ash	Liquefied natural gas (LNG)
Sulfur	Ethyl alcohol
Hydrotreating	Methyl alcohol
	Biodiesel

Review Questions—Chapter 14

Do not write in this text. Place your answers on a separate sheet of paper.

1. Which of the following grades of diesel fuel are recommended for use in high-speed trucks and buses?
   (A) number 1D.
   (B) number 2D.
   (C) number 4D.
   (D) Both A & B.

2. Like gasoline, diesel fuels are blended on a _____ and _____ basis to satisfy anticipated temperature conditions.

3. Heavier grades of diesel fuel usually produce _____ energy than lighter grades.

4. The value(s) used to measure the amount of heat emitted is called:
   (A) British thermal unit (Btu).
   (B) joule.
   (C) flash point.
   (D) Both A & B.

5. Volatility of diesel fuel is indicated when _____ or the fuel is distilled off.
   (A) 75%
   (B) 90%
   (C) 50%
   (D) 100%

6. Sulfur is removed from diesel fuel by:
   (A) distillation.
   (B) hydrotreating.
   (C) filtering.
   (D) None of the above.

7. Which of the following types of tanks should not be used to store diesel fuel?
   (A) Steel.
   (B) Fiberglass.
   (C) Copper.
   (D) Steel clad with fiberglass.

8. Fuel stored above ground will deteriorate _____ than fuel stored underground due to the greater temperature ranges found above ground.

9. Name three ways air commonly enters diesel fuel.

10. *True or False?* Flow improvers help redissolve paraffin crystals that have formed in the diesel fuel.

ASE-Type Questions

1. Technician A says that grade 1D diesel fuel is generally the most refined and volatile diesel fuel available. Technician B says that grade 2D is more widely used in truck fleets due to its greater heat value per gallon, particularly in warm to moderate climates. Who is right?
   (A) A only.
   (B) B only.
   (C) Both A & B.
   (D) Neither A nor B.

2. The _____ of fuel is a general indication of how heat energy is supplied to an engine and how well the engine converts heat energy into work.
   (A) heat value
   (B) viscosity
   (C) flash point
   (D) cetane number rating

3. All of the following are contaminants of diesel fuel EXCEPT:
   (A) water.
   (B) dirt.
   (C) cetane.
   (D) air.

4. All of the following are diesel fuel properties EXCEPT:
   (A) heat value.
   (B) pour point.
   (C) specific gravity.
   (D) octane rating.

5. The temperature at which wax crystals begin to form in diesel fuel is known as the _____.
   (A) cloud point
   (B) pour point
   (C) viscosity point
   (D) specific gravity level

6. The viscosity of diesel fuel affects all of the following EXCEPT:
   (A) the lubrication capability of the fuel at different temperatures.
   (B) the cloud point.
   (C) the burn capability.
   (D) the atomization capability.

7. The _____ of diesel fuel is based on the ability of the fuel to ignite.
   (A) volatility
   (B) specific gravity
   (C) cetane number
   (D) viscosity

8. Technician A says that fungus and bacterial growth in diesel fuel can be eliminated by draining settled water out of the fuel tanks at regular intervals. Technician B says that the only way to eliminate bacterial growth entirely is to treat the fuel system with a biocide. Who is right?
   (A) A only.
   (B) B only.
   (C) Both A & B.
   (D) Neither A nor B.

9. Diesel fuel tanks can be made from _____.
   (A) galvanized copper
   (B) aluminum
   (C) steel clad with neoprene
   (D) fiberglass

10. Technician A says that as the temperature of diesel fuel goes down it will hold less dissolved water. Technician B says that there is no common method now available for removing dissolved water from diesel fuel. Who is right?
    (A) A only.
    (B) B only.
    (C) Both A & B.
    (D) Neither A nor B.

Chapter 15

Basic Fuel Systems

After studying this chapter, you will be able to:
- List and explain the five major jobs of a diesel fuel system.
- Name the major components of a diesel fuel system and describe their use.
- Recognize the various fuel tank arrangements and know how they function.
- Identify the three types of fuel gauges and describe how they operate.
- Describe the three types of fuel lines and know where they are used.
- Explain the general guidelines for replacing a fuel line.
- Name the different types of fuel transfer pumps and understand how each one operates.

The primary job of a diesel fuel system is to move fuel from the fuel tank or tanks to the fuel injectors. The fuel must be delivered at the proper pressure. It must be free of dirt, moisture, wax, and air bubbles. The fuel system must also have return lines to vent any unused fuel from the injector back to the fuel tank. The fuel tank safely stores the fuel until needed and must be properly vented and have a means for draining moisture and sediment. This chapter discusses the various components used in diesel fuel delivery systems.

Fuel System Components

The primary job of a diesel fuel system is to deliver clean fuel to the fuel injectors. The major components of a diesel fuel system include:
- Fuel tank (stores fuel).
- Fuel transfer pump (pushes fuel through the filters and into the injection pump).
- Fuel filters (clean the fuel).

- Injection pump (times, measures, and delivers fuel under pressure to the injectors).
- Injection nozzles (atomize and spray fuel into the cylinders).
- Return line (vents any unused fuel back to the fuel tank).

In addition to these major components, the fuel system may also contain fuel-water separators, fuel heaters, agglomerators, and sedimentation chambers (see **Chapter 16**).

Fuel System Operating Fundamentals

Figure 15-1 illustrates the components of a typical fuel delivery system. While component design and location may differ between manufacturers, most systems operate in a similar manner. As shown in **Figure 15-2,** fuel is pulled from the **fuel tank** and through a primary fuel filter by a **fuel transfer pump.** From the fuel transfer pump, the fuel is pushed through one or more secondary **fuel filters** into the fuel injection pump housing manifold. A bleed orifice allows a constant supply of fuel to pass through the fuel return line and flow back to the fuel tank. This helps keep the fuel cool and free of air.

One or more **fuel injection pumps** takes fuel from the fuel manifold and pushes it under pressure through the fuel lines to the fuel injectors. The fuel injection pump, or metering pump, boosts low and medium fuel pressures to the high pressures needed for injection. Each **fuel injector nozzle** has very small holes in its tip. These holes change the flow of fuel to a fine atomized spray that provides for good combustion in the cylinder. The fuel system also prevents excess air from entering the system, which can cause rough engine operation.

Fuel Delivery System Types

There are several types of fuel delivery systems. Depending on the exact design, segments of the system may operate at low, moderate, or high fuel pressures. For example,

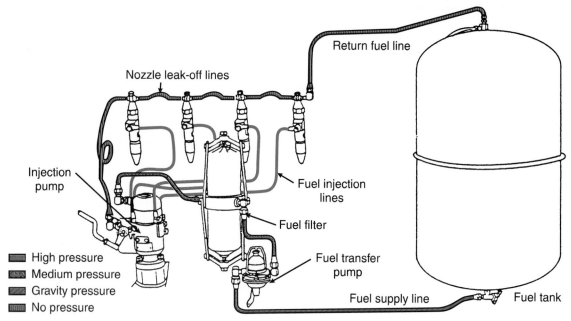

Figure 15-1. *Location of the various fuel pressure lines in a typical fuel supply system.(John Deere Co.)*

fuel lines that link the fuel tank to the fuel transfer pump and the transfer pump to the injection pump operate at low or moderate pressures, **Figure 15-2.** The low and moderate pressure portions of the fuel system also contain the necessary filters, water separators, and fuel conditioning equipment.

The lines running from the injection pump to the injectors are in the high pressure side (high side) of the system. The **return lines** running from the injectors to the fuel tank may operate at very low pressures or with no pressure.

Some systems do not have a high pressure side. Fuel is delivered to the injectors at low or moderate pressure by the transfer pump. Mechanical action inside the injectors raises the fuel pressure to the high levels needed for injection.

Low Pressure Supply System

The low pressure fuel supply system is designed to provide an unrestricted supply of fuel to the transfer pump and the fuel injection pump. Various components within the system must also be designed to comply with all governmental regulations, yet they must be accessible for servicing. In cold conditions, the components should be positioned to take advantage of the heat produced by the engine and the exhaust system. They must also be protected from damage and abuse. If possible, components should not be positioned above electrical equipment; however, if this is unavoidable, the pump should be shielded.

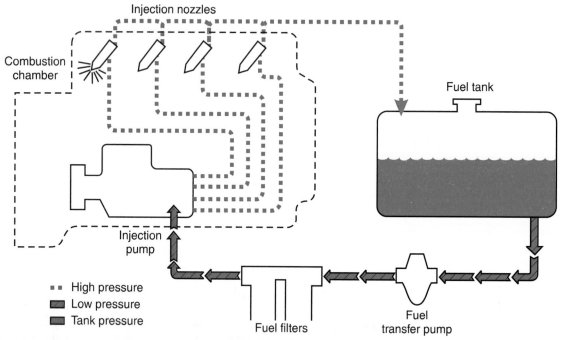

Figure 15-2. *Fuel flow system schematic. (Courtesy of Caterpillar, Inc.)*

In a few applications, using a *gravity-fed* system can eliminate the need for a transfer pump. However, such a setup can cause slow venting unless a hand primer pump is used.

Fuel system feed pressures vary depending on the type of injection pump and fuel circuit used (some inline pump systems require higher feed pressures). Transfer pumps can provide from 3-45 psi (20.9–310.3 kPa). If a standard filter is used, a relief valve should be added to prevent system pressure from exceeding 15 psi (103.4 kPa).

Proper positioning of the low pressure fuel system components helps achieve maximum fuel system and engine performance. For example, sedimenters or agglomerators filters (see **Chapter 16**) must be fitted on the suction side of the transfer pump (as close to the tank as possible) to minimize water freezing in the pipes and to keep water droplets from breaking up. To allow easy draining of the sedimenter, it should be positioned below the fuel level in the tank.

Fuel Heaters

Fuel heaters should be located before the first point at which wax can cause a fuel restriction—normally the primary fuel filter. The final filter in a system employing three or more filters should be mounted away from the engine cooling fan and above the level of the fuel injection pump to prevent siphoning and to assist venting.

Fuel Tanks

While the primary function of the fuel tank is to carry the fuel necessary to keep the engine operating, they also serve other functions. As previously described, the diesel engine requires both draw lines to the injectors and return lines to handle the excess fuel provided by the transfer pump.

Fuel also serves as a coolant for the injectors. Excess fuel, which is hot, must be returned to the fuel tanks for storage and cooling. Fuel tanks, therefore, double as cooling tanks. Hot return fuel sloshing around gives up heat through the tank walls. When tanks are connected by a crossover line, this doubles the area for heat dissipation. When fuel temperatures climb above 115°F (46.1°C), there is a noticeable loss of power and a drop in fuel economy. A vehicle can require a lower gear late in the day when climbing hills and when fuel levels are low.

Fuel tanks are manufactured in many shapes, sizes, and materials. Cylindrical or square shape is more a matter of aesthetics and the space available for the tank. In most cases, the shape of the tank can be altered to fit the space available. In most mobile mounting configurations, the fuel tanks are mounted close to the front axle on conventional cab trucks, sometimes closer to the drive axle(s) on cabovers. The typical mounting installation of the tank mounted on the cab access step is shown in **Figure 15-3**. The tanks on most types of mobile vehicles must be equipped with the following:

- [ ] The tank must have a non-spill vent with a ball-check valve or equivalent.

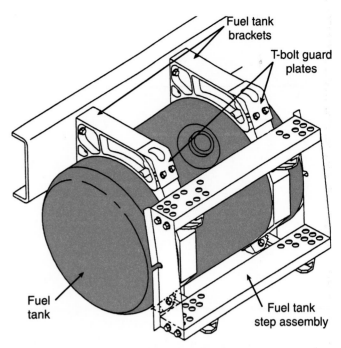

Figure 15-3. *Fuel tank mounting configuration.*

- [ ] The suction line entry fitting must be above the normal full line of the tank.
- [ ] The tank must not extend beyond the vehicle body.
- [ ] The tank must not be high enough to gravity feed the injectors (highway vehicles).

Fuel Tank Material

Fuel tanks can be made from aluminum alloy or steel. Most on-road mobile diesel tanks are of aluminum alloy because of its lighter weight. This is very important to trucking firms. If polished and kept clean, cylindrical aluminum is appealing to the eye, which can pay dividends in driver pride, **Figure 15-4**.

Steel is less costly, but in cases of mobile vehicles, it carries a weight problem. It is used in most industrial diesel installations. Steel is a tough material, and is usually found on off-road applications where withstanding rough treatment is a factor.

 Caution: A galvanized steel fuel tank should never be used for fuel storage because diesel fuel reacts chemically with the zinc coating to form powdery flakes that can quickly clog fuel filters and cause damage to the fuel pump and injectors.

Fuel Tank Construction

Mobile fuel tanks are designed with **baffle plates** formed internally during tank construction. The baffle plates have holes that are large enough to allow fuel to flow freely throughout the length of the tank. However,

Figure 15-4. *Some fuel tanks, when cleaned and polished, can be eye appealing as well as functional. (Andrew Ryder)*

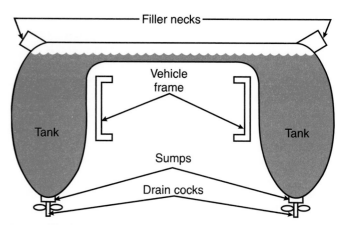

Figure 15-5. *Typical double saddle fuel tank arrangement. (Courtesy of Cummins Engine Co., Inc.)*

must be remembered that fuel weighs up to 7.5 lbs. per gallon. If the truck is carrying over 3500 lbs. (1587.6 kg) of fuel, these pounds might be better devoted to extra cargo.

At least 5% of the total capacity of the fuel tank must be above the *full line* to allow for fuel expansion. It must be remembered that 80% of the fuel flowing through the injectors returns to the fuel tank. This fuel returning to the tank from the engine has been warmed. As the fuel in the tank is warmed by the returning fuel and in some cases, exposure to hot sunlight, it expands. The 5% expansion space required will prevent tank overflow during operation.

these holes are small enough to prevent fuel from sloshing when the vehicle is moving. The fuel inlet and return lines may also be separated by baffles to prevent recirculation of the warm diesel fuel returning from the injectors. The bottom of the fuel tank is angled to collect contaminants and fitted with a drain plug to facilitate contaminant removal. It also allows the fuel to be drained from the tank before removing it for any necessary service procedure.

Fuel Tank Maintenance

Most engine manufacturers recommend that the fuel tank be drained periodically (at least once a year) to remove any water. Some fuel tanks are equipped with a small low-mounted catching basin arrangement that permits easy draining of water from the tank. Remember, the specific gravity of water is greater in weight than the fuel oil and, therefore, sinks to the bottom of the tank.

Tank Capacity

The capacity of a tank must be capable of storing enough fuel to operate the engine for a reasonable length of time and distance, considering the type of operation. For instance, on mobile vehicles, there are usually two or more fuel tanks with total capacity ranging from 100-500 gallons (380-1900 L), **Figure 15-5.** Larger fuel tanks will mean fewer fuel stops, but the extra fuel takes valuable pounds off the vehicle's payload. While a pair of 250 gallon (950 L) tanks can legally carry about 475 gallons (1805 L) of fuel, it

Fuel Tank Ventilation

Fuel circulating in and being drained from the injectors may be under some pressure. This can create air bubbles, which are returned to the fuel tank with the fuel. Therefore, adequate venting is required to allow this air, along with any air that is picked up by agitation from road vibration and as the fuel drops into the tank, to escape.

Vents are usually located at the highest point in the tank so fuel cannot percolate out when air expands elsewhere in the tank. The vents should be positioned to prevent dirt and water from entering the tank. They must also be positioned to prevent clogging.

It is also important to ventilate the tank properly so that air can enter to replace outgoing fuel. Remember that fuel cannot flow unless air is free to enter. Fuel tank vents must breathe freely whenever the engine is operating.

In addition to the vents, three other openings are normally found in the tank: one for filling, one for discharge, and one for drainage. A fourth opening, especially in industrial tanks, is sometimes used to allow leak-off fuel to be returned from the injection system to enter the tank

Fuel Pickup Suction Tube

The end of the fuel pickup or *suction tube* should be positioned near the drain plug to allow access for cleaning. The pickup point should be at least 1″ (2.54 cm), **Figure 15-6,** away from the tank bottom to prevent blockage from sludge or ice. The pipe should not be equipped with a gauze filter.

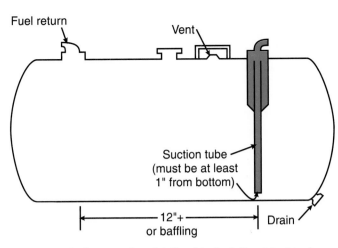

Figure 15-6. *Construction details of typical diesel fuel tank.*

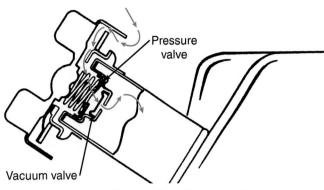

Figure 15-7. *Typical fill cap with both a vent valve and pressure relief valve.*

 Warning: Do not drain fuel near intense heat or an open flame. This is a dangerous practice that could lead to property damage, personal injury, or death.

There are a number of ways in which the suction tube can affect engine operation. Because the suction tube is especially prone to rust (due to condensation in the tank), it is a very common source of contamination. If the suction tube is not adequately braced, vibration may cause the tube to crack or break loose. Replacement tubes should be of the size and material recommended in the engine service manual. If the tube is too small for the fuel flow demands of the engine, engine power can be reduced.

Fuel Return Line

The location of the fuel return line is another important consideration. The fuel return line should return fuel to the open space above "full level," allowing the air bubbles to be vented. It should also return fuel at a point at least 12″ (304.8 mm) away from the fuel pickup point so that the return fuel will not be picked up before the air is vented. If these points are closer, baffling in the tank will be necessary.

As previously mentioned, the return fuel carries varying amounts of air. If this fuel is dropped directly over the pickup point, a low fuel level will permit much of the air to be picked up and recirculated in the fuel system. With mobile diesel engines, this can increase the likelihood of a "soft" or non-responsive throttle pedal when the fuel level is low. However, the return pipe(s) should discharge near the suction tube to assist in wax prevention in cold conditions.

Fuel Tank Filler Neck and Cap

A diesel fuel tank must also prevent fuel contamination. The tank should have a raised filler neck that is positioned to prevent water entry during refilling and normal operation. It should also have a hinged or captive fill cap to prevent fuel loss and to minimize the amount of dirt entering the tank. The filler cap on mobile tanks is generally constructed with a vent and pressure relief valves, **Figure 15-7.** If the tank has

no other ventilation, the filler cap must allow air to enter the tank so that fuel can flow from the tank easily. If necessary, gauze screens (usually 4 or 5 mesh) can be placed in the filler neck. The screens should be positioned deep enough to allow the dispensing nozzle to enter the neck, but they should be easily removable for cleaning.

Dual Fuel Tanks

While a single fuel tank can be used, most heavy-duty highway vehicles employ *dual* or *tandem fuel tanks.* There are several methods of using two tanks in a single fuel system. In one design, the tanks operate independently of one another. Each tank has its own feed and return lines. The only connection between the tanks is a crossover line. This line contains a manual selector valve that is used to control which tank is feeding the fuel system. In this setup, cool fuel may be drawn from one tank while warm fuel is fed back into the second tank. Two independent tanks can be a problem if the operator forgets to switch the selector valve when required.

To overcome this problem, many dual tank installations use an equalizer line to connect the two tanks. This arrangement, **Figure 15-8,** does not require a selector

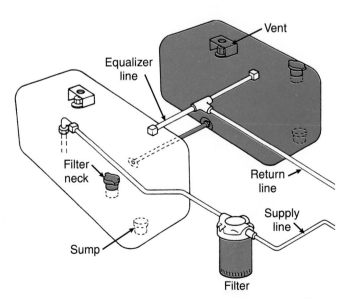

Figure 15-8. *A double saddle tank arrangement with an equalizer line.*

valve. Connecting the tanks in this manner has three basic advantages:

- ❑ The fuel equalizing line allows the tanks to work as one without having to manually switch from one tank to another.
- ❑ The operator of the vehicle can more easily keep track of the amount of fuel remaining, even if the fuel gauge sending unit is located in only one tank. The fuel will settle to the same level in both tanks.
- ❑ Both tanks may be filled from a single opening. However, the equalizer line must be large enough to allow fuel to flow into the second tank rapidly. The crossover line is still needed to help equalize the fuel tank levels so that one tank will not run dry and shut down the engine.

Fuel Tank Installation

In either of the two-tank installations previously described, the tank units on highway vehicles are fastened to the chassis below or just behind the cab. The tanks are usually located below the injector level and the lines are usually protected by a steel rod or angle iron.

In construction equipment or industrial installations, the fuel tank is often higher than the injectors, **Figure 15-9.** To prevent hydrostatic lock of the cylinders in these installations, a check valve is often used in the injector drain line, **Figure 15-9A.** An alternate solution is to use a float

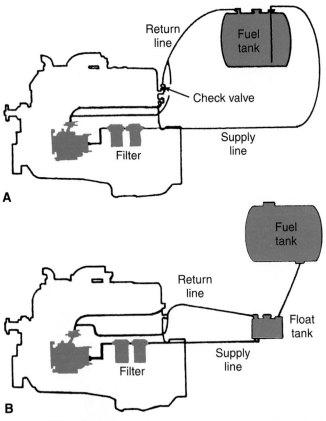

Figure 15-9. Systems designed to prevent hydrostatic lock. A—Check valve system. B—Float tank system. (Cummins Engine Co.)

tank valve. As shown in **Figure 15-9B,** fuel is fed into the float tank and return line, overcoming the possibility of hydrostatic lock. However, float valves occasionally stick, possibly causing a loss of fuel through the tank vent. If leakage from the float vent is noticed, check the float valve for proper operation. Due to the possibility of a sticking valve, the float vent should be located below the level of the engine's injectors. The escape of some fuel through the float vent is more desirable than hydrostatic lock. When the float tank is above the injector level, fuel pump shutdown valve failure or parking the vehicle with the ignition switch on may allow fuel to drain to the engine through the supply line.

Fuel System Problems

Fuel draining from a high fuel tank through a check valve (and through injectors stopped in a position in which drain orifices are uncovered) is a typical cause of hard starting. It can lead to bent connecting rods if the fuel draining into the cylinder causes heavy detonation or hydrostatic lock. Failure of a return line valve in a moving vehicle with a high tank may permit dangerous overfueling when some types of injectors are used.

Air trapped in the fuel lines can prevent fuel from being picked up during starting and is difficult to purge from the lines. Air trapped in high drain loops creates a siphoning condition, which often results in excessive surge during engine idle. Moisture collecting in low loops may freeze and cause enough restriction to create a hard or no-start condition.

When fuel tanks are located above the injector level, it is advisable to have a warning light to prevent the operator from leaving the ignition switch in the on position while the engine is *not* running. In this condition, fuel can drain into the engine, causing a no-start condition due to heavy detonation or hydrostatic lock.

The industrial fuel supply system shown in **Figure 15-10** includes the supply tank, hand fuel prime pump, built-in supply tank in the engine, filter, gauges, and the necessary piping and fittings. The hand–operated pump, when installed, is used to fill the fuel manifolds and deliver it through the filter to the engine inlet. A pressure of about 15 psi (103.4 kPa) is built up and maintained in this fuel center. A relief bypass valve is provided in the piping between the pumps and the filter to protect the system in case of excessive filter clogging. Excessive fuel not used by the injection pumps returns to the supply tank through the clean fuel drain. **Figure 15-11** illustrates a typical marine fuel system installation.

Fuel Gauges Systems

There are three basic types of electric ***fuel gauge*** systems widely used in diesel-powered vehicles: the balancing coil gauge system, the thermostatic gauge system, and the electronic (digital) gauge system. Most gauge systems include a sending unit in the primary fuel tank and a fuel gauge mounted in the cab instrument panel.

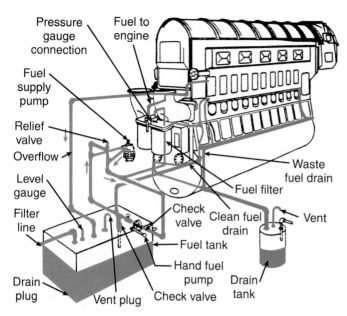

Figure 15-10. *Components of typical industrial marine fuel supply system.*

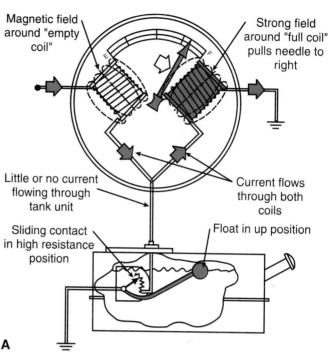

A

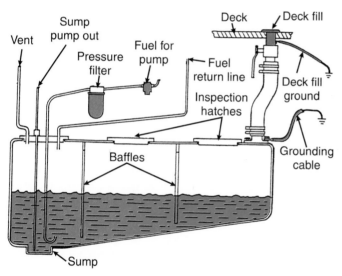

Figure 15-11. *Typical pleasure boat diesel fuel installation. Specifications of the United States Coast Guard must be followed in the USA.*

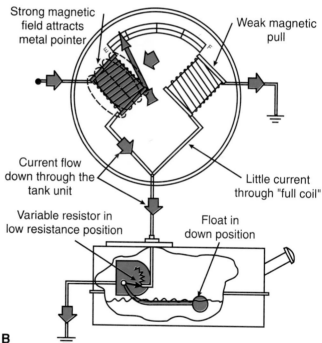

B

Figure 15-12. *Balanced coil fuel gauge. A—Full. B—Empty.*

Balancing Coil Fuel Gauge System

The tank unit for the **balancing coil fuel gauge** system has a sliding contact that moves back and forth as the position of the float changes. The resistance in the unit changes as the contact moves. When the tank is full, current flows through both coils, but the stronger field is around the *full coil* and the needle is pulled to the right. When the float is in the up position, the sliding contact is in the high resistance position. Little or no current is flowing through the tank unit, **Figure 15-12A**. As the tank empties, the float moves downward and the variable resistance also decreases. This permits current to flow easily through the tank unit and to ground. As a result, the magnetic pull of the *full coil* weakens, and the magnetic field around the *empty coil* increases. This pulls the needle to the left towards the empty *(E)* position, **Figure 15-12B**.

Thermostatic Fuel Gauge System

The **thermostatic fuel gauge** system contains a pair of thermostat blades, **Figure 15-13**. Each blade has a heating coil connected in series through the ignition switch to the battery. The tank unit has a float that activates a cam. As the cam moves, it flexes the thermostat blade. Turning on the ignition switch sends current through the heater coils, which, in turn, heat the tank blade. As the tank blade heats up, the dash blade heats up as well, and its movement corresponds to the tank blade movement. Dash blade movement goes through a linkage to the indicator, which moves to the appropriate position on the gauge dial.

When the fuel tank is full and the float is near the top, **Figure 15-13A,** the variable resistance is low and high current flows through the tank unit. This high current heats the thermostatic strip, and pushes the pointer or needle to the full mark. As the tank empties, **Figure 15-13B,** the float moves toward the bottom of the tank, increasing the resistance of the variable resistor. As the current decreases, the thermostatic strip loses some of its heat and the pointer moves toward the empty mark.

Analog and Digital Gauge Systems

Both the balancing coil and the thermostatic gauge systems use *analog dash gauge* displays in which the pointer or

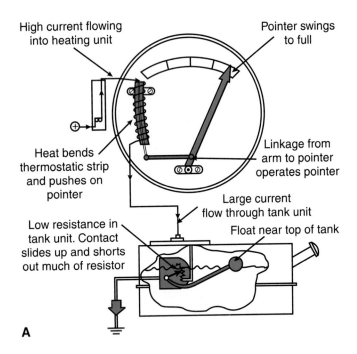

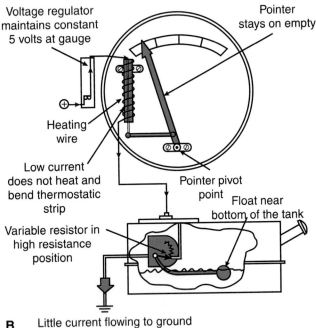

B   Little current flowing to ground

Figure 15-13. Thermostatic fuel gauge. A—Full. B—Empty.

needle swings though an arc formed by a curved row of numbers. The *digital fuel gauge* systems consist of a fuel sensor which reads the amount of fuel in the tank and sends an electrical pulse to a computer or module. The results appear on the dash as a colored light display, **Figure 15-14.** The digital, as well as the analog display types, are used in other gauge systems. Methods of servicing each type of fuel gauge system can be found in the engine service manual.

Fuel Lines

Fuel lines bring the fuel from the tank to the engine and return surplus fuel from the engine to the tank. The three types of fuel lines currently used with diesel engines are:

- ❏ Heavyweight lines: Also called high pressure lines. Used to handle the very high fuel pressures found between the injection pump and the nozzles.
- ❏ Mediumweight lines: Used for low or moderate fuel pressures between the tank and injection pump.
- ❏ Lightweight lines: Used for areas of little or no fuel pressure or for leak-off fuel from the nozzles to the tank or pump.

Engineers responsible for diesel engine and vehicle manufacture select tubing, hoses, and fittings to meet temperature, pressure, chemical, and physical requirements for each part of an installation. Often during repair and replacement, a hose or fitting may be used that looks the same, but is not suitable for the fluid or pressure expected. Therefore, when replacing fuel hose or tubing, hose with equivalent or greater specifications should be used. Hose specifications can generally be found in the appropriate service manual.

Heavyweight Fuel Lines

Heavyweight, or high pressure, fuel lines are usually made of heavy-wall steel tubing and must be capable of

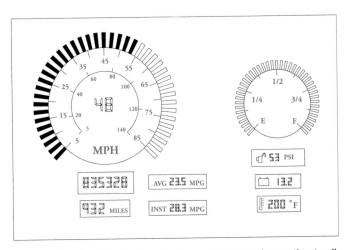

Figure 15-14. This electronic dash display replaces the traditional pointer arrow with digital numbers that indicates the fuel level. Other useful information such as fuel mileage projections are also displayed.

handling fuel pressures as high as 10,000 psi (68 950 kPa) between the metering pump and the injectors. These lines must be properly clamped to the engine and protected from vibration to avoid cracking. Steel tubing of 5/8″ (15.87 mm) outside diameter is suitable for high pressure lines in most mobile installations. Some stationary installations use steel tubing for many lines, but flexible sections are often required for adequate protection.

Some steel tubing is copper-coated to prevent excessive corrosion and to add resiliency. Copper-coated tubing must not be mistaken for copper tubing. Copper tubing is not recommended for any fuel system. This tubing hardens and breaks in a relatively short time in diesel installations. Also, the fittings used with copper tubing are extremely susceptible to leaks.

Note: Hose and tubing must be protected from vibration, pinching, cutting, and abrasion during engine operation and during repair procedures. Anchoring fuel lines and other hose against the engine or solid frame areas will greatly increase their service life.

Flaring Steel Fuel Lines

Steel fuel lines are joined using **flared tubing connections.** A flared connector has three parts: two flare nuts and a threaded flare fitting, **Figure 15-15.** The flare nuts are slid onto the lengths of tubing to be joined. The tubing ends are then flared to increase their end diameters. The angle of the flared end is typically 45°, but 38° flare angles are often used on steel tubing. After the tubing is flared, the flare nuts are threaded onto the fitting from both sides. This pulls the ends of the tubing tight against the fitting to form a leak-free joint.

Tubing is flared using a flaring tool set. For proper flaring, the tube must be cut square and all burrs removed. As shown in **Figure 15-16,** the tube end is clamped into a flaring bar. A special flaring cone is then screwed down into the tubing to expand the tubing end. Flared ends can be single- or double-wall in thickness. Single flare joints are used on medium pressure lines. High-pressure fuel lines typically use double flared joints.

Figure 15-17 shows the steps used in making a single flare. Once the tubing is in place, the cone is tightened

down. A double flare requires two steps, **Figure 15-18.** In the first step a special adapter is pressed down onto the tube end to bend it inward. In the second step the cone is use to fold the bent tubing over onto itself to form the double-wall thickness.

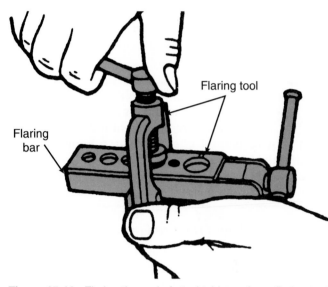

Figure 15-16. *Flaring the end of steel tubing using a flaring tool.*

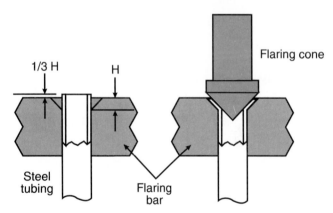

Figure 15-17. *Steps in forming a single wall flared end. A—Place tubing in the flaring bar. B—Tighten down on the flaring cone.*

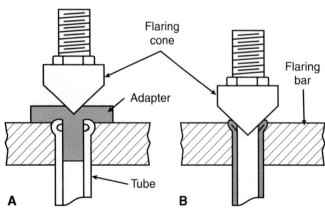

Figure 15-18. *Steps in forming a double-thickness flare. A—Using an adapter to bend tubing inward. B—Using the flaring cone to fold tube over to form the double-thickness end.*

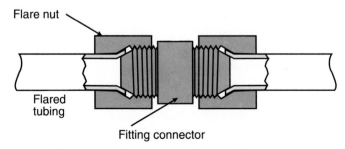

Figure 15-15. *Flared connection used on steel tubing lines.*

Mediumweight Hoses

The construction of mediumweight hose, **Figure 15-19**, is also important. The material used must resist deterioration when exposed to diesel fuel. Most engine manufacturers recommend inner linings of buna-N-synthetic, Teflon®, nylon, or neoprene housed in standard braided-wire hose. Mediumweight hose should handle 250 psi (1723.7 kPa) of pressure safely and should remain pliable at –40°F (–40°C). Temperature resistance up to 200°F (93.3°C) is generally adequate.

Note: Some engine manufacturers are using plastic tubing in fuel systems quite successfully; but replacement tubing specifications require careful checking since some plastics are affected by diesel fuel and by high or low temperatures. Never use rubber hose in a diesel fuel system. Rubber deteriorates when exposed to diesel fuel.

Low Pressure Fuel Line

Low or non-pressure fuel lines may be made of steel tubing, neoprene or mylar hose, or armored neoprene hose. These lines must be large enough in diameter to carry leak-off fuel from the nozzles to the pump or the fuel tank. Fuel drain or leak-off lines with an inside diameter of less than 13/32″ (10.3 mm) may restrict injector drain fuel enough to raise pressure in the injector, reducing engine efficiency.

Fuel Flow Considerations

When the fuel tank is located a considerable distance from the engine, the size of the suction line may need to be increased to overcome flow resistance. Tubing walls themselves create some restriction to flow. The distance through which the fuel must be drawn by the gear transfer pump vacuum also adds to the resistance. If there is excessive length in the system, the resistance can be reduced by increasing line size.

Checking for Resistance

Suction flow resistance checks are taken at the fuel pump inlet fitting by attaching a manometer with the proper fitting (see engine service manual). In all cases, the manometer line should be attached on top of the fitting and lead directly to the manometer. There should be a slight rise (never a drop) between the engine fitting and the manometer inlet fitting to prevent fuel from entering the manometer line and causing a false reading.

A suction line restriction reading is an indication of the amount of vacuum required to pull fuel into the pump. If the ends of the mercury column in the manometer are more than 2″ (5.08 cm) apart (the mercury should rise in the column to which the suction line is attached when measuring vacuum), restriction is above the 4″ (10.2 cm) limit. When measuring pressure, the mercury column to which the line is attached should go down. The difference is read and totaled in the same manner as the vacuum reading. With two manometers and the proper fittings, vacuum and pressure restriction can be read at the same time when checking for fuel system malfunctions. Vertical sections in lines that are 15″ (38.1 cm) high or above, add 1″ (2.54 cm) of mercury restriction to either a vacuum or pressure system.

Fuel Line Inspection

Inspect the fuel lines periodically for loose connections, breaks, or other flaws. Check for any leaks that would pinpoint these problems. Fuel lines must be replaced if they are worn, damaged, or deteriorated. Use the following general guidelines when installing and routing new fuel lines:

- ❑ There should be no loops, sharp bends, or kinks in the lines, since they often add restriction and turbulence. Bends and loops also trap air, increase the possibility of breaks, lead to soft throttle, and cause other engine problems.
- ❑ Any line extending more than 2″ (5.08 cm) below the bottom of the fuel tank must be enclosed in a protective housing.
- ❑ Secure fuel lines to the engine or chassis to prevent damage.
- ❑ Fuel lines should be as short as possible, but long enough to allow movement of any parts to which they are attached.
- ❑ Coat all pipe junctions and tapered threads with an approved sealant.
- ❑ Tighten the line connectors until snug, but avoid overtightening. Overtightening can strip the threads.

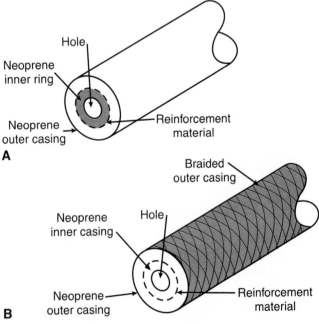

Figure 15-19. *Typical medium pressure fuel line hose construction. A—Hose with neoprene outer cover. B—Hose with braided wire outer cover.*

To prevent metal lines from bending, use only one hand with two wrenches for final tightening, **Figure 15-20.**

Check that new lines and fittings are free of leaks. If air enters through a leak, it can result in a loss of prime in the fuel system.

Drains and other bottom fittings should not extend more than 3/4″ (19.05 mm) below the lowest part of the fuel tank or sump.

When replacing a fuel line, the new line must match the size, shape, and length of the original line.

Normally, it is best to replace a bad diesel injection line with a new factory-formed unit, **Figure 15-21.** Line diameter and length are very critical to ensure correct injection timing and fuel quantity.

Fuel Transfer Pumps (Supply Pumps)

The fuel transfer pump is usually driven by the engine camshaft. It rotates at half engine speed for a four-stroke diesel engine and at engine speed for a two-stroke engine. Many fuel transfer pumps have a hand-prime lever. After the

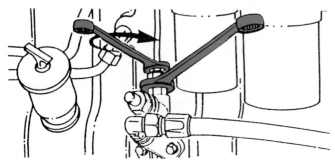

Figure 15-20. When tightening or loosening flare nuts use two wrenches. (Cummins Engine Co., Inc.)

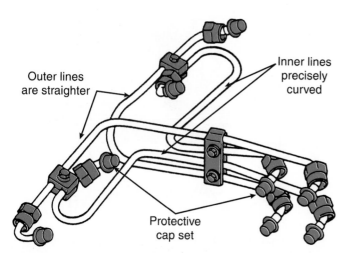

Figure 15-21. OEM drop-in replacement fuel line assembly. (Cummins Engine Co., Inc.)

fuel system has been serviced, the technician can use this lever to manually prime or purge all air from the system before starting the engine. While the function of fuel transfer pumps are basically the same in all systems, there are several types of pumps, including the mechanically-operated diaphragm pump, the electric diaphragm pump, the plunger pump, and the gear pump.

Mechanically-Operated Diaphragm Transfer Pump

The mechanically-operated *diaphragm transfer pump* consists of an actuator arm; a lever-operated, spring-loaded diaphragm; an inlet valve; and an outlet valve, **Figure 15-22.** The engine's camshaft has an eccentric that applies power to the pump's actuator arm.

As the camshaft rotates, the eccentric causes the actuator arm to rock back and forth. The inner end of the actuator arm is connected to a spring-loaded diaphragm located between the upper and lower pump housings. The rocking motion of the actuator arm pulls the diaphragm down and then releases it. The spring, which is located under the diaphragm, forces the diaphragm back up. In this way, the diaphragm moves up and down as the actuator arm moves back and forth.

As the diaphragm is pulled down, a low pressure area is created above it. This causes the higher atmospheric pressure in the tank to push fuel past the inlet valve into the center chamber of the pump. When the diaphragm is forced back up by the spring, the pressure above the diaphragm increases. This pressure closes the inlet valve and opens the outlet valve, forcing fuel to travel through the pump outlet and into the injector pump.

If the needle valve in the injector pump float bowl closes the inlet valve, fuel cannot enter the injection pump. This means the transfer pump can no longer deliver fuel. When this happens, the actuator arm continues to move back and forth, but the diaphragm spring cannot force the diaphragm up. Normal operation cannot resume until the needle valve in the float bowl opens the inlet valve, allowing the spring to force the diaphragm up.

Figure 15-22. Diaphragm-type transfer pump. (Cummins Engine Co., Inc.)

Electric Diaphragm Transfer Pump

The *electric diaphragm transfer pump* is sometimes used in place of the mechanical pump. It has a flexible metal bellows that is operated by an electromagnet, **Figure 15-23.** Turning on the ignition switch connects the electromagnet to the battery. The electromagnet then pulls down the armature, extending the bellows. A low pressure area is created in the bellows, which allows fuel to enter the bellows through the inlet valve.

When the armature reaches its lowest level, it opens a set of contact points. This disconnects the electromagnet from the battery, allowing the spring to push the armature up and collapse the bellows. The fuel in the bellows is forced through the outlet valve to the injector pump. As the armature reaches its highest level, the contact points close. This reconnects the battery to the electromagnet and the cycle repeats itself.

Plunger Transfer Pump

A *plunger transfer pump* is a high pressure pump that is usually mounted on the injection pump housing, **Figure 15-24.** An eccentric on the injection pump camshaft actu-

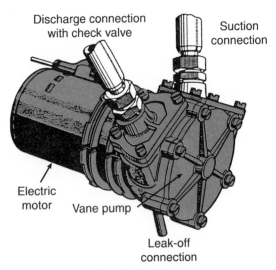

Figure 15-23. *Typical electric diaphragm pump. (AMBAC International)*

Figure 15-24. *Typical plunger fuel transfer pump. (Cummins Engine Co., Inc.)*

ates the pump's plunger. When the plunger moves toward this camshaft, fuel is drawn through the transfer pump's inlet valve, filling the chamber above the plunger.

When the eccentric moves the plunger away from the injector pump camshaft, the outlet valve opens and the inlet valve closes. As the eccentric continues to rotate, the plunger spring pushes the plunger back toward the camshaft, pressurizing the fuel. The plunger then forces the fuel into the injector pump.

When the injection pump receives too much fuel from the supply pump, pressure remains behind the plunger. Both valves are closed, preventing the plunger from returning to its starting point and taking in more fuel. When the injection pump can receive the fuel, the plunger will return to its starting position.

Gear Transfer Pump

The *gear transfer pump* is a positive-displacement pump that is generally driven directly from the end of the governor shaft. A built-in regulating valve prevents damage to the injection equipment in the event of an obstruction in the fuel supply system. An overflow valve returns excess fuel to the supply tank. **Figure 15-25** illustrates a typical gear fuel transfer pump. The fuel pump body is recessed to provide running space for the pump gears. Recesses are also provided at the inlet and outlet cavities, **Figure 15-26.**

The small hole, *A*, permits the fuel oil in the inlet side of the pump to lubricate the relief valve at its outer end and to eliminate the possibility of hydrostatic lock, by providing relief of excess discharge pressure. When the discharge pressure is great enough to move the relief valve back from its seat, fuel reenters the inlet side of the pump through hole *C*. Part of the relief valve may be seen through hole *C*. Cavity *D* provides escape for the fuel squeezed out of the gear teeth on the discharge side of the pump.

In operation, fuel enters the pump on the suction side and fills the space between the gear teeth. The gear teeth

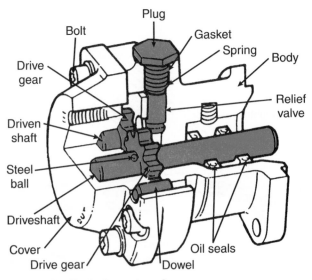

Figure 15-25. *Typical gear transfer pump. (Detroit Diesel Corp.)*

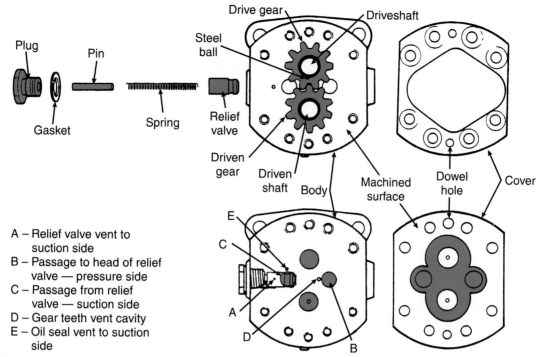

A – Relief valve vent to
 suction side
B – Passage to head of relief
 valve — pressure side
C – Passage from relief
 valve — suction side
D – Gear teeth vent cavity
E – Oil seal vent to suction
 side

Figure 15-26. Internal details of gear-type fuel transfer pump. (Detroit Diesel Corp)

then carry the fuel oil to the discharge side of the pump and, as the gear teeth mesh in the center of the pump, the fuel is forced out into the outlet cavity. Since this is a continuous cycle and fuel is continually being forced into the outlet cavity, the fuel flows from the outlet cavity into the fuel lines and through the engine fuel system pressure.

Inspecting and Cleaning a Diaphragm Transfer Pump

Once the diaphragm transfer pump is removed from the fuel system, visually inspect it for cracks or other damage. Check the weep hole area for signs of fuel leakage, **Figure 15-27.** Clean the pump in solvent, **Figure 15-28,** and blow it dry with compressed air. Carefully examine the camshaft lever and the return spring for excessive wear.

Figure 15-28. After cleaning the transfer pump, blow it dry with clean compressed air. (Cummins Engine Co., Inc.)

> ⚠ **Warning: Check the MSDS and wear the recommended protective equipment while using solvents.**

To complete the inspection of the transfer pump, inspect the diaphragm by blocking the pump inlet with your finger and operating the priming lever, **Figure 15-29.** A good pump will have suction that will not bleed down until your finger is removed from the inlet. With a diaphragm-style fuel transfer pump, repair is usually not practical. If the pump is defective, it must be replaced.

Rebuilding Piston Fuel Transfer Pump

The first step in rebuilding a piston-style fuel transfer pump is to clean any debris from the fuel line fitting and

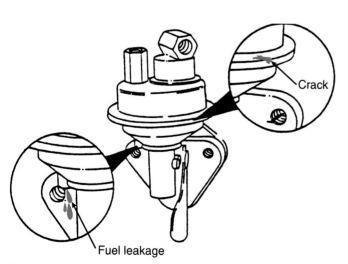

Figure 15-27. A cracked casing and leakage from the weep hole are two signs of a damaged pump.

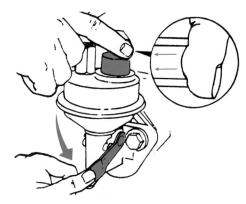

Figure 15-29. Checking a diaphragm transfer pump. (Cummins Engine Co., Inc.)

the fuel transfer pump. Next, secure the pump in a vise, taking care not to damage the housing. Remove the rubber boot from the hand-prime fitting. Then remove the three fittings illustrated in **Figure 15-30A**. Remove all internal pump components, making sure that the check valve gaskets are removed, **Figure 15-30B**.

 Warning: The hand-prime fitting and the inlet fitting are spring loaded. Sudden removal of these fittings can cause personal injury.

Thoroughly flush the pump housing with cleaning solution to remove debris. Then reassemble the pump with the new components supplied in a pump rebuild kit. Torque the fittings to the specifications given in the rebuild

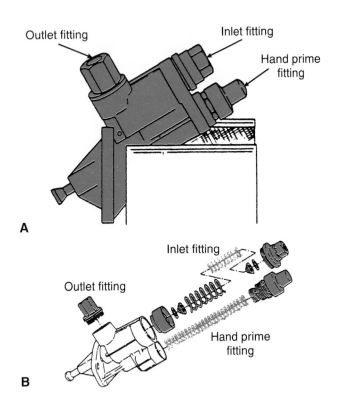

A

B

Figure 15-30. Steps in rebuilding a piston transfer pump. (Cummins Engine Co., Inc.)

kit or the service manual. After reassembly, the pump is ready to be reinstalled in the fuel system.

 Note: Make sure the check valves are installed to open in the direction of the fuel flow. Improper installation of the check valves will result in low power from the engine.

Gear Fuel Transfer Pump Service

When it is necessary to service this type of fuel transfer pump, proceed in the following manner:

1. Clean all parts and dry with compressed air.

Warning: To prevent possible personal injury, wear adequate eye protection and do not exceed 40 psi (276 kPa) air pressure.

2. Inspect the drive coupling and hub for wear or damage. Replace as necessary.
3. Oil seals, once removed, must be discarded and replaced with new seals. Oil seals must be removed whenever the fuel pump driveshaft is withdrawn. Oil seals must be free of nicks, cuts, or cracks.
4. Inspect the drive and idler gear shafts for scoring or wear. Replace shafts as necessary. If the driveshaft is grooved from contact with oil seals, replace the shaft.
5. Check the gear teeth for scoring, chipping or wear. If necessary, replace the gear.
6. The relief valve must be free of score marks and burrs. If the relief valve is scored and cannot be cleaned up with fine emery cloth, it must be replaced. The fuel transfer pump drive is usually non-serviceable. If the bearing is worn, scored, or otherwise unusable, the fuel pump drive assembly must be replaced.

Summary

In a typical diesel fuel system, fuel is pulled from the fuel tank through a primary fuel filter by a fuel transfer pump. From the fuel transfer pump, the fuel is pushed through a secondary fuel filter and into the fuel manifold in the fuel injection pump housing.

Segments of the fuel system may operate at low, moderate, or high fuel pressures. Fuel lines that link the fuel tank to the fuel transfer pump and the transfer pump to the injection pump operate at low or moderate pressures. The fuel injection pump or metering pump boosts low or medium fuel pressures to the high pressures needed for injection. Return lines running from the injectors to the fuel tank may operate at very low pressures or no pressure.

The low pressure fuel system should be designed to provide an unrestricted supply of fuel to the transfer pump and the fuel injection pump. Various components within the system must also be designed to comply with all regulations, yet be accessible for servicing.

Sedimenters, or agglomerators, must be fitted on the suction side of the transfer pump to minimize the chance of water freezing in the pipes and to keep water droplets from breaking up. The sedimenter should also be positioned below the fuel level in the tank to allow easy draining. Fuel heaters should be located prior to the first point at which wax can cause a fuel restriction. The final filter in a system employing three or more filters should be mounted away from the engine cooling fan and above the level of the fuel injection pump to prevent siphoning and to assist venting.

Fuel tanks on many types of mobile vehicles are subject to federal government regulations. They require a non-spill vent and a suction line entry fitting positioned above the tank's full line. The fuel tank should have at least 5% of its total area above the "full line" to provide for fuel expansion.

In addition to the tank air vent, three other openings are normally found in the tank: one for filling, one for discharge, and one for drainage. A fourth opening is sometimes used for leak-off fuel. To prevent fuel from spilling from the tank if the vehicle overturns, most modern diesel fuel systems are equipped with some type of fuel cut-off device. The two most common are the fuel safety shut-off solenoid and the inertia switch.

There are three types of fuel gauge systems used today: the balancing coil gauge system, the thermostatic gauge system, and the electronic (digital) gauge system.

The three types of fuel lines currently used with diesel engines are heavy-, medium-, and lightweight lines. High pressure fuel lines are made of heavy-wall steel tubing. Mediumweight hoses are made of braided wire that can remain pliable at low temperatures. Low or non-pressure fuel lines may be made of steel tubing, neoprene or mylar hose, or armored neoprene hose.

Suction restriction checks are taken at the fuel pump inlet fitting by attaching a manometer with the proper fitting (see engine service manual). Drain restriction is checked in the same way at the head drain fitting. Fuel lines require periodic inspection for loose connections, breaks, or other flaws. Worn, damaged, or deteriorated fuel lines must be replaced.

Important Terms

Fuel tank	Fuel heater
Fuel transfer pump	Baffle plates
Fuel filters	Vents
Fuel injection pump	Suction tube
Fuel injection nozzle	Dual fuel tanks
Return lines	Tandem fuel tanks
Gravity-fed	Fuel gauge

Balancing coil fuel gauge	Flared tubing connections
Thermostatic fuel gauge	Diaphragm transfer pump
Analog dash gauge	Plunger transfer pump
Digital fuel gauge	Gear transfer pump
Fuel lines	

Review Questions–Chapter 15

Do not write in this text. Place your answers on a separate sheet of paper.

1. Which of the following can be used as fuel cutoff devices?
   (A) Inertia switches.
   (B) Electric solenoids.
   (C) Double pole switches.
   (D) Both A & B.

2. Technician A says to use a manometer at the fuel pump inlet fitting to check for fuel line suction restrictions. Technician B says the proper fitting for this test is a head drain fitting. Who is right?
   (A) A only.
   (B) B only.
   (C) Both A & B.
   (D) Neither A nor B.

3. *True or False?* Fuel return lines and fuel pickup lines are usually mounted close to one another in the fuel storage tank.

4. *True or False?* Most over-the-highway vehicles employ dual, or tandem fuel tanks.

5. *True or False?* Neoprene hose is acceptable for high pressure fuel lines.

6. Technician A says that baffles in the fuel tank help mix warm return fuel with the cooler fuel in the tank. Technician B says that fuel expansion is caused by warm fuel being returned to the tank from the injectors. Who is right?
   (A) A only.
   (B) B only.
   (C) Both A & B.
   (D) Neither A nor B.

7. Technician A says that when the fuel tank is located higher than the injectors, precautions must be taken to prevent hydrostatic lock-up of the engine cylinders. Technician B says that you should install a check valve in the injector drain line to prevent hydrostatic lock-up. Who is right?
   (A) A only.
   (B) B only.
   (C) Both A & B.
   (D) Neither A nor B.

8. Technician A says you should install a rubber hose in a low pressure leak-off line from the injector back to the fuel tank. Technician B says you should properly clamp high pressure fuel lines to the engine block to prevent vibration. Who is right?
   (A) A only.
   (B) B only.
   (C) Both A & B.
   (D) Neither A nor B.

9. Any line extending more than _____ below the bottom of the fuel tank must be enclosed in a protective enclosure.
   (A) 1" (2.54 cm)
   (B) 2" (5.08 cm)
   (C) 3" (7.62 cm)
   (D) 4" (10.16 cm)

10. Technician A says you should use two wrenches and one hand when final tightening a connection to avoid bending the lines. Technician B says you should tighten steel line connections as tight as possible to avoid leaks. Who is right?
    (A) A only.
    (B) B only.
    (C) Both A & B.
    (D) Neither A nor B.

ASE-Type Questions

1. In a typical fuel injection system, fuel is pulled from the fuel tank through a primary fuel filter by a _____.
   (A) fuel injection pump
   (B) fuel transfer pump
   (C) hydraulic pressure system
   (D) governor

2. Fuel lines running from the injection pump to the fuel injector are _____.
   (A) low pressure lines
   (B) medium pressure lines
   (C) high pressure lines
   (D) gravity-fed

3. Technician A says that sedimenters or agglomerators should be fitted on the suction side of the fuel transfer pump. Technician B says these devices should be located as close to the fuel tank as possible. Who is right?
   (A) A only.
   (B) B only.
   (C) Both A & B.
   (D) Neither A nor B.

4. The final fuel filter on a system employing three or more filters should be mounted away from the engine cooling fan and _____ the level of the fuel injection pump.
   (A) below
   (B) above
   (C) level with
   (D) None of the above.

5. A fuel tank should have at least _____% of its total area above the full line to provide for fuel expansion.
   (A) 1
   (B) 5
   (C) 10
   (D) 15

6. Balancing coil, thermostatic, and electronic (digital) are all types of modern _____.
   (A) fuel shut-off devices
   (B) fuel gauges
   (C) fuel injectors
   (D) transfer pumps

7. Technician A says the fuel transfer pump is normally driven off the crankshaft. Technician B says it is normally driven off of the camshaft. Who is right?
   (A) A only.
   (B) B only.
   (C) Both A & B.
   (D) Neither A nor B.

8. High pressure fuel lines are constructed of _____.
   (A) wire braided hose
   (B) heavy-wall steel tubing
   (C) PCV plastic tubing
   (D) Both A & B.

9. _____ of the fuel flowing through the injectors returns to the fuel tank.
   (A) 20%
   (B) 40%
   (C) 50%
   (D) 80%

10. Fuel suction tubes are prone to _____.
    (A) air leaks due to tube damage
    (B) rust due to condensation in the tank
    (C) damage due to vibration
    (D) All of the above.

Chapter 16

Fuel Filters and Conditioners

After studying this chapter, you will be able to:
- ☐ Explain the importance of fuel filters.
- ☐ Name several types of fuel filters.
- ☐ Describe the various fuel filtration systems.
- ☐ Change both bolt-on and spin-on fuel filters.
- ☐ Explain the purpose of major types of diesel fuel conditioners.

Delivering clean diesel fuel to the engine's injection pumps is one of the best methods of preventing premature injector and engine wear. Harmful contaminants include dirt, water, and wax. To remove damaging contaminants engine designers typically filter the diesel fuel not once but several times before it reaches the injectors. This chapter discusses the operation and service of various prefilters, filters and fuel conditioners used in diesel fuel delivery systems.

Damage by Dirt and Water

The quantity and size of dirt particles in a diesel fuel system do not have to be large to cause problems. As little as .175 ounces (5 grams) of dirt in 25,000 gallons (94 000 liters) of diesel fuel can cause serious fuel injection pump wear. Abrasive dirt particles that range in size from 5-20 microns are the most damaging. Keep in mind that one micron measures one-one thousandth of a millimeter.

As mentioned previously, dirt is not the only contaminant found in diesel fuel. Water can also do serious damage to the fuel system. Water can corrode injection system components. It can freeze, preventing fuel flow, and it can turn into steam damaging the injection nozzles. Water in a diesel fuel is also responsible for a condition known as *microbial infestation.*

Unlike gasoline, which gives off a vapor barrier that keeps water out, diesel fuel easily absorbs water. Unfortunately, the very design of most diesel engines provides a way for this water to condense out of the fuel and cause problems. As described in **Chapter 15,** excess diesel is delivered to the fuel injectors to help keep them running cool. The warmed diesel fuel is then returned to the fuel tank. As the warm fuel cools, any water held in the fuel condenses out and settles to the bottom of the fuel tank. This water will cause major problems if it is allowed to circulate through the system.

To minimize water in the fuel supply, many fleet owners require that their trucks be filled at the end of the day instead of waiting until morning. Permitting a freshly filled tank to stand overnight allows water and any other contaminants to settle so they can be drained through the tank's drain cock. A full tank also allows no extra air space. Cool air contains moisture that could condense out and settle into the tank. In cold weather, however, it is best to drain any water from the tank in the evening before it has a chance to freeze and damage system components. A drain cock can be used to remove much of the water collecting in the bottom of a diesel fuel storage tank. However, virtually all diesel engines require various combinations of prefilters, filters, and water separators to ensure water and dirt are kept out of injectors and engine components.

Filtration Devices

Prefilters are often the first type of filtration devices located after the fuel tank. They offer little or moderate resistance to fuel flow and have no moving parts. Prefilters are used to collect larger dirt particles and water. The contaminants collect in the base of the unit and are periodically drained off via a *drain cock*. The two most common types of prefilters are sedimenters and agglomerators.

Sedimenters

As their name suggests, *sedimenters* collect large dirt particles and water droplets through a process known as sedimentation. In the sedimenter prefilter shown in **Figure 16-1,**

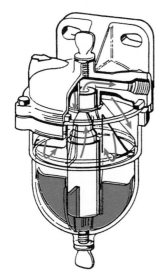

Figure 16-1. *Fuel flow inside a sedimenter type prefilter or filter.*

fuel enters the unit, flows over and around a conical section, and passes through a narrow gap around the periphery of the conical section. It then flows radially towards the center and up and around the unit. The radial flow causes water and heavier abrasive particles to separate from the fuel and settle into a collecting bowl. A small bypass hole ensures that the unit is self venting and that it operates completely full of fuel. Some sedimenters are equipped with an electronic probe that senses the presence of water in the collecting bowl and provides a signal when the unit requires draining. Many sedimenters contain a float that rises with the water level and closes the fuel outlet when the unit requires draining.

Agglomerators

Agglomerators are prefilters that combine water separation and collection with efficient fuel filtration, **Figure 16-2**.

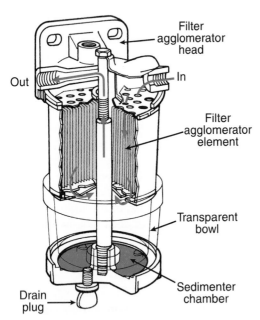

Figure 16-2. *Fuel flow inside a agglomerator type prefilter.* (Lucas CAV Limited)

Fuel enters the agglomerator, travels to the top of the filter element, and then flows down through the element. The fine pores of the paper filter element retain solid particles, while fine water droplets are forced through the pores and gather into larger drops (agglomerate) that are deposited at the base of the unit. Clean fuel then flows up the center tube of the filter element and out of the agglomerator.

When a prefilter is used primarily for water removal, it should be located before the fuel transfer pump. The action of the fuel going through a pump or other restriction before reaching the prefilter lowers the efficiency of the water separation. A typical water separating prefilter can remove approximately 95% of the water in the fuel flow.

However, efficiency may be dramatically lowered if water reaches the 3/4 full level in the housing body. At this point, water in the fuel can pass through the unit without dropping out. This makes it important to maintain a strict maintenance schedule.

To drain a typical water separating prefilter, open the drain valve in the drain line and vent the valve at the top of the unit. Let the water drain until it is all out of the water separator and close both valves. In some installations, an agglomerator unit is placed in series after a sedimenter to provide maximum fuel prefiltering.

Fuel Filters

Diesel **fuel filters** must be capable of trapping extremely small impurities. The porosity of the filtering material will determine the size of the impurities it can remove, **Figure 16-3**. Remember that a typical fuel injector nozzle is set to tolerances that are measured in microns. Therefore, it is necessary to filter impurities in 3 micron diameters or smaller in some fuel filtering systems. Diesel fuel filter elements, like oil filter elements, fall into two common construction categories, depth filters and surface filters.

Depth Filter Elements

Depth fuel filter elements are generally made of a woven cotton sock, **Figure 16-4A**. The most popular material used for depth-type elements is a cotton thread that is

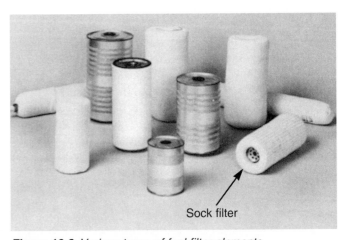

Figure 16-3. *Various types of fuel filter elements.*

blended with a springy supporting material. For special applications, other materials can be used. A depth element can be designed for use in either a shell/base bolt-on assembly or a spin-on filter assembly. These filters are typically used as primary filters and are located between the fuel tank and the transfer pump.

Surface Filter Elements

The most common material used for **surface filter elements** is pleated paper, **Figure 16-4B**. The paper is made of processed cellulose fiber. The fiber is treated with a phenolic resin, which acts as a binder. The physical properties of the paper—thickness, porosity, tensile strength, basis weight, and micron rating can be very closely controlled in the manufacturing process.

Filter Configurations

Like oil filters, there are also two basic styles of fuel filters, bolt-on filters (element or cartridge filters) and spin-on filters. The **bolt-on filters** generally consist of a shell, a cover, and a replaceable filter element, **Figure 16-5.** A shell gasket, a cover nut or bolt, and a cover nut or bolt gasket are used to make the assembly oil tight, **Figure 16-5.** A bolt-on filter can be replaced by a time-saving spin-on filter by the installation of a central stud kit. These kits can be installed in a relatively short time.

Spin-on filters are essentially one-piece filters which reduce downtime and labor costs associated with filter changes, **Figure 16-6A**. Spin-on filters also ensure proper gasket installation on every change. In two-stage filtration systems that utilize spin-on filters, the primary and secondary filters are sometimes designed to use the same head, **Figure 16-6B**.

Filtration Systems

Diesel engine fuel filtration systems may be equipped with one, two, or three fuel filters. A few systems use more than three filters. Single-stage filtration systems contain only

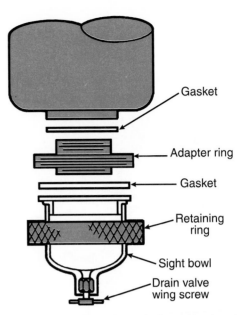

Figure 16-5. *Typical bolt-on filter design with clear bowl and drain valve.*

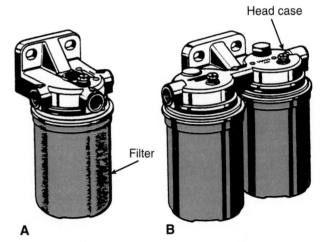

Figure 16-6. *A—Single spin on fuel filter. B—Dual spin on filters that share the same head case.*

one filter and are found on many automotive diesel engines. A low-restriction strainer, found on the suction side of the fuel transfer pump, can be used to stop large particles from entering the system. The main filter, which is located on the pressure side of the fuel transfer pump, must provide efficient operation with low fuel restriction. A surface filter element is generally used in a single-stage filtration system, **Figure 16-7.**

Many diesel engines are equipped with a two-stage filtration system that contains a primary filter and a secondary filter. These filters are not interchangeable and are often marked with the letter *P* or *S* (primary or secondary). In some cases, a color code on the shell is used to identify the filter type. The primary filter (sometimes called a fuel strainer or auxiliary filter) is generally installed between the fuel tank and the transfer pump, **Figure 16-8.** The secondary filter is installed between the transfer pump and the injectors. Some diesel applications may locate the primary and secondary filters side-by-side.

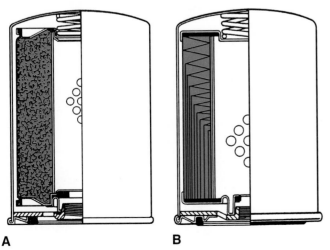

Figure 16-4. *Filter elements used in diesel fuel systems. A—Depth type. B—Surface type.*

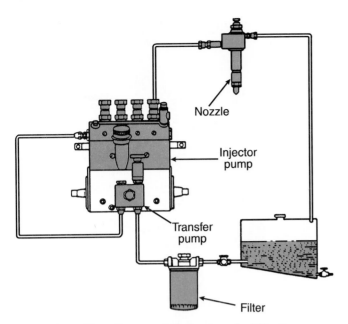

Figure 16-7. *Flow diagram of single-stage fuel filtration.*

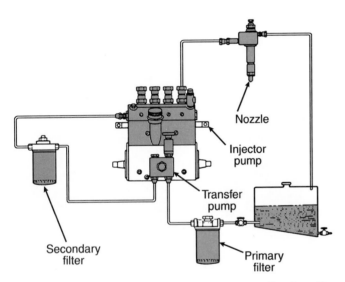

Figure 16-8. *Flow diagram of two-stage fuel filtration. The second filter in this dual arrangement provides additional protection against damage to the pump from dirt, water, and other contaminants in the fuel supply.*

Primary Filters

The *primary filters* offer low restriction to flow because it is mounted on the suction side of the pump, where there is relatively low pressure. This filter has the job of protecting the transfer pump and reducing the load on the secondary filter. An efficient primary filter will greatly increase the life of the secondary filter.

Primary filter elements are usually of the depth construction. Depth filter elements are perfect for primary filters because they offer low restriction to fuel flow and provide the necessary filtering efficiency. In addition, depth filter elements are extremely efficient in removing waxes, gums, and resins without clogging. They are used to remove particles that range in size from 20-30 microns.

Secondary Filters

The *secondary filters* are designed to prevent damaging contaminants from entering the injectors. Secondary filters usually contain surface filter elements. These filters are mounted on the pressure side of the transfer pump. Therefore, restriction is not a determining factor. The filter used must be highly efficient in order to remove fine particles (as small as 3-5 microns) that may damage the injectors. The internal pressure in the secondary filter should not exceed 60 psi (413.7 kPa).

Some diesel fuel systems are equipped with a third filter for added protection against fuel contamination. As shown in **Figure 16-9**, the primary filter is normally located in the line that runs between the fuel supply tank and the transfer pump. The secondary filter and the final filter are located between the fuel transfer pump and the injector metering pump.

Dual Filters

Dual filters are two filters of the same or different type installed one right after the other in the fuel line. Dual filter configurations are normally found on heavy duty vehicles. Dual filters can be configured in series or in parallel. In a series filter arrangement, all of the fuel goes through the first filter and then through the second, **Figure 16-10A.** In a parallel filter arrangement, a portion of the fuel passes through each filter, **Figure 16-10B.**

Series configurations are often used when the engine is exposed to extremely dusty conditions, such as on construction sites. Series filters clean diesel fuel better in these conditions because the second filter will pick up fuel missed by the first filter. However, parallel filters have the advantage of being able to move a larger volume of fuel through the system faster than the series filter.

Fuel Filter/Water Separators

Many diesel engines used in mobile applications are equipped with combination *fuel filter/water separators.*

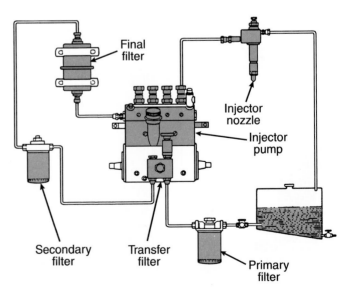

Figure 16-9. *Three-stage fuel-filtering system showing primary, secondary, and final filters. (AMBAC International)*

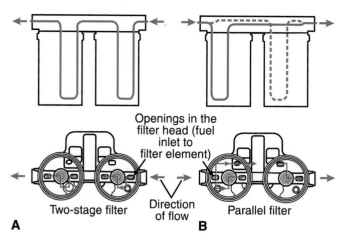

Figure 16-10. Dual fuel filter arrangements. A—Two-stage series. B—Parallel.

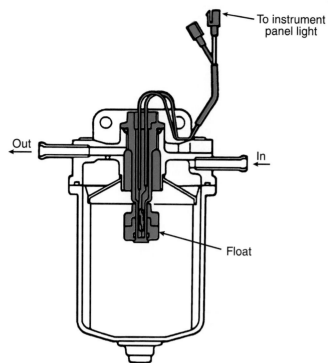

Figure 16-11. Fuel/water flow separator filter with float level sensor. (Lucas CAV Limited)

These units provide both water and particle removal. The most basic filter/separator design uses a full-flow filter to separate both water and solid contaminants from the fuel stream, **Figure 16-11**. The filter/separator shown in **Figure 16-11** is also equipped with a fuel shut-off mechanism. When water in the filter housing reaches its maximum safe level, a valve closes to shut off fuel flow to the injection pump. This prevents water laden fuel from reaching the pump. A warning light circuit is also activated. **Figure 16-12** illustrates a typical warning light/buzzer circuit for water level monitoring.

Three-Stage Filter/Separators

More advanced filter/separators use three separate stages to remove water and solid particles. Each stage removes solids and water droplets of progressively smaller sizes. Fuel flow inside a **three-stage filter/separator** is shown in **Figure 16-13**. The first stage of filtration, known as **separation,** removes contaminants as small as 25-30 microns in diameter. Fuel enters the housing and is directed to the top of a cone shaped baffle located above the filter bowl. As it

passes over the baffle, a device called a *turbine centrifuge* generates centrifugal force that separates solid particles and water droplets from the fuel. These contaminants settle to the base of the bowl while the partially cleaned fuel is directed upward to the second stage of the process.

The second stage of filtration is called **coalescing.** Extremely lightweight droplets in the 10-15 micron size range remain suspended in the fuel. These tiny droplets flow upward with the fuel into the lower section of the filter/separator shell. The droplets begin to bead on the inner walls and on the bottom of the filter element. These beads gradually increase in size and weight until gravity forces them back into the bottom of the bowl.

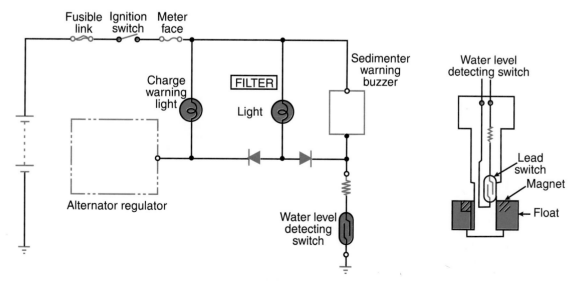

Figure 16-12. Electric schematic for water level warning light and buzzer.

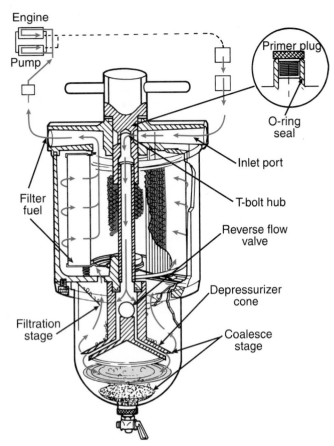

Figure 16-13. Fuel/water flow operation through a heavy-duty filter/water separator assembly. (Baldwin Filters, Clarcor Group)

The third stage of water particle removal is simple filtration. The fuel flows upward through the fine filter element where the tiniest particles are removed. The clean, water-free fuel exits the filter/separator through the outlet port.

Water and dirt are drained from the base of the bowl on a regular basis. A water level sensor may be used to provide maintenance status to the operator. The ultra-fine filter element is also replaceable.

Servicing Filters

To prevent contaminants from reaching the fuel system components, fuel filters must be changed or serviced at specific intervals. During normal engine operating conditions, filters should be changed according to the recommendations found in the service manual. Filters must be changed more often under harsh conditions. When servicing a series filter, the first-stage filter is generally changed more often than the second-stage filter. The second-stage filter is often replaced with every third first-stage filter replacement.

Servicing Bolt-On Filters

In the case of bolt-on filters, it is also important to periodically check the water traps under the filters and drain any water or sediment collected there. The following is a typical procedure for changing the element in a bolt-on filter:

1. If the engine has been running, allow it to cool.
2. Close the fuel shutoff valve.
3. Clean the outside of the filter housing.
4. Place a container under the filter and open the drain cock. Loosen the cover nut or bolt just enough to allow the fuel to drain out freely. Then close the drain cock.

> **Note: The wiring harness, starting motor, and other electrical equipment must be shielded during the filter change. Diesel fuel can permanently damage electrical insulation.**

5. While supporting the shell, unscrew the cover nut or bolt and remove the shell. Remove and discard the filter element, shell gasket, cover nut or bolt gasket, and, if used, the cover bolt snap ring.
6. Wash the shell thoroughly with clean fuel and dry it with compressed air, **Figure 16-14A.**
7. Examine the new element's seat and the retaining ring to make sure they have not slipped out of place. Check the spring by pressing on the element seat. When released, the seat must return to its proper position against the retaining ring.
8. Place a new gasket in the recess of the shell and install a new gasket on the cover nut or bolt. Install the new element, **Figure 16-14B.**

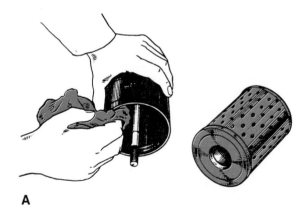

A

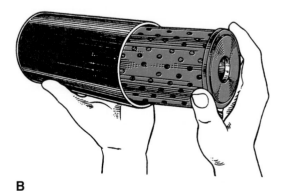

B

Figure 16-14. A—Cleaning the shell of a bolt-on filter. B— Installing the new filter element.

9. Lubricate the new filter element with clean diesel fuel.

10. Install the filter element in the shell.

11. Place the shell and element in position over the cover. Then thread the cover bolt (or nut) into the central stud.

12. With the shell and the gasket properly positioned, tighten the cover bolt or nut just enough to prevent fuel leakage.

13. Remove the pipe plug at the top of the cover and fill the shell with fuel. A fuel system primer may be used to prime the entire fuel system.

14. Start the engine and check the fuel system for leaks.

Replacing Spin-On Filters

The following is a typical procedure for changing a spin-on filter:

1. Clean the outside of the filter housing.

2. Unscrew the filter by hand or with a filter wrench, **Figure 16-15A**. Discard the old filter.

3. Clean the gasket surface on the filter head using a lint-free cloth, **Figure 16-15B**.

4. Clean and reuse any screens (if employed) that are in good condition. Replace the spud gasket or the O-ring seal.

5. Fill the new filter with clean diesel fuel, **Figure 16-15C**. Coat the seal gasket lightly with clean diesel lubricating oil, **Figure 16-15D**.

6. Install the new filter element and gasket, **Figure 16-15E**. Tighten the element one-half turn beyond gasket contact, **Figure 16-15F**. Square-cut sealing gaskets or rings are generally used to form a seal between the mating surfaces of spin-on filters. A large sealing gasket affixed to the filter seals the outside diameter, while a smaller gasket seals the inside diameter (around the threaded support).

 Caution: Overtightening may distort the threads or damage the filter element's seal.

7. Start the engine, allow it to run for five minutes, and check the filter(s) for leaks.

Bleeding the Fuel System

When changing fuel filters, air frequently enters the fuel system. This air must be bled from the system. **Bleeding** can be accomplished by cranking or running the engine and opening the bleeder valves or the injector nozzle lines momentarily to remove the air. Additional instruction on bleeding a fuel system can be found in the engine service manual. The fuel filter(s) usually can be bled as follows:

1. Open the bleed screw, **Figure 16-16A**.

2. Operate the plunger on the lift pump until the fuel flowing from the filter is free of air, **Figure 16-16B**.

3. Close the bleed screw.

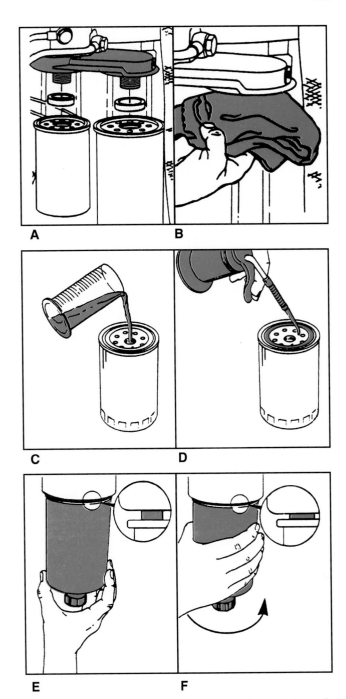

Figure 16-15. *Typical procedure for changing a spin-on fuel filter. (Cummins Engine Co.)*

Air Removal

Air trapped in the diesel fuel system can sometimes cause problems that affect normal operation. Air that is removed from the fuel collects in the fuel filter. This is normal and generally does not slow fuel flow. Nevertheless, it is desirable to eliminate all possible air entry points in the fuel system and to vent the tank adequately to minimize air collection. When air enters the fuel system through leaks or turbulence and does not vent completely at the tank, air collection will occur in the fuel filter—especially if it is a well-made, well-sealed filter.

As long as this air is forced through the filter element at a rate that is relatively close to the rate at which it is

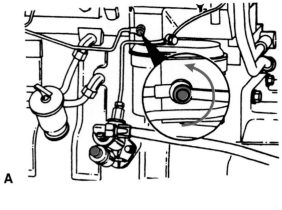

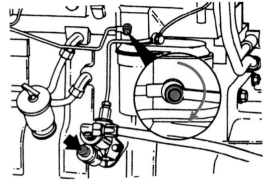

Figure 16-16. *One method of bleeding a fuel system.*
(Cummins Engine Co.)

entering (as will normally happen due to filter operation pressure differences), it will be redistributed in the fuel by the gear pump, and no power loss will be noted unless the system has a suction leak.

In mobile installations, operational conditions sometimes creates large amounts of filter-trapped air, which is forced from the filter. In such cases— particularly when full throttle is suddenly applied after a high-speed, downhill run—air may reach the pump and injectors, resulting in a condition referred to as *soft throttle.*

Large amounts of air in the filter may result in hard starting. A small air leak between the filter head may be uncovered during shutdown (when fuel no longer flows past it). This allows air to completely fill the head and the lines from the pump back to a point in the fuel line that is level with the fuel trapped in the filter. The engine may start and then die when the air is picked up before the fuel. It may be necessary to bleed the lines before the engine can be restarted.

Fuel Heaters

As explained in **Chapter 15,** wax crystals form and settle out of diesel fuel at cold temperatures. These crystals reduce fuel flow and may completely block fuel lines and clog filters. Trying to eliminate wax crystal formation by keeping the engine warm through idling may cause additional problems. An engine that is allowed to idle will cool down faster than an engine that is shut down immediately after use. One hour of idling will cause as much engine wear as 80 miles of driving. The average diesel engine spends 800 hours idling per year, which can shorten the

time between engine overhauls considerably. Fuel cost is another consideration. An idling engine consumes approximately one gallon (3.8 L) of fuel every hour, or 800 gallons (3028 L) a year.

To improve cold engine performance and reduce wear, *fuel heaters* are used. Heating diesel fuel can help eliminate the need for engine idling in cold weather, reducing both fuel costs and engine wear and tear. Both direct and indirect heating methods can help to keep diesel fuel temperatures above the cloud point.

Direct Heating

The diesel fuel can be directly heated to keep its temperature above its cloud point. Heating can be done using electric heating elements powered by the vehicle's storage battery. **Figure 16-17** illustrates a combination fuel filter, water separator, and fuel heater in a single housing. The

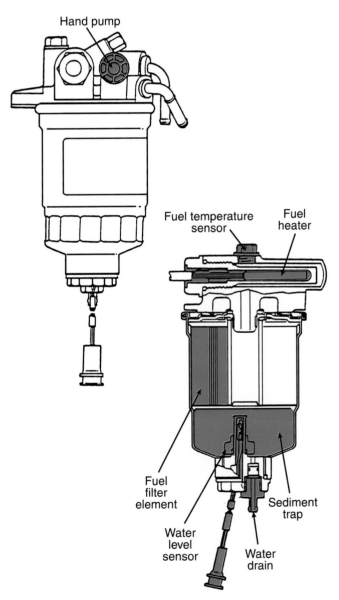

Figure 16-17. *A electric fuel heater that can be used in conjunction with either spin-on or cartridge fuel filters.*
(Lucas CAV Limited)

heating elements are located between the filter head and the filter element. Heaters of this type are generally available in the 100-300 watt power range. The fuel enters the filter head and flows over the heater elements. The heated fuel, now free of wax particles, flows through the element and out of the assembly. A hand operated primer is sometimes included as part of the unit and a full-bowl warning light system is usually available.

The heating element is installed directly into a steel fuel line using a flared connection. As fuel flows through the line, it is warmed as it passes over the heated coil. A flat heat tape works on the same principle. The carbon-based heat tape draws its power from the battery. The electrical draw is highest when the ignition is first turned on and the warm tape begins to heat the cold fuel. The power is automatically reduced as the fuel warms up. In-line heaters can be switched on for several minutes prior to startup or activated while the engine is running, if the operator begins to suspect fuel waxing is occurring in the tank or lines. Warm engine coolant can also be routed through the filter assembly or around fuel lines.

A disadvantage of the coolant system is that the engine must be already started and warmed for the system to be effective. **Figure 16-18** illustrates a filter head heating assembly that uses warm engine coolant to heat the diesel fuel passing into the filter element.

Indirect Heating

The fuel can also be indirectly heated by keeping the engine block warm through the use of immersion block coolant heaters, circulating tank coolant heaters, or oil sump heaters. These electric or propane fired coolant and lube oil heaters are generally considered starting aids, and are discussed in more detail in the chapter on starting systems.

Fuel Coolers

In some stationary and marine diesel installations, a *fuel cooler* is mounted in the raw-water cooling system between the heat exchanger and the raw-water pump, **Figure 16-19.** In this way, the fuel leaving the engine is cooled before it returns to the fuel tank.

Because fuel cycles through the engine and returns to the tank, the fuel in the tank becomes warm after extended operation. Excessive fuel temperatures can affect engine operation. A brake horsepower loss of approximately 2% will occur for each 20° increase in fuel inlet temperature above 90°F (36°C).

Summary

Abrasive dirt particles from 5-20 microns in diameter can damage fuel injectors and wear down engine parts. To remove these damaging contaminants fuel filter components are located throughout the fuel delivery system.

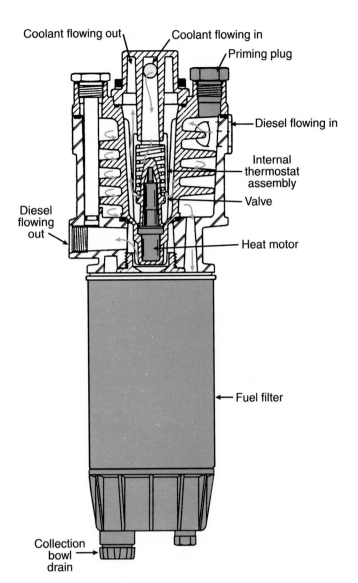

Figure 16-18. *Filter heater that uses warm engine coolant to help heat the fuel.*

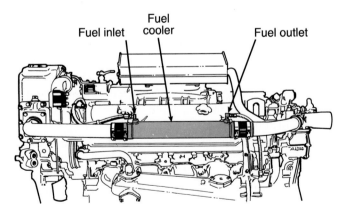

Figure 16-19. *Fuel cooler mounting. (Detroit Diesel Corporation)*

Water can corrode injection system components, freeze and prevent fuel flow, or turn into steam and damage the nozzles. Water in diesel fuel can also cause microbial infestation. When water enters a diesel fuel in the form of condensation, it does not evaporate but settles to the bottom of the fuel tank.

Fuel tank drain cocks are located on the bottom of the tank. Most bolt-on cartridge filters also have drain cocks at the bottom. All drain cocks in the system should be opened and water drained on a regular schedule.

To minimize condensation problems, fill fuel tanks at the end of the day, allow to stand overnight, and drain any water through the tank's drain cock in the morning. In cold weather, it is best to drain any water from the tank in the evening before it has a chance to freeze and damage system components.

Sedimenters and agglomerators are two types of prefilters located in the line between the fuel tank and the injectors. They separate water and contaminants from the fuel.

Large diesel engines often have individual water separators. For efficiency in the action of water separation, the fuel should come directly from the fuel tank and through the unit. A typical water separator can remove approximately 95% of the water in the fuel flow.

Many mobile units use a combination fuel filter/water separators. There are many fuel filter/water separators available for heavy-duty engines. Most are of a two-stage filter design. The first stage of the coalescer filters out solid particles and collects the small drops of water in the fuel and brings them together, coalescing them to form large drops. The water separator, or the second part of the filter assembly, separates this coalesced water from the fuel, providing clean dry fuel to the injectors.

A water-in-fuel detector turns on a warning light to alert the operator when there is water in the fuel tank or filter. Diesel fuel filters must be capable of trapping even the smallest of impurities. The porosity of the filtering material will determine the size of the impurities that it can remove. Some fuel filtering systems must filter impurities as small as 10 microns. Diesel engine fuel filtration systems may use two, three or more fuel filters.

Depth type filters are generally encased in a woven sock or in a perforated metal shell. Surface type filters generally use pleated paper made of processed cellulose fiber treated with phenolic resin binder. Bolt-on cartridge filters consist of a shell, a cover, and a replaceable filtering element. The spin-on filter has a shell, element, and gasket combined into a unitized replacement assembly. The filter covers incorporate a threaded sleeve. Unlike bolt-on filters, no separate springs, seats, or drain cocks are provided. Spin-on filters are manufactured as single, two-stage, or parallel style filters.

Single-stage filtration is commonly found on automotive diesel engines. Many truck and industrial diesel engines are equipped with a two-stage or dual filtration system consisting of a primary and a secondary filter. The primary filter (sometimes called a fuel strainer or auxiliary filter), is generally installed between the fuel tank and transfer pump. The secondary filter is installed between the transfer pump and the injectors.

The primary filter has a low restriction to fuel flow. It protects the transfer pump and lightens the load on the secondary filter. A primary filter is usually of depth construction. It removes the waxes, gums, and resins without plugging rapidly. Secondary filters are designed to offer full protection to the injectors. These surface type filters are mounted on the pressure side of the transfer pump. The media used must be able to remove particles of contaminant as small as 10 microns. Some diesel fuel systems have a third, or final stage filter for added protection against fuel contamination.

During normal operation, filter elements should be changed regularly according to the service manual. These elements are changed more often under extremely harsh conditions. On a series filter, the first-stage filter is changed more often than the second-stage. When changing fuel filters, any air that enters the system must be bled from the system.

When air enters the fuel system through leaks or turbulence, and does not vent completely at the tank, air collection will be noted in the fuel filter. If large amounts of filter-trapped air are forced from the filter all at once, a soft throttle condition can occur. Large amounts of air in the filter may also result in hard starting.

In cold temperatures wax crystals form and settle out of diesel fuel. Heating diesel fuel can help eliminate the need for engine idling in cold weather, reducing both fuel costs and engine wear and tear. The engine block can be kept warm through the use of immersion block coolant heaters, circulating tank coolant heaters, or oil sump heaters. These electric or propane fired coolant and lube oil heaters are generally considered starting aids. The fuel can also be heated directly, usually by electric elements located between the filter head and the filter element or sedimenter bowl. The fuel heater may be controlled by an external thermostat.

Another system combines the fuel heater with the fuel filter and water separator in a single housing. A timer is used to activate fuel system warm-up, which usually takes around five minutes. Another type of fuel heating system relies on a heating element installed directly in the existing fuel line. The carbon-based heat tape draws its power from the battery.

In some industrial stationary and marine diesel installation, a fuel cooler is mounted in the raw-water cooling system between the heat exchanger and the raw-water pump. In this way, the fuel leaving the engine is cooled before it returns to the fuel tank.

Important Terms

Prefilters	Dual filters
Drain cock	Fuel filter/water separators
Sedimenters	Separation
Agglomerators	Three-stage filter/
Fuel filters	separators
Depth fuel filter elements	Coalescing
Surface filter elements	Bleeding
Bolt-on filters	Air
Spin-on filters	Soft throttle
Primary filters	Fuel heaters
Secondary filters	Fuel cooler

Review Questions—Chapter 16

Do not write in this text. Place your answers on a separate sheet of paper.

1. *True or False?* Low resistance to fuel flow is not an important consideration in the design of primary fuel filters.

2. _____ should be bled from the fuel lines whenever fuel filters are changed.

3. A turbine centrifuge _____.
   (A) removes excess air from the diesel fuel
   (B) separates dirt and contaminates from the diesel fuel
   (C) is a type of fuel transfer pump
   (D) None of the above.

4. *True or False?* Engine idling is a practical method of keeping fuel warm in cold conditions.

5. In some fuel systems the _____ is combined with the filter and water separator in a single unit.

6. _____ separate solid impurities and water from diesel fuel.

7. *True or False?* Cartridge type fuel filters should be serviced when the engine is warm.

8. *True or False?* Spin-on fuel filters require considerable tightening force to properly seal their sealing gasket.

9. Most bolt-on cartridge type fuel filters are equipped with _____ that should be drained on a regular schedule.

10. A sedimenter fuel filter relies on _____ to help separate fuel and impurities.
    (A) gravity
    (B) radial fuel flow
    (C) agitation
    (D) Both A & B.

ASE-Type Questions

1. Technician A says a primary filter offers high restriction to fuel flow because it is located on the suction side of the transfer pump. Technician B says primary filter elements normally use depth-type construction. Who is right?
   (A) A only.
   (B) B only.
   (C) Both A & B.
   (D) Neither A nor B.

2. Technician A says that idling a diesel engine overnight to keep it warm causes no damage over the life of the engine. Technician B says that electric heaters are the best method to keep diesel fuel warm in cold weather. Who is right?
   (A) A only.
   (B) B only.
   (C) Both A & B.
   (D) Neither A nor B.

3. Technician A says that in a series filtering system all of the fuel goes through the first filter and then through the second filter. Technician B says that in a parallel fuel filtering system, part of the fuel goes through each of the filters. Who is right?
   (A) A only.
   (B) B only.
   (C) Both A & B.
   (D) Neither A nor B.

4. Technician A says that parallel filtering systems can filter a larger quantity of fuel than series filtering systems. Technician B says that parallel filtering does a better job of removing impurities when compared to series filtering. Who is right?
   (A) A only.
   (B) B only.
   (C) Both A & B.
   (D) Neither A nor B.

5. Technician A says it's best to refill a fuel tank the first thing in the morning. Technician B says that it is best to fill the tank at the end of the day so the full tank will sit overnight. Who is right?
   (A) A only.
   (B) B only.
   (C) Both A & B.
   (D) Neither A nor B.

6. Soft throttle and hard starting are two problems that are associated with:
   (A) air in the fuel system.
   (B) dirt in the fuel system.
   (C) blocked fuel lines.
   (D) All of the above.

7. Coalescer filters _____.
   (A) break down water droplets into tiny particles that are less harmful
   (B) collect small water droplets into larger droplets that can be separated from the fuel
   (C) only filter out solid dirt particles
   (D) only filter out air from the fuel

8. Technician A says you should fill a new filter with clean diesel fuel and coat the seal gasket with clean diesel lubricating oil before installing the filter. Technician B says you should bleed air from the fuel lines after changing the fuel filters. Who is right?
   (A) A only.
   (B) B only.
   (C) Both A & B.
   (D) Neither A nor B.

9. Fuel warmer than _____ begins to slowly reduce the engine brake horsepower output.
   (A) 60°F (15.6 C)
   (B) 90°F (32 C)
   (C) 120 F (49°C)
   (D) None of the above.

10. Fuel coolers are generally found on _____.
    (A) marine diesel engines
    (B) heavy-duty truck engines
    (C) large stationary engines
    (D) Both A & B.

Diesel-powered construction equipment is used to build the roadways traveled by gasoline and diesel-powered vehicles. Here, the concrete roadbed is being poured before asphalt is laid on top.   (CMI Coporation)

Chapter 17

Injection System Fundamentals

After studying this chapter, you will be able to:
- ❑ List the five types of fuel injection systems commonly used today and classify each as either a high pressure or low pressure system.
- ❑ Define and explain the importance of metering, timing, rate control, atomization, injection start and cut-off, and proper pressurization to the fuel injection system.
- ❑ Name the major components of the port and helix metering system and describe its operation.
- ❑ Name the major components of a sleeve metering system and describe its operation.
- ❑ Name the major components of an inlet metering system and describe its operation.
- ❑ Name the three major factors that are used to control fuel delivery in P-T fuel injection systems, and describe the system's basic operating principles.
- ❑ Describe the operation of unit injection systems.

All diesel fuel injection systems must accurately meter or measure the amount of fuel delivered to the combustion chambers. No part of a diesel engine is more important or requires a higher standard of design and manufacturing precision. This chapter discusses the various fuel injection systems used on diesel engines.

Diesel Fuel Injection Systems

The first diesel fuel injection pumps were simple "spill valve" jerk pumps. In the 1930s, Bosch introduced a more advanced helix-controlled injection pump. The Bosch designs were made by several licensees throughout the world. When the licensing agreements began to expire in the 1950s, a number of companies that manufactured Bosch equipment began to introduce their own line of diesel injection pumps. These designs were often similar to the Bosch pumps.

Some engine manufacturers designed and built their own fuel injection systems. Most notable of these were Detroit Diesel's unit injection and Cummins' pressure-time fuel injection systems. In the 1950s distributor type injection pumps were introduced as a more economical pump for light commercial vehicles and farm tractors.

Types of Fuel Injection Systems

Effective combustion depends on the correct operation of fuel injection equipment. Fuel injection systems are becoming increasingly sophisticated to meet the ever more stringent fuel economy, emissions, exhaust smoke, and noise standards. There are five types of fuel injection systems:
- ❑ *Individual jerk pumps* contained in their own housing and serving a single cylinder.
- ❑ *Multiple-plunger inline pumps* that use a number of individual jerk pumps contained in one common housing.
- ❑ *Unit injectors* that time, atomize, meter, and pressurize the fuel within the injector body or unit serving each cylinder.
- ❑ *Pressure-time injection* in which the fuel metering is controlled by engine speed and fuel pressure.
- ❑ *Distributor pump injection* that uses a spinning rotor to distribute pressurized fuel to the individual injectors in the proper cylinder firing sequence.

These systems can be grouped into two classes, depending on the pressures at which they operate. These systems are either high pressure or low pressure injection systems. Their differences are explained in the following sections.

High Pressure Injection Systems

Multiple-plunger inline pumps, individual jerk pumps, and distributor pump injection systems are classified as

high pressure injection systems. Fuel metering and timing is performed inside the injection pump, then delivered to each cylinder injection nozzle through high pressure steel fuel lines, **Figure 17-1.** Delivery pressures to the injectors can be as high as 3000 psi (20 685 kPa). At the nozzle, the pressurized fuel is forced through a series of very small openings that dramatically increase the injected fuel pressure to levels in excess of 17,000 psi (117 198 kPa) on multiple plunger inline systems.

Although distributor pumps are classified as high pressure systems, they do not attain the pressures required to operate larger diesel engines. Their typical spray-in pressure of 12,000 psi (82 728 kPa) is suited for light and medium duty car and truck diesel engines.

Low Pressure Injection Systems

Low pressure injection systems include unit injectors and the pressure/time injection system. Supply pressure in these systems can be as low as 50-75 psi (344-517 kPa). A rocker arm activated plunger then raises this low supply fuel pressure to between 2000-3000 psi (13,788-20,682 kPa). This higher pressure overcomes the spring pressure holding down the injector needle valve and lifts it off its seat. Fuel is then injected into the combustion chamber, **Figure 17-2.** The small opening through which the fuel is forced to pass, raises its pressure dramatically.

For example, the actual spray-in pressure created in a typical unit injector system can exceed 20,000 psi (137 880 kPa). These pressures are actually greater than those produced in high pressure inline pump systems and produce finer atomization, greater fuel penetration into the chamber, and a shorter ignition delay period. There is also no need to maintain a long column of pressurized fuel between the pump and the nozzle as there is on high pressure systems. For these reasons, low pressure mechanical or electronic control unit injector systems have been adopted by many diesel engine manufacturers. In deliv-

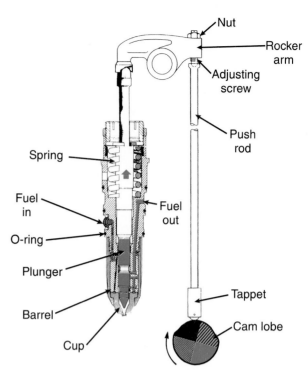

Figure 17-2. In a low pressure fuel injection system, fuel is delivered to the injector at lower pressures, but the mechanical action of a rocker arm actuated plunger attempts to compress the fuel, raising its pressure inside the injector. (Cummins)

ering fuel to the combustion chambers, all diesel fuel injection systems must perform the following functions:

- ❑ Accurately meter fuel to the injectors.
- ❑ Time the injection at the proper point in the combustion cycle.
- ❑ Control the rate of injection.
- ❑ Properly atomize the fuel.
- ❑ Start and stop injection quickly.
- ❑ Generate the proper injection pressure.

Metering

Accurate **metering** or measuring of fuel means that at the same fuel control setting, the same quantity of fuel must be delivered to each cylinder for every power stroke of the engine at a given rpm. Accurate metering allows the engine to operate at a constant speed with a constant power output. It means that the same and equal amount of fuel is delivered to each cylinder. The result is smooth engine operation with an even distribution of load between engine cylinders.

Timing

While it is important to measure the amount of fuel injected, the injection must also be properly timed to ensure efficient combustion in order to maximize the energy output of the fuel. The injection system **timing** must inject fuel at the proper point in the combustion cycle,

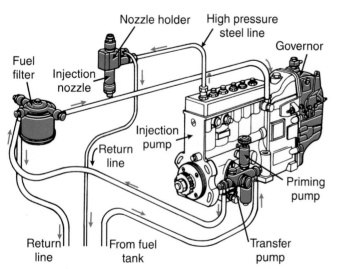

Figure 17-1. In a high pressure injection system, metering and timing take place inside a pump housing and high pressure fuel is then delivered to each injection nozzle.

regardless of the engine's speed and load. If the fuel is injected too early in the cycle, the engine will detonate and lose power, resulting in low exhaust temperatures. If the fuel is injected too late in the cycle, the engine will again lose power, but will have high exhaust temperatures and smoky exhaust. In either case, fuel consumption is high.

Rate Control

The injection system must inject fuel at a rate that controls combustion and the rate of pressure rise in the cylinder. The injection **rate control** should be low enough at first so that excess fuel does not build up in the cylinder during the initial ignition delay before combustion begins. The injection rate should continue so that fuel is injected quickly enough for complete combustion, but not so quickly that there is excessive combustion pressure. A rate that is too high or too low will have the same effect on the engine as improper timing. The result will again be high fuel consumption.

Some systems inject a small amount of fuel 8°-10° of crankshaft rotation before the main injection charge, so that some fuel is already burning when the main charge enters the combustion chamber. This results in smoother combustion and a slower rate of pressure rise in the cylinder.

Atomization

Atomization means the breaking up of injected fuel into small particles that form a mist-like spray as it enters the cylinder. The degree of atomization depends on the design of the combustion chamber; some chambers need only coarse atomization, while others require a very fine spray, **Figure 17-3.** Atomization occurs when liquid fuel under high pressure passes through small openings in an injector or nozzle. The fuel enters the combustion chamber at high speed because cylinder pressure is lower than fuel pressure. Friction resulting from this high velocity fuel passing through the low pressure air causes the fuel to break up into small particles. Proper atomization helps ensure that each particle of fuel will combine with a similar particle of oxygen during the combustion process.

In modern high speed direct injection diesel engines, injection must start and stop abruptly. This helps ensure the fuel is atomized into the finest mist possible, and that all

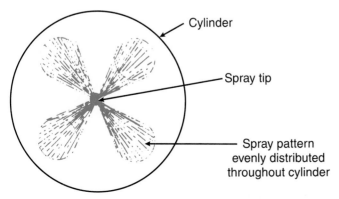

Figure 17-3. *Proper atomization of the fuel is essential for proper combustion.*

droplets are uniform in size. It also controls the duration of the expansion pressure during the power stroke. If the fuel delivery is not stopped abruptly, the last few drops of fuel will not burn properly and will cause heavy black smoke in the exhaust. This condition is frequently referred to as *injector dribble.*

Pressurization

For fuel injection to be effective, fuel pressure must be higher than the compression pressure found in the combustion chamber. High pressure allows the fuel to penetrate the chamber and mix thoroughly with the air to burn efficiently. **Fuel penetration** is the distance the fuel particles travel using their own kinetic energy as they leave the injector or nozzle. Friction between air and fuel in the combustion chamber absorbs the kinetic energy of the fuel particles. Fuel atomization can reduce the size of the fuel particles to the point where they lack penetration. Poor penetration means the particles will ignite before they have fully covered the combustion area.

Atomization and penetration oppose each other in that greater atomization means less penetration. For this reason, fuel injection systems must be designed carefully to achieve the proper balance between fuel atomization and penetration. The fuel pressure needed for both efficient injection and proper dispersion depends on several factors:

- ❏ Compression pressure in the cylinder.
- ❏ The size of the opening leading into the combustion area.
- ❏ The shape of the combustion area.
- ❏ The amount of turbulence in the cylinder.

The fuel spray should not contact the piston crown or cylinder wall as crown burning and/or incomplete combustion could occur.

Fuel Metering

To control a diesel engine, it must run at a desired speed regardless of changes in engine load. The amount of control obtained depends on the engine's performance characteristics and the type of load it drives. In a diesel engine, a varying amount of fuel mixes with a constant amount of compressed air. A full charge of air will enter the cylinder at each intake stroke. It is the amount of fuel injected into the cylinder that controls combustion and, in turn, the speed and power output of the engine. A governor can help regulate the flow of fuel. Governors are discussed in **Chapter 19.** Other devices can be added to either the governor or the engine itself to control overspeed or overload. These are called **fuel metering** devices. Every type of fuel injection system uses one of several basic fuel metering devices.

Port and Helix Fuel Metering

Port and helix fuel metering uses a barrel and plunger unit. The **barrel** is sometimes called a *bushing.* The **plunger** is fitted into the barrel so that it seals even at very high pressures and low speeds. A very slight clearance between components allows enough fuel to enter between mating parts for lubrication.

The pump plunger is machined on its side with the diagonal edge formed on the wall of the plunger known as the **helix.** The helix is sometimes referred to as a *scroll cut* and the system as *scroll metering.* The pressure chamber above the plunger is always connected to the chamber below the helix by the vertical groove in the plunger. The barrel is usually constructed with two ports, one for fuel inlet and one for fuel return after the end of fuel delivery, **Figure 17-4A.** Some barrels have a single port that serves as both fuel inlet and return, **Figure 17-4B.**

If the injection pump is connected to the engine's lubrication oil system, the leakage fuel can cause engine oil dilution. To prevent this, pumping elements used in this type of injection pump are made with a leakage-return duct to the suction gallery. The pump barrel is manufactured with an annular groove that is connected by a port to the injection pump's suction gallery. The leakage fuel expands into this groove and flows back into the suction gallery, **Figure 17-4C.** Due to the precise matching of pump plunger and pump barrel, it is important that only

complete plunger and barrel assemblies be replaced. Never replace one part without replacing the other.

The plunger moves up and down inside the barrel. As it moves, the plunger opens and closes holes or ports in the barrel. The opening and closing of these ports controls fuel flow and pressure inside the unit. This in turn controls the start and end of the injection period. In a port and helix metering system, one of two methods can be used to move the plunger up and down in the barrel to control the start and end of fuel delivery. The first is to use an injection pump camshaft driven off the engine. Lobes on the camshaft move the plungers upward, while a return spring forces them back down, **Figure 17-5.** The second method uses a rocker arm actuated system driven off the engine camshaft. In both cases, the pump plungers operate with a constant stroke.

Port and Helix Metering Operation

Figure 17-6 illustrates the stages of port and helix metering in an inline fuel injection pump. In **Figure 17-6A,** the plunger is at the bottom of its stroke. Both ports in the barrel are open and fuel enters the area of the barrel above the top of the plunger. The delivery valve is closed.

As the plunger starts its upward movement, **Figure 17-6B,** a small amount of fuel spills back through the ports into the reservoir until the plunger body covers the ports completely. This point in the injection cycle is known as **port closing** and is the actual start of injection. The travel of the pump plunger from its bottom dead center position to the start of delivery is known as the **prestroke.**

With the fuel now trapped between the plunger and barrel walls, fuel pressure builds rapidly until the delivery valve at the top of the pumping unit lifts off its seat. The

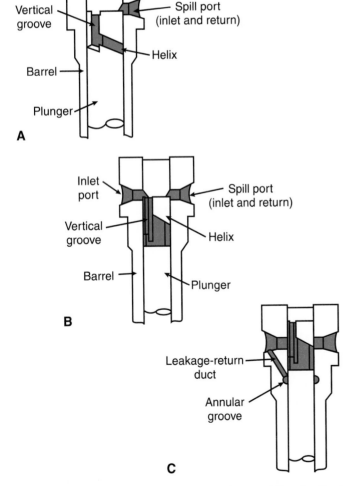

Figure 17-4. Types of plunger and barrel assemblies. A—Dual port with lower helix. B—Single orifice assembly with lower helix. C— Dual port assembly with leakage return duct.

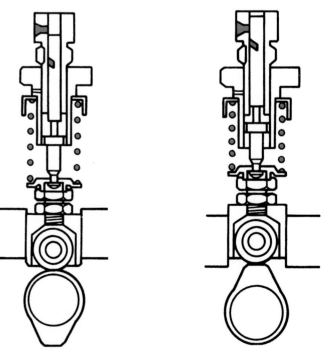

Figure 17-5. The spring-loaded plunger is moved upward by the cam on the injection pump camshaft.

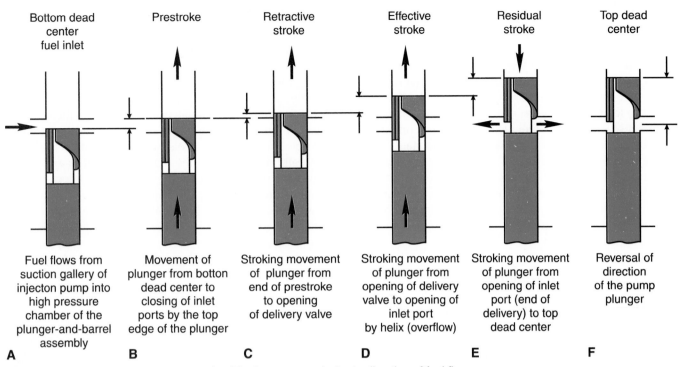

Bottom dead center fuel inlet | Prestroke | Retractive stroke | Effective stroke | Residual stroke | Top dead center

Fuel flows from suction gallery of injecton pump into high pressure chamber of the plunger-and-barrel assembly
A

Movement of plunger from botton dead center to closing of inlet ports by the top edge of the plunger
B

Stroking movement of plunger from end of prestroke to opening of delivery valve
C

Stroking movement of plunger from opening of delivery valve to opening of inlet port by helix (overflow)
D

Stroking movement of plunger from opening of inlet port (end of delivery) to top dead center
E

Reversal of direction of the pump plunger
F

Figure 17-6. Phases of the plunger stroke. The large arrows indicate direction of fuel flow.

travel of the plunger from the end of the prestroke to the opening of the delivery valve is known as the *retractive stroke,* **Figure 17-6C.**

In the case of an inline pump, fuel is now forced through a high pressure fuel line to the fuel injector mounted in the cylinder head. Injection continues until the lower helical land on the plunger uncovers the spill port (also called a control port). Pressurized fuel in the chamber above the plunger can now flow through the vertical groove in the plunger to the chamber below the helix, and out through the spill port. The spill port is opened slightly before the plunger completes its upward stroke. With a drop in fuel pressure, the delivery valve closes and injection ends. The travel of the pump plunger from the end of the retractive stroke to the opening of the spill port by the helix of the pump plunger is know as the *effective stroke,* **Figure 17-6D.**

As the plunger completes its stroke, the remaining fuel is displaced back to the fuel reservoir. The travel between the end of the effective stroke and the top dead center position of the plunger is known as the *residual stroke,* **Figure 17-6E.** The displaced fuel escapes down the relief area of the plunger and out through the control port to the reservoir. Once the plunger completes its total upward stroke and is at its TDC position, **Figure 17-6F,** spring pressure returns it to its original position.

Delivery Valves

The *delivery valve* plays an important role in a spill-metered injection system such as the port and helix system just described. It provides *injection line retraction,* which is the pulling back of a small amount of fuel in the injection line. This lowers the fuel line pressure and allows the injector nozzle to close quickly, abruptly cutting off injection.

Remember, a quick, clean cutoff of fuel to the nozzles is one of the prerequisites for efficient, clean running engines.

A typical delivery valve assembly is made of a valve, spring, and fitting (holder) into which the valve fits, **Figure 17-7.** This assembly is mounted in the fuel pump housing directly above the plunger and barrel assembly. As fuel is trapped above the plunger and the delivery valve is forced off its seat, fuel flows around it and through the fuel line to the nozzle. The valve remains open until fuel flow stops, which occurs as the port is uncovered by the helix and the pumping element stops pumping fuel. When this happens,

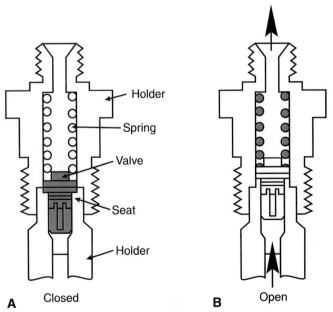

Holder
Spring
Valve
Seat
Holder

A Closed   **B** Open

Figure 17-7. Delivery valve holder with delivery valve. A—Closed. B—Open for delivery.

the pressure that had been holding the valve open is released and the fuel above the valve and the spring pressure seats the valve.

As the delivery valve seats, it moves downward and opens a given amount of area above it. This area is immediately filled with fuel from the line which lowers line pressure. The lowered line pressure allows the injection nozzle valve to seat and stop fuel delivery to the combustion chamber. Once seated, the delivery valve maintains a certain amount of line pressure so that succeeding injection cycles do not have to fill the fuel line with fuel. If the fuel line was not full of fuel, the timing would be retarded, causing the engine to operate poorly. Delivery and retraction all happen very quickly, taking no more than two thousandths of a second.

Return Flow Restrictor

A **return flow restrictor** consisting of a special valve plate or snubber valve may be installed in the delivery valve holder. It is located so that it is between the delivery valve and the injection nozzle. The return-flow restriction reduces the wear on the high pressure fuel system caused by the fast flowing diesel fuel.

As fuel is delivered, the pressure against the valve plate overcomes the spring force and the fuel can flow without resistance to the injection nozzle. At the end of the effective stroke, the closing of the nozzle needle valve creates a pressure wave that moves in the opposite direction of the delivered fuel. This pressure wave can cause damage to the pumping element. To prevent this, the valve plate or snubber valve is made with a special throttling cross-section that dampens the pressure wave, **Figure 17-8.**

Varying Fuel Delivery

The power delivered by a diesel engine depends largely on the quantity of fuel injected. Since the plungers of inline pumps have a constant stroke, some other method must be used to change the quantity of fuel delivered. This is done using the helix and vertical groove cuts on the plunger. If the plunger did not have a helix groove, the plunger's effective

stroke could not be changed and the plunger would pump the same amount of fuel on every stroke.

In addition to moving up and down, the plunger can also rotate in the barrel or bushing. This rotation is possible by the movement of the fuel control rack connected to the throttle or governor linkage, **Figure 17-9.** Rotating the plunger changes the relative position between the helix cut on the plunger and the barrel ports. In turn, this changes the point in the plunger stroke where the ports in the barrel close and open, varying the plunger's effective stroke. The longer the ports are closed, the more fuel is injected.

Plungers can be made with metering lands having a lower helix only, upper helix only, or both lower and upper helixes, **Figure 17-10.** A *lower helix* design produces a constant port closing with a variable ending. A *upper helix* give a variable port closing with a constant ending. When

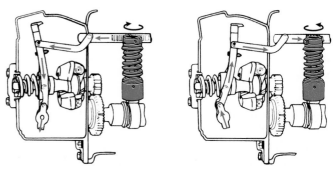

Figure 17-9. The governor lever moves the control rack that meshes with grooves on the control sleeve. As the rack moves it rotates the control sleeve, which in turn rotates the plunger in the barrel. (AMBAC International)

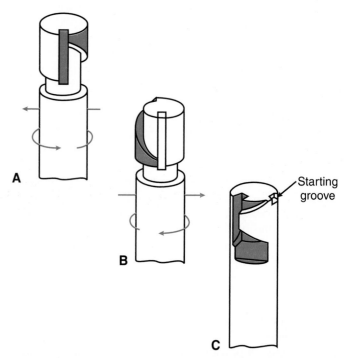

Figure 17-10. Versions of pump plunger designs. A—Lower helix. B—Upper helix. C—Upper and lower helix with starting groove.

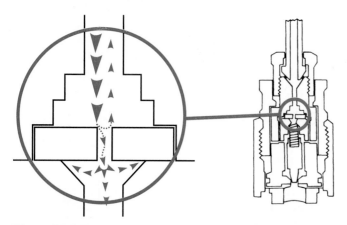

Figure 17-8. The valve plate or snubber valve acts as a reverse flow check valve, breaking down harmful pulsation waves of high pressure fuel. (AMBAC International)

both lower and upper helixes are machined into the plunger, both the beginning and end of injection are variable.

When the vertical groove of the plunger is in line with the spill port, all the fuel is bypassed by the plunger and injection cannot occur, **Figure 17-11A.** If the plunger is rotated into the partial delivery position, port opening takes place sooner, depending on the position of the plunger, **Figure 17-11B.** When the fuel control rack is rotated to the full fuel position, the plunger can travel almost its entire stroke before the helix will uncover the spill port to stop injection, **Figure 17-11C.**

Starting Groove

A diesel engine will start more easily when fuel injection timing is retarded during the cranking process. To provide the retard, the plunger is constructed with a ***starting groove*** at the top. This results in a port closing with a 5°-10° retard. As soon as the engine reaches operating speed, the governor pulls the control rod to normal starting position.

Sleeve Metering System

The ***sleeve metering system*** uses a sliding collar sleeve through which the stroking plunger moves. The sleeve and plunger are lapped together to make a matched set. The position of the sleeve collar controls the length of the plunger's effective stroke, thereby determining the amount of fuel delivered. A control unit or lever connected to the governor is used to control the position of the sleeve collar.

The main components of a fuel injection pump using the sleeve metering fuel system are the barrel, plunger, and ***sleeve.*** The plunger moves up and down inside the barrel and sleeve. The barrel remains stationary, but the sleeve can be moved up and down on the plunger to make a change in the amount of fuel injected.

Sleeve Metering Operation

When the engine is running, pressurized fuel from the transfer pump goes in the center of the plunger through the fuel inlet during the plunger's downward stroke. Fuel cannot go through the fuel outlet at this time because the fuel outlet is blocked by the control sleeve, **Figure 17-12A.** The action of the cam lobes lifts the plunger from its BDC position. As the plunger moves upward in the barrel, it closes the inlet port, **Figure 17-12B,** and injection can begin. This is because the fuel pressure above the plunger increases to the point where it opens the delivery valve and high pressure fuel can flow through the lines to the injection nozzles.

Injection stops when the fuel outlet is lifted above the top edge of the sleeve, **Figure 17-12C.** This is the end of the plunger's effective stroke. With the fuel outlet now open, pressure above the pumping plunger is lost and the fuel above and inside the plunger passes through the fuel outlet and returns to the fuel injection housing.

When the sleeve is raised on the plunger, the fuel outlet will be covered for a longer time, increasing the effective stroke and causing more fuel to be injected. If the sleeve is located low on the plunger, the fuel outlet is covered for a shorter period of time and fuel delivery is decreased. The metering sleeve allows infinitely variable amounts of fuel to be delivered to the injection nozzles.

Figure 17-13 illustrates a sleeve metered multiple plunger inline pump equipped with a mechanical governor. The lever of the governor is connected by linkage and governor springs to the sleeve control shafts. Any movement of the lever causes a change in the position of the sleeve control shafts.

When the lever is moved by governor action to feed more fuel to the engine, the lever acts to compress the governor springs and move the thrust collar forward. As the

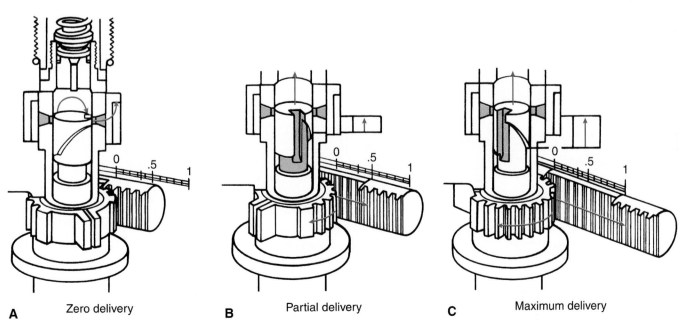

| A | Zero delivery | B | Partial delivery | C | Maximum delivery |

Figure 17-11. *Position of the pump plunger for different quantities of fuel injected. A—Zero delivery. B—Partial delivery C—Maximum delivery.*

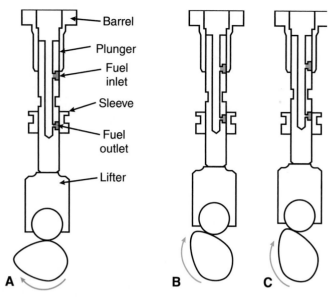

Figure 17-12. *Components and operation of sleeve metering system: A—With the inlet open, fuel fills the plunger body. B—With the inlet and outlet closed, fuel pressure builds and delivery occurs. C—With the outlet open, fuel pressure is relieved and delivery stops. (Caterpillar Inc.)*

Inlet Metering System

Inlet metering systems measure the fuel charge before it reaches the main pumping element. This design was pioneered and developed by Vernon Roosa and is used on distributor (rotor) type fuel injection systems, **Figure 17-14.** The inlet metering system consists of a **metering valve** and a drilled bore into which the valve is fitted. The bore is drilled into a hydraulic head that makes up the main part of the injection pump. A drilled passageway intersects this bore at right angles. Fuel is supplied through this passageway to the metering valve bore. The metering valve is shaped like a round rod measuring approximately .250" (6.35 mm) in diameter and 1" (25.4 mm) in length. It is designed with an angular groove cut on one side to regulate fuel delivery and a slot on top for attachment of the governor control linkage.

Inlet Metering Operation

Fuel is supplied to the metering valve under low pressure from a vane fuel transfer pump, through a drilled hole in the hydraulic head. The delivery valve and its bore are connected to the ring or annulus area of the hydraulic head through a supply drilling. When the metering port and the charging port align, fuel is passed via the metering valve and head to the rotor and pumping plungers. The quantity of fuel is controlled by the governor and the throttle lever.

As the rotor rotates, fuel is isolated in the rotor, **Figure 17-15A.** The amount of fuel delivered to the pumping

thrust collar moves forward, the connecting linkage will cause the sleeve control shafts to turn. The turning movement of the control shafts causes the sleeve levers to lift the sleeves, increasing the amount of fuel sent to the cylinders.

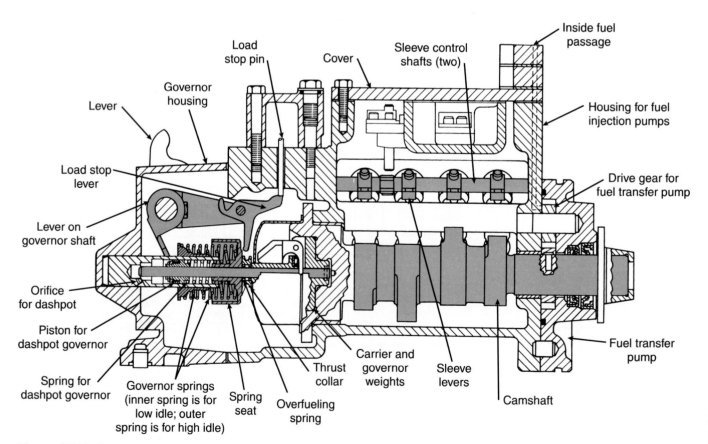

Figure 17-13. *Components of a multiple plunger inline fuel injection pump equipped with sleeve metering and a mechanical governor. (Caterpillar Inc.)*

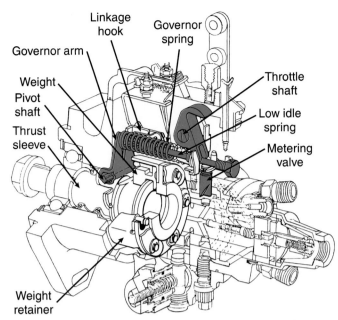

Figure 17-14. *Internal components of an inlet metered distributor pump. (Stanadyne Diesel Systems)*

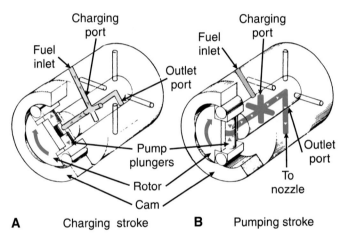

Figure 17-15. *Delivery phases of an inlet metered distributor pump. A—Charging phase. B—Delivery phase. (Lucas CAV Limited)*

plungers depends on the position of the metering valve and its angular groove. If the groove has a large part of the supply drilling from the transfer pump covered, fuel delivery to the pumping plungers during charging is light. If the groove covers very little of the supply drilling, a large amount of fuel will be delivered to the pumping plungers during charging.

As the rotor continues to rotate, one of the distributor ports aligns with one of the outlet ports. At this point in the cycle, the plungers are brought quickly together by the action of the cam ring lobes and pressurized fuel is forced through high pressure piping to the injector, **Figure 17-15B.** The whole cycle is repeated once for each engine cylinder per pump revolution.

Pressure-Time Fuel Metering Systems

The ***pressure-time fuel metering system*** uses fuel injectors that meter and inject the fuel. Metering is based on three factors, the orifice opening of the injector, the time

the injector orifice is open, and the pressure of the fuel supplied to the injector.

Orifice opening size is carefully calibrated and controlled during manufacturing. The time the orifice is open is controlled by engine speed. This is accomplished by using a camshaft-actuated plunger that changes the rotary motion of the camshaft into the reciprocation motion of the injector plunger. The plunger movement opens and closes the injector barrel metering orifice. The period of time the metering orifice is uncovered is the time available for fuel to flow into the injector cup. Metering time in these systems is inversely proportional to engine speed. The faster the engine is running, the less time the metering orifice is open.

Fuel pressure is supplied by a gear-driven, positive displacement low pressure fuel pump. A mechanical governor controls fuel pressure and engine torque throughout the entire operating range. It also controls idle speed and prevents engine overspeeding in the high rpm range. A single low pressure fuel line from the fuel pump serves all injectors. This ensures the pressure and amount of metered fuel for all cylinders is the same.

Pressure-Time Metering Operation

Fuel at low pressure enters the injector and flows through the inlet orifice, internal drillings, around the annular-groove in the injector cup and through the return passage to return to the fuel tank, **Figure 17-16A.** As the injector plunger begins to move upward, the metering orifice is uncovered and fuel enters the injector cup. The amount is determined by fuel pressure. The return passage to the fuel tank is also blocked by the upward moving plunger. This momentarily stops fuel circulation in the injector, isolating the metering orifice from pressure pulsations, **Figure 17-16B.**

As the plunger begins to move downward, it closes the metering orifice and fuel entry into the injector cup is cut off. As the plunger continues down, it forces fuel out of the cup through tiny holes at high pressure as a fine, atomized spray, **Figure 17-16C.** When the return passage is again uncovered, fuel flows back to the fuel tank. After injection, the plunger remains seated until the next metering and injection cycle. Although no fuel is reaching the injector cup, it still flows freely through the injector and returns to the fuel tank. This cools and lubricates the injector and also helps to warm the fuel in the fuel tank, **Figure 17-16D.**

Unit Injector Metering System

In a ***unit injection metering system,*** low pressure fuel from the fuel pump is delivered to the inlet fuel manifold cast internally within the cylinder head and then to the injector through fuel lines. Once the fuel reaches the injector, the injector times, meters, pressurizes, atomizes and delivers the fuel, **Figure 17-17.**

Fuel metering inside the unit injector is accomplished by rotating the unit injector plunger inside its barrel. The plunger is machined with a helical chamber area. As described earlier, under port and helix systems, the rotation of a helix shape will either advance or retard the closing of

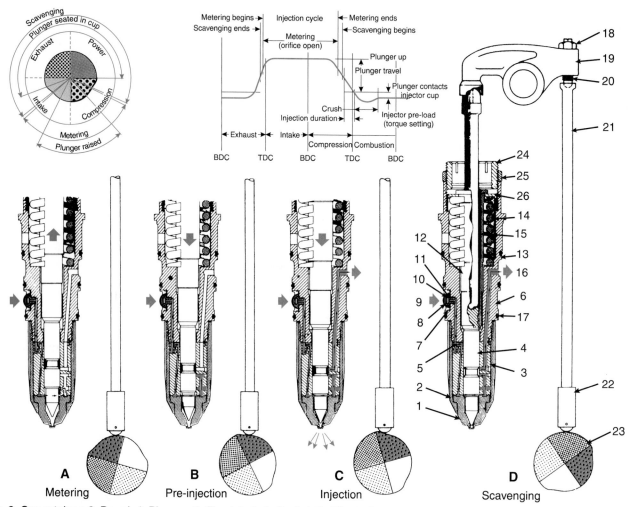

1–Cup, 2–Cup retainer, 3–Barrel, 4–Plunger, 5–Check ball, 6–Gasket, 7–Clip, 8–Screen, 9–Fuel in, 10–Orifice, 11–Orifice gasket, 12–Coupling, 13–Adapter, 14–Spring, 15–Link, 16–Fuel out, 17–O-ring, 18–Nut, 19–Rocker leaver, 20–Adjusting screw, 21–Push rod, 22–Tappet, 23–Camshaft lobe, 24–Top stop, 25–Locknut, 26–Washer.

Figure 17-16. *Plunger movement in a Cummins PT injection system injector is activated by push rod and rocker arm action driven off the engine camshaft. (Cummins Engine Co., Inc.)*

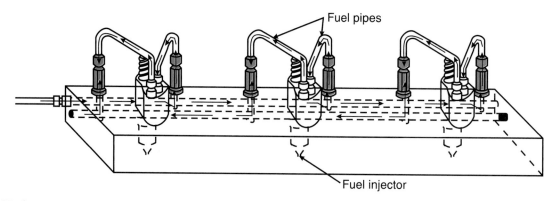

Figure 17-17. *Components of a unit fuel injection system. (Detroit Diesel Corporation)*

the ports in the injector bushing. This alters the start and end of the injection period, changing the effective stroke of the plunger and the amount of fuel injected. A gear on the plunger meshes with a tooth in the fuel control rack.

The **fuel control rack** is connected to a rack control lever mounted on a control tube. The control tube is linked

to the governor. Governor movement rotates the control tube and lever, which in turn moves the control rack and rotates the unit injector plunger. Fuel trapped under the plunger on its downward stroke develops sufficient pressure to force its way past the check or needle valve and enters the combustion chamber. The fuel then passes through small holes in the injector spray tip that increase sthe spray in pressure and atomize the fuel into a fine mist.

Electronic Metering Systems

All major fuel injection systems are now available with electronic controls to time and meter the delivery of fuel much more precisely. **Chapter 25** discusses electronic fuel injection in more detail.

Summary

Multiple-plunger inline pumps, individual jerk pumps, and distributor pump injection systems are classified as high pressure injection systems. Metering and timing of the fuel is done within the injection pump and fuel is then delivered to each cylinder injection nozzle through high-pressure steel fuel lines. Low pressure injection systems include unit injectors and pressure-time injection systems. Fuel is supplied to the injector at low pressure. A rocker arm activated plunger then raises the fuel pressure to high levels needed for injection.

The port and helix fuel metering system uses a barrel and plunger unit. As the plunger moves up and down in the barrel it opens and closes fuel inlet and outlet ports. This controls fuel delivery to the pumping units and to the high pressure fuel lines leading to the injectors. The plunger movement against fuel trapped in the barrel pressurizes the fuel for delivery to the injectors.

The sleeve metering system uses a sliding collar sleeve through which the stroking plunger moves. The position of the sleeve collar controls the length of the plunger's effective stroke, thereby determining the amount of fuel delivered. A control unit or lever connected to the governor is used to control the position of the sleeve collar.

Inlet metering systems measure the fuel charge before it reaches the main pumping element. This design is used on distributor (rotor) type fuel injection systems. The inlet metering system consists of a metering valve and a drilled bore into which the valve is fitted. The bore is drilled into a hydraulic head that makes up the main part of the injection pump. A drilled passageway intersects this bore at right angles. Fuel is supplied through this passageway to the metering valve bore. The metering valve is shaped like a round rod and is designed with an annular groove cut on one side to regulate fuel delivery. It also has a slot on top for attachment of the governor control linkage.

Fuel is supplied to the metering valve from a fuel transfer pump through a drilled hole in the hydraulic head. The delivery valve and its bore are connected to the ring or annulus area of the hydraulic head. When the metering port and the charging port align, fuel is passed via the metering valve and head to the rotor and pumping plungers. The quantity of fuel is controlled by the governor and the throttle lever. As the rotor rotates fuel is isolated in the rotor. The amount of fuel delivered to the pumping plungers depends on the position of the metering valve and its annular groove.

The pressure-time fuel metering system uses fuel injectors that meter and inject the fuel. Metering is based on three factors, the orifice opening of the injector, the time the injector orifice is open, and the fuel pressure supplied to the injector. In a unit injection system, low pressure fuel from the fuel pump is delivered to the inlet fuel manifold cast internally within the cylinder head and then to the injector through fuel lines. Once the fuel reaches the injector, the injector times, meters, pressurizes, atomizes, and delivers the fuel to the combustion chamber.

Important Terms

Individual jerk pumps	Port closing
Multiple-plunger inline pumps	Prestroke
Unit injectors	Retraction stroke
Pressure-time injection	Effective stroke
Distributor pump injection	Residual stroke
High pressure injection systems	Delivery valve
Low pressure injection systems	Injection line retraction
Metering	Return-flow restrictor
Timing	Lower helix
Rate control	Upper helix
Atomization	Starting groove
Fuel penetration	Sleeve metering system
Fuel metering	Sleeve
Port and helix fuel metering	Inlet metering systems
Barrel	Metering valve
Plunger	Pressure-time fuel metering system
Helix	Unit injection metering system
	Fuel control rack

Review Questions—Chapter 17

Do not write in this text. Place your answers on a separate sheet of paper.

1. Which of the following diesel fuel injection system is not considered a high pressure injection system?
   (A) Multiple-plunger inline pump.
   (B) Unit injector.
   (C) Individual jerk pump.
   (D) Distributor pump.

2. In a high pressure diesel fuel injection system, metering and timing of the fuel are done within the _____ and fuel is then delivered to each cylinder.

3. List at least five major functions of any diesel fuel injection system.

4. The power delivered by a diesel engine depends largely on the quantity of _____ injected.

5. *True or False?* It is considered an acceptable service practice to install a new pump plunger into an old barrel assembly, provided the barrel assembly is not visibly pitted or worn.

6. Which of the following is *not* true concerning inlet metering?
   (A) It measures the fuel charge before it reaches the main pumping element.
   (B) It is used in unit injection systems.
   (C) It was pioneered and developed by Vernon Roosa.
   (D) It is used on distributor type fuel injection systems.

7. Which of the following is *not* true concerning sleeve metering?
   (A) It uses a sliding collar through which the stroking plunger moves.
   (B) The barrel remains stationary, but the sleeve can be moved up and down on the plunger to make a change in the amount of fuel injected.
   (C) When the sleeve is raised on the plunger, the effective stroke of the plunger is increased, causing less fuel to be injected.
   (D) Sleeve movement is controlled by governor action.

8. Name the three factors that govern fuel metering in the pressure-time fuel injection system.

9. Once fuel reaches a unit injector, the injector _____, _____ , _____ and _____ the fuel to the combustion chamber.

10. All major types of fuel injection metering systems are now available with _____.

ASE-Type Questions

1. Technician A says that low pressure fuel injection systems include unit injectors and distributor pump injection systems. Technician B says that a rocker arm activated plunger is used in these systems to raise the fuel pressure to high levels needed for injection. Who is right?
   (A) A only.
   (B) B only.
   (C) Both A & B.
   (D) Neither A nor B.

2. All of the following are classified as high pressure injection systems EXCEPT:
   (A) unit injectors.
   (B) individual jerk pumps.
   (C) distributor pump systems.
   (D) multiple-plunger inline pumps.

3. All of the following are functions of a typical diesel fuel injection system EXCEPT:
   (A) time the injection at the proper point in the combustion cycle.
   (B) control the rate of injection.
   (C) properly atomize the fuel.
   (D) pressurize the fuel in the engine cylinder.

4. Technician A says that a multiple-plunger inline injection pump alters the quantity of fuel delivered by varying the length of its individual plunger strokes. Technician B says that this type of pump has a constant plunger stroke length. Who is right?
   (A) A only.
   (B) B only.
   (C) Both A & B.
   (D) Neither A nor B.

5. In port and helix fuel metering systems, rotating the helix groove location in relation to the port _____.
   (A) changes the effective stroke of the plunger
   (B) varies the amount of fuel delivered
   (C) changes the point at which ports in the barrel open and close
   (D) All of the above.

6. All of the following are stages in port and helix fuel metering EXCEPT:
   (A) prestroke.
   (B) effective stroke.
   (C) compression stroke.
   (D) residual stroke.

7. A return flow restriction consisting of a valve plate or snubber may be installed in the _____.
   (A) delivery valve holder
   (B) between the delivery valve and injection nozzle
   (C) fuel injector nozzle
   (D) Both A & B.

8. All of the following are characteristics of sleeve metering EXCEPT:
   (A) uses a fixed collar sleeve.
   (B) the sleeve and plunger are lapped together.
   (C) the plunger moves up inside the barrel and sleeve.
   (D) the sleeve collar position is controlled by a control unit or lever.

9. Technician A says that in a pressure-time injection system, the period of time the metering orifice is uncovered is the time available for fuel to flow into the injector. Technician B says that metering time is decreases as engine speed increases. Who is right?
   (A) A only.
   (B) B only.
   (C) Both A & B.
   (D) Neither A nor B.

10. Technician A says that unit injectors use an inlet metering system. Technician B says that unit injectors rely on port and helix metering systems. Who is right?
    (A) A only.
    (B) B only.
    (C) Both A & B.
    (D) Neither A nor B.

Chapter 18

Injection Nozzles

After studying this chapter, you will be able to:
- ❏ Give the location, name the major parts, and explain the operation of an injection nozzle holder.
- ❏ Describe the difference between open and closed nozzle designs.
- ❏ Describe the construction and operation of hole, pintle, and Pentaux® nozzles.
- ❏ Explain how spray hole design, opening pressure, differential ratio, needle and spindle mass, valve guide diameter, needle lift, fuel sac design, and diameter affect hole nozzle operation and performance.
- ❏ Successfully remove and install a nozzle holder and nozzle.
- ❏ Successfully disassemble, inspect, clean, reassemble, and test nozzle holders and nozzles.
- ❏ Describe the operation and service procedures for pencil injection nozzles.
- ❏ Describe the types of fuel injection tubing and connections used for high pressure injection fuel lines.

As stressed in previous chapters, a diesel engine must be provided with fuel that is properly metered and delivered in the combustion chamber at the proper time. Equally important, the fuel must be atomized into a controlled spray pattern and injected in such a manner that it burns completely without producing excessive black smoke or particulates. Atomization and final fuel delivery is the job of the injection nozzle. Nozzle designs can vary with the type and size of the combustion chamber. This chapter discusses the various types of injector nozzles used in diesel engines.

Nozzle Holders

All *injector nozzles* are enclosed in *nozzle holders*, **Figure 18-1.** The nozzle holders are mounted in the

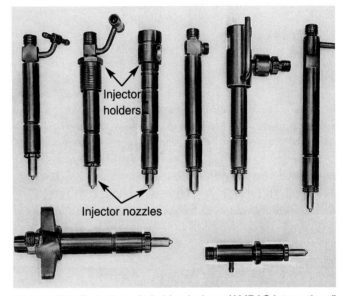

Figure 18-1. *Typical nozzle holder designs. (AMBAC International)*

engine cylinder head. They position the nozzle in the engine cylinder, hold the needed spring and pressure adjustments to provide the proper nozzle valve action, and provide a means for conducting fuel to the nozzle and combustion chamber. A copper gasket is installed between the injector and the cylinder head so that the maximum amount of operating heat possible will be transmitted to the engine coolant.

Holder Construction

A typical holder assembly consists of a body, nozzle spacer, lower spring seat, pressure adjusting spring, spring adjusting spacers, or a screw and a nozzle cap nut, **Figure 18-2.** The holder body contains a high pressure tubing connection at the upper end, a high pressure fuel duct, a leak-off duct, and a leak-off connection. The lower end of the body is machined to accept the pressure adjusting spring and spacers. The lower end of the nozzle holder is provided

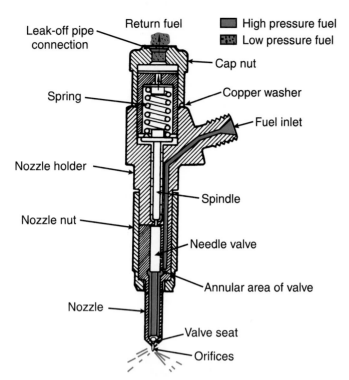

Figure 18-2. *Components of a typical injector holder assembly.*

with a precision ground and lapped surface that forms a leak-proof seal with the lapped surface of the nozzle body. The nozzle is secured to the holder by means of a cap nut.

The lower spring seat provides a seat for the pressure adjusting spring and a piloting surface for the nozzle valve stem. The nozzle spacer contains two locating pins that position the nozzle radially to ensure proper spray pattern orientation within the combustion chamber. Both faces of the spacer are lapped to a fine finish to provide a high pressure seal between the holder and nozzle body faces. The spacer also acts as a valve stop.

Nozzle Fuel Flow

Fuel from the high pressure tubing enters the nozzle holder assembly at the upper end and flows down the high pressure duct through the nozzle spacer duct and into the nozzle. A slight amount of fuel leakage will occur past the nozzle valve's major outside diameter into the adjusting spring chamber. This fuel is used to lubricate the valve spring seat and adjusting spring and then escapes back to the fuel tank through the leak-off duct and connection.

Types of Injector Nozzles

There are two major types of injector nozzle designs: open nozzles and closed nozzles. In **open injector nozzles,** there is no valve used to stop the flow of fuel from the nozzle tip. The amount of fuel passing through an open nozzle is completely controlled by the injection pump. **Closed injector nozzles** are designed with a spring-loaded valve located near the nozzle's exit orifice. Valve opening and closing is controlled in one of three ways: mechanically, hydraulically, or electronically.

One advantage of the open nozzle is that it does not easily become clogged with carbon particles or bits of other solid matter in the fuel stream. The force of the fuel moving through the nozzle opening keeps it clear. However, this type of nozzle is prone to fuel leakage or "dribble" through the tip. **Injector dribble** is a major cause of preignition (knocking) in diesel engines. As fuel runs down and remains on the tip of the nozzle, carbon will form in this area and cause afterburning. Leakage or dribble obviously leads to poor fuel economy and black exhaust smoke. For these reasons, open injector nozzles are not frequently used.

The valve used in closed nozzle injectors eliminates fuel leakage or dribble through the tip. Preignition and afterburning are all but eliminated and fuel economy improves. Modern filtration technology has also reduced clogging of nozzles by removing solids from the fuel before it reaches the nozzle. Closed type injector nozzles are the most popular nozzle type in use today. There are two types of closed nozzles, the hole type and the pintle type. Both types have a valve and seat so the fuel lines are completely closed off when no fuel is being injected.

Hole Injector Nozzles

Direct injection, open combustion chamber diesel engines, such as those used in heavy-duty truck applications, mainly use long-stem **hole injector nozzles, Figure 18-3.** The extended small diameter tip of the long-stem nozzle reduces the amount of mounting space required between cylinder head valves. In this design, the valve guide is located away from the combustion chamber and within the cooling area of the cylinder head. A typical hole nozzle consists of a valve and body that are fitted together to form a mated assembly. Since the valve and body are mated during manufacturing, valves must never be interchanged between nozzle holders. Each hole nozzle is engineered to meet the particular needs of the engine it is serving.

Hole Nozzle Fuel Flow

Fuel under high pressure is generated by the fuel injection pump, flows through the fuel duct in the body, and enters the pressure chamber. When the fuel pressure exerts sufficient force on the differential surface of the valve to overcome the opposing spring preload, the valve is lifted off of its seat, allowing fuel to enter the sac beneath the seat. From the sac, it is directed through the spray holes into the combustion chamber, **Figure 18-3.** The fuel is atomized as it exits the spray holes and enters the combustion chamber. A hole nozzle has one or more spray holes or orifices. These openings are straight round holes through the top of the nozzle body beneath the valve seat.

At the end of the pump's injection stroke, there is a sudden drop in line pressure that results in a rapid pressure drop in the nozzle chamber. Since the pressure adjusting spring is exerting a downward force on the valve and is no longer opposed by fuel pressure in the chamber, the valve immediately reseats itself in the nozzle body, closing the nozzle.

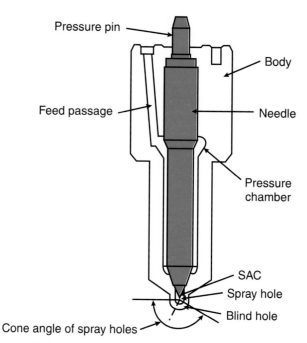

Figure 18-3. *Long-stem hole injector nozzles require less mounting space between cylinder heads.*

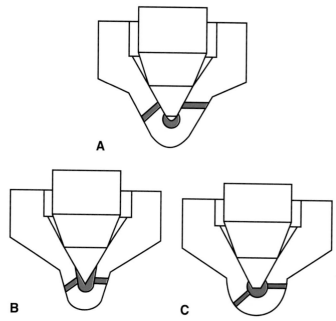

Figure 18-4. *Configuration of spray holes or orifices in typical hole nozzles. A—Seat hole nozzle. B—Hole nozzle with conical fuel sac. C—Hole nozzle with cylindrical fuel sac.*

A slight amount of controlled clearance between the nozzle valve major outside diameter and nozzle body allows a small amount of leakage past the valve into the adjusting spring chamber and nozzle holder body. This leakage provides lubrication for the valve spring seat and adjusting spring. A number of design factors affect the operation and efficiency of hole type nozzles.

Spray Hole Design

The number, diameter, and position of the **spray holes** must be chosen to provide good atomization of the fuel entering the combustion chamber, **Figure 18-4.** This provides optimum air-fuel mixing throughout the engine's full operating range.

Most hole type injector nozzles are set to an opening pressure of between 2500-3500 psi (17 235-24 129 kPa) although some settings may approach 4400 psi (30 333 kPa). This high pressure reduces the injection period and the danger of gas blow back through the holes. In general, opening pressures increase with the degree of engine turbocharging since the injector must be capable of overcoming the increased cylinder pressures generated during turbocharging. Higher opening pressures do increase stress on the nozzle springs and seats. Low speed idling may also be rougher and starting more difficult.

Differential Ratio

The **differential ratio** is the ratio between the diameters of the valve guide and seat. When the valve is closed, fuel pressure acts on the annulus located between the guide and valve seat. When the valve is open, pressure acts on the full guide area. This means the pressure required to open the valve is greater than the pressure required to hold the valve open.

The increase in pressure area once the valve lifts off its seat results in very rapid opening. When the fuel flow rate is low, a condition known as valve overshoot can occur. This is characterized by nozzle buzz, a sound often heard during nozzle bench testing. Most nozzles operate using a differential pressure between 1.7:1-2.7:1. This generates a closing pressure that is between 65-85% of opening pressure. A large differential ratio decreases the amount of stroke-to-stroke variation between injectors, particularly at idle speeds. However, spring forces and the stress placed on these springs during operation are increased.

Needle Mass

The greater the mass of the valve needle, the slower it will move. This is important during the closing cycle of fuel injection when it is necessary to reseat the valve before the fuel pressure below the seat has dropped to the level where combustion gases from the cylinder can blow back through the nozzle opening. A nozzle that is experiencing gas blow back will have a very short valve seat life.

Valve Guide Diameter

The valve guide and nozzle valve can be compared to a cylinder and piston. Like the piston, the nozzle valve moves up and down inside the bore of the guide. This means the **valve guide diameter** has a direct effect on the pushing power of the nozzle and the distribution of the pressures generated. During opening, the rapid expansion of space around the valve seat causes a drop in fuel pressure at the nozzle opening. The opposite effect occurs when the valve closes to cut off injection. A large diameter guide enables the valve to better control this drop in pressure and reduce the danger of combustion gas blow back into the nozzle.

Needle Lift

The valve needle must lift off the seat high enough to provide room for sufficient fuel flow through the seat opening. Too much lift will cause excessive fuel flow and pressure drop. Once again, the danger of gas blow back is increased.

Fuel Sac Volume

There is a direct relationship between the **fuel sac volume** under the seat and the level of hydrocarbon emissions the engine produces. The sac must be as small as practical to reduce emission levels. Fuel remaining in the sac after injection has ended may drip into the combustion chamber as cylinder pressure falls. The fuel then mixes with and dilutes the engine's lubricating oil. Reducing the volume of the injector nozzle fuel sac requires an increase in needle lift setting to obtain the same volume of fuel flow.

Pintle Nozzles

Pintle injector nozzles are used in small-bore, high speed engines. Pintle nozzles are also used in engines having precombustion, divided, air cell or energy-cell combustion chambers. This type of nozzle has a throttling characteristic whereby only a small amount of fuel is discharged at the beginning of injection. The fuel flow rate then increases progressively as the valve lifts higher off its seat.

Pintle Nozzle Construction

The valve of a pintle nozzle is equipped with an extension protruding through the hole in the bottom of the nozzle body, **Figure 18-5.** This protruding extension creates a hollow

cone-shaped spray pattern. The angle of the spray cone is usually between 0-60°, depending on the exact type of combustion chamber used. A pintle nozzle generally opens at a lower fuel pressure than a hole nozzle because fuel flows more readily from the larger hole of the pintle nozzle.

Pintle valves have thin shanks or ends shaped to give the desired spray pattern, **Figure 18-6.** The pintle extends into the nozzle opening so an annular spray (in the shape of a ring) is produced. Varying the shape of the pintle will vary the shape of the spray from a wide angle shape to a hollow cylindrical shape. Cylindrical shaped spray patterns produce the greatest amount of fuel penetration into the chamber.

Pintle Nozzle Operation

Pintle motion can be either up or down, depending on the nozzle design. Upward moving pintles move against fuel pressure, while downward moving pintles move with the flow of fuel. Pintle nozzle bodies have a single hole opening through which the pintle extends. Hole size and angle affect the spray pattern and spray density.

Certain pintle nozzles are classified as **delay nozzles.** This type of nozzle reduces the amount of fuel injected in the early stages of injection. This limits the amount of fuel in the combustion chamber when combustion begins to help reduce preignition and knocking. This "pilot" effect is achieved by providing a parallel portion of the valve pintle that projects through and fits tightly into the nozzle body opening when the valve is closed, **Figure 18-7A.**

As the nozzle begins to open, it produces an initial pilot spray designed to enter the energy cell or precombustion chamber, initiating the combustion process,

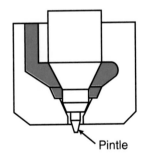

Figure 18-6. *The shape of the pintle nozzle pin affects the shape of the spray cone produced.*

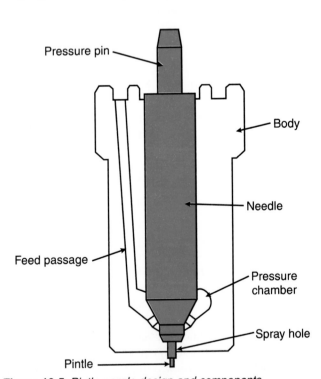

Figure 18-5. *Pintle nozzle design and components.*

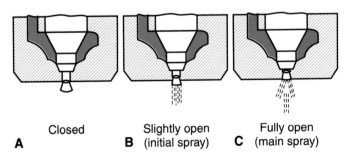

Throttling pintle nozzle (pilot injection)

Figure 18-7. *Throttling pintle nozzle operation.*

Figure 18-7B. The "pilot" characteristics are controlled by the fit of this delay portion, the distance it projects into the body opening bore, the nozzle opening pressure, the differential ratio, and the nozzle spring pressure. The differential ratio and the spring pressure are normally high to reduce the valve opening speed and increase the pilot duration. The pilot is more effective at idling when the valve lift is low. Once the nozzle opens fully, the conical spray pattern characteristic of pintle nozzles is produced, **Figure 18-7C.**

CAV Pentaux® Nozzle

A variation of the pintle nozzle is known as the **CAV Pentaux® nozzle.** This nozzle is designed with an auxiliary spray hole to assist in easy starting in cold weather conditions, **Figure 18-8.** At engine starting speeds, the nozzle valve does not lift sufficiently to clear the standard pintle hole, so fuel must be discharged through the auxiliary hole. Fuel passing through the auxiliary hole is directed at the hot center of the combustion chamber where the chances of ignition are highest.

At normal running speeds, when pressure in the fuel system is higher, the main nozzle valve unseats from the main pintle hole to allow the bulk of fuel to be discharged from it. Under normal operating conditions, no more than 15% of the total fuel volume should pass through the auxiliary hole. This is because the smaller diameter auxiliary pintle hole is prone to blockage from solids in the fuel stream and carbon deposits.

Spray Patterns

Figure 18-9 illustrates the spray patterns of the various nozzle types discussed thus far. To the naked eye, the spray cone generated appears to be solid, but as shown in **Figure 18-10,** the cone actually consists of different regions of air-fuel mixtures. Spray cones are precisely calculated by design engineers to produce maximum fuel efficiency, complete combustion, and minimal emissions. These objective are achieved through the manipulation of nozzle hole size, nozzle hole length, spray angle, and injection pressure.

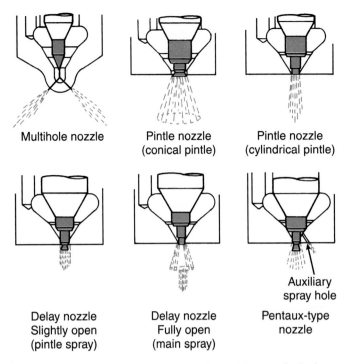

Figure 18-9. Spray patterns produced by various nozzle designs.

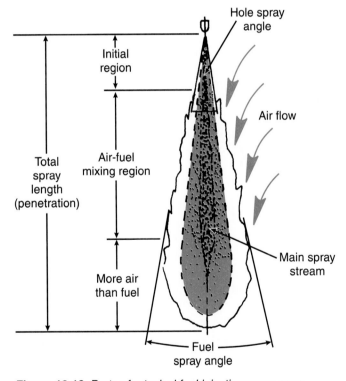

Figure 18-10. Parts of a typical fuel injection spray cone.

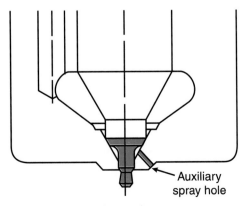

Figure 18-8. CAV Pentaux® nozzle with auxiliary spray hole.

Servicing Fuel Injector Holders and Nozzles

When servicing injectors and nozzles, it is very important to follow the test specifications listed by the injector manufacturer. For example, if the injector opening pressure were mistakenly increased due to poor servicing or adjustment, the start of the injector spray would be delayed and fuel spray velocity would increase. As a result, the length of the spray cone would increase and the spray angle would decrease. Fuel droplet size would decrease and the fuel penetration distance into the combustion chamber would increase. Fuel would now spray to areas in the chamber out of reach of the air stream. This would cause an increase in the delay time, and the engine would operate with increased emissions, a loss of power, high fuel consumption, and noticeable fuel knock.

Injection nozzles are precision units that will operate for very long periods if given proper maintenance. This involves changing all fuel filters at required intervals and using clean, quality diesel fuel. General maintenance also includes a complete cleaning and inspection at manufacturer recommended intervals—usually every 600-1000 hours of operation.

Factors that Can Reduce Injector Life

Several factors can reduce the service life of diesel fuel injectors. **Abrasives** (dirt) in the fuel can wear down fuel carrying components, enlarge the tolerances of the needle valve and bore, damage the valve seat, and enlarge the spray openings. Excessive **heat** will change the expansion rate of the nozzle and accelerate wear. This accelerated wear causes higher leak-by and reduces the amount of fuel available for injection.

Water in the fuel will cause corrosion and can dilute the lubricating oil. Water will also not pass through the spray holes on multi-hole nozzles, and may damage or break off the nozzle tip.

If the **sulfur** content of the fuel is too high, deposits may form that block passages and affect performance. Sulfur is also corrosive. Finally, improper installation can result in insufficient cooling, seizure of the needle valve, or poor valve seating.

Procedures for Injection System Service

When servicing a fuel injection system, always follow a logical step-by-step method. Always perform the following checks and service before turning your attention to the injectors.

❑ Operate the engine to check overall condition.
❑ Check for fuel supply up to the fuel filter(s).
❑ Replace the fuel filter(s).
❑ Check the air supply to the engine.
❑ Check the injection pump timing, adjusting to specifications, if needed.
❑ Check for proper engine speed, adjusting to specifications, if needed.
❑ Check for proper injector advance (if applicable).

If the problem is not isolated or corrected after completing these steps, the nozzles should then be checked. To locate a faulty injector that may be causing incomplete combustion, misfiring, rough running, or excessive smoke

emission, start the engine and check for leaking high pressure connections. When the engine reaches operating temperature, accurately determine engine low idle speed using a tachometer.

> **Warning: The following procedure involves opening high pressure fuel lines during engine operation. Always wear safety goggles and position rags over the connection to catch the fuel stream.**

Crack open the individual fuel injection lines at the injector, one at a time. As each line is cracked open, listen for any change in engine idle. A faulty injector will *not* affect engine idle.

Injector Removal

Cleanliness cannot be overstressed when servicing the fuel injection system. Before removing the injector from the engine, clean the surrounding area. When possible, use compressed air to blow dirt and dust away from the injectors. Be sure to wear safety glasses while using any compressed air. Make sure all lines and connections are absolutely clean before removing the injector.

> **Note: The following injector removal and service procedures are typical for many systems. Each injector manufacturer provides specialized tools and procedures for servicing their injectors. Always consult the proper shop manual before servicing injector holders or nozzles.**

Disconnect the high pressure lines from the injector nozzles, **Figure 18-11A**. Immediately cap all disconnected fuel lines to prevent dirt from entering the system, **Figure 18-11B**. If the injector is held by a flange, remove the hold-down cap screw or the nuts from the stud bolts, **Figure 18-12**. If the injector is held by a clamp, remove the cap screw and slide the clamp out of the injector groove. If necessary, twist the nozzle holder with a wrench placed on the holder flats to help loosen the copper sealing gasket at the bottom of the nozzle.

Using an injector pry bar, pry the injector straight up out of its bore, **Figure 18-13**. Be extremely careful not to bend the nozzle holder and make sure the nozzle sealing washers come out with the nozzles. If the gaskets stick in the bore, remove them with an O-ring pick, **Figure 18-14**. With the injector removed, cap the fuel connection and wrap the nozzle in shop rags to protect it from damage. Cover the cylinder head opening.

> **Caution: Do not use a screwdriver to remove an injector.**

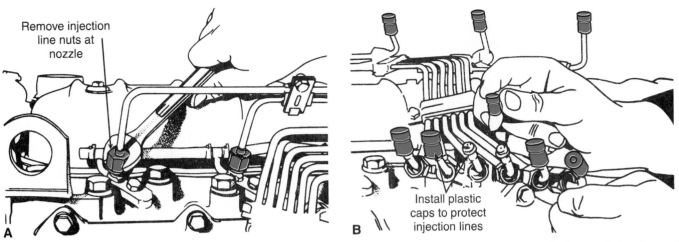

Figure 18-11. *A—Removing injection line nuts at the nozzles. B—Capping the injection lines to prevent dirt from entering the fuel system. (Navistar International Transportation Corp.)*

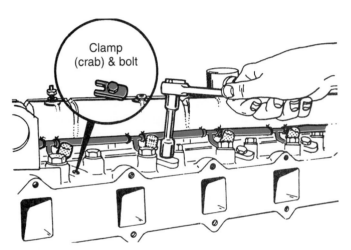

Figure 18-12. *Removing nozzle hold-down clamps. (Navistar International Transportation Corp.)*

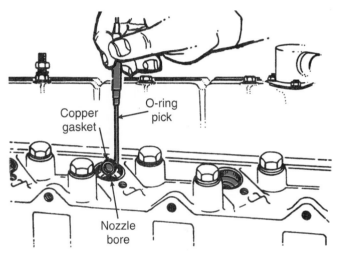

Figure 18-14. *Removing the copper sealing gasket from the injector bore using an O-ring pick. (Navistar International Transportation Corp.)*

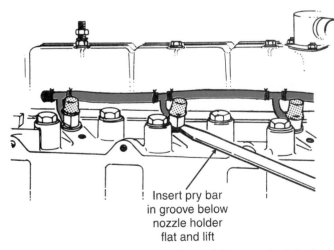

Figure 18-13. *Prying the nozzle straight up out of its bore. (Navistar International Transportation Corp.)*

corroded, or damaged, more extensive serving will probably be required and the unit should be sent to a qualified fuel injection shop.

If the injector appears to be in good shape, clean any external deposits from the nozzle and nozzle holder using a fine-wire brass brush, **Figure 18-15,** and then test injector operation using an injector tester. Pretesting of injectors before disassembly will confirm faulty operation.

⚠️ **Warning: When installing an injector in the tester, position it so that the high-pressure injection spray cannot injure you or others, Figure 18-16. The high pressure of the injected fuel can easily puncture human skin. Follow the injector tester manufacturer's mounting and operating instructions for all tests.**

Injector Testing

Visually inspect the injector to determine if it requires simple cleaning and adjustment or more extensive servicing. If the spray tip, nozzle, or nozzle holder is discolored,

Valve Open and Close Pressure Test

To test the valve *opening pressure* and *closing pressure*, first close the tester's pressure gauge valve and

Figure 18-15. Clean the exterior surface of the injection nozzle with a fine-wire brass brush. The soft metal brass will not damage the surface. (Stanadyne Diesel Systems)

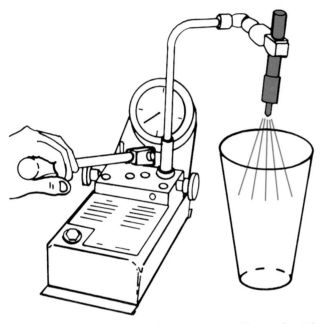

Figure 18-16. Injector mounted in "pop" tester. Always direct the fuel stream into a suitable container. (Mack Trucks, Inc.)

slightly open the fuel supply valve. Rapidly pump the tester hand pump for several strokes to purge any air from the injector. Open the tester pressure gauge valve approximately one-eight turn and slowly operate the hand pump to raise the fuel pressure until the needle valve lifts off of its valve seat and fuel sprays from the nozzle.

Allow the pressure to bleed off and observe when the needle valve reseats. Check the opening and closing pressures noted on the pressure gauge against service manual specifications. In general, the closing pressure should be no more that 200 psi (1375 kPa) less than the opening pressure. If the difference in opening and closing pressure is greater than 200 psi (1375 kPa), the valve may be sticking and will require service.

Valve Seat Test

Test the valve seat seal next. Dry the nozzle tip with a clean, lint-free cloth. Adjust the tester fuel pressure to slightly less than closing pressure. Check the nozzle tip. A

slight wetting after five seconds is permissible, but a droplet of fuel must not form. If a droplet forms after five seconds, the valve seat is leaking.

Back Leakage Test

Checking the amount of leak-off, or **back leakage,** will indicate any problems with the nozzle fit in the bore or the lapped surfaces of the nozzle body and holder. Set the test pressure to within 100 psi (690 kPa) of the opening pressure and time the pressure drop. An average pressure drop of no more than 875 psi (6000 kPa) in six seconds indicates that the mating surfaces are not leaking excessively and that the needle valve is being properly lubricated.

If the pressure drop is higher, the needle valve may fit too loosely in its bore, or the lapped surfaces may not be sealing properly. If the pressure drop is less than 300 psi (2070 kPa) over a six second period, the needle valve is not receiving proper lubrication. Lubricating passages and seams may be clogged with dirt or sulfur deposits.

Spray Pattern Test

To check the **spray pattern** of the injector, close the pressure gauge and operate the hand pump to simulate normal injector operation. On multi-hole nozzles, use long full strokes at a rate of approximately 15 strokes per minute. The spray pattern on multi-hole nozzles should be sharp with a somewhat solid pattern. The angles formed by the individual sprays should be uniform. Usually the spray of a hole type nozzle will be good if the valve seat is tight, the valve free to move, and all spray openings clear. If the spray pattern is poor, **Figure 18-17A,** and cannot be corrected by cleaning, the nozzle should be replaced.

Pintle nozzles should be checked at the stroke rate of about 25 strokes per minute. If the spray is symmetrical and there are no large drops of fuel visible, the nozzle is good. If flags or streamers are visible, **Figure 18-17B,** the unit must be replaced. Pentaux nozzles should be checked at a stroke rate of 1-2 strokes per second. These should be short, quick strokes as opposed to the long, full strokes used with hole and standard pintle nozzles. The spray pattern or cone of each injector used in an engine should be similar in shape, size, and length.

Chatter Test

A slight **chatter** is normally heard as the injector nozzle opens and closes during this test. If there is no audible chatter, the injector valve may be sticking. With a Pentaux nozzle, chatter may or may not occur even though the nozzle is in good operating condition.

Injectors that pass all of the above tests can be reinstalled in the engine, unless it is time for their regularly scheduled cleaning. Before reinstalling any injector, the opening pressure must be adjusted to match service manual specifications as closely as possible. Injectors that fail any of the above tests must be disassembled, cleaned, inspected, serviced with replacement parts as needed, reassembled, and adjusted. The following section outlines a typical disassembly, inspection, and reassembly procedure.

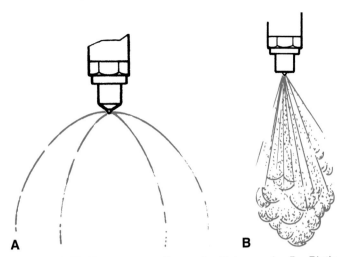

Figure 18-17. *Poor spray patterns. A—Hole nozzle. B—Pintle nozzle.*

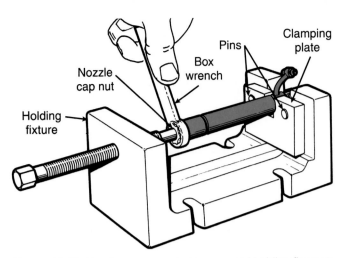

Figure 18-18. *Nozzles are mounted in a special holding fixture to make disassembly easier. (Navistar International Transportation Corp.)*

Disassembly

Complete disassembly of the nozzle holder and nozzle is recommended for inspection of all components. Many nozzle holders require the use of special fixtures and fixture holding plates for safe disassembly. Use the manufacturer's recommended fixtures and tools to prevent damage to components such as the nozzle locating pins. The following sections outline procedures performed by fuel injection service shops:

1. Wash all external dirt, grease, and carbon deposits from the holder assembly with a suitable cleaning agent.
2. Remove the nozzle gasket from the nozzle.
3. Use a brass wire hand brush to remove hardened deposits of dirt or carbon. Do not allow the wire brush to come into contact with the nozzle spray holes.
4. Soak the nozzle holder in a carbon-removing cleaning agent for at least four hours before removing the nozzle cap nut. If this is not done, the nozzle locating pins may be damaged during disassembly.
5. Mount the recommended fixture in a bench vise and assemble the appropriate insert to the fixture holding plate, **Figure 18-18.**

> **Note: Some holders have a mounting flange and do not require an insert. The flange engages the holder bracket pins.**

6. Assemble the holder assembly to the insert and engage the appropriate box-end wrench to the nozzle cap nut.
7. Adjust the fixture clamping plate so that the clamping bolt engages the nozzle tip. Secure the plate in place and tighten the securing bolt to a torque of about 20 ft-lbs. (27 N•m).
8. Loosen the cap nut with a box-end wrench until the nut can be turned by hand. A blow to the end of the wrench is recommended to break any carbon bond between the nozzle and cap nut. On some nozzle holders, it may

also be necessary to loosen the adapter and back off the pressure adjusting screw at the top of the holder.
9. Remove the holder assembly from the fixture.
10. Remove the cap nut, nozzle spacer, and nozzle from the holder body, **Figure 18-19.**
11. Remove the spring seat, spring, and spacers from the holder body. Take care not to lose the spacers.

Cleaning

While keeping each valve with its original nozzle body, soak all parts of each nozzle in a varnish-removing cleaning agent. The disassembled injector may also be cleaned in a 15% **caustic soda solution**, an **air agitated solvent tank**, or an **ultrasonic cleaner**. If the valve nozzle cannot be easily removed after soaking in a carbon solvent,

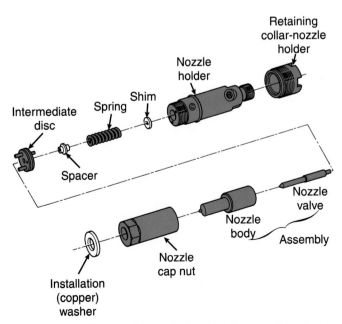

Figure 18-19. *Exploded view of a fuel injection nozzle and components. (Mack Trucks, Inc.)*

a hydraulic nozzle valve extractor should be used. Apply hydraulic pressure from the nozzle test stand to force the valve out of the nozzle body.

 Caution: Do not intermix nozzle valves.

Begin by cleaning the needle valve with a brass wire brush and then use a soft brass tool to remove any carbon from the tip or tip area. Clean the needle valve with mutton tallow placed on a soft cloth or felt pad. Check the valve assembly for pitting or discoloration, **Figure 18-20.** Check the valve seat for wear and replace the nozzle assembly if the seat is excessively worn. Slight irregularities in the seat can be corrected by regrinding using special tools. Injector rebuilding is normally performed in a fully equipped specialty shop.

Clean the outside of the nozzle body with a soft brass brush and check for discoloration and damage to the pressure face and spray holes. The pressure surface of the body that contacts the holder must be clean and free of scratches. The surface of the nozzle body can be lapped using a fine compound such as jeweler's rouge and a lapping plate or block. Use a figure eight pattern when lapping any injector nozzle part, **Figure 18-21.**

Rinse the nozzle body in clean diesel fuel and inspect the lapped surface for smoothness. A small diameter drill bit or wire can be used to clean the internal fuel duct, **Figure 18-22.** Special reaming tools are available for cleaning carbon deposits from cavities and bores. These tools will vary between injector manufacturers. Insert the tool and rotate it to clear out the carbon.

Hard or sharp tools, emery cloth, grinding compounds, or other abrasives must not be used to clean nozzle tips. The pintle nozzle orifice can be cleaned with a toothpick. The small holes in the tip of a hole nozzle can

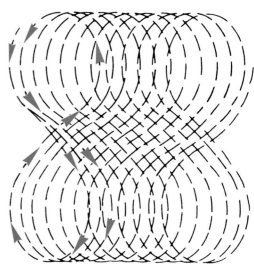

Figure 18-21. *When lapping an injector's parts, it should be moved lightly over the surface of the lapping block in a figure eight pattern. (Detroit Diesel Corporation)*

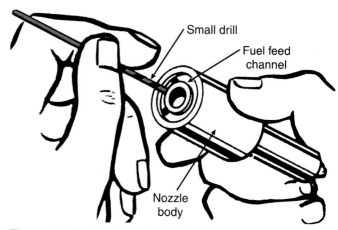

Figure 18-22. *Cleaning the fuel feed channel using a small diameter drill bit.*

be cleaned by probing them with a cleaning wire held in a pin vise or a special tool designed for the job, **Figure 18-23.** The cleaning wire should be slightly smaller than the size of the hole in the tip, **Figure 18-24.** Care must be exercised in cleaning these holes to prevent the wire from breaking in the hole. If this happens, the broken pieces may be impossible to remove. One of the better nozzle tip

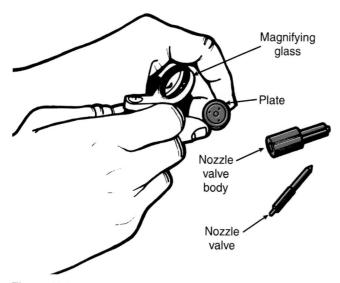

Figure 18-20. *Injector parts must be closely inspected for nicks, scratches, or signs of corrosion. (Detroit Diesel Corporation)*

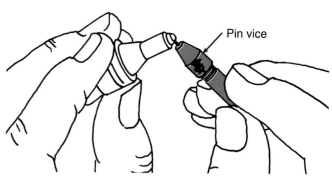

Figure 18-23. *Cleaning spray tip orifices with a special tool. (Detroit Diesel Corporation)*

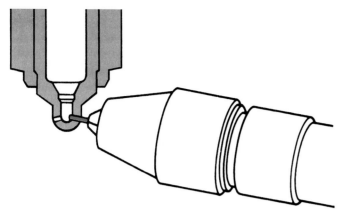

Figure 18-24. Be careful when cleaning nozzle spray orifices. If you break the pin in the nozzle, it may be impossible to recover the broken piece. (Kuboda)

cleaning methods is to use a nozzle-flushing device. The nozzle mounts in the cleaner stand and a special cleaning oil is pumped through the nozzle openings to dissolve and remove the carbon deposits.

Reassembly and Installation

Scrape all loose carbon from the nozzle retaining nut and check for cracks and damaged threads. Clean the sealing surface of the retaining nut using a brass brush. Thoroughly rinse the valve and nozzle body in clean test oil before reassembly and testing. If the inlet fitting was replaced, tighten it to the recommended torque and flush the holder body to remove any metal chips. Assemble any required spacers and insert them into the holder body spring bore. Insert the adjusting spring and spring seat into the spring bore.

Align the nozzle spacer locating pins with the pin holes in the holder body and assemble the spacer into the body. If the spacer has a beveled edge, it should face toward the nozzle. Dip the nozzle valve in clean calibrating or lube oil, and assemble the valve to the nozzle body. The valve must be free in the body. Check by lifting the valve about one-third of its length out of the body. Hold the assembly at a 45° angle and release the valve. It should slide freely back down into the body. If needed, work the valve into the body with clean mutton tallow.

Measure the lift height of the nozzle valve to be sure it is within specifications. Incorrect lift heights will affect liftoff time, fuel flow, and the spray cone and pattern. To measure the lift height, place the spindle of a depth dial indicator against the upper shoulder of the nozzle valve. Zero the dial and then place the spindle against the face of the nozzle body. The difference on the dial reading indicates the lift height of the nozzle valve. If the lift height is greater than specified, you can bring it into specifications by lapping the face of the nozzle body. The nozzle assembly can also be replaced.

Apply a thin even coat of lubricant to the nozzle body seating shoulder and align the nozzle body locating pin holes with the spacer pins. Assemble the nozzle to the

spacer. Assemble the cap nut over the nozzle and tighten the cap nut by hand. Reinstall the injector in the fixture and place the appropriate size box torque adapter wrench onto the cap nut before turning in the fixture securing bolt. Tighten the cap nut to the required torque. Install a nozzle gasket to the nozzle.

Testing and Adjustment

Perform an opening test, chatter test, seat leakage check, spray pattern check, and back leakage test as described earlier in this chapter on all cleaned and reassembled injectors. Injector nozzles that fail these tests after cleaning should be replaced. Set the correct opening pressure as listed in the manufacturer's service manual. When new springs have been installed, the opening pressure should be set about 50 psi (345 kPa) higher than the manual spec to allow for initial setting of the spring.

Reinstallation

The injector bore in the cylinder head must be thoroughly cleaned before the injector is reinstalled. Good cleaning cannot be overemphasized in any phase of fuel injection service. The slightest bit of dirt on nozzle mounting surfaces can cause combustion gas blowby or cocking of the nozzle that can result in sticking. Be sure to use a new nozzle gasket.

Do not use lubricants during installation unless the service manual procedures call for it. Only use the recommended lubricants in the manner described. When lubricants are used improperly, they can turn to carbon under the intense heat of combustion and cause removal problems the next time the injector is serviced. **Figure 18-25** illustrates a typical nozzle installed in the cylinder head. Once the nozzle has been installed, inspect and clean the ends of each injection tubing line before connecting it to the nozzle. The line nut should be left loose to allow air to be bled from the fuel line.

Bleeding the Fuel Injector

Bleed the system by placing the fuel control to the full fuel position and cranking the engine until fuel is visible at all nozzle connections. When bleeding, do not crank the starter motor for more than 30 seconds without a 2 minute cooldown period. After bleeding, tighten all line connections and start the engine. Check for evidence of fuel leakage at all line connections. If a line leak is present, try loosening the connection slightly and allowing fuel to flow from the connection. This may flush out any dirt that is preventing a tight fit. Retighten and recheck.

Pencil Nozzles

The *pencil fuel injection nozzle* is quite different than those discussed to this point. It is a multi-hole, differential

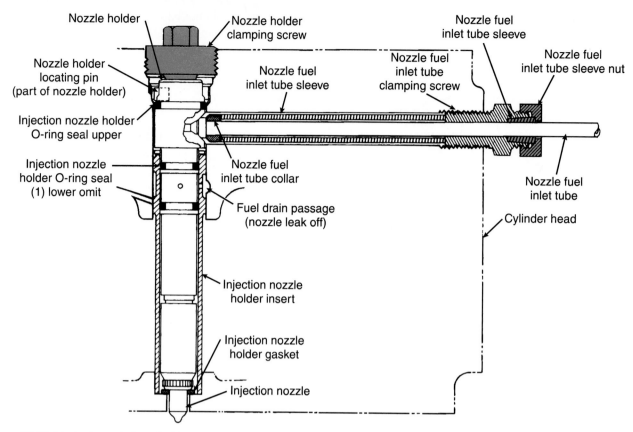

Figure 18-25. Typical nozzle installed in the cylinder head deck. Be sure to leave the fuel line loose to allow air to escape. (Mack Trucks, Inc.)

pressure type nozzle that is much smaller than other nozzle types, **Figure 18-26.** The spray-tip body and inlet fitting are permanently attached to the nozzle body. The needle valve and the spindle are also one unit. The valve is guided by the spray tip body and by the closely fitted valve guide. The lower spring seat rests against the needle valve while the upper spring seat rests against the pressure-adjusting screw. The lift adjustment screw threads into the pressure adjusting screw.

Inlet fittings vary in length and design from engine to engine. Two seals are used: a Teflon® seal to prevent carbon buildup in the cylinder head bore and a nylon seal to prevent loss of cylinder compression and prevent debris from entering the nozzle bore. The nozzle body between the seals is also Teflon coated to prevent carbon buildup and allow for smooth installation and removal.

Pencil Nozzle Operation

The pencil nozzle operates in the same way as the nozzles covered earlier except that its needle valve opens and closes more rapidly. When tested, a good pencil injector will give off a distinct chatter.

Metered fuel under pressure from the injection pump flows through the inlet, the edge filter, and around the valve, filling the nozzle body. When the pressure acting on the differential area overcomes the force of the pressure adjusting spring, the nozzle valve lifts off its seat. As the valve raises to its predetermined lift height, high pressure

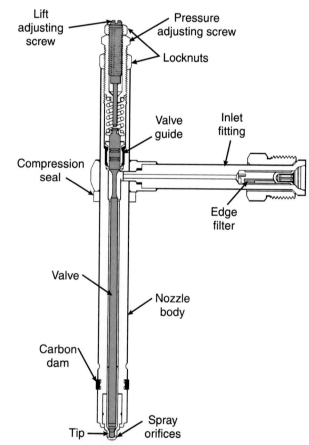

Figure 18-26. Components of a pencil nozzle. (Stanadyne Diesel Systems)

fuel is allowed to flow through the spray orifices in the tip. When delivery to the nozzle ends and injection pressure drops below the preset nozzle opening pressure, the spring returns the valve to its seat. Between injections, positive sealing is maintained by the interference angle, which results in the line contact between the valve and its seat, **Figure 18-27.**

Leak-off fuel that passes through the clearance between the valve and valve guide lubricates the upper half of the injector and returns to the fuel tank through the leak-off connection. The needle valve lift is limited by the lift adjusting screw extension.

Capped Pencil Nozzle

A *capped pencil nozzle,* **Figure 18-28** has no return fuel line like standard pencil nozzles. During injection, fuel accumulates in the nozzle cap rather than returning to the fuel tank. Between injections, the lower line pressure allows the pressurized return fuel in the cap to bleed past the valve to the injection line.

Pencil Nozzle Removal

Figure 18-29 illustrates a typical pencil nozzle installation configuration. Always thoroughly clean the area around the injector before removal. Remove the hold down nut. Disconnect the high pressure fuel lines and remove each nozzle by pulling upward with a slight

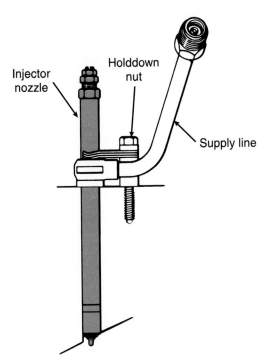

Figure 18-29. *Typical pencil nozzle installation. (Stanadyne Diesel Systems)*

twisting motion. Stubborn nozzles can be removed by using a puller designed for this task.

> **Caution: Never use a pry bar to remove the nozzles as they are easily bent. Plug all holes to prevent contaminants from entering the engine.**

Both the carbon dam seal and the compression seal must be replaced whenever a pencil nozzle is removed from an engine. Failure to use a new carbon dam seal may result in very difficult subsequent nozzle removal. This is due to carbon escaping past the reused carbon dam and accumulating in the clearance between the nozzle body and its bore in the cylinder head. Be sure to cap the nozzle openings and fuel lines to prevent dirt from entering the engine and fuel system.

Disassembly is not recommended for cleaning. The proper cleaning methods involves cleaning the nozzle exterior with solvent and then removing the carbon dam and upper dust seal. Submerge the tips of the nozzles in a parts cleaning solution and soak for at least 30 minutes, **Figure 18-30.** Do not submerge the nozzle completely. This would remove the anti-rust coating on the nozzle body. After soaking, inspect the nozzle holes for corrosion and clogging. Open clogged holes using a proper size cleaning wire. Clean the nozzle tip and seal groove with a brass wire brush.

Pencil Nozzle Testing and Adjustment

The following are the basic steps in testing and adjusting pencil nozzles. Prior to testing, remove the

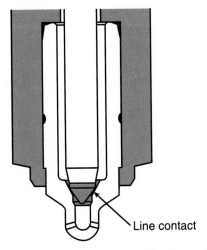

Figure 18-27. *Positive seal is maintained by the contact between the valve and seat. (Stanadyne Diesel Systems)*

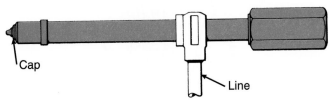

Figure 18-28. *A capped pencil nozzle does not use a return line. Leak-off fuel accumulates in the nozzle cap, and bleeds back into the injection line between injections. (Stanadyne Diesel Systems)*

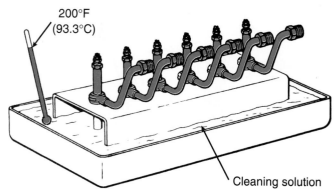

200°F
(93.3°C)

Cleaning solution

Figure 18-30. Nozzles can be cleaned by soaking in a heated cleaning solution. (Stanadyne Diesel Systems)

carbon dam seal from the tip of the nozzle with a razor blade or sharp knife, **Figure 18-31.** Now remove the loose carbon from the tip of the carbon dam seal groove with a brass wire brush. Attach the nozzle to the nozzle tester using a special test pump adapter.

 Caution: Using a standard nozzle adapter will damage the flange of a pencil nozzle.

Opening Pressure

Close the pressure gauge valve and flush the nozzle by operating the test pump rapidly. Open the gauge valve and raise the pressure slowly until the nozzle valve opens. The gauge reading will drop sharply at this point. Check opening pressure against service manual specifications.

 Note: Capped pencil nozzles are tested in the same manner except the cap must be removed.

If the opening pressure is incorrect, remove the nozzle from the test pump and mount in a holding fixture. Loosen the adjusting screw locknut, **Figure 18-32,** and

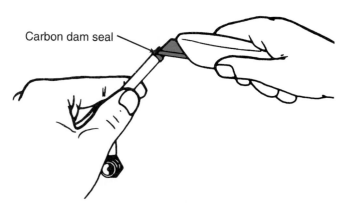

Carbon dam seal

Figure 18-31. Removing the carbon dam seal using a knife. (Stanadyne Diesel Systems)

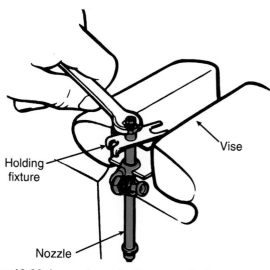

Vise

Holding fixture

Nozzle

Figure 18-32. Loosening adjusting screw locknut with nozzle in holding fixture. (Stanadyne Diesel Systems)

reconnect the nozzle to the tester with the tip facing downward. Loosen the lift adjusting screw two or three turns to prevent interference while adjusting the opening pressure. Turn the pressure adjusting screw clockwise to increase or counterclockwise to decrease the opening pressure, **Figure 18-33.** After setting to specifications, remove the nozzle from the tester and secure it in the holding fixture, tightening the adjusting screw locknut to specifications, **Figure 18-34.**

Valve Lift

Reconnect the nozzle to the tester. Pump fuel through the nozzle and slowly turn the lift adjusting screw clockwise until the valve ceases to open. Do not manually bottom the valve with excessive force as bending may result. Check for bottoming by raising pressure to 200-500 psi (1379-3448 kPa) above nozzle opening pressure. Although some fuel may collect at the top, a rapid dribble should not be apparent.

Now turn the lift screw counterclockwise the amount indicated on the individual nozzle specification sheet.

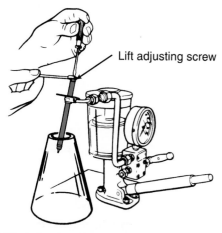

Lift adjusting screw

Figure 18-33. Adjusting the opening pressure by turning the lift adjusting screw. (Stanadyne Diesel Systems)

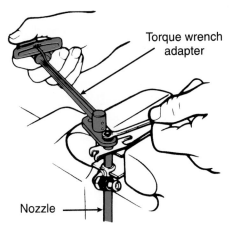

Figure 18-34. *Torquing the pressure adjusting screw locknut. (Stanadyne Diesel Systems)*

Recheck opening pressure. After adjustment is made, the locknut must be torqued to specifications.

 Note: Lift is always set before checking spray pattern and chatter as the amount of lift will affect these characteristics.

Spray Pattern Test

Close the pressure gauge and pump the tester at 30 strokes per minute. The spray pattern should be similar to those illustrated earlier in this chapter for hole nozzles.

Chatter Test

Perform a **chatter test,** pumping the tester handle at a rate of one stroke per second. There should be a distinct chatter. Lack of chatter may indicate a bent or sticking nozzle. If readjustment does not solve the problem, replace the nozzle.

Seat Condition Check

Check valve seat leakage by raising the pressure to 400-500 psi (2758-3448 kPa) below opening pressure and observing for drips and leaks at the nozzle. Check back leakage by raising the nozzle tip up, **Figure 18-35.** Raise the fuel pressure to 1500 psi (10 343 kPa) and count the number of drips from the return in 30 seconds. There should be 3-10 drips. If back leakage is excessive, the nozzle should be replaced.

Nozzle Assembly

Assemble a new compression and carbon dam seal as shown in **Figure 18-36.** A newly installed carbon dam does not return to size immediately, but soaking the nozzle in warm calibrating oil will speed the process, if needed. Cap the nozzle inlet and tip until the nozzle is installed on the engine. On capped nozzles, replace the nozzle cap seal, install the cap, and tighten to specified torque.

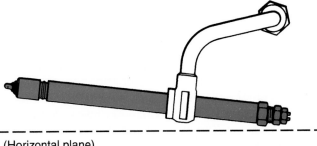

(Horizontal plane)

Figure 18-35. *To check back leakage, the nozzle tip must be higher than the return line. (Stanadyne Diesel Systems)*

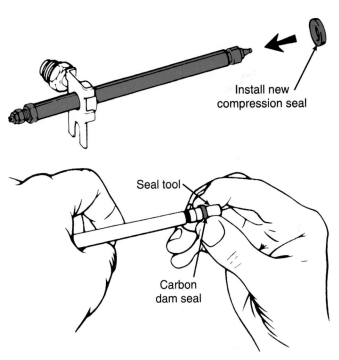

Figure 18-36. *Assembling injector seat. A—Installing compression seal. B—Installing carbon dam. (Stanadyne Diesel Systems)*

Pencil Nozzle Installation

To reinstall the pencil nozzle, first clean the cylinder bore with a bore cleaning tool and clean and check the sealing surface of the cylinder head. Install the nozzles in the cylinder head using a slight turning motion to set them in position. Do not use lubricant. Attach the holding clamps and torque them to specifications. Finally, install the fuel leak-off lines using new plastic boots on the nozzles to prevent leaking.

Summary

Fuel atomization and final delivery is the job of the injection nozzle. Nozzle designs vary with the type and size of the combustion chamber. All injector nozzles are enclosed in nozzle holders. The nozzle holders are mounted in the engine cylinder head.

Direct injection, open combustion chamber diesel engines, such as those used in heavy-duty truck applications, mainly use long-stem hole injector nozzles. A typical hole nozzle consists of a valve and body fitted together to form a mated assembly. Fuel under high pressure generated by the fuel injection pump flows through the fuel duct in the body and enters the pressure chamber. When the fuel pressure exerts sufficient force on the nozzle valve, the valve is lifted off of its seat, allowing fuel to enter the sac beneath the seat. It is directed through the spray holes into the combustion chamber. The fuel is atomized as it exits the spray holes and enters the combustion chamber. A hole nozzle has one or more straight, round spray holes or orifices.

Pintle nozzles are used in small-bore, high speed engines. Pintle nozzles are also used in engines having precombustion, divided, air cell, or energy-cell combustion chambers. With a pintle nozzle, only a small amount of fuel is injected at the beginning of injection. The fuel flow rate then increases progressively as the valve lifts higher off its seat. A pintle nozzle generally opens at a lower fuel pressure than a hole nozzle. Pintle nozzles have thin shanks or pins that extend into the nozzle opening. These pins help create the proper spray cone pattern needed for injection. Pintle nozzle bodies have a single hole opening through which the pintle extends. Hole size and angle affect the spray pattern and spray density.

Delay pintle nozzles reduce the amount of fuel injected in the early stages of injection. This limits the amount of fuel in the chamber when combustion begins to help reduce preignition and knocking. CAV Pentaux nozzles are designed with an auxiliary spray hole to assist in easy starting in cold weather conditions.

Injection nozzles are precision instruments that are very sensitive to dirt and water that can wear and/or corrode nozzle surfaces. Proper precautions must be taken when removing and installing injection nozzles to prevent the entrance of dirt and contaminants into the fuel system. All nozzle cleaning operations must be performed using nonabrasive cleaning tools, such as brass wire brushes.

Injector operation can be tested using a pop tester. Checks include valve open/close pressure, valve seat sealing test, back pressure test, chatter test, and spray pattern checks. Complete disassembly, cleaning, inspection, and reassembly of injection nozzles requires special equipment and training. Most of this work is done in specialized fuel injection service shops. The mounting bore and surrounding area must be thoroughly cleaned prior to reinstalling the nozzle on the engine. Once the nozzle is remounted and its fuel lines reconnected, air must be bled from the fuel lines.

The pencil fuel injection nozzle is a multi-hole, differential pressure nozzle that is much smaller than other nozzle types. The spray-tip body and inlet fitting are permanently attached to the nozzle body. The needle valve and the spindle are also one unit. The valve is guided by the spray tip body and by the closely fitted valve guide. The lower spring seat rests against the needle valve while the upper spring seat rests against the pressure-adjusting screw. The lift adjustment screw threads into the pressure-adjusting screw. The pencil nozzle operates in the same way as hole and pintle type nozzles, but its needle valve opens and closes more rapidly. When tested, a good pencil injector will give off a distinct chatter.

Important Terms

Injector nozzles	Water
Nozzle holders	Sulfur
Open injector nozzles	Opening pressure
Closed injector nozzles	Closing pressure
Injector dribble	Back leakage
Hole injector nozzles	Spray pattern
Spray holes	Chatter
Differential ratio	Caustic soda solution
Valve guide diameter	Air agitated solvent tank
Fuel sac volume	Ultrasonic cleaner
Pintle injector nozzles	Pencil fuel injection
Delay nozzles	nozzle
CAV Pentaux® nozzle	Capped pencil nozzle
Abrasives	Chatter test
Heat	

Review Questions—Chapter 18

Do not write in this text. Place your answers on a separate sheet of paper.

1. *True or False?* Open injector nozzles are the most popular nozzle type in use today.

2. Name the two types of closed injector nozzles in use today.

3. _____ nozzles are used in small-bore, high speed engines.
   (A) Hole
   (B) Pintle
   (C) Open
   (D) None of the above.

4. Direct injection, open combustion chamber diesel engines, such as those used in heavy-duty truck applications mainly use _____ injector nozzles.
   (A) hole
   (B) pintle throttling
   (C) open
   (D) None of the above.

5. The pressure required to open a nozzle delivery valve is _____ than the pressure required to keep it open.

6. In a fuel injector nozzle the fuel sac under the seat must be as _____ as practical to help reduce emission levels.

7. The main advantage of a CAV Pentaux pintle nozzle is _____ .
   (A) higher injection pressures
   (B) easier starting in cold weather
   (C) narrow spray cone for deep penetration
   (D) long service life

8. Which of the following will have an effect on the shape of the injected fuel spray cone?
   (A) Nozzle hole size.
   (B) Spray angle.
   (C) Injection pressure.
   (D) All of the above.

9. Name at least five basic engine checks that should be made before turning attention to the injectors.

10. Which of the following is true concerning pencil injection nozzles?
   (A) The spray-tip body and inlet fitting are permanently attached to the nozzle body.
   (B) Its needle valve opens and closes much faster than a standard nozzle valve.
   (C) A capped pencil nozzle has no return fuel line.
   (D) All of the above.

ASE-Type Questions

1. Fuel injector nozzle holders _____ .
   (A) position the nozzle in the engine cylinder
   (B) contain the needed spring and pressure adjustments to provide the proper nozzle valve action
   (C) provide a means for conducting fuel to the nozzle and combustion chamber of the engine
   (D) All of the above.

2. In a fuel injection nozzle, the slight amount of fuel leakage occurring past the nozzle valves into the adjusting spring chamber is _____ .
   (A) used for cooling
   (B) used to lubricate the valve spring seat and adjusting spring
   (C) unavoidable but causes no harm
   (D) a sign of a defective nozzle

3. Technician A says that the pilot or delay pintle nozzle is used to create a cone shaped spray pattern. Technician B says that this type of nozzle reduces the amount of fuel injected in the early stages of injection. Who is right?
   (A) A only.
   (B) B only.
   (C) Both A & B.
   (D) Neither A nor B.

4. All of the following can be manipulated to affect the injector spray pattern, EXCEPT:
   (A) nozzle hole size.
   (B) spray angle.
   (C) injector pressure.
   (D) spring pressure.

5. Injector nozzles should receive general maintenance _____ .
   (A) every 600-1000 hours of operation
   (B) when a fuel system problem arises
   (C) only if the fuel shows evidence of contamination
   (D) every 600-1000 days of operation

6. All of the following can reduce injector life, EXCEPT:
   (A) abrasives.
   (B) high sulfur content in the fuel.
   (C) frequent fuel filter replacement.
   (D) excessive heat.

7. Technician A says that disabling a faulty injector will not affect engine idle. Technician B says that cleanliness cannot be overstressed when servicing the fuel injection system. Who is right?
   (A) A only.
   (B) B only.
   (C) Both A & B.
   (D) Neither A nor B.

8. Technician A says that high injector opening pressures must be used on turbocharged engines. Technician B says that high nozzle opening pressures may make for rougher low speed idling and more difficult starting. Who is right?
   (A) A only.
   (B) B only.
   (C) Both A & B.
   (D) Neither A nor B.

9. All of the following may be used to clean varnish from injector nozzles, EXCEPT:
   (A) clean diesel fuel.
   (B) 15% caustic soda solution.
   (C) air agitated solvent tank.
   (D) ultrasonic cleaner.

10. All of the following should be done when bleeding the fuel system, EXCEPT:
   (A) place the fuel control to the full fuel position.
   (B) crank the engine until fuel is visible at all nozzle connections.
   (C) tighten all lines after bleeding.
   (D) if line leakage is present, flush out any dirt cranking for 30 seconds with 2 minute cool down periods between.

Freighters and other large ships use multiple diesel engines that must be serviced to ensure dependability and longevity.

Chapter 19

Governors and Acceleration Controls

After studying this chapter, you will be able to:
- ❑ List six reasons why a governor is needed on a diesel engine.
- ❑ Name the different classifications of governors.
- ❑ Understand the terminology of governors.
- ❑ Name the six basic types of governors.
- ❑ Describe a mechanical governor and its operating principles.
- ❑ Know the differences between a hydraulic and a mechanical governor.
- ❑ Understand how a servo governor operates.
- ❑ Describe the underlying principles behind the pneumatic governor.
- ❑ Know the differences between an electric and an electronically controlled governor.
- ❑ Define the terms gain, stability, and dead time compensation.
- ❑ Describe the various devices used on diesel engines to control the generation of black exhaust smoke during acceleration.

This chapter discusses the various speed control devices used on diesel engines. Like gasoline engines, if some form of speed control was not used, the engine would run out of control. A governor is a vital component of a diesel engine's fuel system. It is a speed-sensitive device used to regulate engine speed from idle to maximum full-load (rated) speed. Electronic fuel control systems now perform some of the tasks traditionally handled by a diesel engine governor. Devices such as aneroid/boost compensator controls, fuel ratio controls, and puff limiters are often installed on diesel engines to reduce the amount of black smoke generated during engine acceleration.

Governing Engine Speed

In a diesel engine, the speed at which the engine rotates (rpm) must be controlled or governed. A diesel engine's rotational speed is controlled by varying the amount of fuel injected into the cylinders. Governors can be classified according to the functions they perform, method of operation, and the way they govern the fuel injection system.

As explained in **Chapter 4,** there is more than enough air in the cylinders to support combustion. So under a given load condition, increasing the amount of injected fuel will increase engine power. Pistons will move up and down faster, the time between cylinder firings will decrease, and crankshaft rpms (engine speed) will increase. In the same way, if the amount of fuel injected into the cylinders is decreased, combustion forces will also decrease. Piston movement and engine speed decreases.

A change in engine load can also have a dramatic effect on engine speed. As load increases, engine speed decreases, and more fuel must be injected to maintain a constant engine speed. If the load is suddenly decreased, the amount of fuel injected must also decrease to prevent the engine from racing to dangerously high speeds. On mobile applications, there must be sufficient fuel at idle to prevent stalling.

Monitoring engine speed and making adjustments in the amount of fuel delivered to the cylinders is the job of the engine's *governor* or *fuel governing system.* The governor is often included in the design of the fuel injection pump. However, with the increased use of electronic engine controls, some of the traditional governor functions have been incorporated into the electronic control system. In all cases, the governor or fuel control system is called upon to do the following:
- ❑ Limit the maximum engine speed.
- ❑ Control engine idle speed to prevent stalling.

- ❑ Provide sufficient fuel for starting the engine.
- ❑ Furnish additional fuel when the engine lugs to maximum torque.
- ❑ Shut down the engine if it overspeeds.

To be able to sense engine speed and load changes, the governor must be connected to the engine in some way. This can be a direct mechanical connection or it can be an electronic connection using various speed and engine status input sensors. In nonelectronic controlled inline, distributor, and pressure-time injection systems, the governor is usually an integral part of the fuel injection pump, **Figure 19-1.** On nonelectronic unit injector fuel systems, such as those used on Detroit Diesel's two-cycle engines, the governor may be contained in a separate remote housing. In many cases, the governor is driven at engine speed either directly or through the use of a gear or belt arrangement.

Because governor speed is related to engine speed, any change in engine speed is immediately sensed. The governor can then react accordingly to adjust fuel metering. As shown in **Figure 19-2,** the governor is part of a closed loop system within an engine. The arrows show the direction of feedback to the governor, which reacts to any changes in the steady state condition and then takes corrective action. For the system to operate smoothly, it must be stable, the response swift and accurate, and maintain a constant speed or change the speed to compensate for the load.

Classifications of Governors

Various applications call for different types of engine speed governing. For example, the governing requirements for diesel engines installed in heavy-duty trucks differ somewhat from those needed on stationary generators. The functions designed into a governor depend on the type of load placed on the engine and the degree of control required.

Limiting speed governors have been developed to prevent the engine from stalling at low idle speed and to prevent racing. All speeds in between can be controlled by the operator. This type of governor is generally used in on-highway vehicles.

Variable speed governors are often used on drill rigs, tractors, shovels, locomotives, and marine engines. This

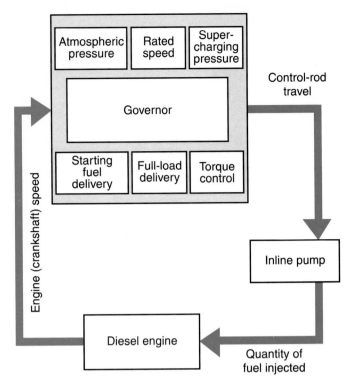

Figure 19-2. *The governor is part of a closed loop system that controls the quantity of fuel injected, and as a result, engine speed.*

type of governor gives the engine automatic speed control and is easily adjusted during operation. It can control fuel flow over a wide range, from low idle to maximum speed. For example, an engine installed in a piece of agricultural equipment may experience many changes in load per minute as it operates over varying terrain and conditions. The operator cannot anticipate these rapid changes, but a variable speed governor can adjust fuel delivery to maintain the proper engine speed during all load levels.

Constant speed governors maintain engine speed at a constant rpm, regardless of the load. These governors are generally used to control engine speed on generators and other stationary applications. ***Overspeed governors*** are designed to prevent the engine from exceeding a specified maximum speed. A ***load-limiting governor*** limits the load that the engine must handle at various speeds. Its purpose is to prevent engine overloading.

Governor Terminology

Before you can understand the operating principles of the various types of governors used on diesel engines, you must become familiar with several terms used to describe the various states of operation.

High and Low Idle Speed

High idle speed or *maximum no-load speed* is the highest rpm at which the governor permits the engine to operate. This speed is achieved when the throttle linkage is moved to its maximum position with no-load acting on the engine.

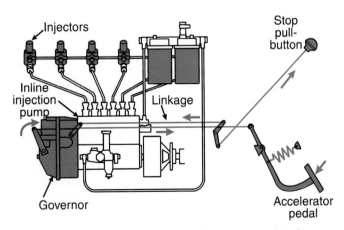

Figure 19-1. *In most cases, the governor mounts directly to the injection pump and is linked to the throttle or accelerator pedal.*

Low idle speed is the rpm at which the engine operates when the throttle linkage is in the closed (released) position. It is the lowest rotational speed the engine can operate at without stalling. For on-highway applications, engine low idle speeds are generally in the 475 to 525 rpm range.

Rated Speed and Maximum Torque Speed

Rated speed or *maximum full-load speed* is the engine rpm at which the engine will generate its maximum horsepower as specified by the manufacturer. **Maximum torque speed** is the rpm at which an engine creates the most torque.

Speed Droop

Speed droop is the change in engine speed caused by a change in engine load. Speed droop is expressed in rpm, or more often, as a percentage of maximum full-load speed. For mobile, marine, and industrial engines, this is usually 5-10%.

Operating Terminology

The engine will experience **under run** if the governor fails to maintain low idle speed whenever the engine rpms drop quickly from higher speeds. If the governor permits the engine to fall below low idle, the engine may stall. **Over run** occurs when the governor allows the engine to exceed its maximum rated speed. Over run is most likely to occur in on-highway applications when the vehicle is descending a steep grade. The vehicle's wheels begin to drive the engine, forcing it above its rated maximum speed.

Every governor has a characteristic **deadband,** which is a narrow speed range during which no measurable speed correction is made by the governor. **Sensitivity** is the change in engine speed necessary before the governor begins making a change in the amount of fuel delivered to the cylinders. Sensitivity is often expressed as a percentage. A governor that responds to a speed change of 3% is more sensitive than one that requires a 6% change before responding. **Promptness** is the length of time required for the governor to move the fuel control mechanism from a no-load to a full-load position.

Governor **hunting** or **surging** is a continuous engine speed fluctuation (slowing down and speeding up) from the desired rpm setting. This condition is most likely to occur at idle. Many governors are equipped with an adjustable buffer screw or bumper spring that is set to minimize this undesirable condition. **Stability** is the ability of the governor to maintain the desired engine speed without fluctuations or hunting.

Governor Designs

Six different types of governors are presently used on diesel engines. They are mechanical, pneumatic, servo, hydraulic, electric, and electronic. A simple method of remembering the six types is to memorize the following mnemonic sentence: *My Poor Sister Hates Everybody Else.*

While electronically-controlled fuel governing systems are being used on nearly all late-model low emissions diesel engines, there are millions of the other governor types still in service. The durability and rebuild capability of diesel engines ensures mechanical and other types of governors will be in service for many years to come.

Mechanical Governors

Mechanical governors are the oldest and most universally used diesel engine governor. This governor relies on centrifugal force and a control spring to regulate the movement of the fuel control mechanism.

The mechanical governor illustrated in **Figure 19-3** is a centrifugal, variable speed type. It is capable of maintaining a constant speed between idle and maximum speed positions. This governor is mounted at the rear of the injection pump housing. The flyweight assembly is mounted on the injection pump camshaft and turns whenever the camshaft turns. The governor is completely enclosed to permit splash lubrication of the working parts, using oil from the injection pump. An operating lever shuts off fuel delivery to the engine by moving the control rack to the stop position.

As the flyweights react to the centrifugal force created by the injection pump's spinning camshaft, they move in (closer to the camshaft) and out (farther from the camshaft). When the engine is off, the governor does not turn and the flyweights are all the way in, **Figure 19-4A.** In this position, the fuel control rod or linkage is in the full-fuel position

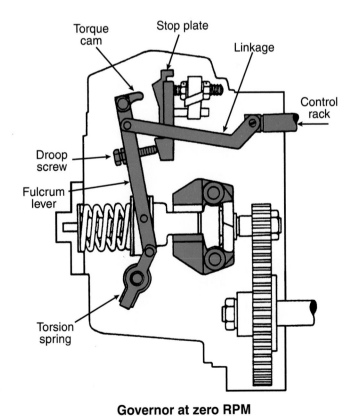

Governor at zero RPM

Figure 19-3. Components of a variable speed mechanical flyweight governor. (AMBAC International)

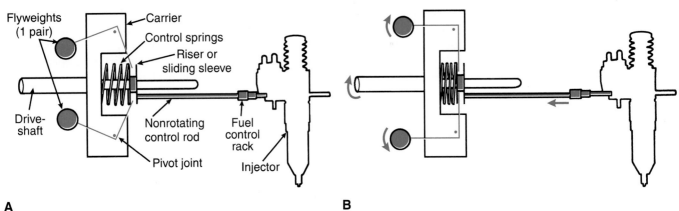

A   **B**

Figure 19-4. Mechanical flyweight governor positions. A—Start-up or full fuel position. B—Balanced running position at constant speed.

and is held by the governor spring. The governor is in this position during engine startup.

As the engine starts, the camshaft begins to turn, carrying the weights along with it. The centrifugal force generated by the turning shaft pushes the flyweights outward. The outward flyweight movement, through a thrust sleeve, compresses the governor spring. The thrust sleeve is also connected to the governor fuel control rod. As the flyweights move outward, the sleeve and fuel control rod both move toward the carrier, compressing the governor spring, **Figure 19-4B.** As the control rod moves, it closes the metering valve in the fuel injector, decreasing fuel delivery. The control rod will continue to move until the flyweight's centrifugal force equals the governor spring's compression force. When these two forces are balanced, the control rod will stop moving, and fuel delivery and engine speed will stabilize.

An important fact to remember is that the force exerted by the governor control spring is always trying to move the control mechanism to an increased fuel setting. The centrifugal force created by the rotating flyweights act against the control spring, compressing it, and moving the fuel control mechanism to a decreased fuel setting.

When a load is placed on the engine, engine speed, injection pump camshaft speed, and flyweight turning speed all decrease. This causes a drop in centrifugal force. The

flyweights move closer to the camshaft, the governor spring decompresses, and the sleeve and fuel control rack all move away from the carrier, **Figure 19-5A.** This movement causes the control rack to open the fuel injector metering valve, increasing fuel deliver to the cylinders. Increased fuel allows the diesel engine to produce the power needed to maintain engine rpm. As before, rack movement continues until the flyweight centrifugal force and the governor spring compression force balance each other out.

If the load on the engine decreases, engine rpm would increase and the flyweights would once more move outward, pulling the sleeve and control rack toward the carrier, **Figure 19-5B.** This would decrease fuel to the injectors, preventing the engine from overspeeding.

Although mechanical governors are the only type to use flyweights to sense engine speed and to move the control rack, all governors operate on the same basic principle. They all have a means of sensing engine speed and controlling the fuel rack movement. Mechanical governor linkage movement under various engine speeds is described in more detail in the following paragraphs.

Starting the Engine

With the engine stopped, the speed control lever is moved into the low idle position and then advanced

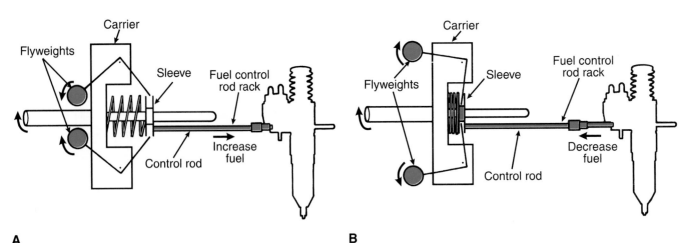

A   **B**

Figure 19-5. Flyweight and spring action in a mechanical governor. A—Increase in engine load and decrease in speed. B—Decrease in engine load and increase in speed.

slightly. The starting spring will then pull the control rack to the excess fuel position, **Figure 19-6A.** At the same time, the tensioning lever will move up against the full-load stop, which moves the guide lever, knuckle, and thrust sleeve forward. The flyweights then come to rest against the thrust sleeve (innermost position).

While the starter is cranking the engine, the injection pump begins supplying excess fuel to the engine. Once the engine starts, the centrifugal force produced by the whirling flyweights overcomes the starting spring tension (even before idle speed is reached). Engine speed increases until the flyweight centrifugal force and the governor main spring are balanced.

Engine Idling

When the engine is idling, **Figure 19-6B,** the governor functions automatically. At this speed, the governor main spring is almost free of tension, and has only a slight effect on the governor linkage. This means that even at low speeds, the flyweights can swing outward with very little resistance. As the control lever and guide lever move, the fuel control rack also moves, increasing governor main spring tension. Since centrifugal force and spring tension are relatively low at idle speed, the torque capsule in the tensioning lever is only slightly compressed. Because the gap between the knuckle and the tensioning lever is greater at low speeds, the tensioning lever will contact the supplementary idling spring and will result in the desired speed regulation.

Engine at Medium Speed

Moving the speed lever above the idle position causes the control rack to move to the maximum fuel delivery position, and the tension lever to move to full-load stop, **Figure 19-6C.** The injection pump delivers more fuel to the engine, resulting in increased speed. As soon as the centrifugal force exceeds the governor main spring force (as determined by the speed control lever position), the governor linkage moves the control rack to a position where the centrifugal force is just equal to the spring force. In this way, the governor maintains a lower fuel delivery rate, which results in a steady engine speed.

Engine at Maximum Speed

Governor operation at maximum speed is similar to medium speed operation, except that the tension lever stretches the governor main spring to its maximum length. The fully stretched main spring causes the tension lever to move against the full-load stop with greater force, and the control rack to move into the maximum fuel delivery position. The torque lever is now compressed and will remain this way until engine speed is reduced enough to slow the flyweight centrifugal force. This moves the fuel control rack into a position where it will provide adequate torque reserve.

Once full-load maximum speed is reached, governor response will regulate fuel delivery between full-load and fast idle to handle any load as long as there is no overload.

Activating the engine shut-off lever moves the stop devices, which then moves the control rack to shut off the fuel

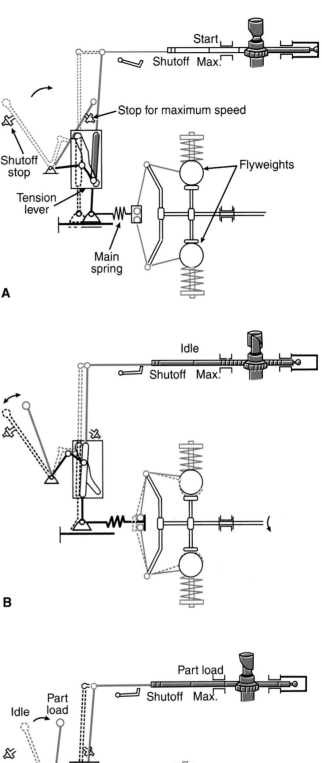

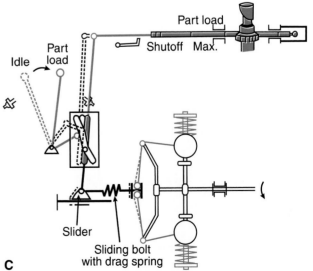

Figure 19-6. *Typical mechanical governor operation. A—Start of engine. B—Engine at idle. C—Engine under load.*

supply to the engine. This movement takes place independently of the flyweight and speed control lever positions. The stop device has a supporting lever that is coupled to the shaft and shut-off lever by pressure springs. This lever continues to pivot until the control rack is in the stop or no fuel delivery position. At this point, the supporting lever stops moving, the pressure springs become tensioned, and the shut-off lever reaches the limit of its travel.

As engine speed decreases, tension on the springs decreases as the flyweights close. This pushes the supporting lever, along with the lower end of the fulcrum lever. The upper end of the fulcrum lever remains nearly stationary to the fuel control rack.

Hydraulic Governors

Hydraulic governors are used on many marine, industrial, and power generator applications, **Figure 19-7.** Hydraulic governors regulate the fuel supply indirectly through oil pressure. Pressurized oil from the engine's lubricating system is supplied to an auxiliary pump in the governor. The auxiliary pump then develops the oil pressure needed to actuate the governor mechanism.

The oil pressure is maintained in the annular space between the undercut portion of the pilot valve plunger and the bore in the ball head. At any given throttle setting, the force of the speeder spring is opposed by the centrifugal force created by the revolving flyweights. When the two forces are equal, the land on the pilot valve plunger covers the lower opening in the ball head, producing a constant speed condition, **Figure 19-7.**

If engine load increases and engine speed decreases, the weights are forced inward by the speeder spring, allowing the pilot valve plunger to uncover the lower port

in the ball head. Pressurized oil now enters the cavity at the lower end of the power piston and forces the piston and the floating lever upward, **Figure 19-8A.** This movement is transmitted through the terminal lever to the fuel rod, and ultimately, the injectors.

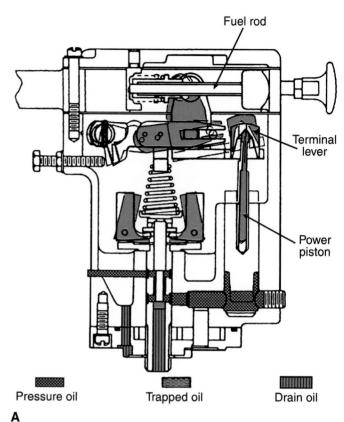

Pressure oil   Trapped oil   Drain oil

A

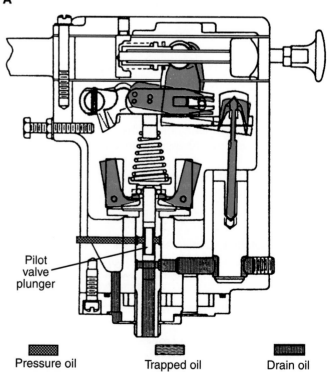

Pressure oil   Trapped oil   Drain oil

B

Figure 19-8. Hydraulic governor operation. A—Increased load and decreased speed. B—Decreased load and increased speed. (Detroit Diesel Corporation)

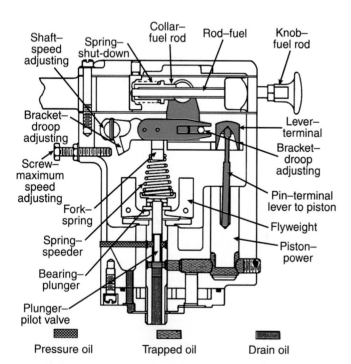

Figure 19-7. Components of a typical hydraulic governor shown in the constant load position. (Detroit Diesel Corporation)

As the flyweights and speeder spring reach a balance, the pilot valve plunger lands will again cover the ports in the ball head, trapping oil under the power piston. With the piston held in its new position, operation is stabilized at a slightly increased speed. **Figure 19-8B** illustrates the governor reaction as load decreases and engine speed increases.

A hydraulic governor also acts as its own automatic shut down device. If the oil supply or oil pressure is cut off or lost, the power piston will drop, allowing the fuel rod to return to the no fuel position, shutting down the engine.

Hydraulic governors are more sensitive than mechanical governors. A hydraulic governor is normally found on larger engines that need more accurate fuel regulation, while mechanical governors are often used on smaller engines that do not require a great amount of fuel regulation.

Servo Governors

Servo governors are similar in design to mechanical governors. However, the servo governor has an additional servo mechanism that assists the mechanical governor, **Figure 19-9.** The servo governor consists of a cylinder, cylinder sleeve, piston, and valve, **Figure 19-10A.**

When the governor moves in the *fuel on* direction, the valve moves in, **Figure 19-10B.** The valve opens the oil outlet. Pressurized oil from the inlet pushes the piston and

the fuel rack to the left. Oil behind the piston goes through the passage, along the valve, and through the oil outlet. When the governor spring and flyweight forces are balanced, the valve stops moving and engine speed is constant. Oil pressure from the inlet pushes the piston until the oil passages are opened. Oil now flows through the oil passages, along the valve, and out through the oil outlet. With no oil pressure on the piston, the piston and fuel rack stop moving.

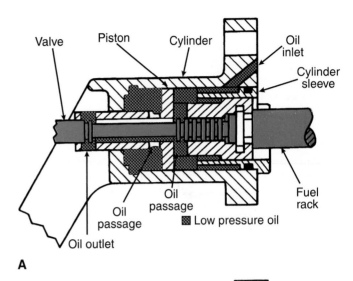

A

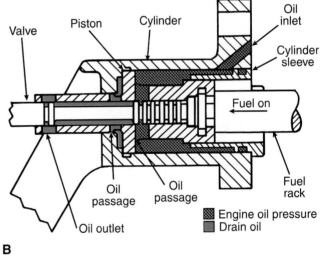

B

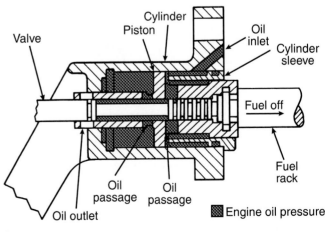

C

Figure 19-10. *Servo governor operation. A—Balanced position. B—Fuel on. C—Fuel off. (Caterpillar, Inc.)*

Figure 19-9. *Components of a servo assisted mechanical governor. (Caterpillar, Inc.)*

When the governor moves in the *fuel off* direction, **Figure 19-10C,** the valve moves out, closing the oil outlet, and opening the oil passage. Pressurized oil from the inlet is now on both sides of the piston. The area and the oil pressure is greater on the left side of the piston, which moves the fuel rack to close.

The **dashpot, Figure 19-11,** gives the governor better control when there are sudden speed and load changes. Dashpot components include a needle valve, oil reservoir, cylinder, piston, dashpot spring, and spring seat. The piston and spring seat are fastened to the dashpot spring. As the spring seat moves, either by a change in load or speed, the dashpot spring moves the piston. The cylinder and oil reservoir are now full of oil. As the piston moves, it moves the oil in or out of the cylinder through the needle valve and oil reservoir. The needle valve restricts oil flow to and from the cylinder, limiting piston and spring seat movement. The faster the governor tries to move the spring seat, the more resistance the dashpot gives to spring seat movement.

Pneumatic Governor

Pneumatic governors do not have flyweights and are not driven by the engine as are mechanical governors. Instead they react to changes in the diesel engine's intake manifold vacuum. Often called *air governors,* pneumatic governors are quite popular for small, high speed diesel engines used in light-duty trucks and farm equipment. Their operating principles make pneumatic governors unsuitable for use on turbocharged engines.

The underlying principle of the pneumatic governor, **Figure 19-12,** is that air enters the system through a tube called a **venturi.** The venturi creates a vacuum in another smaller tube that enters it at right angles. The amount of vacuum created depends on the velocity of the air passing through the venturi.

The pneumatic governor venturi has a throttle-controlled butterfly and is located at the entrance to the intake

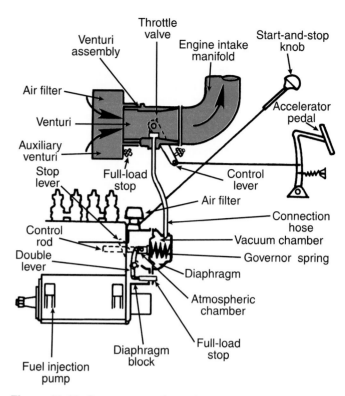

Figure 19-12. Components of a typical pneumatic governor.

manifold. It also has a pressure-tight housing with a spring and flexible diaphragm assembly for actuating the pump rack. As mentioned, air enters the system through the venturi, creating a partial vacuum at the tubing connecting it to the diaphragm chamber. The large spring then pushes the diaphragm and attached pump rack toward full fuel. The partial vacuum acting on the diaphragm counteracts this spring force to move the rack toward fuel shutoff.

When the engine is not running, the governor spring pushes the diaphragm and the control rod to the maximum fuel delivery position. As soon as the engine starts, the incoming air rushes past the nearly closed throttle valve and suction pipe orifice to create a strong vacuum in the diaphragm chamber. The diaphragm and fuel pump control rod then draw back, reducing fuel delivery.

When the engine is operating under load with the throttle valve fully open, air movement is steadier and slower into the open valve, resulting in a weaker vacuum in the diaphragm chamber. The diaphragm and control rod valve are held in this position by the governor spring, resulting in maximum fuel delivery, **Figure 19-13.** Any change in the throttle valve setting or engine load changes the rate at which air enters the throttle valve. This also changes the amount of vacuum in the diaphragm chamber and will change the amount of fuel delivered.

Electric Governors

Electric components and circuitry can be added to mechanical or hydraulic governors to enhance the governor's ability to control the engine. **Electric governors** and *electro-mechanical governors* are often used in applications where constant engine speed must be maintained. A good

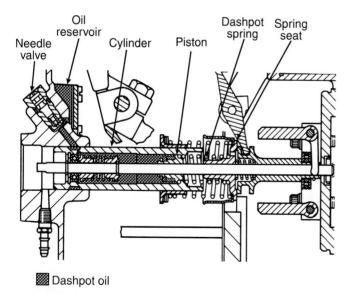

■ Dashpot oil

Figure 19-11. Servo governor dashpot components. (Caterpillar, Inc.)

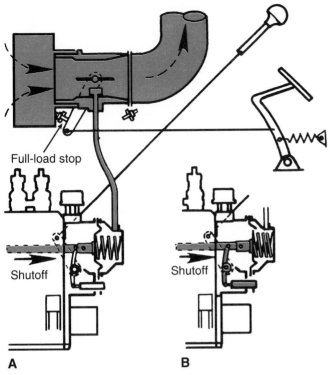

Figure 19-13. *Pneumatic governor full-load speed regulation. A—Nominal speed, full-load. B—High idle speed.*

example is a generator set that uses an electric governor to maintain engine speed within a few rpms of a predetermined setting. Electrically controlled governors are also ideal when engine speed must be controlled from a remote location.

Figure 19-14 illustrates a remote controlled electric governor. The governor body is equipped with a 24-volt reversible electric motor. The motor turns a threaded spindle. A nut that rides on the spindle is connected to the fuel control linkage. When the motor is energized, it turns the threaded spindle. The spindle forces the nut and fuel rack to move in the desired direction for either more or less fuel.

Because the load placed on gen-set motors does not normally fluctuate, increasing or decreasing the fuel has a

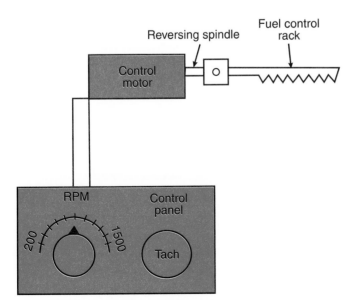

Figure 19-14. *Electric servo-motor controlled governing system.*

highly predictable effect on engine speed. Once the desired engine speed is attained, the motor is de-energized, and the speed remains constant. Holding the engine at one rpm setting helps maintain the frequency output of the generator. Electric governors are not generally used in mobile applications except in large ships.

Electronic Governing

The major difference between mechanical and other non-electronic governors and electronic governing systems is the speed and precision in which the system can make and execute fuel metering decisions. All engine speed governing systems can be subdivided into three separate functions:

- ❑ Sensing
- ❑ Computing
- ❑ Acting

Non-electronic governors typically integrate these functions in a way that cannot be clearly separated. For example, a mechanical governor uses the flyweights for speed sensing; for computation along with the governor spring; and to provide the force to control the fuel delivered by the injection pump. The mechanical governor is very cost effective, but has limited performance and flexibility.

By comparison, electronic governing systems separate the three clearly defined functions. Each component of an electronic governing system is designed to fulfill one specific task, **Figure 19-15.**

- ❑ The sensors detect the actual engine speed and other inputs.
- ❑ The system controller processes the input from the sensors and generates an appropriate output signal to the actuator, controlling the position of the fuel metering mechanism.
- ❑ The actuator moves the fuel metering mechanism to adjust fuel delivery to the engine.

A typical electronic governor speed control system uses a magnetic pickup sensor to detect engine speed. The magnetic sensor is mounted next to a toothed gear that is driven at engine speed. As the gear teeth pass the sensor, a low voltage ac sine wave signal is produced. The signal frequency is proportional to engine speed.

The system controller is really a compact computer. The computer rectifies the ac signal from the sensor to dc current. The computer then counts the number of low voltage pulse signals received per second and uses this figure to accurately calculate engine speed. It compares this value with the desired speed setting. The computer then provides an output voltage to the actuator. The actuator adjusts the position of the fuel control mechanism. In many systems, the actuator is a fast-acting displacement drive, whose motion is proportional to current flow.

Gain and Proportional Control

Gain is the relationship, or error, between the change in fuel and the speed response by the engine. Engine gain varies with load. **Figure 19-16A** shows *high gain* when the

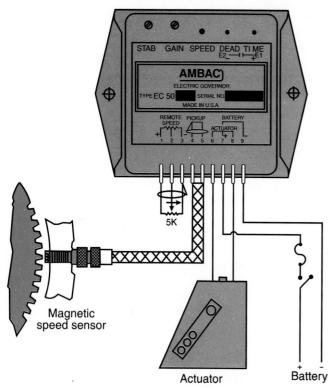

5K

Magnetic
speed sensor

Actuator   Battery

Figure 19-15. *The major components of an electronic governing system are the magnetic speed sensor, controller, and actuator. (AMBAC International)*

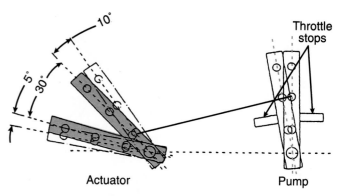

Actuator   Pump

Figure 19-17. *In a non-linear linkage, a large actuator movement does not result in a corresponding large movement of the pump's fuel control mechanism. (AMBAC International)*

engine is carrying no-load. In other words, a small increase in fuel delivery to the engine has resulted in a significant speed increase. **Figure 19-16B** shows *low gain* of the engine when loaded.

Figure 19-17 shows a non-linear linkage between the actuator and the fuel system. This linkage has low gain at

low fuel and/or load and high gain at high fuel and/or load. Linkage gain is matched to engine gain to provide a system which has a relatively constant gain with varying load. Governors with excessive gain become unstable.

Gain is not adjustable on mechanical and other non-electronic governors. To change the gain on a mechanical governor, you must change the governor control spring rate. On a pneumatic governor, it is necessary to alter the shape of the intake port (round ports versus slotted ports).

Electronic governors provide adjustable gain. This minimizes the gain at all load conditions. If the magnitude of the speed error is large, the electronic governor can provide a significant fuel adjustment. If the gain is low, the system can still make a small fuel correction. The ability to operate in this manner is called **proportional control** or proportional gain.

To provide proportional control, the computer's gain circuit amplifies the error signal. When the gain is increased, the error signal size is increased, more correc-

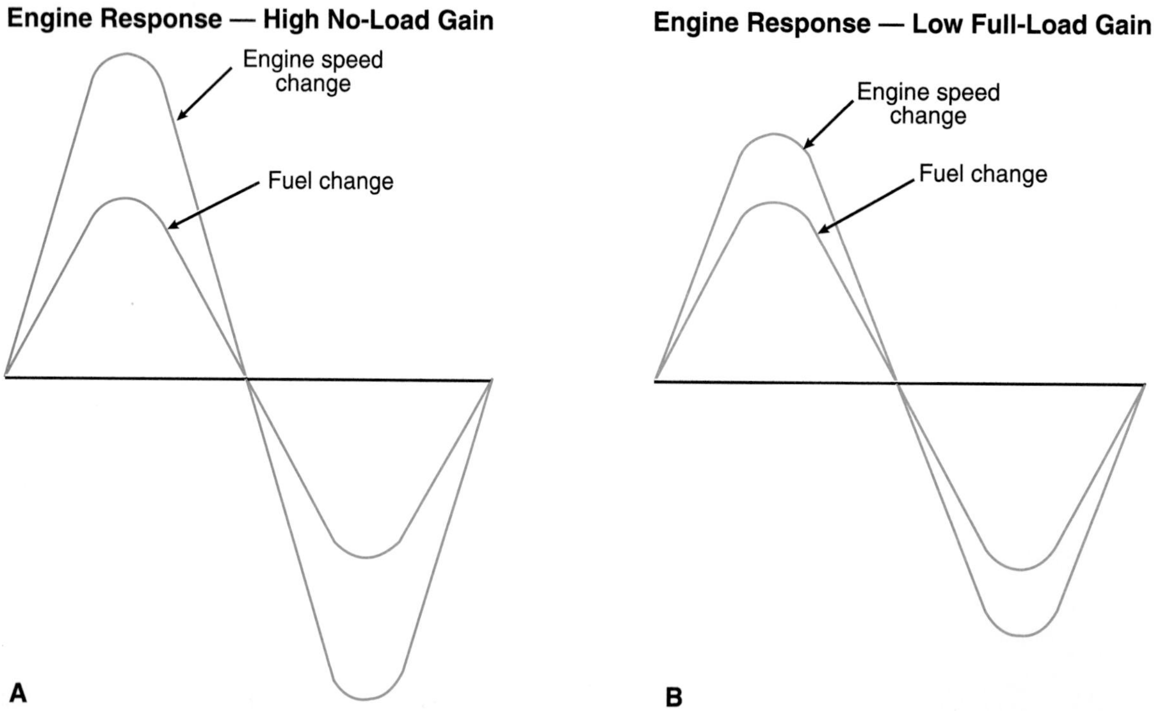

Engine Response — High No-Load Gain

Engine speed change

Fuel change

A

Engine Response — Low Full-Load Gain

Engine speed change

Fuel change

B

Figure 19-16. *Engine response to governor action. A—High no-load gain. B—Low full-load gain. (AMBAC International)*

tive action is taken, and the response is greater. When the gain is reduced, the error signal is reduced, so less corrective action is taken and response is decreased. Adjusting gain in an electronic governing system is achieved by changing the speed error amplification. On some systems, a screw-adjustable potentiometer on the computer is used to set gain. In other systems, a new value is programmed into the engine control module with a hand-held scan tool.

A proportional control governor is referred to as a *droop governor* because it changes fuel delivery with respect to the size of the speed error. The gain cannot be increased until there is no error, since this would require infinite gain. There will always be some change in speed with a change in load when using a simple proportional governor. As explained earlier, this change in speed is called droop and is usually expressed as a percentage of the set speed, **Figure 19-18A**.

Eliminating Speed Droop (Isochronous Governing)

In order to eliminate speed droop, a **stability** function is added to electronic governing systems. This correction is not only proportional to the amount of speed error, but also to the time that the error is present. In this way, the longer the speed error exists, the greater the change in fuel mixture, until the speed error is eliminated and the system runs at the set speed, **Figure 19-18B**. This is called **isochronous governing.** Isochronous governing involves maintaining a con-

stant engine speed, regardless of the load. This means governing with perfect speed regulation or zero speed droop.

Electronic speed controllers provide an adjustable stability function. Increasing the stability amplification will reduce the time needed for the engine to return to the set speed. However, excessive amplification will result in an unstable engine.

Controlling the Rate of Change

The governor corrects fuel delivery as a function of the rate of speed error change. Normally, a rapid deviation from the set speed will result in a rapid response. However, on engines with large actuators, response is slower and speed overshoots can be larger. To correct for this, a setting called **dead time compensation** is increased. This results in a reduction of transient speed error during sudden engine load changes. Dead time compensation is not available on simple flyweight governors.

This control is achieved on electronic governors by changing the value of the dead time compensation capacitors. Excessive dead time compensation will result in high frequency instability. Insufficient dead time compensation will result in speed wandering. The system block diagram is shown in **Figure 19-19**. To summarize electronic governor tuning, the three separate adjustments are as follows:

- ❏ The proportional governor function corrects fuel delivery in proportion to the size of the error. It is adjusted with the gain setting.
- ❏ Integral governor function corrects fuel delivery with respect to the time the error exists. It is adjusted with the stability setting.
- ❏ A governor corrects the fuel delivery as a function of the rate of change of the speed error. It is adjusted with the dead time compensation.

Computer Control of Electronic Governors

A basic sensor/controller/actuator electronic governing system can be installed on diesel engines having no other type of electronic controls. On engines offering compre-

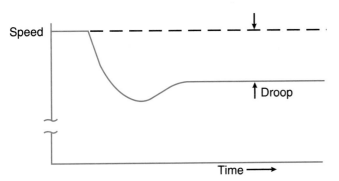

A Speed vs Droop

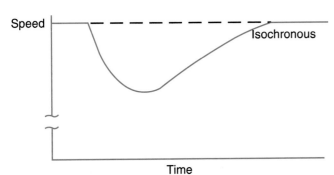

B Speed vs Droop

Figure 19-18. The relationship of speed versus droop between standard and isochronous governors. A—Non-isochronous governor. B—Isochronous governor.
(AMBAC International)

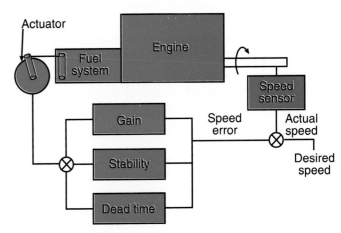

Figure 19-19. Elements of electronic control in diesel fuel governing. (AMBAC International)

hensive electronic engine control systems, many additional operating conditions besides engine speed can be included when calculating the required fuel. Sensors located on the engine continuously update the **engine control module,** or **ECM,** on such factors as air, fuel, and coolant temperature, intake air temperature and pressure, engine oil pressure, throttle position, turbocharger boost pressure, and timing advance status. The ECM looks at all its input data and compares it against the data stored in its memory. The ECM also makes the best decisions in regard to minimizing exhaust smoke and emissions for all loads and speeds.

When the ECM determines the proper fuel metering for the given operating conditions, it sends an output voltage signal to an actuator that increases or decreases the amount of fuel fed to the injectors. In some systems, this actuator consists of a small electric solenoid that moves to open or close a metering valve inside the injection pump or unit injector.

In an electronic governor for inline fuel injection pumps, a linear solenoid actuator with control rod position feedback may be used. The computer reads engine speed and accelerator position along with several correction variables. This input is used to calculate the desired injected fuel quantity or control rod position. The computer then sends the appropriate voltage signal to the control rod actuator solenoid. As the solenoid is energized, it pushes the control rack to the desired metering setting. When the solenoid is de-energized, a return spring moves the rack back to its original position.

Electronic governors and governing systems can respond quickly and accurately to all load conditions placed on the engine. The cruise control and the road speed limiting circuit can also be processed by the computer and used in determining the proper fuel metering setting.

The electronic control system can be programmed to control various engine governing functions such as idle speed control, maximum speed control, speed droop characteristics, vehicle speed cruise control, engine speed cruise control, road speed limit control, and power takeoff speed control. Feedback from these systems is factored into the fuel metering decisions made by the engine control module. Additional information on electrical systems can be found in **Chapter 24. Chapter 25** takes a closer look at electronic engine control systems used on modern diesel engines.

Acceleration Smoke Controls

In addition to particulate control devices required by the Clean Air Act, the amount of exhaust smoke generated by the diesel engine must also comply with federal and local emissions standards. Engine acceleration is a problem area in terms of exhaust smoke generation. This problem is much greater in turbocharged diesel engines. During acceleration, more fuel is injected into the cylinders, but the amount of combustion air supplied by the exhaust gas driven turbocharger lags behind for a few seconds. The result is an excess of unburned fuel, and a large puff of black exhaust smoke upon acceleration. Engine manufacturers use several

methods of controlling fuel injection during acceleration to limit the amount of black smoke produced.

Aneroid/Boost Compensator

An **aneroid/boost compensator** is a pollution control device installed on many governors used on turbocharged engines. It helps to reduce the amount of dense black smoke generated during engine acceleration.

When the engine accelerates, the aneroid/boost compensator limits the fuel supply to the engine until the turbocharger can overcome the temporary lag and begins to supply sufficient boost air to the cylinders. Without aneroid control, any movement of the speed control lever to increase engine speed will immediately move the injection pump rack to the maximum fuel delivery position. However, there is no immediate increase in turbocharger speed to supply more air. This means there is not enough air in the engine cylinders to burn all the injected fuel, and a cloud of dense black smoke is emitted from the engine exhaust.

The aneroid operates under the action of intake manifold pressure on a diaphragm, **Figure 19-20.** An air supply line is installed between the pressure side of the intake manifold (turbocharger outlet) and the top of the aneroid housing. The aneroid fuel control shaft engages the control rack either mechanically or hydraulically.

When the engine stop knob is pulled out, the engine is not running because the fuel pump control rack is moved out to a position where no fuel is injected into the engine. At the

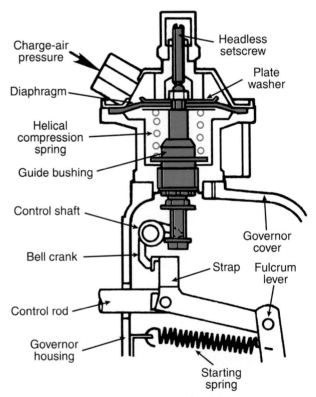

Figure 19-20. Components of an aneroid/boost compensator control mounted on an inline injection pump governor.

same time, the aneroid fuel control link is moved over so that it is no longer contacting the fuel control rack arm. When you advance the engine speed control lever to start the engine, the starter spring moves the fuel control rack to the excess fuel position. As the starter cranks the engine, the injection pump supplies excess fuel to the engine cylinders.

Once the engine starts, centrifugal force from the governor flyweights overcomes starter spring tension—even before idle speed is reached. As the fuel control rack moves from the excess fuel position, the spring on the aneroid fuel control lever shaft moves the control link back to its original position. An arm on the fuel control rack contacts the aneroid fuel control link before the rack has completed 50% of its full travel. This prevents too much fuel from being injected into the engine. The additional pressure will move the lever over to release it.

When the engine is operating, pressurized air from the cold end of the turbocharger is delivered from the air inlet manifold to a chamber in the aneroid/boost compensator. The diaphragm inside the chamber is connected to a push rod that is coupled to a compensator lever. Diaphragm movement is opposed by spring pressure. During acceleration, engine speed gradually increases and the turbocharger spins faster. The air pressure acting on the diaphragm increases to a point where it is greater that the spring pressure holding the diaphragm in position. As a result, the diaphragm and pushrod are pushed down, the compensator lever pivots, and the control rack is forced toward the increased fuel position. The aneroid/boost pressure compensator will react to intake manifold air pressure regardless of governor action.

Hydraulic Activator Operation

The **hydraulic activator** disengages the aneroid from the control rack during engine start-up. During accelera-tion, it engages the aneroid with the control rack. When the engine is shut off, there is no engine oil pressure to the hydraulic activator. Spring pressure moves the piston and control shaft, disengaging the aneroid arms from the control rack, preventing aneroid operation, **Figure 19-21A.**

During initial start-up, the capillary valve temporarily restricts oil pressure movement to the aneroid, delaying aneroid engagement. The length of this delay depends on the ambient temperature and engine oil viscosity. When the oil pressure reaches approximately 9 psi (62 kPa), oil moves through the capillary valve into the aneroid. The piston then moves to the right, forcing the fuel control shaft and aneroid arms to contact the control rack, **Figure 19-21B.** This prevents excess fuel from being injected into the engine. Once again, the arm or lever will hold the rack in position until the turbocharger has increased intake manifold pressure through the diaphragm. This occurs every time the engine is accelerated.

Fuel Ratio Control

The **fuel ratio control** shown in **Figure 19-22** can also be used to limit the amount of fuel entering the cylinders during acceleration. Properly adjusted, it will also minimize the amount of exhaust smoke and soot in the engine. This control restricts the amount of fuel entering the combustion chamber of a turbocharged engine until enough boost has been achieved to ensure clean combustion.

At engine start-up, the governor causes the servo valve and the fuel rack to move into the *fuel on* position. As the servo valve moves, it forces the lever assembly to rotate clockwise against the light tension of the spring. After start-up, the servo valve moves toward the idle rack position, **Figure 19-22A.** As the valve moves, the spring causes the lever assembly to rotate counter-clockwise. The lever assembly rotates until a cross-drilled hole in its sleeve aligns with its piston cavity ports. As engine oil from the governor

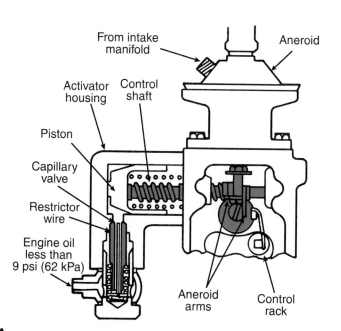

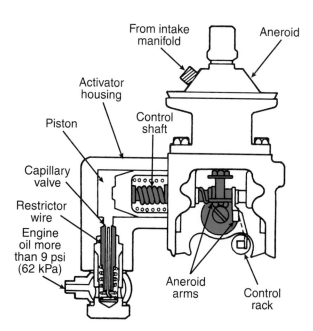

A   **B**

Figure 19-21. *Hydraulic activator aneroid operation. A—Engine off. B—Engine on. (John Deere Co.)*

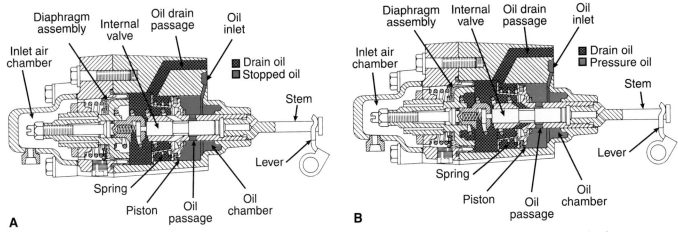

Figure 19-22. Fuel ratio control operation. A—Engine running after start-up. B—Acceleration mode. (Caterpillar, Inc.)

housing feeds into the shaft's center passage, it passes through a cross-drilled hole into a piston cavity in the lever assembly. The pressure of this oil against the piston causes the lever assembly to slide sideways and compress the spring. Lever assembly tangs are overlapped, restricting the assembly's clockwise rotation. The lever assembly, in turn, restricts the *fuel on* travel of the servo valve.

During engine operation, any increase in the intake manifold boost pressure caused by engine acceleration is sent to the chamber through the port. When boost is sufficient, the diaphragm forces the retainer assembly and the rod to the right, **Figure 19-22B.** The lever assembly now rotates clockwise, allowing an increase in fuel as the servo valve moves left.

When the boost pressure drops, the springs return the retainer and the rod to normal position. *Fuel on* movement of the servo valve is restricted until enough boost is achieved. When the engine is shut off, oil pressure is released from the lever assembly's piston cavity. The spring forces the lever assemblies apart so that they are no longer restricted.

Puff Limiter Air Cylinder

The **puff limiter** is an external pneumatic device designed to control the characteristic acceleration smoke puff of turbocharged engines. This system acts directly on the injection pump control rack independent of the fuel injection governor action.

A relay provides an output pressure signal to the air cylinder that decreases in direct proportion to the increase in engine manifold pressure. Therefore, the amount of fuel that can be delivered to the engine is controlled by the puff limiter during low manifold pressure conditions. The relay is mounted directly onto the engine inlet manifold, **Figure 19-23.**

Throttle-Delay Mechanism

On Detroit Diesel mechanical unit injection fuel systems, a **throttle-delay mechanism** is used to reduce exhaust smoke during acceleration, **Figure 19-24.** This devices uses a movable piston and hydraulic oil pressure

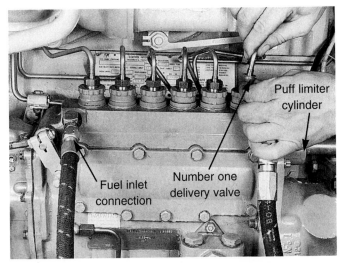

Figure 19-23. Major components of a puff limiter acceleration smoke control cylinder. (Courtesy of Mack Trucks, Inc.)

to slow the movement of the fuel control rod or tube as it moves in the direction of increased fuel delivery.

Pressurized engine oil is fed into a small reservoir located above the throttle delay housing. A check valve in the throttle delay housing allows air to be drawn into a small cavity behind the piston. Whenever the control rack moves toward the no fuel position, oil from the reservoir enters the housing and displaces the air from behind the piston. When the engine is accelerated, the control rack moves forcefully toward the full-fuel position. As it moves in this direction, the piston forces the oil out of the cavity through a small opening. This slows down the rack movement and the rate at which fuel delivery is increased. By preventing excess fuel from entering the cylinders, the throttle-delay mechanism ensures complete fuel combustion and minimizes black smoke during acceleration.

Altitude-Pressure Compensator

At high altitudes, atmospheric pressure decreases and the volumetric efficiency of the engine drops as well. Less air is being drawn into the engine. Unless some method is used

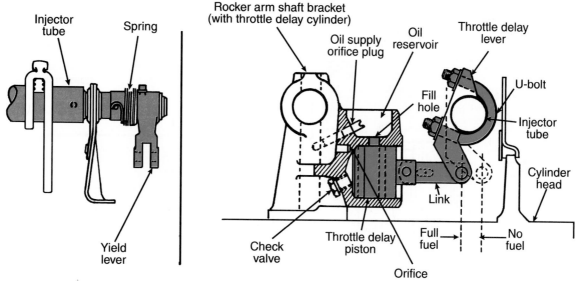

Figure 19-24. *Unit injector throttle delay cylinder and link. (Detroit Diesel Corporation)*

to decrease fuel delivery, problems will occur. This can be a major problem for non-turbocharged engines. Injection pumps used on non-turbocharged engines can be equipped with an *altitude-pressure compensator* as part of the mechanical governor. The altitude-pressure compensator mounts on the governor cover as shown in **Figure 19-25.**

Operation

The altitude-pressure compensator uses an aneroid capsule that can be adjusted to match the highest anticipated operating altitude. The unit's adjustment screw bears against the force of a spring-loaded threaded bolt. The bolt has a fork attached to its lower end. Fork movement acts on a pin, which in turn transfers movement to a swivel mounted cam plate. The cam plate transfers movement to the fuel control rack linkage.

At low altitudes (high air pressures), the force of the coil spring beneath the unit's barometric cell is strong enough to pull up on the cam plate. Maximum fuel delivery can occur when this condition exists.

At higher altitudes (low air pressure), the threaded bolt and fork move downward, causing the pin to follow the curve of the cam plate slot. This pulls the fuel control rod (rack) to a decreased fuel delivery position. The cam plate adjusting screw is used to set full-load delivery. As explained earlier, an aneroid/boost compensator performs the job of the altitude compensator on turbocharged engines.

Summary

A governor is a vital component of a diesel engine's fuel system. It is essentially an engine speed (rpm) sensitive device, designed to maintain a reasonably constant engine speed regardless of load variation.

To be able to sense speed and load changes in the engine, the governor must be connected to the engine in some way. In inline, distributor, and pressure-time injec-

tion systems, the governor is usually an integral part of the fuel injection pump that is being driven at engine speed.

On a unit injector fuel injection system, the governor may be contained in a separate remote housing. It is driven at engine speed through the use of a gear or belt arrangement. Any change in engine speed is immediately sensed by the governor which can then react accordingly to adjust fuel metering.

A governor must be capable of limiting maximum engine speed, regulating engine speed between low and high idle, controlling idle speed to keep the engine from stalling, provide fuel for starting the engine, furnish more fuel when the engine lugs down to maximum torque, and shutting down the engine if it overspeeds.

Constant-speed governors maintain engine speed at a constant rpm, regardless of the load. Variable speed governors give the engine automatic speed control and are easily adjusted during operation over a wide range from low idle to maximum speed. Overspeed governors are designed to prevent the engine from exceeding a specified maximum speed. Limiting speed governors have been developed to prevent the engine from stalling at low idle speeds and to prevent engine racing at full-load speed. A load-limiting governor limits the load that the engine must handle at various speeds.

The six types of governors presently found on diesel engines are as follows: mechanical, pneumatic, servo, hydraulic, electric, and electronic. A simple method of remembering the six types is to memorize the following mnemonic sentence: My Poor Sister Hates Everybody Else.

Mechanical governors use flyweights mounted on a rotating shaft to sense engine speed and generate the force needed to move the control rack. Hydraulic governors regulate the supply through oil pressure. Pressurized oil from the lubricating system is supplied to an auxiliary pump in the governor. The auxiliary pump then develops the oil pressure needed to actuate the governor mechanism.

Servo governors are similar in design to mechanical governors. However, the servo governor has an additional mechanism that assists the mechanical governor in moving

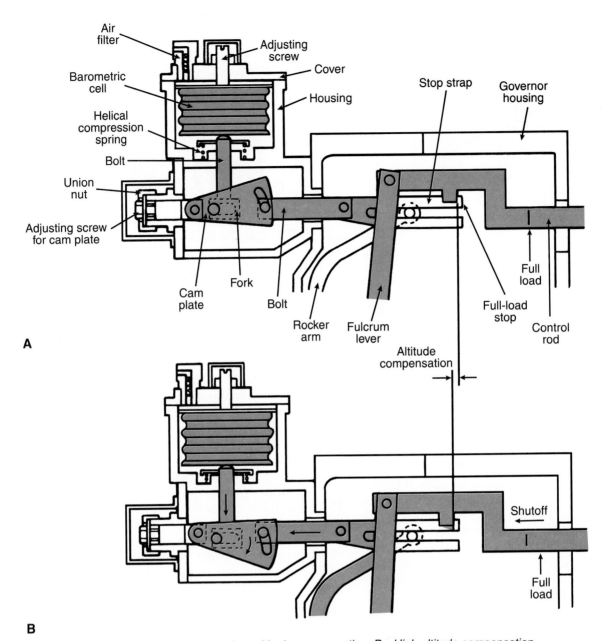

Figure 19-25. *Altitude pressure compensator. A—Low altitude compensation. B—High altitude compensation.*

the fuel rack. A pneumatic governor relies on air entering a venturi to create a partial vacuum. The vacuum acts on a spring controlled diaphragm. Diaphragm movement is transferred to the injection pump fuel control rack.

Electric governors operate using electric relays and solenoids. As a change in engine speed is sensed, a relay closes and voltage is applied to a servo motor or solenoid that changes the fuel control rack position. Response time is very fast, and there is very little droop.

Electronic governors are now used on all diesel engines equipped with electronic engine control systems. Electronic governing systems are computer controlled. This means they can respond quickly and accurately to all load conditions placed on the engine. On electronically controlled engines, input from other engine related systems such as the cruise control and the road speed limiting circuit, can also be processed by the computer and used in determining the proper fuel metering setting.

Most diesel engines used in vehicles and mobile applications are equipped with aneroid/boost compensators, fuel ratio control, throttle delays, or puff limiters to reduce diesel exhaust smoke during engine acceleration. Altitude-pressure compensators are used on non-turbocharged diesel engines to decrease fuel delivery rates at high altitudes. Fuel delivery must be lowered to match the lower levels of combustion air at these altitudes.

Important Terms

Governor

Fuel governing system

Limiting speed governors

Variable speed governors

Constant speed governors

Overspeed governors

Load-limiting governor

High idle speed

Low idle speed
Rated speed
Maximum torque speed
Speed droop
Under run
Over run
Dead band
Sensitivity
Promptness
Hunting
Surging
Stability
Mechanical governors
Hydraulic governors
Servo governors
Dashpot
Pneumatic governors

Venturi
Electric governors
Gain
Proportional control
Stability
Isochronous governing
Dead time compensation
Engine control module (ECM)
Aneroid/boost compensator
Hydraulic activator
Fuel ratio control
Puff limiter
Throttle-delay mechanism
Altitude-pressure compensator

Review Questions—Chapter 19

Do not write in this text. Place your answers on a separate sheet of paper.

1. List the six types of governors presently found on diesel engines.

2. In which type of fuel injection system is the governor not an integral part of the fuel injection pump?
   (A) Inline.
   (B) Distributor.
   (C) Pressure-time.
   (D) Unit.

3. A mechanical governor relies on centrifugal force to move its _____ and _____.

4. Hydraulic governors are more _____ than the mechanical type.

5. Servo governors use a _____ for better speed control during sudden changes in speed or load.

6. Which type of governor gives the engine automatic speed control and is easily adjusted during operation over a wide range of engine speeds?
   (A) Constant speed.
   (B) Overspeed.
   (C) Variable speed.
   (D) Limiting speed.

7. Engines equipped with _____ must use an electronic governor.

8. Altitude-pressure compensators are used on non-turbocharged diesel engines to _____ fuel delivery rates at high altitudes.

9. Aneroid/boost compensators, fuel ratio controls, and puff limiters are all devices used to _____ .
   (A) recirculate exhaust gases back into the air intake
   (B) vent crankcase and fuel tank vapors to the air intake
   (C) reduce diesel exhaust smoke during engine acceleration
   (D) Both A & B.

10. The _____ is an external pneumatic device designed to control the characteristic acceleration smoke puff of turbocharged engines.

ASE-Type Questions

1. Technician A says that the governor must be able to sense and react to changes in engine speed. Technician B says the governor's job is to adjust fuel metering in a diesel engine. Who is right?
   (A) A only.
   (B) B only.
   (C) Both A & B.
   (D) Neither A nor B.

2. Technician A calls the continuous fluctuation of the engine speed away from the desired speed governor hunting or surging. Technician B says this condition is known as under/over running. Who is right?
   (A) A only.
   (B) B only.
   (C) Both A & B.
   (D) Neither A nor B.

3. Which type of governor uses a venturi as part of its operating mechanism?
   (A) Mechanical.
   (B) Pneumatic.
   (C) Servo.
   (D) Hydraulic.

4. Oil pressure for a hydraulic governor is _____.
   (A) by an auxiliary pump in the governor
   (B) by the engine's oil pump
   (C) by the fuel injection pump
   (D) All of the above.

5. With the exception of an additional mechanism that hydraulically assists the movement of the fuel rack, _____ governors are similar to mechanical governors.
   (A) mechanical
   (B) pneumatic
   (C) servo
   (D) hydraulic

6. _____ is the immediate change in engine speed caused by a change in engine load.
   (A) Engine lag
   (B) Speed droop
   (C) Deadpan
   (D) Torque droop

7. All of the following are characteristics of electric governors, EXCEPT:
   (A) they operate using relays and solenoids.
   (B) they use voltages in the 10-12 volt range.
   (C) governor response time is very fast.
   (D) they are used exclusively in over-the-road trucks.

8. Electronic governors operate in the _____ volt range.
   (A) 12-14
   (B) 110
   (C) 2-5
   (D) 22-24

9. _____ involves maintaining the speed of the engine truly constant, regardless of the load.
   (A) Isochronous governing
   (B) Deadpan governing
   (C) Under run governing
   (D) Variable speed governing

10. All of the following are used to control exhaust particulates on acceleration, EXCEPT:
   (A) aneroid/boost compensator.
   (B) servo governor dashpot.
   (C) puff limiter.
   (D) throttle delay mechanism.

Fuel injection pumps come in a variety of sizes for different engines and applications. (AMBAC International)

Chapter 20

Multiple Plunger Inline Injection Pumps

After studying this chapter, you will be able to:

❏ Name the major components in a multiple plunger inline pump fuel injection system.

❏ Name the components of a typical inline fuel injection pump and describe how they operate.

❏ Explain the function and operation of a typical inline injection pump automatic timing device.

❏ Properly remove and install an inline fuel injection pump.

❏ Describe the three methods used to time an inline injection pump.

❏ Properly bleed all air from the fuel lines.

❏ Perform diagnostic tests to check for proper pump operating pressures, relief valve operation, and fuel line restrictions.

❏ Understand the basic pump disassembly, inspection, and calibration work that is performed in injection pump repair shops.

Before the advent of small injection pumps, diesel engines were used in small numbers. The first bulky air compressor driven systems limited diesel engine use to large stationary and marine applications. Inline injection pumps allowed diesel engines to be used in many more applications, not limited by size constraints. This chapter discusses the various inline multiple plunger inline injection pumps.

Inline Injection Pumps

As explained in **Chapter 1,** one of the most persistent problems faced by Rudolf Diesel in the development of his new engine was how to inject fuel into the combustion chamber. In 1927, Robert Bosch, a German engineer and inventor, began mass production of a compact, engine-driven *multiple plunger inline fuel injection pump*. Today,

the Robert Bosch Corporation and its licensees manufacturer over half of all diesel injection equipment.

Several former Bosch affiliates now operate independently. The American Bosch Company is now part of AMBAC International and the British manufacturer, Lucas CAV, is also a former Bosch affiliate. Both Lucas CAV and AMBAC inline pumps are similar to Bosch pumps in overall design, operation, and service, but design differences have evolved over the years. Caterpillar Inc. manufacturers its own line of multiple plunger inline injection pumps for select models in its engine line.

Inline Injection Pump Components

Figure 20-1 illustrates the fuel system layout used with multiple plunger inline pumps. Fuel is drawn from the fuel tank by a transfer or supply pump. The primary fuel filter or fuel filter/water separator is located between the fuel tank and the transfer pump. The transfer pump supplies fuel to the inline pump at pressures between 5-20 psi (35-140 kPa). The fuel passes through the secondary fuel filter on its way to the injection pump. To maintain a set operating pressure inside the injection pump housing, a spring-loaded relief valve may be used. When excess pressure opens the relief valve, warm fuel used for cooling and lubricating the pump is routed back to the fuel tank. Some engines may be equipped with a spring-loaded overflow valve and return line at the secondary fuel filter (not shown). Each injection nozzle is also equipped with its own fuel return line.

Inline Pump Components

Figure 20-2 illustrates the location of the major external components of a Bosch inline injection pump. These include:

❏ An injection pump to produce high injection pressure.

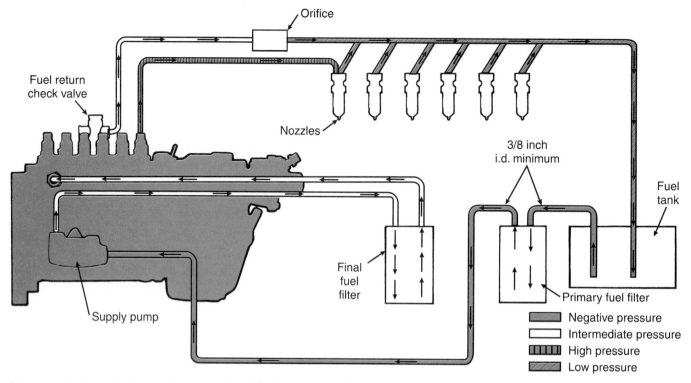

Figure 20-1. *Multiple plunger inline pump fuel injection system. (Navistar International Transportation Corp.)*

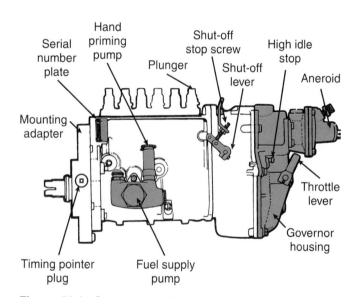

Figure 20-2. *Components of a typical inline injection pump assembly. (Navistar International Transportation Corp.)*

❑ A governor to control the engine speed.

❑ A mechanical transfer/supply pump for drawing in and delivering fuel from the fuel tank to the injection pump suction gallery.

Inline pumps have one plunger and barrel assembly for each engine cylinder. Each plunger and barrel assembly consists of a ***pump plunger*** and ***pump barrel***. The barrel is sometimes called a bushing.

Figure 20-3 illustrates the internal components of an inline injection pump. Its construction and operating principles are typical of many inline injection pumps. The plungers are driven by the pump's camshaft, which is

contained inside a cast aluminum housing. The injection pump camshaft is driven by the diesel engine. In a four-stroke engine, the speed of the injection pump camshaft equals engine camshaft speed (one-half crankshaft speed).

The injection pump camshaft and the engine driving device are connected using a torsion-proof ***coupling element.*** This synchronizes injection pump position with the proper crankshaft position (provided the injection pump is properly installed and timed). The type of driving device used depends on the engine design. The injection pump camshaft is connected to the engine driving device either directly or via a coupling unit or timing device. Torque and rotational speed can be transmitted by the use of toothed gears, plug-in couplings, toothed belts, or timing chains, or gear and shaft arrangements.

For example, on many diesel engines used in mobile applications, the injection pump driven gear directly meshes with the camshaft drive gear. In some cases, there may be an idler gear located between the camshaft drive gear and the injection pump driven gear.

Some engine manufacturers use an auxiliary drive shaft and gear arrangement to provide the drive and support for the fuel injection pump, air compressor, fan drive, tachometer drive, and other engine accessories. A gear on the auxiliary drive shaft meshes with the camshaft gear. This allows the auxiliary shaft to turn at camshaft speed. The injection pump driven gear meshes with another gear mounted at the rear of the auxiliary drive shaft. The auxiliary drive shaft and gears are contained in their own housing. The injection pump mounts to the rear of this housing.

Above each cam on the injection pump camshaft is a roller tappet with a spring seat. ***Roller tappets*** ride directly on the pump camshaft and transmit its motion to the

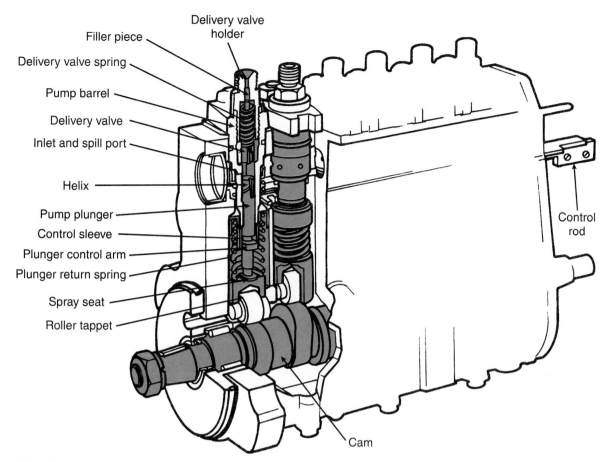

Figure 20-3. Inline injection pump internal components.

plungers. The plunger springs keep the roller tappets pressing on the camshaft. The plunger is guided by the pump barrel. As mentioned in **Chapter 17,** the pump barrel has one or two inlet ports that are connected to the injection pump suction gallery. A delivery valve and holder is positioned above each plunger and barrel assembly. The delivery valves seal the delivery line from the barrel during the intake stroke and relieves the pressure in the injection line to prevent nozzle dribble.

Inline Pump Operation

The **control rack** or rod transmits the governor action to the pumping plunger. Each pump plunger is connected to a control sleeve. The upper end of the control sleeve has a gear segment installed on it. The control rod teeth and gear segment engage one another, **Figure 20-4.** It is possible to adjust the fuel delivery for each plunger by turning the control rod clamping sleeve. Each plunger and barrel assembly must be adjusted for equal delivery as the pump plunger is turned by the control rod, regulating the fuel delivery from shutoff to maximum.

A stop mechanism installed in the pump housing limits the distance the control rack or rod can be moved by the throttle linkage. This limits the maximum full-load fuel delivery to safe levels. Two types of stops are commonly used. The fixed stop design is adjusted using a locknut and screw, **Figure 20-5A.** The adjustment can be made to set

either fuel start-up or full-load delivery, depending on the engine application and the governor. The spring-loaded rack stop shown in **Figure 20-5B** is adjusted by turning the threaded bushing in or out as required.

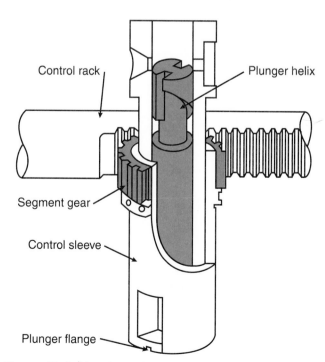

Figure 20-4. Metering principle of control rack and control sleeve. (AMBAC International)

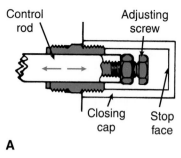

A

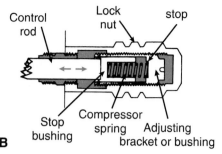

B

Figure 20-5. *A—Fixed control rod stop mechanism. B—Adjustable stop mechanism.*

Each plunger is moved up and down inside the barrel housing by the action of the injection pump camshaft. The camshaft's rotational movement is converted into reciprocating movement by the roller tappet, **Figure 20-6.** The upward movement to top dead center is ensured by the cam's shape. The return spring moves the plunger back to bottom dead center. The return spring pressure also guarantees that the roller tappet does not lift off the cam as the engine accelerates. The cams are arranged so that the injection sequence of the pump matches the engine's firing order.

Most inline injection pumps operate on the port and helix metering principles described in **Chapter 17.** For injection to begin, the pump camshaft must lift the pumping plunger until it closes off the inlet fuel ports. This is known as *lift-to-port closure* and refers to how far the plunger must move to effectively close off both fuel inlet ports in the barrel. **Figure 20-7** illustrates the stages or phases of plunger action during one complete injection cycle.

Pump and Governor Identification

Important information about inline pumps is listed on the pump nameplate. The same is true for the pump governor.

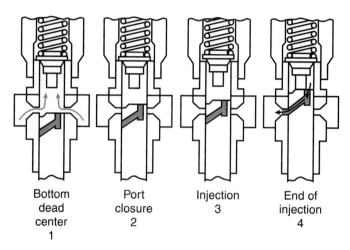

| Bottom dead center 1 | Port closure 2 | Injection 3 | End of injection 4 |

Figure 20-7. *Four principle plunger positions of the injection cycle. (Navistar International Transportation Corp.)*

For a Bosch pump, a typical **code number** can be interpreted as shown in **Figure 20-8.**

The pump can be base mounted onto a support on the engine or bolted to a cradle on the engine. An example of a

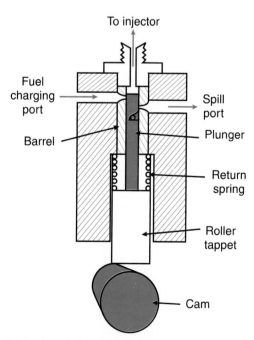

Figure 20-6. *Simplified view of cam and tappet action in an inline injection pump.*

PES 6MW 100/320 RS 1108	
PE	Inline injection pump with enclosed camshaft
S*	Flange mounted
6	Number of pumping plungers (cylinders in the engine)
MW	Pump size
100	Pump plunger diameter (10.0 mm)
/	Pump model
320	Number code for location of feed pump and governor
R	Direction of pump rotation (L = left; R = right)
S1108	Version code number of pump identification and application information

*An S after the PE (PES) indicates that the pump is flange mounted

Figure 20-8. *Injection pump identification numbers can tell a great deal about a particular pump's features, how it is mounted, the engine it is installed on, etc.*

governor code for this inline pump is shown in **Figure 20-9.** Other manufacturers, such as AMBAC and Lucas CAV use similar coding systems for their pumps and governors.

Inline Injection Pump Sizes and Features

Inline injection pump sizes are classified by plunger stroke length and sometimes plunger diameter. A small inline pump, such as the Bosch model M, has a 7 mm plunger stroke, and is used on passenger cars and light duty engines.

Inline pumps with a 8 mm plunger stroke, such as the Bosch A model pump, are found on a number of medium and heavy-duty truck engines. The Bosch MW pump operates on the same basic principle as the A pump, but also uses an integrated flange element at the top of each pumping plunger. Plunger stroke is either 8 mm or 10 mm and is found on many medium- and heavy-duty truck engines as well as some marine applications.

The Bosch P pump can have a plunger stroke of 10-12 mm. It is among the largest inline pumps available and is used on heavy duty truck and other industrial engines. A distinct feature of the P pump is a protective metal cover mounted on top of the pump housing to keep dirt and oil away from the barrel flanges.

Prestroke and fuel delivery rate adjustments on inline pumps must only be performed by authorized service technicians. These adjustments require the use of special tooling and a fuel pump test stand.

Automatic Timing Advance Device

Diesel fuel takes a certain amount of time to ignite and burn. As engine speed increases, the burn time still remains the same. Much of the combustion process would occur after top dead center if some type of timing device were not used to advance the start of fuel injection.

The **automatic timing advance unit** used on most inline injection pumps is classified as an eccentric, fly-weight operated device. It is usually mounted on the pump camshaft at the front of the injection pump, **Figure 20-10.**

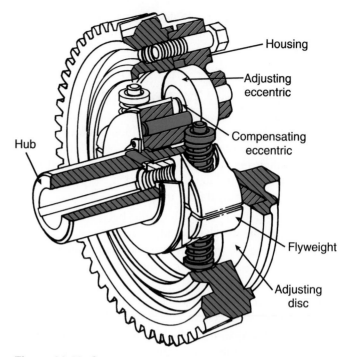

Figure 20-10. *Components of an eccentric timing device.*

In some cases, the timing device may be mounted on an intermediate or auxiliary shaft. The eccentric timing device is available in an open and closed type. Open timing devices are lubricated from the engine oiling system and contain the needed bores and connections for lubricating. Closed timing devices are filled with a lubricant at the time of manufacture and do not require servicing during their service life.

Construction

The open eccentric timing device housing is attached to a toothed gear. An adjusting disc, hub, adjusting eccentric, and compensating eccentric are mounted in the housing and rotate inside it. The adjusting and compensating eccentrics are guided by a pin rigidly connected to the housing. The flyweight pins engage the bore of the adjusting eccentrics. Compression springs fit inside the flyweights.

Operation

The open timing device is driven by a toothed gear housed in the engine's timing gear housing or case. The connection between the drive input and the drive output (hub) is made by interlocking pairs of eccentrics. The eccentrics are located in the adjusting disc bores and are guided by the housing pins. The pins transmit the drive movement from the housing to the hub.

As shown in **Figure 20-11A,** when the engine is stopped, the compression springs hold the flyweights in the starting position. When the engine is running, the centrifugal force increases, causing the flyweights to move outward just as they do in a mechanical governor, **Figure 20-11B.** This outward movement causes the eccentrics to turn. As a result of this movement, the hub is moved against the housing, thus changing the injection timing,

RQV 350 1200 MV 43-5	
R	Flyweight (mechanical) type governor
Q	Fulcrum lever model
V	Variable speed governing (all-speed)
350	Low idle speed
1200	Full load rated speed
MV	Fits on MV size pump
43-5	Application number

Figure 20-9. *Governor identification number. Information on the governor is listed in this number.*

Figure 20-11C. Advance angles of up to 30º with respect to the engine crankshaft are possible using this eccentric timing device.

Inline Pump Testing and Service

Most diesel engine shops do not troubleshoot and repair fuel injection equipment or work on multiple plunger inline injection pumps. Therefore work on injection pumps is limited to the following:

❑ Pump removal from the engine.

❑ Pump installation after it has been serviced at a qualified service center.

❑ Timing the injection pump to the engine so that fuel injection occurs at the proper time in the combustion cycle.

❑ Connecting the injection pump to the engine's lubricating system.

❑ Bleeding air from the injection pump and fuel lines after the service pump has been installed.

❑ System testing for proper relief valve and pump operating pressures and fuel system restrictions.

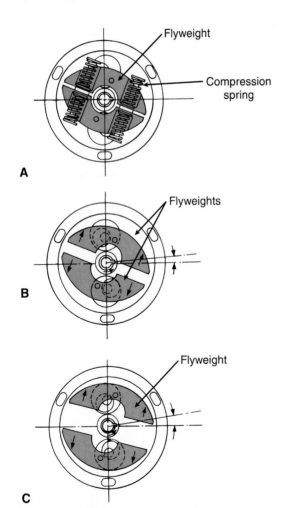

Figure 20-11. Operation of an eccentric timing device. A—At rest position. B—Medium speed position. C—High speed or maximum angle position.

Inline Pump Removal

Removal procedures for inline injection pumps are fairly straight-forward, but slight differences may exist between specific engine manufacturers. The major concern when removing the injection pump is maintaining the proper timing between the injection pump and the engine.

The start-of-delivery markings on the engine and the injection pump are used for timing the injection pump to the engine. Usually, the compression stroke of the number 1 cylinder is used as a basis for timing, but it is possible that other reference points are used. For this reason, check the engine service manual procedures and timing specifications.

The mark for the start-of-delivery is usually located on the flywheel, on the V-belt pulley, or the vibration damper. On the injection pump, the start of delivery for barrel 1 is normally when the mark on the fixed coupling half, or on the timing device is in alignment with the mark on the pump housing, **Figure 20-12.** In the case of flange pumps, the markings are on the drive gear and on the pinion.

If the injection pump is being removed for service, manually rotate the crankshaft to place the number 1 piston on its compression stroke (both inlet and exhaust valves closed), and align the injection pump to engine flywheel or pulley timing marks. The timing mark must be aligned on the compression stroke. To confirm this engine position, remove the rocker cover and check for free play at the number 1 cylinder valve operating mechanism. The presence of lash in both intake and exhaust valves confirms the engine is on its compression stroke. No lash in either valve indicates the engine is on its exhaust stroke and must be rotated 360° to place the piston on its compression stroke.

Some inline injection pumps are equipped with locking or timing pins that are used to lock the fuel injection pump

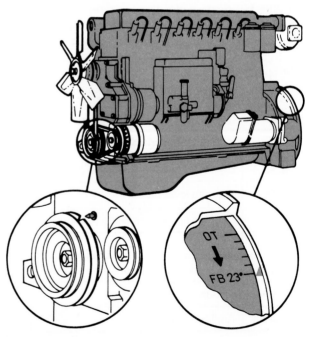

Figure 20-12. Example of reference marks on diesel engine V-belt pulley or flywheel.

prior to removal from the engine. The pin is accessed through a plug in the injection pump housing, **Figure 20-13**. In some injection pumps, the timing pin itself is reversed so that a slot in the pin end fits over the timing tooth, **Figure 20-14**. In other pump designs, the pin is locked into the pump camshaft by reversing the position of a spring inside the pin bore, **Figure 20-15**. The force of the spring pushes the tip of the pin into the locking slot on the camshaft.

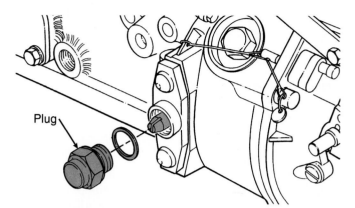

Figure 20-13. Removing the access plug from the injection pump housing to access the camshaft locking pin. (Cummins Engine Co., Inc.)

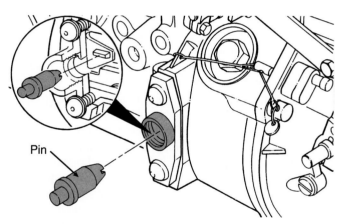

Figure 20-14. A groove in the reversible pin locks into the pump's timing tooth. (Cummins Engine Co., Inc.)

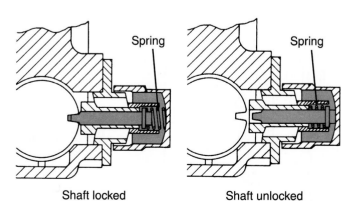

Shaft locked   Shaft unlocked

Figure 20-15. Spring activated locking pin. Repositioning spring forces the pin tip into a locking groove in the camshaft. (Cummins Engine Co., Inc.)

Removal Procedure

The following is a typical inline injection pump removal procedure:

1. Clean all external surfaces of the injection pump, including all line connections and fittings that are to be disconnected. Clean the area around the injection pump gear cover to prevent dirt from entering the crankcase, **Figure 20-16**.
2. Disconnect injection lines from delivery valves and cap all lines so dirt cannot enter the system, **Figure 20-17**.
3. Disconnect the fuel inlet and return lines and cap, **Figure 20-18**.
4. Disconnect the throttle/control linkage return spring and the stop lever linkage, **Figure 20-19**.
5. If equipped, remove the fuel shut-off solenoid by disconnecting the harness at the solenoid connector, removing the bolts and/or brackets that hold the solenoid in place, **Figure 20-20**.
6. Remove the fuel filter and the fuel pump support bracket.

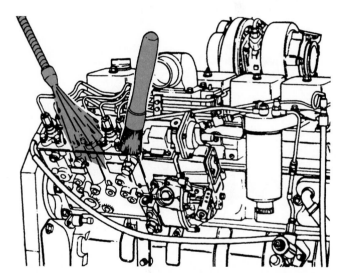

Figure 20-16. Clean the pump area with compressed air and a clean brush. (Cummins Engine Co., Inc.)

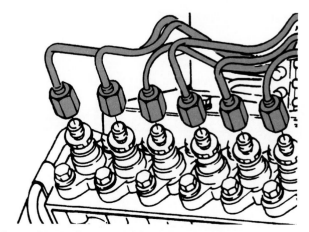

Figure 20-17. Remove injection lines from the delivery valves and cap all lines. (Cummins Engine Co., Inc.)

7. Disconnect the oil supply line from the fuel pump and engine block, **Figure 20-21.**

8. Make certain the number 1 piston is at the top dead center position.

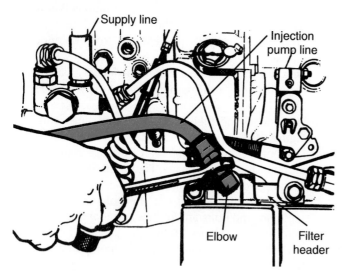

Figure 20-18. Disconnect fuel lines leading to and from the injection pump. (Cummins Engine Co., Inc.)

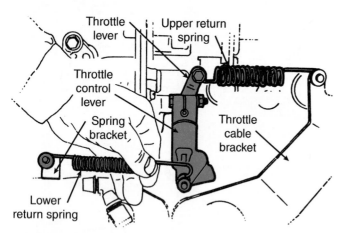

Figure 20-19. Removing the upper and lower throttle control springs. (Navistar International Transportation Corp.)

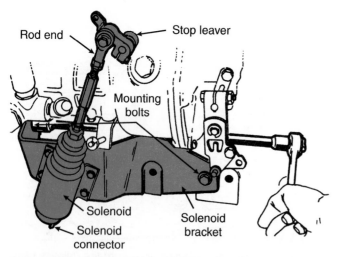

Figure 20-20. Removing the electric shut-off solenoid. (Navistar International Transportation Corp.)

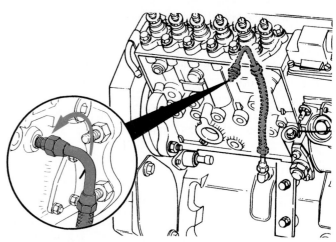

Figure 20-21. Removing the lubricating oil supply line running from the engine block to the injection pump. (Cummins Engine Co., Inc.)

9. If equipped, lock the fuel injection pump in position using the timing or locking pin as described above.

10. On engines where the injection pump is directly driven by the camshaft drive gear or idler gear, the injection pump driven gear must be removed from the camshaft prior to removing the pump from the engine. To do this, remove the gear cover access cap to access the injection pump driven gear. Remove the nut and washer from the pump shaft, **Figure 20-22.**

11. Using pulling equipment, pull the injection pump driven gear from the drive shaft, **Figure 20-23.**

12. Remove the mounting nuts holding the injection pump to the mounting flange and pull the pump straight back away from the mounting flange, **Figure 20-24.**

13. On engines equipped with an auxiliary drive shaft, steps 10-12 are not required. The injection pump mounts to the rear of the auxiliary housing and is simply dismounted by removing the mounting bolts and pulling the pump straight back so the injection pump driven gear disengages from the auxiliary drive shaft drive gear. If required, the injection pump driven gear and hub can be removed and serviced.

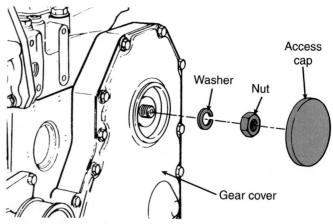

Figure 20-22. Removing the gear cover access cap and the injection pump gear nut and washer. (Cummins Engine Co., Inc.)

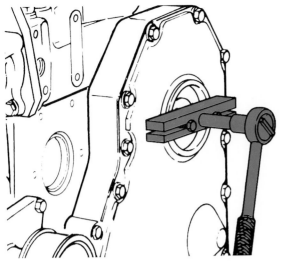

Figure 20-23. *Pulling the pump gear from the drive shaft.* *(Cummins Engine Co., Inc.)*

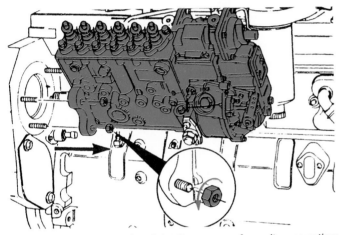

Figure 20-24. *Removing injection pump from its mounting flange. (Cummins Engine Co., Inc.)*

Note: When possible, keep the engine crankshaft in this same position until it is time to reinstall the serviced injection pump. Maintaining this proper crankshaft position will make installation of the injection pump much easier.

If the engine is being overhauled and crankshaft position cannot be maintained, the engine must be reassembled and manually rotated into position prior to installing the fuel injection pump.

Injection Pump Installation

The following is a typical inline injection pump installation sequence:

1. Install a new injection pump gasket or O-ring on the injection pump mounting flange, **Figure 20-25.**
2. If used, install the key in the injection pump shaft keyway, **Figure 20-26.**

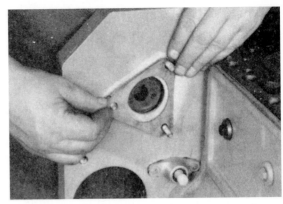

Figure 20-25. *Installing a new injection pump mounting gasket. (Cummins Engine Co., Inc.)*

Figure 20-26. *Install the key in the pump camshaft keyway. (Cummins Engine Co., Inc.)*

3. Recheck that the number 1 cylinder is at the TDC position on its compression stroke.
4. Align the timing marks on the injection pump driven gear with those on the camshaft or idler driving gear and install the gear into the timing cover housing, **Figure 20-27.**

Figure 20-27. *Installing pump driven gear in the timing housing. (Cummins Engine Co., Inc.)*

5. Install the injection pump to the mounting flange by slipping the pump camshaft through the pump driven gear opening. Align the key and keyway as required, **Figure 20-28.**

6. Attach the pump by finger tightening the mounting nuts. The pump should be free to move, **Figure 20-29.**

7. Install the driven gear mounting nut, **Figure 20-30,** but do not torque to specifications at this time. Also be careful not to move the engine crankshaft at this time.

8. Torque the pump mounting bolts to specifications.

9. If equipped, unlock the injection pump camshaft by reversing the locking pin position or otherwise disengaging the timing pin.

10. Tighten the driven gear mounting bolt to specifications.

The following steps describe the installation procedures for mounting a Bosch injection pump to a V-8 engine equipped with an auxiliary drive shaft and housing, **Figure 20-31.**

Figure 20-30. *Installing the driven gear mounting nut. (Cummins Engine Co., Inc.)*

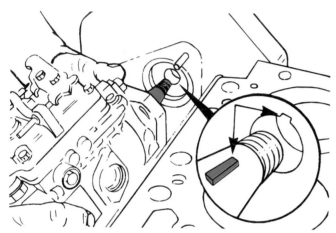

Figure 20-28. *Installing the pump camshaft to the driven gear through the back of the mounting flange. (Cummins Engine Co., Inc.)*

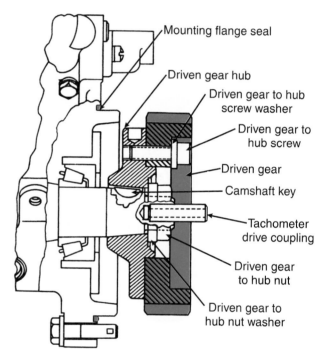

Figure 20-31. *Cutaway of an injection driven gear and hub. (Mack Trucks, Inc.)*

1. Degrease the tapered surfaces of the pump camshaft and driven gear hub with an approved solvent. Install the camshaft gear key and align the keyway in the hub with the key slot in the camshaft key. Install the hub on the camshaft and then install the driven gear hub lock washer and nut. Torque the nut to specifications.

2. Position the injection pump driven gear so that the tapped holes in the hub are in the center of the adjusting slots. Install the gear-to-hub washers and bolts, but do not tighten the bolts completely. They should be loose enough so that the gear can rotate freely.

3. Lubricate and install a new mounting flange O-ring onto the injection pump housing.

4. Position the injection pump hub so that the radially drilled hole is at the top, indicating that the pump is in its port closure position, **Figure 20-32.**

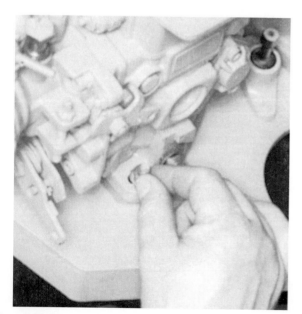

Figure 20-29. *Install the pump mounting nuts. Tighten them finger tight, do not torque. (Cummins Engine Co., Inc.)*

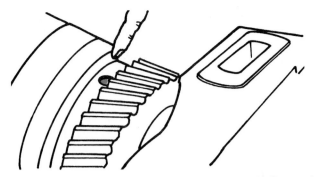

Figure 20-32. *The radial hole at the top of the hub indicates the pump is in its port closure position. (Mack Trucks, Inc.)*

5. Before assembling the injection pump to the engine, make certain that the number 1 piston is on the compression stroke and the damper mark is at the recommended timing degree. If needed, manually rotate the engine into this position.

6. Viewing the injection pump from the front, position the driven gear to its extreme counter-clockwise (CCW) position. Check to be certain that the pump driven gear (as viewed from the front of the engine) is in its extreme counterclockwise (CCW) position before the injection pump driven gear engages the auxiliary shaft gear. Install the injection pump on the engine.

7. If the injection pump driven gear does not mesh with the auxiliary gear, rotate it clockwise (viewing from the front of the engine) until the gears mesh (approximately one tooth) and then move the pump into place.

8. Assemble the pump-to-auxiliary drive housing using bolts and plain washers. Remove the timing plug from the front of the auxiliary drive housing, if installed. Tighten the top hub-to-gear bolts and check driven gear to auxiliary shaft gear backlash. Adjust the backlash to specifications, and then loosen the screws so that the gear can rotate freely.

Static Spill Timing Check

Inline pump installation also involves checking and/or setting the pump-to-engine timing. This is known as a *static spill timing check.* The term *static* is used because the engine is not running during this test. Static spill timing sets the injection pump in time with the engine. This ensures that actual start of injection for each cylinder begins the proper number of degrees BTDC in that cylinder's combustion cycle. The static spill timing specification is stamped on the engine exhaust emission label located on the engine or rocker cover. Proper injection pump to engine timing is the surest method of ensuring optimum fuel economy, engine durability, and proper engine exhaust levels.

The static spill timing check accurately determines exactly when port closure occurs within the injection pump plunger and barrel for cylinder number 1. Three methods can be used to set spill timing, depending on the type of test equipment available:

❑ Fixed timing sensor/light.
❑ Portable high pressure port closure test stand.

❑ Low air pressure port closure test stand.

Always refer to the specific engine manual for all spill timing procedures. The following sections give typical examples of each method, but the procedures will vary between engines.

Timing with Fixed Timing Sensor/Light

Using a fixed timing sensor/light tester is the simplest, most accurate method of setting injection pump to engine timing. The following procedure is recommended for Bosch P7100 series close-coupled injection pumps. This procedure requires the use of a fixed timing sensor/light.

1. Remove the fixed plate plug located at the rear of the pump housing. Retain both the plug and washer for reinstallation.

2. Align and install the fixed *timing probe*, **Figure 20-33**, in the fixed plate opening. Probe alignment should be at the six o'clock position. Install the probe by slowly turning the knurled surface clockwise.

3. Connect the fixed timing sensor tester to ground and press the tester power switch on, **Figure 20-34.**

Figure 20-33. *Fixed timing probe. Install the probe in the injection pump plug opening. (Mack Trucks, Inc.)*

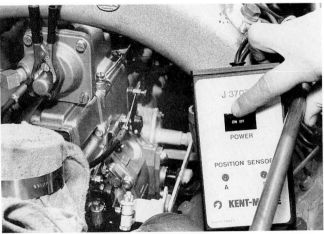

Figure 20-34. *Turning on the fixed timing sensor power switch. Note the properly grounded clip. (Mack Trucks, Inc.)*

4. Insert a 11/23″ (12.1 mm) diameter bar into the radially drilled hub hole and rotate the hub clockwise (viewing engine from front) until the bar touches the side of the auxiliary housing, **Figure 20-35.**

5. Now rotate the hub counter-clockwise. Light A will light first indicating that the injection pump is very near port closure, **Figure 20-36.**

6. Continue rotating the hub until both A and B lamps light at the same time. This indicates that the pump is now positioned at port closure on the number 1 cylinder. Work carefully as there is only a small band in the engine rotation when both lamps will light, **Figure 20-37.**

7. Torque the top hub-to-gear bolt to specifications without disturbing the pump gear-to-hub relationship.

8. Rotate the engine over so the two remaining bolts are accessible through the plug hole and torque to specifications without disturbing the pump gear-to-hub relationship.

9. Recheck the pump to engine timing by rotating the engine clockwise until both lamps are lit. If the timing

Figure 20-37. *Both lights lit indicates the pump is at its port closure position. (Mack Trucks, Inc.)*

indicator does not agree with the desired timing setting on the engine information plate, the injection pump must be retimed.

10. Once timing is correct, turn tester power off, disconnect the tester ground, remove the fixed timing probe, and reinstall the fixed plate plug and washer.

11. Install the timing plug in the auxiliary drive housing, then install the timing hole cover plate and seal.

12. Connect the throttle control linkage, the stop lever linkage, and the boost pressure line. Finally, connect the oil feed line to the injection pump, the fuel inlet and return lines, and all injection lines to their delivery valves.

High Pressure Port Closure Test

The *high pressure port closure test* requires the use of a test stand. This is performed as follows:

1. Cap or connect all injection lines except number 1 cylinder delivery valve outlet. Also cap the valve return and bleed fitting from the nozzle drip line if connected to the overflow valve.

2. Connect the number 6 high pressure line from the portable port closure stand to the fuel pump gallery inlet. Connect the number 4 line from the port closure stand to the number 1 cylinder delivery valve holder on the fuel pump, **Figure 20-38.** Make sure the injection pump stop lever is in the running position, and the throttle lever is secured in the full load position.

3. Rotate the engine in the direction specified in the engine service manual to place the injection pump in a position 10-20° of rotation before port closure is specified to occur.

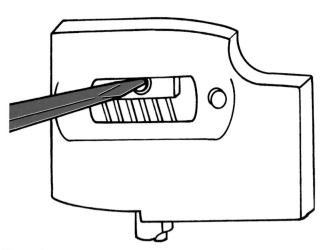

Figure 20-35. *Rotating driven gear hub with bar. (Mack Trucks, Inc.)*

Figure 20-36. *One light lit indicates the injection pump is near its port closure position. (Mack Trucks, Inc.)*

Note: In engines equipped with an auxiliary housing, the pump is placed into position by inserting a 11/32″ (12.2 mm) diameter bar into the radially drilled hub hole, and rotating the hub clockwise until the bar touches the side of the auxiliary housing.

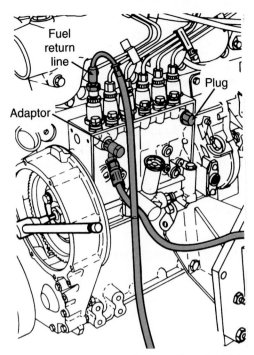

Figure 20-38. Typical test stand connections for high pressure timing of an inline injection pump. (Mack Trucks, Inc.)

4. Activate the portable port closing stand to introduce fuel pressure to the injection pump gallery. A steady stream of fuel should flow from the injection pump delivery valve, **Figure 20-39.**

5. Rotate the engine or injection pump driven gear hub in the direction specified in the engine manual until port closure occurs. This happens when fuel flowing from the test stand changes from a solid stream into a series of drops, **Figure 20-40.**

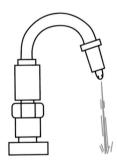

Figure 20-39. Steady fuel flow. Port closure is reached when steady fuel flow falls to a series of droplets. (Mack Trucks, Inc.)

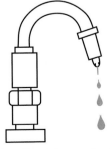

Figure 20-40. Drops from the delivery valve indicating port closure. (Mack Trucks, Inc.)

6. If port closure timing is correct, lock the engine in position with a crankshaft locking tool and torque the hub-to-injection pump gear retaining bolts to specifications. Recheck the pump to engine timing by rotating the engine crankshaft 720°. As the engine approaches the point where the vibration damper timing mark should align with the stationary pointer, slow the rotation speed and note if the fuel flow from the test stand changes from a steady stream to a series of drops. This is the actual engine timing.

If the timing indicator does not agree with the desired timing, the injection pump must be retimed. Proceed as follows:

1. Deactivate the high pressure timing unit and bleed the residual pressure.
2. Manually rotate the engine as required to loosen the two lower pump driven gear-to-hub bolts through the plug hole.
3. With the number 1 cylinder on the compression stroke, manually rotate the engine over until the damper timing mark is at the recommended BTDC position.
4. Loosen the top pump gear hub bolt and insert the 11/32″ (12.2 mm) bar into the hole in the pump gear hub. Rotate the hub clockwise until the bar touches the side of the auxiliary housing.
5. Activate the high pressure port closure timing unit, and slowly rotate the pump hub counter-clockwise until the port closure point is reached (solid fuel flow stops and series of drips begins).
6. Deactivate the high pressure pump and without disturbing the pump gear-to-hub relationship, torque the top hub bolt. Turn the engine manually and torque the two lower bolts.
7. Recheck pump to engine timing, and readjust as needed.
8. Connect the oil feed line to the injection pump.

Assembled Engine Spill Timing Check

If it is necessary to verify the *engine spill timing* of an injection pump that is already installed on an engine, the procedure is slightly different:

1. Connect the high pressure timing unit.
2. Manually turn the engine over in the direction of normal rotation until the number 1 piston is on its compression stroke and the damper mark is 3-4° higher than the recommended timing point listed on the engine information plate or the injection pump nameplate.
3. Activate the high pressure timing unit and manually turn the engine until the correct timing is indicated on the damper. This is the point where the solid fuel stream from the test nozzle slows to a series of drips.
4. If timing is not correct, adjust it as outlined earlier. In order to access the gear-to-hub bolts on a fully assembled engine, break the wire seal and remove both the timing hole cover plate and the plug on the front of the auxiliary housing.

Warning: Breaking the wire seal may void the pump's warranty and in some cases may be illegal.

Low Air Pressure Method

The **low air pressure method** of setting pump to engine timing should only be used when the fixed timing sensor/light or high pressure methods are not available. The major drawback of this method is the need to remove the pumping plunger spring-loaded delivery valve so the low air pressure can move the plunger.

1. Remove the number 1 delivery valve holder from the injection pump. Remove the delivery valve, spring, and spring shim.
2. Install a suitable air line onto the pump gallery *in* fitting. The air supply used must be equipped with a separator and pressure regulator. Moisture-laden air can cause serious damage to the pump components.
3. Attach a high pressure fuel line with an end fitting to the delivery valve holder. Route the end of the fuel line into a container filled with water.
4. Secure the stop lever in the running position and activate the throttle lever several times and secure it in the full-load position.
5. On engines that feature a retard starting device, remove the injection pump control rack cap plug and install the correct timing plug gauge. Engines equipped with a puff limiter and/or torque limiter cylinder do not require timing gauges. However, 80-120 psi (550-8825 kPa) of air pressure must be applied and held to the system air cylinder.
6. Adjust the air line regulator so that a steady stream of bubbles is present in the water-filled container.
7. Manually rotate the crankshaft very slowly in its normal direction of rotation. Watch the air bubbles in the jar. Immediately stop crankshaft rotation when the stream of bubbles stops.
8. Check the flywheel or vibration damper timing indicator position. If properly timed, the indicator will register the correct number of degrees BTDC listed in the engine specifications. (Check manual or engine plate). Recheck to ensure accuracy.
9. If the timing is incorrect, rotate the engine in its normal direction of rotation until cylinder number 1 is on its compression stroke and the timing mark indicates the correct number of degrees BTDC as recommended in the engine timing specification.
10. Loosen the engine to injection pump drive coupling bolts and move the injection pump drive counterclockwise or clockwise, when viewed from the front, depending on the engine and/or manufacturer until the pump drive is at the end of its adjusting slots. Tighten the bolts.
11. Now slowly turn the injection pump drive clockwise or counter-clockwise. Watch the air bubble stream in

the container and immediately stop rotation when the stream stops. This indicates that the pumping plunger of cylinder number 1 is in its port closing position.
12. Check port closing by turning the engine crankshaft counter-clockwise a minimum of one-half turn, followed by clockwise rotation to the desired timing mark on the vibration damper or flywheel.
13. Tighten the two opposite bolts to secure the hub and then rotate the crankshaft to recheck timing.
14. If timing is correct, torque all bolts to specifications, remove the timing plug gauge, replace the control rack plug, and reassemble the delivery valve, spring, and delivery valve holder.

Bleeding the Injection System

Air bubbles in the fuel will impair the injection pump operation or even make operation impossible. It is very important that the system be **bled** whenever the injection pump has been removed from the engine or after changing a fuel filter or opening any fuel line.

Many supply pumps are equipped with a hand primer that can be used to fill the suction and delivery lines, the fuel filter(s), and the injection pump housing with fuel. The bleeder screws on the filter cover and in the injection pump should be kept open until the escaping fuel is completely free of bubbles. The pump control rod should be kept in the shutoff position while the system is being bled. During operation, the fuel system bleeds itself through the overflow valve on the fuel filter.

Basic Troubleshooting

Low power, rough running, or stalling may be an indication that fuel is not reaching the inline pump. A number of simple tests can be performed to pinpoint the cause of fuel starvation.

Relief Valve Pressure Check

The injection pump is equipped with a **pressure relief valve**. This relief valve must open at the correct pressure. To check **relief valve pressure,** disconnect the fuel line that leads to the secondary filter at the outlet side of the transfer pump. Using the proper fitting, connect a fuel pressure gauge at this point as shown in **Figure 20-41**. If possible, use a gauge with clear plastic fuel lines so any air bubbles in the fuel stream can easily be seen.

Start and run the engine. The reading on the fuel gauge indicates the relief valve opening pressure. Check the manual for exact pressure specifications, usually between 20-25 psi (140-175 kPa). A valve that opens at low pressure or sticks open will cause low fuel pressure inside the pump. A valve that opens at high pressure or sticks closed will result in excessive pressures inside the housing and increased fuel temperature that can reduce

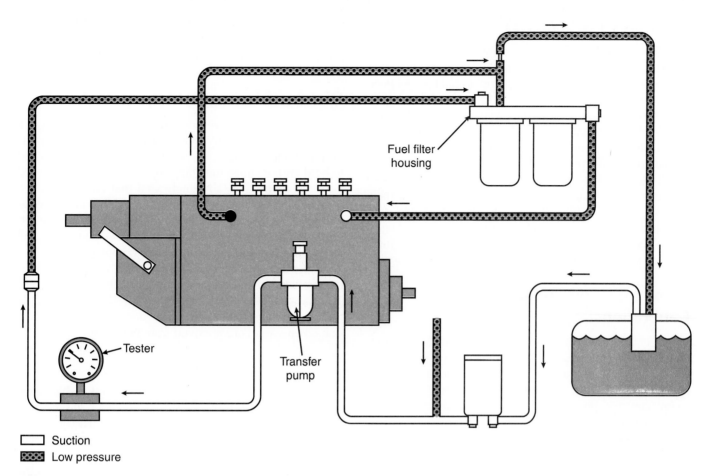

Suction

Low pressure

Figure 20-41. Setup for relief valve and transfer pump pressure testing.

power output. Fuel that is too hot will not properly lubricate and cool injection pump components.

While this test is being performed, inspect for air bubbles in the clear plastic fuel lines. Bubbles in the fuel indicate loose connections or damaged lines. Finally, pinch the fuel line closed downstream of the transfer pump. The reading on the gauge is now the actual operating pressure of the transfer pump. Check this pressure specification; it is typically a minimum of 30 psi (210 kPa).

Transfer Pump Restriction Check

A restriction on the transfer pump suction side will limit fuel flow. Perform a *transfer pump restriction check* at this point in the line using either a vacuum gauge or a manometer. If a manometer is teed into the fuel line, the meter must be positioned higher than the engine to prevent diesel fuel from running back into the manometer when the engine is stopped.

Connect the vacuum gauge or manometer into the fuel line between the primary fuel filter and transfer pump as shown in **Figure 20-42.** Start the engine and note the gauge reading. It should be between 6-12 psi (40-80 kPa) on the vacuum gauge and 12-24 inches on a mercury manometer. This relatively low pressure indicates the lines and connections on the suction side are offering little or no resistance to fuel flow. Pressures above those listed may indicate crimped or damaged lines or a clogged primary fuel filter.

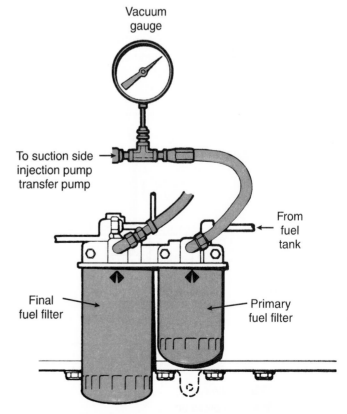

Figure 20-42. Setup for transfer pump vacuum restriction check. (Navistar International Transportation Corp.)

Secondary Fuel Filter Pressure Check

The *secondary fuel filter pressure* should also be checked between the outlet side of the secondary fuel filter and the inlet of the fuel injection pump. Install the fuel pressure gauge as illustrated in **Figure 20-43,** and start the engine. Inspect for air bubbles through the clear fuel lines of the pressure gauge. There will probably be some air that was introduced into the system when the gauge was installed, but this should quickly dissipate. If it does not, bleed the system, and check the fuel filter for a tight seal and fuel lines for loose connections or leaks. Normal system fuel pressure is usually between 14-20 psi (97-140 kPa). Low pressure would indicate a clogged secondary filter. If the pressure reading is within specifications, remove the test equipment, reconnect all lines, start the engine, and bleed all air from the system.

Inline Injection Pump Service and Calibration

Fuel injection pumps are considered precision equipment. Servicing this equipment including disassembly, inspection of parts, reassembly, calibration and testing requires special training, tools, and test equipment. Only authorized technicians may perform this service work. Injection pumps are equipped with special lockwires and tamperproof seals and fasteners that discourage unauthorized disassembly and adjustment. Removing these lockwires or breaking factory seals may void the injection pump warranty and may be considered a form of illegal tampering in some areas.

Every diesel engine technician should have a basic understanding of the work performed by fuel injection specialty shops. Some technicians may choose to specialize in this area of diesel engine service. For these reasons, the following sections present a very general overview of pump

and governor disassembly, inspection, reassembly, and calibration. **Figure 20-44** shows an exploded view of the pump to help identify the components being discussed.

Disassembly

All pump and governor disassembly and repair work is performed in a clean room to protect the precision parts from dirt and dust. Special plastic organizer trays are used to hold and organize the components as they are removed from the pump. It is absolutely vital that all mated parts, such as plungers and barrels are kept together in their original sets. These parts are machined to extremely close tolerances, and any parts that are not properly installed will lead to improper pump operation or complete failure.

To begin, the pump and governor housing is thoroughly cleaned and drained of all lubricating and fuel oils. The pump is then secured in a brass jawed vise. The supply pump, supply pump gasket, inspection cover, and inspection cover gasket are removed from the pump.

Governor Removal

The following components are removed in this order: overflow valve assembly, governor top cover, gasket, governor fastening screws, and lockwashers. The governor housing is then tapped lightly with a plastic mallet to separate it from the pump. As shown in **Figure 20-45,** the governor is pulled back, sliding it to the right to disconnect the control linkage from the control rack.

Pump Disassembly

A drive coupling is attached to the pump camshaft and a special wrench is used to prevent the coupling from

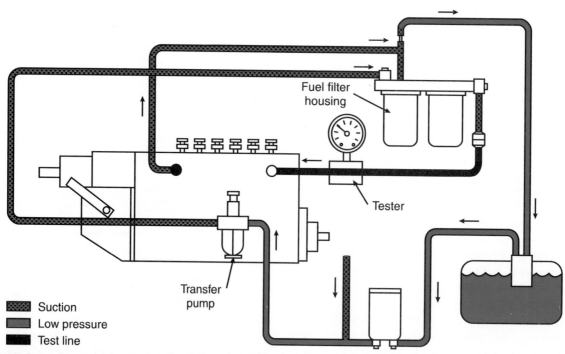

Figure 20-43. Setup for secondary fuel filter pressure check.

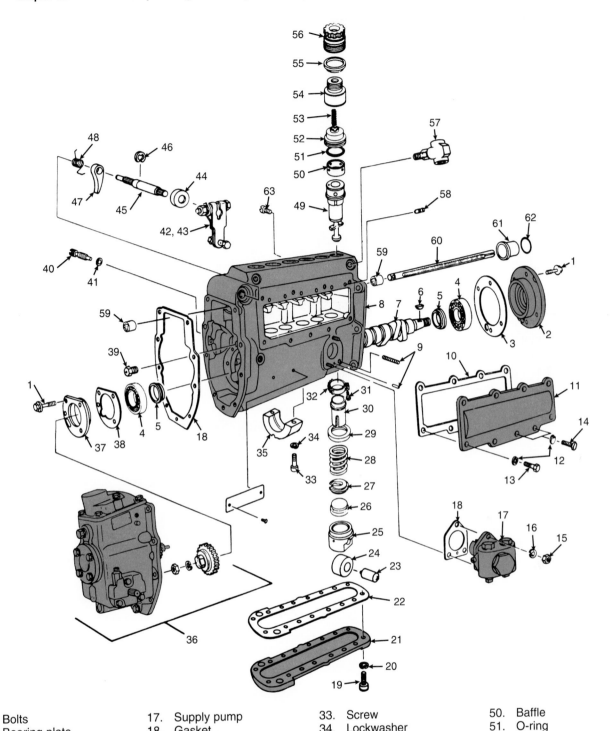

1.	Bolts	17.	Supply pump	33.	Screw	50.	Baffle
2.	Bearing plate	18.	Gasket	34.	Lockwasher	51.	O-ring
3.	Shim	19.	Screw	35.	Center bearing	52.	Delivery valve assembly
4.	Roller bearings	20.	Lockwasher	36.	Governor	53.	Spring
5.	Seals	21.	Base cover	37.	Rear bearing plate	54.	Holder
6.	Woodruff key	22.	Gasket	38.	Shim	55.	O-ring
7.	Camshaft	23.	Pin	39.	Plug	56.	Delivery valve securing nut
8.	Pump housing	24.	Roller	40.	Tappet locating screw	57.	Overflow valve
9.	Supply pump mounting studs	25.	Tappet shell	41.	Gasket	58.	Locating pin
10.	Gasket	26.	Knob spacer	42.	Shutoff lever	59.	Control rack bushing
11.	Inspection cover	27.	Lower spring seat	43.	Lever assembly	60.	Control rack
12.	Lockwashers	28.	Plunger spring	44.	Oil seal	61.	Control rack cap
13.	Screw	29.	Upper spring seat	45.	Shutoff shaft	62.	O-ring
14.	Lockwire screw	30.	Control sleeve	46.	Retaining ring	63.	Control rack locating screw
15.	Nut	31.	Screw	47.	Stop lever		
16.	Washer	32.	Control sleeve segment gear	48.	Spring		
				49.	Plunger		

Figure 20-44. Exploded view of APE 6G multiple plunger inline injection pump. (AMBAC International)

Figure 20-45. Removing governor from the pump housing. (AMBAC International)

turning as the governor drive gear retaining nut is loosened. A puller is attached to the drive gear and used to free it from the shaft, **Figure 20-46.** The camshaft is rotated and a service pin is inserted into a hole as each roller assembly comes up into the inspection area to hold it in place, **Figure 20-47.** The drive coupling is now removed and two union nuts are installed on the center outlets.

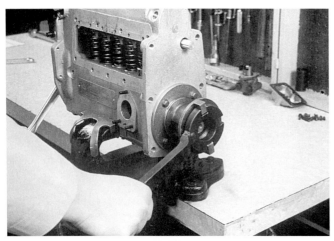

Figure 20-46. Removing injection pump drive gear with puller. (AMBAC International)

Figure 20-47. Lock the roller assembly in place using service pins. (AMBAC International)

The pump is now inverted in the vise with the jaws bearing on the union nuts. The base cover and gasket are removed along with the center bearing screws and lock washers. The rear bearing plate screws are removed and the camshaft tapped lightly on the drive end with a plastic mallet to remove the rear bearing plate and shim pack. The camshaft and center bearing can then be removed from the rear of the housing.

The tappet assemblies are removed from the pump housing using a special compressing tool. Each tappet assembly is compressed, **Figure 20-48,** the loosened service pin removed, and the pressure on the tappet assembly slowly released. Once all service pins are removed, the individual tappet assemblies can be removed, **Figure 20-49.**

Mechanical fingers are used to remove the lower spring seats and plungers which are placed with their corresponding tappet assemblies, **Figure 20-50.** The plunger return springs, upper spring seats, and control sleeve assemblies are the next items removed. After this step, the pump is returned to its upright position in the vise and the delivery valve holders, snubber valves, and springs are

Figure 20-48. Compress the tappet assembly for removal. (AMBAC International)

Figure 20-49. Tappet assemblies must be kept separate. Do not mix these as the parts from each assembly are mated. These are the first mated parts removed from the pump. (AMBAC International)

Figure 20-50. *A—Removing lower spring seats and plungers. B—Keep all parts from each seat and plunger assembly together. (AMBAC International)*

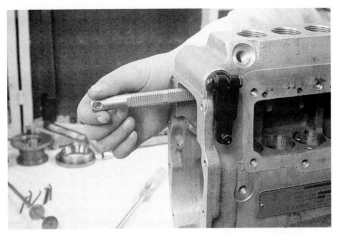

Figure 20-52. *Removing the injector pump control rack. (AMBAC International)*

removed, **Figure 20-51.** A magnetic pencil is used to extract the delivery valve and a puller is needed to remove the delivery valve body. The barrels and baffles are removed and the plungers immediately inserted into their mating barrels to avoid damage.

Finally, the control rack locating screws are removed and the control rack slid from the housing, **Figure 20-52.** This completes the normal pump disassembly procedure. The remaining bushing and bearing assemblies will only be disassembled if they are damaged or work beyond specifications.

Cleaning and Inspection

All parts are cleaned and inspected for damage and excessive wear prior to reassembly. Components are inspected for the following types of damage:

❑ Cracks, stripped screw threads, or burrs on the mating surfaces of the governor housing and cover.

❑ Damage to the governor weight assembly, such as worn bell cranks, damaged weights, stripped threads, worn or damaged springs, etc.

❑ Wear to the adjusting pin or its parts.

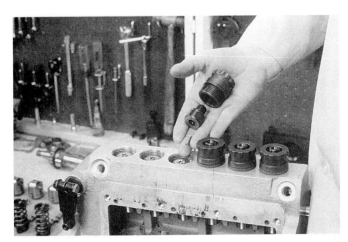

Figure 20-51. *Removing the delivery valve and spring. (AMBAC International)*

❑ Cracks, chips, or leaks in the pump housing including stripped threads.

❑ Scratching or scoring on barrels and plungers, especially their lapped surfaces. Test the set by washing out with test oil and pulling the plunger partially out of the barrel and releasing it. It must slowly fall back into the barrel by its own weight. If one part is damaged, the set must be replaced.

❑ Delivery valve and seat. Any damage to either requires replacement of the set.

❑ Slight pressure marks or grooving on the roller tappets can be smoothed out with a polishing cloth, but more serious damage requires replacement.

❑ Deep wear or grooving of camshaft cams or bearing surfaces requires camshaft replacement.

❑ Damaged roller bearings.

❑ Burrs on the control rack gear teeth or scratches along the sides. Rack may be polished with crocus cloth, but replace if any binding with housing exists.

❑ Damage to the control sleeve and gear teeth segments or to the sleeve slots in which the plunger flange rides.

❑ Worn, bent, or rusted springs and spring keepers should be replaced.

❑ Base plugs can be reused unless they are known to be leaking.

❑ All gaskets, seals, and O-rings must be replaced.

Reassembly

Pump reassembly begins by temporarily installing the camshaft, rear bearing plate, and shim pack, and using a dial indicator to check for proper end play, **Figure 20-53.** End play is adjusted by altering the shim pack thickness. Once the proper shim pack is selected, the control rack is slid into the housing. The slot in the rack is aligned with the locating hole in the pump housing and the locating screw is installed, torqued, and staked in place, **Figure 20-54.**

Ring baffles are then reassembled to their mating barrels with the top edge of the baffle below the top surface of the barrel. The plunger barrels are reinstalled in their original positions with the locating slots in the barrel aligned with the locating pins, **Figure 20-55.** New O-rings are installed on

the delivery valve bodies and the puller is used to install each valve body in its original position, **Figure 20-56.** The delivery valve components are then installed in the valve body. The springs and delivery valve holder assemblies are reinstalled and the delivery valve retaining nuts are installed and tightened to specifications, **Figure 20-57.**

At this point in the reassembly process, union nuts are installed and the pump inverted in the vise. The control rack is positioned at the mid-point in travel and the control sleeve assemblies are positioned over the barrels with the segment gear gap at 90° to the control rack axis, **Figure 20-58.** The rack is moved from end to end to make certain all segment gears are parallel, move freely and have full travel. The upper spring seats and plunger springs are now slid in place over the control sleeves, **Figure 20-59.**

The lower spring seat is assembled and mechanical fingers are used to insert the plunger into its original barrel. The tappet assemblies are then installed in their original bores. Install the compression tool on the tappet and move the control rack to align the plunger flanges with the control sleeve slots. The plunger springs are then compressed and a service pin installed into the tappet hole to hold it in place. This procedure is repeated for each tappet assembly. The camshaft and its bearings are now reinstalled into the housing,

Figure 20-53. Checking for proper injection pump camshaft end play. (AMBAC International)

Figure 20-54. Installing control rack locating screw. (AMBAC International)

Figure 20-56. Installing delivery valve with attached puller. (AMBAC International)

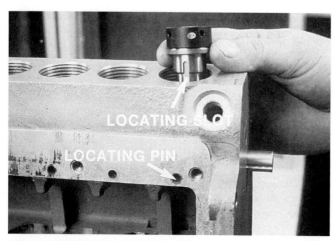

Figure 20-55. Plunger barrel installation. Note the locating slots align with the locating pins. (AMBAC International)

Figure 20-57. Installing springs and delivery valve holder assembly. (AMBAC International)

Figure 20-60. The camshaft center bearings and rear end plate are secured to the housing using screws and lockwashers coated with thread lock compound. The Woodruff key and drive coupling are then assembled to the camshaft.

The camshaft is rotated so that the service pins can be removed from the tappets. The base plate and new gasket are then installed. With the pump in an upright position, the supply pump and new gasket are reinstalled and the governor drive gear retaining nut is tightened to the recommended torque, **Figure 20-61.**

Figure 20-58. Installing barrel control sleeves.
(AMBAC International)

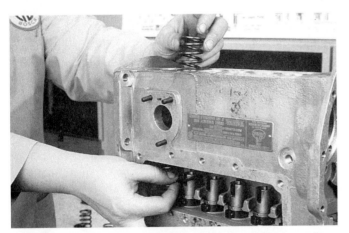

Figure 20-59. Installing upper spring seats and plunger springs.
(AMBAC International)

Figure 20-60. Installing the pump camshaft and bearings.
(AMBAC International)

Figure 20-61. Torquing the governor drive gear retaining nut.
(AMBAC International)

Governor Installation

After installing a new gasket on the pump housing, the governor link pin is engaged with the mating hole at the end of the control rack. The governor housing is mated to the pump housing and the governor screws installed and tightened to specifications. The camshaft must be rotated and the governor linkage moved to ensure that the pump is operating smoothly and that all internal linkages are secured. The injection pump is now ready for timing, pressure testing, and calibration.

Testing and Calibration of Inline Pumps

After servicing and overhaul, an inline injection pump must be mounted on a test stand and checked for lift-to-port closure, calibration, and other basic checks and adjustments that affect performance. Every multiple plunger inline pump manufacturer recommends specific test stand equipment and procedures for testing their injection pumps. As is the case with pump and governor service, the testing and calibration can only be carried out by technicians authorized by the pump manufacturer.

Lift-to-Port Closure

Once lift-to-port closure for the number 1 pumping element is set, the correct setting for the remaining pumps can be easily made. For example, after the correct lift-to-port closure for pump number 1 is set on a six-cylinder pump, the test stand drive hub degree wheel is set to the 0° position. The remaining pumping plungers lift-to-port closure is then adjusted in firing order sequence at 60° intervals on the test stand degree wheel. This is known as phasing the pump.

1. Connect the necessary high-pressure fuel lines on the test stand to the inlet fuel gallery of the injection pump, **Figure 20-62.**
2. Install a dial gauge on the number 1 pumping plunger so that when the pump camshaft is rotated the gauge will register the lift distance. Zero the dial gauge with

Figure 20-62. *Connections to test stand for injection pump lift-to-port closure test. (AMBAC International)*

Figure 20-64. *Some pumps require the installation of shims or spacers under the tappets to adjust plunger lift. (AMBAC International)*

the pump camshaft lobe for the number 1 element at BDC, **Figure 20-63.**

3. Turn on the test stand fuel pump and allow calibration fluid to flow from the top of the number 1 delivery valve holder into a suitable container.

4. Slowly rotate the injection pump camshaft in its normal direction or rotation until the flow of fuel from the test stand tube cuts off to a series of drops. As in spill timing discussed earlier, this indicates that port closure has occurred.

5. Check the measurement on the dial indicator against lift specifications. If it is not within the allowed tolerance, adjust the plunger lift. This is done in a number of ways, depending on the exact injection pump model. Some use shims below the spring seat and tappet, **Figure 20-64.** In others, the tappet locknut is

loosened and an adjust screw is rotated to obtain the desired lift. On pumps with no removable side cover, adjustment is made by loosening the barrel retaining bolts and adding or removing shims under the pump flange, **Figure 20-65.**

6. Once lift-to-port closure is set for cylinder number 1, loosen the test stand drive degree wheel and manually rotate it to the 0° setting.

7. Connect the fuel supply lines from the test stand to the pumping element for the next cylinder in the engine's firing order. Activate fuel flow and rotate the camshaft by turning the test stand drive degree wheel. As it reaches the 60° mark, fuel flow should drop off to a series of drips. This indicates the cylinder's lift-to-port closure setting is correct. As you can see, the dial gauge is no longer needed. If necessary, adjust the plunger lift as described in step 5. Check the next cylinder in the firing order at 120°, number 4 at 180°, number 5 at 240°, number 6 at 300° and adjust as needed.

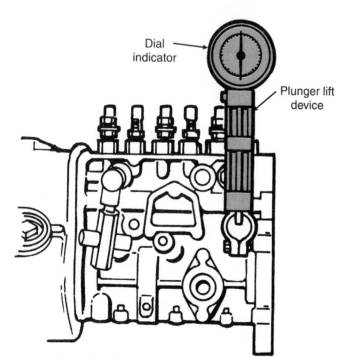

Figure 20-63. *Dial gauge in position on the pump's output to the number 1 cylinder. (Navistar International Transportation Corp.)*

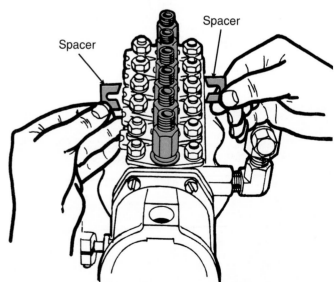

Figure 20-65. *Installing shims under the pump flange to adjust plunger lift. (Navistar International Transportation Corp.)*

Plunger Head Clearance

There must be adequate clearance between the top of the plunger and the delivery valve positioned on top of the barrel. On pumps that use a shim-tappet method of adjusting lift-to-port closure, confirm adequate head clearance by manually rotating the camshaft with the test stand drive bar. If it rotates smoothly after all lift-to-port closure adjustments have been made, there is sufficient head clearance.

Fuel Delivery Rate Calibration

As discussed earlier, the amount of fuel delivery by each pumping plunger must be equal to provide even firing and smooth operation. The amount of fuel delivered by a individual pumping element is controlled by its effective stroke. This is the time during its lift stroke that both inlet ports in the barrel are closed.

Plunger rotation by the fuel control rack causes the helix to either lengthen or shorten the effective stroke. For a given amount of control rack movement, the change in the effective stroke must be the same for all plungers. If it is not, the fuel rate must be adjusted at each plunger.

Measuring the rate of fuel delivery and the effect control rack movement has on it can only be done on a fuel pump test. Test stands measure the delivered volume of fuel from each barrel and plunger assembly through the use of video display monitoring, or by graduated cylinders, **Figure 20-66.** The injection pump is mounted to the stand, and each individual pumping element is checked in the engine's firing order.

Variations between the fuel delivery rates indicate that adjustment is needed to bring all elements into manufacturer's specifications. The exact adjustment procedure varies depending on inline pump design. On some pumps, the fuel rate adjustment is made by loosening the control sleeve on the control rack. By inserting a small punch pin into the radial holes on the sleeve, it can be rotated clockwise or counter-clockwise to change the pumps delivery rate.

On pumps that use bolted barrel flanges on top of the injection pump housing, the adjustment is made by loosening the barrel flange retaining nuts and rotating the flange either clockwise or counter-clockwise. This alters the control fork position. Most pumps provide approximately 10° of barrel rotation for setting the fuel delivery rate.

The test stand must be run at specified speeds per the pump manufacturer's test specifications. Mount the pump on the test stand and be certain there is adequate clearance at the drive coupling. Connect the fuel supply line to the pump inlet and the lines to the valve holders. Also connect the lubrication line from the test stand to the pump. If the pump is not engine oil lubricated, fill the pump sump with the appropriate lubricating oil.

Refer to the appropriate pump manufacturer specification sheet and set the number 1 sleeve gear or control fork to specifications. Some pump specifications call for calibrating the number 6 element first. Select the correct pump drive direction (counter-clockwise or clockwise). Open the fuel supply valve to pressure flow and loosen the pump bleed screw and start the pump drive. Operate the pump until all air has been bled from the system, and then tighten the bleed screw.

The first step is to adjust the maximum fuel delivery from the first element (number 1 or 6 as specified). Move the fuel lever to the maximum speed position. Run the pump at the specified speed and note the delivery from the element for the specified number of shots. Adjust the maximum fuel stop screw until the maximum delivery is as specified. Repeat the test several times to ensure accuracy.

Now adjust the position of the other sleeve gears or control forks so that the maximum delivery rate from their pumping elements is exactly the same as that from the specified element. Check the operation of the excess fuel device and fuel stop control.

Final Checks

Adjust the control rack stops as described earlier in the chapter. Check fuel delivery rates at the excess fuel position and at idle. After calibration is complete, install all plugs, covers, etc., using new gaskets and sealing washers as needed. Install tamper-proof locks and seals as specified.

Figure 20-66. *Graduated cylinders are used to test the pump's fuel delivery rate. (AMBAC International)*

Summary

Multiple-plunger inline injection systems are extremely popular. The injection pump produces the pressure needed for injection using a separate plunger and barrel assembly for each engine cylinder. These plunger and barrel pumping units are contained inside of the injection pump housing.

A transfer or supply fuel pump supplies low pressure fuel to the injection pump. Most inline injection pumps operate on the port and helix metering or sleeve metering principle.

The up-and-down motion of the pump plungers is controlled by cam lobes on the injection pump camshaft.

This camshaft is driven off the diesel engine. The injection pump driving gear can mesh directly with the engine camshaft gear, or an idler gear, or auxiliary shaft gear that is driven off of the camshaft gear.

For injection to begin, the pump camshaft must lift the pumping plunger until it closes off the barrel inlet fuel ports. This is known as lift-to-port closure and refers to how far the plunger must move to effectively close off both fuel inlet ports in the barrel.

A governor linked to a control rod and control sleeve rotates the plungers, varying fuel delivery and controlling the engine speed. The identification plate mounted on the pump housing contains important information concerning size and specifications. An inline fuel injection pump can be base mounted onto the engine or bolted to a cradle on the engine.

The most important aspect of removing and installing an inline injection pump is to properly time the injection pump to the engine. Fixed timing sensor/light, high pressure (hydraulic) port closing, and low air pressure port closure timing methods are used to check injection pump timing.

Low power, rough running, or stalling may be the result of fuel starvation caused by low transfer pump fuel delivery. A number of simple tests can be performed to pinpoint the cause of fuel starvation. These include relief valve and pump pressure checks, pump vacuum restriction checks, and fuel filter pressure checks.

Inline fuel injection pumps are precision assemblies. Servicing this equipment including disassembly, inspection of parts, reassembly, calibration and testing requires special training, tools, and test equipment. Only technicians authorized by the pump manufacturer may perform this service work. Injection pumps are equipped with special lockwires and tamperproof seals that discourage unauthorized disassembly and adjustment.

After servicing and overhaul, an inline injection pump must be mounted on a test stand and checked for lift-to-port closure, calibration, and other basic checks and adjustments that affect performance. Pump calibration ensures that each pumping element is delivering the same amount of fuel at a given fuel rack position. As with disassembly procedures and service calibration, adjustment work requires special equipment and manufacturer's authorization.

Important Terms

Multiple plunger inline fuel injection pump	Automatic timing advance unit
Pump plunger	Static spill timing check
Pump barrel	Timing probe
Coupling element	High pressure port closure test
Roller tappet	Engine spill timing
Control rack	Low air pressure method
Lift-to-port closure	Bled
Code number	

Pressure relief valve

Relief valve pressure

Transfer pump restriction check

Secondary fuel filter pressure

Review Questions—Chapter 20

Do not write in this text. Place your answers on a separate sheet of paper.

1. In a multiple-plunger inline injection pump, a separate _____ and _____ assembly are used to produce the pressure needed for injection into each separate cylinder.

2. An inline injection pump contains its own _____ that is driven off of the diesel engine.

3. Eccentric timing devices used on most automatic timing advance units are available as _____ and _____ types.

4. The mark for the start of injection delivery is on the _____.
   (A) the flywheel
   (B) the V-belt pulley
   (C) the vibration damper
   (D) All of the above.

5. Some inline injection pumps are equipped with _____ that are used to lock the pump in position prior to removal.

6. The installation of an inline pump also involves checking and/or setting the pump to engine timing, known as a _____.

7. List any disadvantages to the low air pressure method.

8. Air must be _____ from the injection pump and fuel lines after the serviced pump has been reinstalled.

9. Name three methods of checking and setting spill timing.

10. Low power, rough running, or stalling may be the result of _____ caused by low transfer pump fuel delivery.

ASE-Type Questions

1. Technician A says that an inline injection pump commonly operates using port and helix fuel metering. Technician B says an inline pump can also use pressure-time metering. Who is right?
   (A) A only.
   (B) B only.
   (C) Both A & B.
   (D) Neither A nor B.

2. Technician A says the distance the plunger must move to close the fuel inlet ports in the barrel is called the plunger's effective stroke. Technician B says that this is known as lift-to-port closure. Who is right?
   (A) A only.
   (B) B only.
   (C) Both A & B.
   (D) Neither A nor B.

3. Technician A says that multiple-plunger inline pumps can be base mounted onto an engine support. Technician B says these pumps can be bolted to a cradle on the engine. Who is right?
   (A) A only.
   (B) B only.
   (C) Both A & B.
   (D) Neither A nor B.

4. All of the following are performed by technicians working outside of diesel injection pump shops, EXCEPT:
   (A) testing lift-to-port closure.
   (B) pump removal from the engine.
   (C) timing the injection pump.
   (D) bleeding air from the injection system.

5. All of the following are simple diagnostic tests that can be performed on inline injection pumps, EXCEPT:
   (A) relief valve and pump pressure check.
   (B) pump vacuum restriction check.
   (C) secondary fuel filter pressure check.
   (D) fuel delivery calibration.

6. Which of the following procedures must be performed by authorized service personnel in a diesel injection pump shop?
   (A) Pump removal and installation on the engine.
   (B) Pump disassembly and overhaul.
   (C) Timing the injection pump to the engine.
   (D) Connecting the injection pump to the engine lubrication system.

7. When removing an inline fuel injection pump from the engine, Technician A says to first align the injection pump to engine flywheel timing marks with the engine rotated to its exhaust stroke. Technician B says to align the marks on the compression stroke before removing the pump. Who is right?
   (A) A only.
   (B) B only.
   (C) Both A & B.
   (D) Neither A nor B.

8. The presence of valve lash confirms the engine is on its _____ stroke.
   (A) exhaust
   (B) power
   (C) compression
   (D) intake

9. Technician A says some inline injection pumps are equipped with locking or timing pins that are used to lock the fuel injection pump in proper position prior to removal. Technician B says this pin is accessed through a plug in the injection pump housing. Who is correct?
   (A) A only.
   (B) B only.
   (C) Both A & B.
   (D) Neither A nor B.

10. The spill timing check accurately determines exactly when _____ occurs within the injection pump plunger and barrel for cylinder number 1.
    (A) port opening
    (B) port closure
    (C) top dead center
    (D) bottom dead center

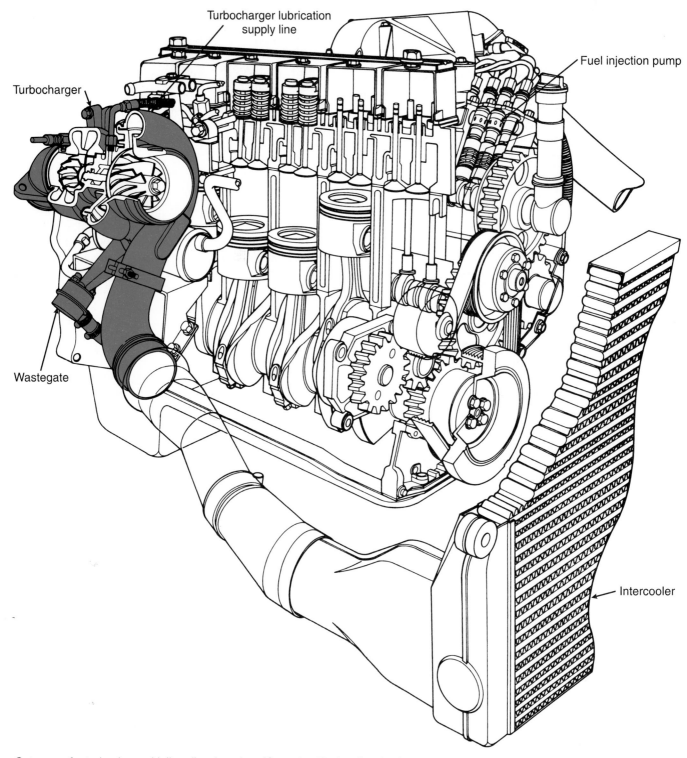

Turbocharger lubrication supply line

Fuel injection pump

Turbocharger

Wastegate

Intercooler

Cutaway of a turbocharged inline diesel engine. (Cummins Engine Co., Inc.)

Chapter 21

Distributor Injection Pumps

After studying this chapter, you will be able to:

❑ Name the major difference between multiple plunger inline injection pumps and distributor injection pumps.

❑ Describe the operating principles and major components of the AMBAC Model 100 distributor pump.

❑ Describe the operating principles and major components of the distributor pump.

❑ Describe the operating principles and major components of opposed plunger, inlet metered distributor injection pumps.

❑ Properly remove, install, and time an opposed plunger, inlet metered distributor pump to an engine.

❑ Describe the operating principles and major components of the Bosch Type VE distributor injection pump.

As explained in **Chapter 20,** an inline fuel injection pump uses a separate pumping element for each cylinder in the engine. A distributor injection pump, however, uses a single pumping element to meter and deliver pressurized fuel to the fuel injectors at all cylinders. This chapter discusses the operation and service procedures for the various distributor pumps used on diesel engines.

Distributor Injection Pumps

Distributor injection pumps are compact pumps designed for applications with relatively broad speed requirements, such as medium to heavy-duty trucks, agricultural equipment, and tractors. Major manufacturers of distributor injection pumps include AMBAC International, Stanadyne, Lucas CAV, and the Robert Bosch Corporation.

The principle of the distributor pump is similar to the distributor in a gasoline engine's ignition system. Just as the distributor in an electronic ignition system sends a high voltage pulse to each spark plug at the precise moment it is needed, a distributor fuel injection pump delivers the precise amount of fuel needed to each cylinder at the proper time in the cylinder's combustion cycle.

Although distributor injection pumps all operate on the same basic principle of a single pumping element, there are differences which vary by manufacturer. This is particularly true in the way the plunger or plungers are activated and synchronized to the combustion cycle. Other distributor pumps activate plunger motion through the use of a camshaft or a cam ring with internal lobes. Other distributor pumps use a cam plate design to synchronize plunger movement. In addition, the manner in which fuel is metered and directed into the fuel lines varies between systems. As with inline pumps, you must have a basic understanding of how each type of distributor pump operates.

The disassembly, overhaul, reassembly, calibration, and testing of distributor pumps requires special training, tools, and test equipment. As with inline and other injection pump systems, service technicians must be authorized by the pump manufacturer to perform service on the pump itself.

Distributor Pump Components

Figure 21-1 is an AMBAC Model 100 single plunger fuel injection pump. One plunger injects fuel into the lines for each cylinder in the correct firing order. The plunger assembly rides on a rotating camshaft. This produces the reciprocating movement needed to pressurize the fuel. At the same time, the plunger is also being rotated by the camshaft through a gear arrangement, **Figure 21-2.** These two actions allow the plunger to accomplish both of its main jobs—pumping and distribution. The precise rotating and reciprocating action of the single plunger distributes a uniform volume of fuel per cylinder for smooth power and fuel economy. There is also no need to make fuel delivery

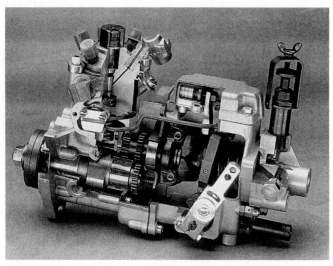

Figure 21-1. *Cutaway of an AMBAC Model 100 distributor fuel injection pump. (AMBAC International)*

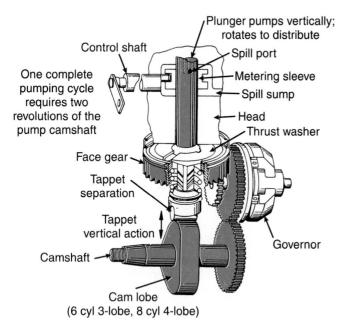

Figure 21-2. *Operational diagram of the camshaft activated AMBAC Model 100 hydraulic head. (AMBAC International)*

adjustments for any one cylinder as in inline pumps. The distributor pump consists of the following major components.

Transfer Pump

The **transfer fuel pump** or *supply fuel pump* is driven by the governor driveshaft. It is a simple gear pump with a self-contained pressure regulator that supplies fuel to the main pump's sump. Pressure is created by a restricted orifice in the return line and controlled by a pressure regulator valve.

Hydraulic Head

The **hydraulic head** is a complete assembly that mounts to the pump housing. It consists of a head core, plunger, metering sleeve, plunger face gear, delivery valve assembly, plunger return spring, spring seat, and plunger

button. The plunger fits into a central bore in the head core. The upper part of the central bore is counterbored and threaded to receive the delivery valve. Drilled **discharge ducts** (one for each cylinder in the engine) extends symmetrically from the plunger bore and end in discharge fittings that assemble into the top of the head core, **Figure 21-3.**

A downward declining duct from the delivery valve bore intersects a mating duct leading to the **distributing annulus.** Drilled holes extend from the sump's outer periphery, through the head and terminate at two horizontal port closing scallops on the same plane near the top of the plunger bore. A horizontal rectangular slot in the lower section of the head accommodates the metering sleeve.

The **plunger** is lapped and mated into the head and metering sleeve, therefore, these three components must always be serviced as a unit. The lower extension of the plunger is machined for positioning on the plunger face gear. This fixes the angular position of the plunger in relation to the face gear. A bronze washer is located between the gear and head to take the thrust of the drive gear. The lower spring and spring seat is fastened to the plunger and the face gear assembly is fastened to the hydraulic head by a spring ring.

The **delivery valve** assembly is positioned at the top of the plunger bore and is held in place by a holder that also serves as the upper seat for the valve spring. A delivery valve cap nut and gasket seal the top of the head. The plunger makes one complete revolution for every two revolutions of the camshaft. The plunger rotates continuously as it moves vertically through the pumping cycle. On an eight-cylinder engine, the four lobe cam actuates the plunger eight times for every two revolutions of the pump camshaft. Pumps for six-cylinder engines use a three lobe

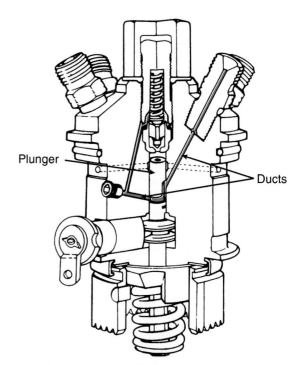

Figure 21-3. *Cutaway view of the AMBAC Model 100 hydraulic head. (AMBAC International)*

cam that actuates the plunger six times for every two revolutions of the pump camshaft.

The pump is engine oil lubricated. Internal ducts supply pressurized oil to the tappet assembly, camshaft bushings, thrust washer faces and governor shaft. All other parts are splash lubricated.

Fuel Pumping

Fuel enters the pump through the housing inlet connection. When the plunger is at the bottom of its stroke, fuel enters the sump and head cavity. As the rotating plunger moves upward, it closes two horizontal galleries that contain the inlet ports, **Figure 21-4**. This distance of plunger movement is the lift-to-port closure distance.

With the **inlet ports** closed, the trapped fuel builds pressure until the spring-loaded delivery valve opens. Further plunger movement forces the fuel through the delivery valve and the intersecting duct to the annulus and distributing slot. As the plunger rotates, the vertical **distributing slot** aligns with the outlet duct. The rotary and vertical movement of the plunger are so placed in relation to the **outlet ports** that the vertical distributing slot overlaps only one outlet duct during the effective portion of each stroke. When the outlet duct aligns with the distributing slot, high pressure fuel flows through the slot, into the outlet duct and to the injection nozzle, **Figure 21-5**.

Fuel Metering

The amount of fuel delivered per stroke is controlled by the **metering sleeve** position in relation to the fixed port closing position. That is the point at which the top of the plunger covers the top of the horizontal galleries, or *fill ports,* as they are called.

Delivery ends when the round metering port in the lower part of the plunger passes the upper edge of the metering sleeve. Since there is a hole drilled from the top

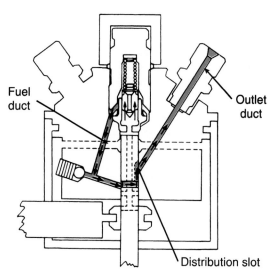

Figure 21-5. *Fuel flow and plunger position in the hydraulic head at the start of fuel delivery. (AMBAC International)*

of the plunger down to the metering hole, the fuel above the plunger is no longer contained under pressure. Its pressure will rapidly drop to match the fuel pressure in the lower pump area, **Figure 21-6**.

The metering sleeve allows variable amounts of fuel to be delivered to the injection nozzles. At the no-delivery position, **Figure 21-7A,** the metering sleeve is at the bottom of its travel. The sleeve is then moved up to the medium delivery position, about halfway through its travel, **Figure 21-7B.** The distance the plunger moves from the moment of port closure until the metering port is uncovered and the pressure is relieved is called the **effective stroke.** When the sleeve is in its highest position, the metering port will remain covered for the maximum effective stroke, resulting in maximum fuel delivery, **Figure 21-7C.**

Delivery Valve Operation

The delivery valve acts as a reverse flow check valve and also controls fuel line pressures between the injections.

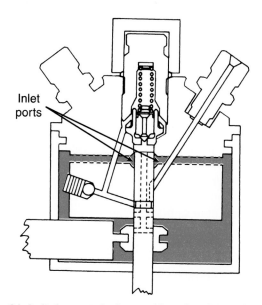

Figure 21-4. *At the port closing position, the plunger blocks the horizontal galleries of the inlet ports. (AMBAC International)*

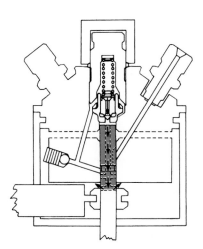

Figure 21-6. *Port opening in the hydraulic head at the end of delivery position. (AMBAC International)*

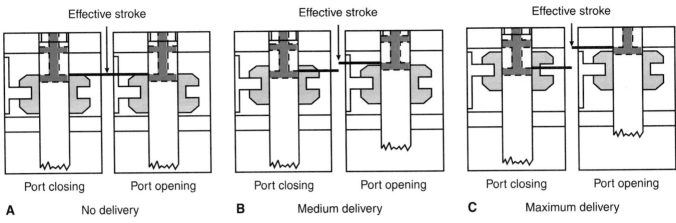

Effective stroke   Effective stroke   Effective stroke

Port closing   Port opening   Port closing   Port opening   Port closing   Port opening

A   No delivery   **B**   Medium delivery   **C**   Maximum delivery

Figure 21-7. Metering sleeve positions. A—No fuel delivery. B—Medium fuel delivery. C—Maximum fuel delivery. (AMBAC International)

Very quickly after port closure, the fuel pressure builds to open the delivery valve. When the plunger metering port is uncovered, fuel pressure rapidly drops and the delivery valve returns to its seat. Since part of the valve stem is actually a piston, it rapidly decreases the injection line pressure as it moves down into the valve bore.

Governor Operation

The governor maintains a specified minimum and maximum rpm, and provides torque control throughout the engine's operating range. It controls fuel delivery by varying the metering sleeve position in the hydraulic head. The movement of the control rod and control unit, with its offset pin, positions the metering sleeve up or down. The control rod is moved by the fulcrum lever, **Figure 21-8.** Lever position is controlled by a flyweight and spring assembly.

The flyweights are turned at engine speed by the gear train. As they spin, the flyweights spread apart, pushing the fulcrum lever in the direction of less fuel. This movement is resisted by the springs, so that at a given rpm, the fulcrum lever is held in a position of equilibrium. As the

engine speed increases, the flyweights move apart, which moves the fulcrum lever in the direction of less fuel delivery. As the engine speed drops, the springs overcome the flyweights to move the fulcrum lever in the direction of increased fuel flow. A stop plate prevents the fulcrum lever from moving beyond the maximum limit set for the engine application.

As engine load increases and speed decreases, the governor springs push the fulcrum lever in the direction of increased fuel delivery. As the fulcrum lever pivots around the torque cam nose, the torsion spring in the fulcrum lever bracket will open and fuel flow increases slightly on each stroke as predetermined by calibration. Decreasing speed will cause the cam nose to slide up the stop plate. This results in increased fuel delivery, which in turn causes increased torque. At peak torque—approximately 500-800 rpm below rated speed, the droop screw contacts the stop plate. At this point, the control rod can move no further and fuel delivery is limited to this upper level.

Excess Fuel and Aneroid Devices

Some distributor pump governors also include an excess fuel and aneroid or puff limiter device as part of the stop plate assembly, **Figure 21-9.** To obtain maximum fuel delivery during engine start-up, the excess fuel device uses a spring to move the governor stop plate to increase fuel flow. When the engine starts, oil pressure against the excess fuel piston overcomes the spring pressure and returns the stop plate to its normal position.

The aneroid device responds to intake manifold pressure to limit fuel flow until enough air is present for efficient combustion. This helps to eliminate the characteristic puff of black exhaust smoke during acceleration. This is the reason the aneroid is called a puff limiter by some truck manufacturers. The aneroid device is spring-loaded and is opposed by the aneroid piston, which is moved by boost pressure in the intake manifold. During acceleration, the aneroid will limit fuel delivery until there is enough air pressure in the intake manifold to allow complete combustion.

Figure 21-8. Control rod, fulcrum lever and other components of the AMBAC Model 100 governor assembly. (AMBAC International)

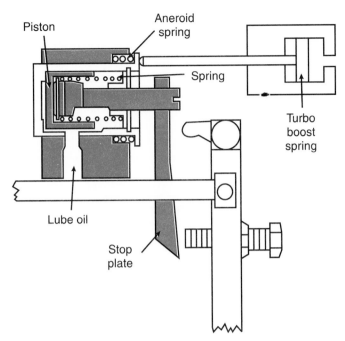

Figure 21-9. *Cutaway of an excess fuel and aneroid device. (AMBAC International)*

Intravance Camshaft

The ***Intravance camshaft*** is the main driveshaft of the AMBAC Model 100 pump, **Figure 21-10.** It is driven by the engine and contains the lobes and gear that moves the plunger and turns the gear train that drives the governor and rotates the plunger. Equally important, the Intravance camshaft automatically varies the fuel injection timing to match specific engine speeds.

The Intravance camshaft has a dual range which establishes one timing advance range for engine cranking, then changes to a second range for normal operation. This variation in timing advance is possible because the outer camshaft twists around the inner driveshaft on a helically splined sleeve, **Figure 21-11A.** If the splined sleeve is moved in one direction, the camshaft is advanced in relation to the driveshaft. If the sleeve is moved in the opposite

Figure 21-10. *Location of the AMBAC Model 100 distributor pump Intravance camshaft. (AMBAC International)*

direction, the camshaft is retarded in relationship to the driveshaft. The sleeve is moved by allowing engine oil to enter or drain from a working chamber behind the sleeve. The oil flow is controlled by a servo valve, which is controlled by an arrangement of flyweights and opposing springs.

Operation

As engine speed increases, the flyweights move apart and move the servo valve inside the Intravance camshaft. This opens the path for oil to enter the working chamber behind the splined sleeve. As high pressure oil flows into the chamber, it pushes the splined sleeve, which in turn advances the camshaft, **Figure 21-11B.**

As the sleeve moves, it also moves a pin and follow-up rod that increases the spring force opposing the flyweights. When the springs and flyweights become balanced, the servo valve is held in balance and there is no oil flow to or from the working chamber. As long as engine speed is maintained, the camshaft will remain advanced in relationship to the driveshaft.

If engine speed drops, the springs will overcome the inertia of the flyweights to move the servo valve. This allows oil to drain from the working chamber. As oil drains from the chamber, fuel pressure from the plunger assembly will move the splined sleeve toward the retarded position. When the springs and opposing flyweights become balanced, the servo valve will again be held in balance, **Figure 21-12.**

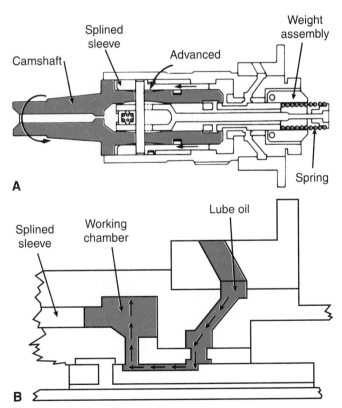

Figure 21-11. *A—Components of the Intravance internal timing device in the advanced position. B—Oil flows through the timing device in the advanced position. (AMBAC International)*

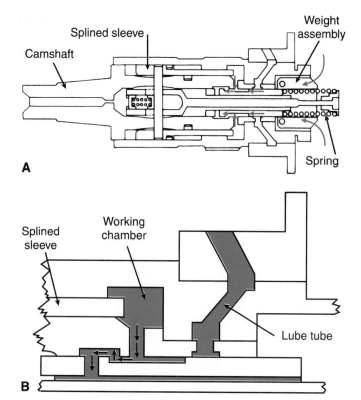

Figure 21-12. *A—Intravance camshaft static position. B—Oil flow in the static position. (AMBAC International)*

Distributor Pump Removal

To remove the distributor pump from the engine, first clean the side of the engine the pump is mounted on. Turn off the fuel supply and rotate the engine in its normal running direction until the timing marks on the pump hub are in alignment. This places the number 1 piston at the end of its compression stroke.

 Note: To ensure the number 1 cylinder is at TDC and not the number 6 cylinder, remove the rocker arm cover and check for valve lash at both valves at the number 1 cylinder. No lash at the number one cylinder indicates the number 6 piston is on its compression stroke. If this is the case, rotate the crankshaft 360°.

1. Remove all fuel inlet and outlet lines, oil supply lines, and air pressure lines (if used). Cap or plug all pump nozzles and line openings immediately after removal.
2. Remove and cap the high pressure injection lines, being careful not to bend the lines.
3. Remove all mounting bolts and clamps that hold the pump to the engine. Slide the pump carefully away from the engine. If the pump is bolted to the drive gear and the drive gear will not come out through the mounting hole, the bolts must be removed and the pump tapped off the gear.
4. Cap all openings on the injection pump.

Distributor Pump Installation

To install the pump on the engine, make certain that the engine is still located on the number 1 cylinder timing mark. If you are not certain, remove the rocker arm cover and confirm that the number 1 cylinder is on its compression stroke. The injection pump must also be in its number 1 firing position. If the side inspection cover is not wire sealed, remove it and align the timing marks in the window. At this time the timing marks on the drive hub should also be in alignment.

Note: If the side inspection cover is wire sealed, the timing marks on the hub should be aligned before the pump is installed. If the engine will not start, the pump is probably timed to the number 6 cylinder rather than the number 1 cylinder. To correct this, simply remove the three bolts securing the drive hub and rotate the engine crankshaft 360° and align the timing marks. Reinstall the bolts and continue with the installation.

1. With the engine in the proper position, mount the pump to the engine and tighten all retaining bolts finger tight.
2. Attach the injection lines, taking care not to bend them. Tighten finger tight at this time.
3. Attach the drive gear bolts and align the pump timing marks precisely. Tighten the pump mounting and drive gear bolts.
4. Attach the inlet and outlet fuel lines, oil supply, and air pressure lines (if so equipped).
5. Using the hand primer, bleed the inlet line and check for fuel leaks at all fittings.
6. Turn on the fuel supply and crank the engine with the throttle in the wide open position until fuel flows from all injection lines. Now tighten all line fittings to specifications and blow or wipe the excess fuel from the pump.
7. Before starting the engine, make certain you have a method of shutting down the engine in the event of a governor failure. A common method is to remove the air intake before starting the engine so that it can be covered with a flat metal plate or board to block off the air supply. Crank the engine with the throttle in the wide open position until the engine starts, then immediately move the throttle to half speed. With the engine running, check for leaks at all connections.

Distributor Pump Calibration

The following points cover the general aspects of distributor pump calibration. They are included here to give a better understanding of how the pump operates. Injection pump overhaul, testing, and calibration work must always be performed by authorized personnel. This procedure is specific to the AMBAC Model 100 distributor pump. Check the service manual for the specific procedure for each pump.

Lift-to-Port Closure Dimension

This setting ensures the plunger's **lift-to-port closure dimension** is correct. With the pump mounted on a test stand, remove the delivery valve from the hydraulic head and install a dial indicator to measure the plunger position. Zero the dial indicator at the bottom of the plunger stroke.

Activate the test stand so oil flows from the number one outlet. Rotate the pump camshaft manually until the fuel flow stops. This is the exact position of port closure. The reading on the dial indicator must be within specifications. If it is not, the lift-to-port closure dimension can be adjusted by changing the plunger tappet button. Tappet buttons of varying thicknesses are available for this purpose.

Rated Speed Adjustment

The **rated speed,** or **full-load delivery adjustment**, is made by moving the stop plate horizontally in the direction of increased or decreased fuel. Remember that at this time, the cam nose should be firmly against the stop plate.

Cam Nose Departure

Cam nose departure is the point where contact is being made by the cam nose on the stop plate, but no force is being exerted in the stop plate. It is adjusted with the high idle screw to be approximately 75 rpm above the rated speed.

Peak Torque Set Point Adjustment

Setting the fuel delivery at the **peak torque set point** is done by moving the cam nose up or down in order to obtain the specified difference in fuel delivery from the rated speed. The peak torque set point is normally 500-600 rpm below rated speed.

After setting the peak torque set point, it is necessary to return to the rated speed and readjust the full-load delivery as needed. Full load and peak torque settings will need to be adjusted until they fall into the required specifications.

Droop Screw Set Point Adjustment

Once the peak torque and the rated speed setting have been completed, the droop screw set point is established by adjusting the **droop screw** to get the specified difference in fuel delivery from the peak torque set point to the droop screw set point. The droop screw set point is usually about 300 rpm below the peak torque set point.

Opposed Plunger Inlet Metered Pumps

The original distributor injection pump, the Roosa Master, was an **opposed plunger inlet metered pump**. This type of injection pump used only one metering valve to control the fuel and either two or four opposed plungers to pump the fuel. One component, the distributor rotor, is used to distribute the metered fuel out through the hydraulic head to the injectors. These pumps have a fuel delivery capacity for engines rated between 10-40 hp per cylinder. These pumps can be found on a variety of car and truck engines.

The original Roosa Master company is now the Diesel Systems Division of Stanadyne Automotive Corporation. Stanadyne Model DB4 and DM4 (four-plunger) and DB2 and DM2 (two-plunger) distributor pumps are highly refined versions of the Roosa Master pump that reflect almost forty years of design evolution and improvement.

In 1956, an agreement was signed which allowed Lucas CAV Ltd., of England to produce a pump based on Roosa Master's Model A distributor pump. Lucas CAV designated the original pump the Model DPA, and has continued over the years to manufacture and refine this and another distributor pump, the ROTO-Diesel. Due to their shared heritage, Stanadyne and Lucas CAV pumps are quite similar in basic design and operation. In this text, they will be discussed together. However, there are differences between specific pump models.

Opposed Plunger Pump Construction

The Lucas CAV pump, **Figure 21-13,** is a single-cylinder, twin-plunger, distributor pump. This pump is lubricated by the diesel fuel that it pumps. Either hydraulic or mechanical governors can be fitted to the pump. The driveshaft, pumping and distributing rotor, and sliding vane transfer pump are an integral unit. The distributor is driven by the driveshaft that couples the rotor to a drive hub located at the end of the pump housing.

The Stanadyne pump shown in **Figure 21-14** is similar in many ways. The major difference is that it incorporates four pumping plungers rather than two. As with the Lucas pump, the driveshaft engages the distributor rotor in the hydraulic head. The rotor holds the four pumping plungers. The plungers are actuated simultaneously toward each other by an internal cam ring through rollers and shoes located in slots at the end of the rotor. The number of cam lobes normally equals the number of engine cylinders.

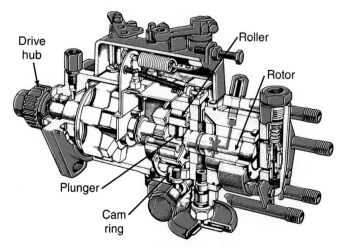

Figure 21-13. Lucas CAV Model DPA inlet metered opposed plunger distributor pump. (Lucas CAV Limited)

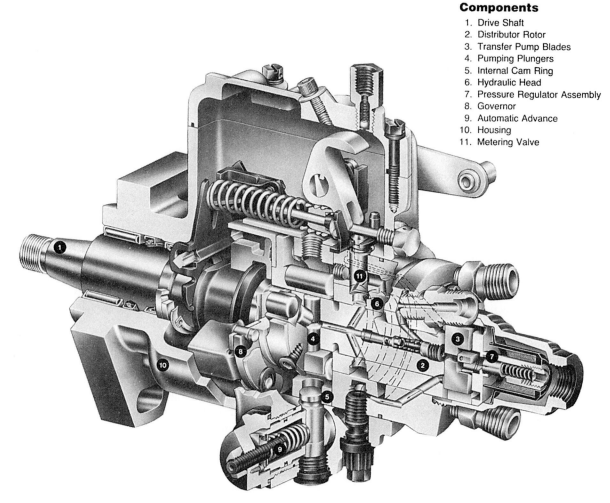

Components
1. Drive Shaft
2. Distributor Rotor
3. Transfer Pump Blades
4. Pumping Plungers
5. Internal Cam Ring
6. Hydraulic Head
7. Pressure Regulator Assembly
8. Governor
9. Automatic Advance
10. Housing
11. Metering Valve

Figure 21-14. *The Stanadyne Model DB4 injection pump uses four opposed plungers. (Stanadyne Diesel Systems)*

The transfer pump is also a positive displacement vane type. It is enclosed in the end cap, which also houses the fuel inlet strainer and transfer pump pressure regulator. The distributor rotor incorporates two **charging ports** and a single axial bore. One discharge port serves all the outlet ports to the injection lines. The hydraulic head contains the bore in which the rotor revolves, the metering valve bore, the charging ports and the head discharge fittings. The high pressure injection lines to the nozzles are fastened to these discharge fittings.

Stanadyne pumps have their own mechanical governor. The centrifugal force of the weights in their retainer is transmitted through a sleeve to the governor arm and to the metering valve. The metering valve can be closed to shut off fuel by an independently operated shut-off lever. The automatic speed advance is a hydraulic mechanism that advances or retards the beginning of fuel delivery. This can respond to speed alone or to a combination of speed and load changes.

Components

The aluminum alloy pump housing of an opposed plunger distributor injection pump contains the driveshaft, distributor rotor, transfer pump blades, pumping plungers,

internal cam ring, hydraulic head, end plate, adjusting plates, transfer pump, pressure regulator assembly, governor, automatic advance, and metering valve.

Driveshaft

The driveshaft connects to the engine drive gear and is supported by a bushing or ball bearing. It supports the governor assembly and drives the distributor rotor and transfer pump. The transfer pump consists of four linear blades. It delivers fuel to the metering valve located in the hydraulic head at low pressure. An end plate acts as a cover for the transfer pump. It also provides a fuel inlet to the pump and contains a pressure regulating valve that controls the transfer pump pressure throughout the speed range.

Hydraulic Head

The hydraulic head is machined with bores and passages that allow fuel to flow from the transfer pump to the metering valve, from the metering valve to the charging ports, and from the discharging ports to the discharge fittings. On the latest designs, hydraulic heads have been fitted with individual delivery valves to maintain residual line pressure and eliminate secondary injection.

Distributor Rotor

The distributor rotor is lap-fitted to the hydraulic head and the governor weight retainer assembly is fastened to its drive end. The plungers are fitted to the rotor and are pushed inward by the rollers and shoes to pump the diesel fuel. The rollers fit into the shoes and contact the cam in a way similar to a cam follower, **Figure 21-15.** Adjusting plates are mounted on the rotor and limit the outward travel of the rollers and shoes to control the fuel delivery rate.

Cam Ring and Metering Valve

A circular cam ring surrounds the rotor base and is located over the shoes and rollers. The number of internal cam lobes equals the number of cylinders. The cam ring forces the plungers toward each other which causes the fuel to be pumped. It can also be rotated back and forth about the rotor to vary the start of injection.

The metering valve contained in the hydraulic head regulates the volume of fuel entering the rotor. A piston valve is used with hydraulic governors. This valve is spring-loaded and controls the fuel according to the valve's axial position. When a mechanical governor is used, the valve is a rotary type, with a slot cut in the periphery. The valve is rotated by the governor arm to regulate fuel injection.

Automatic Advance and Governor

An automatic advance device is located in the bottom of the pump. A hydraulic piston rotates the cam ring against the direction of pump rotation via the cam advance stud. The cam advance stud threads into the cam and connects it to the cam advance mechanism.

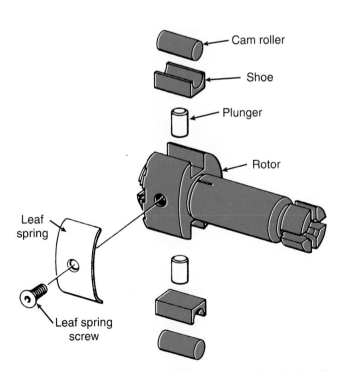

Figure 21-15. Exploded view of a rotor assembly showing the cam rollers and shoes. (Stanadyne Diesel Systems)

The governor weight retainer may be permanently fixed, splined, or bolted to the rotor drive end. Because the fuel metering mechanism can be affected by vibrations and shocks, the retainer often uses a cushioning device to isolate engine vibration and pulsation from the driveshaft. One end of the governor control arm rests against the thrust sleeve and the other end connects to the governor spring, and to the metering valve via a linkage hook. The control lever is connected to the shut-off lever and the fulcrum lever is connected to the governor spring.

Pump Operation and Fuel Flow

The operating principles of a opposed plunger pump can be understood more readily by following the fuel circuit during a complete pump cycle. **Figure 21-16** illustrates the fuel flow for a Stanadyne DB2 two-plunger distributor pump. The fuel flow for the DB4 four-plunger pump is the same with the exception of the charging of two additional plungers. As shown in **Figure 21-16,** the transfer pump pulls fuel from the fuel tank. The fuel passes through a water separator and secondary fuel filter before reaching the transfer pump. Once through the transfer pump, some of the fuel is bypassed to the transfer pump's suction side through the pressure regulator assembly.

Fuel under pressure flows past the rotor retainers and into an annulus on the distributor pump rotor. Some fuel flows through a connecting passage in the head to the automatic advance mechanism. The remaining fuel moves into the charging passage. This fuel flows around the annulus, through a connecting passage, and to the metering valve. The radial position of the metering valve regulates the fuel flow into the charging annulus, which holds the charging ports.

Pressure Regulating Valve Operation

The **pressure regulating valve** is located in the end plate and performs two important functions. When the injection pump is being primed, fuel is forced into the inlet connection through the mesh filter. Fuel enters the regulating sleeve located at the upper port, forcing the regulating piston downward and compressing the priming spring. When the piston has moved down far enough to uncover the lower port in the sleeve, the fuel flows directly into the hydraulic head. The pump is now primed and ready for start-up.

When the engine is running, the pump rotates and fuel is pulled into the end plate by the transfer pump. It passes through the mesh filter and is forced into the hydraulic head and end plate. When the transfer pump builds pressure, it forces the piston upward against the regulating spring, **Figure 21-17.** When the correct pressure is reached, the piston uncovers the regulating port. This bypasses a small amount of fuel back to the inlet side of the transfer pump to maintain fuel pressure at the desired level.

Transfer pump pressure can be adjusted in one of two ways. On some pumps, the spring guide is replaced with one of a different size. This changes the fuel pressure by

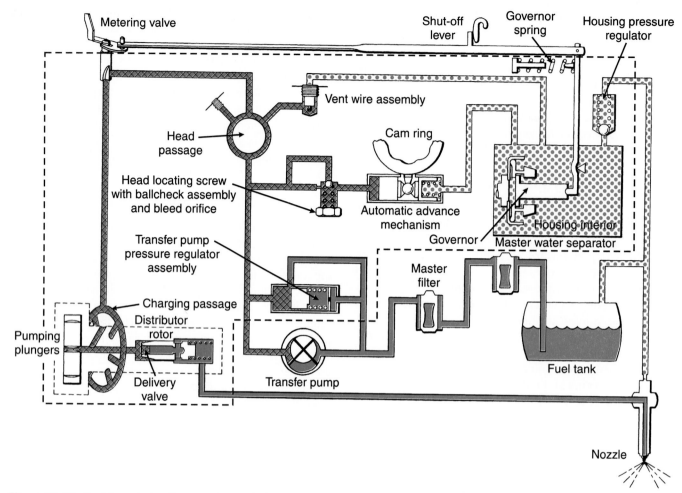

Figure 21-16. *Fuel flow during the pumping cycle in a Stanadyne DB2 distributor injection pump. (Stanadyne Diesel Systems)*

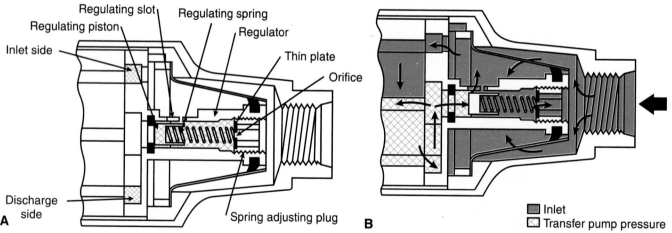

Figure 21-17. *A—Regulator assembly operation in an opposed plunger pump. B—Fuel flow through the regulator assembly. (Stanadyne Diesel Systems)*

altering the amount the regulating spring can be compressed. Other models are equipped with an adjustment device that can be set using a special tool when the pump is running on a test bench.

Charging Cycle

As the rotor revolves, the two inlet passages align with the charging ports in the annulus. Fuel under pressure from the transfer pump and controlled by the metering valve flows into the pumping chamber, forcing the plungers apart.

The plungers move outward a distance proportional to the amount of fuel required for injection on the following stroke. If a small quantity of fuel is admitted into the pumping chamber, the plungers move out a short distance. Maximum fuel delivery is limited by a leaf spring or springs that contact the edge of the roller shoes.

During the charging phase of injection, the angled inlet passages in the rotor are in alignment with the ports in the charging annulus. The rotor discharge port is not in alignment with a head outlet, **Figure 21-18.** The rollers are also off of the cam ring lobes.

Discharging Cycle

As the rotor continues to revolve, **Figure 21-19,** the inlet passages move out of alignment with the charging ports. The rotor discharge port opens to one of the head outlets. The rollers then contact the cam lobes and injection begins. Further rotation of the rotor moves the rollers up the ramps, pushing the plungers inward. During this stroke, the fuel trapped between the plungers flows through the rotor's axial passage and discharge port to the injection line. Delivery to

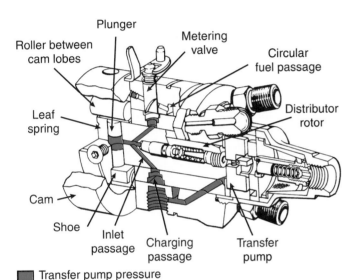

■ Transfer pump pressure

Figure 21-18. *Fuel flow during the opposed plunger pump's charging cycle. (Stanadyne Diesel Systems)*

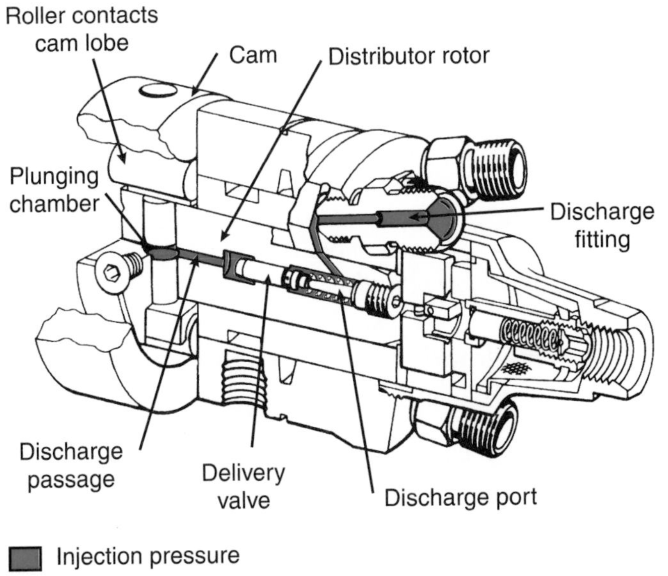

Figure 21-19. *Cutaway showing the opposed plunger pump's discharge cycle. (Stanadyne Diesel Systems)*

the injection line continues until the rollers move past the innermost point on the cam lobe and begin to move outward. The pressure in the axial passage is then reduced, allowing the nozzle to close and ending injection.

Delivery Valve Operation

On some distributor pumps, individual **delivery valves** (sometimes called pressure valves) are installed in the hydraulic head outlets for each cylinder. In other pump models, such as Stanadyne's, a single delivery valve mounted in a bore in the center of the distributor rotor serves all injection lines. The delivery valve(s) keep the lines full of fuel so that a full charge of fuel can be injected at the next cycle for that cylinder.

In addition, the delivery valve rapidly decreases injection line pressure to lower than nozzle closing pressure. This allows the nozzle to snap shut quickly without nozzle dripping or dribble that could cause excessive exhaust smoke.

Lubrication Circuit

The pump is lubricated with diesel fuel supplied by the transfer pump. A lubrication groove runs from the annular ring to the front of the rotor. From this point, the lubricating fuel flows to the main pump housing. At the top front of this housing a return fitting allows fuel to return to the fuel tank, **Figure 21-20.** This fuel lubricates all governor components. It also bleeds any air that may have entered the fuel system.

Advance Mechanism Operation

The opposed plunger distributor pump design permits the use of a direct acting hydraulic **advance mechanism**

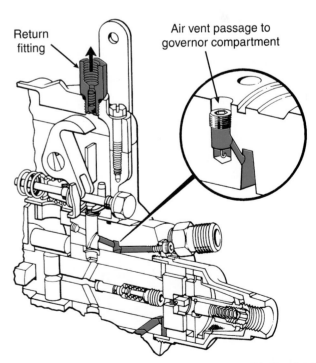

Figure 21-20. *The oil return circuit allows excess oil in the distributor pump to return to the fuel tank. (Stanadyne Diesel Systems)*

powered by pressure from the fuel transfer pump. The pressure is used to rotate the cam and vary delivery timing. The advance mechanism advances or retards the start of fuel delivery in response to engine speed changes.

Fuel from the transfer pump enters the hollow hydraulic head locating screw. Fuel then flows to the advancer piston, **Figure 21-21.** This piston advances the injection timing by pushing the advance cam screw to the cam ring, forcing the cam against rotation, **Figure 21-21.** A check valve prevents the cam ring from retarding by holding the fuel pressure in the chamber. When fuel pressure decreases because of a reduction in engine speed, fuel from the position area drains through the orifice below the ball check valve to allow the cam ring to retard. At engine idle speed, transfer pump fuel pressure is low, and the cam ring is held in the retard position by the spring and roller force. Maximum cam ring movement is limited by the piston length and the size of the piston hole plugs. A *trimmer screw* is provided to adjust advance spring preload which controls the start of the cam movement.

Mechanical Governor Operation

The opposed plunger distributor pump can be equipped with either a mechanical or hydraulic governor. The mechanical governor is shown in **Figure 21-22.** The flyweights transmit force through the thrust sleeve, causing

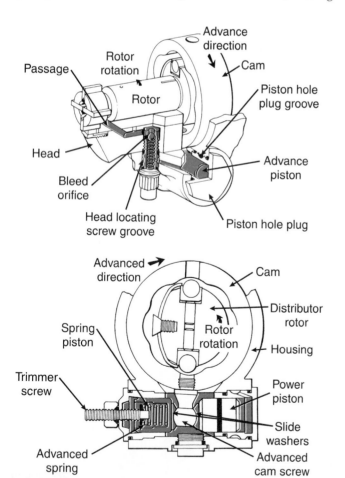

Figure 21-21. Operation of the opposed plunger pump automatic advance mechanism. (Stanadyne Diesel Systems)

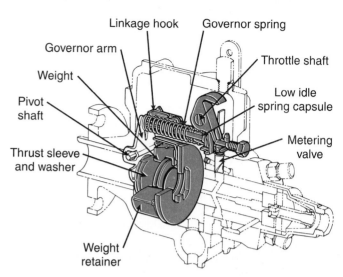

Figure 21-22. Components of DB2 mechanical governor. (Stanadyne Diesel Systems)

the governor lever to pivot. This pivoting movement rotates the metering valve, which reduces or increases the amount of fuel fed to the pumping cylinder. The flyweight force is opposed by the main governor spring and the idle spring. These forces balance each other to maintain a set engine speed. When needed, a mechanical shutoff bar rotates the metering valve to the shutoff position, regardless of the engine speed.

At idle, the flyweights offer little force against the governor lever. However, the main governor spring is slack, so the weights fly outward, causing the metering valve to rotate to the idle position. Within the idling speed range, the idle spring provides sensitive speed control.

At full load speed, the throttle applies maximum spring pressure on the lever. However, because of the load applied to the engine, it can only run fast enough to balance the heavy spring load with the flyweight force. The metering valve rotates to allow the maximum amount of fuel to enter the rotor.

If the load is suddenly removed from the engine, the engine would speed up. The increased force of the flyweights would overcome the spring pressure and rotate the metering valve to the closed position. Once the valve closes completely, governor cutoff occurs, limiting maximum engine speed to safe levels. The engine then slows down until the flyweights and spring reach a balance that allows only a small amount of fuel into the rotor. This condition is known as *high idle speed.*

Hydraulic Governor Operation

The components of a pump hydraulic governor are illustrated in **Figure 21-23.** The metering valve is lap-fitted to the hydraulic head and is center- and cross-drilled. The cross-drilled hole leads to a passage that is in alignment with the metering port. Transfer pump fuel pressure is always exerted onto the lower area of the metering valve. As the valve moves up and down, it varies the area of the metering port opening which controls the amount of fuel

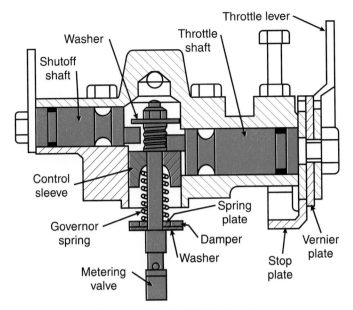

Figure 21-23. Components of a DPA hydraulic governor. (Lucas CAV Limited)

flowing to the pumping cylinder. A rack or control sleeve fits loosely over the extended metering valve stem inside the governor housing.

When the throttle moves to the full fuel position, the throttle lever moves the rack or sleeve and the governor spring moves the metering valve to increase the metering port area. The maximum amount of fuel flows through the metering port. However, fuel pressure increases as engine speed increases. This forces the metering valve back to the left until the metering valve and the governor spring rack are balanced. The metering port opening decreases, and the amount of fuel delivered to the pumping cylinder decreases.

As engine speed decreases, there is a reduction in fuel pressure. Spring force overcomes fuel pressure, and the metering valve opens wider. When the throttle lever is moved to the idle speed position, the idle spring compresses and moves the metering valve to the left until a balance is reached. Adjusting the high and low idle speeds is simply a matter of turning the adjustment screws clockwise or counter-clockwise to increase or decrease the spring force.

Electrical Shut-off

The governor may be equipped with an *electrical shut-off* device housed within the governor control cover. The device is available for 12, 24, and 32 volt systems. The device can either be *energized-to-shut-off (ETSO)* or *energized-to-run (ETR)*.

Energized-to-Shut-off (ETSO)

Energizing the solenoid coil overcomes the force of the shut-down coil spring, pulling the arm in and causing the tab on its lower end to contact the governor linkage hook. This moves the linkage hook against spring tension, rotating the metering valve to its closed position, cutting off the fuel, **Figure 21-24A.**

Energized-to-Run (ETR)

De-energizing the shut-down coil allows the spring to release the shut-off arm. The lower end of the arm moves the governor linkage hook, rotating the metering valve to the closed position and cutting off the fuel, **Figure 21-24B.** Some energized-to-run systems employ a mechanical override device for emergency use if the coil becomes inoperative due to electrical system failure. The override consists of a rod and guide assembly attached to the control cover through a tapped hole at the rear of the cover. It is in a location that aligns the rod with the solenoid arm assembly, **Figure 21-25.** When the rod is pushed into the cover to the limit of its travel, it contacts the solenoid arm assembly, locking it against the solenoid, allowing the linkage hook to operate. Pump shut-off during an electrical failure is accomplished by pulling out the override rod.

Distributor Injection Pump Removal and Installation

To remove the injection pump from the engine, first clean the side of the engine the pump is mounted on.

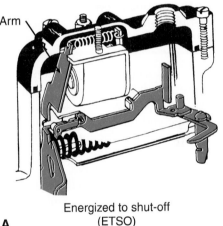

A   Energized to shut-off (ETSO)

B   Energized to run (ETR)

Figure 21-24. Electrical solenoid activated fuel shut-off mechanisms. A—Energized-to-shut-off. B—Energized-to-run. (Stanadyne Diesel Systems)

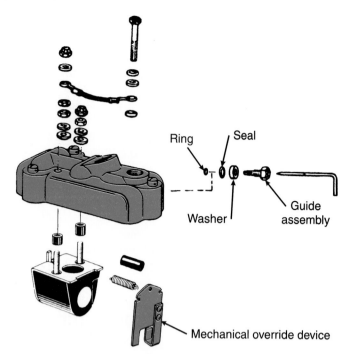

Figure 21-25. *Mechanical override for the electric shut-off.* *(Stanadyne Diesel Systems)*

Caution: Do not steam clean or wash down while the engine is operating. Severe damage to the pump may occur if its temperature is changed radically while running.

1. Turn off the fuel supply and rotate the engine in its normal running direction until the engine timing marks are in alignment. This places the number 1 cylinder at the end of its compression stroke. Check the rocker arm lash on the number 1 cylinder intake and exhaust valves to be sure. Remove the timing line cover from the outboard side of the pump. The timing line on the governor weight retainer hub should be directly opposite the line on the cam. If it is not, rotate the crankshaft 360°.

2. Remove all inlet and fuel return lines and completely remove the lines to the fuel injectors.

Caution: In removing the injection tubing nut from the pump, hold the discharge fitting with a wrench to prevent it from loosening from the hydraulic head. Immediately after the lines are removed, all pump nozzles and line openings should be capped or plugged.

3. Disconnect the throttle and shutoff linkage and electrical connections (if used). In most cases, it will be necessary to remove the pump drive gear from the hub or driveshaft. Refer to the engine service manual for this procedure.

4. Support the pump with one hand and remove the mounting stud nuts and washers on the pump flange. Slide the pump straight back and off of its mounting.

On vertical mounted pumps, lift the pump straight up and out.

To reinstall the pump on the engine:

1. Remove the timing line cover and turn the pump driveshaft in the direction of pump rotation until the timing line on the weight retainer aligns with the line on the cam OD, **Figure 21-26.**

2. Check that the engine is still in its proper timing position.

3. Slide the pump into position over the mounting studs. Assemble washers and the mounting stud nuts and tighten.

4. Install the pump drive gear to the hub or driveshaft.

5. Back off the engine at least 1/2 turn and manually rotate it in the normal direction of rotation to the proper timing mark.

6. Recheck line marks in the pump and correct if necessary. Repeat procedure to ensure proper timing. Install the timing line cover.

7. Remove the caps on the high pressure lines and connect the lines to their respective discharge fittings and tighten to specified torque.

8. Connect the throttle and shut-off linkage and electrical connections if used.

9. Assemble and tighten the fuel return line.

10. Open the bleed screw on the secondary fuel filter and operate the hand primer (if equipped) to bleed air from the system.

Note: Do not open filter bleed if a hand primer is located after the filter. Instead, close the bleed screw and hand prime the system until a quantity of fuel flows air free at the pump inlet line.

11. Connect the inlet line to the injection pump.

12. Start engine as per engine manual instructions.

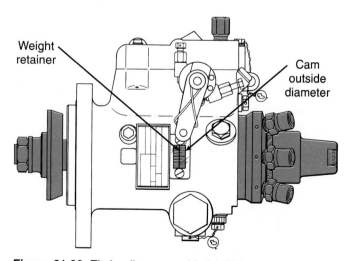

Figure 21-26. *Timing lines on a Model DM pump. Make sure these are aligned during pump removal or installation.* *(Stanadyne Diesel Systems)*

Test Bench Checks

After overhauling, the pump is mounted on an appropriate test bench and operational and pressure checks made. These checks include:

- ☐ Transfer pump pressure check to ensure the pump is capable of generating the required fuel pressure.
- ☐ Pressure regulator adjustment to set transfer pump pressure to specifications.
- ☐ Adjustment of low and high idle speeds and minimum fuel delivery at cranking speed.
- ☐ Adjustment of the trimmer screw to set the automatic advance.
- ☐ Check the electric shut-off (if so equipped).

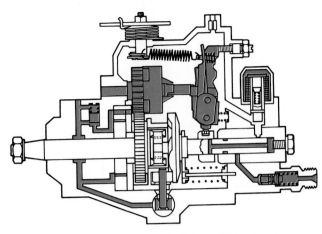

Figure 21-27. Cutaway view of a Bosch VE distributor pump.

Bosch VE Distributor Pump

The Bosch VE distributor pump, **Figure 21-27,** uses a single rotating/reciprocating plunger to meter and deliver fuel. The delivered fuel is fed through a distributor groove to ports that serve the individual engine cylinders. On four-cycle engines the pump is driven at engine camshaft speed. The injection pump's input shaft runs in half-time with the engine's piston motion. Drive connections can be made using a chain, belt, pinion, or helical gears. The pump housing holds the high pressure distribution pump, a vane-type transfer pump, mechanical governor, hydraulic timing device, and a shutoff mechanism.

The distributor pump's input shaft drives the fuel transfer pump and runs on bearings located in the pump housing. This vane transfer pump pulls fuel from the fuel tank and delivers it to the injection pump cavity, **Figure 21-28.** A pressure control valve is used to increase fuel pressure as engine speed increases. The valve screws into a threaded seat in the top of the pump housing, **Figure 21-29.** This spring-loaded slider valve allows the pump housing pressure to be varied, based on the amount of fuel being delivered. If fuel pressure rises above a set value, the valve piston opens the return circuit and fuel flows back to the pump's suction side. If fuel pressure is very low, the return circuit remains closed. Adjusting the spring preload sets the valve's opening pressure. A certain amount of fuel also returns to the fuel tank via an overflow restriction device located on the governor cover, **Figure 21-30.** This provides for pump cooling and self-bleeding.

Plunger Drive Action

A roller ring is located at the end of the distributor injection pump driveshaft. The roller ring is not connected to the driveshaft, but it is held in place inside the pump housing. The drive or input shaft rotation is transferred through a yoke to a cam plate. The cam plate surface rides on rollers mounted in the roller ring. The waves in the cam plate perform the same job as the cams on a camshaft, **Figure 20-31.** The action of the cam plate changes the

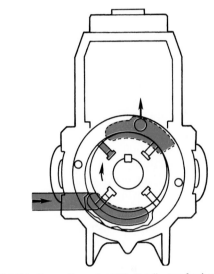

Figure 21-28. Operation of a VE distributor fuel supply pump.

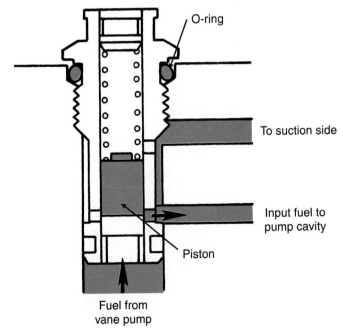

Figure 21-29. Cutaway showing the operation of pressure control valve.

O-ring
To suction side
Input fuel to pump cavity
Piston
Fuel from vane pump

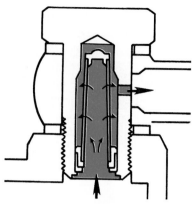

Figure 21-30. *Overflow restriction mechanism for the VE distributor pump.*

input shaft rotation into a reciprocating rotary motion that is transferred to the distributor plunger.

The plunger is set into the cam plate and its position is fixed using a lug. The plunger moves inside the distributor head. The distributor head is bolted to the pump housing and also contains the electric fuel shut off, a screw plug with vent screw, and the pressure valves. If a mechanical fuel shutoff device is used, it is located in the governor cover.

The plunger is pushed to its TDC position by the raised cam plate sections. Once it reaches TDC, two return springs force it back down to BDC. The springs also prevent the cam plate from separating from the roller under high acceleration.

In addition to driving the distributor plunger, the cam plate also affects the pressure and duration of fuel injection. Both the cam stroke and the lift velocity must be individually tailored to the engine's combustion chamber type and configuration. A special cam plate surface is manufactured for each engine type. For this reason, cam plates must never be interchanged with injection pumps from other engines.

Pump Assembly Construction

The distributor head, plunger, and control spool are precisely machined and fitted so that they seal properly,

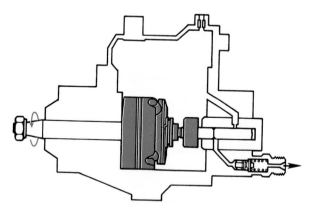

Figure 21-31. *Cutaway of a Bosch VE pump assembly. The surface of the cam plate has raised areas that act like cam lobes.*

even at very high pressures. As with injection nozzles, a small amount of leakage between fitted parts is needed for lubrication. This leakage is designed into the pump. For this reason, always replace the complete pump assembly.

 Note: Never replace the plunger, distributor head, or control spool separately.

Metering Fuel to The Cylinders

The pressure needed for injection is created by the plunger. **Figure 21-32** illustrates the metering of fuel into an engine cylinder. On a four-cylinder engine, the plunger rotates one-quarter of a turn as it moves from bottom to top dead center. On a six-cylinder engine, the plunger rotates one-sixth of a turn.

As the plunger moves from TDC to BDC, a metering slit in the plunger opens the inlet bore in the distributor head. Pressurized fuel in the pump cavity flows through this bore into the high pressure chamber at the end of the distributor plunger. As the plunger moves toward TDC, the plunger closes the inlet bore and opens an outlet bore in the distributor head. Pressure built-up in the chamber and the interior bore now opens the valve. Fuel is forced through the pressure line to the injection nozzle.

The effective stroke is completed when the cutoff bore of the plunger reaches the edge of the control spool. After this point, no more fuel is delivered to the injector and the pressure valve closes the line. Fuel returns to the pump cavity through the cutoff bore as the plunger reaches the TDC position. As the plunger returns, its transverse bore closes at the same time the next control slit opens the fuel inlet bore. The high pressure chamber once again fills with fuel and the injection cycle repeats for the next cylinder.

Pressure Valve

The pressure valve closes the injection line off from the pump. It also relieves the injection line by removing a preset volume of fuel after the delivery phase. The valve is a fluid controlled plunger opened by fuel pressure and closed by a return spring. The pressure valve controls the exact point at which the injector nozzle closes to cut off injection. During delivery, the valve is lifted from its seat by the high fuel pressure. Fuel then flows through longitudinal slits into a ring groove, through the pressure valve body, and into the high pressure fuel lines leading to the injection nozzle.

Governing

A gear set connects the pump's input shaft to the mechanical flyweight governor so it can be driven at camshaft speed. Governor operation influences the position of the control spool on the plunger. The governor cover forms the top of the distributor pump and also contains the full-load adjusting screw, overflow restriction, and engine speed adjustment screw.

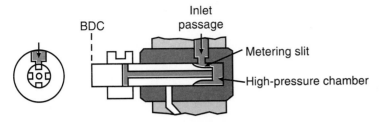

Entry of fuel.
With the plunger at BDC, fuel flows through inlet passage and metering slit into the high-pressure chamber.

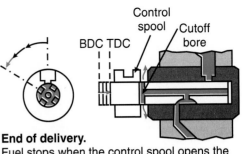

End of delivery.
Fuel stops when the control spool opens the cutoff bore.

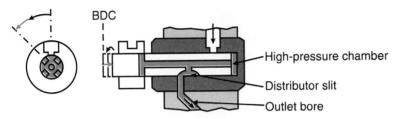

Fuel delivery.
Plunger motion toward TDC closes the inlet passage and pressurizes fuel in the high-pressure chamber. Rotary motion of the plunger causes a distributor slit to open the outlet bore to a cylinder

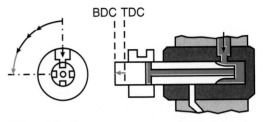

Entry of fuel.
As the plunger returns to BDC its rotary-reciprocating motion closes the cutoff bore and the high-pressure chamber fills again.

Figure 21-32. Stroke and delivery phases of the pump plunger.

Advance Timing Unit

The **advance timing unit** is located in the base of the distributor pump. The unit is driven by fuel pressure. The timing piston is guided by the pump housing and the housing is sealed by a cover on either side. A fuel inlet bore is located on one side of the timing piston. A preloaded spring applies opposing pressure to the other side of the piston. The piston is connected to the roller ring using a sliding block and an actuating pin.

The preloaded spring holds the timing piston in its no-run position. During operation, the fuel pressure applied to the timing piston is proportional to engine speed. This fuel pressure is regulated using a pressure control valve and an overflow restriction.

Fuel pressure increases as engine speed increases. At an engine speed of approximately 300 rpm, fuel pressure exceeds the spring preload pressure and the timing piston begins to move to the left as illustrated in **Figure 21-33.** The piston motion is transmitted through the sliding block and pin to the roller ring, which rotates in bearings. This changes the action of the cam plate and roller ring. The rollers now lift the rotating cam plate at an earlier point. The rollers and ring are turned by a specific angle relative

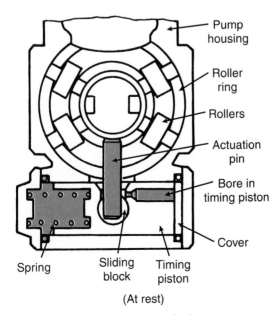

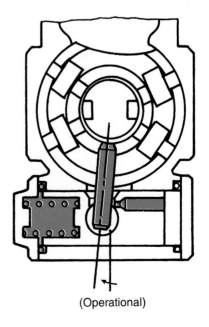

(At rest)   (Operational)

Figure 21-33. Operation of the timing advance device.

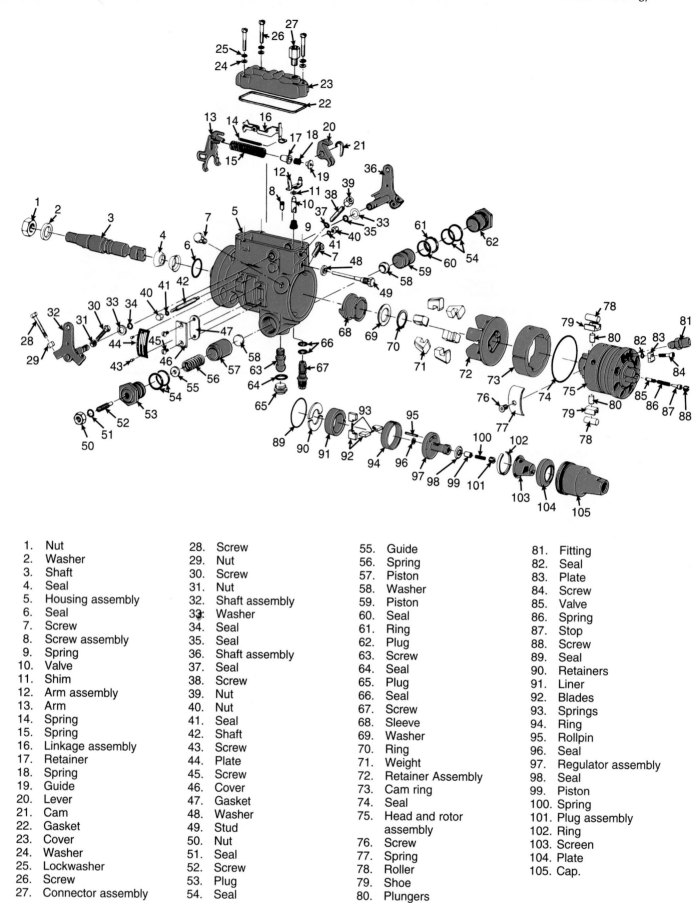

1. Nut	28. Screw	55. Guide	81. Fitting
2. Washer	29. Nut	56. Spring	82. Seal
3. Shaft	30. Screw	57. Piston	83. Plate
4. Seal	31. Nut	58. Washer	84. Screw
5. Housing assembly	32. Shaft assembly	59. Piston	85. Valve
6. Seal	33. Washer	60. Seal	86. Spring
7. Screw	34. Seal	61. Ring	87. Stop
8. Screw assembly	35. Seal	62. Plug	88. Screw
9. Spring	36. Shaft assembly	63. Screw	89. Seal
10. Valve	37. Seal	64. Seal	90. Retainers
11. Shim	38. Screw	65. Plug	91. Liner
12. Arm assembly	39. Nut	66. Seal	92. Blades
13. Arm	40. Nut	67. Screw	93. Springs
14. Spring	41. Seal	68. Sleeve	94. Ring
15. Spring	42. Shaft	69. Washer	95. Rollpin
16. Linkage assembly	43. Screw	70. Ring	96. Seal
17. Retainer	44. Plate	71. Weight	97. Regulator assembly
18. Spring	45. Screw	72. Retainer Assembly	98. Seal
19. Guide	46. Cover	73. Cam ring	99. Piston
20. Lever	47. Gasket	74. Seal	100. Spring
21. Cam	48. Washer	75. Head and rotor assembly	101. Plug assembly
22. Gasket	49. Stud	76. Screw	102. Ring
23. Cover	50. Nut	77. Spring	103. Screen
24. Washer	51. Seal	78. Roller	104. Plate
25. Lockwasher	52. Screw	79. Shoe	105. Cap.
26. Screw	53. Plug	80. Plungers	
27. Connector assembly	54. Seal		

Figure 21-34. *Exploded view of a distributor injector pump. This is a Stanadyne DB2 Opposed plunger injection pump. (Stanadyne Diesel Systems)*

to the cam plate and distributor plunger. This turning advances injection timing by as much as 12° of camshaft rotation or 24° of crankshaft rotation. **Figure 21-34** shows an exploded view of a distributor injection pump.

Electrical shut-off

Energized-to-shut-off (ETSO)

Energized-to-run (ETR)

Advance timing unit

Summary

Distributor pumps are compact injection pumps designed for applications with relatively broad speed requirements, such as medium to heavy-duty trucks, agricultural equipment, and tractors. A distributor injection pump uses a single pumping element to meter and deliver pressurized fuel to the fuel injectors at all cylinders.

Although distributor injection pumps all operate on the same basic principle of a single pumping element, there are design differences between manufacturers. This is particularly true in the way the pumping plunger or plungers are activated and synchronized to the combustion cycle. Some distributor pumps activate plunger motion through the use of a camshaft with cam lobes.

Opposed plunger distributor injection pumps use a cam ring with internal lobes to activate plunger action. Bosch Type VE distributor pumps use a cam plate design to synchronize plunger movement with the engine.

A transfer pump is used to deliver fuel at low pressure to the distributor injection pump housing. Depending on the distributor pump, fuel metering can use port and helix, sleeve metering or inlet metering. Most distributor pumps are equipped with a timing advance device and an automatic shut-down mechanism.

As always, removal and installation require special attention in regard to cleanliness. The distributor pump must also be properly timed to the engine. Disassembly, service, and calibration of distributor injection pumps requires specialized tools and training.

Important Terms

Distributor injection pumps

Transfer fuel pump

Hydraulic head

Discharge ducts

Distributing annulus

Plunger

Delivery valve

Inlet ports

Distributing slot

Outlet ports

Metering sleeve

Effective stroke

Intravance camshaft

Lift-to-port closure dimension

Rated speed

Cam nose departure

Peak torque set point

Droop screw

Opposed plunger inlet metered pump

Charging ports

Pressure regulating valve

Delivery valves

Advance mechanism

Review Questions—Chapter 21

Do not write in this text. Place your answers on a separate sheet of paper.

1. Which of the following can be used to synchronize the pumping plunger of the distributor pump with the timing of the diesel engine?
   (A) Camshaft and cam lobes.
   (B) Cam ring with internal lobes.
   (C) Cam plate.
   (D) All of the above.

2. The simple design of distributor injection pumps requires that they be overhauled using _____ tools and _____ .

3. An injection system equipped with a distributor pump requires the use of a _____.

4. In an AMBAC Model 100 pump, the head core, plunger, metering sleeve and delivery valve assembly are all contained in the:
   (A) governor assembly.
   (B) hydraulic head.
   (C) pump sump.
   (D) auxiliary housing.

5. On an AMBAC 100 distributor injection pump, the _____ automatically varies the timing of the fuel injection to match specific engine speeds.

6. Distributor pumps are compact injection pumps designed for applications with relatively _____ speed requirements.

7. Which of the following is true concerning Lucas DPA and Stanadyne DM and DB distributor fuel injection pumps?
   (A) They are opposed plunger, inlet metered pumps.
   (B) They use only one metering valve to meter the fuel.
   (C) A rotor distributes the metered fuel out through the hydraulic head to the injectors.
   (D) All of the above.

8. Which of the following fuel metering methods can be used with distributor injection pumps?
   (A) Port and helix.
   (B) Sleeve.
   (C) Pressure-time.
   (D) Both A & B.

9. The governor may be equipped with an electrical shut-off device that can operate on either a _____ or _____ principle.

10. Inlet metered, opposed plunger distributor pumps can have either _____ or _____ plungers.

ASE-Type Questions

1. Technician A says that a distributor injection pump uses multiple pumping elements to distribute fuel to all cylinders. Technician B says a distributor pump uses a single pumping element for all cylinders. Who is right?
   (A) A only.
   (B) B only.
   (C) Both A & B.
   (D) Neither A nor B.

2. Technician A says that an AMBAC 100 distributor injection pump plunger is lapped and mated into the head and metering sleeve, so these three components must always be serviced as a unit. Technician B says that in this type of pump, the plunger makes one complete revolution for every two revolutions of the camshaft. Who is right?
   (A) A only.
   (B) B only.
   (C) Both A & B.
   (D) Neither A nor B.

3. All of the following are components of a distributor fuel injection pump EXCEPT:
   (A) hydraulic head.
   (B) distributor rotor.
   (C) driveshaft.
   (D) control rack.

4. Technician A says some opposed plunger inlet metered distributor pumps use individual delivery valves installed in the hydraulic head outlet bolts for each cylinder. Technician B says all opposed plunger distributor pumps use a single delivery valve mount in a bore in the center of the distributor rotor. Who is right?
   (A) A only.
   (B) B only.
   (C) Both A & B.
   (D) Neither A nor B.

5. All of the following adjustments should be done when removing a distributor pump EXCEPT:
   (A) cap the high pressure injector lines.
   (B) perform a peak torque set point adjustment.
   (C) cap all openings on the injector pump.
   (D) place the engine in the number 1 firing position.

6. Technician A says it is safe to steam clean an engine equipped with an opposed plunger inlet metered distributor pump while the engine is running. Technician B says this procedure could severely damage the distributor pump by changing its temperature during operation. Who is right?
   (A) A only.
   (B) B only.
   (C) Both A & B
   (D) Neither A nor B

7. The governor weight retainer in an opposed plunger distributor pump is fixed to the _____.
   (A) hydraulic head
   (B) rotor drive end
   (C) pressure regulating valve
   (D) Both A & B.

8. In an opposed plunger distributor type pump the delivery valve(s) _____.
   (A) keeps the injection lines full of fuel so the next injection cycle for that cylinder can displace a full charge of fuel
   (B) rapidly decreases injection line pressure after injection so the nozzle snaps shut quickly
   (C) actually control the amount of use metered to the injector
   (D) Both A & B.

9. All of the following are bench tests performed on opposed plunger injection pumps EXCEPT:
   (A) transfer pump pressure check.
   (B) lift-to-port closure dimension.
   (C) electric shut-off check.
   (D) pressure regulator adjustment.

10. The opposed plunger inlet metered distributor injection pump is lubricated by _____.
    (A) diesel fuel supplied by the transfer pump
    (B) oil from the engine crankcase
    (C) oil from the pump's own reservoir
    (D) None of the above.

Chapter 22

Unit Injector Fuel Injection Systems

After studying this chapter, you will be able to:
- ❑ Describe the fuel flow routing in an engine equipped with unit injection.
- ❑ Name the four functions of the unit injector.
- ❑ Describe how the unit injector is mounted in the engine and how it is activated by rocker arm action.
- ❑ Explain the function and operation of the control rack and control tube.
- ❑ Properly remove and install a unit injector.
- ❑ Describe the steps required to clean, inspect, and test a unit injector.
- ❑ Locate a faulty injector on two-stroke engines.
- ❑ Perform a fuel spill back check.
- ❑ Perform an engine tune-up on a mechanical unit injection system.
- ❑ List the basic design characteristics of the Detroit Diesel Series 60 engine.
- ❑ Perform valve clearance adjustment and injector timing adjustment on Detroit Diesel Series 60 engines.

Unit fuel injectors are lightweight, self-contained assemblies. They can generate extremely high injection pressures. Their compact design and straight forward operation allow for simplified control and easy adjustment. Unit injectors can be mechanically operated or electronically controlled. This chapter will concentrate on mechanically operated unit injection systems. Refer to **Chapter 25** for information on electronic unit injectors.

Basics of Unit Injection (Two-Stroke Engines)

As mentioned in **Chapter 17,** unit injection is classified as a low pressure fuel delivery system. Fuel is delivered to the unit injectors at pressures typically between 50-75 psi (345-515 kPa).

Unit fuel injection systems were first developed and marketed by the Detroit Diesel Corporation (DDC). Unit injection was offered on the company's first engine, the two-stroke series 71™, in 1937. Other two-stroke unit injection engines, such as the series 53™, series 149™, and series 92™ solidified DDC as the industry leader in both unit injection systems and two-stroke engines. Detroit Diesel also produces four-stroke, unit injected engines, the series® 50 and series® 60. Other major engine and injection pump manufacturers such as Caterpillar, Cummins, and Robert Bosch produce and use unit injectors.

Fuel is drawn from the supply tank, through the fuel strainer, and enters the fuel pump at the inlet side where it is pressurized. Upon exiting the pump, the pressurized fuel is forced through the fuel filter and into the manifold inlet port. To avoid confusion when the fuel lines are installed, some manifolds have the words *in* and *out* cast or stamped on the side of the cylinder head. Once in the manifold, fuel passes through jumper pipes into the inlet side of each unit injector. Fuel flows through the injector where a certain amount is metered and injected into the cylinder's combustion chamber at pressures up to 3300 psi (22 750 kPa), **Figure 22-1.**

Fuel that is not injected circulates through the injector, helping to cool and lubricate it. Fuel then exits the injector through the return jumper line to the return manifold. Fuel flows through the return manifold and out of the restrictor fitting to the fuel return line. The restrictor fitting at the end of the return fuel manifold maintains the manifold pressure inside the cylinder head.

Mechanical Unit Injectors (MUI)

The *mechanical unit fuel injector (MUI),* **Figure 22-2,** is a lightweight, compact unit. Its simple design and operation provides for simplified controls and easy adjustment. The unit injector performs four functions:
- ❑ Creates the high pressures required for proper fuel injection.
- ❑ Accurately times the moment of fuel injection.

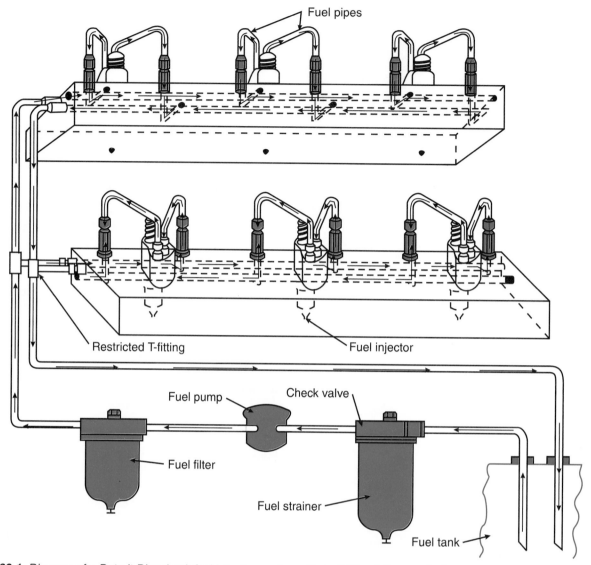

Fuel pipes

Restricted T-fitting

Fuel injector

Fuel pump

Check valve

Fuel filter

Fuel strainer

Fuel tank

Figure 22-1. Diagram of a Detroit Diesel unit fuel injection system. (Detroit Diesel Corporation)

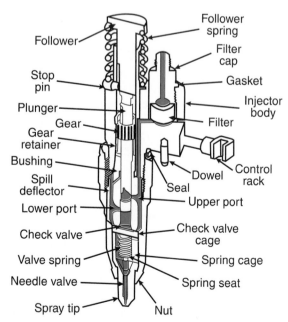

Follower

Follower spring

Filter cap

Stop pin

Gasket

Plunger

Injector body

Gear

Filter

Gear retainer

Bushing

Spill deflector

Control rack

Lower port

Dowel

Seal

Upper port

Check valve

Check valve cage

Valve spring

Spring cage

Needle valve

Spring seat

Spray tip

Nut

Figure 22-2. Cutaway view of a mechanical unit injector. (Detroit Diesel Corporation)

❑ Meters and injects the amount of fuel required to maintain engine speed and to handle the engine load.

❑ Atomizes the fuel for mixing with the air in the combustion chamber.

Metering and timing is accomplished through the use of an upper and lower helix machined in the lower end of the injection plunger. **Figure 22-3** illustrates the phases of injector operation.

Fuel flow through the injector is continuous to prevent air pockets from forming in the fuel system. To vary the engine's output, injectors having different fuel output capacities are used, based on application. The injector's fuel output is governed by the plunger's effective stroke and the spray tip's flow rate. Since the helix angle and the plunger design determines the operating characteristics, it is very important that the injector specified for the engine be used. If more than one type of injector is mistakenly installed in an engine, erratic operation, serious engine damage, or damage to the driver equipment can result. Always check the injector identification data located on the circular disc pressed into the recess of the injector body.

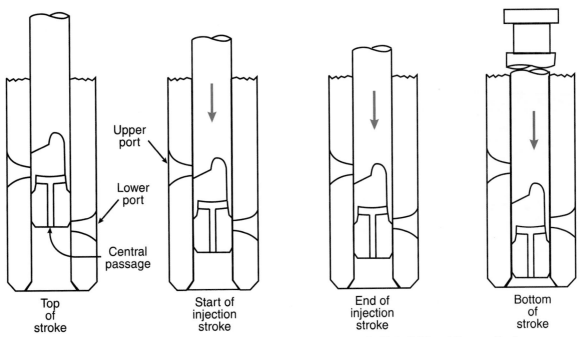

Figure 22-3. *Phases of unit injector operation through the vertical plunger travel. (Detroit Diesel Corporation)*

Each unit injector is equipped with a **control rack.** Control rack movement changes the upper and lower helix positions and alters the plunger's effective stroke. This changes the amount of fuel delivered by the injector. Each injector control rack is actuated by a lever on the **injector control tube.** As the governor reacts to changes in engine speed, its motion is transferred to the control tube and the injector control rack. The levers on the injector control tube can be adjusted independently so that each unit injector delivers the same amount of fuel for a given engine speed. Electronically controlled unit injectors do not require this adjustment.

A roller cam follower, pushrod, and rocker arm are used to operate the injector. The cam follower and pushrod are threaded into the rear of the rocker arm. The threaded pushrod and locknut setup provides for adjustment of the injector follower-to-body height. Fuel is supplied to and from the injector by two **jumper lines.** One jumper line connects to the fuel inlet stud and the other connects to the fuel return stud. The injector's inlet stud is always located directly above the fuel rack. It is on the right hand side of the injector body when viewed from the rack control end. The inlet stud contains a fuel filter underneath it. If the inlet and return lines were mistakenly switched during installation, dirty fuel could enter the injector, causing serious damage.

Unit injectors are accessed by removing the cylinder head cover, **Figure 22-4.** Each injector mounts into a copper tube located in the cylinder head. Coolant circulates around the outside of the tube. To ensure the injector is positioned correctly, a dowel pin on the unit injector slips into a locator hole in the cylinder head. Each injector is secured to the cylinder head using a single bolt and clamp, **Figure 22-5.** The injector control tube is connected to the governor by means of a **fuel rod,** **Figure 22-6.**

Figure 22-4. *The unit fuel injectors are accessed by removing the cylinder head cover.*

Figure 22-5. *Injectors, rocker arm assemblies, and control tube assembly for one bank of a Detroit Diesel series 92 V-8 engine.*

Figure 22-6. *Injector control tube, lever, and fuel rod to the governor.*

Mechanical Unit Injector Operation

Low pressure fuel enters the unit injector through a filter cap and filter positioned over the control rack. From the filter, fuel passes through a drilled passage into the supply chamber. The supply chamber is the area between the plunger bushing and the spill deflector. Fuel also enters the area under the **injector plunger** within the bushing. The plunger moves up and down in the bushing and is supplied with fuel through two funnel-shaped ports in the bushing wall.

The injector rocker arm motion is transmitted to the plunger by the follower. In addition to up and down movement, the plunger can also be rotated around its axis by a gear that meshes with the injector control rack. Rotating the plunger changes the positions of the helix grooves with regard to the bushing ports.

As the plunger moves down under the rocker arm pressure, fuel enters the supply chamber through the lower port in the bushing until it is covered by the plunger. The fuel below the plunger moves through a central passage in the plunger, into the fuel metering recess, and into the supply chamber. The supply chamber continues to receive fuel through the upper port until it is covered by the plunger upper helix. With the upper and lower ports both covered, the remaining fuel trapped under the plunger is subjected to increased pressure by the downward plunger movement.

When enough pressure builds up, it opens the flat check valve. Fuel in the check valve cage, spring cage, tip passages, and tip fuel cavity is compressed until the pressure on the needle valve is sufficient to open the valve. As soon as the needle valve opens, the fuel is forced through the small orifices in the spray tip and atomized as it enters the combustion chamber. When the plunger lower land uncovers the bushing lower port, the pressure below the plunger is relieved, and the valve spring closes the needle valve, ending injection.

The injector plunger is then returned to its original position by the follower spring. As the plunger moves upward, the high pressure cylinder within the injector is again filled with fuel. This constant circulation of fresh, cool fuel through the injector renews the fuel supply in the chamber, cools the injector, and removes any air that could interfere with accurate fuel metering. The injector's fuel outlet opening is directly adjacent to the inlet opening. Excess fuel passes through the outlet and returns to the fuel supply tank.

Needle Valves

Needle valves open at pressures between 2300-3200 psi (15 860-22 750 kPa). The final spray-in pressure inside the combustion chamber is close to 20,000 psi (137 900 kPa). Operating at these high pressures produces a clean, fast closing of the valve. The needle valve also uses a *valve cover orifice nozzle* design, **Figure 22-7.** With this type of nozzle, no fuel remains beneath the valve seat, preventing dribble into the combustion chamber. In the event the valve is accidentally held open by a small particle of carbon or dirt, a check valve directly below the bushing prevents leakage from the combustion chamber into the fuel injector.

Metering

Metering is accomplished by changing the position of the helices. Rotating the plunger retards or advances the port closing and the beginning and end of the injection period. As shown in **Figure 22-8,** when the control rack is pulled out all the way (no injection), the upper port is not closed by the helix until after the lower port is uncovered. With the rack in this position, the fuel is forced back into the supply chamber and no injection takes place. With the control rack pushed all the way in (full injection), the upper port closes shortly after the lower ports are covered. This produces the maximum plunger **effective stroke** and maximum fuel injection. There is an infinite number of fuel metering settings possible, from no injection to full injection.

Matching Fuel Flow Rates

Unit injectors are manufactured to exacting tolerances. Even so, the fuel flow rate can vary slightly for one unit injector model to the next. For each model of injector, Detroit Diesel specifies an acceptable fuel flow range. For example, the manufacturer's acceptable flow for a given model of injector may be between 85-90 cc when tested on a calibrated test jig. This does not mean it is acceptable to install an 85 cc injector and a 90 cc injector on the same engine.

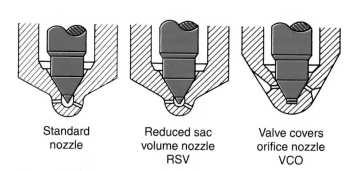

Figure 22-7. *Unit injector valve orifice designs.*

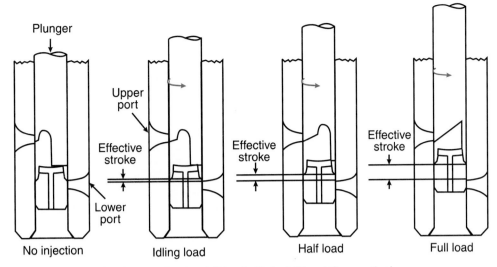

Plunger

Upper
port

Effective
stroke

Effective
stroke

Effective
stroke

Lower
port

No injection   Idling load   Half load   Full load

Figure 22-8. Unit injector fuel metering from no-load to full-load. (Detroit Diesel Corporation)

Matched sets of injectors are available through Detroit Diesel dealers and distributors. In a matched set, all injectors have the same flow range. For example, a matched set of injectors may all flow at an 87 cc rate. Matched injector sets provide the smoothest possible idle and acceleration under load. When obtaining an exchange set of unit injectors, *always specify a matched set.*

Timing Height

When a Detroit Diesel mechanical unit injector is assembled, timing marks on the control rack and the internal gear that the plunger slides through are aligned to set initial timing. After the injector is installed in the cylinder head, a specific *timing height* must be set using a timing pin or a dial gauge.

This timing height determines the exact moment the upper port in the bushing closes and injection begins. If the correct timing height is not set, the engine will run rough and perform poorly under load. Continued operation with incorrect timing can result in engine damage. The timing height specification is listed in the engine service manual and on the engine plate located on the rocker cover.

Isolating a Faulty Injector

On some Detroit Diesel two-stroke mechanical unit injector equipped engines, it is possible to short out each unit injector individually and note its effect on engine idle. To do this, push and hold the injector follower down using a large screwdriver, prybar, or special tool while the engine is idling, **Figure 22-9.** This action prevents the injector lower port from closing, stopping the injector's fuel delivery.

Note any effect shorting out each injector has on the engine. If the engine runs rougher and rpm drops, the injector is good. If there is little or no change in engine smoothness and rpm speed, the injector is defective. Remember, shorting out a good injector adversely affects the engine. Shorting out a defective injector will have little or no effect on the engine.

Injector
follower

Figure 22-9. Shorting out injector by wedging the injector follower using a large screwdriver.

 Caution: Do not short out an injector for an extended period of time.

Another method of isolating a suspect injector is to measure the individual exhaust manifold outlet temperatures using a contact pyrometer. Temperature differences of more than 50°F (10°C) may indicate a faulty injector. Before condemning any injector, however, make certain the cylinder compression is within specifications. Leaking piston rings can create problems similar to a faulty injector.

 Caution: Detroit Diesel four-stroke engine mechanical unit injectors cannot be shorted out using this method. Holding the injector follower down may cause the pushrod to fly out of the engine at high speeds or may result in pushrod damage. Electronically controlled unit injectors for both two- and four-stroke engines can be shut off temporarily using the Detroit Diesel Electronic Control System, referred to as *D-Deck* (DDEC) or an appropriate electronic diagnostic scan tool.

Injector Service (Two-Stroke Engines)

To remove a unit injector from a two-stroke engine. the following procedure should be used:

1. Clean the area around the rocker cover to prevent dirt from entering the engine. Remove the rocker cover and discard the cover gasket.

2. Remove the fuel pipes (jumper lines) from the injector and the fuel connectors, **Figure 22-10.** Cover the injector openings with shipping caps to prevent dirt from entering the injector. Cover all open fuel pipes and connectors to prevent dirt from entering the fuel system.

3. Manually turn the engine over until the flats across the rocker arms (at the pushrod) are all aligned or horizontal. If a wrench is used to turn the crankshaft bolt at the front of the engine, do not turn the crankshaft in a left-hand direction as the bolt could be loosened.

> ⚠️ **Warning: Keep hands and clothing away from the engine's moving parts when manually turning the engine. There is a remote possibility the engine could start.**

4. Remove the two rocker shaft bracket bolts and swing the rocker arms away from the injector and valves, **Figure 22-11.** Remove the injector clamp bolt, washer, and clamp.

5. Loosen the inner and outer adjusting screw or screws and locknut on the injector rack control lever and slide the lever away from the injector.

6. Insert an injector removal tool or small heel bar (also referred to as a ladyfoot) beneath the injector body, **Figure 22-12A.** Do not exert any pressure on the control rack. Lift the injector from its seat in the cylinder head, **Figure 22-12B.**

7. Immediately cover the injector hole in the cylinder head to keep dirt out.

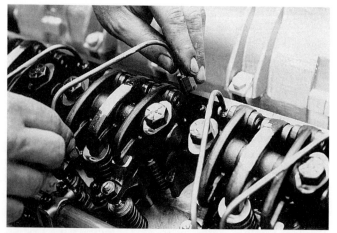

Figure 22-10. *Removing the fuel lines to the injector. Be sure to use a flare-nut crowfoot to loosen the lines.*

Figure 22-11. *Rocker arms swung up and out of way of injectors.*

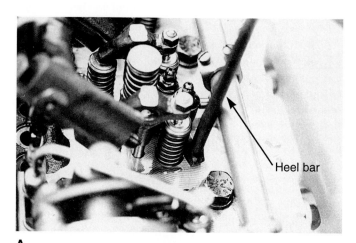

A

B

Figure 22-12. *A—A heel bar can be used to lift the unit injector from its seat. B—Use care when removing the injector from the cylinder head.*

Inspection and Testing

Each unit injector must be cleaned, inspected, and tested to determine if it can be reused or requires a complete overhaul. Many engine shops are equipped with the test equipment needed to perform plunger and rack movement, injector leakage, spray pattern, and fuel flow tests.

Injectors that pass these tests and show no signs of damage or excessive wear can be reused. Unit injectors that fail any of these basic tests or show signs of wear or damage must be overhauled. Clean the exterior of the injector with solvent. Submerge the injector in a container of solvent to wash it. Blow dry the injector using compressed air.

> ⚠ **Warning: To prevent possible personal injury, wear adequate eye protection and do not exceed 40 psi (275 kPa) when blow drying the injector.**

Once the injector is clean, inspect the following parts for wear, rust, corrosion, and abrasion:

- ❑ Follower spring.
- ❑ Injector body.
- ❑ Body nut.
- ❑ Spray tip.
- ❑ Plunger and bushing assembly.
- ❑ Check valve and seat.
- ❑ Spring case.
- ❑ Spill deflector.
- ❑ Injector rack.
- ❑ Filter caps.
- ❑ Follower top.
- ❑ Spray tip orifices.

Plunger and Control Rack Movement Test

The *plunger and control rack movement test* checks the plunger and rack's ability to move freely. Turn the injector upside down and watch the rack. It should fall freely under its own weight. Apply light pressure with the palm of your hand to the top of the injector follower. If the rack falls freely, the injector passes the test. If the rack does not fall freely, several problems may exist. The plunger may be scored or the body, bushing, or nut may be misaligned due to irregular or dirty parts. Disassembly and overhaul may be required.

> ⚠ **Warning: Diesel fuel under pressure can penetrate the skin, causing a serious infection. If injured by escaping fluid, seek medical attention immediately.**

Spray Pattern Test

The *spray pattern test* confirms fuel spray pattern uniformity and atomization. The unit injector is hydraulically clamped into position on the tester and primed with test oil. The injector rack is placed in its full-fuel delivery position. A manually operated pump is operated at 40-80 strokes per minute to simulate injector operation.

Observe the spray pattern to see that all nozzle tip orifices are open and delivering test oil in an even pattern.

Droplets should be finely atomized. When manual pumping stops, injection cut-off should be sharp and clean. Drops of test oil should not form at the nozzle tip after injection cut-off. If the fuel is atomized correctly and no dripping is observed, the injector has passed this test.

Pressure and Leakage Test

A *pressure and leakage test* determines if the plunger body-to-bushing mating surfaces are sealing properly and confirms plunger-to-bushing fit. It should be performed after the spray pattern test.

1. Clamp the injector in the tester as in the spray pattern test.
2. Close the through-flow valve, but do not overtighten it.
3. Set the plunger position lever to its rear horizontal position.
4. Operate the manual pump lever until the gauge reaches 100-200 psi (690-1375 kPa). Inspect for leakage at the injector nut seal ring. Increase pressure to 1500 psi (10 335 kPa) and check for leaks at the filter cap gasket and the body plugs. Note the time it takes for the pressure to drop from 1500 to 1000 psi (10 335 kPa to 6890 kPa). This drop should take 7 seconds or longer. A shorter time indicates the body-to-bushing mating surfaces are not sealing properly.
5. After testing, open the through-flow valve to relieve injector pressure. Carefully unclamp and remove the injector from the tester.

Fuel Output Test

The *fuel output test* is performed by installing each injector into the test stand or calibrator and measuring fuel delivery rate against listed specifications. As mentioned earlier, there is an acceptable fuel delivery range for each model of unit injector. However, perfectly matched injectors provide the finest performance.

Injector Installation

Prior to installing a unit injector into the cylinder head, remove the carbon deposits from the beveled injector tube seat. This ensures correct injector alignment and prevents any undue stress from being exerted against the spray tip.

Clean the seat using the recommended reamer tool or a cylindrical wire brush. Pack the flutes of the reamer with clean grease so that any carbon removed will be caught in the grease and not fall into the engine. Work carefully to remove only the carbon buildup; do not scratch the copper tube surface, which can change the tip protrusion setting.

1. Fill the injector with clean fuel, adding fuel at the inlet until it flows from the outlet.
2. Slide the injector into the tube so the dowel pin in the injector body aligns with the locating hole in the cylinder head. Slide the injector rack control lever over so that it aligns with the injector rack.

3. Install the injector clamp and washer. Make sure the washer's curved side faces toward the injector clamp. Install the bolt and tighten to specifications.

 Caution: Do not overtighten. Make sure that the injector clamp does not interfere with the injector follower spring or the exhaust valve springs. Use a torque wrench.

4. Check the injector control rack for free movement. Excessive torque on the clamp bolts can cause the control rack to stick or bind.

5. Install the exhaust valve bridges over their guides (four valve heads) and onto the valve stem tips. Place the rocker arm assembly in position and secure the rocker arm brackets to the cylinder head. Torque the hold-down bolts to specifications, **Figure 22-13**.

6. Align the fuel pipes and connect them to the injectors and fuel connectors. Do not bend the fuel lines. Avoid overtorquing connections as damage and leakage could result. Fuel leakage from damaged or improperly installed fuel pipes can cause lube oil dilution, which may result in serious engine damage.

 Caution: Do not reuse the original fuel pipes. Install new fuel pipes using a flare-nut wrench crowfoot and torque wrench, Figure 22-14.

If a new set of injectors has been installed, the fuel system should be primed and a complete engine tune-up performed. If only one injector has been removed and replaced, it is only necessary to adjust valve clearance, time the injector for that one cylinder, and re-adjust the injector rack control lever. Depending on the location of the new injector, a complete tune-up may be needed.

Figure 22-13. *Use a torque wrench to tighten rocker arm brackets to specifications.*

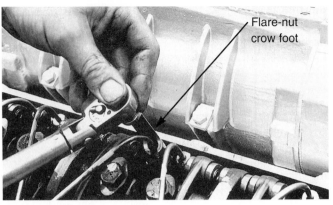

Figure 22-14. *Installing and torquing fuel pipes with a flare-nut crowfoot. Double-check the torque when tightening fuel pipes.*

Fuel Spill Back Check

On DDC unit injection two-stroke engines, a high volume of fuel is delivered to the supply manifold and unit injectors. Very little of the total fuel volume is actually injected into the cylinder as most is returned to the fuel tank. This fuel flows through a restricted fitting in the cylinder head fuel return manifold before entering the fuel return lines. Forcing the fuel through this small opening helps maintain fuel pressure in the system.

 Note: Do not substitute a regular brass fitting for the restricted fitting. Poor engine performance or engine damage will result.

A *fuel spill back* check simply measures the volume of fuel returning to the tank in a given time. A return fuel line is disconnected at an accessible location and routed into a clean bucket or container. The engine is operated at a specified rpm (usually 1200 or 1800 rpm), and the volume of fuel flowing into the container in one minute is recorded and checked against specifications.

By immersing the return line in the fuel collecting in the container and watching for bubbles, you can also determine if air is entering the fuel system on the suction side of the fuel pump. If bubbles are present, check all fuel line connections and filter seals for air tightness.

If the fuel returned is less than specifications, the primary or secondary fuel filters may be clogged. Fuel line diameter also may be too small, particularly if the fuel tank is more than 20′ (6 m) away from the fuel pump. Make sure the fuel inlet line is correctly sized for the engine and application. The relief valve in the fuel pump may be stuck partially open or the pump itself may be defective.

Engine Tune-up Procedure

The *tune-up* sequence outlined in the following sections is typical of two-stroke mechanical unit injection engines. Refer to the engine service manual for all tune-up procedures and specifications. It is important to follow the tune-up sequence in the order listed. On DDC tailored torque governor equipped engines, back off the Belleville washer retainer nut prior to performing the tune-up. On engines equipped with either a mechanical throttle delay mechanism or fuel modulator piston, remove these accessories to prevent any interference when setting the fuel racks and other adjustments.

Exhaust Valve Bridge Check

On four-valve two-stroke engines, the rocker arm pushes down on an *exhaust valve bridge,* which opens two exhaust valves at the same time. When cylinder head work has been performed, the bridges must be reinstalled on their original guides and their settings checked and adjusted as needed.

If the cylinder head has been removed from the block, the valve bridges can be adjusted once the cylinder head has been torqued into position and before the rocker arm assemblies are bolted back down.

Caution: Do not attempt to adjust the bridge while it is still on the guide. This will bend the guide and necessitate guide replacement.

To check and adjust the exhaust valve bridges on an engine in service, proceed as follows:

1. Manually turn the engine over to place the injector follower for that cylinder all the way down.
2. Using two separate .0015" feeler gauges, lift up on the bridge slightly and slip a .0015" feeler gauge between the bridge and both exhaust valve stem tips.
3. Lightly press down on the center of the bridge assembly. Pull on both feeler gauges. An even drag on both gauges indicates the bridge is properly adjusted. If the drag is uneven, adjust as outlined in steps 4-10.
4. Remove the fuel jumper lines and the rocker arm bracket hold down bolts to gain access to the individual cylinder bridges.
5. Place the valve bridge in a soft-jaw vise or appropriate fixture. Loosen the locknut on the bridge adjusting screw, and back out the adjusting screw 2-3 turns.
6. Install the bridge back onto its guide.
7. Firmly apply pressure to the bridge and turn the adjusting screw clockwise until it just contacts the valve stem, **Figure 22-15.** Turn the screw an additional one-eight to one-quarter turn and finger tighten the locknut.
8. Mount the bridge in a soft-jaw vise. Use a screwdriver to hold the adjustment screw as the locknut is torqued to specifications.
9. Lubricate the bridge and guide with engine oil and reinstall it in its original position. Check exhaust valve

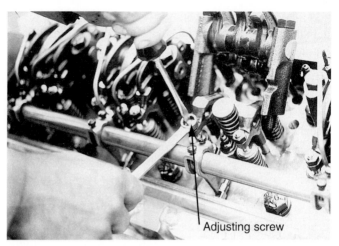

Figure 22-15. *Exhaust valve bridge adjustment. Do not use the wrench to tighten the locknut.*

clearance as outlined in steps 1 through 3. Readjust the screw as needed.
10. Use the same procedure to adjust the remaining valve bridges. When the rocker arm assemblies are brought into position, be sure the valve bridges are properly positioned on the rear valve stems. If they are not positioned properly, the valve and bridge may be damaged during engine operation.

Setting Exhaust Valve Clearance

To adjust the *exhaust valve clearance,* thoroughly clean all dirt from the rocker covers and remove them from the engine. Place the governor speed control lever in the idle speed position, or, if a stop lever is provided, secure it in the stop position. Manually turn the engine until the injector follower is fully depressed on the cylinder to be adjusted. Loosen the exhaust valve rocker arm push rod locknut and place the appropriate size feeler gauge between the valve bridge and the valve rocker arm pallet, **Figure 22-16.** Adjust the push rod to obtain a smooth pull on the feeler gauge. Once the adjustment is satisfactory, remove the feeler gauge

Figure 22-16. *Check and adjust valve clearance using a feeler gauge.*

and hold the push rod with a 5/16" (8 mm) wrench as you tighten the 1/2" (13 mm) locknut. Recheck the clearance to ensure it meets engine manual specifications.

It is recommended that **go/no-go feeler gauges** be used. For example, if setting the required valve clearance is .016", a .015-.017" go/no-go feeler gauge is needed. The initial setting is made using the .017" portion of the feeler gauge. Once the locknut is tightened, the .017" portion of the feeler gauge should not pass through the clearance. However, the .015" portion of the gauge should pass through, indicating the correct .016" clearance has been set. Readjust the push rod if the setting is not correct. Once the correct cold setting is made, run the engine to reach normal operating speed and recheck valve clearance. Hot engine valve clearance is slightly less than cold engine specifications, **Figure 22-17.** Adjust the exhaust valve clearance for all cylinders using this method, **Figure 22-18.**

Injector Timing

To time a unit injector properly, the follower must be adjusted to a definite set height in relation to the injector body. All the injectors must be timed in their correct firing order during one complete crankshaft revolution. After the exhaust valve clearance for each cylinder has been adjusted, time the injector as follows:

1. Place the governor speed control lever in the idle speed position. If a stop lever is provided, place it in the stop position.
2. Rotate the crankshaft manually until the exhaust valves are fully depressed on the cylinder to be timed.
3. Place the small end of the injector timing gauge in the top of the injector body. The flat part of the gauge should be toward the injector follower, **Figure 22-18A.** Using the wrong injector timing gauge can result in serious engine damage.
4. Loosen the injector rocker arm push rod locknut and turn the push rod. Adjust the injector rocker arm until the extended part of the gauge will just pass over the top of the injector follower.
5. Hold the push rod and tighten the locknut, **Figure 22-18B.** Check the adjustment and, if needed, readjust the push rod.
6. Time the remaining injectors in the same manner as outlined above.

Governor Gap Adjustment

Detroit Diesel mechanical governors require the correct **governor gap** linkage adjustment. The purpose of governor gap is to ensure that when the engine throttle is moved from idle speed, engine speed control transfers from the governor to the operator. Governor gap settings are more critical on limiting speed mechanical governors than on variable speed governors. As explained in **Chapter 19,** a limiting speed governor controls only the low and high speed ranges with all intermediate speeds under operator control.

The governor gap divides the travel of the flyweights between the low and high speed spring systems. Governor control is not lost until the engine speed accelerates sufficiently to allow the flyweights to fully compress the low speed spring. When this occurs, there must be a small

Valve Clearance Specifications			
Engine	Cylinder head	Cold Setting	Hot Setting
Series 53	Two-valve Four-valve	.011" (.28 mm) .026" (.66 mm)	.009" (.23 mm) .024" (.61 mm)
Series 71	Two-valve Four-valve	.012" (.31 mm) .016" (.41 mm)	.009" (.23 mm) .014" (.36 mm)
Series 92	Four-valve	.016" (.41 mm)	.014" (.36 mm)
Series 82L	Intake valve Exhaust valve	.012" (.31 mm) .020" (.51 mm)	.011" (.28 mm) .012" (.31 mm)
Series 60	Intake valve Exhaust valve	.008" (.20 mm) .020" (.51 mm)	No Hot Setting

Figure 22-17. Valve clearance specifications for Detroit Diesel engines equipped with unit injection. (Detroit Diesel Corporation)

A

B

Figure 22-18. A—Installing a fuel injector timing gauge. B—After adjustment, tighten the push rod locknut.

enough gap between the low speed spring cap and the high speed spring plunger to allow transfer of control to the operator without resistance from the governor linkage or springs. The speed at which this occurs is usually between 1100-1400 rpm on late model engines, but may be lower (800-1000 rpm) on older engines.

Adjusting Limiting Speed Governors

Detroit Diesel engines use both *single weight limiting speed (SWLS) governors* and *double weight limiting speed (DWLS) governors.* DWLS governors have a gap setting of between .003-.019″ (.07-.48 mm) while SWLS governors have a setting of .2″ (5.08 mm). Always check the engine service manual for gap specifications. Gap adjustment procedures may vary slightly between governor models. It is important to follow service manual instructions.

Begin by ensuring the engine idle speed is set to specifications and adjust if needed. Before making the gap adjustment, remove the governor cover and gasket and turn the buffer screw approximately 5/8″ away from the locknut. On turbocharged engines, back out the starting aid screw. Depending on the engine, loosen the throttle delay or fuel modulator clamp. On DDC tailored torque engines, loosen the Belleville washer retainer adjusting nut to provide a clearance of .060″ (1.52 mm) between the nut and the washers, **Figure 22-19.**

On V-type engines using DWLS governors, a special *wedge tool* can be used to measure governor gap with the engine stopped, **Figure 22-20.** Manually turn the engine until the large low speed weights are in a horizontal position. Insert the governor wedge tool between the weights and the riser shaft. The wedge's tapered face should be against the riser shaft and midway between the flanged ends. Push the wedge down as far as it will go to bottom the weights on the carrier, **Figure 22-21A.** With the wedge tool in place, check the gap between the low speed spring cap and the high speed plunger, **Figure 22-21B.** Specifications for this gap are between .003-.019″ (.07-.48 mm). If the governor gap is not within specifications, loosen the

Figure 22-20. *Governor adjustment wedge tool.*

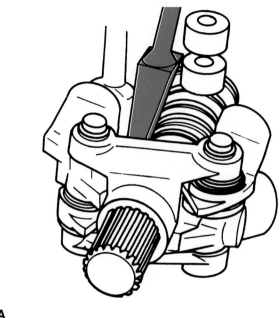

A

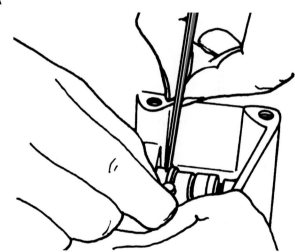

B

Figure 22-21. *A—Install the wedge tool as shown here when setting governor gap. B—When the wedge tool is in place, check the governor gap with a feeler gauge.*

gap locknut and rotate the governor gap adjusting screw clockwise or counter-clockwise until the gap measures .008″ (.2 mm).

If the wedge tool is not available, the governor gap on a DWLS governor V-type engine can be checked by operating the engine at approximately 1200 rpm and checking the gap with the proper feeler gauge. On DWLS inline engines, the weights cannot be accessed to insert the wedge

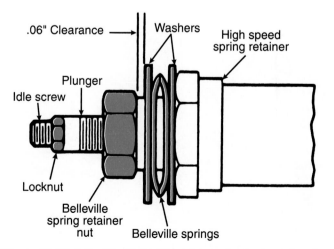

Figure 22-19. *Adjust the Belleville spring pack prior to performing a tune-up on engines with tailored torque governors. (Detroit Diesel Corporation)*

tool, so the running engine/feeler gauge method must be used on these engines. A cutaway governor cover allows the engine to run while checking the governor gap with a feeler gauge. If a cutaway cover is not used, the differential lever and the fuel rod must be removed. Removing these components eliminates the governor's ability to control engine rpm, so you must manually control engine speed by turning the fuel control tube. This is best accomplished by grasping the control tube with a set of locking pliers. Do not use too much clamping pressure as the tube could be damaged. Insert the feeler gauge between the low speed spring cap and the high speed plunger.

On SWLS governors, the gap is set with the engine stopped. The weights are not touched. The gap is set to .17" (4.32 mm) on inline engines and .2" (5.08 mm) on V-configured engines. Recheck and readjust the following settings whenever the governor gap is adjusted: buffer screw, idle speed, maximum no-load rpm, the starting aid screw, injector racks, and the throttle delay or fuel modulators assembly.

Injector Control Racks

The unit injector control racks are connected to the mechanical governor by one or more fuel rods. The injector racks must be set and adjusted to ensure they all react to governor action equally. If adjustment varies between racks, different amounts of fuel will be injected into each cylinder and the engine will perform poorly.

The injector rack closest to the governor is adjusted first. On inline engines, if the governor is located in the front, the number 1 cylinder rack is set first. If the governor is located in the rear (flywheel end), the rear cylinder is set first. On a V-engine with the governor located in front, the number 1 left bank injector rack is set first. If the governor is rear mounted, the injector rack for the last cylinder on the left bank would be set first.

Setting and Adjusting Injector Racks

The following conditions should still exist from setting the governor gap. The throttle delay mechanism should be removed or U-bolt clamp loosened. On engines so equipped, the Belleville spring retainer nut should still show a .060" (1.5 mm) clearance, the buffer screw should be backed out 5/8" (16 mm) from its locknut, and the starting aid screw on turbocharged engines should be backed out 5/8" (16 mm).

Begin by disconnecting any throttle linkage attached to the governor speed control lever. With the locknut against the high speed plunger, turn the idle speed adjusting screw until 1/2" of threads project from the locknut. This reduces the low speed spring tension so that it can be easily compressed. The low speed gap can now be safely closed without bending the fuel rods or damaging the yield link spring mechanism used on throttle delay engines.

 Note: Forgetting to back out the idle adjustment screw will result in an improper fuel rack setting.

Loosen the inner and outer adjusting screws of all injector rack control levers. Newer engines use only one inner adjusting screw with a locknut on it. Setting the first unit injector control rack properly is crucial. The first rack setting is referred to as the master rack setting, and all other injector racks will be set to this position. Once the first rack is set, adjust the remaining unit injector racks.

Adjusting the Number One Injector Rack

 Note: If the engine is equipped with a limiting speed governor, move the speed control lever to the maximum speed position and hold it there with light finger pressure or by securing the lever in position with a spring or length of elastic cord.

With the speed control lever in the full-fuel position, turn down the number 1 injector rack inner adjusting screw (two-screw type engines) or single adjusting screw, **Figure 22-22.** Continue turning the screw until the injector rack is almost against the injector body and the rack just begins to roll up slightly, **Figure 22-23.** An increase in the effort needed to turn the screwdriver will be felt at this point. To make the final adjustment on single screw racks, tighten the screw an additional one-eighth turn and lock the setting by tightening the adjusting screw locknut.

On a two-screw rack, turn the outer adjusting lock screw until it bottoms lightly on the injector control tube. Alternately tighten the inner and outer adjusting screws in one-eighth turn increments until both are snug. This should result in the governor linkage and control tube assembly being placed in the same position they will be in while the engine is running at full load.

Caution: Overtightening the injector rack control lever adjusting screws during installation or adjustment can result in damage to the injector control tube. The recommended torque of the adjusting screws is 24-36 in-lbs (3-4 N•m).

Figure 22-22. *Adjusting the number 1 injector rack control lever. All other injector rack adjustments are based off of this adjustment*

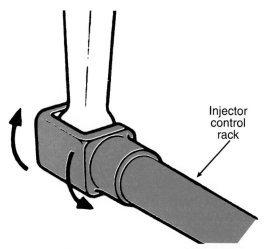

Figure 22-23. Checking the rotating movement of the unit injector control rack. (Detroit Diesel Corporation)

To check the proper rack adjustment, hold the control lever in the maximum speed position and press down on the injector rack with a screwdriver as shown in **Figure 22-24**. Note the rotating movement of the fuel control rack. The rack should tilt downward and spring back upward when the pressure of the screwdriver tip is removed. If the rack does not return to its original position, it is too loose. Back off the outer adjusting screw on two-screw racks very slightly and tighten the inner screw an equal amount. On single-screw racks, loosen the locknut, turn the adjusting screw clockwise very slightly, and tighten the locknut. Recheck the rack for spring back.

To check for an excessively tight setting, move the stop lever to the run position and observe for any tightness before it reaches the end of its travel. To correct this condition on two-screw racks, loosen the inner adjusting screw slightly and tighten the outer adjusting screw slightly. The ball end of the rack leg may experience scuffing on engines in service for an extended period. This can prevent good spring back or bounce when setting the racks. To remove this scuffing, loosen both rack screws and slide the rack leg lever to the side of the injector, and swing it

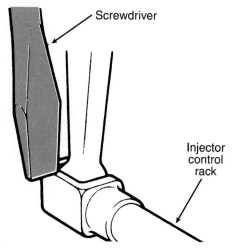

Figure 22-24. Checking the unit injector control rack spring back. (Detroit Diesel Corporation)

upward. Lightly rub the ball end with emery cloth to remove the scuffing.

To adjust the remaining injector racks on each bank, it is necessary to remove the **clevis pins** to disconnect the governor linkage. You can now treat each bank as a separate "inline" control tube and adjust all injector racks to the number 1 master rack setting.

Adjusting Injector Racks on Inline Engines

On inline engines, remove the clevis pin from the fuel rod and the injector control tube lever. This unlinks the governor from the control tube. Use the speed control lever to hold the injector control racks in the full-fuel position. Adjust the number 2 injector control rack, making sure proper rack roll-up and resistance is felt on the screwdriver.

Recheck the number 1 injector rack to be sure that it has remained snug while the number 2 rack was adjusted. If the number 1 injector rack has become loose, loosen the inner adjusting screw slightly on the number 2 injector rack control lever while slightly tightening the outer adjusting screw. On single screw racks, loosen the number 2 rack locknut, turn the adjusting screw clockwise slightly, and tighten the locknut. When the settings are correct, both injectors rack ball ends will be snug on their respective control levers. Adjust the remaining injector rack control levers in the same manner. Reconnect the fuel rod to the injector control tube lever.

Adjusting Injector Racks on V-Type Engines

V-engines have a separate fuel rod and control tube for each bank. Loosen the screws on both banks. Be certain the injector rack control levers are free on the control tube. Once the number 1 injector control rack is set on the left bank of a V-engine, it becomes the master rack setting. The number 1 injector control rack on the right bank is then adjusted to this master setting. To adjust the number 1 right bank control to match the number 1 left bank control rack setting, remove the clevis pin from the fuel rod at the left bank injector control tube lever and install the clevis pin in the fuel rod at the right bank injector control tube lever.

Adjust the number 1 right bank rack by turning down the adjusting screw(s) until proper rack roll up and screwdriver resistance is felt. Verify that the number 1 racks on both the right and left banks are adjusted the same by inserting the clevis pin at the left bank fuel rod and moving the speed control lever to the maximum speed position. The clevis pin drag on both racks should be the same. The bounce or spring back of both number 1 racks must also be the same. If they are not, the number 1 *right bank* injector rack must be readjusted to increase or decrease bounce as described above. Once the number 1 racks on the left and right banks are adjusted to the same setting, they become the master racks for their respective banks and all racks on the bank are adjusted to match them.

Checks After Adjustment

Once all injector racks are adjusted, reinstall the clevis pins in each fuel control rod. Check that each injector control rack has the same amount of bounce or spring back. Also check the drag on the clevis pin at each bank. If they are not the same, further adjustment is needed. Minimize this possibility by working carefully through the entire adjustment procedure.

Turn in the idle screw until the screw projects 3/16" (7 mm) from the locknut. This will allow the engine to start for idle adjustment and other settings. Double check that the racks will move to the no-fuel position when the stop lever is activated. Finally, replace the rocker valve cover or covers before starting the engine.

Setting Maximum No-Load Speed

A *maximum no-load setting* must be made on limiting-speed governors. The maximum no-load speed is matched to engine application and is initially set at the factory. To check it and adjust as needed, first make certain the buffer screw is still backed out 5/8" (16 mm) from the governor housing and locknut. Loosen the spring retainer locknut and back off the high-speed retainer nut five turns, **Figure 22-25A.** Operate the engine until it reaches normal operating temperature. With the engine running under no load, place the speed control lever in the full-fuel position. Turn the high-speed spring retainer nut clockwise until the engine is running at the recommended no-load rpm. To complete the adjustment, hold the high speed spring retainer and tighten the locknut.

Idle Speed

With the maximum no-load speed properly adjusted, the next step is to adjust *idle speed.* With the engine operating at normal operating temperature, and the buffer screw backed out to avoid contact with the differential lever, turn the idle speed adjusting screw, **Figure 22-25B,** until the engine operates at approximately 15 rpm below the recommended idle speed. It may be nec-

essary to use the buffer screw to eliminate engine roll. After the idle is set, be sure to return the buffer screw to its original 5/8" (16 mm) backed out position. When the proper idle is set, hold the idle screw and tighten the locknut. Install the high speed spring retainer cover and tighten the two bolts.

Buffer Screw

With the idle screw properly adjusted, the *buffer screw* can be adjusted to its proper setting. With the engine running at normal operating temperature, turn the buffer screw, **Figure 22-26,** in so it contacts the differential lever as lightly as possible and still eliminates engine roll or hunt. Do not increase the engine idle speed more than 15 rpm with the buffer screw. Recheck the maximum no-load speed. If it has increased by more than 25 rpm, back off the buffer screw until the increase is less than 25 rpm. After proper adjustment is made, hold the buffer screw and tighten the locknut.

Starting Aid Screw Adjustment

On turbocharged engines, a *starting aid screw* is used to limit the amount of fuel injected during start-up to reduce exhaust smoke. To properly adjust the starting aid screw, stop the engine, place the governor stop lever in the run position, and the speed control lever in the idle position. Check the engine emission label for guidance in selecting the proper starting aid gauge size.

Measure the gap between the injector rack clevis shoulder and the injector body at any cylinder location. Turn the starting screw in or out if adjustment is required. When the starting aid screw is properly adjusted, the area between the clevis shoulder and the injector body should have an end clearance of 1/64" (.397 mm). Hold the starting aid screw and tighten the locknut. Remove the gauge, move the speed lever from the idle position to the

A   **B**

***Figure 22-25.** A—Loosen the high speed retainer to adjust maximum no-load speed. B—Adjust engine idle speed by turning the idle adjustment screw.*

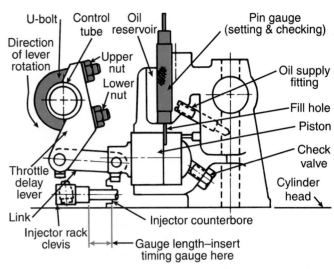

Figure 22-26. The buffer screw adjustment is used to stabilize the engine idle and eliminate roll.

maximum speed position, and back to idle. Recheck the gap and readjust as needed.

Throttle Delay Cylinder Adjustment

As described in **Chapter 19,** a throttle delay mechanism is often used on highway truck applications on both turbocharged and non-turbocharged engines to help reduce emissions during vehicle acceleration. It does this by retarding the fuel control tube and injector rack movement towards the full-fuel injection position when the engine is accelerating. The throttle delay mechanism consists of a rocker arm shaft bracket that uses a throttle delay cylinder, piston, throttle delay lever, connecting linkage, oil supply orifice plug, ball check valve, U-bolt, and a yield mechanism within the governor to prevent fuel rod bending or governor damage caused by accelerating too quickly.

Whenever the piston is removed from the delay mechanism, it must be inspected for burrs or rough edges. If any are found, remove them with a crocus cloth. There are two methods of adjusting the throttle delay mechanism. Late model engines use a throttle delay gauge and a special *go/no-go pin gauge.* The *throttle delay gauge,* **Figure 22-27A,** is inserted between the injector body and the clevis shoulder on the injector nearest the throttle delay cylinder, **Figure 22-27B.** The governor speed lever is then moved to the maximum speed position. The go/no-go pin gauge is now inserted in the cylinder fill hole, **Figure 22-28.**

Figure 22-28. Position of go/no-go pin gauge while adjusting the throttle delay cylinder. (Detroit Diesel Corporation)

The throttle delay lever is then rotated until the piston touches the end of the pin gauge. To finalize the setting, hold the throttle delay piston in this position and tighten the U-bolt nut while exerting a slight amount of torque on the throttle delay lever in the direction of rotation. To check the setting, attempt to reinstall the pin gauge in the cylinder opening. It should not slip in until the injector racks are moved toward the no-fuel position.

On earlier engine models, a larger cylinder opening is used and the go/no-go pin gauge is not used. Instead, the throttle delay gauge is inserted between the injector rack and body, and the piston is pushed manually into its cylinder until it is flush with the housing bore.

Fuel Modulator Adjustment

Some turbocharged DDC two-stroke engines with limiting speed governors are equipped with *fuel modulators* that mount on the cylinder head. Like a mechanical throttle delay, this device limits the movement of the fuel rack to the

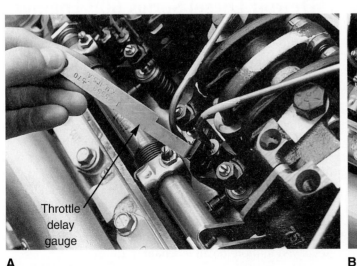

A

B

Figure 22-27. A—Mechanical throttle delay adjustment gauge. B—Position of the throttle delay gauge before adjustment begins.

full-fuel position during acceleration to help control exhaust smoke and emissions.

The modulator is made of a cast housing that holds a cylinder, piston, cam, and spring assembly, **Figure 22-29.** The modulator's lever and roller assembly connects to the injector fuel control tube. Pressurized air enters the housing from a flexible line coming from the turbocharger discharge side. This pressurized air activates the fuel modulator piston. An increase in turbocharger pressure pushes the piston and cam out of the cylinder against the return spring force. This allows the rack control tube to increase fuel supply in direct response to the turbocharger boost.

> **Note: To make this adjustment, it is necessary to bend a 3/8" wide strip of .017" (.43 mm) feeler gauge to take the shape shown in *Figure 22-29.* This gauge will be inserted between the cam and roller during this procedure.**

To adjust the fuel modulator, stop the engine and insert the correct rack gauge between the injector body rack and the rack shoulder, **Figure 22-30.** The placement is similar to that used for injector rack setting and adjustment. Place the rack gauge at a 45° angle forward of the fuel modulator. Hold the governor speed control (throttle) lever in the maximum speed position and the stop/run lever on top of the governor in the run position. The rack gauge should stand up by itself in the 45° position.

Loosen the modulator clamp screw and force the lever assembly forward until the roller contacts the cam. Insert the bent feeler gauge between the cam and the roller. Tighten the bolt only enough to allow the rack gauge to fall under its own weight from the 45° position.

Replace the .017" (.43 mm) feeler gauge with a .004" (.1 mm) feeler gauge and reinsert the rack gauge between

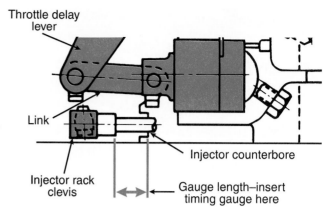

Figure 22-30. *Rack gauge position for adjusting the throttle delay assembly. (Detroit Diesel Corporation)*

the injector body and rack at the 45° angle. Gently tighten the bolt until the rack gauge once again falls under its own weight. To double check the setting, hold the speed control lever in the full-fuel position. When the specified rack gauge is installed between the body and rack, it should be capable of standing at the 45° angle. Now insert a .005" (.13 mm) feeler gauge between the fuel modulator roller and cam. The rack gauge should again fall from its 45° position under its own weight.

Final Tune-up Checks

After completing the tune-up adjustments on any engine, always be sure that the injector racks will move to the no-fuel position when the governor stop lever is moved to the stop position. If the linkage malfunctions during or after the engine is started, overspeeding and possible engine damage could result.

Also check that the governor speed control lever will move to the full-fuel position when the accelerator pedal or throttle lever is fully depressed or activated. If it does not, adjust the throttle linkage.

Detroit Diesel Series 60® Engines

The Detroit Diesel series 60 engine is a six-cylinder, four-stroke engine available in 11.1 and 12.7 liter (L) displacements. This engine features an overhead camshaft, parallel ports for the intake and exhaust valves, and turbocharged air-to-air charged air cooling. The overhead camshaft design eliminates all pushrods and lifters. It also results in stiffer fuel injection and valve operating mechanism that provides better control of the unit injectors and valve action. Unit fuel injector pressures as high as 20,000 psi (137 900 kPa) are attained.

The series 60 fuel system is equipped with a fully electronic controlled governor and fuel injection system. The Detroit Diesel Electronic Control (DDEC) system is reviewed in more detail in **Chapter 25.**

Due to its advanced design and use of DDEC fuel injection and governor control, the number of checks and

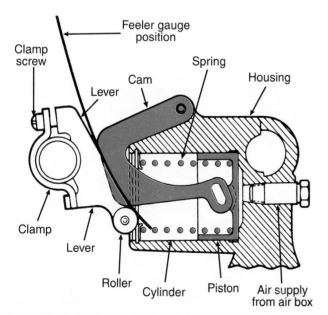

Figure 22-29. *Mechanical fuel modulator components and feeler gauge position for adjustment. (Detroit Diesel Corporation)*

adjustments needed for a series 60 engine tune-up are limited to adjusting the intake and exhaust valve clearances and timing the unit fuel injectors. Intake and exhaust valve clearances and fuel injector height are adjusted using a set screw and locknut located at the end of the rocker arm, **Figure 22-31**.

Note: On engines equipped with a Jake Brake® engine brake, measure valve lash and injector height before removing any brake housings. Only remove the brake housing needed for adjustment.

Valve and Injector Timing

Valve clearance and injector timing adjustments must be made when the engine is cold. Because the series 60 engine is a four-stroke engine, the valves for one cylinder can be checked and adjusted at a given crankshaft position. For example, when the intake and exhaust valves are fully closed on cylinder number 1, the valves for cylinder number 1 can be set. However, the injector timing for its mating cylinder, number 6, can be adjusted at this same engine position. Remember, when number 1 cylinder is at TDC on its compression stroke, its mating cylinder, number 6, is also at TDC at the very end of its exhaust stroke and beginning of its intake stroke. **Figure 22-32** summarizes the valve and injector timing setting sequence for the series 60 engine.

By using this method, it will be necessary to manually turn the engine only two complete revolutions. On each revolution, adjust the valves on three cylinders and adjust the injectors on three cylinders. Never set the valves and

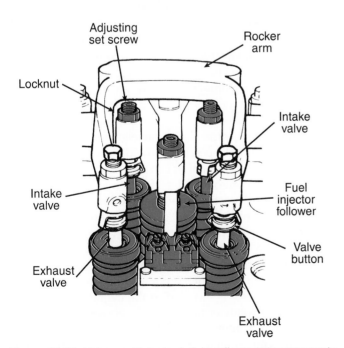

Figure 22-31. *Valve and injector height adjustment components for a Detroit Diesel series 60 engine. (Detroit Diesel Corporation)*

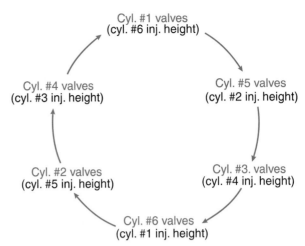

Figure 22-32. *Timing circle chart for Detroit Diesel series 60 engine valve and injector height adjustment. (Detroit Diesel Corporation)*

injectors on the same cylinder at the same time. Always work on companion cylinders. However, it does not matter which cylinder you begin with:

1. Disconnect power from the starting motor or mechanism.
2. Remove the engine rocker cover.
3. Insert a 3/4″ drive breaker bar or ratchet into the square hole in the crankshaft pulley. Turn the engine over in the direction of normal engine operation and observe the fuel injector followers. Stop the engine rotation when any one of the injector followers begins its downward stroke. Note the number of this cylinder.
4. Refer to the timing circle illustrated in **Figure 22-32**. Adjust the valves for this cylinder and set the fuel injector height for the companion cylinder (listed in parentheses).
5. Manually turn the engine in a clockwise direction until the fuel injector follower for the next cylinder in the chart begins its downward stroke. Adjust the valves for this cylinder and the injector height for its companion cylinder. Continue working until the timing circle is complete.

Intake and exhaust valve adjustment is only needed when the clearance between the end of the self-centering valve button at the end of the rocker arm and the valve stem tip are outside of the specifications listed in the engine manual. To check and adjust the valve clearances, either a straight feeler gauge or a go/no-go gauge can be used.

Once the engine has been turned to the proper position for adjustment, insert the feeler gauge between the valve button and valve stem, adjust the clearance by loosening the rocker arm locknut and rotating the adjusting screw with an Allen wrench until a slight drag is felt on the feeler gauge, **Figure 22-33**. When using a go/no-go gauge, turn the adjusting screw until the smaller part of the gauge passes through but the larger part does not. Tighten the locknut and recheck the clearance.

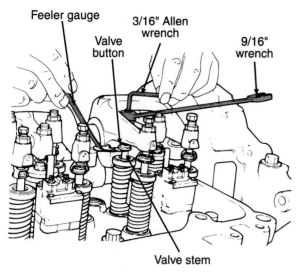

Figure 22-33. *Valve clearance adjustment for a Detroit Diesel series 60 engine. (Detroit Diesel Corporation)*

Injector Timing Height

Injector timing height adjustment is similar to that outlined earlier for DDC two-stroke engines. A special *injector height gauge* is required. On engines equipped with a Jake Brake engine brake, move the handle 90° to the shank. A height gauge pilot hole is provided in the fuel injector body on the machined surface near the solenoid, **Figure 22-34.** The gauge must seat on the machined surface with the tip in the pilot hole. Dirt in the pilot hole or on the machined surface may prevent accurate injector height setting. Use the following procedure to adjust injector timing height on the mating cylinder of the one on which valve clearance has just been set:

1. Make certain the injector hole is clear and clean. Place the gauge into the hole and rotate the gauge until the gauge flat is facing toward the fuel injector plunger, **Figure 22-35.**

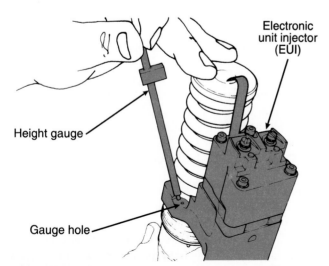

Figure 22-34. *Injector height gauge and location of gauge hole on Detroit Diesel Series 60 engines. (Detroit Diesel Corporation)*

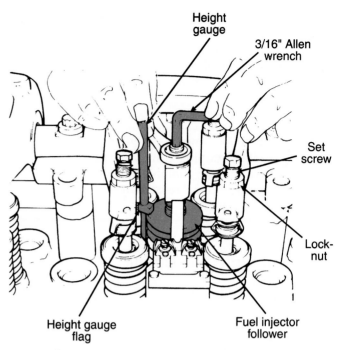

Figure 22-35. *Detroit Diesel Electronic Control unit injector height adjustment with height gauge in place. (Detroit Diesel Corporation)*

2. If adjustment is necessary, loosen the rocker arm locknut and use an Allen wrench to turn the adjusting screw clockwise to decrease timing height and counterclockwise to increase timing height. The gauge flag should just pass over the top of the injector follower.

3. When adjustment is correct, tighten the locknut to specifications and recheck the reading using the height gauge. Readjust if necessary.

Summary

Unit injection is a low pressure fuel system. Fuel delivered to the unit injectors averages 45-70 psi (310-482 kPa). The fuel system includes the fuel injectors, inlet and outlet fuel pipes, fuel manifolds that are an integral part of the cylinder head, a fuel pump, fuel strainer, fuel filter, and the necessary fuel lines.

The unit injector is a lightweight, compact unit. Its simple design and operation provides for simplified controls and easy adjustment. It can be either mechanically or electronically controlled. It performs four functions: times the moment of injection, atomizes the fuel, meters the correct amount of fuel, and generates the high pressure needed for injection.

When a mechanical unit injector is assembled, a timing mark on the control rack and the internal gear that the plunger slides through are aligned to set initial timing. Once the injector is installed in the cylinder head, a specific injector follower timing height above the body must be set using a timing pin dial gauge. This timing height determines the exact moment the upper port in the bushing closes and injection begins.

A mechanical unit injector is equipped with a control rack. Control rack movement changes the plunger's effective stroke, which changes the amount of fuel delivered by the injector. Each injector control rack is actuated by a lever on the injector control tube. The injector control tube is connected to the governor by means of a fuel rod. As the governor reacts to changes in engine speed, its motion is transferred via the fuel rod to the control tube and the injector control rack. The racks must be set and adjusted to ensure they all react to governor action in the same way.

Matched sets of unit injectors are available. These sets all have the same flow rate. Mechanical unit injectors that use a fuel control tube and rod linked to a mechanical governor must also be properly balanced when installed. The fuel rack setting of each injector is adjusted so that all injectors are delivering the same amount of fuel for a given engine speed. Electronically controlled unit injectors are not balanced in this manner.

During servicing, each unit injector must be cleaned, inspected, and tested to determine if it can be reused or requires a complete overhaul. Many engine shops are equipped with the manufacturer's test equipment needed to perform plunger and rack movement, injector leakage, spray pattern, and fuel flow tests.

A tune-up on an engine equipped with mechanical unit injectors includes exhaust valve bridge adjustment, exhaust valve clearance adjustment, injector timing, governor gap adjustment, injector rack setting and adjustment, maximum no-load speed setting, idle speed, buffer and starting screw adjustment, throttle delay adjustment, and fuel modulator adjustment.

To time a unit injector properly, the injector follower must be adjusted to a definite set height in relation to the injector body. All of the injectors must be timed in their correct firing order during one complete revolution of the crankshaft.

A throttle delay mechanism is often used on highway truck applications on both turbocharged and non turbocharged engines to help reduce emissions during vehicle acceleration. It does this by retarding the fuel control tube and injector rack movement towards the full-fuel injection position when the engine is accelerating. Turbocharged engines may be equipped with a fuel modulator that performs the same function. Both must be properly adjusted.

The Detroit Diesel Series 60 engine is equipped with a fully electronic controlled governor and fuel injection system. Mechanically operated control tubes and racks are not used. Due to its advanced design and use of DDEC fuel injection and governor control, the adjustments needed for a series 60 engine tune-up are to adjust the intake and exhaust valve clearances and time the unit fuel injector by adjusting follower height.

Important Terms

Unit fuel injection systems
Mechanical unit fuel injector (MUI)
Control rack
Injector control tube
Jumper lines
Fuel rod
Injector plunger
Needle valve
Metering
Effective stroke
Timing height
Plunger and control rack movement test
Spray pattern test
Pressure and leakage test
Fuel output test
Fuel spill back
Tune-up
Exhaust valve bridge
Exhaust valve clearance
Go/no-go feeler gauges
Governor gap
Single weight limiting speed (SWLS) governors
Double weight limiting speed (DWLS) governor
Wedge tool
Clevis pins
Maximum no-load setting
Idle speed
Buffer screw
Starting aid screw
Go/no-go pin gauge
Throttle delay gauge
Fuel modulators
Injector height gauge

Review Questions—Chapter 22

Do not write in this text. Place your answers on a separate sheet of paper.

1. Unit injection systems can be used on _____ diesel engines.

2. Which of the following is true concerning a unit injection fuel system?
   (A) A low fuel pressure is increased to extremely high pressure through the mechanical movement of the engine's rocker arm.
   (B) Fuel metering can be either mechanically or electronically controlled.
   (C) Mechanical unit injectors that use a fuel control tube and rod linked to a mechanical governor must also be properly balanced when installed.
   (D) All of the above.

3. When a mechanical unit injector is assembled, a _____ on the control rack and internal gear are aligned to set initial timing.

4. When isolating a faulty unit injector, holding the injector follower down with a screwdriver to short out the injector is an acceptable service method on the following Detroit Diesel unit injection engines:
   (A) two-stroke mechanical injection models.
   (B) four-stroke mechanical injection models.
   (C) four-stroke electronic injection models.
   (D) Neither A nor B.

5. List the four main functions of a unit injection system.

6. A _____ measures the volume of fuel returning to the tank in a given time.

7. Remove the _____ or _____ to prevent any interference when setting the fuel racks and other adjustments before beginning the tune-up process on a unit injection engine.

8. In non-electronically controlled unit injection engines, fuel metering in the unit injector is controlled by:
 (A) a unit injector rack connected to the fuel control tube that is linked to the governor via the fuel rod.
 (B) a unit injector rack linked directly to the governor.
 (C) a fuel control tube running directly from the injector to the governor.
 (D) a control rod and sleeve running from the injector to the fuel tube connected to the governor.

9. The overhead camshaft design of the Detroit Diesel series 60 engine eliminates all _____ and _____ .

10. The only checks and adjustments needed for a Detroit Diesel series 60 engine tune-up are to adjust the _____ and time the unit fuel injector by adjusting follower height.

ASE-Type Questions

1. Technician A says that matched unit injectors all have a fuel flow rate that falls within a specific range set by the manufacturer. Technician B says that a matched set of injectors all have the *exact same* flow rate. Who is right?
 (A) A only.
 (B) B only.
 (C) Both A & B.
 (D) Neither A nor B.

2. All of the following are methods of isolating a faulty injector EXCEPT:
 (A) pushing and holding the injector follower with a large screwdriver.
 (B) measuring the individual exhaust manifold outlet temperatures with a pyrometer.
 (C) using an electronic diagnostic tool.
 (D) checking the injector timing.

3. A fuel spill back check indicates that the volume of fuel returning to the tank is less than specified. Technician A says that there may be a clogged fuel filter or the fuel line diameter may be too small. Technician B says that the relief valve in the fuel pump may be stuck partially open or the pump itself may be faulty. Who is right?
 (A) A only.
 (B) B only.
 (C) Both A & B.
 (D) Neither A nor B.

4. Technician A says you should make any needed timing adjustments to the unit injector after setting the exhaust valve clearance to specifications. Technician B says exhaust valve clearance should be the last setting made in the tune-up sequence. Who is right?
 (A) A only.
 (B) B only.
 (C) Both A & B.
 (D) Neither A nor B.

5. Technician A says you should time a unit injector by adjusting the injector follower to a definite set height in relation to the injector body. Technician B says the injectors must be timed in their correct firing order during one complete revolution of the crankshaft. Who is right?
 (A) A only.
 (B) B only.
 (C) Both A & B.
 (D) Neither A nor B.

6. The governor gap divides the travel of the _____ between the low and high speed spring systems.
 (A) throttle
 (B) flyweights
 (C) accelerator pedal
 (D) Both A & C.

7. When the injector rack control levers are properly adjusted, Technician A says the speed control lever will be at the minimum speed position. Technician B says the governor low speed gap will be open. Who is right?
 (A) A only.
 (B) B only.
 (C) Both A & B.
 (D) Neither A nor B.

8. Turbocharged engines use a _____ to limit the amount of fuel injected at start-up.
 (A) starting aid screw
 (B) locknut
 (C) buffer screw
 (D) idle screw

9. Technician A says that the injector rack closest to the governor is adjusted first. Technician B says that on V-configured engines, the right and left cylinder banks are determined when viewing from the rear of the engine. Who is right?
 (A) A only.
 (B) B only.
 (C) Both A & B.
 (D) Neither A nor B.

10. Technician A says that the idle speed should be set with the engine operating at normal operating temperature and the buffer screw backed out to avoid contact with the differential lever. Technician B says the adjustment should be made so the engine operates at about 15 rpm below the recommended idle speed. Who is right?
 (A) A only.
 (B) B only.
 (C) Both A & B.
 (D) Neither A nor B.

Chapter 23

Cummins Pressure-Time Injection Systems

After studying this chapter, you will be able to:

❏ Name the three factors that control fuel metering in a pressure-time fuel injection system.

❏ Describe the functions of the PT fuel injection system fuel pump.

❏ Describe the operation of the PT fuel injection system governor and how it controls rail pressure during engine operation.

❏ Explain the operation of the system's air-fuel control unit, shutdown valve, and cooling kit.

❏ Explain the operating cycle of the PT injector.

❏ Perform basic checks and adjustment to the PT fuel pump.

❏ Check PT fuel injection system for fuel system restrictions and air leaks.

❏ Properly remove and install the PT fuel pump on the engine.

❏ Properly remove and install PT fuel injectors.

❏ Perform a timing check on a PT injection engine using the proper timing fixture.

❏ Perform valve crosshead adjustment, valve clearance adjustment, and injector timing adjustment.

❏ Explain the operating principles of Cummins Step Timing Control and Mechanical Variable Control and properly perform adjustments to both systems.

The last mechanical fuel injection system that will be covered in this text is the pressure-time fuel injection system manufactured by Cummins Engine Company. There are literally hundreds of thousands of engines with mechanically controlled pressure-time injection systems in service. This chapter will concentrate on the operating principles of this system. More information concerning Cummins' electronically controlled systems will be covered in **Chapter 25.**

Pressure-Time (PT) Fuel Injection

For many years Cummins manufactured engines equipped with a mechanically operated **pressure-time (PT) fuel injection system** that remains unique to the industry. To meet increasing stringent emission control laws, Cummins introduced its **PACE system** on its PT-equipped engines. PACE provides electronic control of the PT system through the use of sensors, an electronically activated fuel control valve, and a PT control module (computer). Cummins also offers an electronically controlled unit injector fuel injection system known as *Celect-ECI.* Some Cummins engines also use inline or distributor injection systems, depending on the application.

PT Fuel System Operation

Figure 23-1 illustrates the basic components of the Cummins PT fuel system. The PT fuel injection system derives its name from two of the primary variables affecting the amount of injected fuel per cycle. The *P* refers to the *pressure* of the fuel at the injector inlets. This pressure is controlled by the fuel pump. However, the PT fuel pump is not the same as an inline or distributor injection pump. The *T* refers to the *time* available for the fuel to flow into the injector cup. The time is controlled by engine speed through the camshaft and injection train. Cummins engines are four-cycle engines, so the camshaft is driven at one-half engine speed. However, the fuel pump turns at engine speed, making additional governing of fuel flow necessary. The fuel pump varies pressure to the injectors in proportion to engine rpm.

The final controlling element in the fuel metering process is the size of the fuel injector openings (orifices). The orifice size, also called the **flow area,** is determined by calibration of a complete set of injectors.

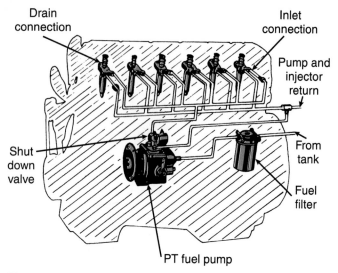

Figure 23-1. Basic components of the Cummins Pressure-Time (PT) fuel injection system. (Cummins Engine Co., Inc.)

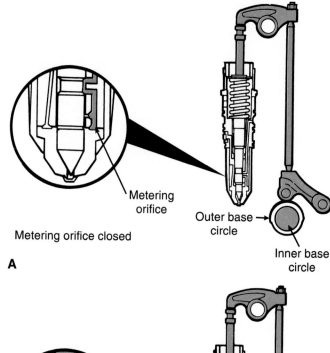

Metering orifice closed

A

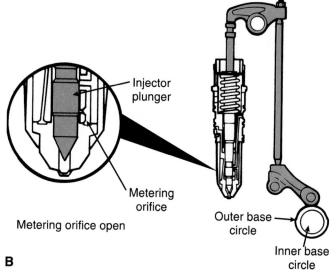

Metering orifice open

B

Figure 23-2. In a PT fuel system, the camshaft actuated inj A—Injector metering orifice closed. B—Here, the injector metering orifice is opened. (Cummins Engine Co., Inc.)

With a given flow area, fuel metering is controlled by **rail pressure** and flow time. The **flow time** or *metering time* is controlled by engine speed through a camshaft-actuated plunger. The camshaft rotary motion is changed into the reciprocating motion of the injector plunger. The plunger movement opens and closes the injector barrel **metering orifice**, **Figure 23-2.** The period of time the metering orifice is open is the time available for the fuel to flow into the injector cup. The greater the engine speed, the less available to meter the fuel.

At any given engine speed, only rail pressure in the PT fuel system controls the quantity of fuel metered to the injectors. Rail pressure regulating takes place inside the fuel pump. The PT fuel pump is designed and calibrated to provide the proper rail pressure during all engine operating conditions. Maximum rail pressure occurs at rated engine speed.

One of the major advantages of the PT system is that unlike other fuel injection pumps, it is not necessary to time the PT fuel pump to the engine. The PT pump supplies fuel at a given flow rate and at a specified pressure setting to a common rail supplying all the injectors. In the PT injection system, the injectors themselves are timed to ensure that the start of injection occurs at the correct time for each cylinder.

Pressure-Time Governed (PTG) Fuel Pump Functions

The pressure-time fuel pump also drives the governor. This assembly is often referred to as the **pressure-time governed (PTG) fuel pump.** The basic functions of the PTG fuel pump include:

- ❑ The transfer of fuel from the supply tank to the engine.
- ❑ Rail pressure to the injectors.
- ❑ Idle and maximum speed governing.
- ❑ Operator control of power output below governed speed.

- ❑ Control of exhaust smoke during acceleration.
- ❑ Shutdown of the engine.

The PTG fuel pump assembly is coupled to the air compressor drive and driven from the engine gear train. The fuel pump main shaft drives the gear pump, governor, and tachometer shaft assemblies at engine speed, **Figure 23-3.** Fuel from the supply tank enters the gear pump inlet and is carried around the outside of the two meshing gears to the gear pump outlet. From this point, fuel flows through a wire mesh magnetic filter to the governor inlet passage. In addition to providing idle speed and maximum speed governing, the PT governor also regulates fuel pressure.

Governor

The **PT fuel system governor** is illustrated in **Figure 23-4.** It is a simple flyweight-operated mechanical governor. The flyweights are driven at engine speed by the fuel

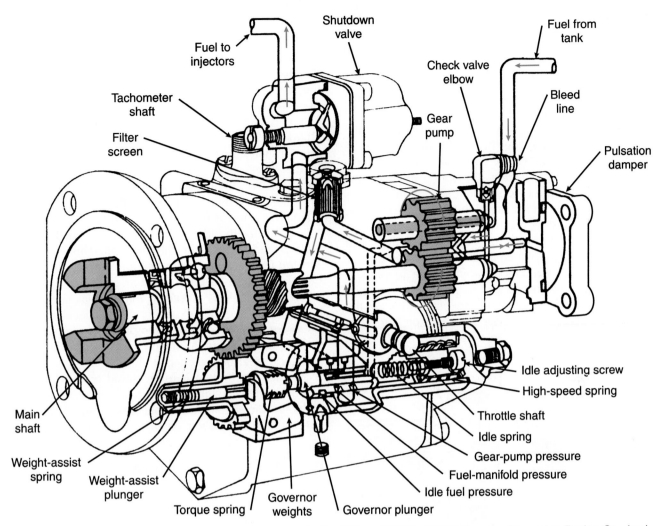

Figure 23-3. Components of the Pressure-Time Governed Air-Fuel Control (PTG-AFC) fuel pump. (Cummins Engine Co., Inc.)

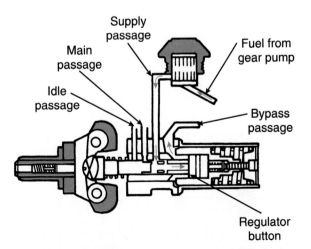

Figure 23-4. Components of a Cummins PT fuel pump mechanical flyweight governor.

speed also increases the force exerted by the flyweights and forces the plunger to move to the right as shown in **Figure 23-5.**

Rail pressure control of the PT fuel system begins by regulating the fuel or **supply pressure** to the governor assembly. Supply pressure control is accomplished by

pump mainshaft. The governor plunger is held between the flyweight feet and rotates with the flyweights. The rotating flyweights pivot on pins, allowing their feet to exert an axial force on the governor plunger. At any given engine speed, the plunger position is determined by the balance between the flyweight and spring forces. Increasing engine

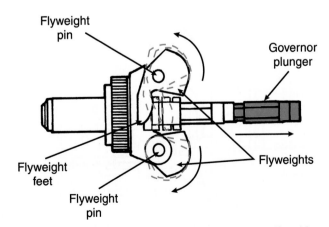

Figure 23-5. As rotational speed increases, the centrifugal force of the governor flyweights force the governor plunger to the right. (Cummins Engine Co., Inc.)

using a bypass pressure regulator inside the governor. The pressure regulator has a button valve designed to unseat when a designated pressure is reached, **Figure 23-6.** In order for the regulator to open, an excess supply of fuel must be delivered to the governor. This ensures that some fuel is always being bypassed, which is necessary for the regulator to maintain control of the supply pressure.

When the fuel pressure exceeds the force holding the button valve and plunger together, fuel is bypassed to the gear pump suction side, **Figure 23-7.** The unseating pressure is determined by dividing the spring force by the button recess area. The recessed area, or **counterbore,** is the area the fuel is pushing against. Increasing the counterbore area will reduce the pressure at which the fuel bypass opens. Decreasing the size of the recessed area will increase the pressure at which the fuel bypass opens.

With a given counterbore, any change in plunger position will change the supply pressure. For example, when engine speed increases, governor action changes the position of the plunger. This raises the supply pressure. The end result is that more fuel is delivered to the injector.

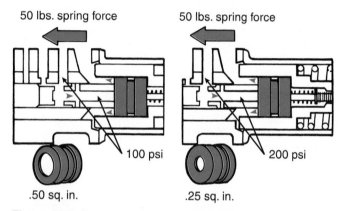

Figure 23-6. *Button valve pressure regulator. Altering the size of the button recess changes the pressure at which the valve will unseat. (Cummins Engine Co., Inc.)*

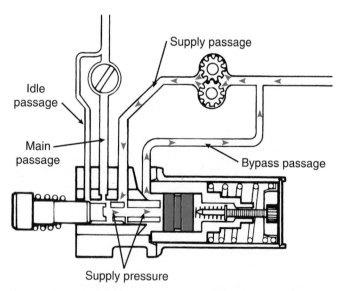

Figure 23-7. *Fuel bypass occurs when the button is unseated. (Cummins Engine Co., Inc.)*

Throttle

The engine's power output can be controlled by the operator, within the established governor limits, through the use of a **throttle.** The throttle shaft is located between the governor and the fuel pump discharge. It allows the operator to control the rail pressure, and therefore, the engine power. It acts as a variable orifice, controlling the amount of fuel exiting the governor main passage. The throttle shaft's travel is limited by two stop screws located in the throttle shaft housing. A fuel adjusting screw is located within the throttle shaft. Its setting determines the maximum flow area when the throttle shaft passage is wide open. This screw is used to adjust the rail pressure.

While the throttle is closed, there is a small amount of fuel flowing through the throttle shaft. This is known as **throttle leakage.** This controlled fuel leakage is needed in the PT system to keep the injectors cooled and lubricated when the throttle is closed. If throttle leakage is set too high, the engine can experience slow deceleration and excessive injector carbon loading. If the throttle leakage is set too low, engine hesitation can result. The injector plunger may also be damaged.

When the throttle is in its closed position, the small amount of fuel flowing to the injectors is not enough to maintain idle speed. Additional fuel is provided from the governor through an idle passage around the throttle shaft.

PTG Air-Fuel Control (AFC)

The **PTG air-fuel control (AFC)** fuel pump used on turbocharged heavy duty truck engines has an exhaust smoke control device built into the pump body. The AFC unit restricts fuel flow in direct proportion to intake manifold pressure during engine acceleration, under load, and lug-down conditions.

The main components of the AFC assembly are shown in **Figure 23-8.** The AFC senses air pressure in the intake manifold. Changes in intake manifold pressure will change

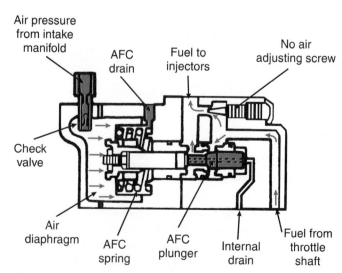

Figure 23-8. *Components and fuel flow through the air-fuel control (AFC) assembly. (Cummins Engine Co., Inc.)*

the piston position, which controls the plunger shoulder position over the AFC inlet passage. This determines the amount of fuel delivered to the injectors during acceleration and other engine operational states. Air pressure is applied to the diaphragm and piston through the cover inlet fitting. Increasing air pressure overcomes spring force, causing the plunger to move in the barrel. As the plunger moves, it uncovers the passage, allowing fuel to flow from the throttle shaft through the AFC. As the air pressure increases, the plunger is pushed even further, eliminating the fuel restriction, and permitting maximum fuel flow through the AFC.

When there is little or no air pressure applied to the AFC diaphragm, maximum fuel pressure and flow is controlled by the *No-air* adjusting screw. The plunger is positioned by the return spring to block the main fuel passage through the AFC. Under these conditions, all fuel flow is around the No-air adjusting screw.

Shutdown Valve

From the AFC assembly, fuel flows to the shutdown valve. Most **shutdown valves** are controlled by an electrically-operated solenoid. In the shutdown mode, a spring washer seats a disc, which prevents fuel flow out of the pump, **Figure 23-9.** When the solenoid is energized, the electromagnetic force overcomes the spring washer pressure. The disc unseats and fuel flows from the pump to the injectors. A single low pressure line from the fuel pump serves all injectors. This means the pressure and quantity of fuel metered to each injector are equal.

Cooling Kit

Many PT fuel pumps are equipped with a **cooling kit elbow.** This elbow, **Figure 23-10,** is designed to prevent fuel pump overheating. Some of the hot fuel inside the pump housing is bled back to the storage tank through the injector drain line. This allows cooler fuel from the tank to circulate through the fuel pump assembly.

The cooling kit elbow contains a spring-loaded check valve. This valve is designed to prevent air and/or fuel from

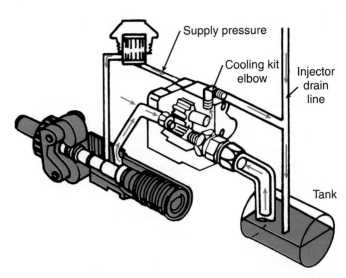

Figure 23-10. *Location of cooling kit elbow in the PT fuel injection system. (Cummins Engine Co., Inc.)*

draining back through the fuel pump assembly when the engine is not operating. During engine operation, gear pump fuel pressure unseats the valve and circulates some fuel back to the tank.

Injector Operation

Figure 23-11 illustrates the **pressure-time type D Top Stop injector** commonly used in Cummins PT fuel injection systems. The injector plunger is actuated by engine camshaft rotation. When the cam follower roller is on the inner base circle, the injector return spring lifts the plunger, uncovering the metering orifice. When the cam follower

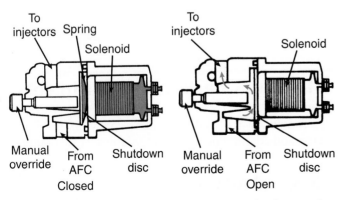

Figure 23-9. *Operation of PT fuel pump shutdown valve. (Cummins Engine Co., Inc.)*

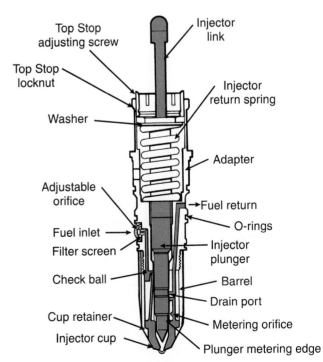

Figure 23-11. *Components of PTD Top Stop Injector. (Cummins Engine Co., Inc.)*

roller is on the outer base circle, the injector plunger's downward movement overcomes the injector return spring, closing the metering orifice, and injecting fuel into the cylinder. The injector plunger is now seated in the injector cup.

Fuel entering the injector flows through a wire mesh filter screen and an adjustable orifice. The size of the orifice determines the injector flow rate and the pressure at the metering orifice. From the adjustable orifice, fuel flows down an internally drilled passage in the injector adapter and barrel. The fuel unseats a check ball, while continuing its flow toward the metering orifice. The check ball prevents the reversal of fuel flow during deceleration and shutdown as the plunger moves downward across the metering orifice, **Figure 23-12.**

When the metering orifice is uncovered, fuel flows into the injector cup. This occurs during the end of the engine's intake stroke and the beginning of the compression stroke. As the cam follower roller travels toward the base circle of the camshaft injector lobe, the injector return spring lifts the plunger, uncovering the metering orifice, **Figure 23-13A.** With the metering orifice open, the plunger also blocks the drain port in the injector.

With continued camshaft rotation, the cam roller travels up the injection ramp. The upward movement of the push rod pushes the injector plunger downward. As the injector plunger moves down on the compression stroke, the metering orifice closes. Shortly after the metering orifice is closed, the drain port opens. At this point, the injector cup contains the proper amount of fuel to be injected, **Figure 23-13B.** The point at which injection begins varies with the level of fuel in the injector cup, **Figure 23-14.** With an increase in the fuel level, the injector plunger contacts the fuel earlier, advancing the start of injection.

Fuel is injected when the pressure exerted by the injector plunger on the fuel is greater than the combustion chamber pressures. Injection ends when the plunger

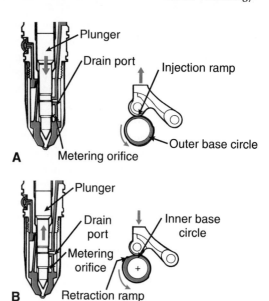

Figure 23-13. Metering of fuel in the PTD Top Stop injector. A—The plunger moves up during the end of the intake stroke. B—The plunger moves down during the first part of the compression stroke. (Cummins Engine Co., Inc.)

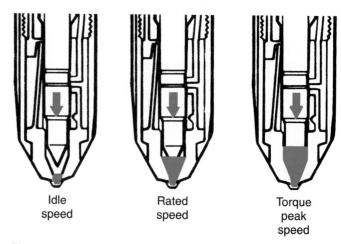

Figure 23-14. The volume of fuel in the injector cup determines when injection will begin. The more fuel that is present in the cup, as at the engine's rated speed or torque peak speed, the sooner injection begins. (Cummins Engine Co., Inc.)

bottoms out in the cup, **Figure 23-15A.** At this point, the drain groove on the injector plunger aligns with drain passages in the injector barrel, permitting fuel to flow out of the drain groove and return to the tank, **Figure 23-15B.**

PT Pump Checks and Adjustments

The following sections outline many of the basic checks and adjustments made to Cummins PT fuel pumps.

Idle Speed Adjustments

Before checking idle speed, run the engine until it reaches normal operating temperature. Connect a

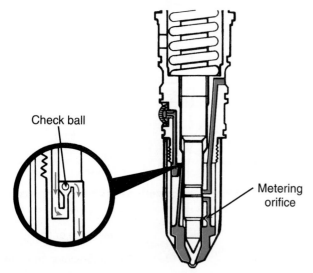

Figure 23-12. Fuel flow into the PTD Top Stop injector. (Cummins Engine Co., Inc.)

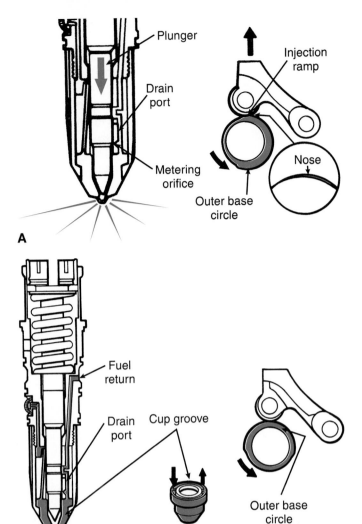

A

B

Figure 23-15. *A—Fuel injection occurs and ends when the plunger reached the bottom of the cup. B—Any fuel not injected returns to the tank. (Cummins Engine Co., Inc.)*

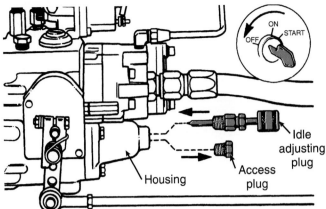

Figure 23-16. *Idle speed adjustment is performed by removing the idle spring access plug and inserting a special adjustment tool. (Cummins Engine Co., Inc.)*

tachometer to the PT pump tach drive or use a digital tachometer to read engine speed. A remote starter is also helpful when making the idle speed adjustment. The exact location of the idle speed adjustment screw can vary between models, but is usually located at the bottom of the pump housing.

To perform the idle speed adjustment on a PTG-AFC model pump, a special idle adjustment tool is needed. Remove the idle speed screw access plug from the housing and install the idle adjustment tool. Thread the tool's fitting into position, **Figure 23-16.** The tool is designed to properly seal the access plug so that no air can enter the pump during adjustment.

Start and run the engine at high idle for about 30 seconds to remove any air from the fuel system. Allow the engine to return to normal idle speed. Make certain the throttle lever is at the low speed position. Turn the adjusting tool clockwise to increase the engine idle speed, or counter-clockwise to lower engine idle speed. The correct idle speed is stamped on the ***control parts list (CPL) data plate.*** It normally ranges from 650-725 rpm, depending on the engine model. When the correct idle

speed is set, shut off the engine, remove the adjusting tool, and reinstall the plug in the pump body. Start and run the engine to remove air that may have entered the fuel system. Operate the engine until it runs smoothly and recheck idle speed.

If the idle adjustment tool is not available, the pump's fuel pressure can be used to set idle. The shop manual lists the pump fuel pressure at idle specification. Knowing this pressure, you can install a fuel gauge at the rail and adjust the idle screw in or out until the recommended pressure registers on the gauge.

To adjust high idle on a PTG-AFC pump, remove the four bolts that fasten the high idle spring pack cover to the bottom of the fuel pump housing. The engine's maximum speed is adjusted by adding or removing shims from behind the high idle spring. Each .001″ (.025 mm) of shim thickness changes engine speed by approximately 2 rpm. Add or remove shims as required and reassemble the spring pact and housing cover.

On PTG-AFC fuel pumps equipped with ***variable speed (VS) governors,*** both low and high idle speeds are set by adjusting separate idle speed screws. The low idle adjustment screw is located on the back of the pump housing, while the high speed adjustment screw is located on top of the housing. Each adjustment screw has a locknut and jam nut threaded onto it. Copper washers are located on either side of the jam nut, **Figure 23-17.**

To make an adjustment, first remove both the locknut and jam nut from the adjustment screw and discard the old copper washers. Install a new copper washer on the screw and then the jam nut. Start and run the engine. When adjusting low idle, hold the VS lever in the idle position. When setting high idle, rotate the lever counter-clockwise to its maximum no-load (high idle) position.

Use an Allen wrench to turn the adjusting screw clockwise to increase or counter-clockwise to decrease engine speed. When the desired speed is set, tighten the jam nut with a wrench, **Figure 23-18.** Place a second new copper washer against the top to the jam nut and install the lock nut onto the adjustment screw. Recheck for correct idle.

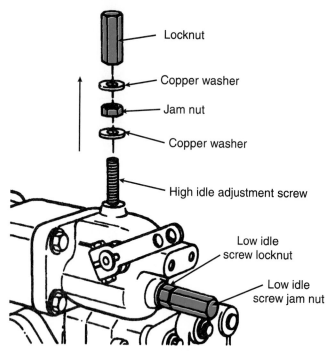

Figure 23-17. *Location of low and high idle adjustment screws on a PTG-AFC pump with a variable speed governor. (Cummins Engine Co., Inc.)*

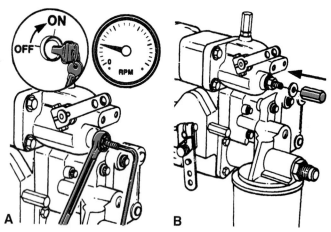

Figure 23-18. *A—Make the adjustment by turning the screw while holding the jam nut in position. B—Install a new copper washer and locknut. (Cummins Engine Co., Inc.)*

Basic Fuel System Checks

Low engine power may be caused by problems in the fuel system. A more complete summary of troubleshooting techniques is covered in **Chapter 29.** The following sections outline several checks that can be made on Cummins PT fuel systems.

Fuel System Restrictions

Restrictions in the fuel system can be caused by dirty fuel filters, bent or kinked fuel lines, or fuel lines that are too small in diameter for the engine. The maximum allowable inlet restriction to the fuel transfer pump on a Cummins PT system is 4″ Hg (mercury) when a new primary filter has

been installed or no more than 8″ Hg when the primary fuel filter has not been replaced.

A vacuum gauge or manometer can be used to check for fuel restrictions. Install the gauge or manometer at the primary filter, **Figure 23-19.** When using a vacuum gauge, be sure to hold it level with the fuel pump to obtain an accurate reading. Start and run the engine at idle and then slowly accelerate until the engine reaches its rated rpm under load, if possible. If the engine cannot be loaded, take the reading at high idle rpm. If the reading is above specifications, replace the fuel filter. If this does not correct the problem inspect the fuel system for damaged lines, excessive turns, and elbows in lines.

The fuel pressure in the return line can also be measured by disconnecting the fuel drain line and installing the pressure gauge. Run the engine at its rated rpm under load or at high idle rpm. The maximum restriction in the return line should not exceed 2.5″ Hg on systems with no check valves and 6.5″ Hg on systems equipped with check valves. If readings are above these levels, check for a plugged filler cap or tank vent, or a kinked, bent, or collapsed fuel return line.

Air Leaks

Air leaks in the fuel pump inlet side will decrease the fuel pressure and feed air to the injectors. The engine will lack power, run rough, and possibly stall. The simplest method of checking for air in the system is to install a temporary fuel return line (flexible hose) and submerge it in a container of diesel fuel, **Figure 23-20.** Run the engine at maximum high idle speed with no load and check for air bubbles in the container. Air bubbles indicate a leak between the fuel tank pickup and the transfer pump inlet. Check all connections, fittings, and seals for signs of leaks.

PT Pump Fuel Rail Pressure Check

The fuel transfer pump inside the PT pump body produces maximum fuel rail pressure at the engine's maximum rated speed. Low pump rail pressure results in a

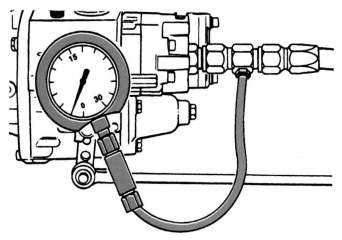

Figure 23-19. *Checking for fuel flow restrictions at the primary filter location. (Cummins Engine Co., Inc.)*

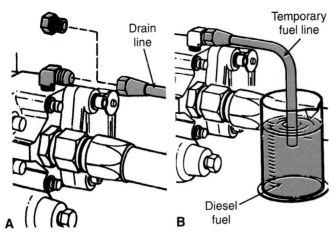

Figure 23-20. *A—Removal of fuel pump suction drain line. B—Install tubing with the end immersed in clean diesel fuel to check for air leaks in the fuel suction line. (Cummins Engine Co., Inc.)*

lower flow rate to each injector and a loss in horsepower. Each fuel pump/engine horsepower setting is assigned a set fuel rail pressure based on the application. Use the fuel pump calibration code number on the CPL data plate to look up the exact rail pressure specification.

To accurately read the maximum fuel rail pressure, the engine must run at its full-load rated speed. If the engine is removed from the vehicle or boat, this can be done on an engine dynamometer. If the engine is installed in the vehicle, a chassis dynamometer can be used or the drive-line disconnected and a portable dynamometer connected to the transmission output shaft. To measure rail pressure, a fuel pressure gauge is installed to a fitting on the shutoff valve, **Figure 23-21.**

Fuel Rate

Each PT fuel pump is factory calibrated to deliver the proper *fuel rate,* measured in pounds of fuel per hour. This rate can be checked on an engine dynamometer using fuel flow metering equipment that is usually available only at fuel injection specialty shops or Cummins service centers.

The fuel rate can be adjusted by removing a tamper-resistant ball and turning the slotted end of an internal adjusting screw within the throttle shaft. Rotating this screw clockwise will lower the fuel rate while turning it counter-clockwise will increase the fuel rate. Fuel rate adjustments on Cummins engines must only be performed by authorized service personnel.

PT Fuel Pump Removal and Installation

Since it is not necessary to time a PT fuel pump to the engine as with other fuel injection pumps, removal and installation of the PT pump is a relatively simple job. Disconnect the batteries from the starting system to prevent accidental engine start-up during pump removal. After cleaning the pump housing and the surrounding area, remove the following components from the fuel pump assembly:

- ☐ Fuel shutoff wire.
- ☐ Throttle lever linkage.
- ☐ Fuel drain line from cylinder head.
- ☐ Gear pump cooling drain line.
- ☐ Gear pump suction line.
- ☐ AFC fuel drain line.
- ☐ Fuel supply line to the injectors.
- ☐ AFC air supply hose to the intake manifold.
- ☐ Tachometer drive cable.

Remove the four retaining bolts that secure the PT pump to the air compressor assembly, **Figure 23-22.** Clean all traces of the old gasket from the pump mounting flange. Inspect the pump drive coupling spider for wear or damage. Install a new drive coupling and new gasket. Reverse the removal procedure for pump installation.

Injector Removal

Injector removal and installation may be required any time the cylinder head is serviced or when injectors require

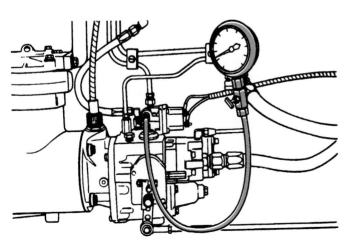

Figure 23-21. *Setup for measuring fuel rail pressure. (Cummins Engine Co., Inc.)*

Figure 23-22. *Removing the PT fuel pump from the rear of the air compressor. (Cummins Engine Co., Inc.)*

inspection and service. After cleaning the engine, remove the rocker cover assemblies. Loosen the injector rocker arm adjusting screw locknut on the cylinder being serviced. If needed, rotate the engine manually to remove the spring tension from the injector rocker lever prior to loosening the rocker lever adjusting screw.

Loosen the adjusting screw sufficiently to allow the injector push rod to be moved to the side or removed from the engine. If the push rod is removed, mark it so that it can be reinstalled in the same cylinder position on the engine.

Rotate the injector rocker lever upward and remove the injector link. Remove the two injector bolts and lift out the hold-down clamp, **Figure 23-23.** Install an injector slide hammer puller on the injector body, **Figure 23-24,** and pull the injector free from its bore. With the injector removed, clean away any carbon from the copper injector sleeve. Use a lint free cloth wrapped around a wooden stick as a cleaning tool. Do not scratch the copper sleeve. If a chip removing unit is available, use it to remove any carbon accumulation from the top of the piston crown.

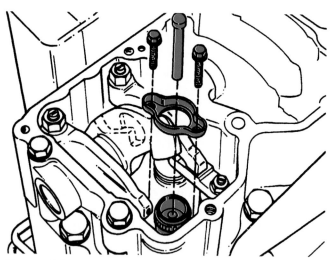

Figure 23-23. Removing injector hold-down clamp. (Cummins Engine Co., Inc.)

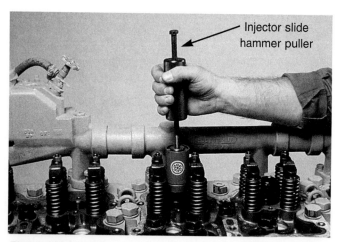

Figure 23-24. Removing injector using slide hammer puller. (Cummins Engine Co., Inc.)

Injector Installation

PT injectors are equipped with three external O-rings on the injector body. These O-rings prevent lube oil from entering the injector bore as well as diesel fuel from leaking into the rocker box area and the combustion chamber. Always replace these O-rings whenever an injector is removed. Roll the O-rings into the retaining grooves on the injector, taking care not to twist the rings. Once the new O-rings are in place, lubricate them with a suitable lubricant such as clean SAE 30 engine oil, **Figure 23-25A.**

Check the bores in the cylinder head for signs of burrs or sharp edges that could nick, cut, or tear the rings as they slide into the bore. Carefully insert the injector into the cylinder head bore, **Figure 23-25B.** Use a clean, blunt object, such as a hammer handle to push the injector into position. Press on the injector adapter body, not on the injector plunger or line, **Figure 23-25C.** Give the injector a quick, hard push to seat it in its bore. A snap should be heard and felt when the injector seats correctly.

Reinstall the hold-down clamp and the two bolts. Alternately tighten the two bolts to proper torque. Install the link into each injector and the push rod for the injector rocker arm. Turn the rocker arm adjusting screw until it seats in the push rod socket sufficiently to hold the push rod in place. Adjust all crossheads, valves, and injectors according to service manual procedures. Examples of these procedures are given later in this chapter.

Engine-to-Injector Timing

Injector timing must be checked and set if the camshaft, timing gears, cam follower box gaskets, cam followers, or cam follower box has been removed or replaced. On a pressure-time injection system, injection must start at a specific amount of degrees BTDC. The injection timing check will confirm that the distance between the injector plunger and the injector cup matches Cummins specifications when the piston is 19° BTDC, **Figure 23-26.** This equals a measurement of .203" or 5.16 mm BTDC.

The injection timing code on the engine data plate can be cross-referenced to manual specifications for all models. The injection timing is directly related to the amount of push rod travel remaining before the injector plunger bottoms out in its cup. If the amount of push rod travel is less than specifications, the timing is advanced or "fast." If the amount of push rod travel is more than specifications, the timing is retarded or "slow." The timing is properly set when the reading on the piston travel dial indicator and the push rod dial indicator matches specifications.

Camshaft Keys

A **camshaft key** is used to provide a method of indexing the camshaft with the cam gear. The size of the keys affect injection timing. Using offset cam keys allow changes in camshaft indexing in relation with the cam gear. Offset keys allow the camshaft profile to be rotated slightly while gear train timing remains stationary, **Figure 23-27.**

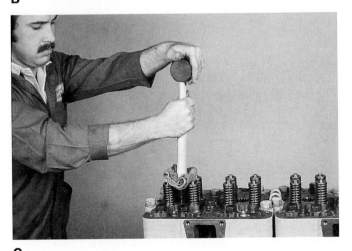

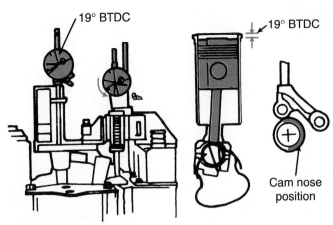

Figure 23-26. *To set injector timing, both piston position and push rod travel is measured using a special dial indicator setup. (Cummins Engine Co., Inc.)*

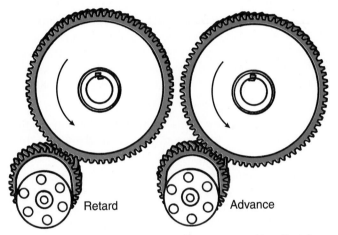

Figure 23-27. *Mounting the camshaft gear with offset keys allows the camshaft profile to be rotated slightly to retard or advance engine timing as required. (Cummins Engine Co., Inc.)*

Figure 23-25. *Installing an injector. A—Lubricate the injector body with clean engine oil. B—Install the injector in the bore. C—Seat the injector using a hammer handle or other blunt tool. (Cummins Engine Co., Inc.)*

Cam keys are available in 3/4″ and 1″ sizes. The 3/4″ offset keys can be interchanged for 1″ straight keys. However, 3/4″ straight keys cannot be interchanged for 1″ offset keys. To determine the required amount of offset and its effect on engine timing, refer to the key data information listed in the Cummins control parts list. To retard injection timing, the top of the offset key always points in the

direction of camshaft rotation. A greater degree of key offset results in greater timing retard.

In addition to installing camshaft keys, cam follower gaskets can also be added or removed to fine tune the timing to specifications, **Figure 23-28.** These gaskets alter the position of the injector cam follower roller in relation to the camshaft lobe. Gasket thickness ranges from .007-.037″ (.177-.939 mm) and can be stacked to achieve a maximum stack thickness of .015″ (.381 mm). Each .007″ (.177 mm) of gasket thickness affects injection timing by approximately .002″ (.050 mm) of dial indicator travel.

Performing Timing Checks

Timing checks on Cummins engines require the use of a timing fixture that can be adapted to most engine models. The engine service manual will indicate the correct adapters needed to mount the timing fixture to each engine.

The fixture is equipped with push rod and piston plunger indicators. The indicators have a total travel of 1″ (25.4 mm). Each complete revolution equals .1″ (2.54 mm) travel on the indicator stem. When zeroing the indicator,

Figure 23-28. Cam follower gaskets stacked between the block and the cam follower box can also be used to fine tune engine timing. (Cummins Engine Co., Inc.)

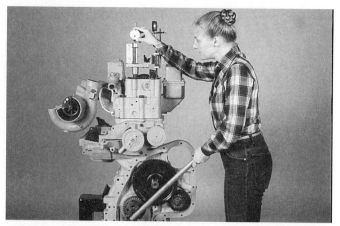

Figure 23-29. The engine is rotated into proper position on the compression stroke before the gauges are zeroed and locked. (Cummins Engine Co., Inc.)

the outer ring is positioned and locked into place. Two reference markers that may be used as a guide when reading indicator travel are located on the outer ring. When checking engine timing, it is necessary to check each cylinder cam follower housing, however, the rocker boxes do not need to be removed. To perform the timing check, proceed as follows:

1. Remove the injector from its bore.
2. Pivot the fixture stems away from their respective plunger rods to prevent indicator damage while installing the timing fixture.
3. Install the timing fixture into the injector bore and tighten the hold-down adapters. Compress the push rod plunger spring approximately .5″ (12.7 mm), line up the vertical scribe mark on the fixture, and tighten the lock nut.

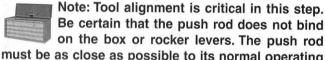 **Note: Tool alignment is critical in this step. Be certain that the push rod does not bind on the box or rocker levers. The push rod must be as close as possible to its normal operating position.**

4. Manually rotate the crankshaft (or accessory drive on some models) in the direction of normal rotation until both indicator plunger rods start moving upward. This indicates the piston is now on its compression stroke, **Figure 23-29.**
5. Continue rotation until the piston plunger rod reaches its upper most travel position.
6. Position the piston indicator over its respective plunger rod. Adjust to its fully compressed position and then raise the indicator .025-.030″ (.630-.762 mm) and lock it in place. Building in this slight overtravel helps prevent damage to the indicator.
7. Turn the engine slowly in the direction of rotation to locate the TDC piston position. Watch the indicator needle. When it begins moving in the clockwise direction, slowly advance the bar with the palm of your hand. The TDC piston position is located when the needle movement stops.

8. Zero the piston plunger indicator. This is a very precise setting and requires careful observation. If the needle reverses direction, you have passed the top dead center position and must repeat the procedure. Never reverse the direction of engine rotation to set the TDC position as backlash will give a false reading.
9. To set the plunger rod dial indicator, rotate the engine until the plunger rod indexes with the 90° NH marking on the fixture. The piston is now 90° ATDC. This positions the cylinder on the cam's outer base circle past the nose crush.
10. Position the push rod indicator and adjust it to its fully-compressed position. Raise the indicator by .025-.030″ (.630-.762 mm) overtravel. Lock it in place and zero the indicator.
11. Turn the engine in the opposite direction of rotation through the TDC position, and continue to 45° BTDC, noting the NH index marks on the fixture. As you turn the engine in the opposite direction, you will notice the push rod indicator travels from 0 to approximately .005″-.015″ (.127-.381 mm) in a clockwise direction. This travel indicates the camshaft lobe nose crush. This step ensures that all gear backlash is removed before the actual setting is checked.
12. Slowly turn the engine in the direction of rotation and stop at a position equal to about 30° BTDC. From this point, gently turn the engine until the piston dial indicator registers .2032″ (5.16 mm), **Figure 23-30.**
13. Note the value registered on the injector push rod dial indicator. This value is read from zero in a counterclockwise direction. Compare the injector dial's value with the specifications listed in the service manual, the CPL data plate on the gear train cover, or in the engine control parts list manual.

Interpreting Readings

If the push rod dial indicator reading is greater that specifications, the engine injector timing is "slow." To advance engine timing, the proper thickness of gaskets

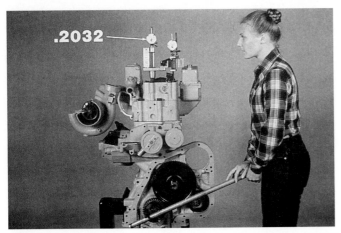

Figure 23-30. Gently nudge the piston into the position .2032" (5.16 mm) from TDC and check push rod indicator reading against engine specifications. (Cummins Engine Co., Inc.)

should be added behind the cam follower housing. If the push rod dial indicator reading is less than specifications, injection timing is "fast," and gaskets must be removed from behind the cam follower housing to retard injection timing.

Figure 23-31 lists the various gasket thicknesses available and the change in push rod travel that will result. If timing adjustment cannot be made using gaskets alone, an offset camshaft key must be installed and the injection timing procedure repeated to determine the gasket thickness now required.

 Note: One Print-O-Seal gasket must be used on each cam follower housing. The sealing end should always be towards the cam follower for effective sealing.

Valve and Injector Adjustments

Once the injection-to-engine timing has been checked, basic tune-up procedures such as adjusting the crossheads, valves, and injector clearances should be completed. These checks and adjustments should be made with the engine cold.

Camshaft Follower Housing Gasket Thickness			
Thickness		Change in Pushrod Travel at 19° BTDC	
Inches	Millimeters	Inches	Millimeters
.006-.008	.15-.20	.040-.050	.0015-.002
.014-.020	.30-.51	.090-.130	.0035-.005
.020-.024	.51-.61	.130-.150	.005-.006
.027-.033	.69-.84	.180-.200	.007-.008

Figure 23-31. Cam follower gasket thickness specifications for Cummins PT injection equipped engines. (Cummins Engine Co., Inc.)

Because the crosshead and valves on one cylinder can be adjusted while the injector is being set on another cylinder, the engine cylinder numbering sequence should be known before beginning any adjustment. You should also be familiar with the *valve set marks* on the engine *accessory drive pulley* and understand which valves and injectors can be checked and adjusted at valve set marks A, B, and C.

The accessory drive pulley valve marks are the most widely used reference markings on Cummins engines. Some older engines may have them on the flywheel or crankshaft pulley. **Figure 23-32** illustrates the accessory drive pulley valve set markings and stationary pointer location on the engine gear cover of one engine. When A is aligned, pistons 1 and 6 are both at 90° ATDC, however, one is on its power stroke while the other is on its intake stroke. When B is aligned, the same holds true for pistons 5 and 2. For C, the pistons affected are 3 and 4.

The respective injector and valve that can be set when the engine is rotated to each mark is shown in **Figure 23-33**. All valves and injectors are set in 720° of engine rotation. It is always easier to perform these adjustments in the firing order of the engine. On Cummins engines, the direction of engine rotation is always determined from the front. Cylinder numbering also starts from the front, with the number 6 cylinder as the closest to the flywheel on an inline six-cylinder engine. On V-configured engines, the left and right bank are determined from the flywheel end. On inline six-cylinder NT and L10 Cummins engines, typical firing order is 1-5-3-6-2-4. On Cummins V-6 engines, the firing order is 1-4-2-5-3-6, and on V-8 engines, 1-5-4-8-6-3-7-2. Left-hand rotation engines will have their firing order reversed. In the case of Cummins V-8 engines, the left-hand firing order is 1-2-7-3-6-8-4-5.

Valve Crosshead Adjustment

Valve crosshead adjustment should always be checked before attempting to set the intake or exhaust

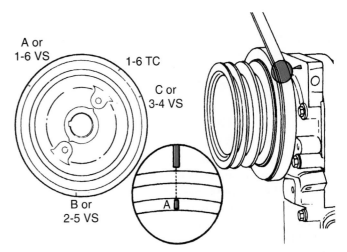

Figure 23-32. Example of accessory drive pulley valve set markings and stationary pointer location on engine gear case cover. (Cummins Engine Co., Inc.)

Injector and Valve Adjustment Sequence (Cummins NT Engines)			
Manually Rotate Engine in Direction of Rotation	Pulley Position	Set Cylinder	
		Injector	Valve
Start	A	3	5
Advance to	B	6	3
Advance to	C	2	6
Advance to	A	3	2
Advance to	B	1	4
Advance to	C	5	1

Figure 23-33. *Injector and valve adjustment sequence. Pulley position in relation to engine rotation is very important. (Cummins Engine Co., Inc.)*

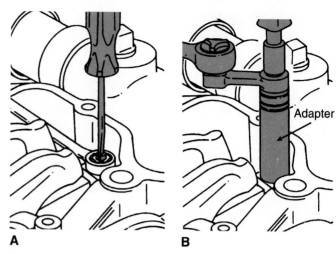

A   **B**

Figure 23-34. *Tightening the valve crosshead locknut. A—Using a wrench/screwdriver to tighten the locknut. B—Using a torque wrench adapter. (Cummins Engine Co., Inc.)*

valve clearance. Before adjusting valve crossheads, the **valve clearance**, or the injector setting, remove the rocker housing cover plates and torque the housing bolts to the proper torque using the sequence recommended in the service manual. Make certain the injector clamp bolt is torqued to specifications. The valve crosshead or bridge is located on both sides of the injector. The valve crosshead allows two intake or exhaust valves to be opened at the same time. There are two crossheads per cylinder. Each must sit square on its guide stud in the cylinder head. To make the adjustment:

1. Use **Figure 23-33** to determine which crosshead to check. For example, in the A position, the valves of cylinder number 5 are fully closed. The crossheads serving this cylinder should be adjusted at this time.

2. Loosen the crosshead screw locknut one full turn and check that there is no binding between the crosshead and its guide.

3. Apply light finger pressure on the crosshead center pallet surface. Gently tighten the adjusting screw until it touches the valve stem. Do not overtighten or the crosshead will raise up off of the other valve stem.

4. Tighten the locknut to proper torque while holding the adjustment screw stationary, **Figure 23-34A.** A special torque wrench adapter is also available to make this job easier, **Figure 23-34B.**

Valve Adjustment

Once the valve crosshead is adjusted, the valve clearance can be set. Any time one set of valve set timing marks is aligned, two separate cylinders, A (1 and 6); B (5 and 2), or C (3 and 4) are positioned at 90° ATDC. However, only one cylinder's valves can be set.

For example, when the B valve set mark is aligned for the first time, either the valves of cylinder 5 or the valves of cylinder 2 can be adjusted, depending on whether the engine is in its first or second 360° of rotation. The other cylinder's valves will be set when the engine is turned into

the B mark position for the second time in the adjustment sequence. The recommended intake and exhaust valve lash specifications are stamped on the engine CPL data plate.

Valve lash can be adjusted by using a torque wrench. Loosen the rocker arm lever locknut, and tighten the adjusting screw to 6 in.-lbs. (.677 N•m). The valve arm adjustment locknut is then tightened to specifications using either the torque wrench adapter or a screwdriver and combination wrench as shown in **Figure 23-35.**

The valve lash can also be set using go/no go feeler gauges. For example, on an engine with a .011″ intake valve clearance specification, a .010″ go, .012″ no go feeler gauge should be used to make the setting. Valve and injector specifications for various Cummins engine models are listed in **Figure 23-36.**

Top Stop Injector Timing Adjustment

Most Cummins PT fuel systems are equipped with Top Stop injectors that have a zero lash setting. Check the

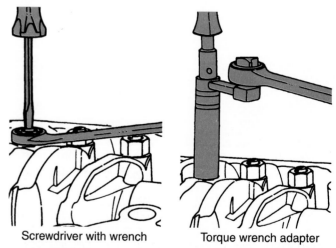

Screwdriver with wrench   Torque wrench adapter

Figure 23-35. *Tightening valve rocker arm adjustment locknut. (Cummins Engine Co., Inc.)*

Valve and Injector Specification Values	
NH/NT	**Adjustment Specifications** in. (mm)
Intake Valve	.011 (.280)
Exhaust Valve	.023 (.580)
Injector Preload (Top Stop)	5-6 in.-lbs. (.564-.677 N•m)
Injector Travel (except Top Stop) Big Cam Small Cam NTE	 .228 (5.79) .170 (4.31) .225 (5.71)
10 Liter Intake Valve Exhaust Valve	 .014 (.350) .027 (.680)
Small V-Type Intake Valve Exhaust Valve Injector	 .012 (.300) .022 (.550) 60 in.-lbs. (6.77 N•m)
V/VT/VTA 903 Intake Valve Exhaust Valve Injector	 .012 (.300) .025 (.630) .187 (4.74)

Figure 23-36. Valve and injector specifications for various Cummins engines. (Cummins Engine Co., Inc.)

engine data plate to confirm the use of Top Stop injectors and zero lash settings. Older engines may use non-Top Stop injectors. In this case, a lash setting other than zero will be listed on the engine plate.

Top Stop Injector Preload

When a Top Stop injector is removed from the engine, ***injector preload*** must be reset using a special adjustment fixture. This procedure sets the proper plunger travel for the injector. Mount the injector in the fixture and preload the internal injector spring by pulling down on the fixture handle, **Figure 23-37.** Change the injector plunger travel by rotating the adjusting nut on the injector top until a torque of 55 ft.-lbs. (73.3 N•m) registers on the fixture gauge. The plunger's upward travel is limited by the position of the adjusting nut. Its downward travel stops when the plunger bottoms in the injector cup.

Once preload is set, the injector can be installed in the engine and its zero lash setting checked:

1. Rotate the accessory drive pulley clockwise to place valve set mark A in position and check the coupled cylinders to determine which one is on its power stroke (valves closed and lash exists).

2. Loosen the injector rocker arm locknut and turn the adjusting screw clockwise until all clearance is removed from the valve train components. This is confirmed by rotating the injector push rod while

Figure 23-37. Setting plunger travel using a Top Stop Injector preload fixture. (Cummins Engine Co., Inc.)

turning the adjusting screw until the push rod can no longer be turned.

3. Tighten the injector adjusting screw one complete turn in order to ensure that the injector link assembly is seated and all oil has been squeezed from the contact surfaces.

4. Slowly loosen the injector adjusting screw, **Figure 23-38,** until the injector spring retainer washer just touches the Top Stop screw. The injector rocker lever is now in an unloaded position.

5. Using a small T-handle torque wrench or a regular torque wrench that does not exceed 12 in.-lbs. (1.35 N•m),

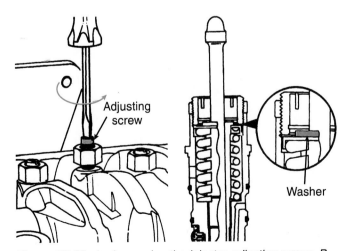

Figure 23-38. A—Loosening the injector adjusting screw. B—The injector retainer washer should just touch the Top Stop screw. (Cummins Engine Co., Inc.)

torque the injector adjusting screw to 5-6 in.-lbs. (.564-.677 N•m). Never overtorque the injector setting. Damage to valve train components and increased fuel consumption and exhaust emissions can result from overtorquing.

6. Tighten the injector adjusting screw locknut to specifications using either of the methods shown earlier.

Step Timing Control (STC)

Step timing control (STC) is a variable injection timing system. This system allows the injection timing to operate in advanced mode during start-up and light loads and in the normal timing mode during medium to high engine load conditions. STC offers many advantages during advanced injection:

❑ Improves cold weather idling.

❑ Reduces cold weather light smoking.

❑ Improves light load fuel economy.

❑ Reduces injector carbon build-up.

During normal injection timing, STC controls cylinder pressures and reduces oxides of nitrogen (NO_x) emissions. Step timing control changes the timing from advanced to normal at a set fuel pressure. This pressure occurs between light and medium engine loads.

Operation

The STC system uses a direct fuel-feed Top Stop injector having two plunger springs and a hydraulic tappet. The Top Stop tappet has both an inner and outer piston. Timing is advanced by filling the tappet with engine oil. This lengthens the tappet and effectively increases the injector plunger length for each degree of camshaft lobe rotation.

During normal timing, **Figure 23-39,** no oil is allowed into the injector tappet and it collapses (the inner piston touches the outer piston) before the injector plunger starts

to move and inject fuel. In the normal mode, the STC injector is similar to a standard injector except that the STC camshaft has a higher lift to reduce the space between the pistons in the tappet.

When the system is in the advanced mode, **Figure 23-40,** the STC control valve directs lube oil pressure to the tappet, filling the space between the two pistons. The inner piston contacts this oil and transmits hydraulic pressure to the outer piston and injector plunger. The plunger begins moving sooner in the injection cycle. Because it started its downward motion early, the injector plunger bottoms in the cup before the cam follower reaches the top of the injection ramp. To allow for this extra camshaft lift, the oil trapped in the tappet is forced out.

Whenever the pressure in the oil manifold is greater than 10 psi (68.95 kPa) the STC inlet check ball moves from its seat and the cavity between the inner and outer pistons is filled with oil. During the injection cycle, oil is held inside the tappet by the inlet and load-cell check balls. At the end of the injection cycle, the injector force increases the oil pressure in the tappet and holds the plunger firmly in the cup. This pressure increase moves the load-cell check ball off its seat and the oil drains through the drain holes in the injector adapter, returning oil to the sump through drain passages in the cylinder head and block. As the oil drains, the continued cam lift causes the inner piston to contact the outer piston, maintaining injector plunger seating force.

Tappet operation is controlled by the STC control valve, which senses fuel pressure and directs lubricating oil to control timing. It also senses engine brake (C-brake) operation and ensures that the engine is in the normal timing mode when the brakes are activated.

STC Control Valve

The *STC control valve* uses fuel pressure and spring force to control the plunger position. The STC control valve is located on the side of the block below the fuel pump. Fuel pressure acts on the plunger. Plunger position determines whether the oil passage to the hydraulic tappets is open or closed. In the advanced timing mode, the spring

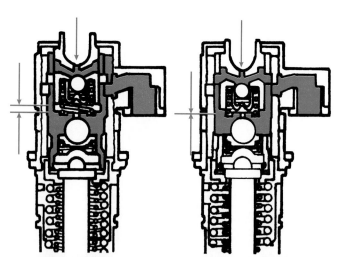

Figure 23-39. *Normal timing STC tappet operation with no oil in the tappet. (Cummins Engine Co., Inc.)*

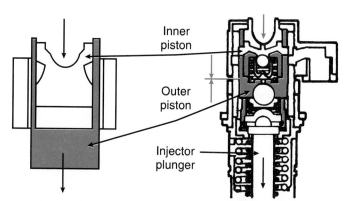

Figure 23-40. *In the advanced timing mode, the STC valve is filled with oil. Motion transfer from the rocker arm to the injector plunger is instantaneous. (Cummins Engine Co., Inc.)*

opposed fuel pressure holds the plunger in the open position, **Figure 23-41.** Pressurized lube oil flows to the tappets and initiates advanced engine timing.

The spring holds the plunger in the open position until the fuel pressure rises above the certified switching pressure. At this point, the higher fuel pressure overcomes spring pressure and shifts the plunger, closing the oil passage. With the oil supply to the tappets interrupted, the engine now operates in the normal timing mode. As fuel pressure decreases, the plunger shifts and opens the passage once more. A specific pressure difference is built into the valve to prevent the plunger from fluttering between the advanced and normal settings.

The control valve supplies oil to the STC external oil manifold. A check valve prevents oil from draining out of the line. The internal oil manifold connects the oil supply to each STC injector in the rocker housing. Fuel pressure to the STC valve is provided by a hose running between the fuel shutdown valve and the STC valve.

C-Brake Activation

The rear **C-brake** is connected to the STC control valve by a hose. When the C-brake is activated, oil pressure in the brake housing is directed through the hose to the STC valve, holding it closed for normal timing. Because the Cummins C-brake can be activated progressively (1-brake, 2-brake, or 3-brake), the STC control valve receives the oil pressure signal only when the system is in the 2-brake or 3-brake mode. When only one C-brake is activated, intake manifold pressure and camshaft loading

are not high enough to require normal timing. On engines not equipped with C-brakes, the C-brake sensing line is vented to the engine crankcase.

> **Note: The oil control valve is calibrated to a specific flow and pressure using a fuel pump test stand. Tampering with the valve or plumbing will result in the loss of both fuel economy and engine durability. Correct valve operation is needed to maintain acceptable cylinder pressures and smoke levels, and to ensure good fuel economy.**

Setting STC Top Stop Injectors

STC Top Stop injectors can be adjusted and set either off or on the engine. Off-engine adjustment consists of two stages: setting base plunger travel and total injector travel. These settings can only be done using the specified Top Stop setting fixture. Once the injector is properly mounted in the fixture, plunger travel is set by rotating the stopscrew with an Allen wrench. Total injector travel is set by rotating the top cap of the injector, **Figure 23-42A.**

On-engine adjustments are made using a tappet extender tool. Insert the locating pin of the extender tool into one of the four "chimney" holes in the tappet assembly. Hold the tappet in its fully extended position and adjust the setscrew with a 5 in.-lb. (.564 N•m) T-handle torque wrench, **Figure 23-42B.** These instructions are only

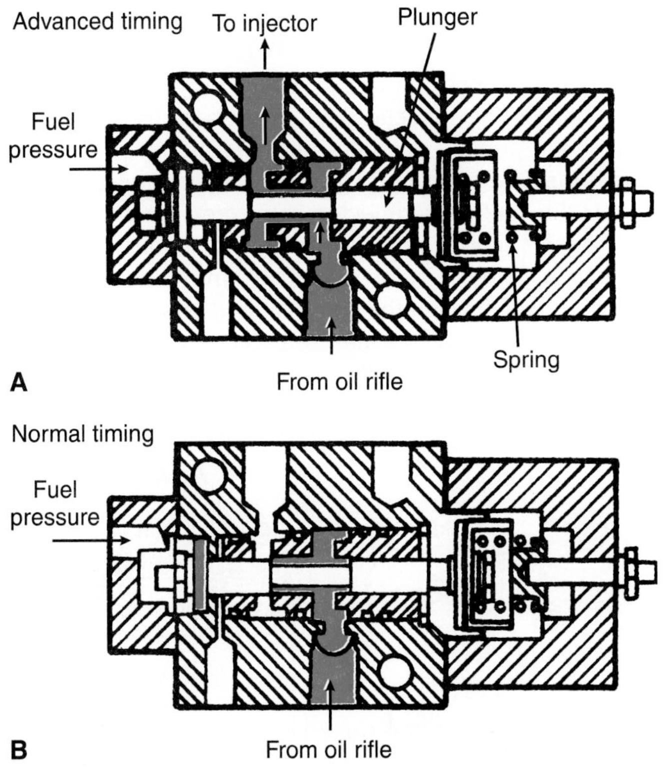

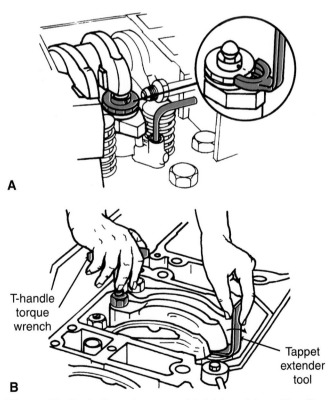

Figure 23-42. Setting plunger and total travel in a Top Stop Injector preload fixture. A—On-engine adjustments are done by using a tappet extender. B—Using a T-handle torque wrench to tighten the setscrew. (Cummins Engine Co., Inc.)

Figure 23-41. STC valve operation. A—Advanced timing. B—Normal timing. (Cummins Engine Co., Inc.)

a summary of the proper setting and adjustment sequence. Always refer to the proper service manual for complete details.

Mechanical Variable Timing

Mechanical variable timing (MVT) is a two-position injection timing control system used on some Cummins engines prior to the introduction of step timing control. MVT is either fully retarded or advanced. It uses an eccentric injector cam follower shaft assembly that is controlled by an air solenoid/actuator mechanism. The actuator is activated by an electrical air pressure sensing device.

When the vehicle air pressure is below 80 psi (552 kPa), the engine starts in the retarded mode, **Figure 23-43A,** and remains there until the pressure exceeds 80 psi (552 kPa). At this pressure, the system automatically shifts to the advanced timing mode, **Figure 23-43B.** The engine shifts back to the retarded position when it reaches 25% of its rated torque. The engine then operates at cruise and full-load conditions in the retarded position.

The injection timing is varied by moving the eccentric mounted injector cam follower assemblies through the use of a rack and pinion gear. The timing is held in the retarded mode by a spring and in the advanced mode by air pressure applied to a piston.

Checking MVT Timing

To check MVT injector timing, place the system in the retarded mode by disconnecting the air solenoid wiring, then starting and operating the engine at low idle for 30 seconds. Remove the actuator cap and watch the spacer move down and seat on the spring retainer. The retarded position may also be obtained by manually turning the crankshaft at least two revolutions in the direction of engine rotation.

1. Insert the proper timing tool in the number 3 cylinder. Be sure the piston is on its compression stroke.

2. Follow the injection timing procedures outlined earlier in this chapter to the point where both indicators have been properly zeroed, and the piston is at .203″ BTDC. Remember that the MVT actuator piston must be down and no air applied to obtain the correct timing readings.

3. Check the injector reading using the timing codes listed on the engine data plate. Check the reading listed on the MVT data plate against the injection timing chart in the control parts list manual or service manual.

4. To change MVT timing, remove the actuator cap and loosen the external locknut. Rotate the spring retainer screw to make the timing adjustment, **Figure 23-44.** Rotate the spring retainer clockwise to retard timing and counter-clockwise to advance timing. One complete revolution of the spring retainer changes push rod travel by approximately .004″ (.10 mm).

5. After completing the necessary adjustments, reconnect the wiring to the actuator/solenoid.

6. Apply 12 volts to the solenoid and 80 psi (552 kPa) of air pressure to the MVT air inlet. The supplied air should allow the actuator piston to move freely to the advanced timing position.

7. Check the injection timing in the advanced position to be sure it is within stated specifications. If specifications are not met, check for incorrect spacer length, binding parts, and misaligned tool placement.

8. After completing spring retainer adjustment, be sure to torque the actuator locknut to specifications.

Summary

The PT fuel injection system derives its name from two of the primary variables affecting the amount of fuel that is metered per cycle. The P refers to the pressure of the fuel at the inlet of the injectors. This pressure is controlled by the fuel pump. T refers to the time available for the fuel to flow into the injector cup. This time is controlled by engine speed through the camshaft and injection train.

Cummins engines are four-cycle engines, so the camshaft is driven from the crankshaft at one-half engine

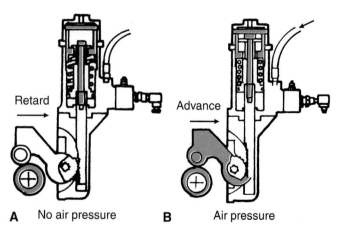

Figure 23-43. *Components of the Mechanical Variable Timing device and its modes of operation. A—No air pressure retards the injection timing. B—Air pressure over 80 psi (552 kPa) advances the timing. (Cummins Engine Co., Inc.)*

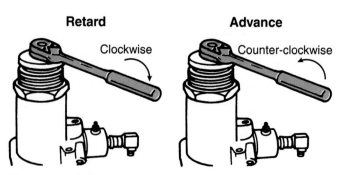

Figure 23-44. *Adjusting the mechanical variable injection timing. (Cummins Engine Co., Inc.)*

speed. The fuel pump turns at engine speed. Because of this, added governing of fuel flow is needed in the fuel pump. The fuel pump varies pressure to the injectors in proportion to engine rpm. The final controlling element in the fuel metering process is the size of the fuel injector openings (orifices). The orifice size, also called the flow area, is determined by calibration of a complete set of injectors.

The PT fuel pump provides fuel transfer from the tank to the engine, rail pressure to the injectors, idle and maximum speed governing, operator control of power below governed speed, exhaust smoke control during acceleration, and engine shutdown. The PT fuel system governor is a simple flyweight mechanical governor driven through gears by the fuel pump mainshaft turning at engine speed.

The PT fuel system begins by regulating the fuel pressure supplied to the governor assembly. This is called the supply pressure. Control of supply pressure is accomplished through the use of a bypass pressure regulator incorporated inside the governor.

The PTG fuel pump used on late-model turbocharged heavy-duty truck engines uses a built-in acceleration exhaust smoke control device. The air-fuel control (AFC) unit restricts fuel flow in direct proportion to engine air intake manifold pressure during engine acceleration, under load, and during lug-down conditions. The PT fuel injector plunger is rocker arm actuated by camshaft rotation.

PT fuel pump checks and adjustments include low and high idle adjustment, checks for fuel system restrictions, checks for fuel pump air leaks, fuel rail pressure checks, and fuel rate checks. Injectors are removed from their mounting bore using a special slide hammer.

On a pressure time injection system, injection must start at a specific number of degrees BTDC. The injection timing check confirms that the distance between the injector plunger and cup when measured with a dial indicator matches specifications when the piston is at 19° BTDC.

A camshaft key can be used to index the camshaft with the cam gear. The size of the key affects injection timing. Using offset cam keys allow changes in camshaft indexing in relation with the cam gear. Offset keys allow the camshaft profile to be rotated slightly while gear train timing remains stationary. Cam follower gaskets can also be added or removed to fine tune the timing to the desired specification. Use of these gaskets alters the position of the injector cam follower roller in relation to the position of camshaft lobe.

Timing checks on Cummins engines require the use of a special timing fixture that can be adapted to most engine models. The fixture is equipped with a push rod indicator and a piston plunger indicator. Basic tune-up checks include valve crosshead adjustment, valve adjustment, and setting injector preload.

Step Timing Control (STC) is a variable injection timing system that allows the engine to operate in the advanced mode of injection timing during start-up and light loads and in the normal timing mode during medium to high engine load conditions. Mechanical variable timing (MVT) is a two position injection timing control used on some Cummins engines prior to the introduction of STC. MVT is either fully retarded or advanced.

Important Terms

Pressure-time (PT) fuel injection system	Control parts list (CPL) data plate
PACE system	Variable speed (VS) governors
Flow area	Fuel rate
Rail pressure	Camshaft key
Flow time	Valve set marks
Metering orifice	Accessory drive pulley
Pressure-time governed (PTG) fuel pump	Valve crosshead adjustment
PT fuel system governor	Valve clearance
Supply pressure	Valve lash
Counterbore	Injector preload
Throttle	Step Timing Control (STC)
Throttle leakage	STC control valve
PTG air-fuel control (AFC)	C-brake
Shutdown valves	Mechanical variable timing (MVT)
Cooling kit elbow	
Pressure-time type D Top Stop Injector	

Review Questions—Chapter 23

Do not write in this text. Place your answers on a separate sheet of paper.

1. In a PT fuel injection system, the fuel pump varies fuel pressure to the _____ in proportion to the engine speed.

2. In a PT fuel system, it is not necessary to time the _____ to the engine as in inline and distributor systems.

3. The PTG-AFC fuel pump is coupled to the _____ drive on most engines.

4. Name at least four functions of a PTG AFC fuel pump in a PT injection system.

5. The PT fuel system governor is operated by _____.

6. If the throttle leakage is set too high, _____ of the injectors will result.
   (A) carbon loading
   (B) engine runaway
   (C) hesitation
   (D) fast deceleration

7. Name two methods of adjusting injection timing in a PT injection system.

8. Which of the following is a truly variable injection timing control system available on PT fuel systems?
   (A) Mechanical Variable Timing.
   (B) Step Timing Control (STC).
   (C) AFC Timing Control.
   (D) PT Timing Control.

9. The PT fuel injector plunger is:
   (A) electronically activated using a solenoid and relay.
   (B) rocker arm actuated by the rotation of the engine camshaft.
   (C) hydraulically activated by fuel pressure.
   (D) Both A & B.

10. On a PT fuel injection system, idle speed can be accurately set by monitoring _____.

ASE-Type Questions

1. Technician A says that control of the fuel rail pressure in the PT fuel system begins by regulating fuel supply pressure to the pump through the use of a bypass pressure regulator inside the governor. Technician B says that the supply pressure is regulated by the use of a control rack and sleeve linked to the governor. Who is right?
   (A) A only.
   (B) B only.
   (C) Both A & B.
   (D) Neither A nor B.

2. Technician A says that on a Cummins pressure time injection system, injection must start at a specific amount of degrees ATDC. Technician B says that injection must start at a specific amount of degrees BTDC. Who is right?
   (A) A only.
   (B) B only.
   (C) Both A & B.
   (D) Neither A nor B.

3. Technician A says timing checks on Cummins engines require the use of a special timing fixture that can be adapted to all Cummins engine models. Technician B says that timing checks on Cummins engines require no special equipment, just a feeler gauge and ordinary shop tools. Who is right?
   (A) A only.
   (B) B only.
   (C) Both A & B.
   (D) Neither A nor B.

4. All of the following is characteristics of camshaft keys, EXCEPT:
   (A) the size of the key affects injection timing.
   (B) offset cam keys will change camshaft indexing in relation with the crankshaft gear.
   (C) the greater the offset, the greater the timing is retarded.
   (D) offset keys allow the camshaft profile to be rotated while the gear train remains stationary.

5. Valve crosshead adjustment should always be checked before attempting to set _____ .
   (A) intake valve clearance
   (B) exhaust valve clearance
   (C) intake valve lift
   (D) Both A & B.

6. All of the following is performed by the STC during advanced injection, EXCEPT:
   (A) improved cold weather idling.
   (B) reduced injector carbon build-up.
   (C) improved heavy load fuel economy.
   (D) reduced cold weather smoking.

7. Technician A says you should remove a PT injector from its mounting bore by using a special slide hammer. Technician B says you should carefully pry the injector out of its bore using a properly sized bar. Who is right?
   (A) A only.
   (B) B only.
   (C) Both A & B.
   (D) Neither A nor B.

8. The AFC unit of a PTG-AFC fuel pump is responsible for _____ .
   (A) filtering fuel entering the pump
   (B) varying engine timing to match operating conditions
   (C) limiting the amount of exhaust smoke generated during engine acceleration
   (D) regulating fuel rail pressure

9. The maximum allowable inlet restriction to the suction side of the fuel transfer pump on a Cummins PT pump-equipped engine is ___ Hg (mercury) when a new primary filter has been installed.
   (A) 2″
   (B) 4″
   (C) 6″
   (D) 8″

10. In a Cummins PT equipped engine, any time one set of valve set timing marks on the accessory drive pulley is aligned with the cast-in stationary pointer on the engine gear cover, two separate cylinders are positioned at _____ ATDC.
   (A) 30°
   (B) 60°
   (C) 90°
   (D) 120°

Chapter 24

Basics of Electricity

After studying this chapter, you will be able to:
- ❏ Explain what electrical current is and what force causes it to flow or move from one point to another.
- ❏ Define and explain how current, voltage, and resistance are measured.
- ❏ Explain conventional theory versus electron theory of current flow.
- ❏ State Ohm's law and describe how it is used to determine voltage, current, and resistance in a dc circuit.
- ❏ Define conductors and insulators.
- ❏ Describe the three types of circuits used in electrical systems.
- ❏ Describe what occurs in closed, open, and shorted circuits.
- ❏ Describe the operation of electrical system components.
- ❏ Describe the proper safety precautions required for servicing, testing, and charging batteries.
- ❏ Perform battery state of charge and capacity tests.
- ❏ Properly recharge batteries using either fast or slow charge methods.
- ❏ Properly jump start vehicles equipped with either 12-volt or 12- to 24-volt electrical systems.
- ❏ Describe the operation of basic semiconductor devices such as transistors and diodes.

All technicians must have a clear understanding of how electricity works. In addition to the electrical circuits and components used in diesel engine starting and charging systems, nearly all late-model high-speed diesels manufactured for light, medium, and heavy-duty trucks are equipped with electronic engine control systems. These systems rely on a network of electronic sensors and actuators controlled by a computer. Troubleshooting these systems requires the technician to take accurate voltage and resistance readings using electronic diagnostic equipment. A simple mistake caused by lack of electrical knowledge can damage expensive components or create a safety hazard.

Principles of Electricity

All material is made up of atoms. Each atom has a nucleus made of **protons** which have a positive charge and **neutrons** which have no charge. **Electrons** are the negatively charged particles orbiting the nucleus of an atom. It is the release of electrons which causes electricity, referred to as current, in a circuit. **Electric current** is the controlled flow of free electrons from atom to atom within a conductor, such as a copper wire. To create this flow of electrons, there must be an outside source of energy acting on the conductor's electrons. In a diesel truck engine, this source of energy is provided by the battery or the alternator.

Figure 24-1 illustrates what happens when a length of copper wire is connected between the positive and negative poles of a battery. The battery's positive pole has excess protons around it, giving it a net positive charge. The positive pole acts like a magnet with a very strong attraction for the negatively charged electrons of the copper atoms in the wire. The attraction is so strong, the battery's positive pole pulls an electron out of the orbit of the copper atom nearest it.

The copper atom now has more protons than electrons. It has lost its electrical balance and is now positively charged. Atoms with an electrical charge are unstable and will try to balance themselves. The only way for an atom to do this is to steal an electron from a neighboring atom. The positively charged copper atom, aided by the pull of the battery's positive pole, attracts an electron from the copper atom next to it. However, on losing an electron, the second copper atom now becomes unbalanced and attracts an electron from the next copper atom. This sets off a chain reaction along the entire length of the wire.

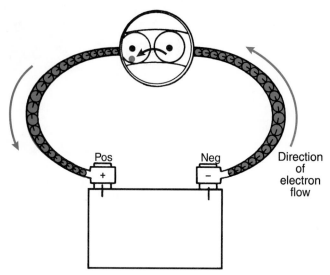

Figure 24-1. *In the electron theory of current flow, electrons are drawn out of the wire at the positive pole of the battery and pushed into the wire at the negative pole.*

The chain reaction continues along the wire to the battery's negative pole. The negative pole has an excess of electrons around it and a net negative charge. It pushes electrons into the copper wire to replace the electrons being pulled out at the positive pole. This chain reaction occurs very fast as electrons move at the speed of light. As long as the wire remains connected, the electron flow will continue for as long as the positive and negative charges at the poles are maintained.

Conventional versus Electron Theory

When scientists first began experimenting with batteries and simple circuits, little was known about the make-up of atoms and the behavior of electrons. From what they observed, scientists believed current was the flow of positively charged particles from the positive pole of a battery to the negative pole. This became known as the **conventional theory** of current flow. Since it was simpler to illustrate and understand, the conventional theory was the only theory used to describe electrical systems for many years. Many electrical diagrams still show current flowing from the positive terminal of the battery, through the circuit to the negative pole.

The **electron theory** of current flow states current moves from negative to positive as described earlier. The electron theory is used in describing the operation of computers and other electronic devices. Because many diesel engines are now electronic and computer-controlled, wiring diagrams using the electron theory of current flow may be found in some service manuals.

Measuring Electricity

The flow of electricity can be measured like the flow of any other substance. Consider the example of water flowing through a hose. A certain amount of water exits the hose in a given time period and can be measured in

gallons or liters per minute. If the hose is kinked, the flow of water meets resistance and less water reaches the end of the hose. Water pressure within the system also affects the volume of water exiting the hose. High pressure results in high water volume exiting the hose. Low pressure will result in a smaller volume of water exiting the hose. These principles of volume, resistance, and pressure can be used to measure electricity.

Current

As mentioned earlier, current refers to the rate at which electrons flow in an electrical circuit. The unit for measuring the rate of current flow is the *amp* or **ampere (A).** An ammeter is used to measure current. In some situations, such as when measuring signals from sensors used in computerized engine control systems, the ampere is too large a unit for measuring current. In such a case, the intensity of current is measured in milliamperes (mA). A milliampere equals one thousandth of an ampere.

Direct and Alternating Current

When electrons flow in only one direction, it is called **direct current (dc).** Direct current is produced by a battery and is the type of current used in automotive and diesel truck circuits. Many electronic sensors and computers also operate using dc current. Direct current appears as a straight line on an oscilloscope and has a square waveform when used in digital circuitry, **Figure 24-2.**

Alternating current (ac) surges back and forth in alternating directions at a fixed rate. Ac current will generate a sine waveform, **Figure 24-3.** Alternating current is produced by the engine's alternator. The ac current is converted, or

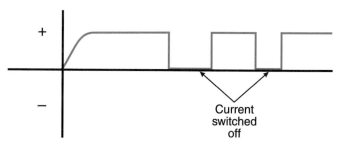

Figure 24-2. *Direct current has a square waveform. Many electronic sensors and computer data will generate this waveform.*

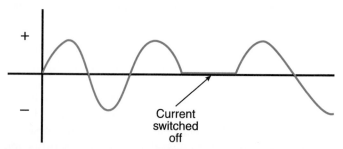

Figure 24-3. *Alternating current generates a sine waveform. This current must be changed to direct current for use in the diesel engine's electrical system.*

rectified, to direct current. This direct current is then used to maintain the battery's charge. If the alternator malfunctions, the battery will quickly lose its charge.

Voltage

Voltage refers to the force or pressure that allows electrons to flow. When two separate electrical points with opposite charges, such as the poles of a battery are joined, voltage is produced. This potential difference is referred to as the *voltage* or **electromotive force (emf).**

A typical truck battery usually has an emf of 12 volts between its positive and negative terminals. Some vehicles use a combination of batteries to provide 24 volts of power to parts of the electrical system, such as starting circuits. A voltage potential exists even when no current-consuming devices are connected to the posts. Voltage can exist without current, but current cannot exist without voltage. Voltage is measured using a **voltmeter.** One volt is the amount of electromotive force needed to move one amp of current through a resistance of one ohm. A millivolt equals one thousandth (.001) of a volt.

Resistance

All conductors and electrical devices offer some amount of **resistance** to current flow. In an electrical system, resistance is measured between two points in a circuit. The basic unit of resistance is the ohm. An **ohm** is defined as the resistance that will allow one ampere to flow when the voltage equals one volt. Resistance is measured using an ohmmeter and is indicated by using the symbol Ω.

Many factors affect the total resistance. The wire and its connections are the main sources of unwanted resistance in a circuit. The material that makes up the wire, plus the wire's length and gage (thickness) all affect resistance.

Ohm's Law

The rate of flow or amps, the pressure or volts, and resistance or ohms, all have a very specific relationship which can be stated in mathematical terms. **Ohm's law** states the relationship between current (I), voltage (E), and resistance (R) in a direct current (dc) circuit. Ohm's law can be applied to the entire circuit or to any part of the circuit. When any two factors are known, the third unknown factor can be found using Ohm's law:

$$E = I \times R$$

The voltage in a circuit is equal to the current (in amps) multiplied by the resistance in ohms. A second method of expressing the same formula is:

$$I = \frac{E}{R}$$

The current (amps) in a circuit equals the voltage divided by the resistance in ohms. Finally,

$$R = \frac{E}{I}$$

The resistance of a circuit in ohms equals the voltage divided by the current in amps. An easy method of remembering the equations of Ohm's law is by using the pie method illustrated in **Figure 24-4.** You may also see this law in triangle form in some manuals. It is extremely important for all diesel technicians to understand Ohm's law. It explains how an increase or decrease in voltage, current, or resistance will affect a circuit.

Conductors and Insulators

Conductors are made of materials which have very low resistance to electron flow. Due to their low resistance, conductors can carry large amounts of current without

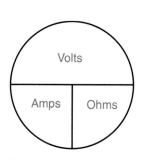

To find volts, multiply across the amps and ohms:

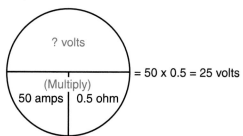

To find ohms, divide volts by amps:

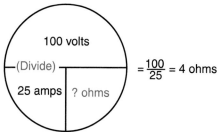

To find amps, divide volts by ohms:

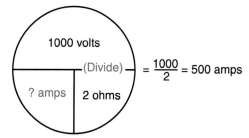

Figure 24-4. *Ohm's law pie chart can be used to find one electrical quantity when the other two are known.*

producing excessive amounts of heat. Good conductors are strong, lightweight, and easy to use. Copper, aluminum, and copper-covered aluminum are the three most common conductors used to make wire and other electrical system components. Gold and platinum are also used for some terminal applications.

> **Warning: Your body is an excellent conductor of electricity. Always remember this when working on any electrical system. Always observe all electrical, battery, and storage safety rules.**

Insulators are materials having very high resistance to electron flow. Plastic, rubber, cloth, glass, wood, and paper are good insulators. Insulators used in automotive and truck electrical systems must be able to resist current flow, withstand heat, moisture, corrosion, and vibrations generated by the engine, and be able to withstand high or continued voltage. Thermoplastics are the most commonly used insulating materials found in automotive and truck

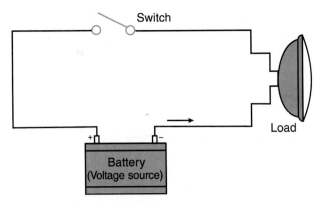

Figure 24-5. Simple series electrical circuit.

electrical systems. Insulators resistant to water and salt are used in marine applications.

Cracked, frayed, or damaged insulation can cause shorts, grounds, and other problems in the electrical system. It can also be dangerous if inadvertently touched or handled. Inspect electrical wiring on a regular basis, and always repair or replace any damaged insulator.

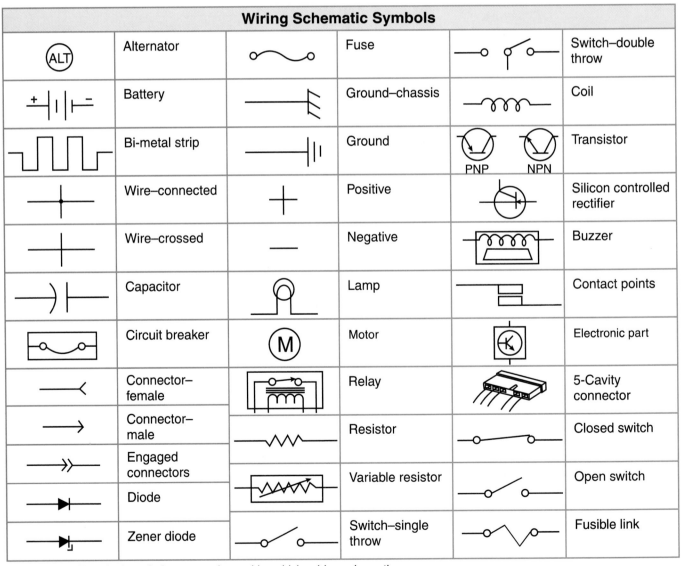

	Alternator		Fuse		Switch–double throw
	Battery		Ground–chassis		Coil
	Bi-metal strip		Ground		Transistor
	Wire–connected		Positive		Silicon controlled rectifier
	Wire–crossed		Negative		Buzzer
	Capacitor		Lamp		Contact points
	Circuit breaker		Motor		Electronic part
	Connector–female		Relay		5-Cavity connector
	Connector–male		Resistor		Closed switch
	Engaged connectors		Variable resistor		Open switch
	Diode				
	Zener diode		Switch–single throw		Fusible link

Wiring Schematic Symbols

Figure 24-6. Electrical symbols commonly used in vehicle wiring schematics.

Electrical Circuits

All electrical circuits in an engine control system and/or vehicle electrical system are powered by the vehicle's batteries. An electrical circuit is simply a continuous path along which current moves. Current moves from the voltage source (battery or alternator), through the conducting wires, through the loads connected to the circuit (such as lights and starter motors), and finally back to the source, **Figure 24-5.**

Figure 24-6 illustrates many of the standard electrical symbols used in automotive and truck electrical diagrams. These diagrams are commonly called *schematics.* There are three basic types of circuits used in electrical systems:

❑ Series circuits.

❑ Parallel circuits.

❑ Series-parallel circuits.

The circuit components are represented by symbols that are fairly standard throughout the industry.

Series Circuits

In all *series circuits,* there is only one path for the current to flow. The series circuit shown in **Figure 24-7** has three resistors placed in the path of current flow. All current supplied by the battery must flow through each of these resistors before it can return to the battery. In a series circuit, total resistance is found by adding the values of all the loads in the circuit. The circuit shown in **Figure 24-7** has a 6 ohm, 4 ohm and a 2 ohm resistor. The total resistance in this series circuit can be found as shown here:

6 ohms + 4 ohms + 2 ohms = 12 ohms

Keep in mind there is a voltage drop each time current passes through a resistor. Use Ohm's law to determine the voltage drop across any resistor. However, use only the amperage and resistance at each particular resistor or load.

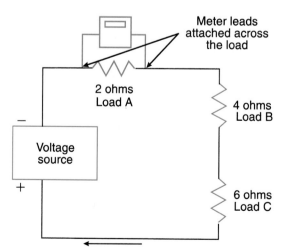

Figure 24-7. *In a series circuit, current can only flow through one resistor or load at a time. The combined voltage drops across resistors and loads in the circuit will equal the terminal voltage of the battery.*

Voltage Drop = Resistance (load) × Amperage (at any one resistor or load).

In testing diesel electrical systems, the amount of voltage required to overcome the resistance in different parts of the circuit is measured using a voltmeter. As shown in **Figure 24-7,** the two voltmeter leads are attached parallel to each resistor. The amount of volts used by each resistor or load is read on the voltmeter. As shown by Ohm's law, high resistance loads require more voltage to push a given amount of current than low loads.

The amount of voltage measured across each of the loads is the voltage drop. The voltage drop for load A in **Figure 24-7** is 2 volts, 4 volts for load B, and 6 volts for load C. All these voltage drops, when added together, equal the total terminal voltage at the battery, in this case, 12 volts.

When resistance is high, the voltage drop will be high. When resistance is low, the voltage required to push this current through (and the voltage drop) will be low. The following points summarize series circuit characteristics:

❑ Total resistance is calculated by adding the resistance of each load:

$R_{total} = R_1 + R_2 + R_3 + ... R_n.$

❑ Current (amps) through each resistor is the same.

❑ The voltage drop across each load will be different if the resistance values are different.

❑ The sum of the voltage drops equals the source voltage.

Parallel Circuits

Parallel circuits provide two or more separate paths for current flow. Each branch of a parallel circuit operates independently of the others, and may have its own resistors and loads connected to it, **Figure 24-8.** If one branch of the total circuit should break or become open, the others will continue to operate. This is why parallel circuits are commonly used in automotive electrical systems.

In a parallel circuit, current can flow through more than one resistor at a time. So total resistance is not additive as in a series circuit. Instead it is determined by the following formula:

$$R_{total} = \cfrac{1}{\cfrac{1}{R_1} + \cfrac{1}{R_2}}$$

So in the parallel circuit shown in **Figure 24-8,** total resistance equals:

$$R_{total} = \cfrac{1}{1/2 + 1/4}$$

When dealing with different fractions, it is necessary to find the lowest common denominator, which is 12 in this case:

$$R_{total} = \cfrac{1}{6/12 + 3/12}$$

$$R_{total} = \cfrac{1}{3/4} = 1.33 \text{ ohms}$$

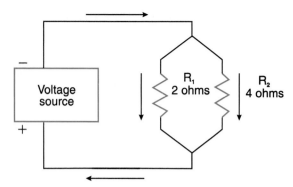

Figure 24-8. *In a parallel circuit, current can flow through several paths.*

In a parallel circuit, total resistance will always be less than the smallest resistor in the circuit. If more resistors are added to the parallel circuit, total resistance will decrease. This is another reason parallel circuits are popular in car and truck electrical systems. The following is a summary of parallel circuit characteristics:

❑ Total resistance is always less than the lowest resistor.

❑ Current (amps) flowing through each resistor is different if the resistance values are different.

❑ Voltage drop across each resistor is the same. This also equals the source voltage.

❑ Total amperage in the parallel circuit is equal to the sum of amperages on the branch circuits.

Series-Parallel Circuits

In a ***series-parallel circuit,*** both series and parallel paths exist. **Figure 24-9** illustrates four resistors connected in a series-parallel circuit. Total resistance in a series-parallel circuit is found by first adding together the resistance in the series portion of the overall circuit:

Series resistance = $R_1 + R_4$
$= 5 + 3 = 8$ ohms

The resistance in the parallel portion of the circuit is then calculated:

$$\text{Parallel resistance} = \cfrac{1}{\cfrac{1}{R_2} + \cfrac{1}{R_3}}$$

$$= \cfrac{1}{1/4 + 1/4}$$

$$= \cfrac{1}{1/2}$$

$$= 2 \text{ ohms}$$

The resistance in the parallel section is added to the series resistance to obtain the total resistance in the entire circuit. So for the series-parallel circuit shown in **Figure 24-9:**

Series-parallel resistance = 8 ohms + 2 ohms = 10 ohms

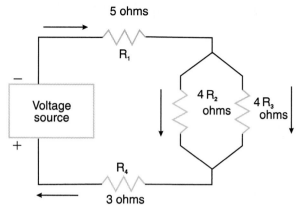

Figure 24-9. *In a series-parallel circuit, both series and parallel circuits exist for current flow.*

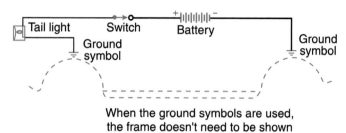

When the ground symbols are used, the frame doesn't need to be shown

Figure 24-10. *In modern electrical systems, the truck frame is used as a ground return circuit to the battery.*

Ground Return Circuit

A good ground is always at zero voltage and can accept an infinite number of electrons. That is why a path for current flow can always be completed by connecting it to ground.

The ground connection can be made at a battery terminal (usually the negative) or it can be made to a metal part of the engine or vehicle frame. The part of an electrical circuit formed by the metal components of the vehicle or its engine is called the ***ground return circuit.*** In most vehicles, the negative terminal of the battery is connected to the frame or engine block. This completes the path or circuit back to the battery. **Figure 24-10** illustrates a simple ground return circuit and the symbol representing ground.

Basic Circuit Conditions

A properly wired and operating circuit has a complete path from its source, through its intended loads, and back to the source through the ground circuit. Problems exist in a circuit when it becomes open, shorted, or improperly grounded.

Open Circuits

An ***open circuit*** occurs when there is a break in the wire or the wire connection at a load. An open circuit is like a circuit with a switch in the open position. Voltage drop across an open circuit is always equal to the voltage of the source. A circuit or portion of a circuit with no opens is said to have continuity. Troubleshooting often calls for checking continuity between two points in a circuit.

Shorted and Grounded Circuits

A **shorted circuit** is a connection of extremely low resistance between two points in a circuit. In a short circuit, the intended path for current flow is changed either accidentally or intentionally. A short circuit is not like an open circuit. A **grounded circuit** is a unwanted connection in a circuit to ground. Grounded circuits can drain the batteries or cause one or more devices in the circuit to operate, even after the operator turns the device off.

Accidental shorts occur when current has strayed from its intended path through a break or crack in a wire's insulation. A short can occur when two exposed wires or terminals of an electrical system accidentally come in contact or when a bare current carrying wire accidentally touches an unintended source of ground such as the metal frame, body, engine, or transmission housing. Both types of shorts are dangerous because they create a condition in the circuit that can lead to unwanted system operation or a fire.

Figure 24-11 shows a short to ground condition. The wire running from the battery to the tail light has become frayed and the bare wire now contacts the truck frame about halfway between the battery and the light location. The current now has two possible paths back to the battery. It can continue on to the tail light and its intended ground circuit, or it can take the shorter route back through the truck frame starting at the point the wire is frayed. Most, if not all of the current will choose the shorter route. As a result, the tail light will either burn dimly or not at all. However, this is not the only problem that exists.

Normally, the tail light bulb filament draws a relatively small amount of current through the circuit wire. Grounding the circuit before it reaches the light bulb filament creates a path that offers very little resistance to current flow. The battery now begins to work hard to pump current into the metal frame and can quickly drain itself. The light gage wire intended to carry a small amount of current to the tail light is now carrying high current. As electrons fight for position inside the small wire, they bump into one another, creating heat. This heat can eventually melt the plastic insulation and likely start a fire.

When troubleshooting engine control systems, the manufacturer's recommended procedures may call for the intentional temporary shorting to ground of a given signal circuit. This is normally done to force an error condition or otherwise check the circuit functionality. It is important to remember the voltages and currents involved in signal circuits are quite low, but not harmless. Follow all testing instructions to ensure safe, accurate results.

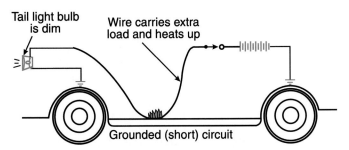

Figure 24-11. *Simple grounded short circuit.*

High Resistance Problems

In addition to opens, shorts, and grounds, circuits are also subject to unwanted resistance caused by loose, corroded, or damaged connectors. A good example of an unwanted high resistance condition is a corroded battery terminal. Dirt and corrosion are the most frequent causes of high resistance in a circuit.

An increase in circuit resistance will cause a decrease in the current supplied to loads such as starter motors, electric fuel heaters, etc. The voltage available to the load will also decrease because there will now be an additional voltage drop across the unwanted resistance. The net result is less available voltage for desired loads such as starter motors, lights, etc.

Electrical System Components

The following sections describe the most common types of electrical devices used in modern heavy-duty truck and diesel powered equipment electrical systems.

Wiring

Electrical **wiring** can be either solid or stranded. Solid wire is made from a single strand of conducting metal. Stranded wires are constructed of many smaller wires twisted together to form a single conductor. Diesel engines equipped with electronic controls will be equipped with shielded cable. This shielding is designed to protect delicate electronic components from the effects of induced voltages (electromagnetic interference), which can adversely affect their operation.

The amount of current a wire can safely carry and the amount of voltage drop that will occur in a wire are determined by two factors: its length and its gage (thickness). The wire gage is determined by the load it has to carry and the distance from the power source to the load.

Wire sizes are established by the **Society of Automotive Engineers (SAE),** which uses the **American Wire Gage (AWG)** system. This system assigns number ratings from 0 to 20 with the smaller numbers indicating the larger gages. Battery cables are typically made of number 2 or 4 gage wire. Most other vehicle wiring ranges from number 10 to number 20 in size. Wires are generally assembled in groups call harnesses.

Printed Circuits

Many of the solid state components used in computerized engine control systems are mounted on **printed circuit boards.** Printed circuit boards are constructed of thin sheets of insulating plastic onto which conductive metal, such as copper, have been deposited. The copper metal coating is then etched with acid to form the necessary circuit paths that link the components mounted to the board. Plug-in connectors allow power and ground wires to be connected to the board.

Circuit Protection

When a short or overload occurs in a circuit, excess heat generated by high current flow can melt the insulation and surrounding materials, posing a potential fire hazard. To provide protection against damage from current flow higher than the wiring was designed to handle, circuit protection devices, such as fuses, fusible links, and circuit breakers are designed into every circuit. These devices can be used alone or in combination with one another.

Fuses

A *fuse* is made by enclosing a strip of low-melting point metal in a glass or plastic housing. If the circuit current exceeds the fuse's amperage rating, the metal element melts, breaking the circuit's continuity and protecting wiring and components located further along in the circuit.

Inline glass cartridge fuses, **Figure 24-12A,** were once the only type of fuse manufactured and are still used in older automotive and truck electrical systems. Most modern trucks and vehicles now use a fuse panel or junction block assembly containing blade-type plug-in fuses, such as the fuse shown in **Figure 24-12B.** These fuses are sometimes called miniature fuses. Larger blade fuses are also used and are referred to as Maxifuses or Pacific fuses.

Fuse amperage ratings are usually keyed to a color code system that conforms to SAE standards, **Figure 24-13.** The rating may also be stamped on the glass or the fuse ends.

Circuit Breakers

Circuit breakers can be mounted in a fuse panel, junction block, or inline in the circuit. Like fuses, circuit breakers are rated in amperes. A circuit breaker is simply a set of electrical contacts connected by a strip or arm made of two different types of metal. If excessive current passes through this bimetal arm, it heats up and the two different metals begin to expand. One metal expands faster than the other, causing the arm to bend, opening the contacts. This breaks the circuit continuity, **Figure 24-14.**

Fuse Color Coding			
Amp Rating	Color	Amp Rating	Color
1	Dark green	14	Black
2	Gray	15	Light blue
2.5	Purple	20	Yellow
3	Violet	25	White
4	Pink	30	Light green
5	Tan	40*	Amber or green
6	Gold	50*	Red
7.5	Brown	60*	Dark blue or yellow
9	Orange	70*	Brown
10	Red	80*	White

*Maxifuses or Pacific fuses

Figure 24-13. *Fuse color coding chart. There may be slight color variations in some cases.*

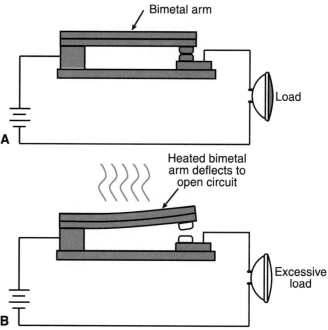

Figure 24-14. *Components and operation of a typical circuit breaker. A—In normal operation, the circuit breaker remains closed, supplying power to the circuit loads. B—If a short occurs or the circuit is overloaded, the excess heat causes the bimetal arm to deflect, opening the circuit breaker.*

Circuit breakers can be designed for automatic or manual reset. In automatic or cycling reset circuit breakers, the bimetal arm flattens out as it begins to cool. In approximately 30-45 seconds, it returns to its original shape and position, and current can again flow between the two contacts of the circuit breaker. Cycling circuit breakers are used in circuits prone to occasional overloads. Automatic recycling circuit breakers are installed in systems vital to safety, such as vehicle headlights.

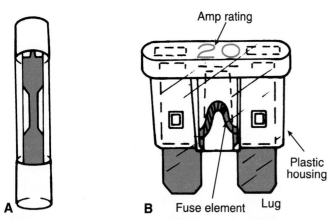

Figure 24-12. *A—Glass cartridge fuses were used in older vehicles. B—Blade fuses have replaced the cartridge fuses in almost all mobile applications.*

Manual reset circuit breakers operate on the same principle except the expanding bimetal arm also pushes a reset button to its open position. The reset button must be pushed in before the circuit breaker can again conduct current. Manual circuit breakers are used primarily on stationary applications.

Fusible Links

In addition to fuses and/or circuit breakers, most vehicles also employ fusible links for added circuit protection in the vehicle's wiring harness. A *fusible link* is a short length of high temperature insulated wire usually three or four gage sizes smaller than the circuit wiring it is designed to protect, **Figure 24-15.**

When excess current heats the fusible link, two things happen. The wire melts, breaking circuit continuity. The heat also blisters and discolors the fusible link insulation, giving the service technician a clear sign the link has been damaged and requires replacement. The gage of the fusible link wire is usually marked on the insulation covering. Color coding on the fusible link flag, insulation, or connector is also used to indicate gage, **Figure 24-16.** Fusible link replacement procedures can normally be found in the shop manual. Maxifuses and Pacific fuses are beginning to replace fusible links in some late-model vehicles.

Switches

A *switch* provides a means of controlling current flow. Switches can be used to turn current on or off or direct it to a specific branch of the circuit. Switches can be either operator controlled, or they can be self-activating. Self-activating switches may be controlled by a condition that exists in the circuit or the vehicle, or by external conditions, such as temperature.

Switches are named using the terms *pole* and *throw*. Simply remember the pole refers to the number of input circuits connected to the switch while the throw refers to the number of output circuits controlled by the switch. So if a switch has one conducting wire running into it and a single wire running out of it, it is called a *single-pole single-throw (SPST) switch.* If the switch has one wire running into it and two circuits running out of it, it is called a *single-pole double-throw (SPDT) switch.* A *double-pole double-throw (DPDT) switch* has two input circuits running to it and

controls current flow to two output circuits. **Figure 24-17** illustrates the electrical symbols of several popular switch configurations.

Mercury Switches

Mercury switches are motion-sensing switches. A small amount of conductive liquid mercury is contained inside a capsule. When the liquid mercury does not submerge the switch contacts, the circuit is open. When the movement such as a door or hood opening changes the capsule's position, the liquid mercury bridges the contacts and the circuit is closed.

Bimetal Switches

Bimetal switches are temperature sensitive switches commonly used as sensors in a vehicle's cooling system. They operate in a manner similar to circuit breakers. When they reach a certain temperature, the bimetal strips move into position to close contacts and complete the circuit to the coolant indicator light, **Figure 24-18.**

Relays and Solenoids

A *relay,* **Figure 24-19,** is a device actuated by a small amount of current. The relay uses this small current to control the flow of high current. A relay relies on the principles of electromagnetism for its operation. When current is

| Fusible Link Color Coding ||
Wire Gage	Color
20	Blue
18	Red
17	Yellow
16	Orange
14	Green

Figure 24-16. Fusible link color coding chart.

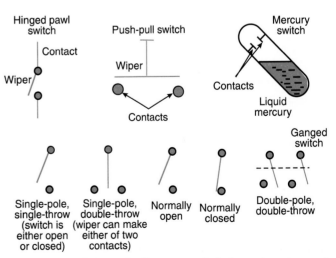

Figure 24-17. Schematic diagram symbols for various types of switches used in electrical systems.

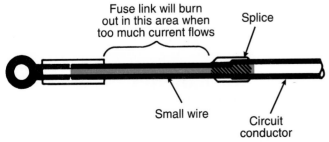

Figure 24-15. Typical fusible link. The link is replaced if it is burned out. Always replace the link with one of the same gage.

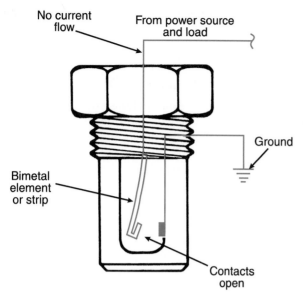

Figure 24-18. *Bimetal switches are sometimes used to sense coolant temperature.*

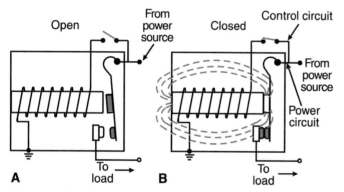

Figure 24-19. *Simple electrical relay construction and operation. A—The relay remains open if no power from the control circuit is applied. B—When the control circuit is closed, the arm inside the relay is attracted to the metal bar, closing the relay contacts.*

passed through a relay's control circuit, it turns a piece of soft iron in the relay housing into an electromagnet. This magnet pulls an armature into position, closing the contacts inside the relay and completing the power circuit. When the low current is switched off in the control circuit, the electromagnet loses its magnetic force, the armature returns to its open position and current is cut off in the relay's power circuit.

A *solenoid,* **Figure 24-20,** is a type of electromagnet used to convert electrical power into mechanical movement. This mechanical movement can also be used to close electrical contacts allowing the solenoid to act as a relay at the same time it performs its primary task.

Resistors

Resistors are electrical components specially constructed to provide a predetermined amount of resistance in a circuit. Their job is to limit current flow in circuits. Resistors can be fixed, or have a series of stepped or variable resistance values.

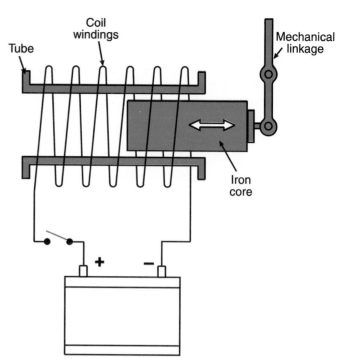

Figure 24-20. *Simple solenoid construction. When power is applied to the coil windings, the iron core is pulled in, operating the mechanical linkage.*

The rating of a fixed resistor does not change. Stepped resistors are designed to provide two or more fixed resistance values. Variable resistors provide a range of resistance available through two or more taps and a control. Potentiometers and rheostats are the two most common types of variable resistors.

Potentiometers have three connections or terminals, one at each end of the resistor and one to a sliding contact connected to the resistor. Turning the potentiometer control or pushing a pedal moves the sliding contact away from one end of the resistance and toward the other end. As resistance increases at one end, it simultaneously decreases at the other end.

A *rheostat* has two connections, one at the fixed end of a resistor and the other to a sliding contact with the resistor. Turning the rheostat control moves the sliding contact toward or away from the fixed end terminal. This increases or decreases the resistance generated.

Thermistors are designed to change resistance value in relation to temperature changes. They are often used as fuel, air, and coolant temperature sensors in electronic engine control systems. Thermistors are also used to provide voltage compensation in components sensitive to temperature change.

Batteries

The *battery* forms the heart of a vehicle's electrical system. It supplies power for lights, glow plugs, fuel heaters, starting and charging systems, and other accessories. The battery also acts as a voltage stabilizer for the electrical system and provides current when electrical demands exceed the charging system output.

The primary job of the battery in a diesel engine electrical system is to start the engine. It must supply high current to the starter motor for a short period of time. The electric starter motor cranks, or rotates the flywheel and crankshaft until the engine starts.

A battery stores chemical energy. When connected to an electrical load, such as a starter motor or glow plug, a reaction occurs in the battery which converts this chemical energy into electrical energy. Once the materials used to produce this chemical reaction are used up, current flow stops. However, these materials can be restored by forcing direct current back into the battery.

Battery Design

Batteries can be grouped into three major types: **conventional, low-maintenance,** and **maintenance-free.** All types share many common design characteristics. All batteries contain grids of positive and negative plates, **Figure 24-21.** Each grid holds the active materials needed for the chemical reaction that produces electrical power. A positive plate is made of a grid filled with lead peroxide (PbO_2) as its active material. A negative plate is made of a grid filled with sponge lead (Pb).

Elements and Cells

A battery element is a group of alternating positive and negative plates, each welded to a post strap. A separator of insulating material such as porous fiberglass or polyethylene is placed between the plates. Each element is placed inside the battery case and immersed in an **electrolyte** solution composed of water (H_2O) and sulfuric acid (H_2SO_4). The sulfuric acid supplies sulfates which chemically react with the lead (Pb) and lead peroxide (PbO_2) on the plates to release electrical energy. The electrolyte also carries the electrons inside the battery between the positive and negative plates.

During this chemical reaction, negatively charged ions accumulate on the negative plates and positively charged ions on the positive plates. The negatively charged ions have an excess of electrons, while the positively charged ions have a lack of electrons. When an electrical circuit is closed, the excess electrons pass through the circuit from the negative to the positive plates. As long as the circuit is closed, current will pass from negative plates to positive plates.

Once a battery element is placed in electrolyte and becomes chemically active, it is called a **cell.** The battery case has a number of individual cell compartments. Cell connectors are used to join battery cells in series with one another. Each cell has an open circuit voltage of approximately 2.1 volts. A 12-volt storage battery with 6 cells connected in series has an actual open circuit voltage of 12.6 volts when fully charged.

The top of the battery is covered by a cell cover. The cell cover may be a one-piece design, or each cell may have its own cover. Conventional and low-maintenance batteries must have vents in their cell covers. The vent holes provide access to the cells for adding water or electrolyte. They also permit the escape of hydrogen and oxygen gases that form during charging and discharging.

> ⚠ **Warning: Hydrogen gases are explosive. Do not generate sparks and flames near a charging battery. Always wear proper eye protection.**

The battery has two external terminals: a **positive terminal (+)** and a **negative terminal (-).** Terminals can be top-mounted tapered posts, threaded studs, or side-mounted internally threaded connectors. Terminals connect to either end of the series of elements inside the battery and are clearly marked either (+) or (-). In many batteries, the positive terminal is slightly larger in diameter to minimize the danger of installing the battery cables in reverse polarity.

Electrolyte Specific Gravity

In a fully-charged battery, the electrolyte consists of about 64% water and 36% sulfuric acid by weight. As the battery discharges, the sulfuric acid is used up, and the percentage of acid in the electrolyte goes down.

The relationship between the amount of water and the amount of sulfuric acid is stated in the electrolyte's specific gravity. **Specific gravity** is the weight of any volume of liquid divided by the weight of an equal volume of water. The specific gravity of water is 1.00. A fully charged battery (64% water to 36% sulfuric acid) has a specific gravity of 1.270 at 80°F (27°C). The battery's state of charge can be determined by measuring the specific gravity of its electrolyte using a hydrometer and correcting for temperature.

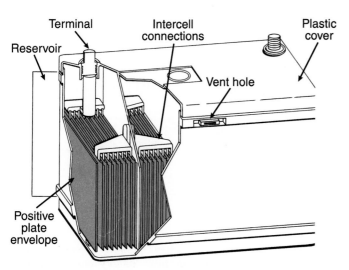

Figure 24-21. *Internal components of a typical 12-volt maintenance-free battery. (Freightliner Corporation)*

There is a direct relationship between specific gravity and the amount of sulfuric acid in the electrolyte solution. In a fully-charged battery, nearly all the sulfuric acid is contained in the electrolyte solution As the battery discharges, the sulfate from the acid combines with the lead in the plates to form lead sulfate ($PbSO_4$). The acid becomes more dilute and its specific gravity goes down as water replaces some of the sulfuric acid in the electrolyte. A fully discharged battery has more sulfate in the plates than in the electrolyte.

As a battery's specific gravity is lowered, the freezing point of its electrolyte becomes higher. This is why discharged batteries will freeze and crack at relatively warm temperatures. **Figure 24-22** lists the freezing points for electrolyte by level of specific gravity and battery charge.

Additional Battery Cell Materials

Antimony is a material used in the battery plate grids to give them strength. It also causes most of the hydrogen and oxygen gassing during battery charging and discharging. In a low-maintenance battery, the amount of antimony used in the grid plates is reduced, so there is less hydrogen and oxygen gas released as the battery operates. Water is lost

from the electrolyte more slowly, and the maintenance interval for adding water to the cell decreases.

A maintenance-free battery replaces the antimony in the grids with either calcium or strontium. This lowers the battery's internal heat level and greatly reduces, but does not eliminate, hydrogen gassing. Oxygen gas is still produced, but is forced to recombine with material on the negative plates. This results in very little water loss during the charging and discharging cycles of a maintenance-free battery. Maintenance-free batteries are sealed batteries with no cell caps for adding water. There are vents for releasing the small amounts of hydrogen produced. Most maintenance-free batteries are equipped with a built-in hydrometer that gives a visual indication of the battery's specific gravity and its state of charge, **Figure 24-23.**

Battery Ratings

Batteries used in heavy-duty trucks and most other diesel powered vehicles are classified as Group 31 batteries. They are rated in two ways: cold cranking performance and reserve capacity.

The cold cranking rating performance is the discharge load in amperes a battery at 0°F (-18°C) can deliver for 30 seconds and maintain a voltage of 1.2 volts or higher per cell. The discharge load is called *cold cranking amperes (CCA).* A standard Group 31 heavy-duty truck battery provides approximately 625 CCA, while a high output Group 31 battery can deliver 900 CCA or more. As shown in **Figure 24-24,** a battery's power is reduced as its temperature decreases. Batteries that are partially discharged will deliver significantly less power than listed in **Figure 24-24.**

The battery's *reserve capacity* is the number of minutes it can produce 25 amps at 10.2 volts with battery temperatures at 80°F (27°C). The battery's reserve capacity indicates how long it can supply power to the electrical system in the event of a charging system malfunction.

Battery Safety

The sulfuric acid in the electrolyte is poisonous and very corrosive. The hydrogen gas produced during charging and discharging is also highly explosive. These factors add up to potentially dangerous situations if batteries are not handled and serviced properly.

Specific Gravity versus State of Charge		
Specific Gravity	**State of Charge**	**Freezing Point**
1.270	100%	-75°F (-59.4°C)
1.225	75%	-35°F (-37.2°C)
1.190	50%	-15°F (-26.1°C)
1.155	25%	5°F (-15°C)
1.120	Discharged	15°F (-9.4°C)

Figure 24-22. *Battery specific gravity as related to state of charge and freezing point.*

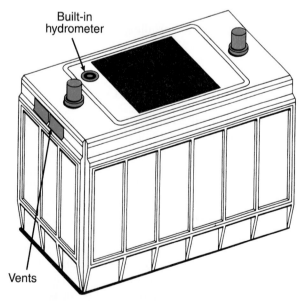

Figure 24-23. *Location of built-in hydrometer on a sealed maintenance-free battery. (Freightliner Corp.)*

Battery Capacity as Related to Temperature	
Temperature	**Percentage of Battery Capacity (Fully Charged)**
80°F (26.7°C)	100%
32°F (0°C)	65%
0°F (-18°C)	40%
-20°F (-29°C)	20%

Figure 24-24. *Relationship between battery cranking power and temperature. The capacity rating assumes the battery is fully charged.*

❑ Always wear safety glasses when working with batteries. Eye injuries caused by splashing electrolyte are the most common accident that occurs in battery service. If electrolyte is splashed into the eyes, rinse with cool, clean water for 15 minutes, and seek medical attention. Do *not* add eyedrops or medication.

❑ If electrolyte is splashed onto your skin, rinse thoroughly with cool water for several minutes. Contact a doctor if burning is evident.

❑ Do not smoke while servicing or charging a battery. Charge batteries in well ventilated areas only. Keep all batteries away from open flames or sparks that can ignite the explosive hydrogen gas.

❑ If the battery is frozen, allow it to reach room temperature before attempting to charge it. Check for cracks and leaks in the battery case before charging.

❑ Never lean over a battery during charging, jump-starting, or other service operations. If the battery should explode, the case and electrolyte will be thrown up and outward.

❑ Always use a battery carrier or strap when lifting or moving batteries. If you must lift a battery for a short distance, lift with your hands positioned under the battery at opposite corners.

❑ Do not wear watches, rings, or other jewelry when servicing batteries or performing electrical system work.

❑ Do not place or drop metal tools or objects on top of batteries. This could short the battery between the terminals or other points, causing sparking and/or damage.

❑ Do not break live circuits at the battery terminals when charging. This can cause sparking which could ignite any hydrogen gas lingering around the battery. The disconnected terminal could also cause a spike that may damage electronic components.

❑ Always disconnect the battery's ground cable when working on the electrical system. This prevents sparks from short circuits, or accidental engine starting.

❑ Always disconnect the battery's ground cable when fast charging a battery in the vehicle, especially if the system is equipped with an alternator. Incorrect connection of charging cables during in-vehicle charging can damage the alternator.

❑ When removing a battery from a vehicle, disconnect the ground cable first and connect the ground cable last when installing.

❑ Never reverse the battery cable connections (reverse polarity). This will damage the alternator and circuit wiring. Generally, all diesel vehicles use a negative terminal ground.

❑ Never use a fast charger as a booster to start a vehicle with a discharged battery.

❑ Do not "short" across battery cables or battery terminals.

Battery Preventive Maintenance

Battery maintenance is simple and can greatly extend its useful life, prevent no-starts, and other electrical system problems. Inspect the outside of the battery for cracks and other signs of physical damage that may allow electrolyte to escape. Replace any damaged batteries.

Inspect the battery, terminal posts, and mounting hardware for corrosion. If terminals are corroded, remove the cables (negative or ground cable first) using a proper size wrench and a cable puller. Remove the corrosion using a solution of baking soda and water.

 Caution: Do not allow the baking soda solution to enter the cells.

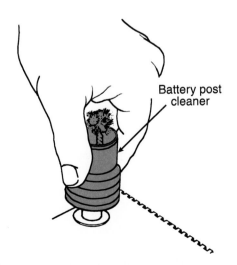

A

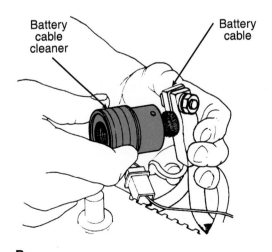

B

Figure 24-25. Proper method of cleaning battery terminals and cables. A—Remove the cable and clean the battery's post. Make sure there is no leakage between the post and the battery cover. B—Clean the cable terminal next. Make sure the cable is not corroded.

Clean the terminals using a wire brush or terminal cleaning brush, **Figure 24-25.** Wipe all dirt from the battery case using a clean cloth or sponge. Coat the battery terminals with terminal grease before reinstalling the cables. Reconnect the negative ground cable last.

On low-maintenance and older battery designs, remove the cell caps and visually check the electrolyte level. Add distilled water as needed to bring the electrolyte level to 1/2″ (13 mm) above the top of the grid plates, (do not overfill). Never add additional electrolyte to a low cell. This would increase the concentration of sulfuric acid to unsafe levels. If a battery appears to be losing a large amount of water between service intervals, it may be leaking or overcharging. Check the battery case, charging system, and voltage regulator.

Battery Testing and Service

Whenever a battery is suspect, perform a state of charge and battery capacity test. These two tests will determine if the battery is charged, requires recharging, or should be replaced. If the electrical system has more than one battery, such as those used in most heavy-duty trucks, disconnect the batteries from one another and perform tests on each battery.

A battery's state of charge can be checked using a hydrometer or by performing an open circuit voltage test. A **hydrometer** is a bulb syringe used to extract a sample of electrolyte from the cell on conventional (unsealed) batteries. A glass float in the hydrometer tube is calibrated to read in terms of specific gravity. The lower the float sinks in the electrolyte, the lower the electrolyte's specific gravity and the battery's state of charge.

Hydrometer Readings

Hydrometer readings can only be taken on batteries having removable cell caps. Many batteries classified as low-maintenance or maintenance-free are designed with removable cell caps. This is done so the batteries can be tested and serviced in the event they are heavily abused. When using a hydrometer, hold the barrel vertically and take the reading with your eyes even with the level of electrolyte in the barrel. Disregard the slight curvature of the liquid where it contacts the barrel sides and float stem, **Figure 24-26.**

Take an electrolyte specific gravity reading from each cell. If the readings are 1.225 or higher and are within .050 of each other for all cells, the battery is sufficiently charged. If the specific gravity readings are less than 1.225 or vary greater than .050 between cells, recharge the battery, then retest.

Never take a hydrometer reading immediately after adding water to the cell or subjecting the battery to prolonged loads, such as heavy starter motor cranking. The best way of mixing water and electrolyte in a recently serviced cell is to charge the battery for a short period of time.

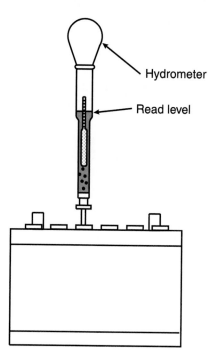

Figure 24-26. *Using a manual hydrometer to measure battery acid specific gravity.*

All hydrometers are calibrated to give accurate specific gravity readings at one temperature only, usually 80°F (27°C). A correction factor of .004 must be added for each 10° the electrolyte temperature is above 80°F (27°C) or subtracted for each 10° the electrolyte temperature is below 80°F (27°C). For example, if the electrolyte temperature is 40°F, and the specific gravity reading is 1.250, it must be corrected by a factor of:

40 ÷ 10 = 4
4 × .004 = .016
1.250 − .016 = 1.234 temperature corrected specific gravity

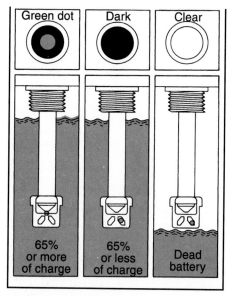

Figure 24-27. *Operation of built-in hydrometer on maintenance-free battery. (Freightliner Corp.)*

Built-in Hydrometers

As mentioned earlier, many "sealed" maintenance-free batteries have built-in hydrometers. These hydrometers are temperature corrected to give accurate visual readings at all times. As shown in **Figure 24-27,** a small colored ball (usually green), floats to various levels in a cage at the end of a tube, depending on the electrolyte's specific gravity. Different ball positions produce a green dot, dark circle, or clear/light yellow circle when viewed straight down from the top.

The green dot indicates sufficient charge for further testing. The dark circle indicates the battery should be recharged before further testing. A clear or light yellow dot indicates the electrolyte level has fallen below the bottom of the cage. This is a sign the battery has reached the end of its service life or is leaking electrolyte. A gas bubble caught in the cage may give a false clear reading. Tap the side of the battery several times to dislodge any bubbles before declaring the battery as "dead."

Open Circuit Voltage Test

Another method of determining a battery's state of charge involves reading the open circuit voltage across the battery terminals using a voltmeter. Connect the voltmeter across the battery terminals as shown in **Figure 24-28** and observe the reading. Compare this reading with the listings in **Figure 24-29** to calculate the battery's state of charge.

Note: If the battery has recently been charged or in vehicle service, it will have a surface charge that must be removed before an accurate reading can be taken. Remove this surface charge by cranking the engine for 15 seconds. Prevent the engine from starting by applying the engine stop control or disconnecting the fuel solenoid valve lead wire.

Battery Load Test

A capacity or **battery load test** determines how well the battery will function under load, such as cranking the starter motor. This test requires the use of an adjustable carbon pile, which is simply a type of adjustable resistor.

The battery should be at or very near its full state of charge for test results to be accurate. Recharge the battery as needed before proceeding. The battery temperature should also be as close to 80°F (27°C) as possible, or test results will be inaccurate. Never perform a capacity test on a sealed battery at temperatures below 60°F (16°C). A capacity test can be performed with the battery in or out of the vehicle.

Note: A voltmeter, ammeter, and carbon pile are often built into one unit called a batter tester.

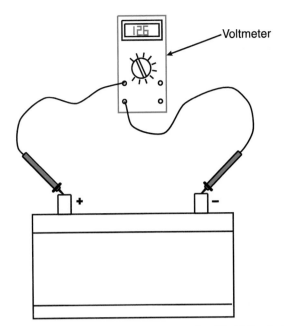
Figure 24-28. *Checking battery open circuit voltage using a voltmeter or multimeter across the battery terminals.*

State of Charge Determined by Open Circuit Voltage Test	
Open Circuit Voltage (Stabilized)	**State of Charge**
12.6 volts or more	100% Charged
12.4 volts	75% Charged
12.2 volts	50% Charged
12 volts	25% Charged
11.7 volts or less	Discharged

Figure 24-29. *State of battery charge as determined by open circuit voltage.*

1. Remove any surface charge from recently charged or in-service batteries by applying a 300-amp load for 15 seconds.

2. Connect the carbon pile and voltmeter to the battery according to equipment manufacturer's instructions. A typical setup is shown in **Figure 24-30.**

3. If the tester is equipped with an adjustment for battery temperature, turn it to the proper setting.

4. Refer to the battery specifications to determine the battery's cold cranking amps (CCA). Turn the load control setting on the carbon pile to draw battery current at a rate equal to one-half the rated CCA. For example, if a battery's CCA is 460 amps, set the carbon pile to 230 amps.

5. Maintain this current load for 15 seconds while watching the voltmeter reading. Turn the control knob off immediately after 15 seconds of current draw.

6. At 70°F (21°C) or above, the voltage reading at the end of 15 seconds should not fall below 9.6 volts. If the tester is not temperature corrected, determine the proper reading by referring to **Figure 24-31.**

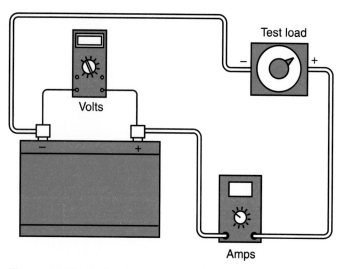

Figure 24-30. *Performing a capacity test on a battery using a carbon pile. Often, both meters and the carbon pile are combined into one tester.*

reason, maintenance-free sealed batteries are normally fast-charged at lower current rates of 25-40 amperes.

"Trickle" or ***slow chargers*** provide lower charging currents of 5-15 amperes and normally require 10-20 hours to recharge a battery. Some high capacity batteries may require as much as 40 hours of slow charging time. General fast and slow charging information for low- and maintenance-free batteries is summarized in **Figure 24-32.**

When applicable, check and adjust the electrolyte level before charging. Charge in a well-ventilated area, following the safety guidelines listed earlier in this chapter. Always be sure to turn the charger off before connecting or disconnecting leads to the battery. Follow the charger manufacturer's directions for operation. When charging a battery in the vehicle, always disconnect the battery cables to avoid damaging the alternator or other electrical/computer system components.

Minimum Load Test Voltages as Affected by Temperature	
Battery Temperature	**Minimum Test Voltage**
70°F (21°C)	9.6 volts
60°F (15.6°C)	9.5 volts
50°F (10°C)	9.4 volts
40°F (4.4°C)	9.3 volts
30°F (−1.1°C)	9.1 volts
20°F (−6.7°C)	8.9 volts
10°F (−12.2°C)	8.7 volts
0°F (−18°C)	8.5 volts

Figure 24-31. *Battery voltage is affected by temperature. These are the minimum load test voltages relative to battery temperature.*

If the voltage reading is higher than the minimum specification by a volt or more, the battery is supplying a sufficient amount of current. Readings at or near specifications indicate the battery may not have sufficient reserve in all operating conditions, particularly cold weather conditions. If the voltage drops below the minimum specifications, the battery should be replaced.

Battery Charging

Batteries can be recharged using either fast-charge or slow-charge equipment. ***Fast chargers*** charge batteries at a higher rate, usually 40-70 amperes. They can recharge a battery in 45 minutes to an hour, but should only be used on newer batteries that have not been subject to excessive sulfation. If done incorrectly, fast charging can result in dangerous heat buildup or gassing. For this

Battery Charging Guide		
6- and 12-Volt Low-Maintenance Batteries*		
Rated Battery Capacity**	**Slow Charge**	**Fast Charge**
75 minutes or less	14 hrs at 5 amps 8 hrs at 10 amps	1.75 hrs at 40 amps 1 hr at 60 amps
75-125 minutes	20 hrs at 5 amps 10 hrs at 10 amps	2.5 hrs at 40 amps 2 hrs at 60 amps
125-175 minutes	28 hrs at 5 amps 14 hrs at 10 amps	3.5 hrs at 40 amps 2.5 hrs at 60 amps
175-250 minutes	40 hrs at 5 amps 20 hrs at 10 amps	5 hrs at 40 amps 3.5 hrs at 60 amps
Above 250 minutes	35 hrs at 10 amps	8 hrs at 40 amps 5.5 hrs at 60 amps
12-Volt Maintenance-Free Batteries*		
Rated Battery Capacity**	**Slow Charge**	**Fast Charge**
75 minutes or less	10 hrs at 5 amps 5 hrs at 10 amps	2.5 hrs at 20 amps 1.5 hrs at 30 amps 1 hr at 54 amps
75-125 minutes	15 hrs at 5 amps 7.5 hrs at 10 amps	3.75 hrs at 20 amps 2.5 hrs at 30 amps 1.75 hrs at 45 amps
125-175 minutes	20 hrs at 5 amps 10 hrs at 10 amps	5 hrs at 20 amps 3 hrs at 30 amps 2.25 hrs at 45 amps
175-250 minutes	30 hrs at 5 amps 15 hrs at 10 amps	7.5 hrs at 20 amps 5 hrs at 30 amps 2.5 hrs at 45 amps
Above 250 minutes	20 hrs at 10 amps	10 hrs at 20 amps 6.5 hrs at 30 amps 4.5 hrs at 45 amps

*Recommended rate and time for fully discharged condition (Initial rate is for standard taper charge)
**Reserve minutes

Figure 24-32. *Fast and slow charging times and rates for low-maintenance and maintenance-free batteries.*

Warning: Never charge a maintenance-free battery that gives a clear/light yellow hydro-meter reading.

Battery Circuits

Most heavy-duty truck electrical systems use multiple batteries connected in series, parallel, or series-parallel to create desired voltage and/or CCA ratings. Most North American trucks use 12-volt as well as 24-volt systems. The 24-volt circuit is used for engine starting, while all the other systems use 12 volts.

Series Hookups

The most common 24-volt system is set-up by connecting two 12-volt batteries in series, **Figure 24-33A.** Another arrangement would be four 6-volt batteries connected in series. When batteries are connected in series, voltages are added together, but the amperage (cold cranking amps) remains the same. However, the system has greater reserve power since two or four batteries are used.

Parallel Hookups

Batteries can also be hooked up in parallel. The result is an increase in amperage while the voltage remains the same, **Figure 24-33B.** For example, when two 12-volt batteries with a CCA of 500 amps are hooked together in a parallel arrangement, voltage remains 12-volts but CCA increases to 1000 amps.

Series-Parallel Hookups

Figure 24-33C illustrates four 12-volt batteries connected in a series-parallel arrangement. System voltage is equal to 24-volts with a capacity of 700 CCA since each battery is rated at 350 CCA.

12- to 24-Volt Systems

Some diesel engines used in extreme heavy-duty or cold weather applications may be equipped with a series-parallel switch that allows the use of a 24-volt starter system with a 12-volt charging system. For starting, the batteries are connected in series through the series-parallel switch to provide 24 volts for cranking. After starting, the charging system automatically switches back to a 12-volt setup so that all accessories can be operated on the conventional 12-volt circuitry.

Jump-Starting Diesel Engines

When jump-starting an engine with a discharged battery, make certain the booster battery and discharged battery are the same voltage (either 6-volt or 12-volt). Make

certain the stalled vehicle and the vehicle containing the booster battery do not touch, or an undesired ground that could lead to sparking when the connections are made can result. Finally, disconnect all engine and brake electronic control equipment to prevent damage from power surges occurring during jump starting.

12-Volt Systems

The following procedure is for systems using a negative ground, **Figure 24-34.**

1. Place both vehicles in neutral and set the parking brakes. Turn off all lights and electrical loads. Make certain the ignition key is turned off in both vehicles.

2. Attach one end of the jumper cable to the positive terminal of the booster battery and the other to the positive terminal of the discharged battery.

3. Attach one end of a second jumper cable to the negative (ground) terminal of the booster battery and the opposite end to a ground at least 12″ (30.48 cm) from the battery of the vehicle being started. The vehicle frame is an excellent ground.

4. Start the engine of the vehicle with the booster battery and allow it to idle for several minutes.

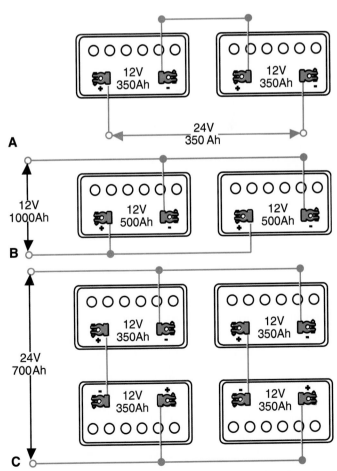

Figure 24-33. *Various battery hook-ups used in diesel applications. A—Series. B—Parallel. C—Series-parallel.*

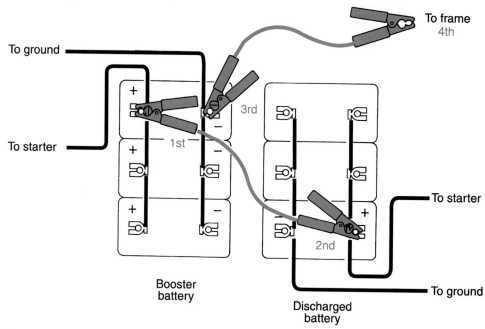

Figure 24-34. *Cable connections and hook-up sequence for jump-starting a 12-volt electrical system. Carefully follow the placement of cables on the batteries. (Freightliner Corp.)*

5. Start the engine in the vehicle with the drained battery. Do not crank the starter motor for longer than 30 seconds without waiting at least 2 minutes between attempts.

6. When the vehicle starts, allow the engine to idle. Disconnect the ground connection from the vehicle with the discharged battery. Then disconnect the opposite end of the cable from the booster battery negative terminal.

7. Disconnect the other jumper cable, removing the clamp from the discharged battery first.

 Note: Always check the appropriate service manual for components that may need to be isolated during jump-starting.

12- to 24-Volt Systems

As mentioned earlier, some heavy-duty diesel starting systems operate on a series-parallel principle where the starting motor operates on 24 volts, but the charging system is only 12 volts. To jump-start a dead 12- to 24-volt system, only use a vehicle equipped with a similar 12- to 24-volt system. Two sets of jumper cables are needed, **Figure 24-35.**

1. Connect the positive terminal of one cable to the battery that connects directly to the booster vehicle's starter. Connect the other end of the cable to the positive terminal of the drained battery that connects directly to its vehicle's starter.

2. Attach a second jumper cable between the negative terminals of the booster and discharged batteries that connect directly to their respective vehicle's series-parallel switches.

3. Attach a third jumper cable between the positive terminals of the booster and discharged batteries that connect directly to their respective vehicle's series-parallel switches.

4. Attach one end of a fourth cable to the negative ground terminal of the booster battery that connects directly to its vehicle's starter. Connect the other end of this fourth cable to a ground at least 12″ (30.48 cm) from the battery of the vehicle being started.

5. Start the booster vehicle's engine and allow it to idle for several minutes.

6. Start the engine in the vehicle with the drained battery. Do not crank the starter motor for longer than 30 seconds without waiting at least 2 minutes between attempts.

7. When the vehicle starts, allow the engine to idle for a few minutes.

8. Disconnect the ground connection in the vehicle with the discharged battery. Then disconnect the opposite end of this cable.

9. Disconnect the positive cables from the discharged vehicle first. Then disconnect the opposite ends.

Semiconductor Devices

Semiconductors are materials which can be either conductors or insulators, depending on voltage level, direction of current flow, and other factors. Semiconductor

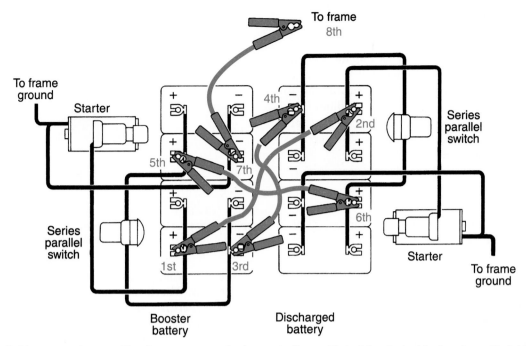

Figure 24-35. *Cable connections and hook-up sequence for jump-starting a 12- to 24-volt electrical system. (Freightliner Corp.)*

material is used to make diodes, transistors, integrated circuits, and other electronic devices.

To understand how a semiconductor device works, you must first understand how semiconductor materials behave. When an element contains exactly four electrons in its valence ring (outermost orbit), it does not easily give up or accept other electrons. Elements with such electron arrangements do not make good insulators or conductors and are referred to as semiconductors. The two most common semiconductor materials used in electronic components are germanium (Ge) and silicon (Si). Each atom shares the four electrons in its outermost orbit with four neighboring atoms. This sharing of electrons is known as a **covalent bond.** It creates a structure with a crystal-like appearance.

Crystals are a very stable electrical structure, **Figure 24-36.** However, some electrons still manage to break free from their covalent bonds. When this happens, it creates a hole in the crystal, **Figure 24-37.** The hole is simply the absence of an electron, but it has the characteristics of a positively charged particle.

When pure semiconductor material is connected to a voltage source, the semiconductor's negatively charged free electrons are attracted to the positive terminal of the voltage source. When free electrons drift toward the positive terminal of the voltage source, positive holes are created in the semiconductor material. As free electrons flow into the positive terminal of the voltage source, an equal number of electrons leave the negative terminal and flow toward the semiconductor material where they are captured or absorbed by the positive holes.

It is important to remember current flow in a semiconductor consists of both electrons and holes. The holes and electrons flow in opposite directions and the number of electron-hole pairs produced within a pure

semiconductor increases as temperature increases. This means semiconductors can carry more current at increased temperatures.

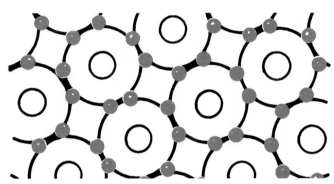

Figure 24-36. *The crystalline structure of semiconductor materials is formed by sharing the four electrons in the valence ring.*

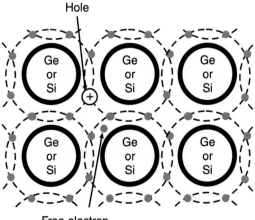

Figure 24-37. *A free-electron/hole pair in a semiconductor material such as silicon or germanium.*

Doped Semiconductors

The conductivity of semiconductors can be dramatically increased by a process known as **doping.** Doping involves adding other materials or impurities to silicon or germanium. There are two types of impurities added to semiconductors. Trivalent impurities such as gallium (Ga), have three valence electrons, **Figure 24-38.** Pentavalent impurity materials, such as arsenic (As), have five electrons in their valence ring, **Figure 24-39.**

Diodes

A **diode** allows current to flow in one direction only. This enables the diode to work as a switch, acting as either

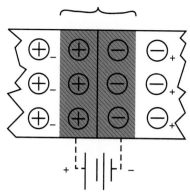

Figure 24-40. Characteristics of a diode junction. Note the depletion zone and barrier voltage created at the junction of the N- and P-type semiconductor materials.

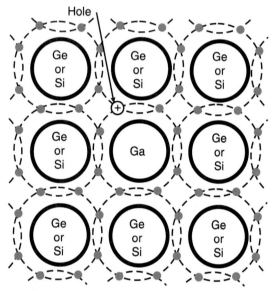

Figure 24-38. N-type semiconductor material. Germanium or silicon doped with pentavalent material such as arsenic.

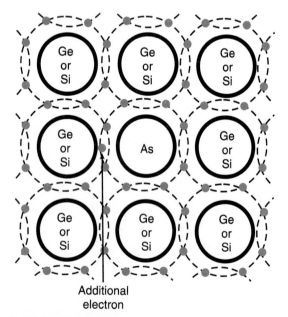

Figure 24-39. P-type semiconductor material. Germanium or silicon doped with gallium.

a conductor or an insulator depending on the direction of current flow. The diode is the simplest of all electronic devices. It is created by joining positive and negative semiconductor material. The point where the two materials are joined is called the **junction, Figure 24-40.**

When the junction is formed, a unique action takes place. The free electrons and holes near the junction are attracted to each other. Some electrons pass over the junction point and fill holes on the positive. This creates a very small electrical imbalance. The region close to the junction is positively charged on the negative side and negatively charged on the positive side. These opposite charges generate a small voltage potential, known as barrier voltage. A diode made of silicon generally has a barrier voltage of approximately .7 volts while germanium diodes have a barrier voltage of about .3 volts. When an outside source of voltage is applied to a diode, it must be greater than the barrier voltage for current to flow.

The switching capability of diodes makes them an important component in the logic circuits of computers. Another common use for diodes is found in modern truck and automotive alternators. In a charging system, diodes in an alternator act as a one-way valve for current flow. In an alternator, current is changed or rectified from ac to dc through the use of diodes. The diodes are arranged so current can exit the alternator as direct current.

Whenever the negative terminal of a battery is connected to the negative side of a diode, current can flow through the diode, provided voltage is sufficient to overcome the barrier voltage at the junction, **Figure 24-41A.** Diodes set up in this manner are said to have a **forward bias.**

If the circuit polarity is changed by connecting the battery positive terminal to the diode's negative side, electrons are drawn into the diode's positive side. These incoming electrons fill the holes in the positive material. This stabilizes the positive side, making it an excellent insulator. At the same time, the pull from the battery's positive terminal attracts excess electrons from the diode's negative side. This makes the negative side an excellent conductor. With the entire diode now acting like an insulator, current cannot pass, **Figure 24-41B.** When the negative terminal of a voltage source is connected to the negative side of the diode, the diode is said to have **reverse bias.**

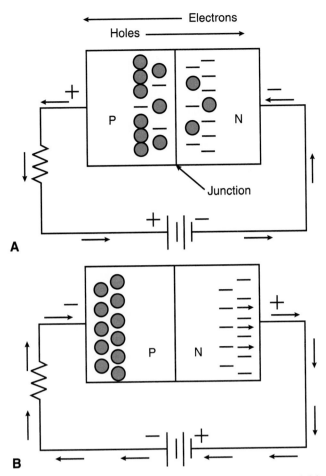

A

B

Figure 24-41. A—Current will flow through a forward bias condition. B—Current will not flow through this reverse bias connection.

Zener Diodes

A *Zener diode* is a type of diode produced using special doping techniques. A Zener diode operates like a standard diode until a certain voltage level is reached. When the circuit voltage reaches this level, the Zener diode's insulating characteristics break down, and the diode can now conduct high levels of reverse bias current. If applied voltage drops below this break down voltage, the Zener diode again acts as an insulator in reverse bias.

This characteristic allows the Zener diode to conduct current in one direction at certain voltage levels and in the other direction at other voltage levels. This is why Zener diodes are used in the construction of electronic voltage regulators.

Transistors

A *transistor* is constructed of semiconductor material just like a diode. However, it uses three alternately doped semiconductor regions instead of two. These three regions can be arranged in one of two ways. An N-region can be sandwiched between two P-regions, forming a PNP transistor, or a P-region can be sandwiched between two N-regions, creating an NPN transistor, **Figure 24-42.** In both types, the center region of the transistor is called

the *base.* The outer regions are called the *emitter* and the *collector.* Each of these three regions has a terminal that allows it to be connected to a circuit.

Transistor Operation

Transistors have insulating and conductive abilities that allow them to act as switches. When the circuit switch is open, **Figure 24-43A,** voltage cannot be applied to the transistor's base. In this condition, the N section of the PNP transistor has a very high resistance. Current cannot pass from the emitter to the collector even though a potential voltage exists between them.

When the circuit switch is closed, voltage is applied to the base and current flows through the base and emitter and back to the voltage source (battery), **Figure 24-43B.** In this condition, the base now becomes a conductor. With no internal resistance inside the transistor, current can flow through the collector to the emitter as shown. In this way, applying a very small amount of current to the base allows a much larger amount of current to flow through the entire transistor.

Integrated Circuits and Computer Chips

When a large number of diodes, transistors, and other electronic components are mounted on a single piece of semiconductor material, the assembly is called an *integrated circuit* or *IC.*

The semiconductor piece, usually silicon, is very small and is called a *chip* or *microchip.* The integrated circuits etched into the surface of the silicon chip can contain as few as two or three, or as many as several hundred thousand electrical circuits which incorporate diodes, transistors, resistors, capacitors, etc. The silicon chip which forms the heart of an engine's electronic control system may be as small as .25" (6.4 mm) square. This chip may contain 10,000 or more separate electronic elements. IC chips used in computers and other electronic devices are normally enclosed in a plastic casing.

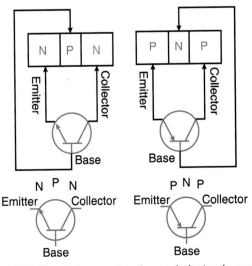

Figure 24-42. Transistor construction and electronic symbols.

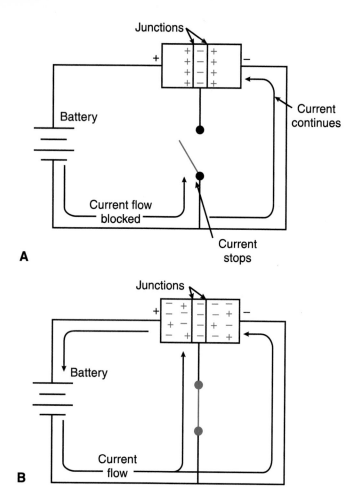

Figure 24-43. *A—PNP transistor acts like an insulator and blocks current flow through the junction. B—When current is applied to base, the PNP transistor conducts current through the circuit.*

The small size and excellent reliability of integrated circuits have produced an electronics revolution in the diesel industry. All diesel engines and diesel fuel injection systems now being manufactured use some type of electronic engine control system to maximize fuel efficiency, engine performance, and minimize exhaust pollutants.

Summary

Electric current is the controlled flow of free electrons from atom to atom within a conductor, such as a copper wire. To create this flow of electrons, there must be an outside source of energy acting on the conductor's electrons. In a diesel truck engine, this source of energy is provided by the battery or the alternator.

The mathematical relationship between current, resistance, and voltage is summarized in Ohm's law, where voltage is measured in volts, current in amperes, and resistance in ohms. There are three basic types of circuits used in diesel engine electrical systems: series, parallel, and series-parallel. Electrical diagrams or schematics use electrical symbols and lines to depict electrical parts and current flow. Schematics are very useful when troubleshooting electrical systems.

An open circuit occurs when there is a break in the wire or the wire connection at a load. An open circuit does not offer a complete path for current flow. A short circuit is a connection of relatively low resistance between two points in a circuit. A short circuit is not like an open circuit. Increased circuit resistance, caused by corrosion and poor connections, can cause a decrease in the current supplied to loads.

Fuses, circuit breakers, and fusible links protect electrical circuits against overloads. Switches control current flow in a circuit. A solenoid is an electromagnet which turns current flow into mechanical movement. The amount of current a given wire can safely carry and the amount of voltage drop that will occur in a wire are determined by two factors, its length and its gage (thickness).

A battery stores chemical energy. When connected to an electrical load, such as a starter motor or glow plug, a chemical reaction occurs in the battery that converts chemical energy into electrical energy or current. The battery has two external terminals: a positive terminal (+) and a negative terminal (-).

In a fully charged battery, the electrolyte consists of about 64% water and 36% sulfuric acid by weight. The relationship between the amount of water and the amount of sulfuric acid in an electrolyte is stated in the electrolyte's specific gravity. They are rated in two ways: cold cranking performance and reserve capacity.

The sulfuric acid in the battery electrolyte is poisonous and very corrosive. The hydrogen gas produced during charging and discharging is also highly explosive. These factors add up to potentially dangerous situations if batteries are not handled and serviced properly.

Most heavy-duty truck electrical systems use multiple batteries connected in series, parallel, or series-parallel to create desired voltage and/or CCA ratings. Most North American trucks use the standard 12-volt or 24-volt system.

When-jump starting an engine with a discharged battery, make certain the booster battery and discharged battery are the same voltage. Make certain the stalled vehicle and the vehicle containing the booster battery do not touch. Disconnect all engine and brake electronic control equipment to prevent damage from power surges occurring during jump-starting. To jump-start a dead 12- to 24-volt system, only use a vehicle equipped with the same 12- to 24-volt system.

Elements containing exactly four electrons in their outer most orbit are called semiconductors and have special electrical properties very useful in electronic devices. Current flow in a semiconductor consists of both electrons and holes. The holes and electrons flow in opposite directions, and the number of electron-hole pairs produced within a pure semiconductor increases as the temperature increases. Doping involves adding other materials or impurities to the semiconductor material to increase its conductivity.

Important Terms

Protons

Neutrons

Electrons

Electric current

Conventional theory

Electron theory

Ampere (A)

Direct current (dc)

Alternating current (ac)

Voltage

Electromotive force (emf)

Voltmeter

Resistance

Ohm

Ohm's law

Conductors

Insulators

Schematics

Series circuits

Parallel circuits

Series-parallel circuit

Ground return circuit

Open circuit

Shorted circuit

Grounded circuit

Wiring

Society of Automotive
    Engineers (SAE)

American Wire Gage
    (AWG)

Printed circuit boards

Fuse

Circuit breakers

Fusible link

Switch

Single-pole single-throw
    (SPST) switch

Single-pole double-throw
    (SPDT) switch

Double-pole double-
    throw (DPDT) switch

Mercury switches

Bimetal switches

Relay

Solenoid

Resistors

Potentiometers

Rheostat

Thermistors

Battery

Conventional

Low-maintenance

Maintenance-free

Electrolyte

Cell

Positive terminal (+)

Negative terminal (-)

Specific gravity

Cold cranking amperes
    (CCA)

Reserve capacity

Hydrometer

Battery load test

Fast chargers

Slow chargers

Semiconductors

Covalent bond

Crystals

Doping

Diode

Junction

Forward bias

Reverse bias

Zener diode

Transistor

Base

Emitter

Collector

Integrated circuit (IC)

Chip

Microchip

Review Questions—Chapter 24

Do not write in this text. Place your answers on a separate sheet of paper.

1. For current to flow between two points, which of the conditions listed below must exist?
    (A) A difference in the quantity of electrons between two locations.
    (B) A path of conductive material between the two locations.
    (C) A magnetic field.
    (D) Both A & B.

2. Explain the difference between the conventional and electron theory of current flow.

3. Your body is an excellent conductor of _____.

4. A series circuit has _____ path(s) for current to flow.
    (A) one
    (B) two
    (C) three
    (D) four or more, depending on the circuit

5. State Ohm's law.

6. If a circuit has 12 volts and 4 ohms of resistance, what is the current in this circuit?

7. The current in a 12-volt circuit is 6 amps. How much resistance is present in this circuit?

8. The amount of current a given wire can safely carry and the amount of voltage drop that will occur in a wire are determined by two factors. Name them.

9. A battery converts _____ into _____ or current.

10. When jump-starting an engine with a discharged battery, the voltage of the booster battery and discharged battery must be _____ .

ASE-Type Questions

1. The unit for measuring the rate of electron flow or current is the _____.
    (A) amp
    (B) volt
    (C) ohm
    (D) watt

2. A _____ circuit offers more than one path to current flow.
    (A) series
    (B) parallel
    (C) series-parallel
    (D) Both B & C.

3. Technician A states voltage drop across an open circuit is always equal to zero. Technician B says a circuit or portion of a circuit with no opens is said to have continuity. Who is right?
   (A) A only.
   (B) B only.
   (C) Both A & B.
   (D) Neither A nor B.

4. Technician A says an increase in circuit resistance will cause a decrease in the current supplied to loads. Technician B says the voltage available to the load will also decrease. Who is right?
   (A) A only.
   (B) B only.
   (C) Both A & B.
   (D) Neither A nor B.

5. Which of the following can be used to protect diesel engine electrical circuits against overloads?
   (A) Fuses.
   (B) Circuit breakers.
   (C) Fusible links.
   (D) All of the above.

6. When diesel engine batteries are suspect, Technician A says state of charge and battery capacity tests should be performed. Technician B says to disconnect the batteries in a multiple battery system and perform these tests on each individual battery. Who is right?
   (A) A only.
   (B) B only.
   (C) Both A & B.
   (D) Neither A nor B.

7. Technician A says a battery must be very near its full state of charge for a battery capacity test to be accurate. Technician B says the battery temperature will have no effect on the results of the capacity test. Who is right?
   (A) A only.
   (B) B only.
   (C) Both A & B.
   (D) Neither A nor B.

8. Technician A says semiconductors carry more current at increased temperatures. Technician B says doping of semiconductors is done to increase their resistance to current flow. Who is right?
   (A) A only.
   (B) B only.
   (C) Both A & B.
   (D) Neither A nor B.

9. Which of the following electronic components can conduct current in one direction at certain voltage levels and in the other direction at other voltage levels, making the component extremely important in the construction of electronic voltage regulators?
   (A) Semiconductor.
   (B) Diode.
   (C) Zener diode.
   (D) Transistor.

10. Which of the following electronic components have insulating and conductive abilities that allow them to act as switches?
    (A) Semiconductor.
    (B) Printed circuit boards.
    (C) Resistors.
    (D) ICs.

Diesel engines use warning lights, such as these, to warn the operator of a condition that could damage the engine. On some engines, one or more of these lights may be computer-controlled.

Chapter 25

Electronic Engine Controls and Fuel Injection

After studying this chapter, you will be able to:
- ❏ Explain the operating principles of basic electronic components.
- ❏ Describe the operating principles of an electronic engine control system including the role of sensors, computers, and actuators.
- ❏ Describe the basic operation of engine control computers including A/D converters, ROM, RAM, PROM, EPROM, EEPROM, and FEPROM memory, interfaces, and clock chips.
- ❏ Explain how on-board engine computers communicate with other control modules, personal computers, and service tools.
- ❏ Define what a software program is, what it does, and give examples of the types of programs used by service technicians and fleet managers.
- ❏ Explain the basic operating principles of electronically controlled multiple plunger inline injection pumps, unit injection systems, and distributor pump injection systems.
- ❏ Define and give examples of factory and customer programmable diesel control system parameters.
- ❏ Give brief overviews of the electronic engine control systems offered by several major manufacturers.
- ❏ Describe the basic steps used when troubleshooting engines equipped with electronic controls.

Modern computer and data communication technologies have changed the way modern diesel engines are designed, operated, and serviced. Advances in computer memory, processing speed, and data sharing technology have dramatically increased the capabilities and value of electronically controlled engine systems. This chapter will discuss electronic diesel fuel injection and the control systems that make them work.

Diesel Engine Control Systems

The first computer-controlled or electronic diesel fuel injection systems were introduced in the mid-1980s. They were designed to reduce engine exhaust emissions from over-the-road trucks to levels set by Federal and state regulations. These systems concentrated on precisely controlling fuel injection timing and volume so fuel would be burned more efficiently.

All electronic diesel engine control systems share a number of common characteristics regardless of the type of fuel injection method used. They all have a network of sensors that report engine conditions to a computer. The computer calculates the needed action based on this input and modifies output device operation to control fuel injection and other vital systems.

Some engine manufacturers use two major controllers, one specifically for fuel injection and a second for other engine and vehicle related functions. These controllers communicate with each other and share sensor input. The use of multiple controllers expands the system capability while reducing the number and length of wires and connectors. Engine control systems can now share data with other on-board computers and communicate with vehicle management and service related control systems such as trip recorders, navigation and driver information systems, security systems, as well as ground and satellite communications networks. Electronic control systems are now used in all newer on-highway diesel engines, as well as some off-highway, marine, and industrial engines.

Electronic Control Module

It is important to remember each manufacturer may have their own name for their computers. In this text, we will refer to any generic computer as an **electronic control** *module* or *ECM.* Depending on the manufacturer of the system, the computer may be referred to as a(n):

- ❏ ECM (Engine Control Module).
- ❏ ECU (Electronic Control Unit).
- ❏ EEC (Electronic Engine Control).
- ❏ CPU (Central Processing Unit).
- ❏ PCM (Powertrain Control Module).
- ❏ VCM (Vehicle Control Module).

The term PCM is gaining in popularity, especially in mobile vehicles.

The electronic control module is the brain for the engine control system. The ECM monitors engine and vehicle operating conditions continuously using sensors located on the engine, transmission, and other major vehicle systems, **Figure 25-1.** The primary job of the ECM is to control fuel quantity and injection timing based on input from engine data sensors. The fuel quantity and injection timing to each cylinder is precisely controlled to obtain optimum fuel economy and reduced emissions in all driving situations.

Other ECM functions may include fan clutch control for engine cooling, engine brake control, and engine protection (oil pressure, coolant temperature, and coolant level) control. In addition, the ECM performs diagnostic

fault code logging, broadcasts data to other control modules for other vehicle systems, and executes password and security access tasks.

The ECM may be mounted to the side of the engine block or located under the instrument panel, under a panel in the cab, or under the seats, **Figure 25-2.** It is contained in a sealed aluminum or plastic housing which may have a protective heat shield and/or a cold plate. Diesel fuel from the transfer pump is routed through the cold plate to help keep the module cool.

Voltage Signals

The ECM uses voltage values as communication signals. The ECM steps down 12 volt battery power to 5–9 volts, which is used as a *reference voltage* by its sensors. Sensors act on the reference voltage in some way. They can send a signal of the same voltage strength back to the ECM. They can change the strength of the signal and send it back, or they can block the voltage signal from returning to the ECM altogether. Based on the nature of the signal returned from the sensor, the ECM can determine changes in the engine's or vehicle's operating conditions. The ECM can then send out voltage signals of its own to actuators such as solenoid valves, relays, and switches. By operating

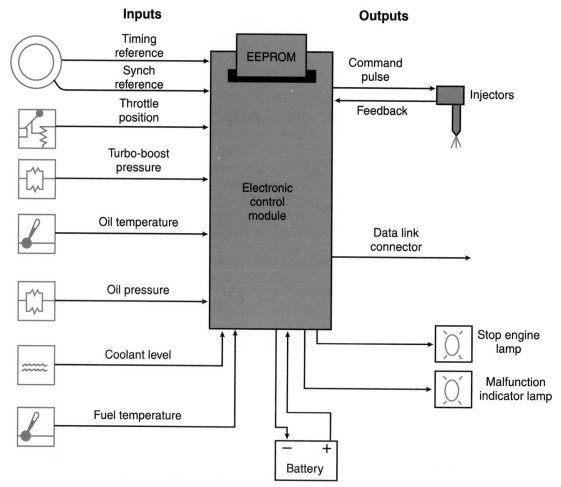

Figure 25-1. Components of an electronic engine control system.

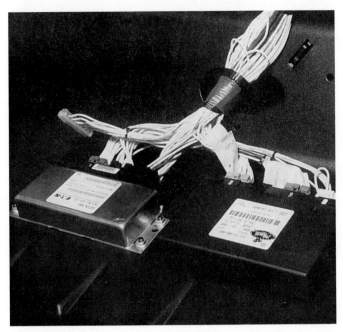

Figure 25-2. Typical engine control ECM.

these actuators, the ECM controls and makes changes to various engine and vehicle systems.

There are two types of voltage signals: analog and digital. As shown in **Figure 25-3A,** an *analog signal* varies continuously over a given range. Analog signals are used to indicate conditions that change gradually and continuously within an established range. Temperature sensors are the best example of sensors that generate an analog signal.

The second type of signal is a digital signal. A *digital signal* is a signal that is either on or off with no range in between, **Figure 25-3B.** Turning a switch on or off generates a digital signal.

Analog-to-Digital (A/D) Converters

All ECMs are digital computers and will only recognize digital or on/off input signals. As explained earlier, most sensors operate using 5-9 volt reference signals. They modify this signal before relaying it back to the ECM. To convert this signal to an on/off mode, the signal is passed through an *analog-to-digital (A/D) converter.* This converter uses diodes to control signal input to the ECM. The converter changes analog signals to digital signals the ECM can understand.

For example, a diode assembly inside the converter may allow sensor signals of 5 volts or less to pass and stop signals greater than 5 volts. The result is the ECM receives either an on or off signal from the sensor. When a precise reading from an analog sensor, such as a temperature sensor is needed, the input signal will pass through a series of diodes, all of which generate on/off signals. Based on the pattern of these on/off signals, the ECM can then determine the exact temperature reading. Very weak voltage signals are passed through an amplifier as they enter the ECM.

The digital-to-analog converter circuit does just the opposite. It converts the ECM's digital signal to an analog signal for any controlled functions that need analog voltage.

Binary Codes

Digital signals are matched to a binary number code, which is the language the ECM uses to communicate. Binary numbers are made up of 1s and 0s. "On" voltage signals are assigned a 1 value, while "off" signals are given a 0 value. Each 1 or 0 equals one bit of data, with eight bits equaling one byte. All information exchanged between the ECM, its memory, and other electronic devices is done in the form of bytes, **Figure 25-4.**

Interface Circuits and Clock Chips

Every ECM has an input/output *interface circuit,* **Figure 25-5.** This interface circuit protects the ECM's electronic components from shorts as well as the higher voltages of the external circuits. The interface also contains the analog/digital signal converters.

In order to maintain an orderly flow of data through the ECM, a quartz crystal is used as a *clock chip* to produce a continuous, consistent time pulse. Data can only be transmitted between pulses. The pulses are timed so one bit of information can be transmitted between each pulse. Without this clock to space out and organize data flow, the ECM would not operate properly.

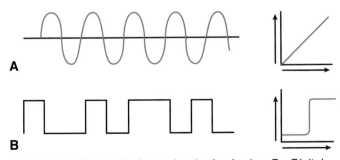

A

B

Figure 25-3. Types of voltage signals. A—Analog. B—Digital.

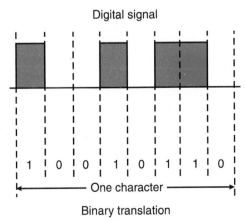

Digital signal

| 1 | 0 | 0 | 1 | 0 | 1 | 1 | 0 |

One character

Binary translation

Figure 25-4. Digital voltage signals are on/off signals that can be converted and read as a binary code.

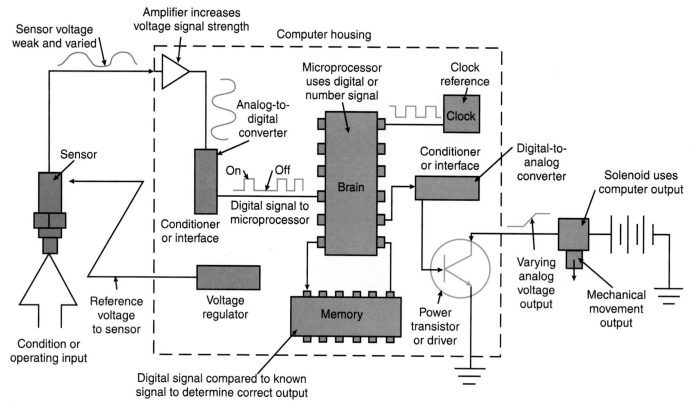

Figure 25-5. *Voltage signal processing in a typical microprocessor. Note the use of conditioners or interfaces on both the input and output sides of the ECM.*

Control System Memory

Memory refers to the internal storage areas of the ECM. For diesel electronic control modules, computer memory is provided in the form of memory chips. In contrast, the memory of personal and laptop computers includes magnetic floppy disks and internal hard disks as well as memory chips.

The ECM's memory chips store all the information or data needed for the engine and vehicle control systems to operate and communicate effectively. The ECM uses memory in two ways. It can pull up, read, and use information stored in system memory, or it can take new or altered data and store it in system memory for future use or viewing. This new data can be information the ECM receives directly from its sensors. It can be data that is the result of calculations or operations the ECM has performed on raw sensor data. It can also be new instructions uploaded into the ECM from an outside source, such as a laptop computer.

During processing, the ECM often receives more data than it can process. Some of this data is temporarily stored in memory until it can be processed. When the ECM is ready to use this data, it retrieves the information. The ECM only takes a copy of the information stored. The original data is kept in memory in case it is needed for future use.

There are a number of different types of memory available in chip form for automotive and truck ECMs. These are random access memory (RAM), read only memory (ROM), and various types of programmable read only memory (PROM), **Figure 25-6.**

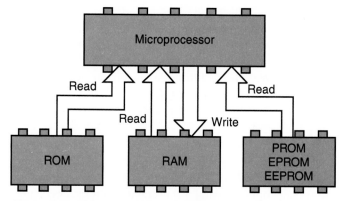

Figure 25-6. *Types of ECM memory and their read/write capabilities to the on-board microprocessor.*

Random Access Memory (RAM)

Random access memory (RAM) refers to read and write memory. The ECM can both write data into and read data from RAM. Data stored in RAM can be erased or changed after it has been read. RAM is the ECM's decision making center. The ECM retrieves data and loads it into RAM where it compares it to sensor input, also stored in RAM. The ECM can read from or write new operational information into RAM memory as needed. Most RAM memory is volatile, requiring a continuous flow of electricity to maintain its stored information. Any information stored in RAM is lost if battery voltage is disconnected from the ECM. Certain data is also erased when the ignition is turned off. In most cases, RAM memory chips are permanently wired into the module and must be replaced with the ECM, if defective.

Read Only Memory (ROM)

As its name states, **read only memory (ROM)** can only be read by the ECM. Computers contain a small amount of read-only memory that holds basic operating instructions for the ECM.

The information stored in ROM cannot be reprogrammed or erased. The ECM can only read and use this information. ROM memory is non-volatile, which means data stored in ROM is not lost if the ECM is turned off or disconnected from the battery. Early electronic control systems relied heavily on ROM memory, which meant they had limited ability to adapt to situations and conditions not originally programmed into ROM by the system manufacturer.

Programmable Read Only Memory (PROM)

A **programmable read only memory (PROM)** chip is a memory chip that can store a particular program or database. However, once the PROM has been programmed, the information on it becomes permanent. It cannot be erased or altered. Programmable read only memory is not erased at system shutdown or when the battery is disconnected.

Because PROM chips were designed to be removed and replaced, they provided the first method for matching engine and vehicle specifications to different engine applications. PROM chips could be factory programmed with application specific data on engine calibration, transmission specifications, vehicle weight, rear axle ratio, etc. Each PROM chip is designed for use in a particular vehicle/engine application. If the engine was placed in a different service environment, the PROM chip could be changed to match the new operating parameters.

 Caution: PROM chips must never be intermixed between ECMs in different vehicles, even though they may be from the same manufacturer.

Eraseable Programmable Read Only Memory (EPROM)

Eraseable programmable read only memory (EPROM) offers a major advantage over basic PROM memory in that data stored on an EPROM can be erased and/or changed by qualified personnel using a device known as an **EPROM burner.** EPROM is a special type of memory that retains its contents until it is exposed to ultraviolet light. The ultraviolet light clears its contents, making it possible to reprogram the memory. Like ROM and PROM memory, an EPROM chip does not require power to retain its data.

Electronically Eraseable Programmable Read Only Memory (EEPROM)

Electronically eraseable programmable read only memory (EEPROM) is a special type of PROM that can be erased by exposing it to an electrical charge. With EEPROMs, a special software program is used to erase data from the EEPROM and then download new information onto the chip. A faster version of EEPROM memory used in some late-model ECMs is called **flash erasable programmable read only memory (FEPROM).** EEPROMs and FEPROMs are non-volatile memory and are fixed to the ECM.

Both EEPROMs and FEPROMs can be completely programmed or partially reprogrammed electronically using a personal computer loaded with the engine manufacturer's programming software. Unlike older style EPROMs, EEPROMs and FEPROMs do not have to be removed from the ECM for programming. The programming information is downloaded from the personal computer to the ECM's programmable memory through a data link connector. This provides control system manufacturers and engine owners with an extremely powerful tool for customizing engine control systems with factory and customer selectable parameters.

A **parameter** is simply a specific engine operating instruction or setting. The ability to program customized parameters means the same model engine can be programmed to respond to unique operating conditions and demands. Modern engine control computers also have the ability to store and transmit large amounts of data. This data can be used to help diagnose engine driveability problems. It can be used by drivers and fleet operators to help maximize engine performance and to coordinate maintenance and service intervals. Additional information on programming system parameters and working with engine control software programs can be found later in this chapter.

EMI and RFI

Any device or system that generates an electromagnetic field has the potential to disrupt the operation of electronic components, devices, and systems in its vicinity. This phenomenon is known as **electromagnetic interference** or **EMI.** Unfortunately, an engine's alternator and arc welding equipment are both sources of EMI that can disrupt the operation of the engine control system if proper precautions are not taken. **Radio frequency interference** or **RFI,** can also disrupt electronic component operation. CB radios, especially those with signal amplification circuitry, can be a source of RFI.

There are three essential elements to any EMI/RFI problem. There must be a source of interference, a receptor that cannot function properly due to EMI/RFI, and a path between them allowing the source to interfere with the receptor. Identifying at least two of these elements and eliminating one of them generally solves EMI/RFI compatibility problems.

Ensuring all electronic equipment is operated with a good electrical ground system can minimize problems with EMI/RFI. In addition, cords and cables connecting the peripherals in an electronic or computer system should, if possible, be routed and/or shielded to help keep signal energy from entering or leaving.

Data Link Connectors and Serial Lines

Data link connectors are used as the standard communication link between on-board computers and off-board service and fleet management equipment. Diagnostic and programming tools, such as hand-held scan tools or personal computers, can communicate to the ECM through these connectors, **Figure 25-7A.** On heavy-duty trucks and other vehicles, the data link connector is usually located beneath the dash next to the steering column, **Figure 25-7B.** Special adapters are sometimes needed between the off-board diagnostic tool and the data link connector, **Figure 25-8.**

The ECM communicates with other on-board control modules as well as external service tools. Several SAE standards are used for ECM on-board and off-board

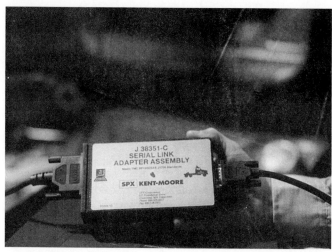

Figure 25-8. *A serial link adapter is used on some systems to establish communications between a personal computer and the truck engine ECM.*

communication. These standards include J1587, J1708, J1922, and J1939. The standards define the methods and protocols by which various electronic devices communicate with one another.

Communication between on-board computers are handled over **serial lines.** These lines also operate fully electronic instrumentation, electronic speedometers, tachometers, and driver displays. The J1587 and J1939 standards outline serial data communication between the ECM and other control modules such as a vehicle control module, anti-lock brake module, and accessory systems.

Remote Data Retrieval Stations

To allow drivers and technicians to perform data downloads from ECM memory without a PC or scan tool, some control systems can be tied into **remote data retrieval stations.** These weather-proof interface boxes can be installed in service bays or fueling islands. A short serial communications cable is installed between the vehicle's communications port and the port on the interface box. Lights on the interface box indicate when the download is in progress and when it is complete. The interface box connects to a centrally located master computer, **Figure 25-9.** The master computer monitors all data transfers and stores and organizes data for easy access by fleet managers and service personnel.

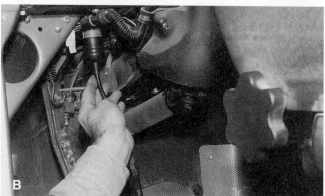

Figure 25-7. *A—The data link connector. B—Connecting to the data link connector, usually located beneath the dashboard. C— Programming engine control parameters using a personal laptop computer.*

Input Sensors

Input sensors are used to monitor mechanical conditions such as movement or position, temperature, or pressure. Sensors used in diesel engine control systems monitor oil pressure, oil temperature, coolant level, coolant temperature, turbocharger boost pressure, air temperature, fuel pressure, vehicle road speed, and throttle position.

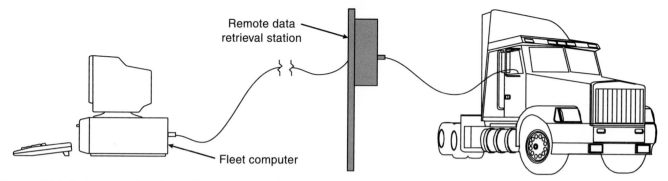

Figure 25-9. *Data retrieval stations allow the engine ECM to interface with a remotely located master computer used by fleet and maintenance personnel.*

Variable Resistor Sensors

A ***variable resistor*** is a very basic sensing device. In this type of sensor, a variable resistor is wired in series with a fixed value resistor located inside the ECM. The ECM sends the sensor a reference voltage. The sensor's variable resistor alters this voltage, depending on the existing conditions at the sensor's location. The modified voltage signal is sent back to the computer, **Figure 25-10.** After reading this signal, the computer knows the conditions that exist at the sensor, and can react accordingly.

The most common type of variable resistor sensor is the thermistor. A ***thermistor*** varies its resistance based on temperature. When the temperature is cold, the sensor resistance is high. As the temperature increases, the sensor resistance decreases. Thermistor sensors include coolant, fuel, oil, and intake air temperature sensors. Potentiometers are another type of variable resistor.

Coolant Temperature Sensor (CTS)

The ***coolant temperature sensor (CTS)*** is located on the engine in one of the top coolant passages. Input from the coolant temperature sensor helps to control the amount of fuel needed. In some control systems, an engine protection feature will be triggered if the coolant temperature exceeds the specified limits.

The ECM monitors the resistance drop across the coolant sensor and uses the signal to calculate fuel injector timing and to evaluate operating conditions that can cause high coolant temperatures such as thermostat failure, fan failure, heavy load, high ambient temperatures, and radiator blockage. This sensor is also used for cold start enhancement.

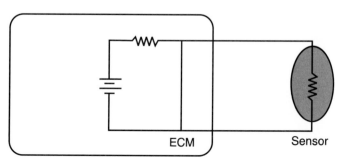

Figure 25-10. *Basic circuitry used in a variable resistor two-wire sensor.*

Oil Temperature Sensor (OTS)

The ***oil temperature sensor (OTS)*** allows the computer to set the correct idle speed during warm-up and control injection timing. This helps improve cold driveability and reduces white smoke. Input from the oil temperature sensor is also used to activate the engine protection system. If higher than normal oil temperature is reported, the ECM will reduce the amount of fuel fed to the engine to slow engine speed. If the problem does not correct itself, an engine shutdown sequence may be started.

Fuel Temperature Sensors (FTS)

The ***fuel temperature sensor (FTS)*** provides a signal to the ECM as to the fuel's temperature. This sensor is normally located in the fuel filter or the fuel inlet line on some engines. The ECM utilizes the fuel temperature signal to help calculate the fuel injection timing. The resulting changes in fuel density, as a function of temperature, allow the engine to maintain peak horsepower and efficiency.

Intake Air Temperature (IAT) Sensor

The ***intake air temperature (IAT) sensor*** monitors and reacts to air temperature within the engine intake manifold or air cleaner assembly. Its input allows the computer to deliver a more accurate air-fuel mixture to the injectors. The ECM adjusts the injector timing to reduce white smoke during warm-up, improve cold starts, and provide engine protection. Intake air temperature information is also used to prevent misfire under light load conditions.

Inductive (Pulse Counting) Sensors

Inductive sensors are used to determine how fast the teeth of a gear are moving past the sensor, **Figure 25-11.** As the gear teeth pass the sensor, the principle of magnetic induction generates an ac voltage pulse that can be read by the ECM. Pulse counting sensors are used to provide input for calculating engine speed (rpm), vehicle speed, and engine position.

All sensors continually update the conditions they monitor. The sensors usually update their signals at least once per second with some updates as often as 10 times per second.

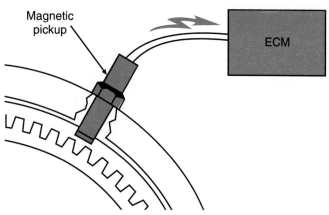

Figure 25-11. *Engine position or engine speed sensor mounted in flywheel housing.*

Engine Speed/Timing Sensor

The **engine speed/timing sensor** is installed near the flywheel. When the engine is running, the flywheel's teeth rotate past the sensor's tip and a pulsed signal is generated. The computer counts the signal pulses generated by the engine speed/timing sensor to calculate the engine RPM. Some engine speed sensors can also determine when the crankshaft is at the top dead center position. The sensor does this by recognizing two "missing" teeth on the ring gear corresponding to top dead center.

Engine Position/Timing Reference Sensor

The **engine position sensor** or **timing reference sensor** is normally located in the outside face of the timing gear cover. It senses the passage of one or more notches on the face of the cam gear. Pulses generated act as trigger signals for the electronic fuel injection sequence. The frequency of the pulses is translated into engine speed and position by the ECM. The ECM uses this signal along with the engine speed sensor signal to synchronize fuel injection.

Vehicle Speed Sensor (VSS)

The **vehicle speed sensor (VSS)** senses the passage of a series of teeth on a gear mounted on the transmission output shaft. It is mounted in the transmission housing, **Figure 25-12A.** The ECM monitors the frequency or pulses of the signal generated by the VSS to calculate the vehicle speed for use with cruise control, vehicle speed limiting, and progressive engine braking. The air gap between the sensor and the toothed gear influences the VSS signal output and should be checked if erratic or inaccurate speedometer readings are reported.

Brake, Clutch, and Cruise Control Switches

The service **brake switch** is located in the brake service line, **Figure 25-12B.** When the brakes are applied, the circuit opens, signaling the computer to disengage the cruise control. The **clutch switch** or sensor is located near the clutch

pedal or linkage, **Figure 25-12C.** When the clutch is engaged (pedal up), the switch is closed to allow cruise control or engine braking operation. Depressing the clutch opens the circuit to the computer, signaling it to deactivate the cruise control and/or engine braking operation.

By monitoring the **cruise control switch,** cruise control set/resume switch, and other inputs, the ECM or an associated control module can maintain the desired vehicle or engine speed. The set/resume switch is normally a spring-loaded switch that allows the operator to resume the preset engine or vehicle speed or accelerate. This is done by pressing the top of the switch. To set or lower engine or vehicle speed, the operator presses the bottom of the switch.

Oil Pressure Sensor (OPS)

The **oil pressure sensor (OPS)** is located in the oil line or on top of the oil filter assembly. It monitors oil pressure during engine operation, **Figure 25-13.** It is a pressure sensitive diaphragm that produces an electrical signal proportional to the amount of diaphragm movement. If the sensor reads a significant or total loss of oil pressure, it will signal the computer to shut the engine down.

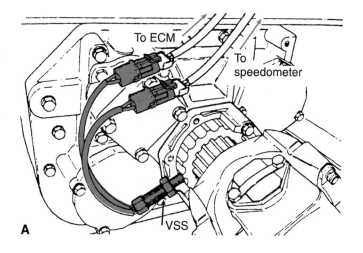

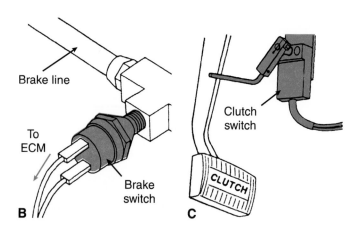

Figure 25-12. *Various sensor locations. A—Vehicle speed sensor. B—Brake switch. C—Clutch switch. (Cummins Engine Co., Inc.)*

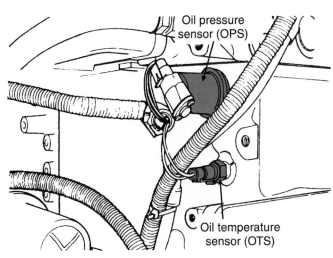

Figure 25-13. *Common position of oil pressure and oil temperature sensors. (Detroit Diesel Corporation)*

Air Pressure Sensor

Some engines are equipped with an **air pressure sensor.** This diaphragm movement sensor provides the computer with data on outside air pressure, which is then used to calculate the altitude at which the engine is operating.

 Note: Do not confuse air pressure sensors with the manifold absolute pressure sensors used on many computer-controlled gasoline engines; they are not the same.

Coolant Level Sensor

The **coolant level sensor** is located in the top section of the radiator or radiator overflow tank, **Figure 25-14.** The sensor detects coolant by measuring the continuity in the coolant level sensor circuit. The sensor contains two electrodes; one is in the probe tip and the other in the probe base. The ECM applies a small current to the coolant level sensor. When the sensor is immersed in coolant, the coolant acts as a conductor between the two electrodes. If the coolant level falls below the level of the two electrodes, circuit resistance increases and the ECM registers a low coolant alarm and switches on a dash-mounted "shut down" warning light. In some systems, the ECM may shut down engine operation. The shut down will usually occur before the engine reaches temperatures that could cause damage.

Turbocharger Boost Sensor (TBS)

The **turbocharger boost sensor (TBS)** is located in the intake manifold on the turbocharger's pressure side. Its signal is used to determine the amount of air entering the engine, so the amount of fuel injected can be increased or reduced as needed. By monitoring turbocharger compressor discharge, this sensor provides air pressure data to the ECM for smoke control during acceleration.

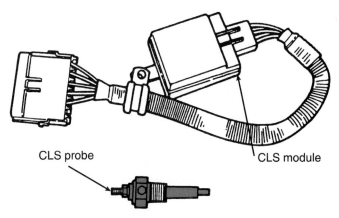

Figure 25-14. *Coolant level sensors are located in the engine radiator.*

Throttle Position Sensor (TPS)

The **throttle position sensor (TPS)** provides the driver's fuel requirement input to the ECM. It is a potentiometer mechanically linked to the accelerator pedal, **Figure 25-15.** As the resistance changes, the sensor signal voltage changes, indicating the accelerator pedal position. At idle (pedal up), voltage output is low, and the computer then knows to deliver a smaller amount of fuel to the cylinders. As the pedal is depressed, the voltage signal from the TPS increases and the computer in turn increases fuel delivery. This "drive by wire" pedal is designed to provide a system that feels similar to the standard type of accelerator pedal and mechanical linkage

Output Devices

After the ECM has calculated a response to the existing conditions or situation, it sends output signals to control devices known as actuators or output devices. **Output devices** are electromechanical devices, such as solenoids and relays that convert electrical current into mechanical action. This mechanical action can then be used to open and close valves in fuel injection units. It can

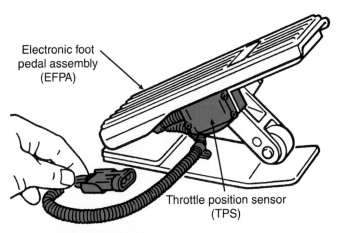

Figure 25-15. *Throttle position sensor. (Detroit Diesel Corporation)*

also be used to operate a switch in the electrical circuit of an engine cooling fan motor, dash-mounted engine warning light, or other electrical device.

Fuel Injection Control

The ECM controls the amount of fuel injected by allowing the injectors to pump fuel only when the injector solenoid is energized. The ECM sends a voltage signal to energize the solenoid. After computing the fuel requirements, the ECM sends a *pulse width signal* to the injector solenoid. The pulse duration determines the length of time fuel is allowed to flow through the injector nozzle. The longer the pulse width time, the greater the fuel volume. As soon as the ECM stops its signal to the solenoid, injection ends.

The ECM carefully monitors the time each injector solenoid is activated. This response time feedback is used by the ECM to monitor and adjust fuel injection timing. This ensures there will be no timing variation at the start of injection. The system's PROM (programmable read only memory) controls the base pulse width for each engine application. This pulse width is altered by the ECM in response to operating conditions and loads. However, the base pulse width cannot be changed without replacing the PROM chip or reprogramming the ECM.

Tachometer/Speedometer Drive Outputs

When the engine is running, the engine speed sensor generates a series of voltage pulses to the electronic control module. The module counts the voltage pulses to calculate engine speed. The ECM sends the calculated engine speed

as a series of signals to the vehicle's tachometer. The *tachometer output* translates the engine speed signal into revolutions per minute (RPM) and displays the result on the tachometer gauge. The ECM also calculates the vehicle speed to operate the speedometer, cruise control and other functions. The *speedometer output* is used to drive the speedometer pointer.

Electronically Controlled Fuel Injection Pumps

All major types of fuel injection pumps are now integrated into electronic control systems. Variations on the unit injector are used in most electronic diesel fuel injection systems.

Electronic Unit Injectors (EUI)

Electronic unit injectors (EUI) have proven to be the most readily adaptable of all existing fuel injection systems to electronic control. The spray-in pressures and fine atomization possible with unit injection systems, coupled with the system's compatibility to electronic control, has led many manufacturers to switch to unit injection. In addition to Detroit Diesel, other manufacturers using electronic unit injection include Cummins, Caterpillar, Mack, and Robert Bosch. To better understand how an electronic unit injector operates, the unit injection system used on the Detroit Diesel Series 60 engine will be shown as an example, **Figure 25-16.**

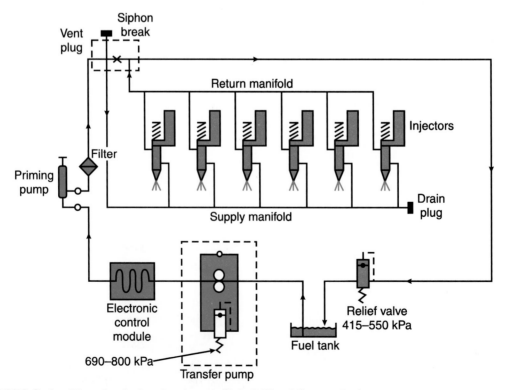

Figure 25-16. *DDEC Series 60 engine fuel system layout. (Detroit Diesel Corporation)*

Electronic Unit Injector Operation

Fuel enters the injector through two filter screens. Fuel not used for injection cools and lubricates the injector before exiting through the return port and returning to the supply tank, **Figure 25-17.**

Figure 25-18 illustrates the actuator components of the unit injector. The electronic unit injection system uses mechanical action to create the pressures needed for injection. As in the mechanical unit injection system, the camshaft pivots the rocker arm through its roller follower.

This forces the injector follower down against its external return spring. This action raises the trapped fuel to a pressure sufficient to lift the injector needle valve off its seat. However, fuel metering is electronically controlled by the ECM based on input signals from various sensors.

Fuel Flow

Fuel flow through the electronic unit injector is illustrated in **Figure 25-19.** Once fuel enters the electronic unit injector, it passes through the inlet filters and flows through a drilled passage to the electronically controlled poppet valve. At this point in the injection cycle, the poppet valve is held open by spring pressure. Fuel flows through the plunger and bushing, into the fuel supply chamber. With the poppet valve open, fuel simply fills the injector.

When the piston is approximately 60° BTDC on its compression stroke, the camshaft begins to lift the injector rocker arm roller follower. The EUI internal plunger moves on its bushing, increasing fuel pressure. However, no increase above fuel pump pressure is possible until the ECM sends out a voltage signal to the electronic distributor unit. The electronic distributor unit (EDU) handles the high current needed to activate the injector solenoid.

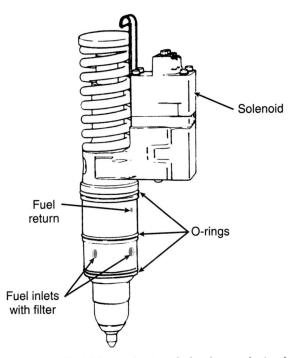

Figure 25-17. Fuel inlet and return holes in an electronically controlled unit injector. (Detroit Diesel Corporation)

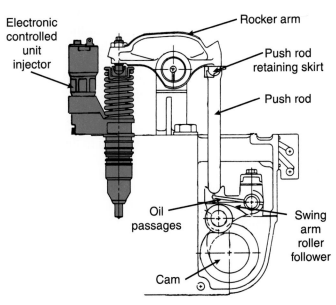

Figure 25-18. Detroit Diesel Series 60 engine electronically controlled/mechanically actuated unit injector. (Detroit Diesel Corporation)

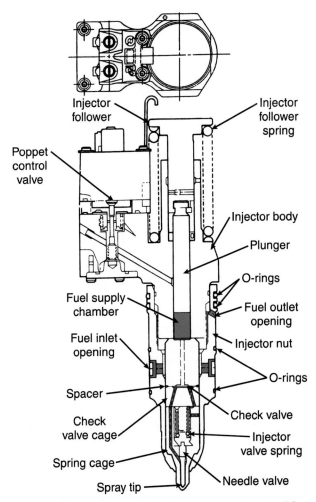

Figure 25-19. Components of Series 60 engine DDEC unit injector. (Detroit Diesel Corporation)

When the solenoid on the EUI is energized, its armature is pulled upward, closing the poppet valve. This traps fuel between the poppet valve and descending plunger. This creates a rapid rise in the pressure within the fuel supply chamber that leads to the spray-tip assembly.

A small check valve located between the plunger base and the spray tip prevents combustion gas blowby from leaking into the injector. During normal operation, the fuel pressure below the plunger increases until it is powerful enough to lift the needle valve from its seat. The strength of the needle valve spring determines when the valve will lift off its seat. Opening pressures of 2800-3200 psi (19 300-22 065 kPa) are common. When the needle valve unseats, fuel flows through the orifices in the injector tip. Forcing the fuel through these small openings increases the pressure to approximately 20,000 psi (138 000 kPa).

The start and duration of injection are controlled by the pulse width signal from the ECM. The longer the EUI solenoid is energized, the longer the poppet valve remains closed and the greater the amount of fuel injected. Holding the poppet valve closed sets the injector plunger effective stroke. The plunger always moves down the same distance on every injection stroke, but the length of time the fuel is pressurized beneath the plunger is controlled by the solenoid.

When the ECM de-energizes the EUI solenoid, spring pressure opens the poppet valve. High-pressure fuel can now flow through the small return passage in the injector body. Pressure is lost, and the force of the needle valve return spring forces the needle valve onto its seat. This results in a clean, quick end to injection. Fuel at pump pressure immediately flows into the EUI through the open poppet valve. The plunger continues to the bottom of its stroke, but it is pushing fuel through the small return passage at low pressure, not through the injector tip at high pressure. When the plunger completes its downward stroke, the follower return spring pulls it up to its original position. The injector is now in position to begin the injection cycle again.

Hydraulic Electronic Unit Injectors (HEUI)

Hydraulic electronic unit injectors (HEUI) were jointly developed by Caterpillar and Navistar. They use high pressure engine oil to provide the force necessary to accomplish injection, **Figure 25-20.** Many of the mechanical drive components found in standard mechanical or electronic unit injection systems, such as cam lobes, lifters, push rods, and rocker arms are not needed in this system.

A solenoid on each injector controls the amount of fuel delivered by the injector. A gear-driven axial pump raises the normal pressure to the levels required by the injectors. The ECM sends a signal to an injection pressure control valve to control pressure, and another signal to each injector solenoid to inject fuel.

Pressure in the engine oil manifold is controlled by the ECM through the use of an injection pressure control valve. The injection pressure control valve or dump valve

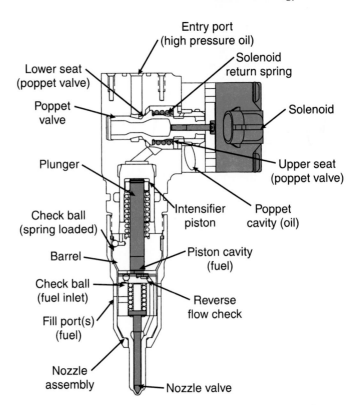

Figure 25-20. *Components of a hydraulic electronic unit injector. (Caterpillar, Inc.)*

controls the injection pump outlet pressure by dumping excess oil back to the sump.

The ECM monitors pressure in the manifold through an injection pressure sensor. The sensor is located in the top of the manifold on the left side of the engine. The ECM compares the pressure sensor signal to the desired injection pressure. Based on this and other sensor inputs, the ECM changes the position of the injection pressure control valve. This adjusts the oil pressure in the high pressure manifold.

High pressure engine oil is routed from the pump to the high pressure manifold through a steel tube. From the manifold, the oil is routed to each injector through short jumper tubes. All injectors have a constant supply of oil while the engine is running. Cutting out an injector disables the electrical signal to the injector solenoid, but does not interrupt the oil flow to the injector.

A variation of the HEUI injector used on small to midsize diesel engines is known as the *split-shot* or *pre-stroke injector.* This injector delivers fuel to the combustion chamber in two stages. A short initial burst is delivered to start combustion slowly with the remainder of the fuel lagging slightly behind. The major advantage of split-shot injection is reduced engine noise and lower exhaust emissions.

Inline Injection Pumps

Electronically controlled inline injection pumps generate fuel injection pressures in the range of 18,000 psi, pressures significantly higher than old style mechanical pumps. These electronically controlled inline pumps also provide precise control over the start of injection through the use of port closing (start-of-delivery) adjustment.

Pump Operation

In a mechanically governed inline injection pump, the start of injection and the quantity of fuel injected are directly controlled by engine speed. **Figure 25-21** illustrates an electronically controlled inline injection pump. In this pump, both the injected fuel quantity and the start of injection are electronically controlled by means of solenoids. This means the start of fuel injection and the fuel quantity can be altered independently, based on inputs from the engine sensors. This particular pump design offers precise control over the start of injection through the use of port closing (start-of-delivery) adjustment.

Port closing is adjusted by changing the position of a sleeve on the pump plunger. The sleeve moves inside a window formed in the pumping element body, **Figure 25-22**. The exact time of port closing is determined by the sleeve's vertical location on the plunger. The quantity of fuel injected is metered by helix control. Electronic inline pumps can produce fuel injection pressures in the range of 18,000 psi (124 100 kPa). Port opening is adjusted by rotating the pump plunger as in mechanical inline pumps.

In the electronic inline fuel injection pump, an electromagnetic actuator combined with a rod-travel sensor, speed sensor, and ECM replaces the mechanical governor. A solenoid adjusts the fuel quantity by moving the fuel injection pump control rod, which in turn adjusts the pump plunger control sleeve. An inductive pickup registers the control rod position. If deviation is detected, an input signal is sent from the governor to the ECM.

Once the input data is received, the ECM calculates the ideal fuel quantity or control rod position. The ECM determines the correct position for the control rod solenoid. As the solenoid is activated, the control rod acts against the force of a return spring and changes the control sleeve position.

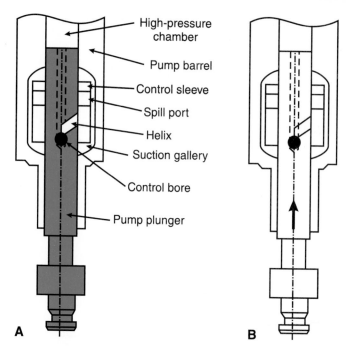

Figure 25-22. Plunger and barrel assemblies with control sleeves.

The control sleeve moves the pump plunger in the direction of delivery, permitting an adjustable prestroke to change the start of injection. When the pump plunger has moved upward a set distance, the control sleeve's bottom edge closes the pump plunger spill port. As delivery starts, pressure builds in the chamber. The pump plunger helix and control sleeve spill port end fuel delivery like a conventional inline pump. The end of fuel delivery (port opening) determines the amount of fuel injected.

Port closing (start of injection) is adjusted by moving the control sleeve in the direction of delivery. A control sleeve position closer to TDC means a large prestroke and a late start of delivery, while a position closer to BDC means a small prestroke and earlier start of injection.

The control rod travel sensor continually updates the ECM as to the control sleeve position. The ECM constantly adjusts the solenoid current to maintain ideal control sleeve position. If the actuator system is de-energized, a restoring spring returns the control rod to the position of zero delivery.

Start of delivery is also adjusted in closed loop. A motion sensor installed in one of the nozzle holders reports the actual port closing position to the ECM. The ECM compares this position and the engine TDC mark with values stored in memory. The ECM then sends the appropriate signal so the actual port closing position and ideal port closing position in the ECM's memory are the same.

Electronically Controlled Distributor Pumps

Electronically controlled distributor pump systems are used on mid-size diesel engines used to power commercial grade pick-up trucks and recreational vehicles.

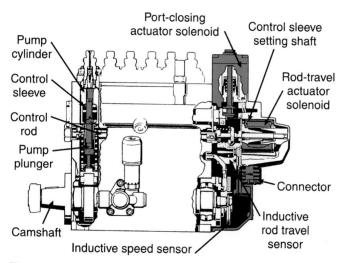

Figure 25-21. Components of an inline injection pump using electronically controlled sleeve metering.

Stanadyne Electronically Controlled Distributor Pump

The Stanadyne electronically controlled distributor pump shown in **Figure 25-23** regulates fuel quantity with a solenoid valve that controls the amount of low pressure fuel entering the high pressure pumping chamber. The solenoid is not pulsed. It is either fully opened or fully closed. The solenoid driver mounted on the side of the pump housing operates the solenoid on command from the ECM. The driver senses when the solenoid is fully closed to tell the ECM when injection has ended.

An *acceleration pedal position(APP) sensor* supplies data on pedal position and movement to the ECM. It then operates the fuel solenoid accordingly. The pedal position sensor contains three separate potentiometers, each with its own 5v reference and distinct return signals. This triple redundancy helps to ensure a signal is delivered to the ECM. If one or two sensors fail, the engine will still run at limited power. If all three fail, the engine will run only at idle. If an APP sensor fails, the ECM will log a trouble code into memory and turn on the *Service Throttle Soon* light. Each APP signal can be checked using a scan tool or an oscilloscope.

An optical/temperature sensor mounts on the pump itself. It consists of a thermistor type fuel temperature sensor and two optical position pick-ups that share a housing and a 5 volt reference signal. The optical pick-ups read the position of tone wheels rotating with the cam ring inside the pump. One pick-up provides a high resolution signal, generating 64 pulses per cylinder firing stroke.

Combined with fuel temperature and the crankshaft position data, this extremely fine position signal makes it possible for the ECM to trim the fuel quantity for each individual combustion stroke.

The sensor's second pick-up has eight slots, and reports pump cam position to locate the start of injection for each cylinder and to index cylinder No. 1. Combined with the crankshaft position signal, this information is used for pump timing, idle speed, and other powertrain control events. Additional inputs include coolant and intake air temperature, crankshaft position, barometric pressure, vehicle speed, and automatic transmission sensors.

Injection pump timing is controlled by a stepper motor mounted to the side of the distributor pump. By changing the position of the cam ring, first movement of the high pressure plunger (plunger lift) can be varied relative to crankshaft position. With mechanical governors, this is a function of hydraulic pressure in the pump housing, increasing with rpm to advance timing as speed increases. On electronically controlled pumps, the ECM operates the stepper motor. Other outputs include the injection pump driver, timing stepper motor, Service Engine Soon and Service Throttle Soon lights, and glow plug relay.

Bosch VP44 Pump

The Bosch VP44 distributor pump uses an electronically controlled governor. A cable running from the accelerator pedal to the side of the pump moves a position sensor, not a mechanical control lever. The accelerator pedal

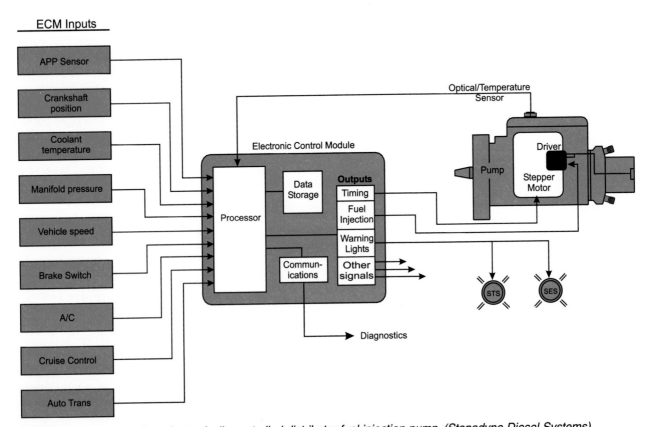

Figure 25-23. An example of an electronically controlled distributor fuel injection pump. (Stanadyne Diesel Systems)

position (APP) sensor uses a single potentiometer similar to a throttle position sensor. The pump itself resembles a standard Bosch VE distributor pump. Fuel quantity is regulated by rotating a governor shaft that positions a control sleeve on a port. This controls the amount of fuel entering the high pressure pumping chamber. The levers, springs, and fly weights used on a mechanical governor are replaced by a electric motor which controls rotation of the shaft.

The ECM handles all engine control functions and starting aids. To run the engine, the ECM must receive signals from the pedal position and crankshaft position sensors and data stream values from the fuel injection pump control module (FPCM) located inside the injection pump. For fuel trim and timing, the ECM uses the pump control data stream as well as inputs from the coolant temperature, intake air temperature, and air pressure sensors.

The ECM also reads signals from two fuel temperature sensors. One is part of the FPCM inside the injection pump. If fuel temperature goes too high, power will be reduced and a trouble code will be set. Another replaceable sensor is located on top of the pump, used for controlling the fuel heater element. Other sensors that input to the ECM are used mostly for diagnostic and/or engine protection purposes. These include battery voltage, a camshaft position sensor (TDC reference), oil pressure sensor, a knock sensor in the FPCM, and a water-in-fuel sensor. On vehicles equipped with a PTO, the ECM will disable the diagnostic functions when the PTO switch is activated.

There is an electric fuel transfer pump on the engine just above the starter motor. This is controlled directly by the ECM through a relay and is operated much like an electric fuel pump on gasoline engines. When the key is first turned on, the pump runs for two seconds to prime, then stops. When the starter is cranked, it runs at 25% duty cycle (7 psi) until the engine starts; then it will run continuously and provide about 10-14 psi. The fuel filter assembly near the injection pump includes the filter, water separator, water sensor, two pressure test ports, a water drain valve, and the fuel heater. The connector on the housing is for the heater, powered by a relay in the power distribution center next to the battery. It is not computer controlled.

While the FPCM does not have final authority over engine control, it receives signals from almost every electric or electronic component on the vehicle and provides a data stream to the ECM. Inputs include transmission sensors, A/C, cruise control, ABS, charging system status, battery temperature, brake light switch, park/neutral or clutch switch, vehicle speed, and fuel level. This information is used by the ECM for fuel trim, engine protection, and diagnostic functions.

Engine Control System Operation

The following sections outline the basic operation of modern diesel engine diagnostic systems. Some systems may not have all the features and options covered.

Open and Closed Loop Operation

When an engine is first started, the electronic control system operates on a set of fixed parameters stored in ECM memory. This is referred to as **open loop** operation. After the engine is started, each sensor monitors the system for which it is designed. Once it detects the engine is at the proper operating temperature, the ECM enters **closed loop** operation. The sensors detect changes in operating conditions and signal the ECM. The ECM calculates the needed response to these conditions and signals the appropriate output devices, **Figure 25-24**. These actuators physically change the engine's operating conditions to match the ECM's commands.

In addition to monitoring engine conditions and issuing output commands, modern engine control system have the memory capacity to record and store substantial amounts of data concerning engine and vehicle operation. This data can be used for engine diagnostics and troubleshooting, engine and vehicle maintenance, or truck fleet management.

Limp Home Mode

In select situations the system will enter a **limp home mode** when a problem is detected. The limp home mode allows the operator to continue vehicle operation. However, the engine control system can substitute a sensor signal value that may result in decreased engine performance. In some cases, the system will continue to function but engine power may be limited to protect the engine and vehicle.

Misfire Monitoring

On some newer diesel engines, the ECM has the capability of detecting the lack of combustion in one or more cylinders. This is referred to as **misfire monitoring.** The misfire monitor can detect which cylinder is misfiring as well as the number of times the cylinder misfired. The ECM will store a code indicating the cylinder misfire. However, if more than one cylinder is misfiring, the ECM will store a code indicating a multiple cylinder misfire, but will not indicate the actual misfiring cylinders.

Engine Diagnostics

Virtually all diesel engine electronic control systems have some type of on-board diagnostic capability. These systems are designed to detect faults or abnormal conditions not within the normal operating parameters of the engine or vehicle. When a fault or abnormal condition is detected, a diagnostic code is logged in the ECM's memory. Diagnostic codes are also referred to as fault codes or trouble codes. Many systems log diagnostic codes into ECM memory using two-digit code numbers, but some systems use longer alpha-numeric codes. Numeric codes offer the advantage of being easily read using a "blink" or flash code system that ties into the engine's *Check Engine* warning light.

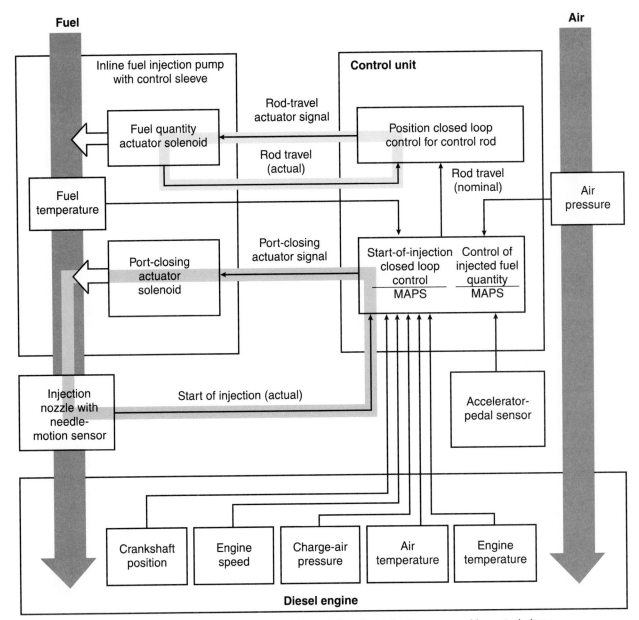

Figure 25-24. *Electronic open loop and closed loop control for an inline fuel injection pump with control sleeve.*

The vehicle's operator will be advised that a fault has occurred by illumination of an electronic *malfunction indicator lamp (MIL)*, usually located on the vehicle dashboard, **Figure 25-25.** This lamp is also referred to as a:

❑ Check Engine lamp.
❑ Service Engine Soon lamp.
❑ Power Loss lamp.
❑ Sensor lamp.
❑ PGM-FI lamp.

The primary purpose for the malfunction indicator lamp is to alert the operator when a system failure occurs, allowing for a quick diagnosis. An *active fault code* is one that is occurring at the present time. An *inactive fault code* is one that has occurred in the past, but may not be presently occurring. In some systems, the malfunction indicator lamp is capable of blinking a two-digit code for every active fault in the system. It does not display inactive fault codes, however, these codes can

be accessed using a personal computer or hand-held scan tool.

 Note: On newer vehicles equipped with a power take-off (PTO), portions or all of the diagnostic system may be disabled while the PTO unit is active. Deactivate any PTO unit before beginning diagnostics.

Engine Protection (Shutdown) System

In addition to the MIL lamp, many systems also feature a *stop engine lamp (SEL)* that warns of potentially damaging engine conditions, such as loss of oil pressure, high oil temperature, or loss of engine coolant. When this lamp illuminates, most systems begin an automatic shutdown sequence. An override button is sometimes available to allow the driver to pull the vehicle off the road.

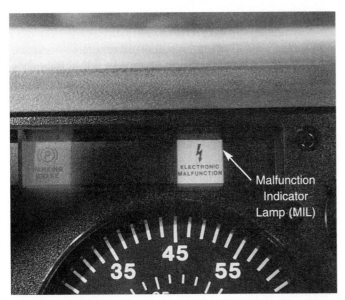

Figure 25-25. Electronic control systems often use dash-mounted warning lights to indicate dangerous operating problems.

Viewing Diagnostic Codes

To help troubleshoot engine and vehicle performance problems, diagnostic fault codes can be read by either counting the blink code via the malfunction indicator lamp or by using a scan tool or personal computer. Specific codes can then be referenced in the engine's service manual that contains detailed testing and troubleshooting procedures designed to pinpoint the cause of the problem.

 Note: Vehicles that use alpha-numeric codes should be accessed using a personal computer or scan tool.

In most cases, one or more codes are logged and stored in the ECM's memory. Active codes indicate that a problem exists requiring immediate attention. Inactive codes are often caused by intermittent problems. Inactive codes stored in memory do not necessarily indicate that a repair is needed, since the problem may have only been temporary or might have been repaired after the time the code was logged.

When diagnosing an intermittent code or condition, it is preferable to use a scan tool or personal computer because access to codes that are not currently active or history codes can only be gained through connection to the serial data line. Some codes cannot be cleared from system memory unless a proper password is entered into the system. Any code that does not need a password to clear is automatically deleted from memory after a predetermined number of ECM operating hours.

Reading Blink Codes

Two-digit blink codes for each detected active fault are displayed on the MIL lamp located on the vehicle's dashboard. The blink code system allows for quick diag-

nosis of active faults, without the need for outside equipment. The following procedure uses the vehicle's cruise control function switches to activate and view blink codes. Each manufacturer uses a slightly different method, so always follow the steps outlined in the service manual.

1. Turn the key *on* and wait until the electronic malfunction lamp's two second power-up test is finished. If there is an active fault in memory, the lamp will remain on after the two second power-up test.

2. Set the speed control On/Off switch to the *off* position and press and hold the Set/Decel switch until the malfunction lamp goes off. The malfunction lamp will remain off for a wait time of approximately one second.

3. Immediately after the wait time, the ECM will begin to flash a two-digit blink code using the malfunction lamp, **Figure 25-26.** The two digits of the code will be separated by a one second idle time (malfunction lamp off). The on and off time between each flash should be one quarter of a second. Count the on flashes of the malfunction lamp to determine the two-digit blink code.

In this system, only one active fault is blinked for each request. Where there are multiple faults within the system, a separate request must be made for each active fault. To display additional faults, hold in the Set/Decel switch until the malfunction lamp goes off. The blinking sequence should reset and begin blinking again after a one second delay. The ECM's internal diagnostic software

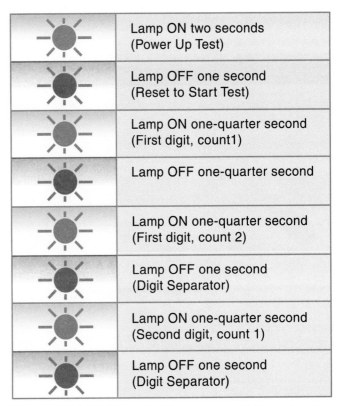

Figure 25-26. Dash-mounted lamps are used for reading two-digit blink codes.

will not provide blink codes for inactive faults. To access the inactive fault tables of the vehicle control unit and engine control unit, either a scan tool or computer must be used.

Scan Tools

Hand-held **scan tools** connect to the vehicle or engine's serial communication port. System specific programming modules can be inserted or downloaded into the scan tool so it can be used with different engine control systems. The hand-held scan tool displays all active faults, all occurrence of inactive faults, and allows monitoring of selected system parameters. The scan tool allows the technician to clear stored fault codes, program customer selectable parameters, and perform certain tasks such as resetting the engine's low idle speed.

Computer

The **computer** is the most advanced of the diagnostic tools available for troubleshooting and interacting with modern engine control systems. An IBM compatible computer will work with most systems, **Figure 25-27**. In addition to performing all the functions of a hand-held scan tool, a personal computer allows for enhanced system diagnostics. It can also be used to program dealer proprietary data and customer specified parameters into the system. Personal computers also allow technicians to use various software programs to manipulate and organize the large stores of data in the ECM's memory. While it is possible to use desktop computers for vehicle diagnostics, laptop computers give the flexibility of being able to take the computer on a road test.

Self-Diagnostics

Most engine manufacturers integrate in their ECMs **self-diagnostics** that allow the detection, logging, and retrieval of diagnostic information. The job of the self-diagnostic software is to aid in troubleshooting the system.

The ECM constantly performs tests and checks of system inputs. When it detects an unusual condition, the programming in the ECM begins a timer to allow the state of this input/sensor to settle back to its normal state. If this condition does not clear within the settling time of that sensor, then a fault will be logged. The fault recognition periods have been selected according to the safety factor of the sensor. These time periods have been specified so as not to detect false faults, while still maintaining engine and vehicle safety.

When a fault or failure is detected, the ECM switches on the electronic malfunction lamp, located on the dashboard. This light will remain *on* for active faults, and will switch *off* when there are no longer any active faults present in the system. A fault message is sent in a standard format on the serial line to alert other devices of this failure. A similar message will be transmitted when the failure is cleared.

The normal transmitted data for this sensor will be replaced with a "bad data" indicator. This signals the other devices on the serial line to ignore the data for this sensor. The control system also updates the fault table in the ECM memory. The fault table lists the number of times the failure has occurred, **Figure 25-28**. The stored codes remain in memory until cleared by an off-board diagnostic tool or computer, or until the information is no longer useful for diagnostic purposes. A default or customer programmable value will be used for circuit failures. This default or customer value allows the engine or vehicle to operate even though the electronic control system does not have any data from the circuit or circuits experiencing the problem.

Figure 25-27. *Laptop PCs and menu-driven software programs have simplified working with engine control systems. (John Deere)*

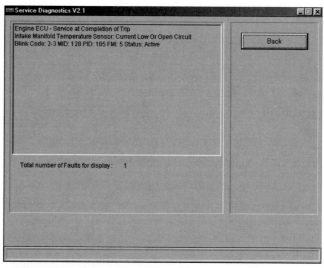

Figure 25-28. *Customer data programming software showing diagnostic fault and information related to the fault. (Mack Trucks, Inc.)*

Software Programs

A *software program* is simply a list of instructions for a computer. Software programs loaded onto a personal computer can extract data from the ECM and organize it into useful reports. For example, a specialized software program may be designed for service technicians. It typically compiles all the basic engine historical information, identifies what is wrong, and can tell the user when a vehicle may not be performing as expected. For example, engine history data can determine if the engine spends too much time at idle or too much time running at high rpms. Examples of specialized software programs manufacturers offer include:

❏ Engine and vehicle specification software that provides a comprehensive performance analysis of one or more selected drive line configurations. Such software may provide charts detailing acceleration, gradeability, startability, shift schedule, vehicle speed in each gear, vehicle power capabilities, and constant speed fuel economy.

❏ Diagnostic and troubleshooting software that includes electronic service manuals and comprehensive code fault log tables.

❏ Software that generates comprehensive trip and summary reports for fleet managers and/or service technicians.

❏ Software that allows for customer and dealer programming of system parameters and features.

Factory and Customer Programmed Parameters

Diesel engines and the vehicles they are installed in can be programmed by the dealer and/or the customer to provide desired performance characteristics, track both vehicle and driver performance, and establish automatic maintenance scheduling and history files.

Parameters can be set in system software to match specific engine ratings and to establish operating guidelines for features such as automatic cruise control, road speed and low gear torque limiting, idle shutdown, and engine protection. Other popular programmable features include optional equipment such as engine brakes, fuel filter monitoring, and optimized idle system that can start and stop the engine to keep the driver comfortable, the batteries charged, and the engine warm for cold weather start-ups. Some systems also provide electronic control for ether injection cold weather starting systems.

Lifetime Totals and Maintenance Indicator Data

Systems designed with built-in comprehensive trip recorders can provide essential information, including trip data as well as cumulative vehicle totals. These totals can

include information such as miles traveled, fuel consumed, and time spent in various operating modes, **Figure 25-29.** The maintenance log feature provides usage data totals in miles, service hours, or fuel gallons for serviceable items such as oil changes, fuel, air, and oil filters, and other customer selectable items, **Figure 25-30.** Dash mounted displays can be programmed to remind drivers when these maintenance items are due, or to warn if battery charge is low.

Security

Newer engine control systems include various security features. For theft deterrence, systems may require the driver or service technician to enter an access code into a dash-mounted display before the engine can be started or the vehicle driven. Authorized personnel can also program

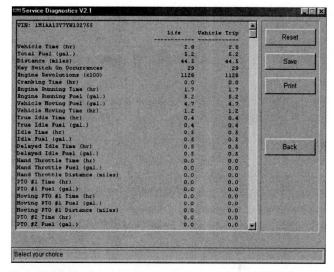

Figure 25-29. *Service diagnostic software listing lifetime and vehicle trip totals for various engine operating states and fuel consumption data. (Mack Trucks, Inc.)*

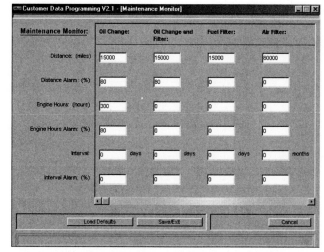

Figure 25-30. *Maintenance monitor software indicating programmable service intervals for specific serviceable items. (Mack Trucks, Inc.)*

the engine into a sleep mode in which the engine simply will not start under any circumstances.

Engine control system software programs also contain multiple encrypted passwords so only authorized personnel can access and change engine data files, vehicle configurations, or customer/fleet programmable settings.

Fuel Injection System Overviews

Despite almost constant evolution, all electronic engine control systems still operate on the sensor/computer/actuator principles outlined earlier in this chapter. Of course, the exact number of sensors, components, actuators, and other features will vary between systems.

In a text of this size, it is not possible to provide detailed service information or operating information for all the electronic engine control systems in use. Many of the electronic fuel injection systems installed on diesel engines have been refined and expanded over the years to take advantage of advancements in electronics, ECM hardware, software development, and wireless communication networks. Always refer to the manufacturer's literature for a full description of the features and associated software programs available for a particular control system.

 Note: Always refer to the manufacturer's service procedures when troubleshooting any electronic or computer-controlled engine.

Detroit Diesel Electronic Controls (DDEC)

All modern Detroit Diesel on-highway engines are equipped with *Detroit Diesel Electronic Controls (DDEC).* First introduced in 1985, there are now four distinct generations of this system. The DDEC fuel injection systems provide precision and flexibility in both fuel injection timing and volume through the use of solenoid operated unit injectors. The system does not use a mechanical governor, so there is no need for control tubes or racks.

System Components and Operation

The major components of DDECs system consist of the electronic control module (ECM), an electronic distributor unit (EDU), the electronic unit injectors (EUI) and various sensors, **Figure 25-31.** The sensors provide information to the ECM regarding various engine and atmospheric conditions. The ECM uses the information to regulate engine and vehicle performance. The DDEC

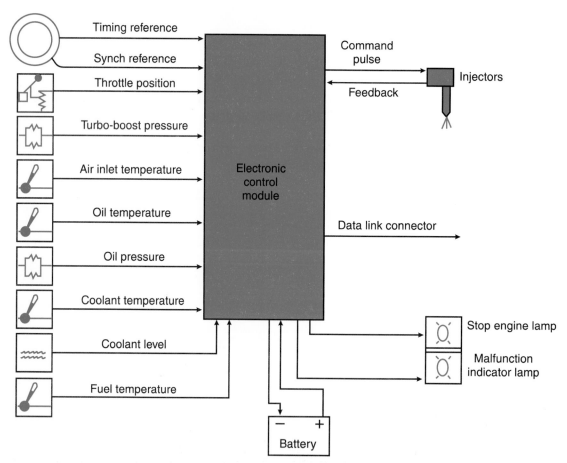

Figure 25-31. Electronic schematic of a DDEC III control system. (Detroit Diesel Corporation)

system also offers on-board diagnostic and engine protection capability.

The DDEC I ECM is separate from the EDU and is fuel-cooled. DDEC II mounts the ECM in the engine compartment and uses an EEPROM. Depending on the application, the ECM may be fuel-cooled. DDEC III has a smaller, faster ECM equipped with more features than the earlier DDEC systems. The most important feature introduced in DDEC III was self-diagnostics. The DDEC IV electronic control module holds more memory, offers increased operating speed, and has a built-in clock and calendar with battery backup. The data recorder and trip file holds information on the most recent 3 months of engine operation.

The DDEC timing reference sensor (TRS) is actually an engine position sensor mounted to the gear case, level with the crankshaft gear's centerline. As the engine runs, a series of evenly spaced teeth cast on the timing wheel rotate past the TRS element, causing the TRS to generate an electrical signal once per cylinder. The signal is used by the ECM to determine the length of time the injector solenoid should be energized.

The synchronous reference sensor (SRS) bolts to the gear case above the air compressor assembly and below the fuel pump drive. It provides a signal to the ECM once per crankshaft revolution. By comparing signals from both the TRS and SRS, the ECM can determine which cylinder is approaching TDC and is ready for injection to begin.

Control Systems

Actual electrical current to the injector solenoid is provided through an electronic distributor unit (EDU) that acts as a relay between the ECM, the vehicle battery, and injector solenoids. In DDEC I systems, the EDU and ECM were separate units. DDEC II and newer ECMs now contain the EDU.

In addition, the DDEC system controls many other subsystems and individual functions such as cruise control functions, idle controls, and advanced transmission interface. DDEC parameters that can be reset include vehicle speed limiting, engine protection, progressive shifting, droop, engine braking, and idle shutdown. The DDEC III and IV systems require no adjustment throughout the engine's life. Using built-in logic, the ECM will compensate for normal engine wear.

Diagnostic Capability

The ECMs in DDEC III and newer systems have the capability of recording and storing trouble codes when engine problems occur. Most problems must exist for at least two seconds before the trouble code will be recorded in the ECM's memory and the MIL lamp on the instrument panel illuminates. If the problem is intermittent, the MIL lamp will turn off when the problem stops occurring, but the trouble code will remain in ECM memory for retrieval at a later date.

In addition to the MIL lamp that warns the driver a problem has been logged into ECM memory, the DDEC

system also has a stop engine lamp. When an engine damaging condition is detected, the system begins an automatic shutdown sequence that gradually reduces fuel to the injectors until the engine shuts off.

Electronically Controlled Ether Start

On DDEC III and IV systems, the ether start system is used to assist a diesel engine in cold start conditions. Ether injection will occur in two modes, preload (before cranking) and block injection (during and after cranking). The mode and duration of injection is determined by the DDEC ECM. Since excessive preloading could be harmful to engine components, the DDEC system will not allow multiple preloads.

ProDriver™ System

ProDriver™ is a dash-mounted display showing real-time engine performance data such as idle time and fuel economy, current average mpg, and active fault codes, **Figure 25-32.** Stored information can be retrieved using a PC-based DDEC software program or through compatible remote retrieval stations or wireless systems.

A full range of software packages are available to complement the DDEC system and expand its capabilities. Such software programs allow vehicle owners to manage their operations better and easily extract useful data from the system concerning the engine and vehicle and the performance of individual drivers.

Cummins Electronic Management Systems

The **Cummins Interact™ System** combines fully integrated electronic engine controls with high-pressure fuel injection and high-strength components. Interact System engines include the heavy-duty high horsepower Signature 600, ISX, ISM, ISC, and ISL engines using electronic unit injection and the medium-duty ISB engine with distributor fuel injection pump.

On all Interact System engines, the ECM coordinates the engine, accessories, ABS brakes, transmission, and other systems. Interact also includes diagnostics with

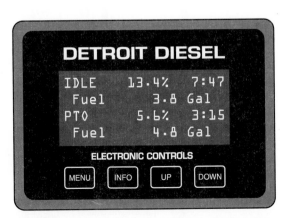

Figure 25-32. ProDriver™ dash-mounted display shows real time data on engine performance. (Detroit Diesel Corporation)

industry-standard data link connectors. Customer and dealer programmable parameters provide the ability to match engine performance characteristics to vehicle specifications and environmental conditions.

The air-cooled ECM connects the vehicle subsystems and accessories, and coordinates all functions from the transmission and ABS to the engine brake, cooling fan, and air compressor. The ECM constantly adjusts engine operating parameters for the best possible vehicle performance. For example, if the engine is equipped with an electronically governed air compressor, it automatically shuts off accessories for maximum acceleration when driving uphill, then turns them back on for extra braking power when going downhill. Idle control features reduce fuel consumption while keeping engine oil warm and batteries charged in cold weather.

Diagnostic Capability

When a fault occurs, the ECM switches on a warning light. Then, the engine protection system automatically derates the engine until the problem comes under control. If the problem worsens, the protection system will shut down the engine to prevent serious damage from occurring. When an indicator light for a fault code comes on, the driver can activate a "snapshot" feature, which records events leading up to the out-of-range conditions. Service technicians can then access this data to look over operating conditions at the time of the logged problem. The system helps to pinpoint the problem area for the service technician. Diagnostics also confirm repairs have been executed properly.

RoadRelay™

Cummins **RoadRelay**™ in-cab information gives drivers the ability to view real-time engine performance data such as fuel consumption, idle time, fault codes, and maintenance reminders. It also provides individual driver ID and antitheft protection.

Cummins Celect™ and Celect™ Plus Systems

The Cummins **Celect**™ and **Celect**™ **Plus** are electronic unit injection systems are in many ways similar to Detroit Diesel and Caterpillar unit injection systems. They are used on Cummins M11, and N14 engines. The Celect System controls engine torque and horsepower curves, air-fuel ratio, low-, and high-, idle speeds, vehicle road speed, cruise control, PTO operation. idle shutdown, and gear down protection.

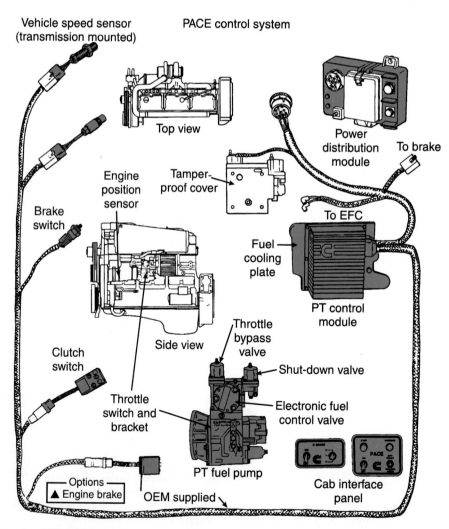

Figure 25-33. Components of PT PACE engine control system. (Cummins Engine Co., Inc.)

The ECM sends voltage signals to control the injectors, engine brake solenoids, fuel shutoff valve, and other actuators. Trouble codes are stored in the ECM and read from a dash-mounted MIL and engine stop lamp system. Electronic unit injector operation is similar to DDEC injection system operation. Once again, the key component to injection start time and metering is a solenoid actuated control valve.

Intelect™ Software

Cummins **Intelect™** software gives service technicians and fleet managers control over engine operating parameters and engine recalibration. It can generate detailed reports about select trip information that can be used for maintenance and fleet management. The types of reports possible include extraction reports, performance, fuel and vehicle reports, driver reports, and service, safety, and route reports. The service and diagnostic software includes drawings and diagrams for troubleshooting and repair. Intelect can extract trip, fault, and maintenance information directly from Signature 600, ISX, ISM, ISL, ISC, ISB, Celect™ and Celect™ Plus electronic control modules (ECM), as well as RoadRelay 4. One portion of the software makes it easy to spot engine tampering.

PT PACE System

The PT PACE system is designed for use on Cummins L10, M11, N14, and Big Cam IV engines equipped with pressure-time (PT) fuel injection systems. The PACE system is an integral part of the PT fuel system. With the exception of the electronic governing features, components and fuel flow in the PACE equipped engine is essentially the same as a mechanically-governed Cummins PT engine. Refer to Chapter 23 for more details on the Cummins PT fuel system.

In the PACE-equipped engine, a pressure-time control module (PTCM) contains the ECM and its related circuitry, **Figure 25-33.** The PTCM receives data from the engine and vehicle sensors, processes this data, and outputs voltage signals that control engine rpm and vehicle speed. The system features a constant bypass valve that enables the engine to continue running in the event of an electronic system failure. The PACE system offers road speed governing, cruise control, PTO control, engine brake control, and built-in diagnostics. Fault codes stored in the PTCM can be read using the dash-mounted MIL and engine stop lamps. They can also be retrieved using a compatible scan tool or PC.

Caterpillar Electronic Engine Control Systems

Caterpillar offers several variations of electronic engine control systems.

Caterpillar ADEM 2000

Caterpillar **Advanced Diesel Engine Management (ADEM) 2000** electronic control system is used on the medium-duty 3126B HEUI engine and heavy-duty

electronic unit injected 3604E, C-16, C-15, C-12, and C-10 engines. The electronic control system includes and provides self-diagnostics; over 100 customizable parameters; and operational, maintenance, and diagnostic data storage. System specific software can tailor this data into reports, histograms, and trend graphs.

The system includes a cold weather startup strategy, electronic idle control, engine monitoring system, cruise control, vehicle speed limiting, and idle shutdown timer. There is maintenance monitoring, password protection, exhaust brake control, adjustable low idle, and programmable PTO functions. The ECM can read up to 140 input/ouput signals to accommodate additional features. Cat Driver Information Display is also an option.

Caterpillar Engine Vision

Caterpillar engines designed for marine applications sometimes come with a software package called **Engine Vision.** Engine Vision is a touch-screen monitoring and display system that allows the boat or ship operator to monitor engine operation, **Figure 25-34.** It has the capability of monitoring up to three engines. It can provide data on engine speed, temperature, idle time, fuel consumption, and load. It can be linked to a global positioning system to provide vessel position, heading, and speed. It operates on a PC platform and can be customized to the owner's preferences.

Caterpillar HEUI Injection Control

The hydraulic electronically controlled unit injection fuel system (HEUI) provides total control over fuel injection for better emissions without sacrificing economy or performance. It is currently used on model 3116 HEUI and 3126B truck engines. The Caterpillar HEUI engine control

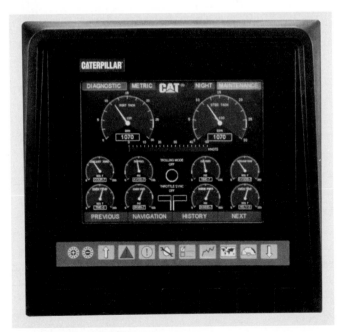

Figure 25-34. Caterpillar Vision software used in marine applications. (Caterpillar, Inc.)

system consists of the electronic control module (ECM), sensors, injection pressure control valve, and vehicle interface, **Figure 25-35.** The ECM controls the 3100 HEUI engine and a personality module stores information that defines power rating, torque curves, rpm, etc., which controls how the ECM behaves.

The ECM on the HEUI engine serves as the governor. Based on actual and desired conditions, the ECM determines when and how much fuel to deliver to the cylinders, as well as injection pressure. The desired conditions are typically the position of the accelerator pedal, desired vehicle speed when in cruise control, or desired engine rpm. The actual conditions are based on current operating conditions such as coolant temperature, load conditions, etc.

Once the ECM has determined how much fuel is required, it must determine when to inject fuel. Injection timing is determined by the ECM using input from the coolant temperature sensor, intake air temperature sensor, and boost pressure sensor. The ECM "knows" the cylinder position from the engine speed/timing sensors. The ECM decides when injection should occur relative to each cylinder's top dead center position and provides the signal

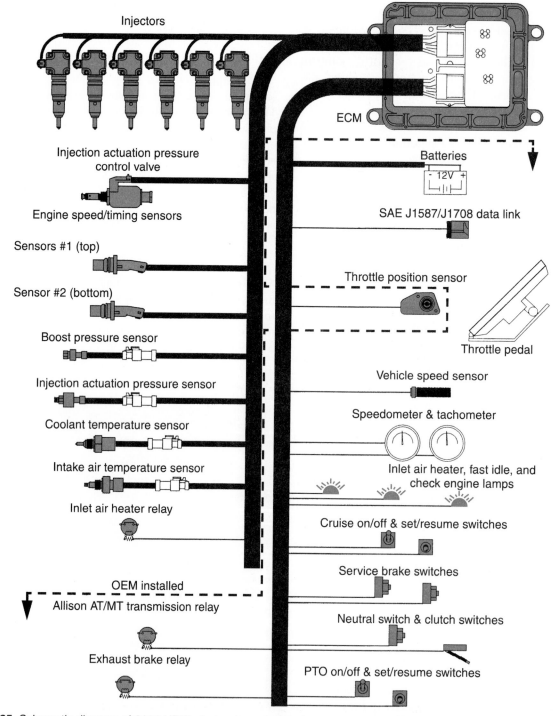

Figure 25-35. *Schematic diagram of 3100 HEUI electronic control system components. (Caterpillar, Inc.)*

to the appropriate injector at the desired time. It also adjusts timing for best engine performance, fuel economy, and smoke control.

The ECM controls the amount of fuel injected by varying signals to the injectors. The personality module inside the ECM sets certain limits on the amount of fuel injected. *Fuel ratio control (FRC)* fuel position is a limit used to control the air-fuel ratio based on boost pressure. When the ECM senses a higher boost pressure (more air into cylinder), it increases the FRC fuel pos limit (allows more fuel into the cylinder). Rated fuel position is a limit based on the engine's power rating and rpm. It provides power and torque curves for a specific engine family and rating. These limits are programmed into the personality module at the factory and cannot be changed in the field.

Certain parameters affecting HEUI operation are programmable. The parameters are stored in the ECM, and are protected from unauthorized changes by passwords. These parameters are either system configuration or customer parameters. System configuration parameters are set at the factory and affect emissions or power ratings. Factory passwords must be obtained and used to change a system's configuration.

Customer specified parameters allow the truck owner to tell the ECM how to fine tune engine operation to accommodate typical vehicle operation, as well as matching powertrain specifications. Engine monitoring can be programmed to three different modes (off, warning, or derate) according to customer parameters. Customer parameters vary and can be used to set limits on cruise control, vehicle speed, progressive shifting, and rpm/power ratings within the factory limitations. A customer-specified password is needed before any changes are made to customer-specified parameters.

The HEUI system has a maintenance indicator which notifies the driver when maintenance is needed. Also, the ECM maintains engine lifetime total data for engine running time, distance traveled, PTO time and fuel use; and idle time and idle fuel use.

Caterpillar 3176 Electronic Unit Injection

The Caterpillar 3176 engine is equipped with an electronically controlled unit injection system similar in operation to the Detroit Diesel system described earlier. **Figure 25-36** illustrates the layout for the Caterpillar 3176 electronic unit injection system. The high pressure needed for injection is generated by the camshaft operated pushrod. The start of injection and the amount of fuel metered is controlled by a solenoid actuated valve as in the DDEC unit injection system, **Figure 25-37.**

System features include self-diagnostics; serial data link compatibility; cold weather startup strategy and electronic idle control functions; and system operational, maintenance and diagnostic data storage. Customer programmable parameters include cruise control options, vehicle speed limiting, idle, exhaust brake, and various PTO functions. The system also has maintenance monitoring and password protection.

Caterpillar Programmable Electronic Engine Control (PEEC) Fuel System

Caterpillar's Programmable Electronic Engine Control (PEEC) fuel injection system was introduced in 1987 on their 3406B four-stroke, inline six-cylinder truck engine. This engine is equipped with an inline fuel pump. The major components of the PEEC system are an electronic control module, a timing advance unit, a transducer module, and an electronically controlled actuator, **Figure 25-38.**

The electronically controlled actuator moves the inline injection pump fuel rack in response to system conditions. A vehicle speed buffer located on top of the

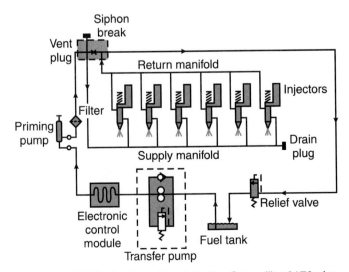

Figure 25-36. Fuel system layout for the Caterpillar 3176 electronically controlled unit injection engine. (Caterpillar, Inc.)

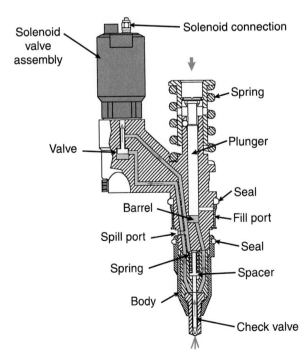

Figure 25-37. Details of the Caterpillar 3176 electronically controlled, mechanically actuated unit injector. (Caterpillar, Inc.)

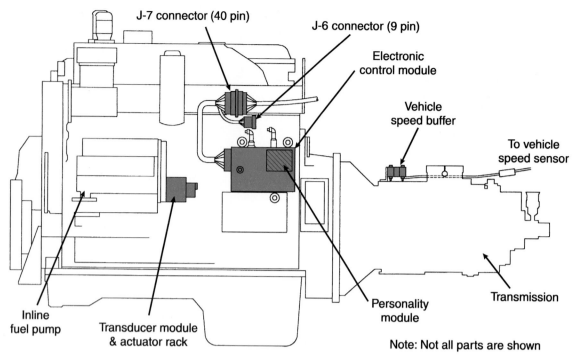

Figure 25-38. *Components of the Caterpillar PEEC system. (Caterpillar, Inc.)*

transmission amplifies and modulates the signal from the vehicle speed sensor. This modified signal is sent to the ECM and other devices requiring vehicle speed input.

The PEEC system provides electronic governing and control of the air-fuel ratio as well as torque-rise shaping. *Torque-rise shaping* is the ability to tailor the engine's torque characteristics at different rpm rates. In addition, horsepower, altitude and power compensation, injection timing control (advance/retard), self-diagnostics, PTO governing, and other parameters can be programmed.

Cruise control and PTO governing are activated by the driver using dash-mounted switches. The remaining functions are performed automatically based on parameters programmed into the ECM. The system is accessed by an on-board trip recorder, dash-mounted display, or using a scan tool.

Timing Advance Unit

The *timing advance unit* mounts on the front of the engine. This housing contains an electronically controlled, brushless torque motor. This motor controls a hydraulic servo which in turn controls the flow of oil to the drive carrier. Movement of the drive carrier advances or retards the engine timing, **Figure 25-39.**

Rack Actuator and Transducer Module

The *rack actuator* also mounts to the rear of the injection pump. This unit houses the engine speed sensor, rack position sensor, rack solenoid, and shut-off solenoid, **Figure 25-40.** The rack actuator is powered by engine oil pressure. It provides the oil (hydraulic) pressure needed to move the rack. The brushless torque motor is spring-loaded and will move the rack to the fuel shut-off position if a

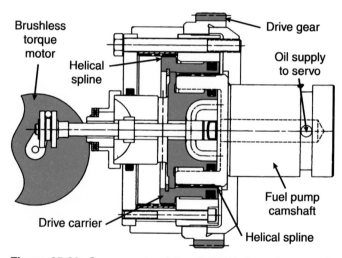

Figure 25-39. *Components of the PEEC timing advance unit. (Caterpillar, Inc.)*

problem occurs. If the brushless torque motor is unable to move the injection pump rack to the fuel shutoff position when required, the electronically activated shut-off solenoid performs this task. There is also a manual fuel shut-off valve.

Mack V-MAC III® Control System

The third generation of Mack's electronic *Vehicle Management and Control (V-MAC)* system, V-MAC III is available on the Mack E-Tech™ electronic unit pump (EUP) E7 engine.

The system consists of a vehicle electronic control unit and a separate engine control unit, which communicate through a high-speed serial line. With the use of the

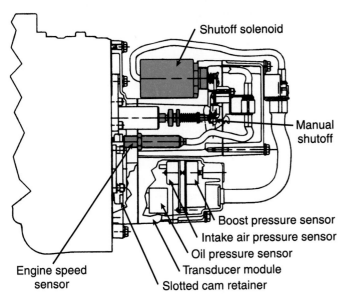

Figure 25-40. PEEC rack actuator and transducer module. (Caterpillar, Inc.)

serial line, sensors and control outputs can now be localized to the appropriate control unit. This means significantly fewer wires are needed between the cab and engine compartment, which improves serviceability.

The system's electronic architecture is based on two primary functions: engine control (engine ECM), and vehicle management (vehicle ECM). Advanced engine control functions and programmable features include electronic fuel and timing controls, idle speed settings, road speed and lower gear road speed limiting, idle shutdown, and accessory relay control. The V-MAC III system also offers numerous customer programmable and vehicle management control features, such as adjustable low idle, torque limiting, road speed limiting, shutdown, and multiple PTO controls.

The V-MAC III system also includes a comprehensive built-in trip recorder called DataMax™, which provides essential trip recording information on miles traveled, fuel consumption, idle time, and time spent in various operation modes. The trip recorder can also be used to program customer selectable features. It can also generate a comprehensive maintenance log.

Co-Pilot™

The Mack Co-Pilot™ information center is an in-cab dashboard display that provides drivers with real-time monitoring of engine and vehicle performance. Working in conjunction with the V-MAC III electronic engine control system, the display enables a driver to monitor fuel mileage, trip data, fault alerts, and to adjust driving practices to optimize vehicle performance.

The fuel economy button displays current and average trip miles-per-gallon, as well as customer-defined fuel economy goals. The display provides access to trip information, such as distance, fuel used, idle time, and a comparison of fuel used while idling versus total

fuel usage. The Co-Pilot dash display can also be used to monitor numerous system sensor outputs to supplement dash instrumentation or aid in problem diagnosis. For example, with the sensor display, an operator can easily monitor engine temperature, pressure, load, or protection status. V-MAC systems equipped with the Co-Pilot™ display monitor have the ability to deter theft by limiting vehicle use to only those with authorized access codes. Four different access codes are available. The engine sleep mode feature, which can only be activated by a scan tool or personal computer, ensures the engine will not start under any conditions.

Enhanced Overspeed Tracking

The V-MAC III system provides an overspeed fault feature, as well as a record of occurrences and total number of times the engine and vehicle speeds exceeded customer defined parameters. The system will also record overspeeds when fuel was actually added. This helps distinguish speeding on a downgrade from speeding resulting from a heavy throttle.

The system also reduces fan clutch cycling by monitoring air conditioning pressure and intake air temperature as part of its overall control strategy. Fan override from the dash is programmable so the customer can preset the time and conditions for override, including eliminating the override entirely. This feature also improves fuel economy by limiting unnecessary fan cycling.

On-Command Fast Idle

With the **on-command fast idle** feature, the engine speed will increase to a customer defined level, provided the park brake is set and the power take-off (PTO) is turned off, if so equipped. This is useful for those applications requiring remote engine speed control. Operators can tie into the system using a connector in the wiring harness. The switch can be positioned anywhere on the chassis.

The J-Tech™ compression brake control is fully integrated into the vehicle's V-MAC III electronic control system. V-MAC III provides a seamless transition between the engine's powered and brake modes. It offers such features as two-position progressive braking (one or both cylinder banks can be activated) and brake delay in cruise control mode. Both of these improve fuel economy by eliminating unnecessary engine brake applications while cruising.

Service Diagnostics and Software

Mack's PC-based service diagnostic software program includes all the tools needed for dealer and customer parameter programming, maintenance monitoring, troubleshooting, and diagnostics, **Figure 25-41.** An electronic unit pump diagnostic software routine allows the service technician to deactivate each fuel injection unit pump individually to aid in engine troubleshooting.

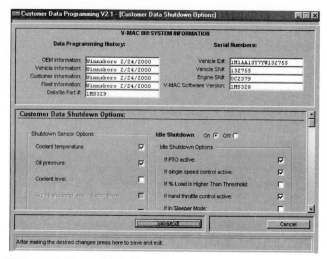

Figure 25-41. *V-MAC III software showing customer data options. (Mack Trucks, Inc.)*

Mack Inline Pump Control Systems

Mack Truck uses earlier versions of their V-MAC system on their E7 engines equipped with multiple plunger inline fuel injection pumps. On inline pump equipped engines, the *fuel injection control (FIC) module* contains two computers that process data from the engine and vehicle sensors. Output signals from the FIC positions the injector rack to control fuel metering. The mechanical governor is replaced by a rack actuator located at the rear of the fuel injection pump. A solenoid is used to provide positioning force that works against spring pressure. In the event of an electronic system failure, the spring ensures fuel cutoff.

The FIC also monitors rack position, and will activate a fuel shutoff solenoid if it detects an error in fuel rack positioning. The V-MAC system's timing advance mechanism, known as Econovance, is an integral part of the fuel injection pump drive. It is located between the engine cylinder block and the injection pump. The ECM controls timing changes by commanding the Econovance to rotate the injection pump camshaft relative to the engine crankshaft.

Engine rpm is the primary parameter used to determine injection timing. The FIC uses crankshaft position data along with timing event mark information to calculate the current injection timing. The timing event marker is located at the rear of the injection pump and senses the timing finger passage in the injection pump rack actuator. The timing finger passage is adjusted to occur at the number six plunger's port closure. The same finger is used to orient the injection pump camshaft as the pump is assembled to the engine and initial injection timing is set.

The system runs an automatic self-check each time the engine is started. Engine trouble codes for active and intermittent problems are stored in the ECM's memory. These trouble codes can be accessed through a flash code system using dash warning lights (active codes only), by using a scan tool, or a personal computer system loaded with the applicable software.

Troubleshooting Electronic Controls

This text cannot discuss all the troubleshooting methods used to diagnose problems in electronic diesel engine control systems. Each manufacturer offers detailed service manuals outlining corrective procedures and test methods. To use service manuals effectively, you must understand how the system works and follow a logical troubleshooting process.

Troubleshooting electronic systems often involves the process of elimination. Some electronic components are completely sealed and cannot be individually tested. To determine if such a part is faulty, all related parts must be checked first. If all related components are ok, then the sealed part in question must be defective. This troubleshooting method requires first visually inspecting for such basic problems as loose or corroded connections and proper grounds. If a visual inspection does not reveal the problem, then testing using diagnostic equipment should be performed.

When troubleshooting an engine with electronic controls, always check all non-electronic engine functions for possible problems first before turning to the control system. For example, a weak battery or a problem in the charging system may result in faulty sensor readings. Always start with the simplest possible cause and work your way to the more difficult and less likely causes of failure.

Service manuals often outline this logical process in troubleshooting flow charts that list service and test procedures in a series of definite steps, **Figure 25-42.** Skipping steps or ignoring instructions can lead to a misdiagnosis, resulting in a good and often very expensive component mistakenly being replaced.

Many electronic engine control problems are the result of a faulty or malfunctioning sensor or actuator. The best troubleshooting procedure is to check input sensors and wiring, actuators and wiring, and then the ECM itself. Keep in mind the ECM is a reliable, but expensive part. Computers do fail, but far too often, good ECMs are mistakenly replaced.

Always use the control system's self-diagnostics as an aid to troubleshooting problems. A problem in any sensor, actuator, or in the ECM itself is logged into the ECM memory as a numerical fault code.

Troubleshooting Procedures

To accurately diagnose a problem, it is helpful to follow the troubleshooting outline given in the following paragraphs. Failure to follow this outline may result in an incorrect diagnosis, unnecessary replacement of good components, and loss of time.

Step 1—Verify the Complaint

Before beginning any testing procedure, make certain the problem actually exists. If possible, talk to the operator, driver, or the person who noticed the problem. Try to

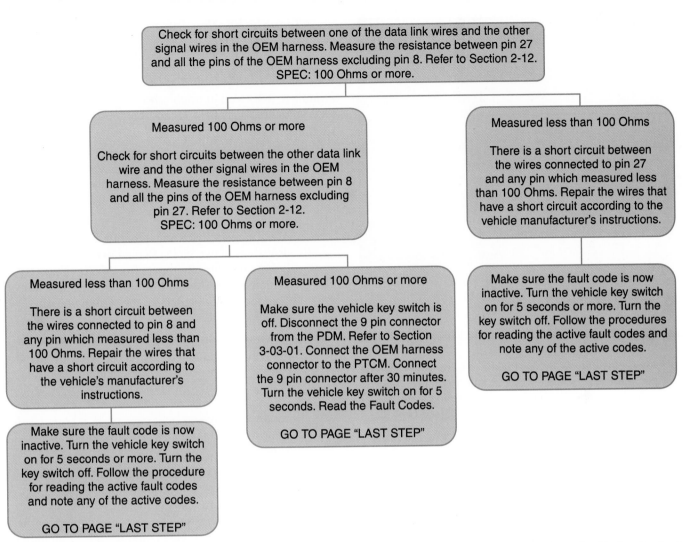

Figure 25-42. *Example of an electronic engine control troubleshooting flow chart for a particular fault code. (Cummins Engine Co., Inc.)*

obtain as much information as possible. In some cases, there will only be a verbal complaint instead of a fault registered by the system.

Step 2—Perform a Visual Check

One of the most important checks that must be done before any diagnostic activity is a careful visual inspection of the suspect system, its components, and related systems. Visual inspection involves looking for obvious problems such as broken, cracked, corroded, or burned parts. It also means checking for clean, secure electrical connections at all parts. This involves pulling and flexing connectors to ensure there are no poor electrical connections. Other visual inspection techniques include lightly tapping on parts to see if vibration disrupts engine operation.

The visual inspection is very important and must be done carefully and thoroughly. Many times, a problem can be found during the visual inspection, saving time performing needless tests. Replace any damaged components and repair wiring before continuing the diagnosis.

Step 3—Perform a System Check

Check which of the features of an improperly working system are truly not working correctly. A system check can help to isolate the problem area. It may also help to define what the problem is not. It may also help identify a condition or factor (engine warm, occurs only with cruise on, etc.) that may be contributing to the malfunction. This step is similar to step 1, verifying the complaint. Before proceeding with any diagnostic testing on the system, you should be familiar with any test precautions as outlined in the service manual.

Step 4—Check for Active or History Codes

Check the computer for codes, **Figure 25-43.** If the code is currently active, refer to the diagnostic procedures in the service manual and follow the step-by-step test procedures in the correct order. If the code is not active or is intermittent, try to induce the code to set by moving or flexing the connectors and wires associated with the suspect system or component. If the code cannot be induced, perform the associated step-by-step test procedures while

Figure 25-43. *Always check the computer for active codes or faults before beginning any diagnostic tests.*

moving the connectors and wires associated with the suspect system or component.

Keep in mind many intermittent problems are caused by poor terminal connections, not failed components. Be sure to inspect carefully the connector terminals for defects and deformities caused by meter probes. Check both sides of a connection for deformity and tightness.

Step 5—Repair

Once the suspect component has been isolated as the source of the problem, carefully disconnect the old component and inspect its connections. Clean and repair the component connections and then reconnect the component to ensure the component was at fault, not the connection itself. Clear all codes stored in the system at this time.

Step 6—Verify the Repair

After the repair has been made, check to be sure the problem was corrected. Perform a complete system check of the newly repaired system under a variety of conditions. Make sure all related systems are operating properly as well. After testing the engine or vehicle, recheck the computer for codes.

Step 7—Documentation

Once the repair is complete, operate the engine or drive the vehicle to ensure the problem has been corrected. When you are satisfied with the repair, be sure to document on the repair order the following:

❑ The problem or complaint.
❑ The cause of the problem.
❑ What you did to correct the problem.

Good documentation is important as it serves as a history for the engine. If the vehicle should return with the same problem, it gives you or the technician working on the engine a place to start. Be sure to write down all codes stored in the computer.

Troubleshooting Intermittent Problems

Intermittent problems can be very frustrating to troubleshoot. In many cases, there will not be a diagnostic code or an inactive fault logged in the electronic control module to assist you in troubleshooting. The following service guidelines can help locate the cause of intermittent problems.

❑ Always make certain the system's electronic malfunction and engine shutdown lamps work properly. In most systems, both lamps should light for two seconds when the ignition key is turned on.

❑ Always check system memory for inactive faults as well as active faults. Remember many handheld scan tools retain data from the previous vehicle it was used to troubleshoot. Be sure the scan tool is configured for the system you are working on.

❑ When checking a wiring harness, always refer to the wiring diagrams in the service manual. Check each wire between the component and control unit for a short to ground, a short to power, and for continuity with other wiring. Also, check for continuity between both ends of each wire. Use a known good ground when ground check testing.

❑ Inspect sensors, connectors, and other components for broken pins, loose terminals, corrosion and dirt, high resistance, moisture, and bad grounds. If you suspect a sensor or component is faulty, disconnect, clean, and reconnect all terminals to that sensor or component. Retest to determine if the problem has been eliminated.

❑ Isolate suspected sensors and components from the system where possible. Remember that disconnecting a faulty component from the system will not change the engine's performance. If the engine's performance does not change with the component disconnected, the component or related system is probably at fault.

❑ Before disconnecting the harness from the control module, be certain the ignition key is in the *off* position. If the key is not in the *off* position, internal electrical damage to the module can occur. If an electronic control module failure is suspected, verify there is power from the circuit breaker to the suspect ECM.

❑ *Do not* open an electronic control module. *Do not* use an ohmmeter to test the ECM. *Do not* make any electrical measurements at the ECM pins unless specifically instructed to by the service literature. Visually inspect the pins for repairable problems. Eliminate all other possible faults and repeat the procedure before replacing a control module.

- ❑ When a sensor, output device, or ECM has been replaced, check to be sure the new unit has corrected the original problem. Never reprogram a replacement ECM until it has been proven the new ECM has corrected the problem.
- ❑ If the problem lies outside the computer control system, be sure to clear all codes from the ECM when you are finished working.

Multimeter Checks

A digital multimeter is used to make the required measurements in the diagnostic testing procedures. These tests will include measuring voltage and resistance, and checks for short circuits and open circuits.

Since there are many types of digital multimeters available, specific instructions are not given here. Basic knowledge of multimeter operation is assumed. It is often helpful to check the manufacturer's instructions for further specific information.

Voltmeter Tests

A *voltmeter* can be used to check voltage drop to and from sensors and actuators. It can also be used to check for excessive voltage drops in the circuitry. Since voltmeters do not pose the risk to computer circuitry that an ohmmeter does, voltmeters are preferred for conducting tests on an engine's computer control system.

Continuity Tests

Continuity tests are performed using a ohmmeter or self-powered test light. Lack of continuity indicates an open circuit due to a break in wiring or a bad connection. Make certain the ignition key is *off* before performing any continuity checks. Remember, continuity checks cannot be performed if there is current flowing.

 Caution: Disconnect the wiring harness from the ECM before making continuity tests to any part of the computer control wiring harness. Damage to the ECM will result if an ohmmeter or continuity tester is used to test an ECM.

Ohmmeter Tests

An *ohmmeter* is useful in checking the resistance of sensors and actuators. For example, an ohmmeter can be used to check a thermistor used as a coolant or air temperature sensor for proper resistance. Actuators such as coils and electric motors can also be checked using the ohmmeter.

 Caution: Use of an analog ohmmeter may damage certain delicate electronic components. Only use a high impedance digital meter (at least 10 meg ohms).

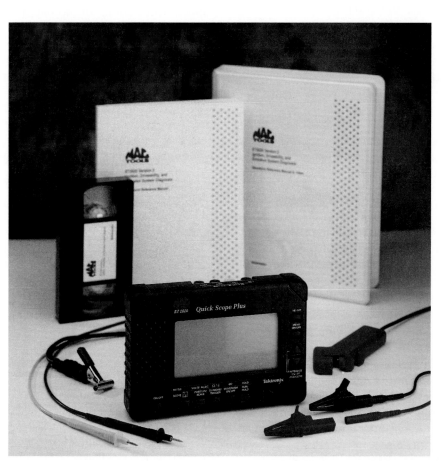

Figure 25-44. *New hand-held oscilloscopes have greatly simplified the task of electronic fuel injection diagnostics. (Mac Tools)*

Oscilloscopes

Oscilloscopes have been used in diagnosing gasoline engines for years. With the addition of electronic controls on diesel engines, they can be used as a diagnostic tool to check the waveform patterns from sensors and outputs from the ECM. The newer oscilloscopes are small, hand-held devices that can be taken on road tests, **Figure 25-44.** With additional probes and adapters, they can be used for a variety of diagnostic tasks. The latest scopes have on-board memory functions, which can be used to capture and store waveform patterns for comparison to good patterns.

Substitution

The absolute last step in testing is *substitution.* Due to the cost of procuring replacement parts, substituting a new component for a suspected faulty component should always be the last method used. Before turning to substitution, remember many parts stores do not accept returned electronic components.

Service Precautions

Electronic components are very voltage sensitive, and problems can arise if they are not handled and serviced correctly. The following are several precautions that must be followed when handling electronic components.

Static Electricity

Static electricity can build up in your body and/or clothing simply by sliding across a seat. It can generate sufficient voltage to damage electronic parts. Discharge any static electricity by touching a good ground, such as the engine block or vehicle frame, before touching the control ECM case or other components, such as PROM chips. Avoid touching the terminals and metal legs of electrical components. This can cause static voltage damage. The oil from your fingers can also corrode terminals and lead to poor contacts. If available, you should wear an anti-static wristband connected to a good ground any time you must handle the ECM or its components.

Plug-in Connectors

Modern electronically controlled engines use dozens of *plug-in connectors.* Many of the operational and troubleshooting procedures outlined in service manuals will begin by directing you to check a specific electrical connector. Intermittent electrical problems are often the result of poor connections.

> **Caution: Never probe the back of a connector or pierce the wiring for the purpose of taking voltage, resistance, or continuity readings. This can cause intermittent faults and system failures.**

Always check for active diagnostic codes before separating any connector. Turn the engine and ignition off before disconnecting or reconnecting harness connectors. Always apply pulling force to the connectors, not to the wires that extend from them. The simple step of disconnecting and reconnecting a connector can sometimes eliminate a code. This indicates the problem is in the connector. Loose pins, improperly crimped pins, and corrosion are likely causes.

On locking connectors, be certain the connector is properly locked (clicked) together and cannot be pulled apart. If any locking mechanism is cracked, broken, or will not lock, replace the connector. If connectors use locking bolts, such as at the ECM, tighten the bolt to the recommended torque, **Figure 25-45.** Do not overtighten. Perform a pull test to ensure the wire was properly crimped in the pin or socket and properly inserted into the connector. Some connectors use small wedges to lock the pins in place. Be sure these wedges are not missing. Pins should always be crimped onto the wires using a crimping tool.

Make certain the connector seals and sealing plugs are in place. If any seals or plugs are missing, replace them or the connector, if necessary. Inspect the wiring harness to be sure it does not make a sharp bend. This condition will deform the connector seal and allow moisture to penetrate the connector.

If moisture or corrosion is found in a connector, locate the path of the moisture entry and repair it. Look for missing or improperly installed seals, nicks in exposed insulation, or unmated connectors. Moisture can also travel from one connector through the inside of a wire to the ECM and is known as *wicking.* If moisture is found in the ECM connector, thoroughly check all connectors and wires on the harness that connect to the ECM. Remember, the ECM is not the source of the moisture. If moisture is present, repair the source of the moisture and replace any wires that may have moisture trapped inside their insulation.

Repair or replace damaged wiring and connectors as needed. Ensure all seals are in place and connectors are completely mated. Verify the repair eliminated the problem

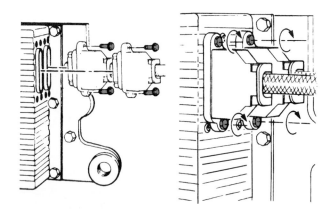

Figure 25-45. *Connector bolts must be tightened to the specified torque. (Cummins Engine Co., Inc.)*

by running the engine for several minutes and then checking again. Clean corrosion from pins, sockets, or the connector using only denatured alcohol. Use a cotton swab or a soft brush. Do not use any cleaners that contain trichloroethylene as this may damage the connector.

Checking Wiring and Harnesses

Carefully inspect each wire for signs of abrasion, nicks, or cuts. Likely locations for such problems are anywhere the insulation is exposed or anywhere the wire rubs against the engine or a sharp point. Check all harness hold-down clamps to confirm the harness is properly clamped and the clamps are not compressing the harness. Pull the wire looms away to inspect for flattened wires in the area where the clamp holds the harness.

Inspect the connector pins and sockets. Be sure the pins and sockets are not damaged, **Figure 25-46.** Be sure the pins are properly aligned. Check for individual pin retention in each socket. Use a new pin and insert it into each socket one at a time to check for a good grip on the pin by the socket. Repeat for each pin on the mating side of the connector, using a new socket for each test. The terminal contact (pin or socket) should stay held as shown in **Figure 25-47.**

Improper Grounds

When test procedures require that electrical power or ground be supplied to the circuit being tested, avoid unintentional grounding with metal tools or by touching live leads to metal parts. This can lead to personal injury and/or components being destroyed. Use jumper wires with circuit protection devices such as circuit breakers.

Protecting the ECM from Adverse Conditions

Remove the ECM and other computer modules from the vehicle when performing any welding, hammering, grinding, or other body work in the area of the ECM. Be careful not to damage terminals and connections when removing the ECM or other components. Store the ECM or parts in a cool dry location. Place them in a sealable plastic bag or wrap them in plastic.

Figure 25-46. *If connector pins are bent or heavily corroded, the connector must be replaced. (Cummins Engine Co., Inc.)*

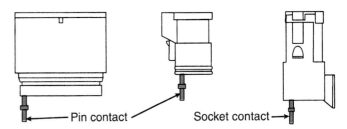

Pin contact   Socket contact

Figure 25-47. *Test the fit of both pin and socket contacts by inserting good contacts into the connector. (Caterpillar, Inc.)*

Summary

Most diesel engines now being manufactured for the heavy-duty truck market feature electronically controlled fuel injection systems. These electronic controls reduce engine emissions to levels that meet U.S. Environmental Protection Agency standards. Electronically controlled diesel engines are also being used in marine and other applications.

A typical electronic engine control system is made up of sensors, actuators, an ECM, and related wiring. ECMs are electronic decision-making devices. Input devices called sensors feed information to the ECM. The ECM processes this data and sends signals to output devices. Output devices or actuators are electromechanical devices that convert current into mechanical action. Solenoids and electric motors are two examples. There are several types of ECM memory used in engine control systems. ROM, RAM, PROM, EPROM, FEPROM, and EEPROM. EEPROM and FEPROM memory are fixed, electronically reprogrammable, and allow for both factory and customer specified parameters to be used to control engine performance.

The ECM can communicate with other control units and with external devices such as personal computers and service tools through the use of a data link connector. The personal computer has become a powerful programming, diagnostic, and fleet management tool. Specialized software programs allow fleet owners and service technicians to customize engine operation, track vehicle performance and maintenance histories, troubleshoot engine problems, and transform raw data into numerous useful reports.

When troubleshooting an engine with electronic controls, always check all other nonelectronic engine functions for possible problems before turning to the control system. Always start with the simplest possible cause and work your way to the more difficult and less likely causes of failure. Service manuals often outline this logical process in troubleshooting flow charts that map out the service or test procedure into a series of definite steps. Skipping steps or ignoring instructions can lead to the wrong service conclusion. This wastes time and can result in a good, and often very expensive component, being mistakenly replaced.

Always use the control system's self-diagnostic ability as an aid to troubleshooting problems. Electronic components are very sensitive, and precautions must be taken to avoid static electricity. All connections must be clean and secure. All circuits must be properly grounded.

Important Terms

Electronic control module (ECM)

Reference voltage

Analog signal

Digital signal

Analog-to-digital (A/D) converter

Interface circuit

Clock chip

Memory

Random access memory (RAM)

Read only memory (ROM)

Programmable read only memory (PROM)

Eraseable programmable read only memory (EPROM)

EPROM burner

Electronically eraseable programmable read only memory (EEPROM)

Flash erasable programmable read only memory (FEPROM)

Parameter

Electromagnetic interference (EMI)

Radio frequency interference (RFI)

Data link connectors

Serial lines

Remote data retrieval stations

Variable resistor

Thermistor

Coolant temperature sensor (CTS)

Oil temperature sensor (OTS)

Fuel temperature sensor (FTS)

Intake air temperature (IAT) sensor

Inductive sensors

Engine speed/timing sensor

Engine position sensor

Timing reference sensor

Vehicle speed sensor (VSS)

Brake switch

Clutch switch

Cruise control switch

Oil pressure sensor (OPS)

Air pressure sensor

Coolant level sensor

Turbocharger boost sensor (TBS)

Throttle position sensor (TPS)

Output devices

Pulse width signal

Tachometer output

Speedometer output

Electronic unit injectors (EUI)

Hydraulic electronic unit injectors (HEUI)

Split-shot

Pre-stroke injector

Acceleration pedal position(APP) sensor

Open loop

Closed loop

Limp home mode

Misfire monitoring

Malfunction indicator lamp (MIL)

Active fault code

Inactive fault code

Stop engine lamp (SEL)

Scan tools

Computer

Self-diagnostics

Software programs

Detroit Diesel Electronic Controls (DDEC)

ProDriver™

Cummins Interact™ System

RoadRelay™

Celect™

Celect™ Plus

Intelect™

Advanced Diesel Engine Management (ADEM) 2000

Engine Vision

Fuel ratio control (FRC)

Torque-rise shaping

Timing advance unit

Rack actuator

Vehicle Management and Control (V-MAC)

On-command fast idle

Fuel injection control (FIC) module

Voltmeter

Continuity tests

Ohmmeter

Oscilloscopes

Substitution

Static electricity

Plug-in connectors

Wicking

Review Questions—Chapter 25

Do not write in this text. Place your answers on a separate sheet of paper.

1. The first computer-controlled or electronic diesel fuel injection systems were introduced in the _____.

2. Name the three major components of an electronic control system.

3. Which of the following are examples of variable resistance sensors having a variable resistor in series with a fixed value resistor in the control ECM?
   (A) Throttle and clutch pedal position sensors.
   (B) Coolant and intake air temperature sensors.
   (C) Engine and vehicle speed sensors.
   (D) Both A & C.

4. Which of the following are examples of three-wire sensors having base voltage, ground, and variable wiper terminals?
   (A) Throttle and clutch pedal position sensors.
   (B) Coolant and intake air temperature sensors.
   (C) Engine and vehicle speed sensors.
   (D) Both A & B.

5. Which type of ECM data storage memory cannot be erased by turning off the engine or removing the ECM?
   (A) RAM
   (B) EPROM
   (C) EEPROM
   (D) Both B & C.

6. The engine position sensor or timing reference sensor is used to trigger the operation of _____.

7. _____ has proven to be the most readily adaptable of all existing fuel injection systems to electronic control.

8. What is the primary job of the electronic control module?

9. Give three examples of specialized software programs manufacturers may include with their engine control systems.

10. List the seven basic steps for effective troubleshooting of engine problems.

ASE-Type Questions

1. Technician A says some diesel engine control systems provide computer generated trouble codes that can be accessed and read through lights on the vehicle dashboard. Technician B says in most cases these codes can also be read using a personal computer equipped with the proper software program. Who is right?
   (A) A only.
   (B) B only.
   (C) Both A & B.
   (D) Neither A nor B.

2. Which of the following are examples of pulse counting sensors?
   (A) Throttle and clutch pedal position sensors.
   (B) Coolant and intake air temperature sensors.
   (C) Engine and vehicle speed sensors.
   (D) Both B & C.

3. Technician A says engine control computers are digital computers that can only understand on/off input signals. Technician B says analog (variable voltage) signals are generated by some types of sensors, but must be converted to digital signals before entering the ECM. Who is right?
   (A) A only.
   (B) B only.
   (C) Both A & B.
   (D) Neither A nor B.

4. The Caterpillar 3100 HEUI electronic engine control system contains all of the following components, EXCEPT:
   (A) a mechanical governor.
   (B) an ECM.
   (C) injection pressure control valve.
   (D) vehicle interface.

5. Technician A says you should follow the manufacturer's troubleshooting sequence when working on electronic control systems. Technician B says you should use past experiences to cut out many test procedures unlikely to reveal the problem, speeding up the repair process. Who is right?
   (A) A only.
   (B) B only.
   (C) Both A & B.
   (D) Neither A nor B.

6. Technician A says you should discharge static electricity from your body by touching a good ground before touching any electronic component. Technician B says you should check the condition of all connectors and wires before assuming a component is faulty. Who is right?
   (A) A only.
   (B) B only.
   (C) Both A & B.
   (D) Neither A nor B.

7. When troubleshooting an engine with electronic controls, Technician A says you should check all non-electronic engine functions for possible problems before turning to the control system. Technician B says you should start with the simplest possible cause and work back to the most difficult and least likely cause of failure. Who is right?
   (A) A only.
   (B) B only.
   (C) Both A & B.
   (D) Neither A nor B.

8. Technician A says an engine speed sensor generate electric pulses when the flywheel ring gear teeth pass close to the sensor. Technician B says the engine speed sensor can also determine when the crankshaft is at the top dead center position by recognizing two "missing" teeth on the ring gear that correspond to top dead center. Who is right?
   (A) A only.
   (B) B only.
   (C) Both A & B.
   (D) Neither A nor B.

9. Technician A says PROM memory can be erased by exposing it to ultraviolet light. Technician B says EEPROM memory can be erased and reprogrammed electronically. Who is right?
   (A) A only.
   (B) B only.
   (C) Both A & B.
   (D) Neither A nor B.

10. Technician A says ensuring all electronic equipment is operated with a good electrical ground system can minimize problems with EMI. Technician B adds that cords and cables connecting the peripherals in an electronic or computer system should be shielded to keep unwanted RF energy from entering or leaving. Who is right?
    (A) A only.
    (B) B only.
    (C) Both A & B.
    (D) Neither A nor B.

11. Technician A says serial data lines are used for communication between the ECM and other control units such as a vehicle control unit, anti-lock brake system module, and accessory systems. Technician B says diagnostic and programming tools, such as hand-held scanners or personal computers communicate to the ECM through the data link connector. Who is right?
    (A) A only.
    (B) B only.
    (C) Both A & B.
    (D) Neither A nor B.

12. Technician A says inactive diagnostic codes stored in memory always indicate that a repair is needed. Technician B says inactive codes may show probable cause for intermittent problems and usually are not viewable using blink codes. Who is right?
    (A) A only.
    (B) B only.
    (C) Both A & B.
    (D) Neither A nor B.

13. Technician A says personal computers are one of the most advanced diagnostic tools available for troubleshooting and interacting with modern engine control systems. Technician B says personal computers can also be used to program of dealer proprietary data and customer specified parameters into the system. Who is right?
    (A) A only.
    (B) B only.
    (C) Both A & B.
    (D) Neither A nor B.

14. When troubleshooting intermittent problems, Technician A begins troubleshooting at the sensor, switch, or other device. Technician B replaces sensors or components thought to be faulty. Who is right?
    (A) A only.
    (B) B only.
    (C) Both A & B.
    (D) Neither A nor B.

15. All of the following are likely causes of intermittent problems, EXCEPT:
    (A) loose pins
    (B) corrosion.
    (C) poorly crimped pins.
    (D) open wire.

Chapter 26

Diesel Engine Charging Systems

After studying this chapter, you will be able to:
- ❏ List the basic components of a charging system.
- ❏ Describe the principles of magnetism as they relate to charging systems.
- ❏ Explain the basic differences between alternators and generators.
- ❏ List the major components of an alternator and describe its operation.
- ❏ Describe the principles of ac rectification.
- ❏ Describe the operation of an electronic voltage regulator.
- ❏ List the service precautions that must be taken when working on alternators and charging systems.
- ❏ List common causes of overcharging and undercharging.
- ❏ Perform checks and adjustments on the charging system.
- ❏ Remove and install an alternator on a typical diesel engine.

As you learned in **Chapter 24,** the battery provides the power to start and operate the diesel engine and vehicle. However, the engine and vehicle needs a source of power to keep the battery charged as well as to operate electrical systems. This is the job of the charging system, which we will look at in this chapter. This chapter will also prepare you for **Chapter 27,** which covers starting systems.

Charging System

The **charging system** includes the battery, alternator, voltage regulator, connecting wiring and cables, and all associated electrical loads in the system. The job of the charging system is to recharge the batteries as needed and to provide the current needed to power the electrical loads in

the system. The charging system does this by converting part of the engine's mechanical energy into electrical energy.

As electrical loads drain the battery, its voltage drops. When voltage falls below a certain level, an electrical switch called a **voltage regulator** turns on the alternator. The **alternator** produces electric current and sends it to the battery to restore voltage to the required levels, usually around 12.6 volts. The voltage regulator also ensures the battery is not overcharged.

This process happens very fast and often. A typical voltage regulator may turn the alternator on and off up to 10 times per second. When current demand is high, the alternator will stay on longer. When current demand is low, the alternator will freewheel and use less engine power. Freewheeling during periods of low demand is desirable, as the alternator will use 6 to 7 horsepower during operation.

Principles of Magnetism

To generate electricity, alternators use **magnetism.** Passing current through a wire creates a magnetic field around that wire. Moving a wire through a magnetic field creates a current flow in the wire. Alternators and old style dc generators use this principle of electromagnetism in their operation. **Electromagnetism** is used in starting systems and in the operation of many actuators, such as solenoids and relays. Magnets have the ability to attract substances that contain iron, such as steel, nickel, and cobalt. These are known as **ferrous metals.**

Every magnet has two points, or **poles.** These poles are located at the ends of the magnet, with one being called the North pole and the other the South pole, **Figure 26-1.** Opposite poles attract one another while similar poles repel. A magnetic field, called a **flux field,** exists around every magnet. This flux field is made of invisible lines along which magnetic force acts. The lines of force emerge from the North pole, travel to the South pole, and return to the North pole through the body of the magnet.

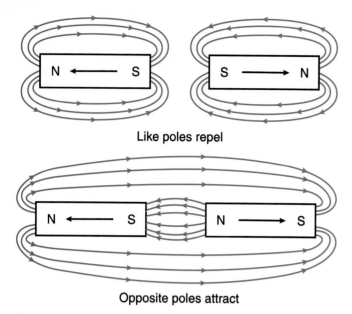

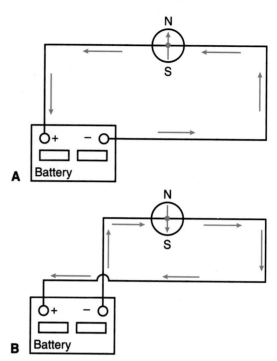

Figure 26-1. *Every magnet has a North and South pole. A—Opposite poles attract one another. B—Similar poles repel one another.*

Figure 26-2. *A—A compass will align itself a right angles to a current carrying wire. B—Changing the direction of current flow reverses the magnetic field polarity and the compass rotates 180°.*

All lines of force form a complete loop and never cross or touch one another.

Magnets do occur in nature, but most are created artificially. Artificial magnets are manufactured by inserting a bar of ferrous material inside a coil of wire and passing a strong direct current through the coil. Artificial magnets can be temporary or permanent magnets. When a malleable material such as soft iron is used, it magnetizes very easily but loses its magnetic qualities very quickly once the current is removed from the coil. Permanent magnets are made from steel, nickel, or cobalt. These materials are difficult to magnetize, but retain their magnetic qualities for very long periods after the magnetizing current is removed.

The earth's North and South poles are magnetic poles. This is why the metal arm of a compass will always align itself in a north-to-south direction. A compass can be used to illustrate the relationship between electricity and magnetism. When a compass is suspended over a current carrying wire, it aligns itself at right angles with the wire, **Figure 26-2A.** This proves the existence of magnetic force around the wire. Also, the polarity of a wire's magnetic field reverses itself when the current flow changes direction, **Figure 26-2B.**

The lines of magnetic force form concentric circles around the wire. The density of these lines is very heavy near the wire and decreases as the distance from the wire increases, **Figure 26-3.** The *magnetic flux lines* do not actually move. The lines are always at a right angle to the conducting wire.

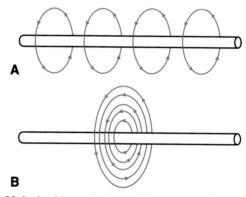

Figure 26-3. *A—Magnetic lines of force around a current carrying conductor. B—Flux density is greatest at the conductor surface.*

Coils and Flux Density

Increasing current flow will increase *flux density* around the wire. Positioning two wires that are carrying equal amounts of current side by side in the same direction will also double the flux density around the wires, **Figure 26-4.** Adding additional wires increases the flux density proportionately.

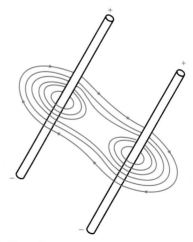

Figure 26-4. *Flux density doubles around two wires carrying current in the same direction.*

Looping the wire into a coil also concentrates the lines of magnetic force inside the coil, **Figure 26-5.** The magnetic field generated equals the sum of all single loops added together. This is a simple and effective method of generating a strong magnetic field with relatively small amounts of current and conducting wire, **Figure 26-6.**

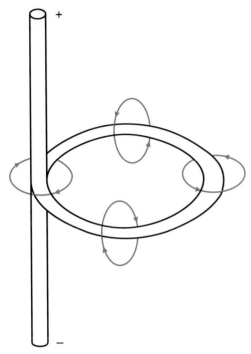

Figure 26-5. Looping the wire concentrates the magnetic lines of force inside the loop.

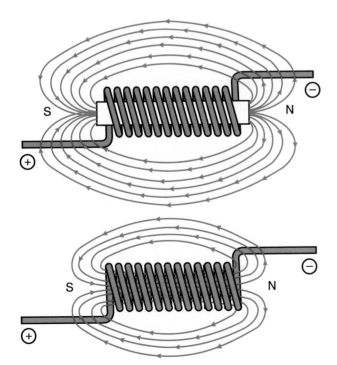

Figure 26-6. Forming a wire coil is an effective method of generating a strong magnetic field. Placing an iron core inside the coil also increases the field strength dramatically.

Reluctance

The lines of magnetic force created by a magnet can only occupy a complete circuit or path. The resistance this circuit or path offers to the line of flux is called **reluctance.** Reluctance can be compared to electrical resistance. Every material has a certain reluctance value. For example, air has a very high reluctance, while soft iron has a very low reluctance.

When a wire is looped into a coil, the air inside the coil offers a very high reluctance. This limits the strength of the magnetic field produced. However, when a bar of soft iron is placed inside the coil, reluctance is dramatically reduced, and the strength of the magnetic field increases tremendously.

When a coil of conducting wire is wound around a soft iron core, an **electromagnet** is produced. This is the operating principle used in solenoids, relays, and other actuators used in starters and electronically controlled fuel injectors and governors. The magnetic attraction or repulsion of the electromagnet is used to perform work. The strength of the electromagnet is directly proportional to the number of wire turns and the amount of current passing through the wire.

Induced Voltage

Using current to generate a magnetic field is one way electromagnetism is used in automotive and truck electrical systems. In an alternator, the opposite principle is employed. Magnetic fields are used to generate electrical current. This is the principle of **induced voltage.**

If a wire is moved across a magnetic field, a voltmeter will indicate a small voltage reading. Voltage has been induced in the wire. The wire must cut across the lines of flux in order for a voltage to be induced. If it moves in the same direction as the lines, nothing will happen.

When voltage is induced in a conducting wire, the wire takes on a distinct polarity with a positive and negative end. This polarity is based on the direction the wire is moving, **Figure 26-7.** If the wire changes direction, and cuts back across the lines of flux, the polarity of the induced voltage is reversed, **Figure 26-7.** This is the reason an alternator produces alternating current.

The strength of the induced voltage depends on a number of factors. Stronger magnetic fields induce higher voltages. The speed at which the lines of flux are cut also affects the amount of voltage produced. Increasing the speed a wire cuts across the lines of flux increases induced voltage. Increasing the number of conductors also increases the induced voltage. Finally, the closer the conductor and magnetic field are to right angles with one another, the greater the induced voltage will be. All of these points are considered in the design and operation of alternators.

Alternators

In **Figure 26-7,** the wire or conductor is the moving element and the magnetic field remains stationary. This was the design principle used in the construction of older automotive and truck **dc generators.** These belt-driven

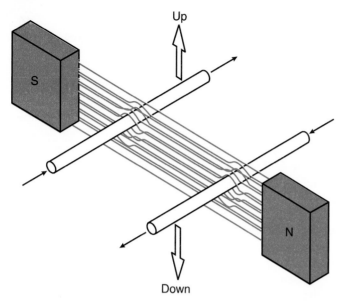

Figure 26-7. *Moving a wire across the lines of force in a magnetic field will induce a current flow in the wire. Reversing the direction the wire cuts across the lines of force will also reverse the direction of current flow.*

generators operate like an electric motor. An armature (conductor) spins inside a stationary magnetic field, **Figure 26-8A.** This induces an output voltage in the armature that is used to recharge the battery. Dc generators produce a limited voltage output, especially when the engine is operating at low speed or idle. This was the main reason for the development and universal acceptance of the alternator.

Alternator Design and Components

Alternators use a design that is essentially the opposite of a dc generator. In an alternator, a spinning magnetic field

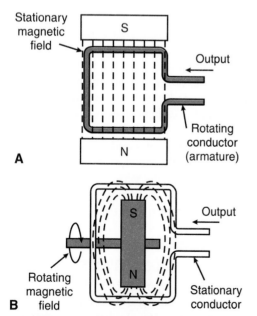

Figure 26-8. *A—Old style generators rotated a conductor in a stationary magnetic field. B—Alternators rotate a magnetic field inside a stationary conductor.*

rotates inside stationary conductors, **Figure 26-8B.** As the rotating magnetic flux field passes the conducting wires, voltage is induced, which creates alternating current flow, first in one direction and then in the opposite direction. Because the battery and all electrical loads in the system operate on direct current, alternating current must be changed or *rectified*.

When compared to dc generators, alternators offer many advantages including:

❑   Lighter and more compact design.
❑   More power produced at all speeds.
❑   Fewer moving parts and less maintenance.
❑   Built-in electronic voltage regulators (most newer models).
❑   Easy adjustment and troubleshooting.
❑   Less wear and longer life.

Figure 26-9 illustrates the components used in a typical alternator.

Rotor

The **rotor** assembly is the only moving part in the alternator. It consists of a drive shaft, coil, and two pole pieces. The rotor drive shaft is connected to the crankshaft through the use of pulleys and a belt, or by gears. A **field winding** or coil made of insulated wire wound around a soft iron spool is housed inside the rotor

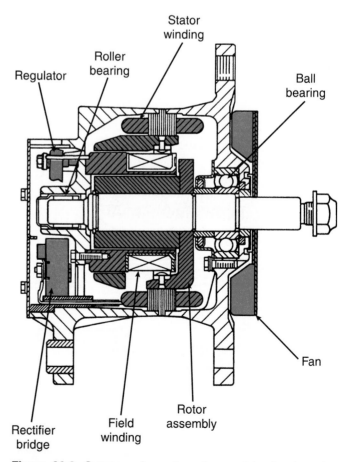

Figure 26-9. *Cutaway of an alternator used in diesel engine applications (Caterpillar, Inc.)*

assembly, **Figure 26-10.** When a small amount of current is passed through this coil, a strong electromagnetic field is generated.

The spool is located between the two pole pieces. When current is passed through the coil, the pole pieces assume magnetic polarity. One pole piece has a north polarity and the other south polarity. Extensions on the pole pieces, known as *fingers,* form the actual magnetic poles. A typical alternator has 12-14 poles with half being north polarity and half being south polarity.

The amount of current supplied to the rotor field coil has a direct effect on the strength of the magnetic field and overall alternator voltage output. Increasing current will increase voltage output. Decreasing current to the field coil will reduce voltage output. Monitoring and regulating current flow to the field coil is the job of the voltage regulator.

Current to the field coil can be supplied from the battery or from the alternator itself when the engine is running. This current passes through the voltage regulator before it reaches the coil. Most alternators use two slip rings mounted directly on the rotor shaft to conduct this current. These slip rings are insulated from the shaft and each other. Each end of the field coil connects to one of the slip rings. A carbon brush mounted on each ring conducts current to and from the field coil. Current passes through the field coil, second slip ring, and carbon brush before it returns to ground, **Figure 26-11.** Some heavy-duty truck applications now use brushless alternators.

Stator

The **stator** is the stationary conductor that the magnetic field cuts across to induce voltage. A stator is a series of wire loops arranged in such a way as to allow magnetic lines of force from the spinning rotor to cut across the wires, **Figure 26-12.**

Stators also have a laminated iron frame and three stator or output windings wound into the frame slots. Three windings are used to produce the required amperage output. These windings are staggered around the iron

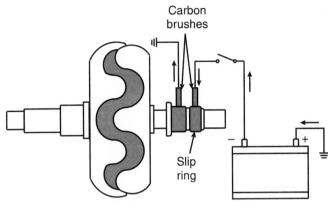

Figure 26-11. *Slip rings and brushes are used to conduct current to the rotating field coils mounted in the rotor.*

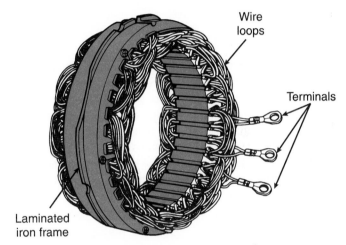

Figure 26-12. *Basic alternator stator construction.*

frame and wired together to form a circuit that resembles either a **delta** (triangle) or **wye, Figure 26-13.** The ends of each winding are attached to separate pairs of diodes. One diode is positive while the other is negative.

The rotor fits inside the stator's diameter and turns freely. A small air gap exists between the rotor and stator. If this air gap is not maintained, the alternator will short to ground and be damaged.

As the North and South poles of the spinning rotor pass each loop in the stator, they induce voltage and current

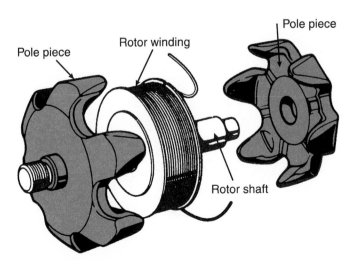

Figure 26-10. *Exploded view of a rotor assembly. Current flowing through the winding magnetizes the pole pieces.*

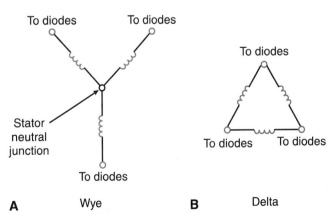

Figure 26-13. *Stator wiring arrangements: A—Wye circuit. B—Delta circuit configuration.*

flow in the windings. As a North magnetic pole passes the winding, current will flow in one direction. When a South magnetic pole passes the same winding, current will reverse itself and flow in the opposite direction. **Figure 26-14A** illustrates this alternating current flow as an oscilloscope sine wave reading. The positive side of the sine wave is produced by the North poles while the negative side is produced by the South poles.

The reading from a single stator loop represents what is known as *single-phase voltage.* Remember that the alternator stator contains three separate windings and that all three are energized at the same time. This creates a *three-phase voltage pattern* as shown in **Figure 26-14B.** This overlapping operation ensures that both positive and negative voltage is being produced at all times. This ensures consistent current flow.

Brushless Alternators

Some engines are equipped with *brushless alternators.* They do not need brushes to deliver start-up current to the rotor coil field windings. This is because the brushless alternator's rotor is equipped with a permanent magnet. This permanent magnet does not need a current supply to become active. When the rotor spins during engine start-up, the permanent magnet induces a small voltage and current flow in the stator. Part of this current is fed back to the

field windings. This increases the strength of the magnetic field around the permanent magnet, which in turn increases the stator current output. This cycle continues until the alternator reaches its rated output.

End Frame

The alternator *end frame* assembly is a two-piece cast aluminum housing that holds the bearings for the rotor shaft. The drive pulley and fan are mounted to the rotor shaft, outside of the end frames. Each end frame is equipped with ducts so that air from the fan can circulate freely through the alternator, cooling the alternator.

A rectifier bridge or diode holder is attached to the rear end frame. It houses three positive diodes. The negative diodes are often mounted in the end frame itself, **Figure 26-15.** The alternator end frames are bolted together and then the assembly is bolted directly to the engine block, which provides a good electrical ground path.

Ac Rectification

Before it can charge the battery or power any electrical load, the alternating current must be changed or rectified into direct current. Old-style generators do this using a mechanical commutator and brushes. Alternators perform the job of ac *rectification* electronically, through the use of diodes.

Diodes act like an electrical check valve, allowing current flow in only one direction. When a diode is connected into an ac circuit, one-half of the ac current is blocked. As shown in **Figure 26-16,** current can flow from A to B, but not from B to A. When the voltage reverses itself, current can pass from A to B but not back. Current is only available one-half of the time. This arrangement is known as *half-wave rectification.*

Alternators must have *full-wave rectification,* so that current can be supplied to the system at all times. Full-wave rectification is reached by adding more diodes to the circuit. **Figure 26-17A** illustrates how current flows from A to B during one-half of the full rectification cycle.

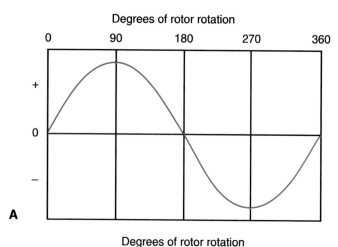

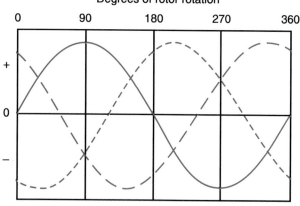

Figure 26-14. Comparison of alternating current waveforms. A—Single phase alternating current. B—Three-phase alternating current.

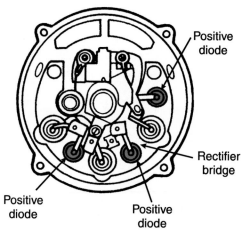

Figure 26-15. Alternator end frame. Note the position of the positive diodes mounted in the rectifier bridge.

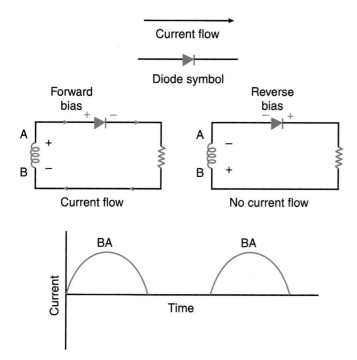

Figure 26-16. Halfwave rectification. Note how the diode prevents current flow in one direction.

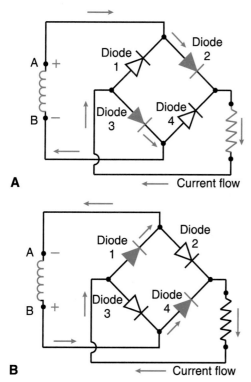

Figure 26-17. By adding additional diodes to the circuit, current can now flow. The result is full rectification of both halves of the sine wave. A—Current flow from A to B. B—Current flow from B to A.

Beginning at A, current flows through diode 2, through the resistor, through diode 3 and on to B. **Figure 26-17B** shows what happens when current direction reverses itself. Current flows from B, through diode 4, through the resistor, through diode 1 and back to A. In both cases the current passed through the load moving in the same direction.

Even during full rectification, there are still very brief moments when no current flows. This is why alternators use three separate windings and six diodes to produce overlapping current pulses. In this system, current output is never zero.

Figure 26-18 illustrates the typical three-phase alternator circuits for delta and wye arrangements using six rectifying diodes. At any time during operation, two of the windings will be in series and the third will be neutral. Depending on the combination of windings and the direction of current flow, one positive diode will rectify current flowing in one direction while one negative diode rectifies current flowing in the opposite direction. This is true for both wye and delta configurations.

Delta stators are common in truck and auto applications. They produce maximum current output at high speed ranges. Wye stators produce more current at lower speed ranges and are used in applications such as electric generating sets.

Voltage Regulators

The voltage regulator controls the amount of current generated by the alternator, which in turn controls the voltage level in the charging circuit. If the system did not have a voltage regulator, the battery would overcharge and the voltage level in the electrical system would rise to levels that would burn out fuses, trip circuit breakers, and possibly damage electronic parts. Precise voltage control is very important in electronic engine control systems. Computers, sensors, and relays are easily damaged by high

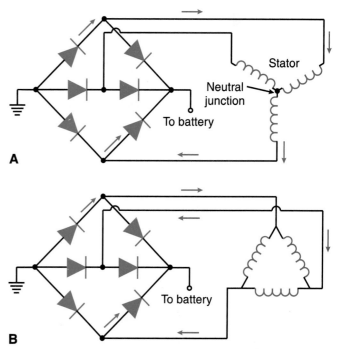

Figure 26-18. Typical current flow during full rectification. A—Current flow in a Wye circuit. B—Current flow in a Delta winding arrangement.

voltage levels and momentary voltage spikes. Most voltage regulators are mounted either on or inside the alternator housing. Some older systems may have the voltage regulator mounted separately.

The voltage regulator controls alternator output by switching the field coil current on and off. The regulator receives a sensing voltage that informs it of the battery's voltage level. When the battery voltage is below a preset level, usually around 13.5 volts, the regulator will switch the field coil circuit *on*. The coil will be energized, its magnetic field strengthened, and current output produced. This dc voltage output appears at the alternator output or *Bat* terminal. Most of the current is used to meet load demands in the electrical system. The remaining current is fed back to the battery to recharge it and to the alternator's field circuit to sustain its magnetic field and current output.

When the battery voltage level reaches 13.5 volts, the voltage regulator switches the field current circuit *off*. With no field current, the magnetic field in the rotor collapses, and current output stops.

This on/off switching is performed by **Zener diodes** and transistors. **Figure 26-19** illustrates the simplified circuit of an electronic voltage regulator during no-charge and charge operation. When charging system voltage rises above the preset limit, the Zener diode conducts current to the base of transistor 1. Transistor 1 switches *on*, which in turn switches transistor 2 *off*. Transistor 2 controls the circuit carrying current to the alternator field coil. With transistor 2 *off*, no field current can flow, and the alternator output drops to zero, **Figure 26-19A**.

When sensing voltage falls below the preset level, the Zener diode does not conduct, transistor 1 switches *off*, and transistor 2 switches *on*. Current is then supplied to the field coil and the alternator produces an output current, **Figure 26-19B**.

Electronic voltage regulators contain capacitors and resistors in various configurations. The capacitors smooth out voltage surges in the alternator circuitry that could damage the diode trio or rectifier bridge, as well as suppress radio noise. Resistors are used to adjust voltages at the diodes and to prevent current leaks at high operating temperatures. A thermistor is placed in the regulator circuit to set the ideal battery charging rate based on battery temperature. This means that during cold operating conditions, the regulator will operate at a higher voltage setting to provide the needed voltage to maintain a good battery charge.

Modern regulators are also designed to provide **load response control** or **LRC.** For example, if the engine is running at a low rpm, but is experiencing a high electrical load, the load response control will gradually increase field current, boosting alternator output without the need for increasing engine speed.

Reverse Polarity Protection

It is very important to maintain the correct polarity when connecting the alternator to the vehicle battery. If the battery is connected backwards, such as connecting the positive battery terminal to ground or connecting any negative

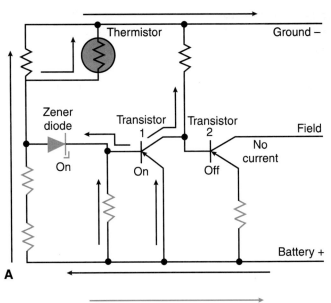

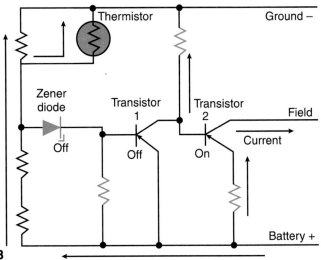

Figure 26-19. *Voltage regulator operation. A–No-charge. B–Circuit charging.*

connection on the alternator to the positive battery terminal, the internal diodes in the alternator, which have very little resistance, would be subject to a tremendous current flow. These diodes would be instantly destroyed and the alternator would need a major overhaul or replacement.

To prevent such damage, some alternators are equipped with a **polarity diode** and fuse in their circuitry. If polarity is accidentally reversed, the fuse will blow, protecting the other diodes in the alternator.

A and B Field Circuits

The field circuit in the alternator is one of two types. In **A-type field circuits,** the voltage regulator is on the ground side of the rotor. Battery voltage is picked up by the field circuit inside the alternator. The regulator switches the field circuit on and off by controlling its ground. A-type field circuits are used in electronic voltage regulators.

In a **B-type field circuit,** the voltage regulator is positioned on the input side of the alternator. Battery voltage passes through the voltage regulator to the field circuit. The field circuit is then grounded in the alternator. The voltage regulator switches the field circuit on and off by controlling current flow from the battery to the field circuit.

Service Precautions

There are certain precautions that apply to alternator and charging system service. The following is a general list. Always refer to the service manual for specific precautions and testing instructions:

- ❏ Be absolutely sure of polarity before connecting any battery or alternator for starting.
- ❏ Never ground the alternator output or positive battery terminal.
- ❏ Never ground the field circuit between the alternator and voltage regulator.
- ❏ Never operate an alternator on an open circuit with the field circuit energized.
- ❏ Keep the regulator cover in place when taking voltage readings.
- ❏ Place the ignition switch in the *off* position before removing the voltage regulator cap or cover.
- ❏ Use insulated tools when adjusting the voltage regulator setting.
- ❏ Never restrict air flow around or through the alternator housing. Do not route hot exhaust pipes near the alternator.
- ❏ Never weld near an alternator. Remove it from the vehicle.
- ❏ Never disconnect the alternator or regulator while the engine is operating. Never remove the regulator ground lead while the regulator is operating.
- ❏ Never disconnect the alternator output lead while the alternator is operating.
- ❏ Never disconnect the battery while the engine is in operation.
- ❏ Never remove an alternator from a vehicle without disconnecting the battery ground cable first, followed by the battery positive cable.

Charging System Problems

Problems in the charging system will result in either an undercharged or overcharged battery. Undercharging will lead to a loss of electrical power and hard or no-start conditions. Battery electrolyte will have a low specific gravity. Undercharging is not always the result of a faulty alternator. In fact, many undercharge conditions are simply the result of:

- ❏ Defective batteries.
- ❏ Loose, broken, or corroded connections at the battery or alternator.
- ❏ Undersize wiring between the battery and alternator.

- ❏ Loose or slipping alternator drive belts.

Overcharging will cause rapid water loss in the battery, leading to hardened plates and the inability to accept a charge. Electronic components may also be damaged due to high voltage levels. Overcharging is caused by problems such as:

- ❏ Defective batteries.
- ❏ Defective or poorly adjusted voltage regulators.
- ❏ Bad sensing lead contact to the voltage regulator or rectifier assembly.

Testing the battery's state of charge is the first step in checking out any undercharge condition. The battery must be at least 75% charged before the alternator will perform to specifications. Some manufacturers state that the battery should be fully charged before testing the alternator. Recharge the battery if its charge is low. If the battery will not accept a charge, replace it before conducting further tests.

A quick visual inspection should uncover any obvious problems. Check for loose drive belts. Tighten belts to specifications and secure any loose hardware. Inspect for broken and loose wiring, corroded battery terminals, etc. Clean and repair as needed before moving on to specific alternator testing.

Alternator Problems

Undercharging can be caused by a number of defects in the alternator's field and generating circuits. These include:

- ❏ Shorted, open, or grounded field coils.
- ❏ Damaged or worn slip rings or brushes.
- ❏ Poor contact between the voltage regulator and brushes.
- ❏ Poor contact between the slip ring assembly and the field coil leads.
- ❏ Defective diode trio.
- ❏ Shorted, open, or grounded stator windings.
- ❏ Grounded rectifier bridge or shorted or open rectifier diodes.
- ❏ Loss of residual magnetism in the rotor.

If the battery is fully charged and functional and the wiring and belts appear in good condition, testing the charging system is the next step. Always check the vehicle or engine manual for specific alternator and charging system tests. Some manufacturers may not specify the following tests, while others may outline additional tests. The main difference in most cases is in the test connection points and alternator specifications. Always take care to match your test readings to the specifications listed in the service manual.

Voltage Regulator Adjustment

Many voltage regulators are equipped with an adjusting screw beneath the regulator cap. This screw can be used to set the charging rate. To check the charging rate:

1. Install a voltmeter across the battery terminals to read system voltage.

2. Turn off all system electrical loads and operate the engine at approximately 2500 rpm to produce maximum alternator output.

3. Note system voltage. It should be between 13.5–14.5 volts on a 12-volt system. If the voltmeter reading is either high or low, the charging system may be defective or the voltage regulator may require adjustment.

To adjust the voltage regulator, remove the cap from the regulator and turn the adjustment screw to bring system voltage to specifications. If a low output voltage cannot be corrected, either the alternator, regulator, or diode trio may be defective. If a high output condition cannot be corrected through adjustment, the voltage regulator is the most likely problem.

Alternator Output Test

An *alternator output test* determines the charging system's maximum output. This test is conducted as follows:

1. Disconnect the ground battery cable from the alternator.

2. Connect an ammeter in series with the alternator output terminal and the ground battery cable. (If an inductive clamp ammeter is used, simply clamp it onto the ground battery cable. There is no need to disconnect the cable.)

 Note: Use a high amperage ammeter. High amperage ammeters are normally part of a charging system tester.

3. Install a voltmeter and a carbon pile across the terminals of the battery, **Figure 26-20.**

4. Start the engine and run it at 2000 rpm or above to attain maximum alternator output.

5. Turn on all loads in the electrical system, such as lights, wipers, air conditioning, etc., and increase the load on the carbon pile until the voltage across the battery terminals reads less than 12.6 volts. This will cause the voltage regulator to send full field current to the rotor, maximizing the alternator output.

6. Compare the maximum output reading on the ammeter with the alternator manufacturer's specifica-

tions. If the output is within 10% of specifications, the alternator is good. If the alternator output is more than 10% below specifications, full-field test the alternator as described in the next section.

Full-Field Testing

When the system is experiencing undercharging, the voltage regulator may be defective. A *full-field test* can determine if this is the case. In this test, full battery voltage is applied directly to the field winding in the rotor. The exact procedure for full-field testing an alternator depends upon whether it has an external or internal voltage regulator.

 Caution: When output testing a full fielded alternator, monitor the rise in system voltage very carefully. Bypassing the voltage regulator means current output is not regulated, and battery voltage can quickly rise to unsafe levels. The battery can overheat and sensitive electronic components can be damaged. Do not allow system voltage to exceed 15 volts during this test.

External Regulators

When the field circuit is controlled through an external voltage regulator, disconnect the regulator from the field terminal and connect a jumper wire from the field terminal to the battery's positive terminal. This bypasses the voltage regulator completely, supplying full battery voltage to the field circuit. If the external regulator connects to the alternator by a wiring harness, remove the connector from the voltage regulator and connect a jumper wire between two terminals in the harness connector. Refer to the service manual to determine which terminals should be used to bypass the regulator.

Once the regulator is bypassed, retest alternator output as outlined earlier. With the load applied, check if alternator output now reaches its rated amperage. If it does, the voltage regulator is defective. If the alternator still does not reach its rated output, further testing is needed.

Internal Regulators

Alternators with internal voltage regulators are full-fielded in one of two ways. Some alternators are equipped with a hole in their casing through which the field circuit can be accessed. This test requires the use of a short jumper with insulated clips and a stiff piece of thin wire or a 1/32″ drill bit.

 Note: Many alternators with internal regulators must be removed from the vehicle and disassembled before the regulator can be isolated and the coil full-field tested.

1. Turn off the engine and all electrical loads.

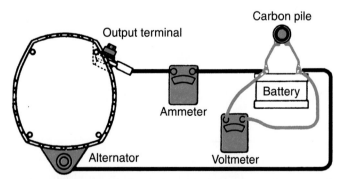

Figure 26-20. Circuit setup for an alternator output test.

2. Measure and record system voltage across the battery terminals.

3. Start the engine and run it at 2000 rpm or higher if needed, to generate full alternator output

4. Connect one end of the short jumper wire to the negative output terminal of the alternator and the other to the stiff wire or drill bit.

5. Insert the wire or drill bit into the full field access hole as far as it will go and make a note of the voltage reading. With the full battery voltage now applied directly to the field windings, the alternator output should climb to within 10% of its rated output. If the alternator is now functioning properly, the undercharge condition is likely caused by a faulty regulator or diode trio.

 Caution: Do not leave the alternator full-fielded for more than 10-15 seconds.

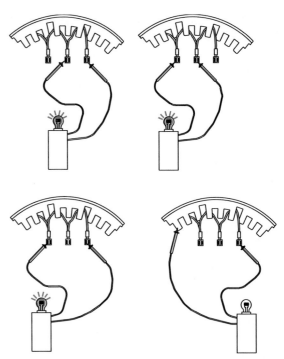

Figure 26-21. *Checking for improper stator winding grounds using a self-powered test lamp. The lamp should light whenever there is continuity. There should be no continuity between the stator windings and the frame*

Stator Winding Test

If the regulator passes the full-field test, the problem lies within the alternator. On many alternators, the unit must be disassembled before the stator windings can be tested for shorts and opens. On some alternators, the stator windings can be checked by connecting a self-powered test lamp, continuity tester, or ohmmeter across the alternator's ac terminals 1 and 2, 1 and 3, and 2 and 3. If the lamp fails to light or if the ohmmeter reads high resistance when connected between any pair of stator terminals, there is an open circuit in the windings, **Figure 26-21.** If the test lamp lights, or the ohmmeter reading is low when connected from any stator lead to frame, the windings are grounded.

Shorts in the stator windings are difficult to locate without special test equipment. However, if all other electrical checks are normal and the alternator fails to supply its rated output, suspect a defect in the stator windings. If the windings are defective, replace them.

Testing Diodes and Brushes

Undercharging can also be caused by problems in the regulator's diode trio. If one or more diodes are bad, full-field voltage will not be delivered to the rotor coil, reducing the alternator's output. Some alternators have a diode trio that can be removed from the vehicle without removing the alternator. To test the diodes:

1. Connect a digital ohmmeter to the single connector and to one of the three diode connectors. Note the resistance reading.

2. Reverse the ohmmeter lead connections to the same two connectors and note the reading.

3. A good diode will give one high resistance reading and one low resistance reading. If the readings are the same, replace the diode.

4. Repeat the same test for the other two diodes.

Undercharging may also be caused by corroded or worn carbon brushes (if so equipped). On some alternators, the voltage regulator can be removed to provide access to the brushes and the contact pads. If the brushes appear damaged or burned, or are worn to a length less than 3/16" (4.76 mm), they should be replaced. Check for a broken shunt lead inside the brush spring. Clean the contact caps on the brushes and the contact pads on the voltage regulator using a crocus cloth or fine grade sandpaper.

Regulator Circuit Test

The voltage regulator must keep charging system voltage within a set range, usually between 13.5–14.5 volts on a 12-volt system and 27–29 volts on a 24-volt system. To test the charging system's voltage regulation:

1. Connect a volt-amp tester to the battery as described earlier under the alternator output test.

2. Start the engine and run it at 2500 rpm to produce maximum alternator output.

3. Check both the voltage level and current output. As the voltage level approaches the system's specified level (13.5 volts for example), the alternator output should slowly decrease. When the battery is fully charged, alternator output should be low.

If alternator current output remains high after the system has reached its specified operating voltage, the system will begin to overcharge. System voltage will rise above 15 volts, and the battery will overheat and experience violent gassing.

To test the ground of an externally mounted voltage regulator, connect a voltmeter between the regulator case and battery ground and start the engine. With the engine running, the voltmeter should read 0, indicating there is no voltage difference between the two points.

If the regulator ground is good, check the level of the sensing voltage being read by the voltage regulator. If there is a voltage drop in the supply or sensing circuit to the regulator, resistance in the circuit is high and the regulator is not receiving an accurate signal. This problem could be caused by corroded or loose connections or an intermittent open in the circuit. Clean and repair as needed. If there is no excess voltage in the sensing circuit, the overcharge problem is being caused by either the alternator or its regulator.

Charging Circuit Resistance Test

Excess resistance in the wiring between the alternator and battery or in the alternator ground circuit can also cause problems. Resistance in the circuit between the alternator and battery will result in the battery not receiving full charging current. Under peak loads, the battery will not become fully charged and damage to the battery can result. To test for voltage drop in this circuit:

1. Connect the volt-amp tester and load the battery to achieve full alternator output.
2. Connect a voltmeter across the alternator's output terminal and the positive battery terminal.
3. Read the voltage drop in the system. In 12-volt systems, the voltage drop should be no more than .2 volts; in 24-volt systems the drop should be no more than .5 volts. Voltage drops higher than these indicate high resistance caused by corrosion, loose connections, or intermittent opens in the circuit.

Check the ground side of the alternator circuit as well. Install the voltmeter leads between the alternator casing and the battery ground terminal. The voltage drop should be 0. If a voltage drop does exist, the alternator is not properly grounded. Check for loose alternator mounts or corrosion between the casing and the mounting bracket.

Alternator Removal and Installation

The following section is a general description of alternator removal and installation. Check the service manual for the specific procedure for each application. To remove the alternator from a vehicle:

1. Disconnect the negative and positive terminals at the batteries.
2. Disconnect and tag (if needed) the wires and wiring harness from the alternator.
3. Remove all surrounding components needed to gain access to the alternator such as vacuum hoses and oil hoses.
4. Loosen the alternator adjustment lock bolt or nut and remove the belt or belts. Check the belts for excessive wear.

5. Remove the alternator retaining bolts(s) and remove the alternator adjusting bolt or nut.
6. Lift out the alternator. If needed, remove the alternator pulley retaining nut and remove the pulley.

Installation is the reverse of removal:

1. Install the alternator fan and pulley, if needed, and install the alternator.
2. Install the alternator retaining bolt(s) and adjusting bolt or nut.
3. Install the belt(s), adjust to specifications, and tighten the adjustment bolt or nut.
4. Reinstall all surrounding components that were removed to gain access to the alternator.
5. Reconnect the wires and wiring harness to the alternator.
6. Reconnect the battery's positive and negative terminals.

Restoring Rotor Magnetism on Brushless Alternators

On brushless alternators, the rotor normally retains a certain level of magnetism to provide voltage build-up when the engine is started. This magnetism may be lost during alternator disassembly or service and must be restored before the alternator can be returned to service. To magnetize the rotor:

1. Connect the alternator to the battery in the normal manner.
2. Momentarily connect a jumper wire lead from the battery positive post to the alternator's relay R terminal.

Summary

Modern diesel engine electrical systems use an alternator to produce electrical current in the charging system. Diodes in the alternator rectify alternating current into the direct current needed for battery charging and circuit operation.

In an alternator, a spinning rotor rotates inside the stator. As the rotating North and South poles of the magnetic flux field pass the stator windings, they induce voltage that creates current flow first in one direction and then in the opposite direction.

A voltage regulator keeps charging system voltage higher than battery voltage to ensure that current flows into the battery. Modern voltage regulators are solid state devices that can be an integral part of the alternator or mounted to the back of the alternator housing.

Voltage regulators operate by controlling the amount of current flowing to the field coils of the alternator. This varies the strength of the magnetic field generated by the alternator, which in turn varies the current output.

The field circuit in the alternator is one of two types. In A-type field circuits, the voltage regulator is on the ground side of the rotor. Battery voltage is picked up by the field circuit inside the alternator. The regulator switches the field circuit on and off by controlling a ground. In B-type field circuits, the voltage regulator is positioned on the feed side of the alternator. Battery voltage is fed through the voltage regulator to the field circuit. The field circuit is then grounded in the alternator. The voltage regulator switches the field circuit on and off by controlling the current.

Many voltage regulators are equipped with an adjusting screw beneath the regulator cap. This screw can be used to set the charging rate. Some new alternator designs are brushless. They do not need brushes to deliver start-up current to the rotor coil field windings. This is because the rotor of a brushless alternator is equipped with a permanent magnet instead of a soft iron core. Worn or loose belts, corroded or loose electrical connections, and battery problems can all result in charging system troubles.

Tests for pinpoint alternator and voltage regulator problems include an alternator output test, full-field test, stator winding test, diode resistance checks, voltage regulator circuit tests, and a charging circuit resistance check.

Important Terms

Charging system	Delta
Voltage regulator	Wye
Alternator	Single-phase voltage
Magnetism	Three-phase voltage pattern
Electromagnetism	Brushless alternators
Ferrous metals	End frame
Poles	Rectification
Flux field	Half-wave rectification
Magnetic flux lines	Full-wave rectification
Flux density	Zener diodes
Reluctance	Electronic voltage regulators
Electromagnet	Load response control (LRC)
Induced voltage	Polarity diode
Dc generator	A-type field circuits
Rotor	B-type field circuit
Field winding	Alternator output test
Stator	Full-field test

Review Questions—Chapter 26

Do not write in this text. Place your answers on a separate sheet of paper.

1. A spinning magnetic field rotating inside a stationary loop of wire conductors is a simple definition of a(n)_____.
   (A) generator
   (B) alternator
   (C) solenoid
   (D) voltage regulator

2. Looping the wire into a _____ concentrates the lines of magnetic force.

3. A brushless alternator does not need a _____ to become active.

4. Name three advantages of alternators.

5. If alternator output rises to specifications when the alternator is full-fielded it is highly likely that the _____ is faulty.

6. Most alternators with _____ must be removed from the vehicle before full-field testing can be performed.

7. Name three alternator defects that can cause undercharging.

8. Name three causes of overcharging.

9. A voltage drop of _____ in 12-volt systems and _____ in 24-volt systems indicates high resistance in either the circuit or ground wiring of the charging system.
   (A) .1 v / .3 v
   (B) .2 v / .5 v
   (C) .2 v / .2 v
   (D) .5 v / 1 v

10. Explain the procedure used to remagnetize a brushless alternator rotor once it is returned to service.

ASE-Type Questions

1. All of the following are parts of an alternator, EXCEPT:
   (A) diodes.
   (B) brushes.
   (C) solenoid windings.
   (D) stator coil.

2. What component of an alternator produces the rotating magnetic field?
   (A) Rotor.
   (B) Stator.
   (C) Slip rings.
   (D) Zener diodes.

3. Technician A says that any physical contact between the rotor and stator will cause an internal short in the alternator. Technician B says that most stators contain three separate windings so they can produce both positive and negative voltage at all times. Who is right?
   (A) A only.
   (B) B only.
   (C) Both A & B.
   (D) Neither A nor B.

4. Technician A says alternators use commutators and brushes to rectify alternating current into direct current that can be used by the vehicle's electrical system. Technician B says that voltage rectification in alternators is performed by diodes. Who is right?
   (A) A only.
   (B) B only.
   (C) Both A & B.
   (D) Neither A nor B.

5. Which type of alternator construction is most common in trucks and vehicle applications?
   (A) Delta wound stator.
   (B) Wye wound stator.
   (C) Tri wound stator.
   (D) All of the above.

6. Which type of stators produce more current at lower speed ranges and are used in applications such as standby diesel electric generating sets?
   (A) Delta wound stator.
   (B) Wye wound stator.
   (C) Tri wound stator.
   (D) All of the above.

7. When the battery voltage drops below a preset level, usually around _____ volts, the voltage regulator will switch the field coil circuit *on* and begin outputting current to the battery and electrical loads.
   (A) 9.6
   (B) 12.6
   (C) 13.5
   (D) 14.5

8. _____ are used in the alternator circuits to smooth out voltage surges that could damage diodes in the diode trio or rectifier bridge and to suppress radio noise.
   (A) Transistors
   (B) Resistors
   (C) Capacitors
   (D) Semiconductors

9. Modern alternators can _____ .
   (A) adjust output according to temperature conditions to maintain proper battery charge
   (B) provide load response control to boost output when needed, even at low engine speeds
   (C) provide self-diagnostic information through their own computer
   (D) Both A & B.

10. Which of the following are examples of a damaging reverse polarity condition?
   (A) Connecting the positive battery cable to ground.
   (B) Connecting the negative cable to the battery last.
   (C) Connecting the negative battery cable to a metal frame member.
   (D) Both A & B.

Chapter 27

Diesel Starting Systems

After studying this chapter, you will be able to:
- ❏ Name the five major components of a diesel engine electrical starting system.
- ❏ Describe the two major circuits used in an electrical starting system and how they operate.
- ❏ Understand how an electric starting motor operates.
- ❏ Perform starter system tests.
- ❏ Describe no-load testing of the starter system.
- ❏ Explain the basic operating principles of an air-powered starting system.
- ❏ Describe the operation of a hydraulic starting system.
- ❏ Name and describe the operating principles of the cold weather starting aids used on diesel engines.

The diesel engine requires tremendous torque in order to start. This task is performed by the starting system. Depending on the engine, there are three possible starting systems that can be used. This chapter discusses the various starting systems and starter motors used on diesel engines.

Starting System

The job of the **starting system** is to turn or "crank" the diesel engine over at a sufficient speed to begin combustion. Once the cylinders fire, the engine can run on its own and the starting system ceases operation. Most diesel engines use an electric starter motor. When the starter motor is energized, it spins rapidly. This mechanical motion is transferred through a drive mechanism to the engine's flywheel. As the flywheel and crankshaft turn, the pistons move up and down in the cylinders, compressing and igniting the initial fuel charges.

Some diesel engines are equipped with pneumatic (air) or hydraulic starter motors. Pneumatic starter motors are powered by compressed air from the vehicle's air system. Hydraulic starter motors are driven by pressurized fluid stored in a special steel shell accumulator. As with electrical starter motors, both air and hydraulic starter motors use a drive pinion to transfer motion to the flywheel.

Electric Starter System Components

The basic **electric starting system** has five major parts and two separate electric circuits to control their operation. As shown in **Figure 27-1**, these circuits and components are:
- ❏ Starter circuit.
- ❏ Control circuit.
- ❏ Battery or batteries.
- ❏ Starting switch.
- ❏ Relay switch.
- ❏ Battery cables.
- ❏ Starter motor.

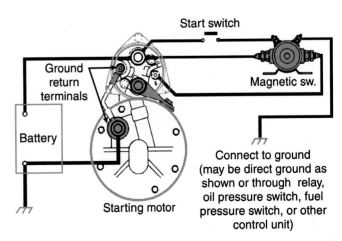

Figure 27-1. Components of an electrical starting circuit. (Detroit Diesel Corporation)

The *starter circuit,* **Figure 27-2,** carries the high current needed to energize the starter motor. Some starter motors draw up to 400 or more amperes. This high current is carried through heavy gage cables from the battery to the starter motor. Current flow in the starter circuit is controlled by the use of a starting switch and a relay switch.

These two components and their wiring make up the control circuit in the starting system. The control circuit allows the operator to use a very small amount of current to control the high current between the battery and starter motor. When the starter switch is placed in the "start" position, the control circuit is closed, and a small amount of battery current flows to the magnetic relay switch. This current magnetizes the coil in the relay, closing the contacts. With the relay in the closed position, current now flows from the battery, through the remote relay, and to the starter motor. Current at the starter motor closes a second solenoid activated relay, completing the circuit. Full battery current now flows to the starter motor.

The starting system on heavy-duty trucks and equipment is usually equipped with a neutral safety switch circuit. This *neutral safety switch* ensures that the vehicle can only be started when the transmission is in neutral or park (automatic transmissions) or the clutch pedal is fully depressed (standard transmissions).

Starter Motor Design and Operation

The *starter motor* is the component that actually performs the task of starting the engine. It must produce high torque for the period of time that the starter circuit is closed. It must also have a way of engaging its pinion gear with the flywheel ring gear during this time, and a way of disengaging once the engine is operating. If the pinion gear does not disengage when the engine started, the flywheel would begin driving the pinion gear at speeds that would quickly damage the starter motor.

There are two basic designs of starter motors: direct-drive and gear reduction. *Direct-drive starters* do not use any type of internal gearing to reduce the speed of the starter motor armature and pinion gear. The speed of the pinion gear limits the amount of torque it can generate at the flywheel.

Large diesel engines require sufficient torque at the flywheel to start the engine, especially in cold operating conditions. To accomplish this, *gear reduction starters* use an additional gear inside the starter housing. This gear reduces the speed of the pinion gear. This drop in speed translates to an increase in starting torque at the flywheel.

Motor Operating Principles

Electric starter motors use the principles of magnetism to generate rotational motion. As shown in **Figure 27-3,** when a current carrying conductor is placed between the ends of a horseshoe shaped magnet, two separate magnetic fields are generated. The first field exists between the poles of the horseshoe magnet. The second field exists around the current carrying conductor. As the lines of force travel from the North to the South pole, they pass the conductor in a clockwise direction. On the left side of the conductor, the magnetic lines of force from the horseshoe magnet and conductor fields are moving in the same direction.

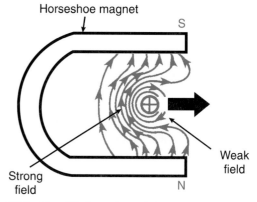

Figure 27-3. *Simplified version of the magnetic lines of force generated in an electric motor.*

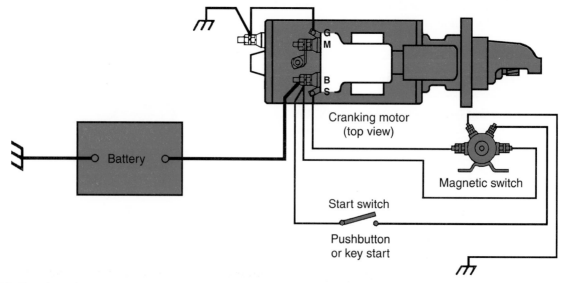

Figure 27-2. *Top view of the starter motor and cranking circuit. (Freightliner Corporation)*

On the right side of the conductor, the lines of force move in the opposite direction. These opposing forces cancel each other out, and a very weak magnetic field exists in this area. This imbalance forces the conducting wire to move to the right. Magnetism has now been used to generate motion. As the strength of the magnetic fields generated by the horseshoe magnet and conducting wire increases, the force also increases.

In a starter motor, **Figure 27-4,** an armature serves as the conductor. It is formed into a loop and its ends are connected to semicircular bars known as commutators. To form the equivalent of the horseshoe magnet, heavy wire is wound around a piece of iron to create an electromagnet. The windings are called field coils, while the iron bar is known as a pole shoe.

One field coil is wound around the pole shoe in a clockwise direction, and the other in a counter-clockwise direction. When current is passed through the field coils, the clockwise wound coil produces a North magnetic pole and the counterclockwise wound coil produces a South magnetic pole. Current is supplied to the armature through spring-loaded brushes placed onto the commutator. When the starter circuit is energized, current from the battery flows through the right field coil, to the left field coil, and through the left brush. It continues through the armature and right brush, back to the battery ground terminal. This current flow creates:

- ❏ A magnetic field between the North and South field coils.
- ❏ A magnetic field with a counter-clockwise direction around the right hand leg of the armature.
- ❏ A magnetic field with a clockwise direction around the left hand leg of the armature.

The interaction of these fields forces the armature right leg downward and the left leg upward, **Figure 27-5.** The armature assembly rotates clockwise. As the legs reach their highest positions, magnetic force acting on them is reduced, but there is sufficient inertia to pull the legs through this static position.

At this point in rotation, the commutator bars interchange position with the brushes, causing the current to flow through the armature in the opposite direction. This changes the direction of the magnetic lines of force. The top

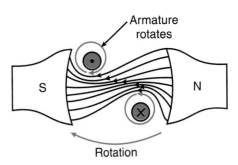

Figure 27-5. Lines of magnetic force that cause armature rotation.

armature leg is forced downward and the lower one is forced upward. Clockwise rotation continues uninterrupted.

Starter Motor Components

Starter motors consist of a housing, field coils, armature, commutator and brushes, end frames, and solenoid-controlled drive pinion, **Figure 27-6.** The iron *starter housing* protects and supports the field coils, brush holders and brushes, and pole shoes. It also acts as a conductor for the lines of magnetic force running between the field coils.

Field Coils

The *field coil windings* are made of heavy copper ribbon capable of handling high current flow. A four pole, four coil starter motor consists of two right-hand and two left-hand wound field coils. The field coils and pole shoes attach to the inside of the housing in an alternating North/South mounting pattern and are insulated from the housing. They are connected to a terminal that passes through the housing wall. The field coils are wired in series with the armature windings through the starter brushes. This setup allows all current passing through the field coils to pass through the armature as well.

Armature

The *armature* is located between the pinion drive and commutator end frames and the field windings. Starter motor armatures use many windings to ensure smooth uniform rotation and strong torque. These coils are often called segments. As one armature segment rotates past the field coil's strong magnetic field, another armature segment moves into position to take its place. This means there are always several armature segments receiving a strong push from the field coils at any one time.

The armature assembly consists of a shaft, laminated iron core, commutator, and armature segments or coils. The shaft has machined bearing surfaces and drive splines for mounting the commutator and a laminated core section. In most heavy-duty starters, the armature coils are not made of individual strands of wire. Instead, flat strips of copper capable of conducting heavy current loads without overheating and melting are used. The sides of the coils fit into slots in the iron core. Each slot contains the sides of two coils, however, the coils are insulated from the core and shaft.

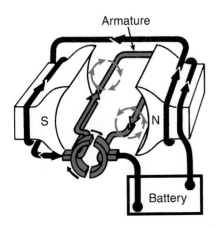

Figure 27-4. Simplified starter motor having a single armature.

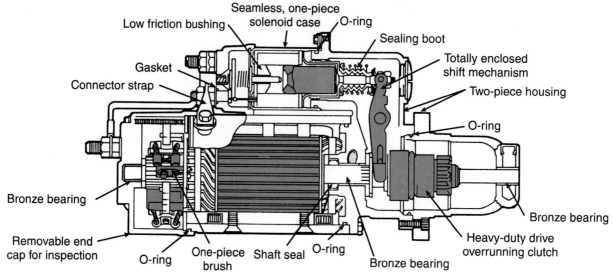

Figure 27-6. Cutaway of a typical electrical starter motor. (Detroit Diesel Corporation)

The armature coils connect to each other and the commutator. This allows current to flow through all armature coils at the same time. This current flow generates a magnetic field around the armature coils that interacts with the magnetic field between the field coils. The resulting force causes the armature to turn.

Commutator

The ***commutator*** assembly is pressed onto the armature shaft. It consists of heavy copper segments insulated from each other and the armature shaft. The commutator segments connect to the ends of the armature. Starter motors have four to twelve stationary carbon brushes that press against and ride on the rotating commutator segments, **Figure 27-7.** These brushes conduct current from the stationary field coils to the rotating armature coils. The brushes are held in position by holders. One-half of the brushes serve as "live" brushes while the other half serve as "ground" brushes.

As the armature segment turns through a half revolution, the commutator segments reverse positions. The segments that were in contact with live brushes now contact ground brushes. This reverses the direction of current flow in the armature loops, and ensures that the magnetic forces continue to push or spin the armature assembly in only one direction.

Internal Circuits

The motor's field coil windings are wired in combinations of series and/or series/parallel circuits to increase the starter motor torque or speed. Torque can also be increased by adding more field coils or increasing the windings on the field coils.

Straight series circuits are used when top rotational speeds are desired, **Figure 27-8.** One or more shunt coils may be included in the circuit to limit the motor's top speed and prevent possible damage. A shunt coil is not wired through the brushes and commutator to the armature. Instead, it has its own ground, **Figure 27-9.** This decreases overall current flow to the armature, limiting the strength of its magnetic field.

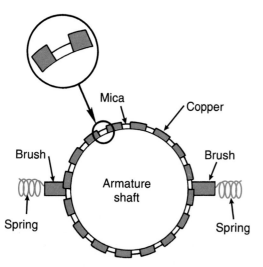

Figure 27-7. Cross-section of the commutator and brushes mounted on an armature shaft.

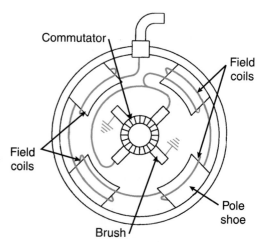

Figure 27-8. Typical series cranking motor circuit.

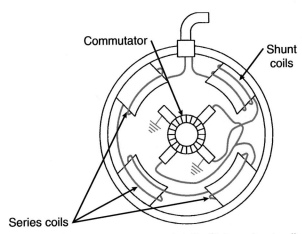

Figure 27-9. *Series starter motor circuit with two shunt coils.*

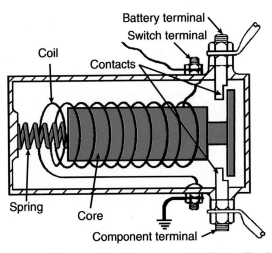

Figure 27-10. *Starter solenoid components. (Caterpillar Inc.)*

Torque is increased by boosting current flow to the field coils, which in turn increases the strength of the magnetic fields. To do this, field coil current can be directed into two, three, or four separate paths. Each separate field coil circuit has its own live/ground brush, except when the cranking current is split into four separate paths. In this case, two field coils share one live and one ground brush.

All live brushes are connected to each other by a number of jumper leads. This equalizes voltage to prevent arcing between the brushes and commutator that can lead to damage. Starter motors used in marine applications or those used with series-parallel switches have their ground brushes connected to the insulated ground circuit by a jumper lead.

Thermal Protection

Some heavy-duty starter motors are equipped with a **thermal protection circuit** that acts like a circuit breaker to protect against overheating. Overheating can be caused by extended cranking or a problem in the motor's internal circuits. The circuit is usually mounted to the starter housing and attached to the field coils. When it senses high internal heat levels, the circuit opens and the current to the field coils is cut off. This stops operation until it cools and its contacts close, which usually takes from one to five minutes.

Solenoid

The **solenoid** is an electromagnetic coil with a hollow center. A fitted metal plunger is located inside the center of the coil. The solenoid coil has two separate windings, a pull-in winding and a hold-in winding, **Figure 27-10.** The pull-in winding produces the magnetic force needed to pull the plunger into the coil and the hold-in winding holds the plunger in position.

When the starter switch is activated, a small amount of battery current passes through the remote starter relay to the field or *F* terminal on the solenoid. Both the pull-in winding and hold-in winding are energized. The plunger is pulled into the coil and its contact disc touches the solenoid terminals. This cuts the current flow to the pull-in winding, but the hold-in winding remains energized.

As the plunger disc contacts the solenoid terminals, it acts like an electrical relay, completing the feed circuit from the battery to the starter motor. Full cranking current is delivered to the field coils and armature. As this electrical connection is made, the moving plunger transfers its mechanical motion to a shift lever. The shift lever pushes the drive pinion into mesh with the ring gear. If the pinion gear teeth are not in mesh with the ring gear teeth, a spring located behind the pinion compresses, allowing the plunger to complete its stroke. When the starter armature begins to rotate, the pinion teeth will quickly line up with the ring gear teeth and spring pressure will force them to mesh.

Starter Motor Drives

Every starter motor requires some type of drive mechanism to engage and disengage its pinion gear with the flywheel ring gear. The starter drive must be able to do this because there is no way the operator can release the starter switch quickly enough to disengage the pinion at the proper time. The two main types in starter drives are the overrunning or sprag clutch design and the inertia or Bendix drive.

Overrunning Clutches

The **overrunning clutch drive** is used on nearly all modern electric starting systems. This type of drive requires the use of a solenoid switch and a shift mechanism. These components are housed in a sealed case mounted on top of the starter motor case, **Figure 27-11.**

The clutch housing is splined to the armature shaft and the drive pinion turns freely on the shaft. When the clutch housing is being driven by the armature shaft, a series of spring-loaded rollers are forced into the small ends of matching tapered slots. The rollers are wedged tightly against the drive pinion barrel. This action locks the pinion and overriding clutch housing firmly together. Thus, the armature shaft, clutch housing, drive pinion, and flywheel ring gear can all turn together and the engine is cranked.

When the engine starts, the flywheel becomes the driving gear, turning the pinion gear and clutch housing. As

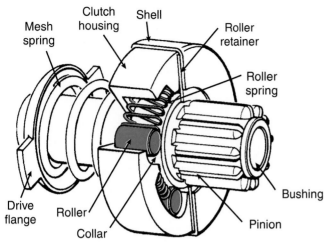

Figure 27-11. Typical starter pinion with overrunning clutch.

the clutch housing begins to spin more rapidly, the centrifugal force pushes the rollers out of the tapered slots. This releases the pinion from the clutch housing and armature shaft. The spinning pinion drive now overruns the armature shaft until the operator releases the starter switch and the ring gear is pulled out of mesh.

Some overriding clutches use a series of **sprags** rather than rollers between the clutch housing and sleeve. The operating principles of the sprag clutch are the same as the roller clutch described earlier.

Inertia Drives

The **inertia drive** or **Bendix drive** is no longer used on electric starting systems, but can be found on some air-powered starter motors. The drive housing is splined to the armature shaft and contains multiple clutch plates that are spring-applied. A screw shaft is fastened to the outside clutch hub and the pinion gear is mounted on this shaft.

When the armature shaft begins to spin, the pinion gear tries to remain stationary. Its inertia forces it to move along the spiral-threaded shaft until it engages the larger ring gear on the flywheel. When the engine starts, the pinion gear is thrown back along the threaded shaft, disengaging the gears.

Series-Parallel Starter Motor Switch

As explained in **Chapter 24,** many heavy-duty trucks and other diesel powered vehicles use two 12-volt or four 6-volt batteries. They are equipped with a **series-parallel switch.** This switch provides 24 volts for cranking and 12 volts for electrical system operation. Older starter switches were solenoid-controlled, but new designs, such as the one shown in **Figure 27-12,** uses diodes to control current flow between the alternator, batteries, and cranking motor. They are referred to as **dual voltage control units (DUVAC).**

When the engine is not running, the control rectifier is open, and current cannot flow between the battery and alternator. The batteries are connected in series and when the starting circuit is energized, 24 volts is delivered to the starting motor.

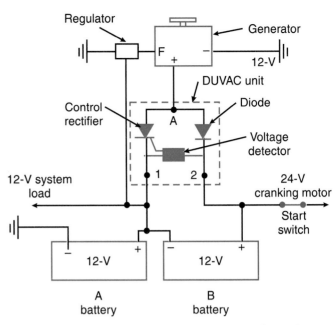

Figure 27-12. Electronic dual voltage control unit is used to provide 12-volts for charging and 24-volts for cranking the engine.

When the engine starts, the alternator begins to charge the batteries. Charging current flows from the alternator, through terminal A of the voltage control unit, through the diode, and out of terminal 2 to the positive terminal of battery B. It then passes to battery A and finally to ground. The B battery becomes fully charged before the A battery.

When the voltage detector in the DUVAC senses the B battery has reached its full charge, it switches the control rectifier on. Current can now only flow to the A battery; the B battery has been effectively removed from the electrical circuit. The system operates as a single 12-volt battery system until the engine is turned off. This opens the control rectifier, and both batteries are again in series to deliver 24 volts of cranking power to the starter motor.

Electric Starting System Tests

Be sure the batteries are fully charged and the charging system is in good working order before performing any starting system tests. Inspect system wiring for damage and be sure all connections to the starter motor, solenoid, relay, ignition switch, ground, and neutral safety switch are clean and tight. Loose or corroded connections can increase system resistance and reduce current to the starter motor.

Be sure the diesel engine is in good working order. The starter motor drive pinion should also be properly adjusted. When the starter is cool, check the resistance between the two thermal protection terminals on the motor. The ohmmeter should read zero. A high resistance indicates an open in the thermal protection circuit. Remember, the circuit is designed to open when it gets hot from overuse.

Figure 27-13 details a systematic approach to starter system troubleshooting. Starter system tests are either area tests that check current or voltage throughout the entire

Condition: Engine will not crank		
Possible Cause	Action	Solution
Battery discharged	Perform battery open circuit voltage, load voltage, and capacity tests	Recharge or replace battery
Faulty cable connections	Visually inspect and test cables	Replace as needed
Open in control circuit	Perform control circuit test to locate open circuit or high resistance	Repair or replace components
Seized engine	Inspect for loose, corroded or frozen parts, manually turn engine	Repair, replace, clean, and tighten as needed
Bad starter relay	Perform relay bypass test	Replace relay if starter turns
Internal starter motor problem	Perform starter relay bypass test	Replace if starter does not turn
Starter loose on mount	Visually inspect	Tighten as required
Condition: Engine cranks slowly		
Possible Cause	Action	Solution
Low battery charge	Perform battery open circuit voltage and load voltage tests	Recharge or replace battery
Bad starter circuit connections	Visually inspect for loose and/or corroded parts	Replace, clean, and tighten as needed
High resistance in starter circuit	Pinpoint areas of high resistance using cranking current, insulated circuit, and ground circuit tests	Replace and repair components as needed
Internal starter motor high resistance problem	Perform cranking current test. If pass, proceed with no-load test	Replace or repair as needed
Overheated starter solenoid	Inspect for missing or damaged heat shields	Replace or repair as needed
Heavy engine oil	Check engine oil specifications	Change to proper weight oil
Internal engine problems	Perform engine compression tests	Repair engine as needed
Poor starter drive-to-flywheel engagement	Visually inspect	Replace damaged components
Starter loose on mount	Visually inspect	Tighten as required
Condition: Starter spins, but engine does not crank		
Possible Cause	Action	Solution
Worn/damaged pinion gear or defective starter drive	Visually inspect	Replace damaged components
Worn/damaged flywheel gear	Visually inspect	Replace damaged components
Condition: Starter disengages early, noisy, or does not operate		
Possible Cause	Action	Solution
Battery discharged	Perform battery open circuit voltage, load voltage, and capacity tests	Recharge or replace battery
High resistance in starter circuit	Pinpoint areas of high resistance using cranking current, insulated circuit, and ground circuit tests	Replace and repair components as needed
Open in solenoid or open in movable pole shoe winding	Bypass solenoid, test movable pole shoe windings	Replace solenoid or movable pole shoe starter
Starter loose on mount	Visually inspect	Tighten as required
Solenoid unable to overcome return spring pressure	Visually inspect	Replace worn solenoid
Defective starter motor	Perform no-load test	Replace motor

Figure 27-13. Troubleshooting chart for starting system problems.

system, or pinpoint tests which concentrate on one particular component or branch of the wiring circuit.

Cranking Current Test

This test measures, in amperes, the amount of current the starter motor is drawing. To perform the test:

1. Turn off all lights and accessories.
2. Clamp the multimeter or volt/amp tester inductive pickup around the circuit wire to be measured, **Figure 27-14.**
3. Crank the engine while observing the ammeter reading. Prevent the engine from starting by using the fuel cutoff switch or holding the exhaust brake button while cranking (if so equipped).
4. Note the speed of the starter motor while cranking.
5. Check the amperage reading against manual specifications.

Low current draw indicates an undercharged or defective battery or high circuit resistance due to faulty components or connections. If current draw is high, there may be a short in the starter motor or mechanical resistance in the engine or starter due to binding, misaligned, or failed components.

Available Voltage Check

If the battery, switches, and wiring appear in good order, test for available voltage at the starter motor solenoid while cranking the engine, **Figure 27-15.** Place the red lead of a voltmeter on the solenoid's *BAT* terminal and the black lead to the starter ground terminal. Crank the engine and note the voltage reading. At room temperature, at least 9 volts should be available on a 12-volt system. On a 24-volt system, voltage should be 18 volts or more.

If the voltage drop is excessive, check the interconnecting cables by measuring the terminal voltage of each battery while cranking the engine. Differences in battery terminal voltage should be no more than .5 volts. If the difference is greater than .5 volts, check the battery cables and replace as

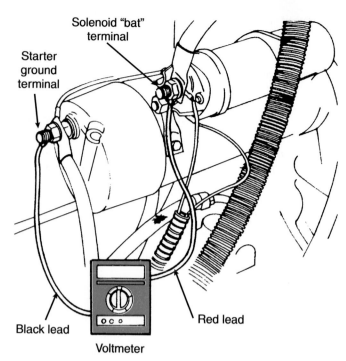

Figure 27-15. *Setup for starter motor available voltage test. (Detroit Diesel Corporation)*

needed. If the voltage is satisfactory, the starter motor may be defective or an internal engine problem may exist.

Relay Bypass Test

If the starter motor does not operate, determine if the remote starter relay is at fault by performing a ***relay bypass test.*** Connect a temporary jumper cable around the starter relay. If the starter motor cranks with the jumper in place, the relay is faulty and should be replaced. If the starter motor is still not receiving current, test the starter.

The relay can also be checked using a voltmeter. Connect the meter across the relay winding from the ignition or push button connection to one of the relay's mounting bolts. With the ignition or push button starter activated, the meter should read battery voltage. Zero voltage indicates an open in the circuit. A reading less than battery voltage indicates excess circuit resistance or a faulty relay. Look for causes of high circuit resistance and retest. If the voltage is still less than battery voltage, replace the relay.

Control Circuit Tests

The ***control circuit test*** checks all wiring and components used to operate the starter relay and solenoid. High resistance in the control circuit will reduce current flow to the relay or solenoid coil windings. Solenoid switch contacts may also be burnt, resulting in high resistance in the circuit. Control circuit components include the ignition or starter button switch, neutral safety switch, remote starter relay, and starter motor solenoid. To test this circuit:

1. Connect a voltmeter between the positive battery terminal and the starter switch terminal on the relay or solenoid.

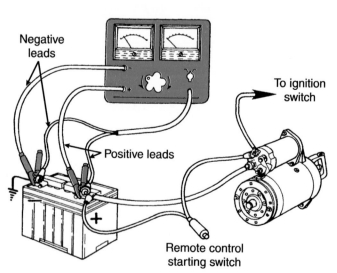

Figure 27-14. *Setup for performing cranking current test. (Navistar International Transportation Corp.)*

2. Prevent the engine from starting using the exhaust brake button or fuel cutoff switch. Crank the engine and note the meter reading.

3. A reading below .5 volts indicates acceptably low circuit resistance. If the reading is greater than .5 volts, perform individual voltage drop tests across each system component. A voltage drop greater than 1 volt across any connection, component, or stretch of wire usually indicates corrosion or a loose connection.

If the neutral safety switch indicates high resistance, adjust the switch according to the service manual and retest. Clutch operated safety switches are nonadjustable and must be replaced if defective.

Starter Circuit Resistance Tests

If the control circuit is operating properly, accurate voltage drop tests can be performed on the starter circuit to pinpoint any sources of excess resistance. Use a voltmeter that can take readings in the 2-3 volt range. One of the voltmeter leads should be equipped with a clean, sharp probe that can be inserted directly or "jabbed" into the battery post. By inserting the probe directly into the post, the voltage drop across the terminal post and cable clamp will be measured along with the drop in the cable itself. When connecting the voltmeter to the relay or motor, connect it to the stud rather than the terminal itself. This will ensure that you are measuring the drop across the connection.

 Note: Prevent the engine from starting by holding the exhaust brake button or by shutting off fuel to the engine.

Begin with a *ground circuit resistance test*. Connect one voltmeter lead to the battery ground terminal and the other to the starter base, **Figure 27-16A**. Crank the engine and read the meter. If the meter indicates battery voltage, the ground circuit is open somewhere between the batteries and the starter. The ground circuit voltage drop should not exceed .2 volts. If the voltage drop exceeds .2 volts, excessive resistance exists in the ground circuit and additional voltage drop checks should be performed to isolate the cause. Test points should include:

❑ The ground cable connections at the battery and engine.

❑ The starter bolt connection to the engine.

❑ The contact between the starter frame and its ends.

❑ Any connection between the ground terminal of the battery and the starter base.

Clean and tighten all connections requiring service.

If the starter ground circuit is in good condition, test the positive leads for excessive resistance. Begin by connecting the voltmeter's positive lead to the battery's positive terminal. Connect the second voltmeter lead to the starter terminal and crank the engine, **Figure 27-16B**. A voltmeter reading higher than .05 volts indicates high resistance in the cable connec-

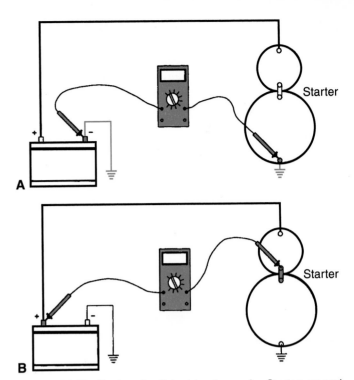

Figure 27-16. Starter circuit test hookups. A—Starter ground circuit. B—Starter positive circuit.

tion at the battery or starter solenoid; in the cable itself; or in the starter solenoid. Isolate the excess resistance by performing individual voltage drop tests across these points. Repair or replace connectors, cables, or parts as needed.

Removing and Installing the Starter Motor

If the battery, wiring, and relay switch appear to be operating properly and the engine is in sound mechanical condition, remove the starter motor from the vehicle for further testing. Even if you are certain the starter is defective, you should perform further tests to verify your diagnosis and to isolate the defective part, if you are repairing the starter.

1. Disconnect the negative (ground) terminal or ground strap from the battery(s).

2. Remove the battery cables and lead wires from the starter motor, tagging each lead to ensure proper reinstallation.

3. Remove the mounting bolts and carefully lift the starter up and out of the engine compartment. Starter motors can be extremely heavy.

To reinstall the starter motor when you are finished, perform the steps as follows:

1. Carefully place the starter motor in position and install the starter mounting bolts.

2. Reconnect the battery cables and lead wire from the starter. Remove any tags that were attached during removal.

3. Reconnect the negative battery terminal or ground strap.

Conducting a No-Load Test

The **no-load test** can help pinpoint internal starter motor problems prior to disassembly. The starter should be clamped in a bench vise during this test. As shown in **Figure 27-17,** a voltmeter is connected across the motor terminal and frame. An rpm indicator is mounted to the end of the motor drive shaft to measure armature speed. The starter motor, an ammeter, and an open switch are connected in series with a fully charged battery.

Close the switch to perform the no-load test. Note the voltage and ampere readings plus the armature speed (rpm). It is not necessary to perform the test at exact system voltage. If the test voltage is slightly higher than system specifications, the armature rpm will increase proportionately. Open the switch and disconnect the test equipment. No-load test results are summarized in **Figure 27-18.** If the motor does not perform to specifications, it requires further disassembly and testing. Refer to the service manual for the specific starter motor.

Air Powered Starting Systems

Air starting systems are simple to operate and maintain. A typical system consists of an air tank, two control valves, a cranking motor, and the required connecting air lines. Air starting systems can reach higher cranking speeds than most electric systems (100 rpm or more). This extra speed is needed to start large displacement diesel engines. Air starters also crank longer and do not generate heat like electrical starting systems.

System Components

Figure 27-19 illustrates a typical remote control air starter system. The system has its own air supply tank connected through a one-way valve to the air brake system reservoirs. A relay valve links the system's air tank, starter motor, and push button starter valve. In some systems, the relay is housed inside the starter motor. The starter motor also has an automatic lubrication system.

Operation

The operator depresses the push button valve to activate air flow to the starter motor. Air from the tank is directed to the relay valve and in some systems, to the automatic oiler. This air pressure activates the oiler system and relay valve.

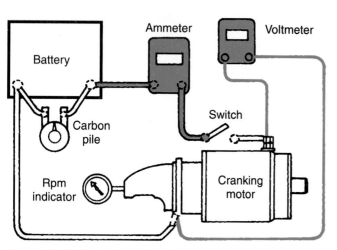

Figure 27-17. Setup for performing starter no-load test.

Test Results	Condition or Problem	Action
Rated current draw and no-load speed	None	Starter OK, reinstall on engine
Will not operate; no current draw	1. Open field circuit 2. Open armature coils 3. Broken brush springs, worn brushes, poor contact between commutator and brushes	1. Disassemble, inspect terminal connections, trace circuit with test lamp 2. Disassemble, inspect commutator for burnt bars 3. Disassemble, replace faulty components, ensure good contacts
Will not operate; high current draw	Direct ground in terminal or fields. Locked bearings	Disassemble, inspect, and test circuits
Low speed operation; high current draw	Excessive friction due to worn or dirty bearings, bent armature shaft or loose pole shoes Short in armature Grounded armature or fields	Disassemble, inspect, and test components
Low speed operation; low current draw	High internal resistance	Disassemble and check for poor connections, bad leads, dirty commutator
High speed operation; high current draw	Shorted field	Replace field coil assembly

Figure 27-18. Chart showing the conditions and corrections for problems found during starter load tests.

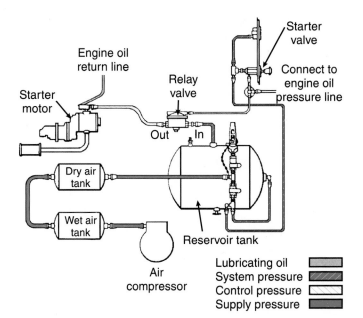

Figure 27-19. *Components of a typical air-operated starting system. (Stanadyne Diesel Systems)*

Air pressure acts on either a diaphragm or piston allowing air pressure to reach the air motor's inlet port.

The air motor consists of a vane rotor, a reduction gear, and either an inertia drive or overrunning clutch. The motor's rotating parts are supported by ball or roller bearings. A cylinder sleeve is pressed into the motor housing to decrease wear and to seal the rotor vanes.

As pressurized air enters the motor's inlet port, it is forced into one or two pockets created by the rotor vanes, end covers, and cylinder sleeve. Air is also directed through a small passage onto a servo piston, forcing it to move. This movement engages the drive pinion gear with the flywheel ring gear.

At this point, there is sufficient air pressure pushing on the vanes to turn the rotor on its bearings. This rotation aligns the next pocket with the inlet port. This pocket also fills with pressurized air and is forced to rotate. As the individual pockets rotate, they pass an exhaust slot and air pressure inside the pocket is released to the atmosphere. The pockets are now ready to be refilled with air the next time they pass the inlet port.

This fill/relieve process occurs very quickly as the rotor rapidly spins, driving the engaged drive pinion and ring gear as it turns. Cranking continues until air flow from the relay valve is stopped. Lightweight motor oil is normally used for lubrication. Lubrication is an important factor in maintaining air starter motor life and may be stored in a separate reservoir.

Troubleshooting Air Starting Systems

Air starting systems are relatively simple to troubleshoot and repair. Problems with the push button valve can usually be traced to air leaks caused by damaged O-rings, worn seats, or valves. The relay valve can be damaged by oil or water in the air supply system. Water and oil can cause the valve to stick or damage the inlet and exhaust valves and seats. Replacement kits are available for rebuilding the relay valve. Be sure to clean the valve thoroughly and use lubricant specified by the manufacturer on all moving parts.

Most problems with the air motor are caused by lack of lubrication. Poor lubrication increases wear on the motor vanes, cylinder sleeve, end plates, and bearings. The motor can be rebuilt by replacing the vanes, honing the sleeve, and refinishing the end plates as needed. If damage is severe, the components must be replaced.

Hydraulic Starting Systems

Hydraulic starting systems use fluid energy to crank the starter motor. Like air starting systems, hydraulic starters can produce higher cranking speeds than electric starters. However, cranking time is limited to approximately 3 seconds per accumulator charge. Components of a hydraulic starter system include a fluid reservoir or tank, a hand pump, an engine-driven pump, an accumulator, a hydraulic cranking motor, and hydraulic lines, **Figure 27-20.**

The *reservoir* cleans and stores the hydraulic fluid. It is connected to the inlet ports of both the hand and engine-driven pumps. Return lines from the hydraulic motor and the engine driven pump are also connected to the reservoir.

Pressurized hydraulic fluid is stored in a steel *accumulator,* closed by two end caps. A moveable piston divides the accumulator's interior into two halves that contain fluid on one side and nitrogen gas on the other side. The end cap on the fluid side is connected to the directional valve and to the hand and engine pumps. The end cap on the nitrogen side is charged at a pressure of approximately 1500 psi (10 343 kPa). Nitrogen is used as a charging gas because it is nonexplosive.

The engine-driven pump may be belt-driven or directly driven by the engine. When the engine runs, the pump forces hydraulic fluid into the accumulator against the pressurized nitrogen gas. When the accumulator is charged, the pump routes excess fluid back through the return lines to the reservoir tank. The hand pump is only used to prime the accumulator with fluid in the event it becomes completely drained.

The hydraulic cranking motor can be a vane, swashplate, or bent axis design, with most new models using an overrunning clutch drive mechanism. The control valve may be bolted directly to the motor's housing or it can be remotely mounted between the accumulator and the motor.

Operation

Hydraulic starter motor operation is similar to that of an air driven motor. The operator activates the system by moving a starter lever or handle, opening the directional spool valve. Pressurized fluid from the accumulator enters the motor's inlet port and acts against the vanes or swash plates, forcing the rotor to spin, and engaging the motor's

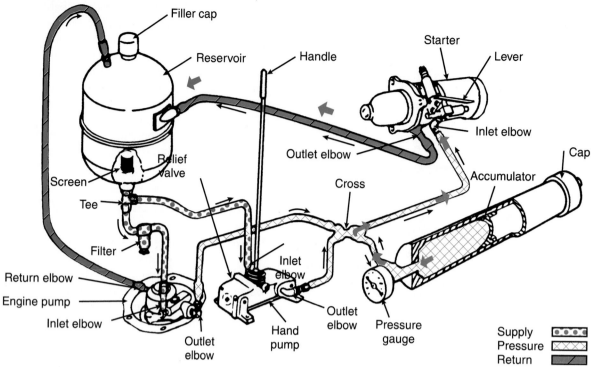

Figure 27-20. *Components of a hydraulic starting system. (Detroit Diesel Corporation)*

drive pinion with the flywheel ring gear. As the rotor spins, individual pockets of pressurized fluid align with the outlet port. Pressure is relieved and fluid flows back to the system reservoir. When the operator returns the starter lever or handle to its neutral position, fluid flow to the motor is cut, cranking stops, and the overriding clutch drive is taken out of engagement.

Troubleshooting Hydraulic Starting Systems

Slow or very brief cranking can be caused by low fluid pressure, a faulty directional valve, a seized starter, or diesel engine. No-start conditions can also be caused by low accumulator pressure, a worn pump, low hydraulic fluid level, air in the fluid lines, fluid leaks, restricted directional valve movement, or a damaged or worn starter motor. Because of the close tolerances involved, most hydraulic starter motor rebuilding and repair work is done by specialty shops.

Cold Starting Aids

Starting a cold diesel engine can be somewhat frustrating. The heat generated by compression tends to dissipate through the cylinder and head metal. Clearances when the engine is cold may be such that much of the compressed air escapes past the piston rings. Other problems may include the effect of cold on lubricating oil and diesel fuel viscosity. The spray pattern at the injectors coarsens and the drag of heavy oil between the engine's moving parts increases friction.

It is important to remember that starting has three distinct phases. During the initial phase, the breakaway torque needed to start cranking can be substantial, since the engine's rotating parts have settled to the bottom of their journals and are only partially lubricated. The second phase occurs during the first few revolutions of the crankshaft. In most cases, the first few crankshaft revolutions are free of heavy compressive loads. Cold oil pumped to the journals collects and wedges between the bearings and the shafts. As the shaft rotates, this oil is heated by friction. As the oil thins, drag reduces significantly, and cranking speed increases dramatically. Engine compression also begins to build during this phase.

The third phase begins as the engine reaches firing speed. Compression levels within the cylinders are sufficient to initiate combustion. The time between the initial breakaway phase and the firing speed depends on the capacity of the starter and battery, the mechanical condition of the engine, lubricating oil viscosity, ambient air temperature, flywheel inertia, and the number of cylinders in the engine.

To ensure long engine life, if the ambient temperature has dropped below 25°F (-4°C), it is advisable to warm any diesel engine before initial start-up. Warming the engine will allow it to start quickly and reduce wear on starting system components. There are several methods of warming an engine in cold weather. These include coolant heaters, lube oil heaters, glow plugs, intake air heaters, and battery heaters.

Coolant Heaters

There are two types of *coolant heaters:* circulating heaters and immersion heaters. *Circulating coolant heaters* are located outside the engine, **Figure 27-21.** Cold coolant leaves the bottom of the engine block and enters the heater tank where it is warmed by a copper heating element, **Figure 27-22.** The warm coolant is then pumped back into the engine block. Circulating heaters require 120 or 240 volts ac and operate at over 3500 watts. They are available with or without thermostatic control. The main advantage of circulating coolant heaters is that they warm the engine in a very short period of time. While most circulating coolant heaters are electric, some are propane fired.

Immersion coolant heaters are installed directly into the engine block in place of a freeze plug. Large engines and V-type blocks usually require the use of two immersion heaters. Immersion coolant heaters do not circulate the coolant. Heat can only move through the cylinder walls by convection, so warm-up time is increased when compared to circulating coolant heaters. Immersion heaters operate on 120 or 240 volts ac, and are available up to 2500 watts.

Lubricating Oil Heaters

Most lubricating **oil heaters** are electric powered immersion heaters that are installed in the oil sump through the drain plug or dipstick opening, **Figure 27-23.** A thermostat can also be installed as part of the heating unit. This type of heater is designed to keep the oil pan warm and allow heat to flow up into the block, warming the entire engine.

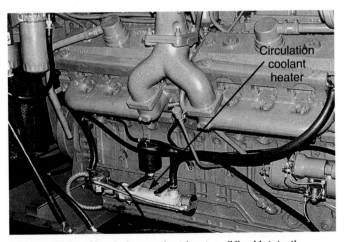

Figure 27-21. *Circulating coolant heater. (Kim Hotstart)*

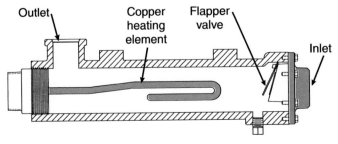

Figure 27-22. *Cutaway of a circulating coolant heater. (Kim Hotstart)*

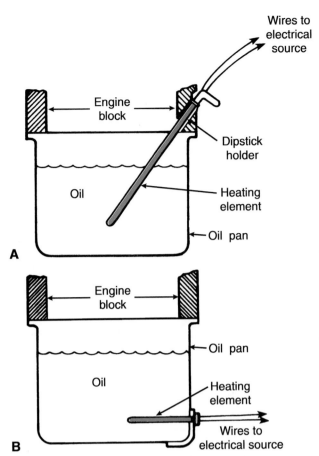

Figure 27-23. *Typical lubricating oil heating systems A—Dipstick. B—Engine block.*

Another oil heater warms the oil using an external heater mounted below the sump. You can heat the oil before start-up, although this method is highly impractical for all but a few stationary applications.

Glow Plugs

Glow plugs are threaded into the individual cylinders and are heated by battery current. Operating temperatures of 1500°F (816°C) can be reached in seconds. Once the engine is running, the glow plugs are turned off.

 Note: Do not confuse glow plugs with spark plugs used in gasoline engines. They are not the same in either function or operation.

Glow plugs may be wired in parallel or series and have a fairly high current draw. If a glow plug is suspect, run a continuity check with a test light or voltmeter while cranking the cold engine to check if current flows from the wire to the plug. If current is present at the plug, check the resistance between the insulated glow plug terminal and the ground with an ohmmeter. Infinite resistance indicates an open circuit and the glow plug must be replaced, **Figure 27-24.** Normal resistance in a glow plug is very low, usually around 1.5 ohms.

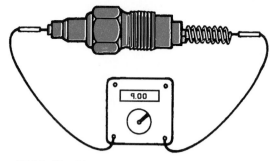

Figure 27-24. Checking glow plug resistance with an ohmmeter.

Glow plugs offer a method of warming the engine without the need for an ac hookup or alternate fuel fired heat source. Intake air heaters operate on the same principle. A small electrical coil energized from the vehicle batteries is mounted in the diesel engine's intake manifold.

Fluid Starting Aids

 Warning: Starting fluids must never be used if the engine is equipped with either glow plugs or an electric coil air intake heater.

For many years, highly combustible *ether* was used as a starting aid for diesel engines. Technicians would pour a spoonful of ether onto a rag, and place it over the air intake. If the technician guessed correctly, the engine might start. But if too much ether was used, severe detonation or an explosion could result in broken pistons and rings and bent connecting rods. The force of these explosions destroyed many engines, injuring (and sometimes killing) the technician working on the engine. *Do not make the same mistake.*

Use of spray cans of ether is discouraged as they can be dangerous. The only method of safely using ether is with a closed dispensing system. Starting aid systems of this type normally consist of a cylinder of pressurized ether, a metering valve, tubing, and an atomizer installed in the intake manifold. The valve is tripped only once during each starting attempt, to prevent a build-up of ether in the intake manifold that could lead to an explosion or hydraulic lock.

Diesel engines can also be fitted with another one-shot starting device consisting of a holder and needle. A capsule containing ether is inserted into the device. The needle pierces the capsule, releasing the premeasured dose of ether into the intake manifold. Regardless of the system used, starting fluid should only be introduced into the intake manifold while the engine is cranking and then very sparingly.

 Caution: Diesel engines can become "dependent" upon ether as a starting aid, even to the point that an engine will not start without it.

Air Shutdown Systems

Some engines use an *air shutdown system* to close off the engine's air intake. This gives the operator the means to shutdown the engine if an abnormal condition causes the engine to run out of control. Air shutdown systems are used primarily on two-stroke engines. In most cases, the shutdown system is simply a valve located at the engine's air intake. This valve should move free and completely close the intake when tested. Do not use the air shutdown system except in an emergency. Repeated use will cause damage to the blower.

Battery Warmers

Battery warmers use an electric heating pad beneath or around the battery to keep it at a temperature that allows for full cranking current to be delivered to the starter motor. Battery warmers can be used alone, but should be coupled with another cold starting aid, such as a coolant or lube oil heating system.

Summary

The starting system turns the engine over until it can operate under its own power. While the majority of diesel engine starting systems are electrically powered, some diesel engines are equipped with starter motors that are air or hydraulically driven.

An electric starting system has two distinct circuits, the starter circuit and the control circuit. The starter circuit carries high current flow from the battery or batteries through heavy cables to the starter motor. The control circuit uses a small amount of current to operate a magnetic switch that opens and closes the starter circuit.

The starter motor must produce a high cranking torque for a very short period of time when the starter circuit is energized. Direct-drive starters do not use any type of internal gearing to reduce the speed of the starter motor armature and pinion gear. A gear reduction starter uses an additional gear inside the starter housing to increase turning torque for large engines or cold weather operation. This gear reduces the pinion gear speed.

Starter motors consist of a housing, field coils, an armature, a commutator and brushes, end frames, and a solenoid-controlled drive pinion. The motor's field coil windings are wired in combinations of series and/or series/parallel circuits to increase the torque or speed of the starter motor.

Every starter motor requires some type of drive mechanism to engage and disengage its pinion gear with the flywheel ring gear. The two main types in use are the overrunning or sprag clutch design and the inertia or Bendix type drive.

Many heavy-duty trucks and other diesel powered vehicles use two 12-volt or four 6-volt batteries. They are equipped with a special series-parallel switch. This switch provides 24 volts for cranking and 12 volts for electrical system operation. Electrical starting system tests include cranking current test, relay bypass testing, control and starter circuit resistance checks, and no-load testing.

Air starting systems are simple to operate and maintain. They can reach higher cranking speeds than most electric systems, an aid in cranking high displacement diesel engines. Air starters also crank longer, and do not generate heat like electrical starting systems.

Hydraulic starting systems use hydraulic energy to crank the starting motor. Like air starting systems, hydraulic starters can produce higher cranking speeds than electric starters. However, cranking time is limited to approximately 3 seconds per accumulator charge.

In cold conditions, warming the engine will allow it to start quickly and avoid wear and tear on starting system components. Methods used to warm an engine include coolant heaters, lube oil heaters, glow plugs, intake air heaters, starting fluids, and battery heaters.

Important Terms

Starting system	Relay bypass test
Electric starting system	Control circuit test
Starter circuit	Ground circuit resistance test
Neutral safety switch	No-load test
Starter motor	Air starting systems
Direct-drive starters	Hydraulic starting systems
Gear reduction starters	Reservoir
Starter housing	Accumulator
Field coil windings	Coolant heaters
Armature	Circulating coolant heaters
Commutator	Immersion coolant heaters
Thermal protection circuit	Oil heaters
Solenoid	Glow plugs
Overrunning clutch drive	Ether
Sprags	Air shutdown system
Inertia drive	Battery warmers
Bendix drive	
Series-parallel switch	
Dual voltage control units (DUVAC)	

Review Questions—Chapter 27

Do not write in this text. Place your answers on a separate sheet of paper.

1. Name the two distinct electrical circuits that make up a diesel engine electrical starting system.

2. The control circuit of an electric starting system is used to:
   (A) regulate the amount of current passing through the alternator.
   (B) open and close the system's starter ciruit.
   (C) prevent battery overcharging.
   (D) Both A & B.

3. _____ do not use any type of internal gearing to reduce the speed of the armature.

4. To form the equivalent of a horseshoe magnet, heavy wire is wound around a piece of iron to create an _____ .

5. Which of the following is not a component in an electric starter motor?
   (A) Pole shoes.
   (B) Field coil windings.
   (C) Armature.
   (D) Neutral safety switch.

6. The _____ can pinpoint internal starter motor problems prior to disassembly.

7. *True or False?* Most problems with the air motor starting systems are caused by lack of lubricant.

8. _____ is used as a charging gas in hydraulic starting systems.

9. List at least four methods commonly used to warm diesel engines for better cold weather starting.

10. Starting fluids should never be used on diesel engines equipped with _____ .

ASE-Type Questions

1. Technician A says that both hydraulic and air starter are capable of reaching higher cranking speeds than electric starting systems. Technician B says that hydraulic starters are capable of continuous cranking for longer periods of time than electric or air driven systems. Who is right?
   (A) A only.
   (B) B only.
   (C) Both A & B.
   (D) Neither A nor B.

2. A series-parallel switch in many diesel engine electrical systems allows _____ .
   (A) 24 volts for both starter cranking and electrical system operation
   (B) 24 volts for starter cranking and 12 volts for electrical system operation
   (C) 12 volts for starter operation and 24 volts for electrical system operation
   (D) 12 volts for both systems but much greater capacity

3. Which of the following starter motor designs provides the most torque at the pinion to flywheel ring gear connection?
   (A) Direct drive.
   (B) Gear reduction.
   (C) Inertia drive.
   (D) Overriding clutch drive.

4. Technician A says that straight series wired starter motors generate the most torque possible. Technician B says that wiring in series-parallel will increase the motor's torque output. Who is right?
   (A) A only.
   (B) B only.
   (C) Both A & B.
   (D) Neither A nor B.

5. Which of the following is true concerning starter motor drive mechanisms?
   (A) The overriding clutch design is the most common in use today.
   (B) The inertia drive is the most popular in use today.
   (C) The drive is moved into position by the action of a solenoid.
   (D) Both A & C.

6. Technician A says that you should check the starter motor thermal protection circuit for high resistance when the engine is hot. Technician B says that the thermal protection can only be accurately tested when the engine is cool. Who is right?
   (A) A only.
   (B) B only.
   (C) Both A & B.
   (D) Neither A nor B.

7. Technician A says that low current draw during the starter cranking test is an indication of a short in the starter motor or some type of resistance in the engine or starter due to binding, misaligned or failed components. Technician B says that low starter current draw is a sign of battery problems or high resistance in the circuit. Who is right?
   (A) A only.
   (B) B only.
   (C) Both A & B.
   (D) Neither A nor B.

8. Technician A says that you should check the starter relay by bypassing it in the circuit with a jumper wire. Technician B says that you should check the relay using a voltmeter connected across the relay winding from the ignition or push button connection to vehicle ground. Who is right?
   (A) A only.
   (B) B only.
   (C) Both A & B.
   (D) Neither A nor B.

9. Total starter system control circuit voltage drop should not exceed _____ volts for the entire system and _____ volts across any one connection or component.
   (A) .3 / .1
   (B) .5 / .1
   (C) 1.0 / .1
   (D) 1.0 / .2

10. Technician A says low free speed and high current draw during a starter motor no-load test may indicate dirty, worn, or sticking internal bearings, a shorted armature, or a grounded armature or fields. Technician B says an open field circuit or open armature coils can also be the problem. Who is right?
    (A) A only.
    (B) B only.
    (C) Both A & B.
    (D) Neither A nor B.

Chapter 28

Engine Reassembly and Installation

After studying this chapter, you will be able to:

❏ Outline the procedures for reassembling a completely overhauled diesel engine.

❏ Explain the proper methods of engine run-in using engine dynamometers, chassis dynamometers, and in-vehicle run-in procedures.

❏ List the situations when sealer should be applied to fittings, plugs, and fasteners.

❏ Explain the procedures for prelubrication and preparing the engine for startup and run-in.

❏ Outline the steps for reinstalling the engine in the vehicle.

This chapter will review the general reassembly procedure as a whole and outline the basics steps taken for the safe start-up and break-in of an overhauled engine. More detailed reassembly procedures for major diesel engine systems and components such as the crankshaft and bearings, cylinder liners, pistons, and connecting rods were covered in more detail in the chapters dealing with those subjects. These instructions are very general and are *not* intended to be a substitute for the manufacturer's service manual. Always follow the manufacturer's recommended procedures.

Establish Good Work Habits

Keep the work area, tools, equipment, and all engine components clean at all times during the reassembly process. Remember, dirt in the engine may have been the reason the engine required repair. Do not contribute to the engine's next overhaul through careless work habits. If parts have been stored prior to reassembly, reinspect them for rust and cleanliness, and clean as needed.

Modern engines are equipped with many accessory items and specialized electronic systems. Examples include electronic control modules, air compressors, air inlet heaters, fan drives, power steering pumps, and turbochargers. These items were removed and are normally installed in a specific order.

Late-model engines will have an extensive electronic wiring system linking its sensors to the control module. The location and wiring for these items is usually engine specific, so check the service manual. Install all sensors and wire connectors in the order outlined by the manufacturer. Be sure all **wiring harness straps** are secure and any ground straps are reconnected to the block or another good ground point.

In addition to general shop and service manuals, many manufacturers provide specialized assembly/disassembly manuals for their engines. These manuals outline the recommended service procedures for a particular engine model. They list all important assembly data such as proper nut, bolt, and stud torque, component clearances, and end play specifications. The manual will also outline the proper use of sealants, anti-seize compounds, thread lock, primers, cleaners, adhesives, and lubricants in the reassembly process.

Always use new gaskets, O-rings, filters, and other components. Apply fresh engine oil to wear surfaces of mating parts. Finally, always use the installation and special tools specified by the engine manufacturer. The exact assembly sequence will vary from engine to engine. This is particularly true of such components as:

❏ Crankshaft front and rear seals.

❏ Idler and auxiliary gears.

❏ Timing covers.

❏ Valve lifter assemblies.

❏ Fuel system components.

❏ Sensors and engine control equipment.

❏ Auxiliary engine-driven equipment.

❏ Front and rear engine **lifting plates** and engine supports.

The following assembly steps are typical for a complete overhaul. Always follow the manufacturer's service manual procedures and recommendations.

Cylinder Block, Sleeves, and Crankshaft Assembly

The cylinder block is the foundation for the entire engine. Specific procedures for it were covered in **Chapter 5.** Crankshaft removal and checking procedures were covered in **Chapter 6.**

Mount the cylinder block on a rebuild stand before beginning reassembly. After applying Loctite 277 or equivalent to the main oil gallery bore plugs, cooling gallery end plugs, miscellaneous pipe plugs, and water drain plugs, install them into their respective bores. Install the plugs flush or slightly below the block's finished surface. Small diameter plugs are often installed using a hex head wrench.

If the engine uses wet cylinder sleeves, carefully clean the O-ring grooves in the lower O.D. of each sleeve. Lubricate and install a new O-ring into each groove. Coat the cylinder bores in the block and the sleeve's O-ring surface with fresh vegetable oil. Install each sleeve into its bore by hand, using a push-fit to seat each sleeve, **Figure 28-1.** Some sleeves have to be driven in by force. Once they are in place, install retainers to secure the sleeves in position. Check sleeve protrusion using a dial indicator. If needed, install shims to meet the manufacturer's specifications.

The crankshaft gear is installed onto the crankshaft end prior to installation. An alignment key is often used to position the gear. This key is placed in a keyway cut into the crankshaft. The gear is then heated to approximately 400°F (205°C) and driven or pressed onto the shaft. If used, the crankshaft external counterweight is installed onto the crankshaft in the same manner.

Installing Bearings and Crankshaft

Install the main bearing upper halves following the order established during disassembly, **Figure 28-2.** Align the lube holes in each bearing shell with its mating port in

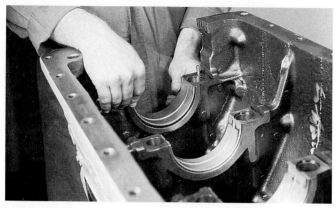
Figure 28-2. Installing upper main bearing shells. (Cummins Engine Co., Inc.)

the cylinder block. Apply a film of engine oil to each bearing wear surface and crankshaft journal. Do not lubricate the back of the bearings.

Make certain the crankshaft is clean prior to installing it in the cylinder block, **Figure 28-3.** Carefully lift and install the crankshaft into the cylinder block using a hoist and slings positioned to balance the weight. Slide the thrust washer upper halves into the cylinder block at the appropriate locations. Be sure the lubrication grooves in each bearing half are positioned to face toward the crankshaft wear surface. Install the main bearing lower halves onto their respective bearing caps, observing the order established during disassembly. At this point in the reassembly, a bearing clearance check using Plastigage may be performed.

Apply a film of engine oil to the wear surface of each bearing shell. Install the main bearing caps according to the order observed during disassembly. Install the bearing cap mounting bolts and washers. Snug the bolts, but do not tighten as the manual may specify a specific tightening sequence. After checking the service manual, tighten the main bearing cap bolts to specifications.

Slowly turn the crankshaft one revolution. Use a **dial indicator** to check crankshaft end play against manual specifications, **Figure 28-4.** If the crankshaft **end play** is out of specifications, correct by installing the proper oversize or undersize main thrust washer.

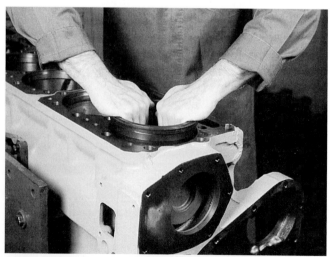
Figure 28-1. Installing a cylinder sleeve into its bore. (Cummins Engine Co., Inc.)

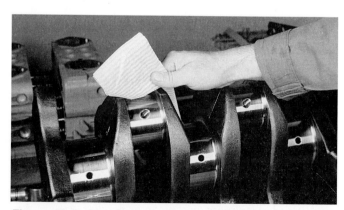
Figure 28-3. Cleanliness is the most important factor when reassembling an overhauled engine. Do not allow dirt to ruin an overhauled engine. (Cummins Engine Co., Inc.)

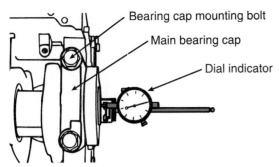

Figure 28-4. During reassembly, a dial indicator is used to check crankshaft end play. (Caterpillar Inc.)

Gear Housing

On many engines, the gear housing mounts to the front of the engine block. It provides protection for the crankshaft drive gear and any gears driven off the crankshaft, **Figure 28-5.** The housing body often provides a mounting surface for components such as the fuel injection pump, air compressor, fan drive, tachometer drive, and power steering pump (when applicable).

Install new bushings for any idler or auxiliary drive gears as needed into the housing. Mount the drive shaft in the housing, and install the drive gears on each end of the shaft. To install the reassembled gear housing on the cylinder block, begin by thoroughly cleaning the mounting surfaces. Install a new O-ring in the engine block-to-auxiliary drive housing oil supply passage.

Apply a bead of sealant to the mounting surface and install the housing on the block. Snug the mounting bolts but do not tighten. When the timing gear cover is installed later in the reassembly process, some mounting bolts screw into the front face of the cover. Once the timing cover is in place, the bolts used to mount the housing to the block are torqued to specifications.

Installing Pistons and Connecting Rods

The pistons and connecting rods transfer the power generated from combustion to the crankshaft. Service procedures for both pistons and rods were discussed in **Chapter 7.**

It is very important to properly install compression ring(s), sealing rings, and oil control rings. The engine will burn oil or run hot if the rings are damaged. With the rings in place, the piston can be installed, **Figure 28-6.** Begin by installing a piston pin snap ring or retainer disk on one side of the piston. Install the connecting rod to the piston by lubricating the piston pin and pressing the pin through the piston and connecting rod bores, **Figure 28-7.**

Install a second snap ring to lock the piston pin in the bore, **Figure 28-8.** On two-cycle engines, install the other pin retainer disk and vacuum check for leakage. Lightly oil the piston and rings and compress with a piston ring compressor.

Install the upper bearing on the connecting rod. Lightly lubricate the bearing wear surface with clean engine oil. With the crankshaft at the bottom dead center position, lubricate the I.D. of the cylinder sleeve and install the piston and rod assembly in the bore. Using a firm steady push, slide the piston out of the piston ring

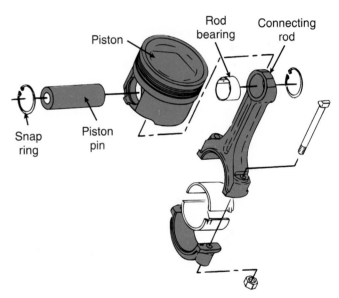

Figure 28-6. Piston and connecting rod assembly. (Caterpillar Inc.)

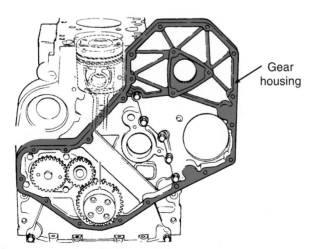

Figure 28-5. Many engines have a protective gear housing mounted to the front of the engine. (Cummins Engine Co., Inc.)

Figure 28-7. Assembling a connecting rod to a piston by inserting a piston pin. (Cummins Engine Co., Inc.)

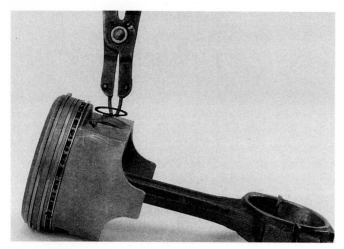

Figure 28-8. Install any bearings and snap rings when assembling the piston and connecting rod.

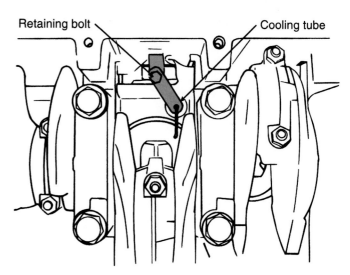

Figure 28-10. Installing piston cooling tubes. (Caterpillar Inc.)

compressor and into the bore, **Figure 28-9.** Install the connecting rod end on the crankshaft journal.

Install the connecting rod bearing caps and bolts. Snug the bolts but do not tighten. Torque the bearing cap bolts to specifications. Reinstall the piston cooling tubes in the cylinder block, if so equipped. Be sure each tube is seated properly and is not bent or damaged when tightening the tube mounting bolt, **Figure 28-10.**

To check the crankshaft installation, manually turn the crankshaft slowly and feel for free rotation. Use a *thickness gauge* to check connecting rod side-play. Finally, use a dial indicator to check piston protrusion above the cylinder block deck.

Oil Pump and Oil Pan

Install any needed O-rings onto the oil pump, coat them with clean oil, and install the oil pump into the cylinder block. Make sure that the oil ports are aligned. Torque the mounting bolts to specifications. Install the oil pump inlet and outlet tubes to the oil pump housing. Coat the new pan gasket with sealing compound and position on the engine's mounting flange. Install the oil pan and torque the bolts to specifications, **Figure 28-11.**

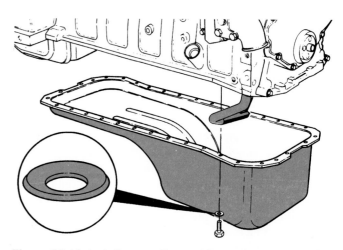

Figure 28-11. Installing an oil pan with gasket to the bottom of the cylinder block. Note the shape of the washer on the pan bolts. (Cummins Engine Co., Inc.)

Flywheel and Crankshaft Rear Seal

Install the crankshaft rear seal housing and gasket onto the guide pins. Torque the rear seal housing bolts to specifications. Mount the flywheel housing onto the guide pins. Use sealing compound on the mounting flange and torque to specifications.

Install the crankshaft rear *oil seal.* If needed, replace the starter ring gear on the flywheel. Using a hoist, install the flywheel onto its guide pins and fasten with new retaining bolts, **Figure 28-12.** Lock the crankshaft and torque all bolts to specifications. Install a new pilot bearing in the flywheel bore (where necessary) and remount the starter motor to the block.

Camshaft Installation

Install all camshaft bearings using the proper tool or driver. Make sure the oil hole in each bearing is aligned with the oil feed hole in the block. Install the front camshaft

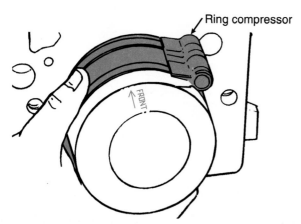

Figure 28-9. Installing the piston into the cylinder block using a ring compressor. (Cummins Engine Co., Inc.)

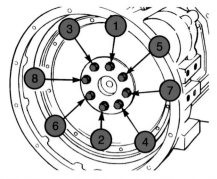

Figure 28-12. The flywheel is installed inside the flywheel housing which bolts to the rear of the engine. Torque the mounting bolts in the order specified in the engine manual. (Cummins Engine Co., Inc.)

bearing last, aligning each bearing oil passage with the corresponding passage in the cylinder block, **Figure 28-13.**

With the bearings in place, the camshaft can be reinstalled. If a Woodruff key is used to secure the drive gear, install it in the camshaft. Heat the camshaft drive gear, usually around 600°F (315°C). Use a heating oven; never use a torch. Install the drive gear and Woodruff key onto the camshaft end and allow the gear to cool.

Camshaft timing is very important. When installing the camshaft assembly, follow the guidelines in the service manual. Usually, the number one cylinder must be at top dead center (TDC) on the compression stroke before timing can be set. Timing marks on the camshaft drive gear, idler gear, and crankshaft gear must be in alignment.

Install the camshaft end play stop plate. Apply fresh oil or cam lube and carefully install the camshaft assembly into the cylinder block, **Figure 28-14.** Engage the camshaft and crankshaft gears, making sure the timing marks are aligned properly. On some engines, it may be necessary to install an idler gear and shaft. Install the drive gear onto the oil pump shaft and secure with a washer and retaining nut.

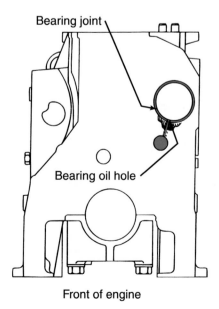

Figure 28-13. Proper positioning of the camshaft bearing oil hole is important for engine lubrication. (Caterpillar Inc.)

Inline or Distributor Injection Pumps

Both inline and distributor fuel injection pumps are driven off the camshaft gear. In some cases, the injection pump is not installed until the cylinder heads have been installed, so check the service manual.

To install the pump on the engine, make certain that the engine is still located on the number 1 cylinder timing mark. The injection pump must also be in its number 1 firing position. With the engine in the proper position, install a new gasket or O-ring on the injection pump mounting flange. If used, install the key in the injection pump shaft keyway. Align the timing marks and install the injection pump gear into the gear housing. Be careful not to move the engine crankshaft during the pump installation.

Install the injection pump to the mounting flange by slipping the pump camshaft through the opening in the pump driven gear, **Figure 28-15.** Align the key and keyway as required. Attach the pump by finger tightening the mounting nuts. The pump should be somewhat free to move. Install the driven gear mounting nut, but do not torque to specifications at this time. Torque the pump mounting bolts to specifications. If equipped, unlock the injection pump camshaft by reversing the locking pin position or otherwise disengaging the timing pin. Now tighten the driven gear mounting bolt to specifications. Attach the inlet and outlet fuel lines, oil supply, and air pressure lines (if so equipped).

Figure 28-14. Installing the camshaft into the engine block. (Cummins Engine Co., Inc.)

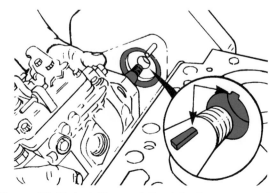

Figure 28-15. Installing an inline injection pump to the engine mounting flange. (Cummins Engine Co., Inc.)

Timing Gear Cover

Install a camshaft thrust bearing onto the timing gear cover inside face and install and tighten the mounting bolts. Coat a new timing cover gasket with sealing compound and install onto the mounting plate. Do not delay assembly as sealing compounds usually have a limited working life. Carefully install the *timing gear cover* onto the mounting plate guide pins. Torque the cover mounting bolts to specifications and check camshaft gear end play.

Install the new crankshaft front oil seal into position using a special driver tool. Install the oil seal bushing onto the crankshaft, making sure the bushing is properly oriented. Install the pulley-damper assembly onto the crankshaft.

Lifter Assemblies (Cam Roller Followers)

On some engines, lifter arm or cam roller followers mount into openings on the side of the cylinder block, **Figure 28-16.** The lifter arms contact the camshaft lobes. As the camshaft turns, the lifter arms move up and down to operate the pushrods and rocker arm assemblies. The lifter arm is usually mounted on a support shaft built into the side cover assembly. Installing the side cover using a gasket of proper thickness positions the lifter arms over the camshaft.

Cylinder Head Reassembly and Installation

At this point in the reassembly, the engine foundation has been completed. The next major task is to install all valve-related components into the cylinder head and remount the head on the block.

Install new *freeze plugs* if they were removed to flush the coolant passages. When new valve seats are installed, they are cooled prior to installation. Cooling shrinks the seat slightly so they can be installed in the head. As the

seats warm, they expand, providing a very tight fit. Once installed, the seats are ground to specifications.

The valve guides are then pressed into the cylinder head using a suitable driver. The valve guides normally protrude above the head surface a specified distance, **Figure 28-17.**

Oil the valve stems and install the valves into the cylinder head. Install the seat washer and spring seat. Dip the valve stem seal in engine oil and install the seal on the valve guide. Install the springs and retainer, compress the valve springs using a spring compressor, and install the valve keepers.

Mount Cylinder Head to Block

Remove the cylinder sleeve retainer (if used) and double-check that the cylinder head deck surfaces and cylinder block are clean. If the fuel injector sleeves were removed, install them in the cylinder head. Dry fit a new cylinder head gasket on the cylinder block. Use any guide pins in the block to help align the gasket; do not use any type of sealant, **Figure 28-18,** unless specified by the manual.

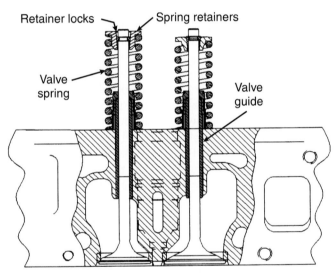

Figure 28-17. *Valve assemblies installed in the cylinder head. (Caterpillar Inc.)*

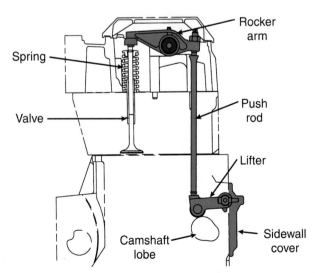

Figure 28-16. *Location of lifter assemblies and cylinder side covers. Other valve train components are shown for reference and may not be installed at this time. (Caterpillar Inc.)*

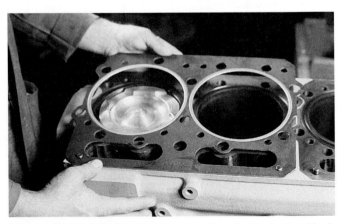

Figure 28-18. *Installing cylinder head gasket onto the block. (Cummins Engine Co., Inc.)*

Using a hoist, carefully lower the cylinder head assembly into position on the cylinder block. Lightly lubricate the cylinder head bolt threads with engine oil. Install and snug the bolts, being careful not to overtighten them. Once all the bolts are snug, tighten them in sequence to the correct torque, **Figure 28-19.**

Unit Fuel Injectors

Unit fuel injectors are normally installed before the air inlet manifold and the rocker arm assemblies, **Figure 28-20.** Inspect the cylinder head sleeve to ensure that there is no damage to the sleeve and that the injector sealing surface is clean. Clean the sleeve seat with a reaming tool if it is dirty. Install new O-ring seals on the injector and lubricate them with clean engine oil. Position the fuel injector and push it down into the bore. Be certain the injector is fully seated in the bore and that the O-ring has not become twisted. On electronic unit injectors, reconnect the wiring harness to the injector.

Air Inlet Manifold

The air inlet manifold installs onto the cylinder head assembly, **Figure 28-21.** Be sure the mating surfaces are

Figure 28-19. Installing the cylinder head bolts. (Cummins Engine Co., Inc.)

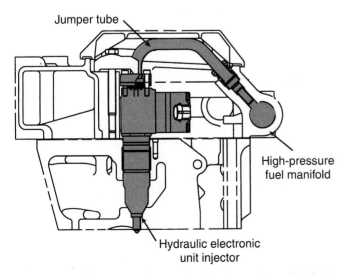

Figure 28-20. Details of a hydraulic-electronic unit injector installation. (Caterpillar Inc.)

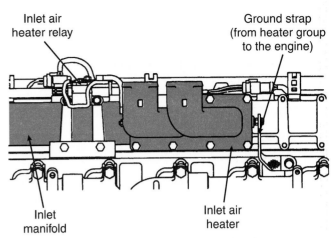

Figure 28-21. Air inlet manifold and related components. (Caterpillar Inc.)

clean and use a new gasket. Install the air temperature sensor into the bore in the inlet manifold body. Ether starting aid nozzles are usually fitted into the top of the air inlet manifold. If the engine is equipped with an *air inlet heater,* it is normally installed at this time. Be certain to reconnect any ground straps and wire assemblies.

Push Rods and Rocker Arms

All parts of the rocker arm assemblies must be clean. Lightly coat the push rods with clean engine oil and install them in their original locations, **Figure 28-22.** Be sure they are seated in the lifter arm assemblies. Assemble and install the rocker arm shaft assemblies. Check for the absence of binding in the assemblies and then torque the mounting bolts to specification.

On engines equipped with unit injectors, install the jumper lines between the fuel manifold and the individual unit injectors. Be sure to use the required washers and O-rings. Adjust the valve lash to specifications.

Fuel Injectors

On engines equipped with inline or distributor injection pumps, the fuel injectors are installed in the cylinder

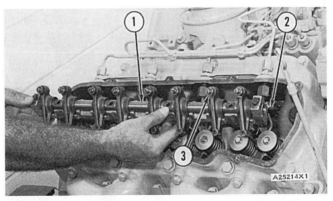

Figure 28-22. Installing the rocker arm assemblies. 1—Rocker-shaft 2—Frame 3—Adjusting screw. (Caterpillar Inc.)

head, **Figure 28-23.** The bore must be thoroughly cleaned before the injector is reinstalled. Be sure to use a new nozzle gasket as well as any special installation tools recommended by the manufacturer. Once the injector has been installed, inspect and clean the ends of each injection line before connecting it to the nozzle. The line nut should be left loose to allow air to be bled from the fuel line.

Valve Cover

Be sure the valve cover and air inlet manifold mating surfaces are clean. An adhesive is normally applied to the mating surface of the valve cover between each mounting bolt hole. Position a new valve cover gasket in the valve cover and install, using bolts and washers, **Figure 28-24.** Torque all bolts to specifications.

Vibration Damper, Crankshaft Pulley, and Crankshaft Front Seal

Make sure the crankshaft flange is clean before installing the vibration damper, crankshaft pulley, and crankshaft front seal, **Figure 28-25.** Once the pulley is installed, drive belts for components such as the water pump and cooling fan may be installed. The vibration

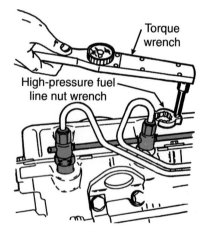

Figure 28-23. Installing fuel injectors and fuel lines. (Navistar International Transportation Corp.)

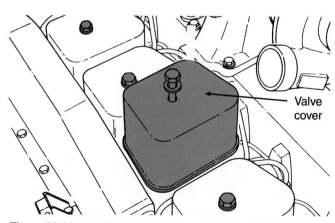

Figure 28-24. Installing the valve cover. (Cummins Engine Co., Inc.)

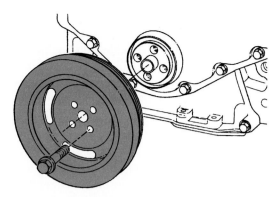

Figure 28-25. Installing the vibration damper on the crankshaft. (Cummins Engine Co., Inc.)

damper is installed last. On rubber dampers, double-check the marks for proper alignment. Replace the damper if it is out of alignment.

Exhaust Manifold

The exhaust manifold mounts to the cylinder head, **Figure 28-26.** A new gasket must be used and manifold sealing compound applied to the mating surfaces. The mounting bolts are normally coated with anti-seize compound prior to installation. Torque the mounting bolts in the proper sequence. Once the exhaust manifold is in place, the turbocharger (if used) can be reinstalled on the manifold using a new gasket and lock bolts.

Fuel Transfer Pump

Check the condition of all parts of the fuel transfer pump. If any parts are worn or damaged, replace them. Use new mounting gaskets or O-rings if so equipped. Carefully install the inlet and outlet lines, **Figure 28-27.**

Electronic Control Module

Position the electronic control module and install the mounting bolts, washers, and any special spacers used. Connect all wiring harness connectors to the ECM. The end

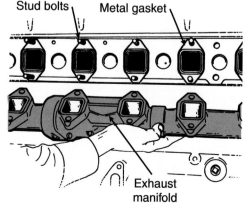

Figure 28-26. Exhaust manifold installation. (Navistar International Transportation Corp.)

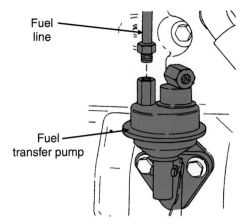

Figure 28-27. Installing the fuel transfer pump and fuel line connections. (Cummins Engine Co., Inc.)

of the connector receptacle may contain a bolt that must be in place and tight to secure the connection.

Fuel and Oil Filters

Reinstall the fuel filter base to the engine block and connect the tube assemblies. Install the breather assembly and primary and secondary fuel filters to the base, and reconnect the fuel supply line. Reinstall the oil level guide tube and install the oil level gauge. Reinstall the oil cooler (if equipped) and oil filter base to the engine block. Make certain to install the oil filter bypass valve if so equipped. Connect all tube assemblies to the oil filter base and install a new oil filter.

Power Steering Pump and Water Pump

Remount the power steering pump to the block if the engine is equipped and reconnect the hose assemblies to the pump inlet and outlet ports. Reinstall the water pump using a new gasket. Make sure that sealant is applied to any bolts that pass through a water jacket.

Air Compressor

A hoist may be needed to lift and position the air compressor for mounting to the engine. Use a new mounting gasket and torque all bolts to specification. Carefully reconnect all hose assemblies to the compressor. These hose connections are often held by clips.

Final Assembly

Final engine assembly involves remounting components such as the starter motor, alternator, fan drive, and batteries. Remount the belt tensioner and install all drive belts. Reinstall other items such as the thermostat and fuel pressure regulating valve.

Rebuilt Engine Run-In Procedures

After a complete overhaul or a major repair involving the installation of piston rings, pistons, cylinder liners/sleeves, or bearings, the engine must be properly **run-in** before being released into service. The durability and service life of a rebuilt engine is directly affected by the quality of the run-in procedure.

Engine Preparation

An engine can be run-in using one of three methods: an engine dynamometer, a chassis dynamometer, or a highway run-in. The exact procedure for run-in will vary depending on the type of method used, but in all cases the rebuilt engine must be properly prepared before starting for the first time. The following is a typical engine preparation procedure for run-in. Always follow the exact procedure outlined in the engine service manual.

Lubrication System

The oil film on the engine's rotating parts and bearings may not be sufficient for proper lubrication when the engine is started for the first time, therefore, the engine should be prelubricated using a pressure lubricator, **Figure 28-28.** If a pressure lubricator is not available, hold the fuel injection pump or governor in the stop or no-fuel position and crank the engine until oil pressure is indicated. Connect the pressure lubricator to the main oil gallery, and prime the engine. Remove the oil level dipstick and check the crankcase level. If needed, add oil to reach the full mark, but do not overfill.

Turbocharger

Add about one pint of clean engine oil in the turbocharger inlet line to lubricate the bearings for initial startup. To make sure the oil pump is providing oil to the

Figure 28-28. Pressure lubrication of a rebuilt diesel engine. (Caterpillar Inc.)

engine, disconnect the wires leading to the transfer pump. Disable the fuel system and crank the engine for 30 seconds with 2 minute cool-down periods until pressure registers on the oil pressure gauge and reconnect the wires.

Air Intake and Cooling Systems

Check the integrity of the air intake system and inspect all piping, connections, and the air cleaner element. Replace components as required. Fill the cooling system and bleed any air from the engine.

Fuel System

Inspect all fuel lines and connections. Prime the system with clean diesel fuel and bleed all air from the lines. Next, check valve clearance and be sure injection pump to engine timing is set to engine specifications.

Final Checks

Walk around the engine and visually inspect for any external oil, fuel, or coolant leaks. Recheck the oil level after several minutes. If it has suddenly risen, coolant or fuel is probably leaking into the crankcase.

Engine Dynamometer Run-In

An **engine dynamometer** is a piece of test equipment that enables the rebuilder to fully load-test the overhauled engine before it is reinstalled in the vehicle, boat, or generator set location, **Figure 28-29.**

With an engine dynamometer, the engine can be fully checked and fine-tuned before it is placed into service. The engine is mounted in the stand and the fuel lines, cooling system hoses, exhaust system pipes, oil lines, and all available instrumentation connected. A typical dynamometer will include the following gauges and meters for monitoring run-in conditions:

❑ Fuel and oil pressure gauges.
❑ Coolant temperature gauge.
❑ Tachometer for engine rpm.
❑ Pyrometer for measuring exhaust gas temperature.

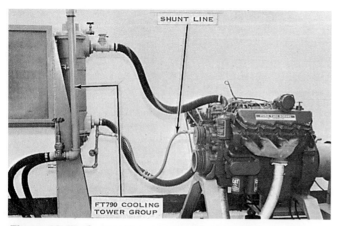

Figure 28-29. Connections to an engine dynamometer. (Caterpillar Inc.)

❑ Barometer for measuring air intake pressure.
❑ Intake air temperature gauges.

Most dynamometers will have one gauge upstream and one downstream of the aftercooler to indicate how well it is doing its job.

Be extremely observant during any run-in procedure to detect possible problems that may develop. Monitor all gauges and readings on the dynamometer, vehicle dash display, or engine control panel. Look for the following problems:

❑ Any unusual noises such as knocking, scraping, etc.
❑ Any significant drop in engine oil pressure.
❑ Any significant rise in coolant temperature above 200°F (93°C).
❑ Any significant rise in oil temperature that exceeds 240°F (116°C).
❑ Any exhaust temperature that exceeds the maximum acceptable limit for the specific engine involved.
❑ Any oil, coolant, fuel, or air inlet system leaks.

If the engine develops any of these problems, it must be immediately shut down. Investigate and correct any problem before continuing with the test.

Dynamometer Check

The following steps are typical of a dynamometer run-in procedure.

1. With the engine stop engaged, crank the engine until oil pressure registers on the gauge. Release the stop and start the engine.

2. Run the engine at 800-1000 rpm with no load for 3-5 minutes to check for proper oil pressure, unusual noises, leaks, etc.

3. Shut down the engine and check oil and coolant levels, refilling as needed.

4. Increase engine speed to 1600 rpm and set the dynamometer to 1/4-1/2 load for the engine being tested. Operate for 15 minutes at this rpm/load or until the thermostat opens and the oil temperature reaches 140°F (60°C) minimum.

5. Set the throttle at full load and adjust the dynamometer load to obtain the proper governed speed. Operate the engine at this speed/load for 30 minutes and record all data from dynamometer gauges and readouts.

6. With the throttle at full load, adjust the dynamometer to obtain peak torque. Operate the engine at this speed/load for 30 minutes.

Note: On heavy-duty truck engines that use dual-speed, variable-speed governors (such as Mack Maxidyne® engines), this should be performed with the throttle at full load and the engine at intermediate speed (1600 rpm) for 15 minutes, and then throttle at full load and engine at peak torque for 15 additional minutes.

7. Check the engine's high and low idle speed.

8. Retorque heads, adjust valves, tighten hose clamps, manifold nuts/bolts, etc.

9. Prepare the vehicle for chassis installation.

Engine Installation into Vehicle

Reinstallation is essentially the reverse of removal as covered in **Chapter 5.** Lift the engine using a spreader bar and lifter chain.

1. Install the hoist and unbolt the engine from the repair stand. Raise the engine slowly, taking care to keep plenty of room around the engine.

 Warning: When removing or installing an engine, serious personal injury or death can occur if an improper lifting method is used. If any part of the lifting apparatus fails, allow the engine to fall.

2. Carefully lower the engine into the engine compartment, making sure that it does not bind or damage related components. Support the engine with floor jacks under the engine compartment. Install all engine mounts.

3. Mate the engine to the transmission and install the engine mounting bolts. Install the bolts that hold the transmission bell housing to the engine flywheel housing.

4. Install the radiator, air conditioning condenser, and bumper, and support, if they were removed.

5. Reinstall the air cleaner-to-engine hose and air control linkage (if so equipped).

6. Install all electrical leads, fuel lines, linkages, oil lines, and coolant hoses. Refill the cooling system with the proper coolant/water mixture

7. Install the fan shroud, brackets, fan, and hardware.

8. Reinstall the aftercooler as described in the manufacturer's service manual.

9. Finally, reinstall the battery.

Chassis Dynamometer Operation

A *chassis dynamometer* allows for a controlled vehicle run-in without the need of a road test. The truck's drive axles are placed on a set of in-ground rollers and the vehicle is operated through a specific range of speeds and loads to monitor performance, **Figure 28-30.**

A chassis dynamometer must be properly calibrated. On vehicles with two bogie axles, the dynamometer must be adjusted so there is no more than a 5 mph (8 kph) difference between axle speeds during testing. On vehicles equipped with a power divider lockout control, the lockout should be engaged during testing. Vehicles without lockout control may require a specially fabricated lockout device. Finally, it may be necessary to disconnect the front pro-

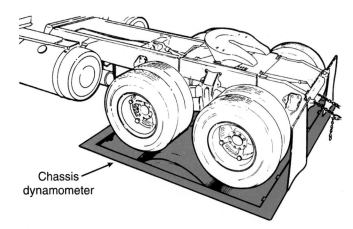

Chassis dynamometer

Figure 28-30. *Setup for chassis dynamometer engine run-in. (Mack Trucks, Inc.)*

peller shaft from the transfer case. Check the engine manual for these and other precautions prior to testing.

To perform the test, proceed as follows:

1. Check the engine oil and coolant levels. Check the gear oil level in the transmission and carrier. Check tire pressure and lug tightness. Inspect for shifted axle or loose drive lines and make certain the chassis rolls readily with the brakes released.

2. Start the engine and operate it for at least 15 minutes, preferably by a controlled road test as this will bring the transmission to operating temperature as well.

3. Place the vehicle on the chassis dynamometer, chock the front wheels, and install the safety chains. Be sure the governor is properly set to prevent the engine from overspeeding. Check and record no-load rpm at neutral and full throttle. Be sure the pump lever is wide open at full throttle. Unload the dynamometer, and record the mileage on the chassis and dynamometer odometers.

4. Hold full throttle and apply the dynamometer load slowly and evenly until the maximum rpm is reached. Allow indicators to stabilize and record wheel horsepower, rpm, and miles per hour.

5. Operate the vehicle on the dynamometer in direct gear with no load at 1600 rpm for approximately 15 minutes or until the thermostat opens and the oil temperature reaches 140°F (60°C) minimum.

6. Select both an rpm and gear range so that the dynamometer can operate for 15-30 minutes at or near full load governed speed. Record all applicable data after this run-in period.

 Note: In some cases, it may be necessary to perform the run-in in 5-10 minute increments to allow the rollers and tires to cool.

7. Start at the engine's governed idle speed and run the vehicle to full-load in direct gear through the entire speed range. Increase engine speed at 200 rpm inter-

vals for approximately 3 minutes in each speed range. Record all appropriate data.

8. Check the engine high and low idle.

9. Retorque heads, adjust valves, tighten hose clamps, manifold nuts/bolts, etc., as needed.

10. Ready the vehicle for road testing.

Highway Run-In Procedures

When a dynamometer or chassis dynamometer is not available, the engine run-in must be performed under controlled highway driving conditions.

1. With the engine stop engaged, crank the engine until an oil pressure reading registers on the gauge. Release the stop and start the engine.

2. Run the engine at 800-1000 rpm with no load for 3-5 minutes to check for proper oil pressure, unusual noises, leaks, etc.

3. Shut down the engine and check the oil and coolant levels, refilling as needed.

4. Use a loaded trailer or body and operate the truck through all gear ranges for approximately 30 minutes with the rpm not exceeding 1800 except as needed to make gear changes.

5. Enter a limited access highway and operate the truck at or near governed speed for 30-60 minutes, monitoring all instrumentation for signs of problems. Pull over and stop the engine at any sign of pressure or temperature problems, noises, grinding, etc.

6. Check the engine high and low idle speed.

7. Retorque heads, adjust valves, tighten hose clamps, manifold nuts/bolts, etc., as needed.

Summary

Keep the working area, tools, equipment, and engine components clean at all times during the reassembly process. If parts have been stored prior to reassembly, reinspect the parts for rust and cleanliness, and clean if needed. Refer to the engine service manual for proper nut, bolt, and stud torque, clearances between components, and any end play or other specifications. Apply a sealer to any plug, fitting, or fastener (including studs), that intersects with a through hole and comes in contact with lubricating oil, coolant, or diesel fuel.

The exact assembly sequence will vary slightly from engine to engine. Always follow manufacturer service manual procedures and recommendations. Always use new gaskets and seals. Lubricate press-fitted parts using vegetable-base lubricant before assembly.

Always use the installation and protective tools specified by the engine manufacturer. The durability and service life of a rebuilt engine is directly affected by the quality of the run-in procedure. An engine can be run-in using one of three methods: an engine dynamometer, a chassis dynamometer, or a highway run-in. The exact procedure for run-in will vary depending on the type of method used.

Because the lubricating oil film on the rotating parts and engine bearings may not be sufficient for proper lubrication when the engine is first started, the engine should be prelubricated using a pressure lubricator filled with the recommended oil.

Important Terms

Wiring harness straps	Timing gear cover
Lifting plates	Freeze plug
Dial indicator	Air inlet heater
End play	Run-in
Thickness gauge	Engine dynamometer
Oil seal	Chassis dynamometer

Review Questions—Chapter 28

Do not write in this text. Place your answers on a separate sheet of paper.

1. A _____ should always be replaced, even if it shows no sign of wear.

2. Install all piston rings in the piston grooves using a _____.

3. When installing the flywheel, a _____ and _____ should be used.

4. List the three major methods of performing engine run-in.

5. Engines should always be prelubricated using a _____ before being started for the first time after rebuilding.

6. What are the proper fuel system checks prior to beginning the engine run-in procedure?

7. The _____ should be lubricated with about one pint of _____ prior to initial startup after rebuilding.

8. If the oil level suddenly rises after sitting for several minutes, _____ or ____ is leaking into the crankcase.

9. List at least five conditions that require the engine dynamometer run-in procedure be stopped immediately.

10. On some vehicles, it may be necessary to disconnect the _____ from the transfer case.

ASE-Type Questions

1. Technician A says keeping the work area and components clean is vital to a successful overhaul and engine reassembly. Technician B says that slight amounts of dust entering the engine will flushed out in the lube oil and cause no major problem. Who is right?
   (A) A only.
   (B) B only.
   (C) Both A & B.
   (D) Neither A nor B.

2. During reassembly, Technician A says you should use a clearance setting based on past experience with other engine models by the same manufacturer. Technician B says you should look the clearance up in the service manual for that specific engine. Who is right?
   (A) A only.
   (B) B only.
   (C) Both A & B.
   (D) Neither A nor B.

3. During reassembly, Technician A says you should lubricate all press fit parts using heavy-duty engine oil. Technician B says you should use no lubrication at all on these parts. Who is right?
   (A) A only.
   (B) B only.
   (C) Both A & B.
   (D) Neither A nor B.

4. Technician A says any plug, fitting, or fastener (including studs), that intersects with a through hole and comes in contact with lubricating oil, coolant, or diesel fuel should have a sealer applied to it. Technician B says these components should have a locking compound applied to them before installing. Who is right?
   (A) A only.
   (B) B only.
   (C) Both A & B.
   (D) Neither A nor B.

5. Wear surfaces of mating parts should be coated with _____ before reassembly.
   (A) vegetable-based oil
   (B) clean engine oil
   (C) diesel fuel
   (D) nothing

6. All of the following should be done when installing the injection pump, EXCEPT:
   (A) reuse the old O-ring.
   (B) ensure the pump timing marks are centered with the port.
   (C) clean injection pump camshaft.
   (D) bleed the fuel system.

7. All of the following final checks should be made before initial engine start-up, EXCEPT:
   (A) walk around the engine and inspect for external leaks.
   (B) recheck the oil level after several minutes.
   (C) connect all instrumentation.
   (D) adjust valve lash to specifications.

8. During an engine dynamometer test, Technician A begins by operating the engine at 800-1000 rpm with no load for 3-5 minutes to check for proper oil pressure, unusual noises, leaks, etc. Technician B retorques the cylinder head, adjusts valves, and tightens hose clamps, manifold nuts/bolts, etc., after the test is completed. Who is following proper procedure?
   (A) A only.
   (B) B only.
   (C) Both A & B.
   (D) Neither A nor B.

9. On vehicles with two bogie axles, the chassis dynamometer must be adjusted so there is typically no more than a __ mile per hour difference between axle speeds during testing.
   (A) 2
   (B) 5
   (C) 7
   (D) 0, both must turn at the same speed.

10. On a typical road run-in, use a loaded trailer or body and begin by operating the truck through all gear ranges for approximately 30 minutes with rpm not exceeding _____ except as needed to make gear changes.
    (A) 1000 rpm
    (B) 1200 rpm
    (C) 1500 rpm
    (D) 1800 rpm

Troubleshooting is an easy task, if you follow a logical routine and use information contained in service manuals. (Lloyd Wolf, VICA)

Chapter 29

Preventive Maintenance and Troubleshooting

After studying this chapter, you will be able to:
- ❑ Explain the importance of setting up a workable preventive maintenance program and keeping accurate maintenance records.
- ❑ Describe how new diesel engine technology has affected PM procedures.
- ❑ Outline the types of tests and checks possible with modern oil analysis and the problems the results may indicate.
- ❑ Describe the steps in taking a proper oil sample suitable for analysis.
- ❑ Name the number one cause of diesel engine failure and steps to prevent it.
- ❑ Outline the procedures for winterizing the engine's coolant, electrical, starting, and fuel systems.
- ❑ Name the qualities of a successful diesel engine troubleshooting technician.
- ❑ Outline the general procedure for troubleshooting engines with electronic control systems.
- ❑ Perform the following troubleshooting tests: locating a misfiring cylinder, cylinder compression check, manometer pressure checks including crankcase pressure, exhaust back pressure, air inlet pressure, air leakage checks, and air box and/or turbocharger boost pressure checks.
- ❑ Perform primary engine checks.
- ❑ Use troubleshooting charts and service information to pinpoint the source of engine problems.
- ❑ Properly prepare a diesel engine for short- and long-term storage.

A well-planned and executed preventive maintenance program is probably the most important step in extending engine life and reducing the number of breakdowns in service. A good PM program reduces overall maintenance costs by eliminating repairs and breakdowns caused by dirt, water, and contaminants entering the engine. Engines run more efficiently with greater fuel economy and reduced emissions. Preventive maintenance is absolutely necessary for engines to meet today's stringent emission standards.

Setting Up a Preventive Maintenance (PM) Program

No one **preventive maintenance (PM) program** can apply to all operations. Determining the right PM service interval to maximize vehicle performance and engine life while minimizing down time requires a combination of art and technology. Creating the PM schedule is usually accomplished by merging the service manual recommendations with the unique operating conditions of the fleet. A typical engine manufacturer's PM schedule is illustrated in **Figure 29-1.**

The cooperation of all service technicians and driver/operators is needed to ensure all servicing and inspections are performed correctly, on time, and that accurate records are maintained. Inspections must be more than filling in a little paperwork; performing the fluid level checks and careful visual inspections is required. A quick glance under the hood will not do the job.

Intervals for maintenance procedures are based on fuel consumption, distance driven, or service hours, whichever comes first. The engine manufacturer's maintenance schedule assumes normal operating conditions. Severe operating conditions may require more frequent maintenance. While all scheduled maintenance is important, particular emphasis must be placed on emission-related components, air cleaner, oil, oil filter, fuel quality, and fuel filter maintenance.

Daily Walk-Around Inspection

Perform a thorough inspection prior to starting the engine. Check for oil or coolant leaks in the radiator, air-to-air aftercooler, water pump, and related hoses and lines.

DAILY	DAILY	PM 1	PM 2	PAGE
WALK-AROUND INSPECTION–LEAKS AND LOOSE CONNECTIONS	I			90
ENGINE CRANKCASE OIL LEVEL	I			92
COOLING SYSTEM COOLANT LEVEL	I			93
AIR CLEANER INDICATOR (If Equipped) (NOTE 2)	C			94
WATER SEPARATOR (If Equipped) (NOTE 2)	D			95

PM LEVEL 1–EVERY 3800 L (1000 U.S. GALLON) of FUEL or 250 HOURS (NOTE 3*)	DAILY	PM 1	PM 2	PAGE
SCHEDULED OIL SAMPLING S•O•S		S		96
ENGINE OIL AND FILTER(S) (NOTE 1)		R	R	97
FUEL SYSTEM–WATER & SEDIMENT (Tank)		D	D	102
COOLING SYSTEM–SUPPLEMENTAL COOLANT ADDITIVE (NOTE 4)		T	T	106
ATAAC AFTERCOOLER (NOTE 2)		I		108
RADIATOR FINS (NOTE 2)		I		110
CRANKCASE BREATHER (every other oil change)		CL	CL	111
ALTERNATOR, FAN & ACCESSORY DRIVE BELTS (NOTE 2)		I		112
HOSES & CLAMPS		I		114
FAN DRIVE BEARING (If Equipped) (NOTE 2)		L	L	115
BATTERY (NOTE 2)		CL,C	CL,C	116
AIR COMPRESSOR AIR FILTER (NOTE 2)		R	R	117

PM LEVEL 1–BETWEEN FIRST AND FOURTH OIL CHANGE	DAILY	PM 1	PM 2	PAGE
VALVE LASH		check/adjust		118

PM LEVEL 2–EVERY 28 500 L (7,500 U.S. GALLON) of FUEL or 80 000 km (48,000 MILES) or 2000 HOURS or 2 YEARS* (NOTE 3)	DAILY	PM 1	PM 2	PAGE
ENGINE			CL	119
TURBOCHARGER			I,C	120
CRANKSHAFT VIBRATION DAMPER			I,C	122
ENGINE MOUNTS			I,C	123
AIR CLEANER ELEMENT (ANNUAL)			R	124
COOLING SYSTEM–COOLANT (NOTE 2,4)		CL,R	CL,R	124
THERMOSTAT		I,R	I,R	128

PM LEVEL 3–EVERY 56 850 L (15,000 U.S. GALLON) of FUEL or 160 000 km (96,000 MILES) or 3000 HOURS or 4 YEARS* (NOTE 3)	PM 1	PM 2	PM 3	PAGE
AIR COMPRESSOR		(NOTE 2)	I	130
TURBOCHARGER			I	131
WATER PUMP			I	132
WATER PUMP BELT TIGHTENER & FAN DRIVE			I,C	133
STARTING MOTOR & ALTERNATOR (NOTE 2)			I	134
VALVE LASH			C,A	118

EVERY 240 000 km (150,000 MILES) OR TWO YEARS	PM 1	PM 2	PM 3	PAGE
COOLING SYSTEM–FOR LLCA ONLY	ADD EXTENDER			135

EVERY 480 000 km (300,000 MILES) OR FIVE YEARS (MAXIMUM)	PM 1	PM 2	PM 3	PAGE
COOLING SYSTEM–FOR LLCA ONLY	CLEAN, FLUSH, REPLACE			136

EVERY 113 000 L (30,000 U.S. GALLON) OF FUEL	PM 1	PM 2	PM 3	PAGE
OVERHAUL				139

NOTE:– *FIRST PERFORM PREVIOUS SERVICE HOUR ITEMS.

NOTE 1 – REFER TO THE OIL CHANGE INTERVAL GRAPH FOR THE CORRECT OIL CHANGE INTERVAL FOR YOUR ENGINE.

NOTE 2 – REFER TO THE SERVICE MANUAL INFORMATION FOR THE TYPE OR MODEL INSTALLED ON YOUR ENGINE AND FOLLOW THE MANUFACTURER'S RECOMMENDATIONS.

NOTE 3 – UNDER SEVERE OR DUSTY CONDITIONS, OR WHERE AVERAGE OPERATING SPEEDS ARE LESS THAN 32km/h (20 mph), SUCH AS INNER CITY GARBAGE COMPACTORS, DELIVERY TRUCKS, OR DRILL RIGS, THE MAINTENANCE INTERVAL SHOULD BE REDUCED BY ONE-HALF THE DISTANCE INTERVAL.

NOTE 4 – THIS COOLING SYSTEM MAINTENANCE IS FOR ENGINES USING STANDARD TYPE ANTIFREEZE ONLY. LONG LIFE COOLANT/ANTIFREEZE (LLCA) MAINTENANCE IS PERFORMED AT 240 000 km (150,000 MILES) OR TWO YEARS AND 480 000 km (300,000 MILES) OR FIVE YEARS MAXIMUM INTERVALS ONLY.

CODES

R – REPLACE   C – CHECK   A – ADJUST   CL – CLEAN   D – DRAIN   L – LUBRICATE   S – SAMPLE

I – INSPECT, SERVICE, CORRECT, REPLACE AS NECESSARY   T – TEST/REPLENISH AS necessary

Figure 29-1. Typical engine manufacturer's maintenance schedule. (Caterpillar Inc.)

On some water pumps, it is normal for a small amount of leakage to occur as the engine cools and parts contract. However, excessive coolant leakage at the pump may indicate a leaking water pump seal.

Inspect for frayed or cracked fan and accessory drive belts, loose connections, and trash build-up. Remove trash build-up and repair leaking or worn components. Check fluid levels and wipe all fittings, caps and plugs before performing maintenance to reduce the possibility of contaminating internal engine areas.

Inspect air inlet system hoses and elbows for cracks and loose clamps. A hose failure or system leak will allow enough dirt and debris to cause engine wear. Check the lubrication system for leaks at the front and rear crankshaft seals, oil pan, oil filters, and valve covers. Check the fuel system for leaks, loose fuel line clamps and fittings, and loose or worn hoses. Do not overtighten fuel line clamps. Excessive torque can cause the clamps to deform, reducing clamping force. The result is fuel line leakage and eventual failure.

Check engine wiring and harnesses for loose connections and worn or frayed wires. All engine grounds must have good connections. Be sure all guards and harness locks are in place. Check the engine oil and coolant levels. Check the oil level with the engine off.

> **Note: Some manufacturers are now using a long-life coolant in their engines. Do not mix standard and long-life coolant. Check the service manual to see if the engine uses long-life coolant.**

Inspect the air cleaner service indicator if the engine is so equipped, **Figure 29-2.** This type of indicator visually displays the restriction of air flow through the air cleaner element. Clean or replace the air cleaner element when the yellow diaphragm enters the red zone, the red piston locks in the visible position or 25″ of water (6.25 kPa) registers on the restriction gauge. If the fuel system is equipped with a water separator, depress the spring-actuated drain valve until all water is drained from the separator.

Level One Maintenance

Level one maintenance usually includes such procedures as changing oil and oil filters, draining water from the fuel tank, changing fuel filters, testing coolant additive levels, and inspection and cleaning of the air-to-air aftercooler. Tips on performing these procedures include:

❑ Drain the old oil with the engine at normal operating temperature, engine off, and the vehicle parked on a level surface.

❑ Replace the filters at every oil change. The replacement intervals are affected by such factors as the fuel's sulfur content (high), grade of oil used, and the capacity of the lubrication system (sump and filters). **Figure 29-3** illustrates an engine manufacturer's oil change interval graph.

❑ Cut open the old filter element and inspect for metal debris. If an excess amount of debris is present, have an oil analysis performed.

❑ If equipped, use the rotation index marks on the filters as a guide for proper tightening.

❑ If the engine is equipped with an auxiliary oil filter or system, extra oil must be added when filling the crankcase, however, do not overfill. Do not load or accelerate the engine until oil pressure registers on the in-cab gauge. Stop the engine and allow the oil to drain into the sump for five minutes. Inspect for leaks and recheck the oil level.

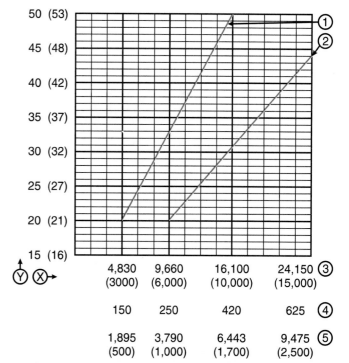

Y = Sump and filter(s) oil capacity in liters (quarts.)
X = Distance, time, or fuel consumed between oil changes.
1 = EITHER: CG-4/CF-4 oils used with 0.5 to 1.0 percent fuel
      sulfur, OR, CE oils used with 0.5 percent or less fuel sulfur.
2 = CG-4/CF-4 oils used with 0.5 percent or less fuel sulfur.
3 = Kilometers (miles) between oil changes.
4 = Service hours between oil changes.
5 = Liters (US gallons) of fuel consumed between oil changes.

Figure 29-2. Air cleaner indicator used to indicate if the air cleaner element needs to be replaced. (Caterpillar Inc.)

Figure 29-3. Graph indicates oil change intervals, dependent on oil quality, distance, time, and engine oil capacity. (Caterpillar Inc.)

❑ Drain water and sediment from the fuel tanks at every oil change interval (more frequently in high humidity conditions). Prime the fuel system with the hand pump, **Figure 29-4,** before attempting to start the engine. Never crank the engine for more than 30 seconds without allowing for starter motor cool-down.

The coolant system must be tested for proper coolant additive concentration (usually 3-6%). Clean and flush the cooling system before the recommended maintenance interval if the coolant is heavily contaminated, the engine overheats frequently, foaming occurs in the radiator, or if incorrect additives or coolant products have been added. Make sure the radiator cap and gasket are in good condition.

Remove and clean the crankcase breather, **Figure 29-5.** Use non-flammable solvent and allow the breather to dry before reinstallation. A clogged breather can result in excessive crankcase pressure and premature crankshaft seal leakage.

Inspect drive belt condition and check belt tension on engines not equipped with an automatic belt tensioner, **Figure 29-6.** When correctly adjusted, the belt should deflect 3/8-5/8 inches (9-15 mm). If new belts are installed, check the belt tension after operating the engine for 30 minutes.

Inspect all hoses for leaks due to cracking, softness and loose clamps. Be sure the end fittings are not damaged, leaking, or displaced. Replace the hose if the wire reinforcing is exposed or the outer covering is ballooning in spots. Heating and cool-down can loosen hose clamps over time. Use of constant torque hose clamps will help

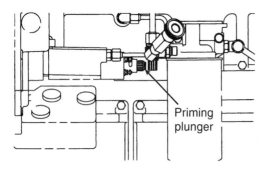

Figure 29-4. *Manual hand primer used for priming the fuel system after filter service. (Caterpillar Inc.)*

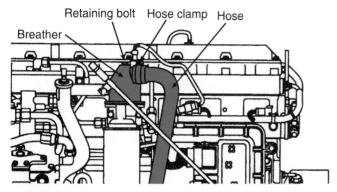

Figure 29-5. *The crankcase breather requires regular cleaning and maintenance. (Caterpillar Inc.)*

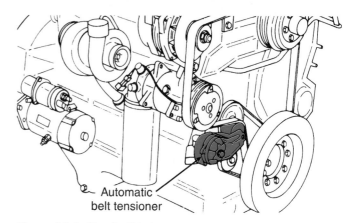

Figure 29-6. *Check all belts and adjust belt tension as needed. (Caterpillar Inc.)*

eliminate this problem. Inspect the fan drive pulley assembly and check bearing end play and radial play against specifications.

Servicing the Engine Air Inlet System

Reduced power, response, or an increase in exhaust temperature may indicate an air leak. Do not operate the engine with an air leak as dirt may enter the air stream. Clean the front of the aftercooler using soapy water and a stainless steel brush to remove insects, dirt, and other debris. Remove dirt from between the aftercooler and radiator cores. Inspect the air ducts and gaskets at each oil change interval. Use constant torque hose clamps to secure ducting. Check all welds and mounting brackets for cracks and damage. Use compressed air to clean core blockage due to dirt and debris.

If the turbocharger has failed or developed an oil leak, the aftercooler core should be removed and cleaned. Back flush with a solvent to remove oil and finish with a non-caustic soap solution. Dry the core with compressed air, blowing in the reverse direction of normal flow. Perform a valve lash adjustment, if specified in the PM schedule.

Level Two Maintenance

Level two maintenance is performed at more extended intervals such as after 7500 gallons of fuel consumption or 2000 service hours. These procedures commonly include steam cleaning the engine, inspection and service of the turbocharger, vibration damper, and engine mounts, replacement of the air cleaner element, and cooling system flush.

Grease and oil accumulation on an engine presents a fire hazard. Steam cleaning the engine at level two intervals eliminates this hazard. It is also easier to detect fluid leaks on a clean engine. A clean engine also transfers heat more efficiently.

Turbocharger bearing failures can result in a large quantity of oil entering the air inlet and exhaust systems. Loss of engine lubricant can result in serious engine damage. Regularly scheduled turbocharger inspections will reduce the chance of a major bearing failure. Remove the exhaust outlet and air inlet piping from the turbocharger and visually check for oil leaks. Turn the turbine and compressor wheel by hand; the assembly should rotate freely. The turbine and compressor wheels should not contact the housing. Check the compressor wheel for cleanliness. If only the blade side of the wheel is dirty, dirt and/or moisture is passing through the air filtering system. If only the back side of the wheel is dirty, the turbo oil seal may be leaking. This can be caused by running the engine at idle for extended periods or by an air inlet restriction or plugged air element.

Nicks in the wheel indicate dirt is entering the system. Clean the turbocharger with standard shop solvents and a soft bristle brush. Check the end play and bearing clearance on the turbine wheel and shaft. Finally, if the unit is equipped with a wastegate boost line, be sure it is connected properly. A loose boost line will raise heat rejection, increase turbocharger speed, and peak cylinder pressure, which will reduce engine performance.

If the engine is equipped with a viscous vibration damper, inspect it for dents, cracks, and fluid leakage. Check engine mounts for deterioration and retorque mounting bolts to specifications. Loose mounts can result in vibration and harshness.

Replace the air cleaner elements at the recommended intervals. The intervals may be decreased in extremely dusty or dirty operating conditions, such as off-road construction sites. Even in extremely clean operating environments, the element should be replace annually.

Coolant System Service

Level two maintenance includes cleaning or *flushing* the coolant system to remove rust, corrosion by-products, solder bloom, scale, and other deposits. With the coolant removed, the thermostat may be replaced. Some service intervals call for the thermostat to be replaced every 3000 hours or two years of operation. Test the radiator cap opening pressure before flushing the system, **Figure 29-7.**

A typical cooling system flush involves running the engine to reach normal operating temperature, draining the existing coolant, and refilling with the fast acting flushing solution. The engine is then operated at normal operating temperature, usually 180°F (82°C) for as little as 5 minutes or as long as 90 minutes, depending on the type of flushing solution used. Once the compound has had time to loosen

scale and dirt, the radiator is drained and refilled with fresh coolant and water mixed to the proper strength. If core clogging (indicated by cool spots on the core) is evident, the system may require pressure flushing using a special nozzle powered by compressed air. If the core is extremely clogged, it may need to be removed and cleaned.

Note: Some slow-acting coolant system cleaners are designed to remain in the cooling system for up to 120 hours of operation before they are drained. With some fast-acting cleaners, the manufacturer may state that the vehicle should not be driven during the cleaning process. Always read the direction of the flushing/cleaning product you are using. If the engine has aluminum parts, do not use tri-sodium phosphate cleaners.

Once the cooling system and radiator are cleaned and serviced, fill the radiator with antifreeze/coolant mixed to the proper strength. Normally, a 50/50 mix of antifreeze and water provides protection against freezing to -34°F (-37°C). **Figure 29-8** lists typical antifreeze concentrations and protection levels.

Do not use water that is classified as hard or has a high mineral content. Ideally, distilled or deionized water should be used to mix the antifreeze solution. If potable water must be used, its mineral content should be analyzed by a private laboratory or an agricultural agent. Hardness should not exceed 300 parts per million (ppm). Sulfur and chlorine content must not exceed 100 ppm.

Most coolants contain inhibitors and additives that resist the formation of scale within the radiator. However, supplemental coolant additives may be required. If supplemental additives are required, they are mixed in as part of the coolant/water mix. The concentration of the scale inhibitor can be checked using a simple chemical litmus test paper. Dipping the paper in a coolant sample triggers a color change in the paper, which is then matched to a color chart keyed to protection levels. Check additive concentration at each oil change and at the intervals recommended for additive charging.

Add coolant to bring the system to the proper level. Start and run the engine with the filler cap removed. Allow the coolant to warm, the thermostat to open and the

Figure 29-7. Test the radiator cap before flushing the cooling system. Replace the cap if it does not pass this test. (Kubota)

Antifreeze Concentrations (Glycol)	
Protection Temperature	**Concentration**
Protection to -5°F (-15°C)	30% antifreeze and 70% water
Protection to -10°F (-23°C)	40% antifreeze and 60% water
Protection to -34°F (-37°C)	50% antifreeze and 50% water
Protection to -60°F (-51°C)	60% antifreeze and 40% water

Figure 29-8. Typical antifreeze/coolant concentrations. (Caterpillar Inc.)

coolant level to stabilize. Add coolant mixture as needed to bring the system to the proper level.

Used coolant should be recycled or disposed of safely. Follow the engine manufacturer's recommendations for recycling coolant. Some engine manufacturers specify only fully-distilled coolant if recycled coolant is used.

Level Three Maintenance

Level three maintenance tasks may include such items as inspection and rebuilding of the engine's turbocharger and air compressor (if equipped), replacement of the water pump seal, and inspection of the fan drive assemblies, **Figure 29-9.** Check the starter motor and alternator for unusual noise, loose fasteners, and noisy or loose pulley bearings.

Records and Forms

Neat and accurate **documentation** is essential to the success of any preventive maintenance program. Service technicians must work with **PM forms, repair orders, vehicle files, oil analysis reports, major component histories,** and other records. These records are important for several reasons. No maintenance manager could develop a systematic, preplanned PM program without documentation. Permanent records are invaluable from a performance standpoint as well. By studying the types of repairs performed and their frequency, managers and technicians can spot trends and tendencies and take corrective action. They can also confirm components and service procedures that have performed satisfactorily.

From a safety and emission standpoint, accurate records and documentation of service and testing are often required by the Department of Transportation and other government agencies. In the event a breakdown or failure results in an accident, a service file contains written proof that all maintenance and required inspections were

performed. This record of repairs may be invaluable in the defense of legal action.

A preventive maintenance form, such as the one illustrated in **Figure 29-10** provides the service technician with an orderly listing of items to be checked, inspected, serviced, or replaced. Although space limitations on the form may only allow for general statements, such as "inspect cooling system", technicians must be aware of all steps involved in this inspection. They may include pressure testing the system for leaks, checking thermostat operation, and other tasks. The service manager or shop foreman should review all inspection and maintenance procedures with the service technicians and perform periodic checks to see that all work is being carried out in a consistent manner.

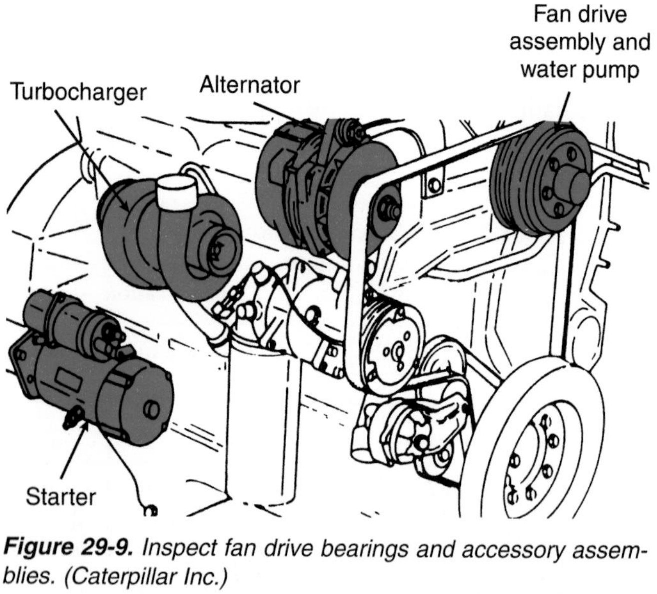

Oil Changes

In the past decade, the "one-million mile" engine has become a reality, and its arrival at the fleet level has created new maintenance demands and benefits. The need to meet increasingly strict emission regulations has accelerated the natural evolution of design changes in diesel engines.

For example, engine manufacturers have modified the design of their engines to meet emission standards by placing the rings higher on the pistons. This adjustment improves the air-fuel mix at the top of the bowl. This increases power and burns fuel more efficiently, giving the added benefit of better fuel economy. However, this change in ring location raises the temperature in the ring area, placing greater demands on the lubricating oil. This means that **oil change** intervals must come more often.

Some manufacturers lists oil change intervals based on fuel usage, which translates directly into miles driven for most long-haul fleet operations. A 15,000 mile (24 000 km) oil change interval on new engines is now quite common. Certain conditions will alter oil change intervals. For example, when a bypass filter is used, the added capacity may allow the change interval to be increased. If a high sulfur diesel fuel (.05% or more) is being burned, the interval must be decreased.

Oil Analysis

Oil consumption in modern engines designed to meet emission standards has greatly decreased. With less oil being burned, less oil is being replaced, and the level of oil contamination in the sump increases. This is why it is critical to change the oil at the established interval. Extended oil change intervals will lead to high soot levels and possible premature engine failure.

Regular laboratory **oil analysis** is a highly useful diagnostic tool. Oil analysis allows lubricating oil and filters to be changed when they have exhausted their effectiveness; not before or after. Impending failures can be detected in the early stages and engines can be scheduled for proper maintenance without unexpected and excessive down time.

Figure 29-9. Inspect fan drive bearings and accessory assemblies. (Caterpillar Inc.)

ENGINE DIAGNOSTICS
INTERNATIONAL® DTI/DTA/DT-466C
DIESEL ENGINES WITH R. BOSCH MODEL MW
INJECTION PUMP
ENGINE PERFORMANCE ANALYSIS GUIDE

REGION			DEALER		ALTITUDE	AMBIENT AIR TEMP.	DATE
CHASSIS MODEL	ENGINE MODEL	ENGINE HP	RATED SPEED	COMPLETE ENGINE SERIAL NUMBER	TRUCK MILEAGE	HOURS (if equipped)
PUMP SERIAL NUMBER 1ST CHECK	PUMP SERIAL NUMBER 2ND CHECK	VIN	MILEAGE ON ENGINE	HOURS (if equipped)

TESTS

1 SUFFICIENT CLEAN FUEL
FREE OF WATER – ICING & CLOUDING. CORRECT GRADE OF FUEL.
METHOD   1st CHECK   2nd CHECK
VISUAL CHECK ☐

2 EXTERNAL LEAKAGE
☐ FUEL   ☐ OIL   ☐ AIR   ■ ☐ COOLANT
RECORD LOCATION OF LEAKS IN "COMMENTS"
METHOD   1st CHECK   2nd CHECK
VISUAL CHECK ☐
PRESSURE CHECK ☐

ACCELERATOR LINKAGE
3
A. ADJUSTED TO OBTAIN OVERRIDE AT FULL PEDAL DEPRESSION
B. ADJUSTED SO LINKAGE HITS HIGH IDLE STOP.
C. ADJUSTED SO LINKAGE RETURNS TO LOW IDLE.
See Illustration – Reverse Side
METHOD   1st CHECK   2nd CHECK
VISUAL CHECK ☐

4 SHUTOFF CABLE OR ELECTRIC SHUTOFF See Illustration – Reverse Side
ADJUSTED TO ALLOW FULL RUN POSITION
METHOD   1st CHECK   2nd CHECK
VISUAL CHECK ☐

5 LOW IDLE (RPM)*
A. MANUAL TRANSMISSION - NEUTRAL
B. AUTOMATIC TRANSMISSION - DRIVE
INSTRUMENT   GUIDELINE DATA†   1st CHECK   2nd CHECK
MASTER
TACHOMETER

6 HIGH IDLE RPM (NO LOAD)*
THROTTLE IN OVERRIDE POSITION AND AT MAXIMUM (HIGH IDLE) STOP
See Illustration – Reverse Side.
INSTRUMENT   GUIDELINE DATA†   1st CHECK   2nd CHECK
MASTER
TACHOMETER

7 AIR CLEANER MAXIMUM RESTRICTION*
A. MEASURE AT HIGH IDLE (NO LOAD).
B. VEHICLES WITH DUAL ELEMENT AIR CLEANERS:
VISUALLY INSPECT THE RESTRICTION INDICATOR
INSTRUMENT   GUIDELINE DATA†   1st CHECK   2nd CHECK
WATER MANOMETER
OR MAGNAHELIC
GAUGE
*ENGINE MUST BE AT NORMAL OPERATING TEMPERATURE

COMMENTS

TESTS

8 TRANSFER PUMP PRESSURE (MEASURE AT HIGH IDLE) (NO LOAD)*
A. 1. HEADER W/VENT PLUG, TEST @ FINAL FUEL FILTER VENT.
2. HEADER W/O VENT PLUG, TEST @ FINAL FUEL FILTER OUTLET.
See Illustration – Reverse Side.
INSTRUMENT   GUIDELINE DATA†   1st CHECK   2nd CHECK
PRESSURE GAUGE
(180 PSI RANGE)
IF PRESSURE REMAINS LOW AFTER REPLACING FUEL FILTERS, PERFORM STEP B
B. TEST TRANSFER PUMP INLET RESTRICTION:
MEASURE AT HIGH IDLE (NO LOAD).
See Illustration – Reverse Side.
INSTRUMENT   GUIDELINE DATA   1st CHECK   2nd CHECK
VACUUM GAUGE   6 in Hg (max.)

9 INJECTION PUMP INITIAL TIMING (ENGINE STOPPED)
A. TIMING VERIFICATION.
See Illustration – Reverse Side.
INSTRUMENT   GUIDELINE DATA†   1st CHECK   2nd CHECK
VISUAL POINTER
INSPECTION

10 CRANKCASE PRESSURE* See Illustration — Reverse Side
1. Measure @ breather tube with orifice restrictor tool, SE-4039.
2. Measure @ High Idle (no load) RPM.
INSTRUMENT   GUIDELINE DATA†   1st CHECK   2nd CHECK
WATER MANOMETER
OR MAGNAHELIC
GAUGE

11 INTAKE MANIFOLD PRESSURE*
A. MEASURE AT FULL LOAD AND AT SPECIFIED SPEEDS.
B. SELECT TRANSMISSION GEAR TO ACHIEVE FULL LOAD.
See Illustration – Reverse Side.
INSTRUMENT   GUIDELINE DATA†   1st CHECK   2nd CHECK
PRESSURE GAUGE
(30 PSI RANGE)

READ THIS INSTRUCTION:
STOP!
IF GUIDELINE DATA WAS OBTAINED DURING THE FIRST 11 TESTS,
ENGINE OPERATION IS SATISFACTORY
THE FOLLOWING TESTS ARE **NOT** REQUIRED – STOP!

TESTS

12 ANEROID DIAPHRAGM
PRESSURE TEST DIAPHRAGM (20 PSI MAX – FOR 30 SECONDS)
NO LEAKAGE PERMISSIBLE.
See Illustration – Reverse Side.
INSTRUMENT   GUIDELINE DATA†   1st CHECK   2nd CHECK
D-100 PRESSURE KIT
(SE-2239)

13 TEST INJECTION NOZZLES
A. SPRAY CONDITION
B. VALVE OPENING PRESSURE
C. LEAKAGE
INSTRUMENT   GUIDELINE DATA†   1st CHECK   2nd CHECK
HAND TEST PUMP
(NOZZLE TESTER)
(SE-4045)

14 EXHAUST BACK PRESSURE
A. MEASURE AT POINT 3 TO 6 IN. AFTER ELBOW.
B. MEASURE AT FULL LOAD & RATED SPEED.
See Illustrations – Reverse Side.
INSTRUMENT   GUIDELINE DATA†   1st CHECK   2nd CHECK
WATER MANOMETER
OR MAGNAHELIC
GAUGE

15 MEASURE SMOKE INTENSITY*
MEASURE AT FULL LOAD & RATED SPEED (RPM)
INSTRUMENT   GUIDELINE DATA†   1st CHECK   2nd CHECK
SMOKE SAMPLING   Bosch No.
KIT (SE-2580)

16 INTAKE AND EXHAUST VALVE CLEARANCE
ENGINE OFF – HOT OR COLD
INSTRUMENT   GUIDELINE DATA†   1st CHECK   2nd CHECK
FEELER GAUGE

-IMPORTANT-
† All operating data is obtained from "PERFORMANCE DATA
GUIDELINE" for specified application listed in "DIESEL ENGINE
DIAGNOSTIC MANUAL" CGES-450 (or later).

SERVICE TECHNICIAN   DATE
    / /

DEALER SERVICE MANAGER   DATE
    / /

Figure 29-10. A detailed engine analysis guide for a specific engine model. (Navistar International Transportation Corp.)

Minor repairs can often prevent major overhauls. By pointing out a minor problem, such as the need for an adjustment or replacement of a single part, the knowledge gained from oil analysis reports can prevent eventual catastrophic damage. In addition, oil analysis taken after a major repair or overhaul can confirm the success of the work.

The most limited, and least effective use of oil analysis is a single sample taken on a onetime basis from an engine on the verge of failure. The next level of analysis is **selective sampling.** It involves selecting representative vehicles from the fleet for regular oil analysis and using the test results to establish oil and filter change intervals for the entire fleet. Samples are then taken on a planned basis to include all vehicles in the fleet, on a rotating basis.

Complete sampling involves testing all engines in the fleet on a regular basis, such as during scheduled oil changes or at an established mileage or hourly rate. Ideally, oil sampling should be scheduled frequently enough to provide a continuous, accurate record of engine conditions, performance, and to detect potential problems. This will vary depending on application and operating conditions. For example, equipment that operates in hostile environments such as quarries and mines will require more frequent sampling intervals than on-highway trucks. Engine oil analysis is recommended at every oil change, even on on-highway trucks that operate in a relatively clean environment.

Taking Samples

To obtain an accurate laboratory oil analysis, special care must be taken when drawing the oil sample. Samples are normally collected from one of three spots: the crankcase, sump, or reservoir drain; a petcock or other sampling valve installed upstream from the oil filter; or the oil dipstick tube or other service opening. Always take the sample from the same location on a particular engine.

1. Take the sample with the engine at or near operating temperature.
2. Always take samples before adding oil.
3. Only use the containers supplied by the laboratory to store the oil sample. Never use contaminated containers.
4. Always take oil samples in a manner prescribed by the laboratory and supply a new, unused sample of the same oil for comparison when requested.

Analysis Data

The results of a complete oil analysis will usually include:
- ❏ Viscosity.
- ❏ Fuel dilution.
- ❏ Diesel soot.
- ❏ High solids.
- ❏ Coolant contamination (Glycol).
- ❏ Water.
- ❏ Large particles analysis (in ppm).
- ❏ Silicon (dirt).
- ❏ Spectrographic analysis.

- ❏ Total base number.
- ❏ Total acid number.
- ❏ Wear metals (in ppm).
- ❏ Oxidation.
- ❏ Any special test(s) requested.

Viscosity

A decrease in **viscosity** can be caused by fuel dilution, additive or polymer shear, improper oil grade, or oil transfer. The results are reduced oil film strength, increased wear, reduced sealing ability, and reduced dispersant properties. An increase in viscosity can be the result of solid materials mixing with the oil, contamination with coolant or water, oxidation, or improper oil grade or oil transfer.

Fuel Dilution and Soot

Fuel oil entering the crankcase can be caused by poor combustion, leaking injectors, bad timing, or worn pistons and rings. The presence of fuel in quantities greater than 5% may lead to **fuel dilution** and rapid engine wear.

Measuring the amount of **fuel soot** in the oil is an excellent indicator of fuel efficiency. There is a direct relationship between the soot level in the oil and carbon deposits on pistons, lacquer formation, and oil oxidation rates. Soot particles can also neutralize some oil antiwear additives. Soot is the result of improper air-fuel ratio, which can be caused by a plugged air filter, a restricted air intake manifold, excessive idling or lugging, and low compression.

While soot accounts for most of the solids found in the oil, products of oxidation and particles of worn components and dirt may also be present in various levels. These abrasives can wear metal surfaces, thicken oil, and plug filters.

Glycol and Water

Glycol contamination from the cooling system is one of the leading causes of premature engine failure. Glycol increases the oil viscosity, forms sludge, and interferes with the additive's performance. Trace amounts of glycol may enter an engine during an overhaul or from minor leaks in the coolant system. If trace amounts are found, and a second analysis shows no decrease, a repair should be scheduled.

Water contaminating the lubricating oil may be the result of poor handling and storage, or the byproduct of combustion that forces its way pass the piston rings and into the crankcase. Water droplets in the oil will create hot spots and "welding" as they pass between bearings, journals, and other close tolerances. Combustion byproducts can also make this water highly acidic and corrosive. Water will also plug oil filters and deplete oil additives.

Dirt Contamination

Dirt entering the engine is the *number one cause* of engine failure. Engine manufacturers have stated that their engines will last for 10,000 hours or 10 ounces of dirt, whichever comes first. Dirt most often enters an engine through a leak in the air filter or in the air inlet ducting. Dirt can also enter during regular servicing, maintenance, and overhaul procedures if the technician does not practice clean servicing techniques.

Spectrographic Analysis

Metals found in the oil sample are identified through the use of a **spectrometer** which analyzes the light frequency of metal when it is burned. For example, when sulfur burns, it produces a light that is high in the yellow spectrum bands.

In oil analysis, an abnormal condition can be spotted by establishing a pattern of normal wear metal levels through a series of samples taken on a regular basis. A sharp rise in iron and chromium levels from the last analysis indicates accelerated cylinder liner and ring wear. A high lead reading in a diesel engine might indicate bearing wear, especially when coupled with high aluminum or copper readings.

Spectrometer testing can also analyze the oil additive package. For example, using different brands of oil in the same engine may result in different additives actually working against one another. The spectrometer can also detect the presence of silicon (dirt) in the oil, which may indicate an air cleaner or breather problem.

Oil Total Base and Acid Numbers

As mentioned in **Chapter 10,** *total base number (TBN)* is a measurement of an oil's ability to neutralize acids formed during combustion. Overextended oil drain intervals are the most common cause of low TBN readings. The *total acid number (TAN)* test is a companion to the TBN test and measures the acidity of an oil. High oil acidity may contribute to internal corrosion of parts, oil thickening, and accelerated wear patterns.

Wear Metal Ferrography

Wear metal is separated by *ferrography* from other particles in the oil and categorized by size, shape, and the ratio of small-to-large particles. The particle's shape helps identify its source, such as from normal rubbing wear, cutting wear, bearing wear, rolling contact wear, and fatigue wear. Particle size helps determine whether the wear is normal or accelerated and is a good indicator of problems in a suspect engine. The ratio of large-to-small particles indicates the engine's wear. The ratio should remain constant except during initial break-in and at the start of severe wear, prior to a catastrophic failure.

Oil Oxidation

Oil oxidation gives an indication of the oil's condition. Although most oils contain anti-oxidants, they are gradually depleted as hydrocarbons in the oil combine with oxygen. Heat, pressure, and the abundance of air speed up the oxidation process. Other common causes of

oxidation include extended oil drain intervals, incorrect oil for the application, excessive combustion blowby, and the use of high sulfur fuels. As oil oxidizes, corrosive acids can form and deposits may accumulate on critical engine parts, inhibiting operation and accelerating wear.

 Note: Oil analysis alone is not a substitute for an effective preventive maintenance program. Always use oil analysis along with visual inspections, knowledge of the engine's application, operating conditions, and maintenance history to make informed judgments regarding serviceability.

Winterizing Engines

Winterizing a diesel engine is a very important aspect of any preventive maintenance program. Areas of concern include:

- ❑ Cooling systems.
- ❑ Batteries and electrical systems.
- ❑ Starting systems.
- ❑ Fuel systems.

Winterizing maintenance must be scheduled in midautumn before the onset of extremely cold weather. Do not wait until the last minute to winterize.

Coolant System

The coolant system must be properly winterized to support cold weather operation. Visually inspect all components such as the radiator, water pump, hoses, etc. Look for leaks or dried inhibitor deposits, which will appear as gel or crystals. Perform any needed repairs or adjustments. Use a cooling system pressure tester, **Figure 29-11,** that will test the system at 15-18 psi (105-120 kPa) after turning on the heat control valves. Repair any leaks evident during pressure testing.

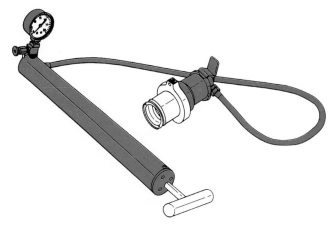

Figure 29-11. *Coolant system pressure test gauge and hook-up. (Caterpillar Inc.)*

Starting Aids

As described in **Chapter 27,** there are several types of engine starting aids available. These systems must be checked and tested prior to the onset of cold weather.

1. Check the mounting, hose condition, and hose routing of circulating coolant heaters. Heaters with manual shut-off valves should be turned on and tested prior to cold weather.

2. Check ether starting systems by removing the charge cylinder and activating the valve. Be sure the charge cylinder contains sufficient ether for the winter season.

3. Electrical resistance heaters, such as glow plugs, oil pan heaters, immersion block heaters, battery warmers, etc., must have all electrical plugs and wiring inspected for wear and damage. Check the resistance of the heating elements using an ohmmeter.

Batteries and Electrical System

Problems with batteries, starting systems, and charging systems are more likely to occur during winter months, when the efficiency of these systems are reduced by the cold. Avoid potential problems by performing a full range of battery, starting, and charging system service and checks as outlined in **Chapters 26** and **27.** Start the winter season with all batteries fully charged, and all electrical system connections clean, tight, and sealed.

Fuel System

Water in the fuel system is always a problem, but during cold weather it can also freeze, causing added blockage problems. Follow all fuel handling guidelines given in **Chapter 14.** When using a fuel-water separator during cold weather, mount the unit in a warm protected area, such as under the hood. The unit can also be wrapped in insulation.

If a separate diesel fuel warmer is used with the fuel-water separator, mount the warmer upstream so that the separator will only receive warm fuel. Do not use 90° fuel line fittings, as these are likely freeze-up points in cold weather. Drain the separator frequently so there is always a minimum of water. If this water turns to ice after shutdown, blockage will occur. Melting may require a substantial amount of time.

Troubleshooting

Successful **troubleshooting** of engine problems requires a thorough understanding of how the particular engine operates, plus a precise analytical approach to the problem using all the resources available to you.

Troubleshooting can be difficult. Troubleshooting tables and charts can only give an indication of where a possible problem might be and what repairs are necessary.

Normally, additional repair work is needed beyond the recommendations on such a chart. Keep in mind that a problem is not normally caused only by one part, but by the relation of one part with other parts.

Obtain as much information as possible concerning the problem. Talk to the vehicle driver or engine operator and ask him or her to describe the problem and symptoms as accurately as possible. Drivers may be quick to offer their opinion as to the repair or service needed, but do not accept any conclusions except your own. A driver's questionnaire, such as the one illustrated in **Figure 29-12** may be helpful in accurately recording driver input on a given problem.

Be observant and use all of your senses. Look at the color of the exhaust smoke. Inspect for fuel, oil, or coolant leaks, damage to air ducts, fuel lines, and corroded electrical connections. Listen for unusual noises, such as air or exhaust leaks, particularly on turbocharged engines. Listen for rough running or misfiring cylinders. A mechanics stethoscope is ideal for listening to the injectors, fuel pump plunger, valve train, and bearing noise. Your sense of smell should alert you to burning lube oil, fuel, coolant, or melting wire insulation. High pressure fuel lines that vibrate under the touch of your fingertips may indicate a malfunctioning injector or pumping plunger in the injection pump.

Electronic Diagnostics

Modern electronically controlled diesel engines offer the service technician another highly accurate tool for troubleshooting work. **_Electronic diagnostics,_** when used properly, can provide a better understanding of engine and fuel system conditions, particularly intermittent problems that may not be evident on a test drive or spot inspection. However, they can also lead you down the wrong path if you do not know how to use or access them properly.

Electronic sensors act as another set of eyes or ears. They can take and record temperature and pressure readings that must be taken manually on non-electronic controlled engines, or may not be taken at all. Coolant temperature, coolant level, engine oil pressure, intake air temperature, engine timing and speed, fuel injector rack position, and other key parameters are constantly monitored, and any readings outside the normal range are stored in the computer memory as fault codes, **Figure 29-13.**

These codes can be read using electronic diagnostic tools, a flashing light readout on the vehicle dash, or by interfacing the engine computer with a personal computer programmed with the appropriate software. Each trouble code is provided with its own troubleshooting flow chart that can help you pinpoint the cause of the problem in a logical, analytic way.

It is important to remember, however, that an electronic diagnostic system is not a solution to all engine troubleshooting. Although it is a highly accurate and reliable tool, which every service technician should learn to take advantage of, it is simply another tool at the service technician's disposal. Many times, the system's trouble codes and charts will pinpoint the same problem that may have also

Questions	Yes	Comments
1. Did the Check Engine Lamp come ON during or after the time the problem occurs? – How long does/did it stay ON?		_____ _____
2. How often does the problem occur? – Can you (the driver) take the truck and demonstrate the problem within one hour?		_____ _____
3. Has the truck been to other shops for the same problem? – If so, when and where?		_____ _____
4. Did the engine completely shut down, requiring a restart using the ignition key switch? – How long do you have to wait before you can restart it after the shutdown?		_____ _____
5. Did the engine hesitate/burp/misfire **without** a complete shutdown?		_____
6. Did the radio, dash gauges, or lights momentarily turn OFF when the problem occurred? – Any other observations about truck components?		_____ _____
7. Does the problem occur only at specific engine loads? – If so, at what load (light, medium, or heavy)?		_____ _____
8. Does the problem occur at a specific **engine** operating temperature? – If so, at what **engine** temperature?		_____ _____
9. Does the problem occur only when above or below specific outside temperatures? – In what temperature range?		_____ _____
10. Does the problem occur during other conditions (during or after rain, or spray washing)?		_____
11. Did the problem occur at a specific vehicle speed? – If so, at what vehicle speed?		_____ _____
12. Did the vehicle kickout of cruise control or PTO mode?		_____
13. Does the problem only occur when using the accelerator, not in cruise control or PTO mode?		_____
14. Does the problem occur at specific engine rpm? – If so, at what engine rpm?		_____ _____

Figure 29-12. A typical questionnaire used to accurately record driver's comments concerning an engine problem. (Caterpillar Inc.)

been evident during your visual inspection and/or test drive. On other occasions, the flow charts will lead to a problem that would have been extremely difficult or impossible to locate without the use of electronic monitoring.

Follow the troubleshooting sequence outlined in the service manual. As you will see, virtually every manufacturer of electronic controlled systems stresses the need to perform the standard checks and tests that have always been used in

Diagnostic Flash Code	Effect on Engine Performance				Suggested Driver Action		
	Engine Misfire	Low Power	Engine Speed Reduced	Engine Shutdown	Shutdown Vehicle	Service ASAP	Schedule Service
Diagnostic Flash Code/Engine Performance Relationship–3100 HEUI Diesel Truck Engine							
01 - Idle Shutdown Override							
15 - Injection Actuation Pressure Sensor Fault	X	✓	✓			✓	
17 - Excessive Injection Actuation Pressure Fault[2]		✓	✓			✓	
18 - Inj. Actuation Press. Control Valve Driver Fault	X	X	X	X		✓	
19 - Inj. Actuation Pressure System Fault	X	✓	✓			✓	
21 - Sensor Supply Voltage Fault[1,2]		✓				X	✓
25 - Boost Pressure Sensor Fault[1]		X					✓
27 - Coolant Temperature Sensor Fault[1,2]	X					X	✓
28 - Check Throttle Sensor Adjustment			✓			✓	
31 - Loss of Vehicle Speed Signal			✓				✓
32 - Throttle Position Sensor Fault			✓			✓	
34 - Engine RPM Signal Fault				X		X	X
35 - Engine Overspeed Warning							
36 - Vehicle Speed Signal Fault			✓				✓
38 - Inlet Air Temperature Sensor Fault[1,2]	✓	✓					✓
41 - Vehicle Overspeed Warning							
42 - Check Sensor Calibrations		X					✓
47 - Idle Shutdown Occurrence				✓			
49 - Inlet Air Heater Driver Fault[1]							✓
51 - Intermittent Battery Power to ECM	✓			X		✓	
55 - No Detected Faults							
56 - Check Customer/System Parameters		X	X			X	✓
59 - Incorrect Engine Software				✓		✓	
61 - High/Very High Coolant Temperature Warning		X	X			✓	
71 - Cylinder 1 fault	✓	✓				✓	
72 - Cylinder 2 fault	✓	✓				✓	
73 - Cylinder 3 fault	✓	✓				✓	
74 - Cylinder 4 fault	✓	✓				✓	
75 - Cylinder 5 fault	✓	✓				✓	
76 - Cylinder 6 fault	✓	✓				✓	

NOTE: – An "X" indicates that the effect on engine performance WILL occur if the code is active. A ✓ (check mark) indicates that the effect on engine performance MAY occur if the code is active, depending upon the exact failure.

[1] – These Diagnostic Flash Codes may affect the system only under specific environmental conditions, such as engine start-up at cold temperature, cold weather operation at high altitudes, etc.

[2] – These Diagnostic Flash Codes reduce the effectiveness of the Engine Monitoring feature when active.

Shutdown Vehicle: Drive the vehicle cautiously off the road and get immediate service. Severe engine damage may result.

Service ASAP (As Soon As Possible): The driver should go to the nearest qualified service location.

Schedule Service: The driver should have the problem investigated when convenient.

Figure 29-13. In an electronically controlled engine, the diagnostic flash codes can give an indication of needed PM. (Caterpillar Inc.)

diesel engine troubleshooting. Once the simple causes of a problem are eliminated as a possible source of trouble, you can then turn to the more elaborate troubleshooting procedures. Ignoring these recommendations or performing checks and tests out of their recommended sequence may lead to a misdiagnosis and wasted time. Troubleshooting diesel engine problems is no time to take short cuts. Approach each repair in the same organized, logical manner each time.

Primary Engine Checks

The primary engine checks consist of a fairly quick and easy method of identifying many common engine performance related problems in a minimum amount of time. Always make these checks before beginning more involved troubleshooting procedures.

1. Make sure the air filter elements are in good condition and are not plugged.

2. Check oil and coolant levels.

3. Check accelerator and governor linkage for proper adjustment.

4. Test the dash mounted tachometer for accuracy. A bad tachometer can cause operation at the wrong engine rpm and give the feel of low power.

5. Test fuel transfer pump pressure and other applicable pressures in the fuel injection system. Check for kinked and damaged fuel lines.

6. Disconnect the fuel return line and check for excessive air in the fuel. A few tiny air bubbles in the fuel may be acceptable. Excessive air bubbles indicate a problem.

7. Check the fuel API gravity rating if there is a complaint of low power.

8. Check fuel heater operation. Check the fuel temperature. At fuel temperatures above 95°F (35°C), power loss is experienced due to fuel expansion. If the engine appears starved for fuel, check the pressure at the secondary fuel filter. If you suspect the engine is using too much fuel (poor mileage), check the fuel flow rate using the appropriate test equipment.

9. Check the color of the exhaust smoke.

10. Measure exhaust temperature with a pyrometer and check against engine specifications. High temperatures may indicate a low power output condition.

11. Check the air inlet system and measure air temperature with a thermistor-thermometer. Air must be cool for engine to develop full horsepower.

12. On mobile equipment check for locked brakes, improper tire size, incorrect gear ratios, misaligned axles, or transmission or driveline problems that may produce symptoms that can be mistaken for low engine power output.

13. Check for restrictions in the fuel system and kinked or damaged fuel lines.

14. Check the operation of the engine brake (if so equipped).

15. Check for proper engine timing, valve adjustment clearance, and injection pump timing. If the injector pump or injector timing is off, problems such as smoky exhaust, low power, high fuel consumption, and internal engine damage can result. Improper engine timing can lead to an increase in exhaust gas temperature.

16. Check for hard starting that can indicate low compression. This problem is usually accompanied by white exhaust smoke. Electrical problems may also exist.

General Test Procedures

Proper compression, fuel injection, and air delivery are essential for efficient operation. If the engine does not have these three things, any preventive maintenance procedures performed will be useless. Several basic tests can be performed to verify operation of these systems.

Locating a Misfiring Cylinder

An engine that runs rough or expels large amounts of black exhaust probably has one or more malfunctioning injectors. A misfiring injector can be located by disabling one injector at a time and noting the engine response. The exact method of disabling the injector depends on the type of fuel injection system the engine uses.

On multiple plunger inline injection systems or distributor pump systems, cylinder injectors are disabled by loosening the line nut for each cylinder at the fuel injection pump. This reduces or stops the fuel flow to the cylinder. After loosening the fuel line nut about 1/2 turn, listen closely to the sound of the engine. If you have disabled a good injector, the engine will run rougher. If you have disabled a bad injector, the sound and rough idle will not change.

A different method is used on Cummins PT fuel injection systems. Warm the engine to its normal operating temperature of 160°F (70°C). Remove the rocker arm cover and install a special rocker actuator tool over the injector rocker arm. Operate the engine at low idle and hold the injector plunger down with the actuator tool to stop fuel to the injector. No change in operating conditions means you have located the faulty injector. If the engine speed drops, the injector is good.

On Detroit Diesel two-cycle engines with unit injectors and no electronic controls (non-DDEC), remove the rocker cover and with the engine at idle speed, use a large screwdriver to wedge the injector follower. Doing so prevents fuel from being injected into the combustion chamber. If a faulty injector is present, there will be no change in operating conditions on that cylinder.

⚠️ **Warning: Do not use this procedure on Detroit Diesel four-cycle 8.2 L V-8 engine unit injectors. Unlike the two-cycle DDC engine, the injector pushrod of the 8.2 L four-cycle engine is not threaded into the rocker arm. Consequently, it may be ejected from the engine or damaged if the injector is shorted out as described earlier. Injury or death may occur.**

The proper method of disabling an injector on the four-cycle 8.2 L is to individually push the injector fuel rack into its full-fuel position. This is called flooding the cylinder. If the injector in question is firing properly, the engine should pick up speed when the cylinder is flooded. No change in engine speed indicates the injector is faulty. On Detroit Diesel's electronically-controlled DDEC engines, the cylinder can be shorted out using a diagnostic scan tool.

Regardless of the method used to disable a diesel fuel injector, always remember that this test only confirms the injector's operation. There could be a problem with the fuel pump delivery valve or elsewhere in the fuel system.

Cylinder Compression Check

Worn piston rings, scored cylinders, damaged pistons, and leaking valves can all cause low cylinder pressure. Low engine power, high amounts of white, blue, or black exhaust smoke, or hard starting may indicate that a *cylinder compression test* should be performed. Before proceeding, make sure the fuel, air inlet, or exhaust system are not the cause of the compression problem.

The compression test is made using a dummy injector mounted in the cylinder. A special adapter connects the dummy injector to a pressure gauge capable of reading the high pressures inside the diesel cylinder. Complete injector test kits are available from engine and tool manufacturer.

 Note: Many shops and technicians often make their own tester using an old injector, adapter, and pressure gauge.

The exact procedure and equipment used to perform the test will vary slightly between engine manufacturers, so check service manual procedures. The following is a typical test sequence:

1. Start the engine and allow it to warm to its proper operating temperature.
2. Stop the engine and disconnect the high pressure fuel line from the injector at the cylinder.
3. Remove the injector as a unit.
4. Direct the end of the high pressure fuel line into a suitable container so the fuel can be caught and retained. On unit injectors, an old fuel pipe can be connected between the fuel inlet and return manifold connections. On two-cycle unit injector engines, you can use an old rocker cover with suitable holes cut in it to minimize oil throw-off during the test.
5. Hold the injector pump control lever in the "stop" position to cut fuel flow and crank the engine for several revolutions to blow any loose carbon from the injector bore.
6. Place a copper gasket beneath the dummy injector and mount it in the cylinder bore, tightening it in place with the nozzle hold-down nut.
7. Start the engine and run it at the speed recommended in the service manual (usually between 600-1000 rpm). Operate the engine until the needle on the test pressure gauge has risen to its maximum reading.
8. Record the compression pressure and compare it to manual specifications. Typical readings at sea level may range from 475-575 psi (3275-3965 kPa).
9. Remove the dummy injector from the cylinder and reinstall the operating injector.

 Note: Inspect and clean the injector as needed prior to reinstallation.

10. Start the engine at idle speed and loosen the fuel line nut at the injector to bleed all air from the line.
11. Tighten the nut to proper torque.
12. Repeat the procedure for all cylinders to be tested.

Difference in compression pressure readings between the individual cylinders should not exceed 25-50 psi (172-345 kPa). If the turbocharger is not disconnected from the intake manifold during the compression check, add 10-20 psi (70-140 kPa) to the minimum compression readings.

Pressure Checks

The following are the major *pressure checks* used to detect air leakage or restriction problems in a diesel engine. A set of manometers is one of the most versatile and effective of all engine troubleshooting tools. It is used to measure either a pressure or vacuum reading on the engine. The five major checks a manometer is used for are:

- ❑ Crankcase pressure test.
- ❑ Exhaust back pressure test.
- ❑ Air inlet restriction pressure test.
- ❑ Air induction system pressure test.
- ❑ Checking air box pressure (two-cycle engines) or turbocharger discharge pressure (two- and four-cycle engines).

These tests can be made using a water or mercury *manometer,* **Figure 29-14A,** or pressure tester equipped with a *magnahelic gauge* capable of reading up to 60" of water, **Figure 29-14B.**

A manometer consists of a U-shaped tube that is filled with either water or mercury. Water-filled manometers are capable of accurately recording lower system pressures

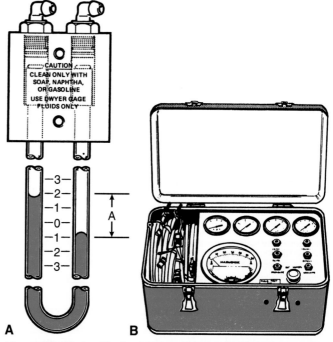

Figure 29-14. A—Slack tube manometer. A dimension indicates total fluid column. B—Pressure tester equipped with magnahelic gauge. (Navistar International Transportation Corp.)

such as those found in crankcases or air inlet systems. When pressure or vacuum is applied to the tube, the water or mercury moves within the tube or is displaced. A special scale, calibrated in either U.S. Customary or metric measurement, is used to read the amount of displacement. The displacement can then be compared to service manual specifications. **Figure 29-15** lists various manometer conversion factors for water, mercury, and psi readings.

Water manometers use a colored dye to make reading the displacement easier. The tubes of mercury manometers tend to discolor over time, but holding a bright light behind the tube will make reading the mercury level easier. The scale of the tool is adjustable so it can be set at zero at the start of each test. This sliding scale may be calibrated on one or both sides depending on the type of manometer used.

Most manometers are calibrated so that the displacement on each side of the tube can be read. When a reading is taken, displacement on both sides must be read and added together to obtain the final pressure or vacuum reading. The reason for this is that there may be slight differences in the internal diameter of the pliable plastic tubing used in this type of manometer. By splitting the displacement measurement between two readings, an accurate total displacement reading is obtained.

Solid tube, cabinet mounted manometer sets have very precise and consistent internal diameters so their scales need only be calibrated on one side. **Figure 29-16** shows the proper sighting line when reading either mercury or water manometers. To take a reading using a manometer:

1. Open both valves at the top of the tube about one turn.
2. Shake the tool slightly and allow the water or mercury to settle.
3. Sight the level horizontally and move the sliding scale on the manometer until the zero on the scale is equal to the fluid level.
4. Connect one end of a rubber hose or tube to the engine test point and the other end to the manometer tube valve. If the manometer scale is calibrated on both sides, the tube or hose can be connected to either tube valve. If the manometer is calibrated on only one side, connect the tube or hose to that side.
5. Operate the engine at the specified rpm and note the fluid displacement in the tube(s). Compare this displacement against service manual specifications.

Crankcase Pressure

A small amount of internal *crankcase pressure* (5-10" of water) in an engine is required to help keep dirt and dust suspended in the lubricating oil. Without this slight crankcase pressure, solid contaminants could settle out of the oil onto internal engine parts.

To take the manometer reading, a hose or tube can be easily adapted to the dipstick tube and then connected to the manometer tube valve or magnahelic gauge. In large bore engines, a special crankcase pressure plug is often located in the engine block slightly above the oil pan rail. The test can also be performed by mounting a special orifice restrictor tool onto the lower breather tube, **Figure 29-17**. Once the manometer or pressure gauge is connected, start and run the engine at the recommended speed. Note and record the pressure reading.

A higher than specified crankcase pressure indicates a problem in the engine. Continued loss of oil through the engine breather tube, crankcase ventilator, or dipstick tube hole are indications the crankcase pressure is too high.

Manometer Pressure Conversion				
Units of measurement	Inches water	Inches mercury	psi	kPa
1 inch water	1	.0735	.0361	.248
1 inch mercury (Hg)	13.6	1	.491	3.85
1 psi	27.7	2.036	1	6.985

Figure 29-15. Manometer conversion factors.

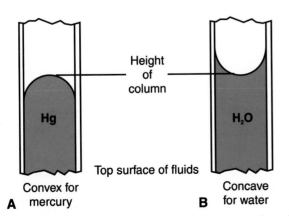

Figure 29-16 Sight line when taking manometer reading. A—Mercury manometer. B—Water manometer. (Detroit Diesel Corporation)

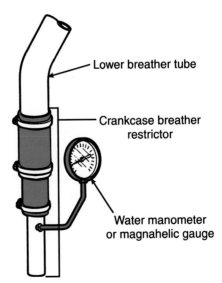

Figure 29-17. Setup for crankcase pressure test. (Navistar International Transportation Corp.)

Causes of high pressure include:

- ❏ Crankcase overfilled.
- ❏ A plugged crankcase breather tube or ventilator.
- ❏ Leaking cylinder head gasket.
- ❏ High exhaust back pressure.
- ❏ Worn or broken piston rings.
- ❏ Loose piston ring retainers (two-cycle engines).
- ❏ Worn blower oil seals.
- ❏ Leaking cylinder block-to-end plate gasket.
- ❏ High air box pressure.
- ❏ Blower or turbocharger air pressure leakage into an oil drain passage.
- ❏ Damaged piston or liner.

Exhaust Back Pressure

Since exhaust gases leaving the engine are under pressure, the exhaust system will induce a slight amount of **back pressure.** This back pressure is used to drive the turbocharger (on engines so equipped). A slight back pressure is also desirable in cold weather operating conditions, light-load operation, and excessive idling conditions. However, excessive exhaust back pressure can increase operating temperature significantly, leading to lack of power, smoke-filled exhaust, burned exhaust valves, poor fuel economy, or shortened engine life.

The exhaust back pressure can be measured using a mercury manometer or magnahelic pressure gauge. Non-turbocharged engines are equipped with a small pipe plug located slightly downstream of where the exhaust manifold companion flange bolts to the exhaust piping. On turbocharged engines, the pressure plug is located after the turbocharger. If it were located before the turbocharger, a higher than normal pressure reading would be obtained.

If the engine is not equipped with back pressure plugs, one can be made in either location mentioned earlier by drilling a small hole in the exhaust piping and tapping it to accept a small brass plug, **Figure 29-18.** Check the exhaust back pressure in both stacks of a V-type engine. If the V engine has only one stack, take the reading about 4-6" (10-15 cm) from the Y-connection toward the muffler or silencer.

Connect and zero the mercury manometer. Start the engine and run it at the specified rpm for the test. Record the pressure reading at both no-load and full-load conditions. Higher than specified exhaust back pressure may be caused by a crushed exhaust pipe or muffler, a rain cap mounted at the end of the exhaust stack, carbon blockage within the system, or a muffler that is too small. Other causes include excessive exhaust pipe runs, pipe diameter too small or obstructed, or too many bends or elbows in the system.

Air Inlet Restrictions

A diesel engine that is not receiving enough air will lack horsepower, have poor fuel economy, produce smoky

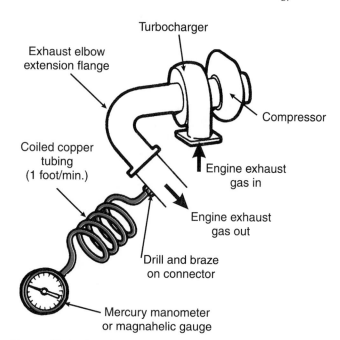

Figure 29-18. *Test for measuring exhaust back pressure by tapping into the exhaust manifold.*
(Navistar International Transportation Corp.)

exhaust, or high exhaust temperature. The air inlet system air cleaner, the diameter and length of ductwork, and the number of bends and turns in the air inlet ductwork all contribute to restricting air flow to the engine. A water manometer check is a simple method of measuring air inlet pressure as an indicator of air flow. The lower the pressure reading, the greater the air flow to the engine.

On a naturally aspirated engine, the manometer or magnahelic pressure tester is used to measure air pressure within the intake manifold in relation to atmospheric pressure. On a turbocharged engine, the same type of reading is taken, but only on the suction side of the turbocharger, **Figure 29-19.** An air inlet restriction check should take into account the restrictions placed on the system by the air cleaner and element, as well as the piping and ductwork to the engine.

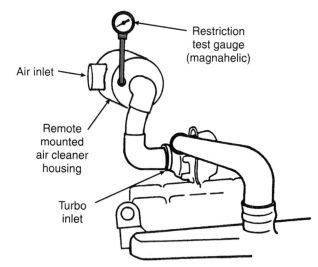

Figure 29-19. *Setup for air inlet restriction test.*
(Navistar International Transportation Corp.)

Many heavy-duty diesel air cleaners have a small plug installed in the outlet to the engine. The plug can be removed and an adapter installed for connecting the manometer or test gauge. Some air cleaners are equipped with their own air restriction indicator gauge. This gauge is usually a small, clear glass window that reads red when air restriction at the cleaner reaches 25" of water. Other air cleaners are connected to a dash-mounted indicator gauge through the use of small bore tubing. Many modern turbocharged engines, aftercooled constant horsepower, and high torque rise diesel engines may have air flow restriction pressure specifications around 20" of water.

To check air restriction within the entire inlet system, a plug is removed from the air inlet manifold on naturally aspirated engines. This plug is normally located a short distance from the air inlet housing or manifold. On turbocharged engines, the plug is always a short distance upstream of the suction (inlet) side of the turbocharger.

Connect the manometer or pressure gauge to the plug using the proper adapter and tube. Start the engine and take the pressure reading with the engine running at the manufacturer's recommended rpm for this test. If needed, use conversion factors to calculate the system air pressure and compare it to service manual specifications.

Remember, high system pressure (insufficient air flow) can be caused by a plugged air cleaner, or by using a filter that is too small to handle the air flow. When used, a plugged precleaner or turbocharger inlet screen may also be the problem. Also, crushed or kinked inlet piping or ductwork that is too small in diameter to handle the air flow, or has multiple bends and elbows may be the cause. In all engine applications, the air inlet ductwork should be kept as short and straight as possible, although this is not always practical.

Air Induction System Leak Check

As mentioned earlier, the most common cause of engine failure is dirt. Leaks in the air induction system will draw dirt into the engine. An air induction leak on the suction side, no matter how small, can cause an eventual engine failure.

Test for leakage in the air induction system using a manually regulated compressed air supply and a pressure gauge capable of reading 0-15 psi (0-103.4 kPa). Block off the air inlet pipe or mask off the outer diameter of the air cleaner inlet with duct tape. Disconnect the air cleaner restriction indicator or tubing at the air cleaner. Install a T-fitting and connect the indicator or tubing and the pressure test gauge. Remove the pipe plug from the intake manifold, connect the manually regulated air supply, and adjust to 5 psi (34.47 kPa). Coat the ductwork and tubing connections with a soap solution and watch for bubbles at the joints. Repair any leakage found.

Dust Sight Window

If the air inlet system is equipped with a clear plastic **dust sight window,** inspect it for signs of clouding over. This is an indication that dirt and dust are entering the engine. Accumulation of dust and dirt indicates that there is a hole either in the primary filter or elsewhere in the air induction system.

Air Box and/or Turbocharger Boost Pressure

Many engines use an exhaust-driven turbocharger to draw air into the inlet system. Once in the inlet system, it compresses the air to pressures greater than atmospheric, and then forces it into the engine. In addition to a turbocharger, two-cycle diesel engines also employ a gear-driven air blower or supercharger to supply air to the engine for efficient combustion and scavenging. The turbocharger delivers its air charge to the blower, which then passes it on to the engine.

On a four-cycle engine, turbocharger boost pressure is normally around 24.5 psi (50" Hg). It is checked by removing a small plug from the intake manifold and connecting the manometer at this point. The engine is then started and placed under load while the reading is recorded. The engine must be under load so the turbocharger can receive maximum exhaust gas flow and deliver maximum air pressure. Compare the reading against manual specifications.

On a two-cycle engine equipped with an air box, turbocharger and air box pressure can be taken individually, but most service manuals only list acceptable air box pressure. The manometer is connected to either an air box drain tube or to a fitting in the air box manifold. As with a four-cycle engine, the test is performed with the engine under load so maximum air pressure is generated. Low turbocharger boost or air box pressure can be caused by the following conditions:

❏ Reduced air supply due to restrictions or other problems in the air inlet system.
❏ Leaking O-ring seals between the inlet and the turbocharger.
❏ Leaking air inlet manifold-to-cylinder head gasket.
❏ Leaking blower-to-engine block gasket.
❏ Damaged blower rotor.
❏ Plugged or restricted aftercooler.
❏ Clogged or plugged blower inlet screens (if equipped).
❏ Leaking hand hole cover plate gaskets (if equipped).
❏ Leaking end plate to engine block gasket (if equipped).

In some cases, turbocharger boost or air box pressures may be too high. High air box pressure will reduce blower efficiency, causing reduced scavenging, poor combustion, increased emissions, and higher exhaust temperatures. Causes could include:

❏ High exhaust back pressure.
❏ Carbon build-up in the cylinder liner port area. This is caused by light load operation or periods of extended idling.
❏ Excessive fuel delivery, causing engine and turbocharger overspeeding.

Storage of Diesel Engines

When a diesel engine is removed from service and placed in storage for an extended period of time, special preparations and precautions must be taken to ensure the engine does not rust or otherwise deteriorate. All rust or corrosion must be removed from exterior engine surfaces and a rust prohibitive compound applied.

Short Storage Preparation

When storage time is relatively short (30 days or less), the following steps should be taken.

1. Drain all lubricating oil from the crankcase and refill with fresh oil of the recommended grade and viscosity.
2. Fill the fuel tank with the recommended grade of fuel and run the engine for 2 minutes at 1200 rpm and no load.
3. Check the air cleaner and service as needed.
4. If temperatures will drop below freezing during the storage period, be sure the radiator is filled with an antifreeze mixture. If a raw water cooling system is used, drain it completely and leave the drain cocks in the open position.
5. Clean the entire engine exterior except the electrical system using fuel oil and dry it with compressed air.
6. Seal all engine openings with a waterproof, vapor-proof material that is strong enough to resist breakage from the expansion of air trapped inside the engine.
7. The engine can be restored to service by removing all seals and checking coolant, fuel, lubricating oil, and transmission oil levels and priming the raw water pump (if used).

Extended Storage Preparation

When an engine is to be removed from service for an extended period of time, the preparation is more extensive.

1. Drain and flush the cooling system. Refill with sufficient strength antifreeze fortified with rust inhibitor additives.
2. Remove, inspect, and service the fuel injectors as needed to ensure they will be operational on start-up.
3. Circulate the coolant through the engine by operating the engine until it reaches its normal operating temperature.
4. Stop the engine and completely drain the lubricating oil from the engine crankcase. Install new oil filters and fill the crankcase to the proper level with a 30 weight preservative oil.
5. Drain the fuel tank and refill with enough rust preventive fuel oil and/or a biocide to run the engine for approximately ten minutes. Drain the fuel filter and strainer. Clean the fuel filter shell with diesel fuel and install a new element. Fill the cavity between the shell and element 2/3 full with rust preventive fuel oil.
6. Run the engine for 5 minutes to circulate the rust inhibitor throughout the engine.
7. Service the air cleaner as needed.

8. For marine gears, drain the oil from the gearing, change the strainer and filter element, and refill with fresh oil of the proper specifications. Start and run the engine at 600 rpm for ten minutes to coat all internal gearing with lubricant.

 Note: Step 8 is not needed on torque converter units.

9. On engines equipped with torque converters, change oil and filters and prepare for storage as outlined in the torque converter service manual. A special preservative oil may be specified. Seal the converter breather with moisture-proof tape.
10. If equipped, lubricate the power take-off unit with all purpose grease and drain, flush, and refill the gear box with proper grade oil. Replace the filter element.
11. Seal the turbocharger air inlet and turbine outlet connections with moisture resistant tape.
12. Apply a non-friction rust preventive compound to all exposed parts.
13. Disengage the clutch from the flywheel to prevent sticking over time. Never apply oil, grease, or wax to the cast iron flywheel as these substances can be absorbed by the porous iron and cause slipping during operation.
14. Drain the engine cooling system.
15. Clean and service the battery. Coat all electrical connections with corrosive preventive compound. Protect all electronic components from excessive heat or moisture.
16. Insert heavy paper strips between all pulleys and belts to prevent sticking.
17. Seal all engine openings including the exhaust outlet using moisture resistant tape.
18. Clean and dry the exterior surfaces of the engine. Spray the surfaces with a rust preventive compound, a synthetic resin varnish, or a liquid auto body wax. Cover with a weather-resistant tarpaulin (outdoor storage) or clear plastic sheeting (indoor storage). Whenever possible, store the engine indoors in a dry, heated location.

When restoring the engine to service, remove all sealing tape and paper strips. Wash off all rust preventive compound using clean solvent or a high pressure washer. Drain the preservative oil from the crankcase and fill with lubricating oil. Fill the fuel tank with clean fuel. Fill the cooling system with a suitable coolant rust inhibitor mix. Install and connect the charged batteries and make all electrical system connections. Service the air cleaner. Drain the preservative oil from the torque converter or marine gear (if so equipped) and fill with proper grade lubricant. Bleed any air from the fuel system.

The engine can now be started. The exhaust may be smoky for the first several minutes due to the rust preventive compound remaining in the fuel system.

Summary

A well-planned and executed preventive maintenance program will extend engine life and reduce the number of breakdowns in service. A good PM program greatly reduces overall maintenance costs which helps the engine to run better with improved fuel economy and reduced emissions. Preventive maintenance is absolutely necessary for engines to meet today's more stringent emission regulations.

Determining the right PM service interval to maximize vehicle performance and engine life while minimizing down time requires a combination of art and technology. The optimum PM schedule is usually a combination of the engine manufacturer's service manual recommendations and the unique operating conditions of the fleet. For a PM program to be successful, PM inspections must be thorough and accurate records must be maintained.

Regular laboratory oil analysis is a highly useful diagnostic tool for preventive maintenance. Oil analysis allows lubricating oil and filters to be changed when they have exhausted their effectiveness; not before or after. Impending failures can be detected in their early stages and engines scheduled for proper maintenance without unexpected, expensive, and disruptive down time.

Winterizing a diesel engine is a very important aspect of any preventive maintenance program. Areas of concern include the cooling system, batteries and electrical system, starting system, and fuel system.

Successful troubleshooting of engine problems requires a thorough understanding of how the engine operates, plus a precise analytical approach to the problem. Obtain as much information as possible concerning the problem. Talk to the vehicle driver or operator. Be observant and use all your senses. An electronic diagnostic system is not a solution to all engine troubleshooting. It is simply another tool at the service technician's disposal.

Always perform primary engine checks on filters, operating pressures, exhaust smoke condition, fuel system functions, engine and fuel injection pump timing, etc., before undertaking any major troubleshooting procedure.

A misfiring injector can be located by disabling one injector at a time and noting engine response. Disabling a faulty injector will have no noticeable effect on engine operation.

A cylinder compression check is made using a dummy injector that is mounted into the cylinder. A special adapter connects the dummy injector to a pressure gauge that is capable of reading the high pressures that exist inside the cylinder.

When a diesel engine is removed from service and placed in storage, special preparations and precautions must be made and followed to ensure that the engine does not rust or otherwise deteriorate. This includes draining and flushing the coolant system, draining the lube oil and filling the crankcase with preservative oil, and servicing the fuel injectors and air cleaning system.

Important Terms

Preventive maintenance (PM) program
Level one maintenance
Level two maintenance
Flushing
Level three maintenance
Documentation
PM forms
Repair order
Vehicle files
Oil analysis report
Major component history
Oil change
Oil analysis
Selective sampling
Viscosity
Fuel dilution
Fuel soot
Glycol contamination
Spectrometer
Total base number (TBN)
Total acid number (TAN)
Ferrography
Oil oxidation
Troubleshooting
Electronic diagnostics
Cylinder compression test
Pressure checks
Manometer
Magnahelic gauge
Crankcase pressure
Back pressure
Dust sight window

Review Questions—Chapter 29

Do not write in this text. Place your answers on a separate sheet of paper.

1. Newer piston designs produce higher ring area _____ and places greater demands on the _____ .

2. Which of the following are records essential to the success of any PM program?
   (A) Repair orders.
   (B) Vehicle files.
   (C) Oil analysis reports.
   (D) All of the above.

3. Although newer engines burn less oil, the oil in a newer engine crankcase is just as likely to be _____ than oil in an older engine.

4. Name at least three benefits of an oil analysis program when it is properly instituted and maintained.

5. Fuel oil entering the crankcase can be caused by:
   (A) poor combustion.
   (B) leaking injectors.
   (C) bad timing or worn pistons and rings.
   (D) All of the above.

6. _____ from the cooling system is one of the leading causes of premature engine failure.

7. Name the four areas of primary concern when winterizing a diesel engine.

8. Modern _____ can aid the technician in engine troubleshooting.

9. Fuel injectors should be inspected and serviced _____ the engine is placed into long-term storage.

10. Long-term storage (more than 30 days) requires the use of _____ in the engine's fuel system.

ASE-Type Questions

1. A good PM program can reduce overall maintenance cost by _____ .
   (A) eliminating repairs caused by dirt, water, and other contaminants
   (B) eliminating paperwork
   (C) increasing technician awareness
   (D) reducing paperwork

2. All of the following are reasons to keep accurate records and documentation, EXCEPT:
   (A) required by some state Departments of Transportation.
   (B) provides written proof that all required inspection and maintenance procedures were performed.
   (C) record for defense of legal action.
   (D) to ensure the technician does his or her job.

3. If a high sulfur diesel fuel (.05% or more) is being burned, the oil change interval must be _____.
   (A) increased
   (B) decreased
   (C) kept the same
   (D) set by the engine manufacturer

4. Technician A says that an oil analysis sample should be taken with the engine at or near operating temperature and before adding any oil. Technician B says that you should always take the sample from the same engine location and use the containers supplied by the laboratory. Who is right?
   (A) A only.
   (B) B only.
   (C) Both A & B.
   (D) Neither A nor B.

5. Studying an oil analysis report, Technician A says that an excess of large wear particles in the oil when compared to small wear particles is an indication of accelerated wear and possible catastrophic engine failure. Technician A says this is normal for diesel engines. Who is right?
   (A) A only.
   (B) B only.
   (C) Both A & B.
   (D) Neither A nor B.

6. All of the following are primary engine checks, EXCEPT:
   (A) check oil and coolant levels.
   (B) check the color of exhaust smoke.
   (C) holding a unit injector follower down with a screwdriver at idle speed.
   (D) check fuel system pressure from the tanks to the injectors.

7. Technician A says you should begin troubleshooting a rough running engine by removing the injectors and looking for signs of fouling. Technician B says you should perform primary engine checks on filters, operating pressures, and engine timing before disassembling any major system. Who is right?
   (A) A only.
   (B) B only.
   (C) Both A & B.
   (D) Neither A nor B.

8. A cylinder compression check should be performed when the following problems exist, EXCEPT:
   (A) excessive white, blue, or black smoke.
   (B) hard starting.
   (C) no cranking.
   (D) low power.

9. Technician A says that short-term engine storage (less than 30 days) does not require the filling of the crankcase with 30 weight preservative lubricating oil. Technician B says any storage period over 14 days requires the use of this special oil. Who is right?
   (A) A only.
   (B) B only.
   (C) Both A & B.
   (D) Neither A nor B.

10. Low turbocharger boost or air box pressure can be caused by _____.
    (A) restrictions in the air inlet system
    (B) leaking air inlet manifold to cylinder head gasket
    (C) clogged or plugged blower inlet screens
    (D) All of the above.

Chapter 30

Career Opportunities

After studying this chapter, you will be able to:
- ❑ List areas in which you can obtain a position servicing diesel engines.
- ❑ Name the various specializations in the diesel field.
- ❑ Discuss ASE certification.
- ❑ List the types of questions and test categories in the ASE test battery.

After you complete your study of diesel engines, you are probably asking yourself this question. "What are my chances of employment?" The answer is "very good." The diesel industry offers many career opportunities in the general categories of automotive, medium-heavy duty trucking, marine, railway, and industrial (including agriculture, construction, and utilities). Many careers are available in the diesel field for people of diversified skills. These jobs provide good pay and a rewarding experience.

The Diesel Field

The proliferation of diesel engines in the American infrastructure is easily overlooked at first glance, but they are a vital and inseparable part of commerce and everyday life. Look around wherever you may be reading this text. If you are in a building, consider this: the ground where the building is sitting was most likely cleared by diesel powered vehicles. The foundation was laid and all the other construction materials—steel, concrete, block, etc., used to construct the building were transported by diesel powered vehicles. The roads and bridges used by the diesel trucks to transport the steel structure and other materials were also constructed by diesel power.

Indeed, if all the diesel engines in the world were suddenly silenced, there would be no lumber, food, minerals,

ores, petrochemicals, or many other goods and services. No matter how well these powerful engines are designed and constructed, they must be maintained and repaired to keep them on the job. That is where diesel technicians provide a most valuable service to their employer, their community, and to society as a whole.

Certainly there are cleaner, easier, and more glamorous vocations available, but without diesel technicians, the engines of industry would eventually break down and cease to operate. The diesel service and repair profession, even with its occasional moments of frustration and adverse working conditions, offers one of the most rewarding careers available today.

Over the past several years, the diesel field has grown tremendously. Reasons for this growth include the increasing demand for diesel engines in utility operations, **Figure 30-1,** and the continuing use of diesels in the transportation and construction industries. Also, once the engine is put into operation, it must be properly maintained to last for years of extensive use. This increases the demand for highly skilled service and repair technicians.

Figure 30-1. *Power plants need diesel technicians to keep engines operating.* (Diesel Progress)

Diesel Engine and Equipment Manufacturers

Companies that manufacture and recondition diesel engines employ thousands of workers. A limited knowledge of diesel engines is necessary for the manufacturing positions. Assembly line jobs provide one answer for entrance into the industry. Companies that recondition diesel engines require technicians with more in-depth training, **Figure 30-2.** However, there are very few openings for these positions and further study and experience are necessary for advancement.

Diesel Technician

Diesel engine service and repair is the area that requires the largest number of employees. Persons working in this area are generally called *diesel technicians.*

Diesel technicians do repair and maintenance work on construction equipment such as earthmovers, bulldozers, graders, and cranes, **Figure 30-3.** They also service transportation vehicles including ships, locomotives, heavy trucks, and buses, **Figure 30-4.** Stationary equipment, such as pumps and generators, require the skills of diesel technicians for periodic service and repair.

Diesel technicians have many trade responsibilities. They do engine inspections and troubleshooting; preventive maintenance, repair, and adjust engine components. These people usually are referred to as *general technicians,* since they do all types of diesel engine repair, **Figure 30-5.**

Specialization

Specialization has brought about a number of specific job responsibilities. Some diesel technicians specialize in the repair of fuel injection systems or other technical jobs such as engine rebuilding, reconditioning cylinder heads, repairing turbochargers, or starter motors. Experience and

Figure 30-3. Diesel construction equipment, such as bulldozers, need the care of diesel technicians. (Komatsu Trading International, Inc.)

Figure 30-4. Transportation services on land or water are powered by diesel engines. (W. Scott Gauthier)

Figure 30-2. Working on the assembly line of a diesel engine manufacturer or rebuilder. (Babcox Publications)

Figure 30-5. Work in a general diesel engine shop is the most likely place to find employment after your studies. (Lloyd Wolf, VICA)

further training often lead to more specific work and job titles. Some examples of diesel specializations include:

- ❏ *Heavy equipment technicians* specialize in bulldozers, earthmovers, and other construction equipment.
- ❏ *Truck technicians* are assigned to maintenance and repair of truck fleets.
- ❏ *Marine technicians* repair and maintain the diesel engines on boats and ships of all sizes. It is often considered by some to be the glamour position of the diesel engine trade. See **Figure 30-6.**

Obviously there are a number of other specialty technician jobs in the workplace, whether it be agriculture or the stationary generation of electricity. Although not highly visible, most utilities, hospitals, and public safety organizations have diesel-powered emergency generators to supply electricity during outages and natural disasters. It is of paramount importance to keep these diesel applications well maintained and in a constant state of readiness, since many times, the prevention of loss of life will depend on them.

Shop Supervisors

Shop supervisors are responsible for assigning jobs to diesel technicians. They must know the skills and abilities of the technicians they supervise as well as be able to work closely with the service manager. When problems arise, the shop supervisor should be able to assist the technician in locating the cause and make suggestions for testing and repair. Scheduling training classes and ordering supplies,

Figure 30-6. Marine diesel technicians have to work in confined spaces, but it is considered the glamour position of the industry. (Diesel Progress Engines & Drives)

tools, and equipment may also be the responsibility of the shop supervisor.

Shop supervisors are usually technicians who have become familiar with the workings of the shop. They must combine their experience with the supervisory skills necessary to manage shop personnel and keep the work moving smoothly through the shop. Although technical skills are a required part of the shop supervisor's background and training, other skills such as handling records, dealing with customers, scheduling work, and monitoring safety conditions are all necessary skills.

Service Manager

Many large companies sell and service their own equipment. In such cases, a **service manager** is employed to coordinate the service department. Responsibilities include taking customer service orders, seeing that time schedules are met, and working with the shop supervisor and other managers. Often, the service manager is responsible for the inventory of parts and equipment. Qualifications for this position include those of the shop supervisor plus a number of business, management, and/or marketing courses.

Self-Employment

Although nearly all technicians begin their employment career working for someone else, many have a desire to eventually work for themselves. These technicians in particular should strive to become proficient in as many different phases of the job as possible. Many technicians normally avoid such non-mechanical duties as bookkeeping, bill collection, and others associated with most business enterprises. However, without these skills, it is difficult for an independent entrepreneur to succeed, no matter how technically skilled he or she may be.

Educational Requirements

If you are interested in a career as a diesel technician, there are many avenues open for you to secure the necessary skills and knowledge to enter this field.

Many start as helpers to experienced automotive or diesel technicians. Usually in 3-4 years, they have gained the necessary background to do all types of general service and repair. If they are employed by companies that utilize or service diesel powered equipment, they are often given a year or more of additional training.

Formal **apprenticeship programs** are often operated by companies that manufacture diesel engines and equipment as well as labor unions. These programs usually consist of 2-4 years of practical work experience in diesel repair combined with classroom study. Many of these programs are located in local vocational-technical schools and community colleges.

Another method of entry into the diesel field is through **full-time study.** Some vocational-technical schools and community colleges offer comprehensive programs in diesel engine service and repair. Most provide 1-2 years in specialized areas of study. Upon graduation, the individual usually needs considerable on-the-job training before becoming a skilled diesel technician.

The armed forces also needs personnel that can service and repair diesel engines. Most tanks, personnel carriers, trucks, some ships, and other military vehicles are diesel powered. The armed forces will also provide the necessary training that can lead to a successful career in diesel mechanics, either in uniform or civilian life.

All companies in the diesel industry prefer to employ apprentices and trainees who have at least a high school education. High school courses in auto and diesel technology (if available), electricity, welding, drawing, and machine trades, as well as courses in science, language arts, and mathematics, are highly desirable. With the wide usage of computerized diagnostic equipment, computer literacy is becoming a necessity, not only to perform well as a technician, but to function in an increasingly complex, computer-driven world.

It is important to remember that you will probably be required to attend additional training on new engines and systems. This training is meant to supplement your experiences in the field. In many ways, your training in diesel technology is never complete as you are faced with new challenges each day.

Employment Availability and Wages

Employment opportunities for diesel technicians are expected to increase rapidly. This is primarily due to the expansion of industries that use diesel engines. Diesel engines are replacing gasoline engines on a wide variety of power equipment. Some of the equipment that is now powered by diesel engines includes agricultural machinery, heavy industrial machinery, power plants, and transportation vehicles.

The outlook for employment appears attractive. New industries and service repair facilities are on the increase. Therefore, job opportunities for skilled diesel technicians will be readily available.

Wages for diesel technicians, supervisors, and service managers are equal to or better than those of other comparable service industries. Most technicians earn a good living. With dedication to their job, they may advance to positions of responsibility and respect in their chosen career.

Teaching

A teaching position is a highly rewarding experience, both in salary and working conditions. For those who are qualified, teaching positions are available in high schools, trade, technical and vocational schools, factory and apprentice training programs, community colleges, and universities. See **Figure 30-7.**

Educational requirements generally include from 2-5 years of occupational experience plus a bachelor's or master's degree in a technical program at a college or uni-

Figure 30-7. There is a demand for teachers of diesel technology because of various retraining programs.

versity. Information on teaching may be obtained by writing for college catalogs.

ASE Certification

The **National Institute for Automotive Service Excellence,** referred to simply as **ASE,** offers a voluntary certification program for technicians. It is recommended and in some cases, required by most diesel engine manufacturers in the United States. ASE was founded in 1972 as a non-profit, independent group dedicated to improving the quality of automotive service and repair through voluntary testing and certification of technicians, **Figure 30-8.** Certified technicians promote customer trust and improve the image of the industry as a whole.

There are eight tests in the **Medium/Heavy Truck Technician certification series.** A technician who passes one or more tests and has at least two years of relevant work experience can become certified as an **ASE Truck**

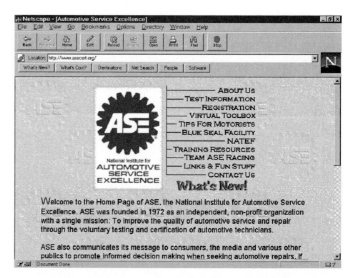

Figure 30-8. More information on ASE and its certification programs can be found at their World Wide Web site on the information superhighway. (ASE)

Technician. A technician certifying in the Medium/Heavy Truck series has the option of taking either the Diesel or Gasoline Engine test (many technicians become certified in both). By passing either the Diesel Engine or Gasoline Engine test and the other six tests in the series, a technician can earn the title and certificate of ***Master Truck Technician.***

There are many advantages to becoming ASE certified. Most employers prefer to hire technicians who are ASE certified. ASE certification is often a prerequisite for entering many state inspection and manufacturer sponsored training classes. Almost all automotive and diesel training programs require master technician certification for their instructors. ASE tests and certifies technicians in the following Medium/Heavy truck areas:

- ❏ Gasoline Engines
- ❏ Diesel Engines
- ❏ Drive Trains
- ❏ Brakes
- ❏ Suspension and Steering
- ❏ Electrical/Electronic Systems
- ❏ Heating, Ventilation, and Air Conditioning Systems
- ❏ Preventive Maintenance Inspection (PMI)

The preventive maintenance test is an additional certification, however, it does not certify you to inspect vehicles in accordance with federal vehicle inspection requirements.

ASE Test Content

Each ASE test is comprised of multiple choice questions with only one correct answer. The types of knowledge and skills needed to pass these tests are as follows:

- ❏ Basic technical knowledge of a system's components and how they work, as well as the proper procedures and/or precautions that should be followed when making repairs or adjustments.

- ❏ Correction or repair knowledge and skill to apply generally accepted inspections, repair procedures, and precautions in assembly, disassembly and reconditioning operations, and adjustments. Also tested is the ability to use service information and precision tools.

- ❏ Testing, diagnostic knowledge, and the ability to use generally available measurement and test equipment to diagnose problems. Also the ability to trace the effects of a particular condition and find the cause of a particular set of symptoms.

The questions on these tests are written by technicians, instructors, and other service industry experts familiar with all aspects of medium/heavy duty-truck repair. All questions asked on ASE tests are entirely job-related. They are designed to test the skills that the technician needs to know in servicing and repairing trucks; theoretical knowledge is not tested. There are no questions on operational theory and no manufacturer-specific questions.

ASE provides a task list to allow technicians to focus their preparation on those subjects which need attention. The diesel engine test contains 70-80 questions. The task

list for the diesel engine test and the typical number of questions asked in each of these categories are as follows:

- ❏ General Engine Diagnosis
- ❏ Cylinder Head and Valve Train Diagnosis and Repair
- ❏ Engine Block Diagnosis and Repair
- ❏ Lubrication and Cooling System Diagnosis and Repair
- ❏ Air Induction and Exhaust Systems Diagnosis and Repair
- ❏ Fuel System Diagnosis and Repair
- ❏ Starting System Diagnosis and Repair
- ❏ Engine Brakes

ASE also offers certification in the areas of automotive repair, engine machinist, parts specialist, school bus maintenance and repair, collision/paint, and specialized areas such as advanced automotive engine performance and compressed natural gas (CNG) light vehicles. Other specialty tests developed by ASE, include an advanced test for electronically controlled diesel engines.

Many certified engine and truck technicians also take the engine machinist tests since much of the information overlaps. The three tests are Cylinder Head Specialist, Cylinder Block Specialist, and Engine Assembly Specialist.

To help the prospective technician prepare for these types of examinations, please note that the questions at the end of each chapter in this text are asked in the same format as ASE test questions. For further information on the ASE certification program, as well as for registration booklets and preparation guides, write:

National Institute for Auto Service Excellence
13505 Dulles Technology Drive, Suite 2
Herndon, VA 20171-3421

Taking ASE Tests

When you receive the registration booklet, fill out the information in the application form, **Figure 30-9.** After mailing it in with your payment, you will receive an admission ticket, which has instruction as to where to go and what time the test begins. Arrive at the test center early, as no one will be admitted once testing begins. Be sure to bring a photo ID and several Number 2 pencils.

Take your time and answer each question as accurately as you can. If you find yourself taking a great deal of time on a particular question, skip it and come back to it later. You will have plenty of time to take the test, however, keep in mind that it is timed and you will have to eventually turn in the test, whether you are done or not. When you are finished, turn in the test booklet and answer sheet.

Your test results will arrive from ASE in approximately 6-8 weeks. You will first receive a confidential pass/fail letter, which will state whether you passed each test or if more preparation is needed. A more in-depth test diagnostic report will follow within 1-2 weeks, including a breakdown by area of how many questions you answered correctly. If you passed one or more tests and you are eligible for certification, this package will also contain a

Registration Form
ASE Tests

ASE

If extra copies are needed, this form may be reproduced.
Important: Read the Booklet first. Then print neatly when you fill out this form.

1. a. Social Security Number:

1. b. Telephone Number: (During the day)
Area Code   Number

1. c. Previous Tests: Have you ever registered for any ASE certification tests before? ☐ Yes ☐ No

2. Last Name: **First**   **Middle Initial**

3. Mailing Address:

Number, Street, and Apt. Number

City

State   ZIP Code

4. Date of Birth:
Month   Day   Year

5. Sex: ○ Male ○ Female

6. Race or Ethnic Group: Blacken one circle. (For research purposes. It will not be reported to anyone and you need not answer if you prefer.)

1 ○ American Indian   3 ○ Caucasian/White   4 ○ Hispanic/Mexican
5 ○ Oriental/Asian   6 ○ Puerto Rican   7 ○ Other (Specify)

7. Education: Blacken only one circle for the highest grade or year you completed.

Grade School and High School (including Vocational)   After High School: Trade or Technical (Vocational) School   College

○7 ○8 ○9 ○10 ○11 ○12   ○1 ○2 ○3 ○4   ○1 ○2 ○3 ○4 ○More

8. Employer: Which of the below best describes your current employer? (Blacken one circle and fill in subcodes from pp. 19-20 as appropriate.)

1 ○ New Car/Truck Dealer/Distributor, Domestic—
2 ○ New Car/Truck Dealer/Distributor, Import— } Enter code for make of car or truck . . . .

3 ○ Service Station—Enter code . . . .
4 ○ Military—Enter Code . . . .
5 ○ Indp't Repair Shop— Enter code for
6 ○ Fleet Repair Shop— vehicle type }
7 ○ Specialty Shop—Enter code . . . .
8 ○ Government/Civil Service
9 ○ Student
10 ○ Educator/Instructor
11 ○ Manufacturer
12 ○ Independent Collision Shop

13 ○ Volume Retailer—Enter code . . . . . .
14 ○ Tire Dealer—Enter code . . . .
16 ○ Utility
17 ○ Lift Truck Dealer/Repair Shop
18 ○ Machinist Facility—Enter code . . . . .
19 ○ Leasing & Rental Shop—Enter code
15 ○ Other

9. Test Center: (See pp 21-22.)
Center Number   City   State

10. Tests: Blacken the circles (regular) or squares (recertification) for the test(s) you plan to take. (Do not register for a recertification test unless you passed that regular test before.) Note: Do not register for more than four regular tests on any single test day.

Regular	Recertification	Regular	Recertification
May 7	**May 14**	**May 9**	**May 14**
○ A1 Auto: Engine Repair	☐ A	○ A2 Auto: Automatic Trans/Transaxle	☐ E
○ A8 Auto: Engine Performance	☐ B	○ A3 Auto: Manual Drive Train and Axles	☐ F
○ A4 Auto: Suspension & Steering	☐ C	○ A6 Auto: Electrical/Electronic Systems	☐ G
○ A5 Auto: Brakes	☐ D	○ A7 Auto: Heating & Air Conditioning	☐ H
○ F1 Alt. Fuels: Lt. Vehicle CNG			
○ S4 School Bus: Brakes		○ L1 Adv. Level: Auto Adv. Engine Perf. Spec.	
○ S5 School Bus: Suspension/Steering		○ T1 Med/Hvy Truck: Gasoline Engines	☐ I
May 14	**May 14**	○ T2 Med/Hvy Truck: Diesel Engines	☐ J
○ T3 Med/Hvy Truck: Drive Train	☐ L	○ T6 M/Hvy Truck: Elec./Electronic Systems	☐ K
○ T4 Med/Hvy Truck: Brakes	☐ M		
○ T5 Med/Hvy Truck: Suspension/Steering	☐ N	○ S6 School Bus: Elec./Electronic Systems	
○ T8 Med/Hvy Truck: PMI		○ B2 Body: Painting & Refinish	☐ P
○ M1 Machinist: Cylinder Head Specialist	☐ Q	○ B3 Body: Nonstructural Analysis	
○ M2 Machinist: Cylinder Block Specialist	☐ R	○ B4 Body: Structural Analysis	
○ M3 Machinist: Assembly Specialist	☐ S	○ B5 Body: Mechanical & Electrical Components	

11. Fees: Number of **Regular** tests (except L1) marked above_____ x $18" = $_____

If you are taking the L1 Advanced Level Test, add $40 here + $_____

If you marked **any** of the **Recertification** tests, add $20 here + $_____

Registration Fee + $ **20**

Important: School Bus Technicians, see page 8 for a money-saving offer! If you qualify, check here ☐ and adjust your fees (Item 11, first line) accordingly.

TOTAL FEE = $_____

☐ MasterCard   ☐ VISA

Credit Card #   Expiration Date
Month   Year

Signature of Cardholder _____

12. Fee Paid By: 1 ○ Employer   2 ○ Technician

13. Experience: Blacken one circle. (To substitute training or other appropriate experience, see page 3.)

1 ○ I certify that I have two years or more of full-time experience (or equivalent) as an automobile, medium/heavy truck, school bus or collision repair/refinish technician or as an engine machinist. (Fill out #14 on opposite side)
2 ○ I don't yet have the required experience. (Skip 14.)

14. Job History: Required. Provide job history and job details on the opposite side of this form.

15. Authorization: By signing and sending in this form, I accept the terms set forth in the *Registration Booklet* about the tests and the reporting of results.

Signature of Registrant   Date

Figure 30-9. *Registration form used to apply to take ASE tests. (ASE)*

certificate and an ASE shirt patch. You can retake any test you fail to pass as many times as you like.

All information submitted to ASE, as well as your test results, are considered confidential. ASE will not provide test results to anyone, even if your employer has paid your test fees. The only information ASE will release without your permission is to confirm that you are certified in a skill area.

Association of Diesel Specialists

Trade associations, such as the *Association of Diesel Specialists (ADS)* are a good source of up-to-date information. ADS is an international organization with members in the United States, Canada, and 35 other countries. Its primary focus is the service and technology of diesel fuel injection, governors, and turbocharger systems. For member companies and individuals, ADS means worldwide recognition by diesel owners and operators. To their customers, ADS signifies a network of professionals dedicated to the achievement of high-caliber service on diesel fuel injection, governor, and turbocharger systems.

Members include service shops, manufacturers, distributors, and associates who work together and share their

expertise in order to provide diesel users with the best possible service, equipment, and parts. For further information on this organization, write:

The Association of Diesel Specialists
9140 Ward Parkway
Kansas City, MO 64114

A Few Final Words

There is little information available concerning the human behavior of technicians, particularly young technicians who are new graduates beginning their professional life. In some cases, it is assumed that they will "learn the hard way." There is a transition period from working on classroom diesel engines that disassemble easily to diesel engines in the field where bolts break, threads strip, and mating surfaces refuse to part without great effort and aggravation. For many, this transition period is difficult, and is too much for some.

In most cases, technicians who do not succeed in the diesel industry fail because of inadequate personal traits and insensitivity to the workforce diversity, not from a lack of training or ability. Diesel technicians are a diverse group of people, and the ability to get along and work harmoniously cannot be overemphasized. Yet, the students who most need to realize the truth of this statement do not believe it applies to them or only applies to other technicians. There is nothing more loathsome to a seasoned technician than a new technician in the shop who is convinced they know everything.

Working in the Field

In many cases, you will probably be working under the guidance of an experienced, certified technician. Listen carefully to any instructions any experienced technician or shop supervisor gives you. While your training has given you many skills to begin your career, they are only a fundamental base that will be built upon and tempered by years of work experience. As in many other professions, you will only prove your worth by performance on the job.

There is a premium on technicians that have the ability to get things done. Most technicians quickly realize the importance of being a self-starter, but what many miss is the other half of the equation: being a sure-finisher. The extra effort is always worth the payoff, however, do not expect to be recognized or rewarded every time that extra effort is expended. The true value is in the self-esteem gained by knowing your own self worth, and the respect you earn from your coworkers and managers.

At first, you will be assigned simple tasks, then more complex jobs as your experience increases. These duties, such as scraping gaskets and washing parts, will at first seem mundane and unworthy of your training. But however trivial these first assignments may seem, give them your best efforts. Remember, they are as important as properly assembling a cylinder head or other complex job. It may not be apparent, but these initial assignments will be closely monitored and checked. Do the best job you know how to do.

Every shop, large and small, has its share of housekeeping chores that must be done. In most shops, you are responsible for the upkeep and cleanliness of your work area. Floors must be swept and mopped, air compressor tanks drained, and spills cleaned. It is natural that the newest employee in a shop inherit many of these chores. Every technician in the shop, at one time, has performed their share of these chores. Do these chores willingly, graciously, and with a minimum of grumbling.

Finally, regard your integrity as one of your most important assets. The automotive service profession has suffered tremendous criticism and contempt from the general public. Unfortunately this criticism has been leveled because of a few dishonest and unscrupulous mechanics out for a "quick buck." Uncompromising integrity must be worked at constantly. Remember that everything you do reflects on the industry as a whole.

The profession of diesel service and repair, even with its occasional moments of frustration and adverse working conditions, offers one of the most rewarding careers available today. Make your career choices carefully and deliberately, and success is sure to follow.

Summary

Many careers are available in the diesel field for people of diversified skills. Many of these jobs pay well and provide a rewarding professional career. Diesel engines are a vital part of commerce. Although many of these jobs are not highly visible or widely publicized, they are numerous and important.

Although many people work in the manufacture and assembly of diesel engines, the majority of technicians service, repair, and maintain diesel engines. Technicians may serve as general technicians or specialized technicians, and may eventually rise to various levels of management.

There are many avenues open to those interested in working as a diesel technician. Formal apprenticeship programs are offered by labor unions or companies that manufacture or distribute diesel equipment. Many vocational schools and technical colleges offer part- or full-time programs in diesel service and repair. Most companies prefer to hire apprentices and trainees with a minimum of a high school diploma or equivalent. Experience in the field coupled with a bachelor's or master's degree from an accredited college or university is usually necessary for those who wish to enter the diesel teaching profession.

The National Institute for Automotive Service Excellence, referred to simply as ASE, offers a voluntary certification program recognized throughout the industry. There are eight tests in the Medium/Heavy Truck Technician certification series. A technician who passes one or more tests and has at least two years of relevant hands-on experience can become certified as an ASE Truck Technician. By passing the Diesel Engine Test (or the Gasoline Engine Test) and the other six tests in the series, a technician can earn the certificate of Master Truck Technician.

The types of knowledge and skills needed to pass these tests are basic technical knowledge, correction/repair procedures, and testing/diagnostic knowledge and skills. A number of other certification areas are available, but the most commonly pursued area after Master Truck Technician is that of Master Engine Machinist. Other trade associations, such as the Association of Diesel Specialists (ADS) are a good source of updated information on diesel servicing.

Important Terms

Diesel technicians	Full-time study
General technicians	National Institute for Automotive Service Excellence
Specialization	
Heavy equipment technicians	ASE
Truck technicians	Medium/Heavy Truck Technician Certification Series
Marine technicians	
Shop supervisors	ASE Truck Technician
Service manager	Master Truck Technician
Apprenticeship programs	Association of Diesel Specialists (ADS)

Review Questions—Chapter 30

Do not write in this text. Place your answer on a seprate sheet of paper

1. Name the categories in which employment is available in the diesel industry.

2. *True or False?* Diesel engines must be maintained and repaired, no matter how well they are constructed.

3. Companies that manufacture diesel engines employ _____ of workers

4. _____ has brought about a number of specific job responsibilities.

5. Scheduling training classes and ordering supplies, tools, and equipment is the responsibility of the _____.

6. Which diesel industry positions involves the most direct interaction with technicians on a day-to-day basis?
   (A) Shop supervisor.
   (B) Service manager.
   (C) Vocational school instructor.
   (D) None of the above.

7. Many technicians start as _____.
   (A) master technicians
   (B) helpers
   (C) entrepreneurs
   (D) shop supervisors

8. By passing either the Diesel Engine or Gasoline Engine test and the other tests in the Medium/Heavy truck series, a technician can earn the title and certificate of _____.

9. *True or False?* All information submitted to ASE is confidential.

10. What is the only information that ASE will release without your permission?

ASE-Type Questions

1. Diesel engines are used in all of the following applications, EXCEPT:
   (A) motorcycles.
   (B) bulldozers.
   (C) cranes.
   (D) ships.

2. All of the following are job duties of the service manager, EXCEPT:
   (A) taking customer service orders.
   (B) seeing that schedules are met.
   (C) assist technicians in diagnosing problems.
   (D) working with the shop supervisor and other managers.

3. A voluntary certification program is offered by _____.
   (A) SAE.
   (B) ACT.
   (C) SAT.
   (D) ASE.

4. The questions on ASE tests are written by _____.
   (A) service industry experts
   (B) guidance counselors
   (C) principals
   (D) school administrators

5. The ASE Master Truck certification requires passing _____ of the eight test areas in the Medium/Heavy Truck test battery.
   (A) four
   (B) five
   (C) seven
   (D) six

6. All of the questions on ASE tests are _____ .
   (A) true or false
   (B) theory
   (C) manufacturer specific
   (D) job-related

7. Technician A says a percentage of ASE tests cover operational theory. Technician B says that certain questions are manufacturer specific. Who is right?
   (A) A only.
   (B) B only.
   (C) Both A & B.
   (D) Neither A nor B.

8. Technician A says that passing one or more ASE tests is all you need to become ASE certified. Technician B says that in addition to passing the tests, you need at least two years of relevant work experience before you can become ASE certified. Who is right?
   (A) A only.
   (B) B only.
   (C) Both A & B.
   (D) Neither A nor B.

9. Technician A says that all information provided to ASE is confidential. Technician B says employers that pay a technician's ASE test fees will be provided test results. Who is right?
   (A) A only.
   (B) B only.
   (C) Both A & B.
   (D) Neither A nor B.

10. If you fail to pass an ASE test, you _____.
    (A) can retake the test as many times as you wish
    (B) can retake the test only once
    (C) can retake the test twice in a five year period
    (D) cannot retake the test

Troubleshooting Engine and Fuel System Problems

This Appendix will allow you to look up the causes and corrections for various diesel engine problems. Other diagnostic charts are located throughout this text. Diagnostic charts are also included in engine and vehicle service manuals.

Low Power Output

Possible Cause	Correction
Low engine rpm	Adjust governor linkage to rest against the high idle stop when accelerator pedal is depressed.
	Adjust high idle rpm with adjustment screw. If proper setting cannot be made, disassemble, inspect, and service governor.
Restriction in air inlet system	Inspect air cleaner element and ductwork for damage and excessive bends and turns. Install new air cleaner element.
Water in fuel	Test for fuel in water. Drain fuel tank as needed to remove water. Install new fuel filters and fill tanks with clean fuel.
Fuel with low specific gravity	Test fuel API gravity. Fuel should have 35 API gravity at 60° F (16°C). Replace fuel if API gravity is greater than 38.
Low fuel pressure	Check fuel pressure at fuel pump housing.
Fuel rack setting incorrect	Check rack setting and adjust if needed.
Fuel too hot due to blocked constant bleed valve.	Check by removing fuel return line and checking for sufficient fuel flow from valve. Replace valve if blocked.
Faulty shut-off solenoid or linkage	Remove solenoid and start engine to see if it has full power. If there is full power, replace solenoid and/or adjust linkage.

Misfiring and Rough Running

Possible Cause	Correction
Air in fuel system	Bleed air from lines.
Valve adjustment incorrect	Adjust intake and exhaust valves to service manual specifications.
Fuel injection timing incorrect	Check and make necessary adjustments.
Automatic timing advance malfunction	Check for correct timing. If a timing light is not available, check for smooth acceleration from low to high idle. Replace timing mechanism if defective.

Defective fuel nozzles	Locate misfiring injector; clean and service as required.
Engine valve leaking, worn, or damaged pistons and/or rings, worn or damaged cylinder liners	Check with cylinder leakage tester. Replace components as required.
Cylinder head gasket leakage	Check with cylinder leakage tester and replace gasket as needed.
Incorrect camshaft timing	Check timing. If timing marks are correct, check for a broken drive dowel in the camshaft drive gear.
Fuel leakage at nozzle nut or adapter	Tighten nut to specifications.

Black or Gray Exhaust Smoke (Engine Runs Smooth)

Possible Cause	Correction
High altitude operation at 2500 ft. (762 m) or greater	Adjust fuel rack position.
Engine used in lug condition	Operate engine in a gear where engine can have an increase in speed under load.
Dirty air cleaner	Replace dirty air cleaner element.
Air inlet system restriction	Inspect ductwork for damage and/or excessive turns and bends. Check for damaged rain cap or cap that is pushed too far onto the inlet pipe. Adjust or repair as needed.
Exhaust system restriction	Inspect system piping and muffler. Perform exhaust back pressure test. Locate and correct source of any restriction.
Fuel injection timing incorrect	Check and make necessary adjustments.
Fuel rack setting incorrect	Check rack setting and adjust to specifications as needed.
Fuel with low specific gravity	Test fuel API gravity. Fuel should have 35 API gravity at 60°F (16°C).
Valve adjustment incorrect	Adjust intake and exhaust valves to service manual specifications.
Leaking fuel nozzles	Locate misfiring injector and clean and service as required.

Black or Gray Exhaust Smoke (Engine Runs Rough)

Possible Cause	Correction
Misfiring cylinders	See *Misfiring and Rough Running.*
Automatic timing advance malfunction	Check for correct timing. If light is not available, check for smooth acceleration from low to high idle.
Air in fuel system	Check for leaks and bleed air from system.

White Exhaust Smoke

Possible Cause	Correction
Cold outside temperatures	Normal condition until engine warms to operating temperature, as not all fuel is burning. Minimize by using number 1 diesel fuel.
Long idle periods	Reduce unnecessary idling or use fuel heating system.
Fuel with low specific gravity	Test fuel API gravity. Fuel should have 35 API gravity at 60°F (16°C).
Air in fuel system	Check for leaks and bleed air from system.
Fuel injection timing incorrect	Check and make necessary adjustments.
Automatic timing advance malfunction	Check for correct timing. If timing light is not available, check for smooth acceleration from low to high idle.

Valve adjustment incorrect   Adjust intake and exhaust valves to service manual specifications.

Bad fuel nozzles   Locate misfiring injector. Clean and service as required.

Misfiring cylinders   See *Misfiring and Rough Running.*

Blue Exhaust Smoke

Possible Cause	Correction
Engine oil level too high	Do not overfill crankcase. Check oil for fuel dilution.
Damaged PCV valve	Inspect for damage and replace as needed.
Worn valve guides	Check for maximum permitted wear and service as needed.
Worn piston rings and/or cylinder walls.	Check cylinder compression and leakage. Visually inspect piston rings and cylinder walls. Measure and compare against specifications. Service as needed.
Worn or damaged pistons	Check piston ring-to-groove clearance. Make sure return holes under oil control rings are open.

Excessive Fuel Consumption

Possible Cause	Correction
Air inlet system restriction	Inspect ductwork for damage and/or excessive turns and bends. Check for damaged rain cap or cap that is pushed too far onto the inlet pipe.
External fuel system leakage	Check fuel system external piping and tubing for signs of leaks. Repair as needed.
Fuel injection timing incorrect	Check and make necessary adjustments.
Leaking fuel nozzles	Locate leaking injector and service as required.
Fuel injection pump calibration incorrect	Remove injection pump and nozzle assemblies from engine. Check calibration and adjust.
Internal engine wear	Overhaul engine

Excessive Oil Consumption

Possible Cause	Correction
External oil leaks	Check engine for visible signs of leaks. Replace defective parts.
Clogged crankcase breather pipe	Remove obstruction.
Exhaust system restriction	Inspect system piping and muffler. Perform exhaust back pressure test. Locate and correct source of any restriction.
Worn valve guides	Replace valve guides.
Failure of seal rings in turbocharger	Check inlet manifold for oil and make any needed repairs.
Internal engine wear.	Overhaul engine.

Low Lubricating Oil Pressure

Possible Cause	Correction
Oil leak or low oil level	Check for leaks sand add oil as needed.
Wrong oil viscosity	Drain oil, replace filter, and fill with proper oil.
Defective pressure gauge	Check gauge operation and replace as needed.

Plugged oil filter	Replace filter. Clean or install new oil cooler core. Change oil.
Fuel oil in crankcase	*See listing above.*
Defective oil pump relief valve	Check valve seat condition, and valve spring for sticking and proper tension. Make needed repairs or install new relief valve.
Incorrect meshing of oil pump gears	Check mounting arrangement. If the oil pump has been rebuilt, check for proper gear ratio combination. Check for correct oil pump cover gasket.
Excessive clearance between crankshaft and bearings	Overhaul engine.

Erratic Engine Speed

Possible Cause	**Correction**
Air leaks in fuel system	Check for air leaks and make needed repairs.
Throttle linkage loose or out of adjustment	Check and adjust linkage.
Injection pump governor failure	Check injection pump for damaged or broken springs or other components. Check fuel rack for free travel. Check for correct governor spring. Install new parts as needed and recalibrate injection pump.

Engine Stalls at Low Speeds

Possible Cause	**Correction**
Idle speed too low	Adjust idle as required.
Fuel tank vent plugged	Check vent arrangement and make needed repairs.
Low fuel supply	Check tank for fuel. Check fuel lines for sharp bends and restrictions. Check fuel pressure. If pressure is low, replace fuel filters. If pressure is still low, replace transfer pump. Bleed fuel system.
Defective fuel injection nozzle or pump	Inspect, test, and replace nozzle parts or injection pump.

Engine Does Not Reach No-Load Governed Speed

Possible Cause	**Correction**
Air in fuel system	Check for leaks and bleed air from lines.
Accelerator linkage loose or misadjusted	Check linkage and adjust as needed.
Restricted fuel lines/stuck overflow valve	Check for restrictions and for defective spring, poor valve setting, or sticking. Make all necessary repairs.
High idle adjustment set too low	Check setting and adjust as needed.
Fuel injection pump calibrated incorrectly	Remove injection pump and nozzle assemblies from engine and recalibrate.
Internal fuel pump governor wear	Remove injection pump from engine and make all necessary repairs.

Difficult Starting (Crankshaft Turns)

Possible Cause	**Correction**
Cold outside temperatures	Use starting aids.

Air in fuel system	Bleed air from lines.
Water in fuel	Test for water in fuel. Drain fuel tank as needed to remove water. Install new fuel filter and fill tanks with clean fuel.
Low fuel pressure	Check fuel pressure at fuel pump housing.
Fuel injection timing incorrect	Check and make necessary adjustments.
Valve adjustment incorrect	Adjust intake and exhaust valve to service manual specifications.
Bad fuel nozzles	Locate misfiring injector and clean and service as required.
Low compression	See *Misfiring and Rough Running.*
No exhaust smoke visible while starting	No fuel in tank. Tank valves inadvertently closed. No fuel from injection pump due to cold weather waxing or fuel line restriction. Faulty fuel injection pump shut-off solenoid. Repair as needed.
Exhaust system blockage	Loosen exhaust pipes at manifolds. If engine starts, check for blockage or restrictions in exhaust piping.

Difficult Starting (Crankshaft will not Turn)

Possible Cause	Correction
Low or no battery voltage	Recharge battery or replace.
Faulty relay switch or wiring connection in starter circuit.	Check for voltage at switch connection on starter solenoid. Repair as needed.
Bad starter solenoid	Replace starter solenoid.
Bad starter motor	Repair or replace.
Transmission or PTO problem prevents crankshaft from turning.	Disconnect transmission and/or PTO and see if crankshaft will turn. If it does, troubleshoot transmission or PTO.
Internal engine problem	Remove fuel nozzles and check for fuel in the cylinders while turning crankshaft. If no fuel is evident, disassemble engine to check for bearing and/or piston seizure, valve contact with pistons, etc. Repair or replace engine as needed.

Difficult Starting (Crankshaft Turns Slowly)

Possible Cause	Correction
Low battery voltage	Check voltage and recharge/replace battery as needed.
Poor electrical connections	Inspect, clean, and tighten electrical connections.
Lubricating oil too thick	Use proper grade oil for temperature conditions.
Bad starter motor	Remove and test. Repair or replace as needed.
Extra outside loads	Check transmission and PTO for damage.
Internal engine problem	Disassemble and troubleshoot engine.

Cooling System (Above Normal Heating)

Possible Cause	Correction
Low coolant level	Fill to capacity. Check for leaks and repair as needed.
Bad temperature gauge	Check with known good gauge. Repair or replace gauge.
Clogged radiator	Clean and flush radiator core.
Loose fan or water pump belts	Tighten to specifications.
Bad hoses	Check for leaks and collapsed hoses. Change hoses as needed.
Shunt line restriction	Check for restrictions from the radiator top tank to the engine front-cover.

Shutter not opening	Check opening temperature of shutters. Shutters must be closed below the water temperature regulator's fully open rating.
Bad temperature regulator	Test thermostats and replace as needed.
Bad water pump	Check for loose water pump pulley or impeller. Remove pump and check for impeller damage and proper impeller clearance.
Air in cooling system	Air can enter when coolant level is low. Combustion gases can also enter through cracks or a bad cylinder head gasket. Perform tests to determine if exhaust gas is leaking into the coolant. Note: Be sure this test is in manual or book.
Undersize fan, fan and/or shroud in incorrect position	Fan must be large enough to send air through most of the radiator area. Fan and shroud must be positioned as per truck manufacturer's specifications
Undersize radiator	Make certain radiator is sized to manufacturer's specifications.
Air restriction in engine compartment	Make certain air flow through radiator comes out of engine compartment and is not blocked by filters, air conditioners, etc.
High outside ambient temperature	Insufficient temperature difference between cooling air and coolant. To get better cooling, operate vehicle in lower gear.
High altitude operation	Cooling capacity goes down at higher altitudes. Be sure application is equipped with sufficient cooling capacity.
Operating in lug condition	Lugging occurs when the truck is operated in a gear too high for rpm to rise as the accelerator is pressed. Low rpm reduces air flow through radiator. To correct, operate truck in lower gear.
Air inlet system restriction	Inspect ductwork for damage and/or excessive turns and bends. Check for damaged rain cap or cap that is pushed too far onto the inlet pipe.
Exhaust system restriction	Inspect system piping and muffler. Perform exhaust back pressure test. Locate and correct source of any restriction.
Fuel injection timing incorrect	Check and make necessary adjustments.
Transmission problems	Slippage or power shifting automatic transmissions cooled by the radiator. Service transmission as needed.

Cooling System (Below Normal Heating)

Possible Cause	Correction
Long idling periods	Limit idling period.
Very light loads	Install shutters to correct problem.
Defective water temperature regulator	Regulator stuck open; replace.

Loss of Coolant (Outside Leaks)

Possible Cause	Correction
Leaks in hoses or connections	Inspect hoses, clamps, and connections and repair as needed.
Leaks in radiator or expansion tank	Pressure test radiator and tank to check for leaks.
Leaks in heater	Pressure test heater and check for leaks.
Leaks in water pump	Check for leaks from pump with engine off and engine running. Install new pump as needed.
Cylinder head gasket leak	Inspect for leaks along the surface of the cylinder head gasket. Install new gasket as needed.

Loss of Coolant (Leaks at Overflow Tube)

Possible Cause / Correction

Possible Cause	Correction
Defective pressure cap	Check sealing surfaces of cap. Pressure test cap. Replace if needed.
Engine runs hot	See *Cooling System/Above Normal Heating.*
Expansion tank too small	Use properly sized tank and check for proper installation.

Loss of Coolant (Inside Leakage)

Possible Cause / Correction

Possible Cause	Correction
Cylinder head gasket leakage	Repair as needed.
Cracks in cylinder head	Repair or replace cylinder head.
Cracks in cylinder block	Repair or replace as needed.

Fuel in Crankcase Oil

Possible Cause / Correction

Possible Cause	Correction
Fuel injection pump camshaft seal leaking	Replace seal.
Fuel injection pump drain line restricted	Remove restriction.
Loose injector nuts	Inspect and replace O-ring, if needed, and tighten injector nuts.
Leaking fuel nozzles	Rebuild or replace fuel nozzles.

Excessive Engine Vibration

Possible Cause / Correction

Possible Cause	Correction
Loose vibration damper hub nut	Check damper and tighten nut.
Defective or damaged vibration damper	Replace damper.
Fan blades not in balance	Replace fan assembly.
Loose engine supports	Tighten all bolts and replace any defective parts.
Misfire or rough running engine	See *Misfiring and Rough Running.*

Conversion Charts

Some Common Abbreviations			
U.S. Customary		**Metric**	
Unit	Abbreviation	Unit	Abbreviation
inch . . . . .in.		kilometer . . . . .km	
feet . . . . .ft.		hectometer . . . . .hm	
yard . . . . .yd.		dekameter . . . . .dam	
mile . . . . .mi.		meter . . . . .m	
grain . . . . .gr.		decimeter . . . . .dm	
ounce . . . . .oz.		centimeter . . . . .cm	
pound . . . . .lb.		millimeter . . . . .mm	
teaspoon . . . . .tsp.		cubic centimeter . . . . .cm³	
tablespoon . . . . .tbsp.		kilogram . . . . .kg	
fluid ounce . . . . .fl.oz.		hectogram . . . . .hg	
cup . . . . .c.		dekagram . . . . .dkg	
pint . . . . .pt.		gram . . . . .g	
quart . . . . .qt.		decigram . . . . .dg	
gallon . . . . .gal.		centigram . . . . .cg	
cubic inch . . . . .in³		milligram . . . . .mg	
cubic foot . . . . .ft³		kiloliter . . . . .kL	
cubic yard . . . . .yd³		hectoliter . . . . .hL	
square inch . . . . .in²		dekaliter . . . . .dL	
square foot . . . . .ft²		liter L	
square yard . . . . .yd²		centiliter cL	
square mile . . . . .mi²		milliliter . . . . .mL	
Fahrenheit . . . . .F		square kilometer . . . . .km²	
barrel . . . . .bbl.		hectare . . . . .ha	
fluid dram . . . . .fl. dr.		are . . . . .a	
board foot . . . . .bd. ft.		centare . . . . .ca	
rod . . . . .rd.		tonne . . . . .t	
dram . . . . .dr.		Celsius C	
bushel . . . . .bu.			

Measuring Systems

U.S. Customary	**Metric**

Length

12 inches = 1 foot	1 kilometer = 1000 meters
36 inches = 1 yard	1 hectometer = 100 meters
3 feet = 1 yard	1 dekameter = 10 meters
5,280 feet = 1 mile	1 meter = 1 meter
16.5 feet = 1 rod	1 decimeter = 0.1 meter
320 rods = 1 mile	1 centimeter = 0.01 meter
6 feet = 1 fathom	1 millimeter = 0.001 meter

Weight

27.34 grains = 1 dram	1tonne = 1,000,000 grams
438 grains = 1 ounce	1 kilogram = 1000 grams
16 drams = 1 ounce	1 hectogram = 100 grams
16 ounces = 1 pound	1 dekagram = 10 grams
2000 pounds = 1 short ton	1 gram = 1 gram
2240 pounds = 1 long ton	1 decigram = 0.1 gram
25 pounds = 1 quarter	1 centigram = 0.01 gram
4 quarters = 1 cwt	1 milligram = 0.001 gram

Volume

8 ounces = 1 cup	1 hectoliter = 100 liters
16 ounces = 1 pint	I dekaliter = 10 liters
32 ounces = 1 quart	1 liter = 1 liter
2 cups = 1 pint	1 deciliter = 0.1 liter
2 pints = 1 quart	1 centiliter = 0.01 liter
4 quarts = 1 gallon	1 millileter = 0.001 liter
8 pints = 1gallon	1000 millileter = 1 liter

Area

144 sq. inches = 1 sq. foot	100 sq. millimeters = 1 sq. centimeter
9 sq. feet = 1 sq. yard	100 sq. centimeters = 1 sq. decimeter
43,560 sq. ft. = 160 sq. rods	100 sq. decimeters = 1 sq. meter
160 sq. rods = 1 acre	10,000 sq. meters = 1 hectare
640 acres = 1 sq. mile	

Temperature

Fahrenheit		*Celsius*
32°F	Water freezes	0° C
68°F	Reasonable room temperature	20°C
98.6°F	Normal body temperature	37°C
173°F	Alcohol boils	78.34°C
121°F	Water boils	100°C

Useful Conversions

When you know:	Multiply by:	To find:
Torque		
pound-inch	0.11298	newton-meters (N•m)
pound-foot	1 .3558	newton-meters
Light		
foot candles	1.0764	lumens/meters^2 (lm/m^2)
Fuel Performance		
miles/gallons	0.4251	kilometers/liter (km/L)
Speed		
miles/hour	1.6093	kilometers/hr (km/h)
Force		
kilogram	9.807	newtons (n)
ounce	0.278	newtons
pound	4.448	newtons
Power		
horsepower	0.746	kilowatts (kw)
Pressure or Stress		
inches of water	0.2491	kilopascals (kPa)
pounds/sq. in.	6.895	kilopascals
Energy or Work		
btu	1055.0	joules (J)
foot-pound	1.3558	joules
kilowatt-hour	3600000.0	joules

Conversion Table
Metric to U.S. Customary

When you know:	Multiply by: * = Exact		To find:
	Very Accurate	**Approximate**	
Length			
millimeters	0.0393701	0.04	inches
centimeters	0.3937008	0.4	inches
meters	3.280840	3.3	feet
meters	1.093613	1.1	yards
kilometers	0.621371	0.6	miles
Weight			
grains	0.00228571	0.0023	ounces
grams	0.03527396	0.035	ounces
kilograms	2.204623	2.2	pounds
tonnes	1.1023113	1.1	short tons
Volume			
milliliters	0.20001	0.2	teaspoons
milliliters	0.06667	0.067	tablespoons
milliliters	0.03381402	0.03	fluid ounces
liters	61.02374	61.024	cubic inches
liters	2.113376	2.1	pints
liters	1.056688	1.06	quarts
liters	0.26417205	0.26	gallons
liters	0.03531467	0.035	cubic feet
cubic meters	61023.74	61023.7	cubic inches
cubic meters	35.31467	35.0	cubic feet
cubic meters	1.3079506	1.3	cubic yards
cubic meters	264.17205	264.0	gallons
Area			
square centimeters	0.1550003	0.16	square inches
square centimeters	0.00107639	0.001	square feet
square meters	10.76391	10.8	square feet
square meters	1.195990	1.2	square yards
square kilometers		0.4	square miles
hectares	2.471054	2.5	acres
Temperature			
Celsius	*9/5 (then add 32)		Fahrenheit

Conversion Table
U.S. Customary To Metric

When you know: ⬇	Multiply by: * = Exact		To find: ⬇
	Very Accurate	**Approximate**	
Length			
inches	* 25.4		millimeter
inches	* 2.54		centimeter
feet	* 0.3048		meters
feet	* 30.48		centimeters
yards	* 0.9144	0.9	meters
miles	* 1.609344	1.6	kilometers
Weight			
grains	15.43236	15.4	grams
ounces	* 28.349523125	28.0	grams
ounces	* 0.028349523125	.028	kilograms
pounds	* 0.45359237	0.45	kilograms
ton	* 0.90718474	0.9	tonnes
Volume			
teaspoons	* 4.97512	5.0	milliliters
tablespoons	* 14.92537	15.0	milliliters
fluid ounces	29.57353	30.0	milliliters
cups	* 0.236588240	0.24	liters
pints	* 0.473176473	0.47	liters
quarts	* 0.946352946	0.95	liters
gallons	* 3.785411784	3.8	liters
cubic inches	* 0.016387064	0.02	liters
cubic feet	* 0.028316846592	0.03	cubic meters
cubic yards	* 0.764554857984	0.76	cubic meters
Area			
square inches	* 6.4516	6.5	square centimeters
square feet	* 0.09290304	0.09	square meters
square yards	* 0.83612736	0.8	square meters
square miles	* 2.589989	2.6	square kilometers
acres	* 0.40468564224	0.4	hectares
Temperature			
Fahrenheit	* 5/9 (after subtracting 32)		Celsius

Dimensional and Temperature Conversion Chart

Inches	Decimals	Millimeters	Inches to Millimeters in.	Inches to Millimeters mm	Millimeters to Inches mm	Millimeters to Inches in.	Fahrenheit & Celsius F	Fahrenheit & Celsius °C	Fahrenheit & Celsius °C	Fahrenheit & Celsius °F
1/64	.015625	.3969	.0001	.00254	0.001	.000039	-20	-28.9	-30	-22
1/32	.03125	.7937	.0002	.00508	0.002	.000079	-15	-26.1	-28	-18.4
3/64	.046875	1.1906	.0003	.00762	0.003	.000118	-10	-23.3	-26	-14.8
1/16	.0625	1.5875	.0004	.01016	0.004	.000157	-5	-20.6	-24	-11.2
5/64	.078125	1.9844	.0005	.01270	0.005	.000197	0	-17.8	-22	-7.6
3/32	.09375	2.3812	.0006	.01524	0.006	.000236	1	-17.2	-20	-4
7/64	.109375	2.7781	.0007	.01778	0.007	.000276	2	-16.7	-18	-0.4
1/8	.125	3.1750	.0008	.02032	0.008	.000315	3	-16.1	-16	3.2
9/64	.140625	3.5719	.0009	.02286	0.009	.000354	4	-15.6	-14	6.8
5/32	.15625	3.9687	.001	.0254	0.01	.00039	5	-15.0	-12	10.4
11/64	.171875	4.3656	.002	.0508	0.02	.00079	10	-12.2	10	14
3/16	.1875	4.7625	.003	.0762	0.03	.00118	15	-9.4	-8	17.6
13/64	.203125	5.1594	.004	.1016	0.04	.00157	20	-6.7	-6	21.2
7/32	.21875	5.5562	.005	.1270	0.05	.00197	25	-3.9	-4	24.8
15/64	.234375	5.9531	.006	.1524	0.06	.00236	30	-1.1	-2	28.4
1/4	.25	6.3500	.007	.1778	0.07	.00276	35	1.7	0	32
17/64	.265625	6.7469	.008	.2032	0.08	.00315	40	4.4	2	35.6
9/32	.28125	7.1437	.009	.2286	0.09	.00354	45	7.2	4	39.2
19/64	.296875	7.5406	.01	.254	0.1	.00394	50	10.0	6	42.8
5/16	.3125	7.9375	.02	.508	0.2	.00787	55	12.8	8	46.4
21/64	.328125	8.3344	.03	.762	0.3	.01181	60	15.6	10	50
11/32	.34375	8.7312	.04	1.016	0.4	.01575	65	18.3	12	53.6
23/64	.359375	9.1281	.05	1.270	0.5	.01969	70	21.1	14	57.2
3/8	.375	9.5250	.06	1.524	0.6	.02362	75	23.9	16	60.8
25/64	.390625	9.9219	.07	1.778	0.7	.02756	80	26.7	18	64.4
13/32	.40625	10.3187	.08	2.032	0.8	.03150	85	29.4	20	68
27/64	.421875	10.7156	.09	2.286	0.9	.03543	90	32.2	22	71.6
7/16	.4375	11.1125	.1	2.54	1	.03937	95	35.0	24	75.2
29/64	.453125	11.5094	.2	5.08	2	.07874	100	37.8	26	78.8
15/32	.46875	11.9062	.3	7.62	3	.11811	105	40.6	28	82.4
31/64	.484375	12.3031	.4	10.16	4	.15748	110	43.3	30	86
1/2	.5	12.7000	.5	12.70	5	.19685	115	46.1	32	89.6
33/64	.515625	13.0969	.6	15.24	6	.23622	120	48.9	34	93.2
17/32	.53125	13.4937	.7	17.78	7	.27559	125	51.7	36	96.8
35/64	.546875	13.8906	.8	20.32	8	.31496	130	54.4	38	100.4
9/16	.5625	14.2875	.9	22.86	9	.35433	135	57.2	40	104
37/64	.578125	14.6844	1	25.4	10	.39370	140	60.0	42	107.6
19/32	.59375	15.0812	2	50.8	11	.43307	145	62.8	44	112.2
39/64	.609375	15.4781	3	76.2	12	.47244	150	65.6	46	114.8
5/8	.625	15.8750	4	101.6	13	.51181	155	68.3	48	118.4
41/64	.640625	16.2719	5	127.0	14	.55118	160	71.1	50	122
21/32	.65625	16.6687	6	152.4	15	.59055	165	73.9	52	125.6
43/64	.671875	17.0656	7	177.8	16	.62992	170	76.7	54	129.2
11/16	.6875	17.4625	8	203.2	17	.66929	175	79.4	56	132.8
45/64	.703125	17.8594	9	228.6	18	.70866	180	82.2	58	136.4
23/32	.71875	18.2562	10	254.0	19	.74803	185	85.0	60	140
47/64	.734375	18.6531	11	279.4	20	.78740	190	87.8	62	143.6
3/4	.75	19.0500	12	304.8	21	.82677	195	90.6	64	147.2
49/64	.765625	19.4469	13	330.2	22	.86614	200	93.3	66	150.8
25/32	.78125	19.8437	14	355.6	23	.90551	205	96.1	68	154.4
51/64	.796875	20.2406	15	381.0	24	.94488	210	98.9	70	158
13/16	.8125	20.6375	16	406.4	25	.98425	212	100.0	75	167
53/64	.828125	21.0344	17	431.8	26	1.02362	215	101.7	80	176
27/32	.84375	21.4312	18	457.2	27	1.06299	220	104.4	85	185
55/64	.859375	21.8281	19	482.6	28	1.10236	225	107.2	90	194
7/8	.875	22.2250	20	508.0	29	1.14173	230	110.0	95	203
57/64	.890625	22.6219	21	533.4	30	1.18110	235	112.8	100	212
29/32	.90625	23.0187	22	558.8	31	1.22047	240	115.6	105	221
59/64	.921875	23.4156	23	584.2	32	1.25984	245	118.3	110	230
15/16	.9375	23.8125	24	609.6	33	1.29921	250	121.1	115	239
61/64	.953125	24.2094	25	635.0	34	1.33858	255	123.9	120	248
31/32	.96875	24.6062	26	660.4	35	1.37795	260	126.6	125	257
63/64	.984375	25.0031	27	690.6	36	1.41732	265	129.4	130	266

Capacity Conversion U.S. Gallons to Liters

Gallons	0 Liters	1 Liters	2 Liters	3 Liters	4 Liters	5 Liters
0	00.0000	3.7853	7.5707	11.3560	15.1413	18.9267
10	37.8533	41.6387	45.4240	49.2098	52.9947	56.7800
20	75.7066	79.4920	83.2773	87.0626	90.8480	94.6333
30	113.5600	117.3453	121.1306	124.9160	128.7013	132.4866
40	151.4133	155.1986	158.9840	162.7693	166.5546	170.3400

Millimeter Conversion Chart

mm In.								
	15 = .5905	30 = 1.1811	45 = 1.7716	60 = 2.3622	75 = 2.9527	90 = 3.5433	105 = 4.1338	120 = 4.7244
.25 = .0098	15.25 = .6004	30.25 = 1.1909	45.25 = 1.7815	60.25 = 2.3720	75.25 = 2.9626	90.25 = 3.5531	105.25 = 4.1437	120.25 = 4.7342
.50 = .0197	15.50 = .6102	30.50 = 1.2008	45.50 = 1.7913	60.50 = 2.3819	75.50 = 2.9724	90.50 = 3.5630	105.50 = 4.1535	120.50 = 4.7441
.75 = .0295	15.75 = .6201	30.75 = 1.2106	45.75 = 1.8012	60.75 = 2.3917	75.75 = 2.9823	90.75 = 3.5728	105.75 = 4.1634	120.75 = 4.7539
1 = .0394	16 = .6299	31 = 1.2205	46 = 1.8110	61 = 2.4016	76 = 2.9921	91 = 3.5827	106 = 4.1732	121 = 4.7638
1.25 = .0492	16.25 = .6398	31.25 = 1.2303	46.25 = 1.8209	61.25 = 2.4114	76.25 = 3.0020	91.25 = 3.5925	106.25 = 4.1831	121.25 = 4.7736
1.50 = .0591	16.50 = .6496	31.50 = 1.2402	46.50 = 1.8307	61.50 = 2.4213	76.50 = 3.0118	91.50 = 3.6024	106.50 = 4.1929	121.50 = 4.7885
1.75 = .0689	16.75 = .6594	31.75 = 1.2500	46.75 = 1.8405	61.75 = 2.4311	76.75 = 3.0216	91.75 = 3.6122	106.75 = 4.2027	121.75 = 4.7933
2 = .0787	17 = .6693	32 = 1.2598	47 = 1.8504	62 = 2.4409	77 = 3.0315	92 = 3.6220	107 = 4.2126	122 = 4.8031
2.25 = .0886	17.25 = .6791	32.25 = 1.269	47.25 = 1.8602	62.25 = 2.4508	77.25 = 3.0413	92.25 = 3.6319	107.25 = 4.2224	122.25 = 4.8130
2.50 = .0984	17.50 = .6890	32.50 = 1.2795	47.50 = 1.8701	62.50 = 2.4606	77.50 = 3.0512	92.50 = 3.6417	107.50 = 4.2323	122.50 = 4.8228
2.75 = .1083	17.75 = .6988	32.75 = 1.2894	47.75 = 1.8799	62.75 = 2.4705	77.75 = 3.0610	92.75 = 3.6516	107.75 = 4.2421	122.75 = 4.8327
3 = .1181	18 = .7087	33 = 1.2992	48 = 1.8898	63 = 2.4803	78 = 3.0709	93 = 3.6614	108 = 4.2520	123 = 4.8425
3.25 = .1280	18.25 = .7185	33.25 = 1.3091	48.25 = 1.8996	63.25 = 2.4901	78.25 = 3.0807	93.25 = 3.6713	108.25 = 4.2618	123.25 = 4.8524
3.50 = .1378	18.50 = .7283	33.50 = 1.3189	48.50 = 1.9094	63.50 = 2.5000	78.50 = 3.0905	93.50 = 3.6811	108.50 = 4.2716	123.50 = 4.8622
3.75 = .1476	18.75 = .7382	33.75 = 1.3287	48.75 = 1.9193	63.75 = 2.5098	78.75 = 3.1004	93.75 = 3.6909	108.75 = 4.2815	123.75 = 4.8720
4 = .1575	19 = .7480	34 = 1.3386	49 = 1.9291	64 = 2.5197	79 = 3.1102	94 = 3.7008	109 = 4.2913	124 = 4.8819
4.25 = .1673	19.25 = . 7579	34.25 = 1.3484	49.25 = 1.9390	64.25 = 2.5295	79.25 = 3.1201	94.25 = 3.7106	109.25 = 4.3012	124.25 = 4.8917
4.50 = .1772	19.50 = .7677	34.50 = 1.3583	49.50 = 1.9488	64.50 = 2.5394	79.50 = 3.1299	94.50 = 3.7205	109.50 = 4.3110	124.50 = 4.9016
4.75 = .1870	19.75 = .7776	34.75 = 1.3681	49.75 = 1.9587	64.75 = 2.5492	79.75 = 3.1398	94.75 = 3.7303	109.75 = 4.3209	124.75 = 4.9114
5 = .1968	20 = .7874	35 = 1.3779	50 = 1.9685	65 = 2.5590	80 = 3.1496	95 = 3.7401	110 = 4.3307	125 = 4.9212
5.25 = .2067	20.25 = .7972	35.25 = 1.3878	50.25 = 1.9783	65.25 = 2.5689	80.25 = 3.1594	95.25 = 3.7500	110.25 = 4.3405	125.25 = 4.9311
5.50 = .2165	20.50 = .8071	35.50 = 1.3976	5050 = 1.9882	65.50 = 2.5787	80.50 = 3.1693	95.50 = 3.7598	110.50 = 4.3504	125.50 = 4.9409
5.75 = .2264	20.75 = .8169	35.75 = 1.4075	50.75 = 1.9980	65.75 = 2.5886	80.75 = 3.1791	95.75 = 3.7697	110.75 = 4.3602	125.75 = 4.9508
6 = .2362	21 = .8268	36 = 1.4173	51 = 2.0079	66 = 2.5984	81 = 3.1890	96 = 3.7795	111 = 4.3701	126 = 4.9606
6.25 = .2461	21.25 = .8366	36.25 = 1.4272	51.25 = 2.0177	66.25 = 2.6083	81.25 = 3.1988	96.25 = 3.7894	111.25 = 4.3799	126.25 = 4.9705
6.50 = .2559	21.50 = .8465	36.50 = 1.4370	51.50 = 2.0276	66.50 = 2.6181	81.50 = 3.2087	96.50 = 3.7992	111.50 = 4.3898	126.50 = 4.9803
6.75 = .2657	21.75 = .8563	36.75 = 1.4468	51.75 = 2.0374	66.75 = 2.6279	81.75 = 3.2185	96.75 = 3.8090	111.75 = 4.3996	126.75 = 4.9901
7 = .2756	22 = .8661	37 = 1.4567	52 = 2.0472	67 = 2.6378	82 = 3.2283	97 = 3.8189	112 = 4.4094	127 = 5.0000
7.25 = .2854	22.25 = .8760	37.25 = 1.4665	52.25 = 2.0571	67.25 = 2.6476	82.25 = 3.2382	97.25 = 3.8287	112.25 = 4.4193	
7.50 = .2953	22.50 = .8858	37.50 = 1.4764	52.50 = 2.0669	67.50 = 2.6575	82.50 = 3.2480	97.50 = 3.8386	112.50 = 4.4291	
7.75 = .3051	22.75 = .8957	37.75 = 1.4862	52.75 = 2.0768	67.75 = 2.6673	82.75 = 3.2579	97.75 = 3.8484	112.75 = 4.4390	
8 = .3150	23 = .9055	38 = 1.4961	53 = 2.0866	68 = 2.6772	83 = 3.2677	98 = 3.8583	113 = 4.4488	
8.25 = .3248	23.25 = .9153	38.25 = 1.5059	53.25 = 2.0965	68.25 = 2.6870	83.25 = 3.2776	98.25 = 3.8681	113.25 = 4.4587	
8.50 = .3346	23.50 = .9252	38.50 = 1.5157	53.50 = 2.1063	68.50 = 2.6968	83.50 = 3.2874	98.50 = 3.8779	113.50 = 4.4685	
8.75 = .3445	23.75 = .9350	38.75 = 1.5256	53.75 = 2.1161	68.75 = 2.7067	83.75 = 3.2972	98.75 = 3.8878	113.75 = 4.4783	
9 = .3543	24 = .9449	39 = 1.5354	54 = 2.1260	69 = 2.7165	84 = 3.3071	99 = 3.8976	114 = 4.4882	
9.25 = .3642	24.25 = .9547	39.25 = 1.5453	54.25 = 2.1358	69.25 = 2.7264	84.25 = 3.3169	99.25 = 3.9075	114.25 = 4.4980	
9.50 = .3740	24.50 = .9646	39.50 = 1.5551	54.50 = 2.1457	69.50 = 2.7362	84.50 = 3.3268	99.50 = 3.9173	114.50 = 4.5079	
9.75 = .3839	24.75 = .9744	39.75 = 1.5650	54.75 = 2.1555	69.75 = 2.7461	84.75 = 3.3366	99.75 = 3.9272	114.75 = 4.5177	
10 = .3937	25 = .9842	40 = 1.5748	55 = 2.1653	70 = 2.7559	85 = 3.3464	100 = 3.9370	115 = 4.5275	
10.25 = .4035	25.25 = .9941	40.25 = 1.5846	55.25 = 2.1752	70.25 = 2.7657	85.25 = 3.3563	100.25 = 3.9468	115.25 = 4.5374	
10.50 = .4134	25.50 = 1.0039	40.50 = 1.5945	55.50 = 2.1850	70.50 = 2.7756	85.50 = 3.3661	100.50 = 3.9567	115.50 = 4.5472	
10.75 = .4232	25.75 = 1.0138	40.75 = 1.6043	55.75 = 2.1949	70.75 = 2.7854	85.75 = 3.3760	100.75 = 3.9665	115.75 = 4.5571	
11 = .4331	26 = 1.0236	41 = 1.6142	56 = 2.2047	71 = 2.7953	86 = 3.3858	101 = 3.9764	116 = 4.5669	
11.25 = .4429	26.25 = 1.0335	41.25 = 1.6240	56.25 = 2.2146	71.25 = 2.8051	86.25 = 3.3957	101.25 = 3.9862	116.25 = 4.5768	
11.50 = .4528	26.50 = 1.0433	41.50 = 1.6339	56.50 = 2.2244	71.50 = 2.8150	86.50 = 3.4055	101.50 = 3.9961	116.50 = 4.5866	
11.75 = .4626	26.75 = 1.0531	41.75 = 1.6437	56.75 = 2.2342	71.75 = 2.8248	86.75 = 3.4153	101.75 = 4.0059	116.75 = 4.5964	
12 = .4724	27 = 1.0630	42 = 1.6535	57 = 2.2441	72 = 2.8346	87 = 3.4252	102 = 4.0157	117 = 4.6063	
12.25 = .4823	27.25 = 1.0728	42.25 = 1.6634	57.25 = 2.2539	72.25 = 2.8445	87.25 = 3.4350	102.25 = 4.0256	117.25 = 4.6161	
12.50 = .4921	27.50 = 1.0827	42.50 = 1.6732	57.50 = 2.2638	72.50 = 2.8543	87.50 = 3.4449	102.50 = 4.0354	117.50 = 4.6260	
12.75 = .5020	27.75 = 1.0925	42.75 = 1.6831	57.75 = 2.2736	72.75 = 2.8642	87.75 = 3.4547	102.75 = 4.0453	117.75 = 4.6358	
13 = .5118	28 = 1.1024	43 = 1.6929	58 = 2.2835	73 = 2.8740	88 = 3.4646	103 = 4.0551	118 = 4.6457	
13.25 = .5217	28.25 = 1.1122	43.25 = 1.7028	58.25 = 2.2933	73.25 = 2.8839	88.25 = 3.4744	103.25 = 4.0650	118.25 = 4.6555	
13.50 = .5315	28.50 = 1.1220	43.50 = 1.7126	58.50 = 2.3031	73.50 = 2.8937	88.50 = 3.4842	103.50 = 4.0748	118.50 = 4.6653	
13.75 = .5413	28.75 = 1.1319	43.75 = 1.7224	58.75 = 2.3130	73.75 = 2.9035	88.75 = 3.4941	103.75 = 4.0846	118.75 = 4.6752	
14 = .5512	29 = 1.1417	44 = 1.7323	59 = 2.3228	74 = 2.9134	89 = 3.5039	104 = 4.0945	119 = 4.6850	
14.25 = .5610	29.25 = 1.1516	44.25 = 1.7421	59.25 = 2.3327	74.25 = 2.9232	89.25 = 3.5138	104.25 = 4.1043	119.25 = 4.6949	
14.50 = .5709	29.50 = 1.1614	44.50 = 1.7520	59.50 = 2.3425	74.50 = 2.9331	89.50 = 3.5236	104.50 = 4.1142	119.50 = 4.7047	
14.75 = .5807	29.75 = 1.1713	4475 = 1.7618	59.75 = 2.3524	74.74 = 2.9429	89.75 = 3.5335	104.75 = 4.1240	119.75 = 4.7146	

Tap/Drill Chart

Coarse Standard Thread (N.C.) *Formerly U.S. Standard Thread*					Fine Standard Thread (N.F.) *Formerly S.A.E. Thread*				
Sizes	Threads per inch	Outside diameter at screw	Tap drill sizes	Decimal equivalent of drill	Sizes	Threads per inch	Outside diameter at screw	Tap drill sizes	Decimal equivalent of drill
1	64	.073	53	0.0595	0	80	.060	1/64	0.0469
2	56	.086	50	0.0700	1	72	.073	53	0.0595
3	48	.099	47	0.0785	2	64	.086	50	0.0700
4	40	.112	43	0.0890	3	56	.099	45	0.0820
5	40	.125	38	0.1015	4	48	.112	42	0.0935
6	32	.138	36	0.1065	5	44	.125	37	0.1040
8	32	.164	29	0.1360	6	40	.138	33	0.1130
10	24	.190	25	0.1495	8	36	.164	29	0.1360
12	24	.216	16	0.1770	10	32	.190	21	0.1590
1/4	20	.250	7	0.2010	12	28	.216	14	0.1820
5/16	18	.3125	F	0.2570	1/4	28	.250	3	0.2130
3/8	16	.375	5/16	0.3125	5/16	24	.3125	1	0.2720
7/16	14	.4375	U	0.3680	3/8	24	.375	Q	0.3320
1/2	13	.500	27/64	0.4219	7/16	20	.4375	25/64	0.3906
9/16	12	.5625	31/64	0.4843	1/2	20	.500	29/64	0.4531
5/8	11	.625	17/32	0.5312	9/16	18	.5625	0.5062	0.5062
3/4	10	.750	21/32	0.6562	5/8	18	.625	0.5687	0.5687
7/8	9	.875	49/64	0.7656	3/4	16	.750	11/16	0.6875
1	8	1.000	7/8	0.875	7/8	14	.875	0.8020	0.8020
1 1/8	7	1.125	63/64	0.9843	1	14	1.000	0.9274	0.9274
1 1/4	7	1.250	1 7/64	1.1093	1 1/8	12	1.125	1 3/64	1.0468
					1 1/4	12	1.250	1 11/64	1.1718

Bolt Torquing Chart

Metric Standard						SAE Standard / Foot Pounds					
Grade of Bolt	5D	8G	10K	12K		Grade of Bolt	SAE 1 & 2	SAE 5	SAE 6	SAE 8	
Min. Tensile Strength	71,160 P.S.I.	113,800 P.S.I.	142,200 P.S.I.	170,679 P.S.I.		Min. Tensile Strength	64,000 P.S.I.	105,000 P.S.I.	133,000 P.S.I.	150,000 P.S.I.	
Grade Markings on Head	5D	8G	10K	12K	Size of Socket or Wrench Opening	Markings on Head					Size of Socket or Wrench Opening
Metric					Metric	U.S. Standard					U.S. Standard
Bolt Dia.	U.S. Dec. Equiv.	Foot Pounds			Bolt Head	Bolt Dia.	Foot Pounds				Bolt Head / Nut
6mm	.2362	5	6	8	10	10mm	1/4	5	7	10	10.5	3/8	7/16
8mm	.3150	10	16	22	27	14mm	5/16	9	14	19	22	1/2	9/16
10mm	.3937	19	31	40	49	17mm	3/8	15	25	34	37	9/16	5/8
12mm	.4720	34	54	70	86	19mm	7/16	24	40	55	60	5/8	3/4
14mm	.5512	55	89	117	137	22mm	1/2	37	60	85	92	3/4	13/16
16mm	.6299	83	132	175	208	24mm	9/16	53	88	120	132	7/8	7/8
18mm	.709	111	182	236	283	27mm	5/8	74	120	167	180	15/16	1.
22mm	.8661	182	284	394	464	32mm	3/4	120	200	280	296	1 1/8	1 1/8

Physical Properties of Metals

Metal	Symbol	Specific Gravity	Specific Heat	Melting Point*		Lbs. per Cubic Inch
				°C	°F	
Aluminum (cast)	Al	2.56	.2185	658	1217	.0924
Aluminum (rolled)	Al	2.71	–	658	1217	.0978
Antimony	Sb	6.71	.051	630	1166	.2424
Bismuth	Bi	9.80	.031	271	520	.3540
Boron	B	2.30	.3091	2300	4172	.0831
Brass	–	8.51	.094	–	–	.3075
Cadmium	Cd	8.60	.057	321	610	.3107
Calcium	Ca	1.57	.170	810	1490	.0567
Chromium	Cr	6.80	.120	1510	2750	.2457
Cobalt	Co	8.50	.110	1490	2714	.3071
Copper	Cu	8.89	.094	1083	1982	.3212
Columbium	Cb	8.57	–	1950	3542	.3096
Gold	Au	19.32	.032	1063	1945	.6979
Iridium	Ir	22.42	.033	2300	4170	.8099
Iron	Fe	7.86	.110	1520	2768	.2634
Iron (cast)	Fe	7.218	.1298	1375	2507	.2605
Iron (wrought)	Fe	7.70	.1138	1500-1600	2732-2912	.2779
Lead	Pb	11.37	.031	327	621	.4108
Lithium	Li	.057	.941	186	367	.0213
Magnesium	Mg	1.74	.250	651	1204	.0629
Manganese	Mn	8.00	.120	1225	2237	.2890
Mercury	Hg	13.59	.032	–39	–38	.4909
Molybdenum	Mo	10.2	.0647	2620	47.48	.368
Monel metal	–	8.87	.127	1360	2480	.320
Nickel	Ni	8.80	.130	1452	2646	.319
Phosphorus	P	1.82	.177	43	111.4	.0657
Platinum	Pt	21.50	.033	1755	3191	.7767
Potassium	K	0.87	.170	62	144	.0314
Selenium	Se	4.81	.084	220	428	.174
Silicon	Si	2.40	.1762	1427	2600	.087
Silver	Ag	10.53	.056	961	1761	.3805
Sodium	Na	0.97	.290	97	207	.0350
Steel	–	7.858	.1175	1330-1378	2372-2532	.2839
Strontium	Sr	2.54	.074	769	1416	.0918
Tantalum	Ta	10.80	–	2850	5160	.3902
Tin	Sn	7.29	.056	232	450	.2634
Titanium	Ti	5.3	.130	1900	3450	.1915
Tungsten	W	19.10	.033	3000	5432	.6900
Uranium	U	18.70	–	1132	2070	.6755
Vanadium	V	5.50	–	1730	3146	.1987
Zinc	Zn	7.19	.094	419	786	.2598

*Circular of the Bureau of Standards No.35, Department of Commerce and Labor

Machine Screw and Cap Screw Heads

	Size	A	B	C	D
Fillister Head	#8	.260	.141	.042	.060
	#10	.302	.164	.048	.072
	1/4	3/8	.205	.064	.087
	5/16	7/16	.242	.077	.102
	3/8	9/16	.300	.086	.125
	1/2	3/4	.394	.102	.168
	5/8	7/8	.500	.128	.215
	3/4	1	.590	.144	.258
	1	1 5/16	.774	.182	.352
Flat Head	#8	.320	.092	.043	.037
	#10	.372	.107	.048	.044
	1/4	1/2	.146	.064	.063
	5/16	5/8	.183	.072	.078
	3/8	3/4	.220	.081	.095
	1/2	7/8	.220	.102	.090
	5/8	1 1/8	.293	.128	.125
	3/4	1 3/8	.366	.144	.153
Round Head	#8	.297	.113	.044	.067
	#10	.346	.130	.048	.073
	1/4	7/16	.183	.064	.107
	5/16	9/16	.236	.072	.150
	3/8	5/8	.262	.081	.160
	1/2	13/16	.340	.102	.200
	5/8	1	.422	.128	.255
	3/4	1 1/4	.526	.144	.320
Hexagon Head	1/4	.494	.170	7/16	
	5/16	.564	.215	1/2	
	3/8	.635	.246	9/16	
	1/2	.846	.333	3/4	
	5/8	1.058	.411	15/16	
	3/4	1.270	.490	1 1/8	
	7/8	1.482	.566	1 5/16	
	1	1.693	.640	1 1/2	
Socket Head	#8	.265	.164	1/8	
	#10	5/16	.190	5/32	
	1/4	3/8	1/4	3/16	
	5/16	7/16	5/16	7/32	
	3/8	9/16	3/8	5/16	
	7/16	5/8	7/16	5/16	
	1/2	3/4	1/2	3/8	
	5/8	7/8	5/8	1/2	
	3/4	1	3/4	9/16	
	7/8	1 1/8	7/8	9/16	
	1	1 5/16	1	5/8	

Drive Belt Tension

SAE Belt	Belt Tension Gauge Part Number		Belt Tension New		Belt Tension Used*	
Size	Click-Type	Burroughs	ft.-lbs	N	ft.-lbs	N
.380 in	3822524		140	620	60-110	270-490
.440 in	3822524		140	620	60-110	270-490
1/2 in	3822524	ST-1138	140	620	60-110	270-490
11/16 in	3822524	ST-1138	140	620	60-110	270-490
3/4 in	3822524	ST-1138	140	620	60-110	270-490
7/8 in	3822524	ST-1138	140	620	60-110	270-490
4 rib	3822524	ST-1138	140	620	60-110	270-490
5 rib	3822524	ST-1138	150	670	60-120	270-530
6 rib	3822525	ST-1293	160	710	65-130	290-580
8 rib	3822525	ST-1293	200	890	80-160	360-710
10 rib	3822525	3823138	250	1110	100-200	440-890
12 rib	3822525	3823138	300	1330	120-240	530-1070

* A belt is considered used if it is has been in service for more than 10 minutes.

Pipe Plug Torque Values

Size		Torque In Cast Iron or Steel Components		Torque In Aluminum Components	
Thread (in)	Actual Thread OD (in)	ft.-lbs	N•m	ft.-lbs	N•m
1/16	0.32	10	15	45 in.-lbs	5
1/8	0.41	15	20	10	15
1/4	0.54	20	25	15	20
3/8	0.68	25	35	20	25
1/2	0.85	40	55	25	35
3/4	1.05	55	75	35	45
1	1.32	70	95	45	60
1-1/4	1.66	85	15	55	75
1-1/2	1.90	100	135	65	85

Metal Thickness Gage Chart

Gage	Brown and Sharp Gage (inch)	Steel Sheet Mfrs. Std. (inch)	Gage	Brown and Sharpe Gage (inch)	Steel Sheet Mfrs. Std. (inch)
6-0	0.5800	—	17	0.0453	0.0538
5-0	0.5165	—	18	0.0403	0.0478
4-0	0.4600	—	19	0.0359	0.0418
3-0	0.4096	—	20	0.0320	0.0359
2-0	0.3648	—	21	0.0285	0.0329
0	0.3249	—	22	0.0253	0.0299
1	0.2893	—	23	0.0226	0.0269
2	0.2576	—	24	0.0201	0.0239
3	0.2294	0.2391	25	0.0179	0.0209
4	0.2043	0.2242	26	0.0159	0.0179
5	0.1819	0.2092	27	0.0142	0.0164
6	0.1620	0.1943	28	0.0126	0.0149
7	0.1443	0.1793	29	0.0113	0.0135
8	0.1285	0.1644	30	0.0100	0.0120
9	0.1144	0.1495	31	0.0089	0.0105
10	0.1019	0.1345	32	0.0080	0.0097
11	0.0907	0.1196	33	0.0071	0.0090
12	0.0808	0.1046	34	0.0063	0.0082
13	0.0720	0.0897	35	0.0056	0.0075
14	0.0641	0.0747	36	0.0050	0.0067
15	0.0571	0.0673	37	0.0045	0.0064
16	0.0508	0.0598	38	0.0040	0.0060

Wire Case Selection Chart

Total Approx. Circuit Amperes 12V	Total Circuit Watts 12V	Total Candle Power 12V	Wire Gage (For Length in Feet)											
			3′	5′	7′	10′	15′	20′	25′	30′	40′	50′	75′	100′
1.0	12	6	18	18	18	18	18	18	18	18	18	18	18	18
1.5		10	18	18	18	18	18	18	18	18	18	18	18	18
2	24	16	18	18	18	18	18	18	18	18	18	18	16	16
3		24	18	18	18	18	18	18	18	18	18	18	14	14
4	48	30	18	18	18	18	18	18	18	18	16	16	12	12
5		40	18	18	18	18	18	18	18	18	16	14	12	12
6	72	50	18	18	18	18	18	18	16	16	16	14	12	10
7		60	18	18	18	18	18	18	16	16	14	14	10	10
8	96	70	18	18	18	18	18	16	16	16	14	12	10	10
10	120	80	18	18	18	18	16	16	16	14	12	12	10	10
11		90	18	18	18	18	16	16	14	14	12	12	10	8
12	144	100	18	18	18	18	16	16	14	14	12	12	10	8
15		120	18	18	18	18	14	14	12	12	12	10	8	8
18	216	140	18	18	16	16	14	14	12	12	10	10	8	8
20	240	160	18	18	16	16	14	12	10	10	10	10	8	6
22	264	180	18	18	16	16	12	12	10	10	10	8	6	6
24	288	200	18	18	16	16	12	12	10	10	10	8	6	6
30				18	16	16	14	10	10	10	10	10	6	4	4
40				18	16	14	12	10	10	8	8	6	6	4	2
50				16	14	12	12	10	10	8	8	6	6	2	2
100				12	12	10	10	6	6	4	4	4	2	1	1/0
150				10	10	8	8	4	4	2	2	2	1	2/0	2/0
200				10	8	8	6	4	4	2	2	1	1/0	4/0	4/0

Acknowledgments

The production of **Diesel Technology** would not have been possible without the contribution of these outstanding companies and individuals. The authors and publisher would like to give their thanks to the following:

Advanced Thermal Products, Inc.; Association of Diesel Specialists; AXE Equipment, Inc.; Babcox Publications, Inc.; Brent Black; Caterpillar, Inc.; Central Tools, Inc.; Chicago Pneumatic Tool Company; Clayton Industries, Inc.; Colt Industries, Inc.; Consolidated Metro, Inc.; Cummins Natural Gas Division; Cummins Engine Co., Inc.; Dana Corporation; Davco Mfg. Corp.; Deere & Company; Detroit Diesel Corp.; Deutz Corp.; Diesel Progress Engines & Drives; Donaldson Company, Inc.; Dresser Industries, Inc.; Engelhard Corporation; Federal Mogul Corp.; Fel-Pro Gaskets; John Fluke Corporation; Ford Motor Company; Ford New Holland, Inc.; Freightliner Corporation; Garrett Corp.; Gates Rubber Company; Corey Glassman; Goodson Shop Supplies; Harrier, Inc.; Hartridge Equipment Corp.; Hatz Diesel of America, Inc.; Marsh Hayward; Heavy Duty Trucking; Hercules Engines, Inc.; Houston Industrial Silencing; ICC Federated, Inc.; Ingersol-Rand Corp.; Ironite Products Company, Inc.; Jacobs Manufacturing Company; Kawasaki Motors Corp. U.S.A.; Kent-Moore Division; Kenworth Truck Co.; Kiene Diesel Accessories, Inc.; Kohler Company; Komatsu Trading International, Inc.; Kubota Corporation; Kwik-Way Manufacturing Co.; L.S. Starrett Company; Mac Tools Corporation; Mack Truck Company; Magnaflex Corporation; MAN Aktiengesellchaft Historical Archives; Mercedes Benz of North America, Inc.; Mercury Marine Corp.; National Institute for Automotive Service Excellence; Navistar International Corp.; Nelson Corp.; Neway Manufacturing, Inc.; Norwegian Cruise Lines, Inc.; Onan Corp.; OTC Tool Division; Parker Hannifin Corporation; Penn Diesel; Perkins Group of Companies; Peterbilt Trucks, Inc.; Petters, Ltd.; Peugeot Engines, Inc.; Philips Temro, Inc.; Racor Division; Robinair Corp.; Rogers Machine Corp.; Sealed Power Corp.; Sioux Tools, Inc.; Snap-on Tools Corp.; SPX Corporation; Stanadyne Automotive Inc.; Stanley Proto; Stewart-Warner Corp.; Stork-Werkspoor, Ltd.; Storm Vulcan, Inc.; Sunnen Products Co.; Teledyne Total Power Group; Toyota Industrial Engine Operation; TRW, Inc.; Viking Yachts, Inc.; Volkswagen of Americas, Inc.; Volvo Penta of America; Waukesha Engine Division; Winona Van Norman Machine Co.; Wix Corporation; Young Radiator Company.

Special thanks to Sears, Roebuck and Company—Orland Park retail store, Orland Park, Illinois for the many tools illustrated in **Chapter 3** and to Jody M. Janssen, Caterpillar, Inc., for supplying the artwork used for the cover.

This textbook is dedicated to the memory of Bob and Mary Scharff. Their encouragement, guidance, and friendship helped make this book possible.

Glossary

A

Abbreviations: Letters or letter combinations that stand for words.

ABDC: Abbreviation for *After Bottom Dead Center.*

Above-ground storage tank (AGST): A tank that is used to store extra diesel fuel above ground. Required in some areas due to environmental concerns.

Abrasion: Wearing or rubbing away of a part.

Acceleration: When the velocity of a body is increased it is said to have accelerated.

Accelerator: A driver-operated pedal used to control engine speed.

Access covers: Removable plates that allow access to parts of an engine, such as the camshaft or lifter gallery.

Accumulator: Cylinder used to store pressurized fluid. Used in diesel hydraulic starting systems.

Acetylene: Gas commonly used in oxyacetylene welding or cutting operations.

ACT: *American College Testing* program.

Actuator: Any output device controlled by a computer.

Additives: Any material added to automotive oils to improve properties such as viscosity, cetane, pour point, and film strength

ADS: *Association of Diesel Specialists.*

Advance: To move the timing of the injection pump or injectors to an earlier injection point so that combustion begins earlier, or more degrees before top dead center.

Advance mechanism: Advances or retards the start of fuel delivery in response to engine speed changes.

Advance timing unit: Fuel pressure driven unit located in the base of the distributor pump. During operation, fuel pressure is regulated using a pressure control valve and an overflow restriction.

AEA: *Automotive Electronic Association.*

AERA: *American Engine Rebuilders Association.*

Aftercooler: A device used on turbocharged engines to cool air which has undergone compression.

Aftermarket device: A device or accessory installed by the customer or manufacturer's dealer after taking delivery from the factory.

Air box: The passage that conducts intake air to the cylinders in two-cycle engines. The air box surrounds the cylinders and, in many engines, is built into the block.

Air cleaner: A device for filtering, cleaning, and removing dust, dirt, and foreign debris from the air being drawn into an engine, air compressor, etc.

Air compressor: A device used to increase air pressure.

Air-cooled system: An engine or other object that is cooled by passing a stream of air over its surface.

Air dam: A metal or plastic shroud placed beneath the front of a vehicle. Designed to channel airflow over the radiator and aftercooler.

Air filter: A pleated paper element installed inside the air cleaner. Used to filter dirt, dust, and foreign debris from the air stream entering the engine.

Air lock: The presence of air in a pump or pipes which prevents the delivery of liquid.

Air pollution: Any release of harmful substances into the atmosphere by engine operation or other causes.

Air starting systems: Starting system that uses compressed air to power the starter motor. Normally used on large displacement diesel engines.

Air-to-air aftercooler: Heat exchanger that cools the intake air after the turbocharger before going to the intake manifold, by using ambient air.

Align: To bring various parts of a unit into their correct positions with respect to each other or to a predetermined location.

Allen wrench: A hexagonal wrench, which is usually "L" shaped, designed to fit into a hexagonal hole.

Alloy: A mixture of two or more metals, usually to produce improved characteristics. An example is bronze, which is a mixture of copper and tin.

Alternating current (ac): An electrical current that moves in one direction and then the other.

Alternator: A generator in which alternating current is first generated, then changed into direct (dc) current.

Altitude-pressure compensator: Any sensor or device that automatically compensates for changes in altitude.

Aluminum: A metal known for its lightweight and good heat dissipation characteristics. It is often alloyed with other metals.

Ambient temperature: The temperature of the air surrounding an object.

American Trucking Association (ATA) data link connector: A two-wire electrical connector used for communication with on-board computers such as trip recorders, electronic dashboards, powertrain controls, and maintenance systems. The data link is also the serial communication medium used for programming and troubleshooting with many OEM engine manufacturer's devices.

American Wire Gage (AWG): A measure of the diameter and the current carrying ability of electrical wire. The smaller the AWG number, the larger the wire.

Ammeter: Instrument used to measure the flow of electric current in a circuit in amperes. Normally connected in series in a circuit.

Ampere (A): The unit of measurement for the flow of electric current. An ampere is defined as the amount of current that 1 volt can send through 1 ohm of resistance.

Anaerobic sealer: A chemical sealer that cures in the absence of air.

Analog: A signal that continually changes in strength.

Aneroid: A device, such as a barometer bellows, that does not contain or use a liquid.

Aneroid/boost compensator: A pollution control device installed on turbocharged engine governors. It helps to reduce the amount of dense black smoke generated during acceleration.

Annealing: A process used to soften metals. Metal is heated and cooled gradually over a period of time.

Anode: The positive pole in an electrical circuit.

Antifreeze: A liquid, usually ethylene glycol, that is added to water in a vehicle's cooling system. This mixture serves as the engine's coolant while lowering the freezing point.

Antifriction bearing: A bearing that uses balls or rollers between a journal and bearing surface to decrease friction.

API: Abbreviation for American Petroleum Institute.

Apprentice program: Any educational program designed to teach a trade through a combination of on-the-job training and classroom study.

Apprentice technician: A beginner who is learning under the direction of one or more experienced technicians.

Arcing: Electricity leaping the gap between two electrodes.

Armature: The revolving part in an alternator or motor. The moving part in a relay or horn.

ASE: Abbreviation for *National Institute For Automotive Service Excellence.*

ASIA: *Automotive Service Industry Association.*

ASME: *American Society of Mechanical Engineers.*

Aspirated: To draw by suction.

ATA: *American Trucking Association.*

ATDC: Abbreviation for *After Top Dead Center.*

Atmospheric pressure: The pressure exerted by the earth's atmosphere on all objects. Measured with reference to the pressure at sea level, which is 14.7 psi (101 kPa).

Atom: A tiny particle of matter made up of electrons, protons, and neutrons. The electrons orbit around the center or nucleus, made up of protons and neutrons. Combinations of atoms make up molecules.

Atomization: The process of breaking a liquid, such as diesel fuel, into tiny drops that can easily mix with air.

Automatic advance device: A sensor regulated device that changes the pump-to-engine timing according to engine speed.

Automatic Transmission Fluid (ATF): Mineral oil with special additives to make it compatible with clutch and band materials and to prevent foaming and thermal breakdown.

Auxiliary oil cooler: A system designed to remove excess heat from engine oil by either passing it through a heat exchanger exposed to the air or by channeling engine coolant through an oil filter adapter mounted on the engine.

AWG: Abbreviation for *American Wire Gage.*

B

Backfire: A condition where unburned fuel in the intake or exhaust manifold ignites.

Backlash: The clearance or "play" between two parts, such as the teeth of two gears.

Backpressure: The exhaust system's resistance to the flow of exhaust gases.

Baffle: An obstruction used to check or deflect the flow of gases, liquids, sounds, etc.

Ball bearing (Antifriction): A bearing consisting of an inner and outer hardened steel race separated by a series of hardened steel balls.

Barometric pressure sensor: A sensor that measures atmospheric pressure.

Battery: A device containing one or more cells that produce electricity through electrochemical action.

Battery cables: Heavy wires used to connect the battery to the vehicle's electrical system.

Battery capacity: The amount of current a battery is capable of delivering.

Battery charging: The process of restoring a battery's charge by passing current through it in a reverse direction (positive to negative).

Battery plate: Battery components made of lead peroxide in sponge form and porous lead.

Battery rating: Standardized measurement of a battery's ability to deliver an acceptable level of energy under specified conditions. Standards established by the Battery Council International (BCI).

Battery temperature sensor: Sensor installed in some batteries to control overcharging.

Battery voltage: Determined by the number of cells in the battery. Each cell is capable of producing approximately 2.1v.

BBDC: Abbreviation for *Before Bottom Dead Center.*

BCI: *Battery Council International.*

BDC: Abbreviation for *Bottom Dead Center.*

Bearing: A part on which a shaft, journal, or pivot turns.

Bell housing: Metal cover installed around the flywheel and clutch assembly.

Bendix drive: A self-engaging starter drive gear. It is mounted on a screw shaft attached to the starter motor armature. Also used as a general term to describe the overrunning clutch on other starters.

Biocide: An additive designed to eliminate hydrocarbon utilizing microorganisms (HUM).

Biodiesel: A fuel oil that is a blend of diesel fuel and another, often recycled, oil. A typical biodiesel blend is number 1 diesel fuel and soybean oils.

Bleed: The process of removing air or fluid from a closed system, such as the fuel lines.

Block: The part of the engine containing the crankshaft, pistons, and cylinders. It is made of cast iron or aluminum and is the foundation for the entire engine.

Block heater: A heating device installed in place of one of the block's freeze plugs. Used to heat an engine during cold weather.

Blowby: Any leakage or loss of compression past the piston rings. Often accompanied by oil from the crankcase passing through into the combustion chamber.

Blower: A low-pressure air pump used on diesel engines to increase the amount and pressure of the air coming into the engine. Sometimes referred to as a *supercharger.*

BMEP: Abbreviation for *Brake Mean Effective Pressure.* The difference between compression and combustion pressure.

Boiling point: The temperature at which a liquid vaporizes.

Boost pressure sensor: This sensor measures intake manifold air pressure and sends a signal to the ECM.

Boost pressure: A measure of positive air pressure provided by a supercharger or turbocharger.

Bore: The diameter of an engine cylinder. Sometimes used to refer to the cylinder itself.

Boring bar: A machine that uses a bar equipped with cutting bits to cut engine cylinders to specific size or to a new diameter.

Boss: A rib or enlarged area designed to strengthen a portion or area of an object.

Bottled gas: Petroleum gas compressed into liquefied form and contained in strong metal cylinders.

Bourdon tube: A circular hollow piece of metal used in some instruments. Pressure on the hollow end causes it to attempt to straighten. The free end moves the needle on the gauge face.

Brake horsepower (BHP): The measurement of usable horsepower delivered at the crankshaft. Usually computed by placing the engine on an engine dynamometer.

Braze: To join two pieces of metal together by heating the edges to be joined and then using brass or bronze on the area as a solder.

Break-in: The process of wearing in two or more parts that have been replaced or reconditioned.

Breather pipe: A pipe opening into the crankcase to assist ventilation.

British thermal unit (BTU): Measurement of the amount of heat required to raise the temperature of one pound of water by 1°F.

Broach: To finish a metal surface by forcing a multiple edge cutting tool across its surface.

Bronze: An alloy consisting of tin and copper. Normally used in bushing and bearing material.

Brown and Sharpe (B & S) Gage: A standard measure of wire gage, similar to AWG standard.

Brushes: A bar of carbon, copper, or other conductive material that rides on the slip rings of a motor or alternator.

BTDC: Abbreviation for *Before Top Dead Center.*

Bull gear: A large gear directly driven by the crankshaft timing gear that directly or indirectly drives all the gear-driven engine accessories.

Burnish: To bring the surface of a metal to a high shine by rubbing it with a hard pad or other smooth object.

Bus: Pathway for data inside a computer. Can also be used to refer to a circuit used to connect two or more on-board computers.

Bushing: A smooth, removable liner used as a bearing for a shaft, piston pin, etc.

Butane: A petroleum hydrocarbon compound which has a boiling point of approximately 32°F. Sometimes referred to as *liquid petroleum gas.* Often combined with propane.

Bypass: To move around or detour from the normal route taken by a flowing substance, such as electricity, air, or fluid.

Bypass circuit: A circuit, usually temporary, to substitute for an existing circuit, typically for test purposes.

Bypass filter: An oil filter that only strains a portion of the oil flowing through the engine lubrication system.

Bypass valve: Valve used to permit coolant flow when the thermostat is closed. Also used in lubrication systems to permit oil flow if the filter becomes clogged.

C

Calibrate: An adjustment made to produce a desired effect, such as an electronic adjustment of a sensor signal.

Calibration: An electronic adjustment of a sensor signal.

Calipers: An adjustable measuring tool placed around or within an object and adjusted until it just contacts. The tool is then withdrawn and the distance between the contact points is measured.

Calorie: Metric unit of measurement for heat. The amount of heat needed to raise the temperature of one gram of water by 1°C.

Cam: Offset part that as it turns, will impart motion on another object.

Cam ground piston: Piston ground slightly egg-shaped. When heated, it becomes round.

Cam lobe: The offset portion of a shaft that, as it turns, will impart motion to another part.

Cam nose: The high point of a camshaft lobe.

Camshaft: A shaft with offset lobes used to operate valves.

Camshaft gear: A gear that is used to drive the camshaft.

Capacitance: The property of a capacitor or condenser that permits it to receive and retain an electrical charge.

Capped pencil nozzle: An injector nozzle that has no return fuel line like standard nozzles.

Carbon (C): Hard or soft black deposits found in combustion chamber, valves, etc. An excellent conductor of electricity.

Carbon dioxide (CO_2): One of the products of combustion. Also a dry chemical mixture that is an excellent fire retardant. Compressed into solid form this material is known as dry ice, and remains at a temperature of -109°F.

Carbon monoxide (CO): A deadly colorless, odorless gas that is formed when fuel is not burned completely.

Carcinogen: Any substance, such as asbestos and carbon tetrachloride, that can cause cancer.

Case-harden: A process of heating a piece of steel to harden its surface while the inside remains relatively soft.

Castellated nut: A nut that has a series of slots cut into it, into which a cotter pin or safety wire may be passed to secure the nut. The top of the nut resembles a castle battlement.

Castle-top piston: A piston crown that has valve reliefs that resemble the battlements of a castle.

Cathode: The negative pole of an electrical circuit.

CAV Pentaux® nozzle: Nozzles that are designed with an auxiliary spray hole to assist in easy starting in cold weather conditions.

Cell: The compartments in a battery where plates are immersed in electrolyte. The plates are separated into positive and negative sections.

Celsius: Metric unit of measurement for temperature. Under standard atmospheric conditions, water freezes at 0°C and boils at 100°C.

Center of gravity: The point on an object on which it could be balanced.

Central Processing Unit (CPU): A microprocessor inside an electronic control unit responsible for controlling the ECM's operations.

Centrifugal advance mechanism: A series of weights used to advance engine timing as engine speed increases.

Centrifugal force: A force that tends to impel an object outward from the center of rotation.

Cetane: A colorless liquid ($C_{16}H_{34}$). Used as a basis to test the performance characteristics of diesel fuel.

Cetane number: A measure of the ignition quality of diesel fuel. it influences both ease of starting and combustion quality of the fuel.

CFM: Cubic Feet per Minute. Used as a measurement for the amount of air entering an engine's intake.

Chamfer: To bevel an edge on an object at the edge of a hole.

Change of state: Any condition where a substance changes from a solid to liquid or from liquid to vapor.

Charge: The electrical rate of flow that passes through the battery to restore it to full power.

Chase: Process of using a tap to repair slightly damaged threads.

Chassis dynamometer: A machine used to measure engine power at the vehicle's wheels.

Chatter: A condition where a clutch disc vibrates severely as it accelerates.

Check engine lamp: Used to alert the operator to the presence of a problem. Also see *Malfunction indicator lamp*

Check valve: A valve that permits flow in only one direction.

Chilled iron: Cast iron possessing a hard outer surface.

Chromium: A substance added to steel to help resist oxidation and corrosion.

Circuit: A source, resistance, and path to carry the flow from the source to the resistance and back to the source. An example is an electrical circuit. The wires (path) allow electricity from the battery (source) to flow to a headlamp (resistance) and back to the battery.

Circuit breaker: Protective device that will open and close when current draw becomes excessive. Unlike a fuse, it does not blow out.

Clearance: A given space between two parts such as a piston and cylinder.

Clockwise rotation: Rotation in the same direction as the movement of the hands of a clock.

Closed loop: Engine ECM operation mode used when an engine has reached its optimum operating temperature and conditions. When in closed loop, the ECM relies on the sensors to determine if engine operating conditions need to be changed.

Cloud point: The temperature at which wax crystals start to form in diesel fuel. Also referred to as the *wax appearance point.*

Clutch: A device for connecting and disconnecting an engine from the device being driven.

Clutch switch: Switch used to monitor clutch position. Normally OEM supplied and installed. It is typically a limit switch mounted near the pedal and is usually adjustable. The switch is normally closed with the pedal released. Depressing the clutch will open the circuit.

Coefficient of friction: The amount of friction developed when two objects are moved across each other. Coefficient of friction is calculated by dividing the force needed to push a load across a given surface.

Coil spring pressure plate: A type of pressure plate used with diesel engines. Contains coil springs to soften the clutch's impact.

Cold cranking amps: Measurement of cranking amperes that a battery can deliver over a period of 30 seconds at 0°F (-18°C).

Cold mode: A mode of engine operation where the timing is retarded and the low idle may be raised for engine protection, reduced smoke emissions, and faster warm-up time.

Combustion chamber: The area above a piston with the piston at TDC. Measured in cubic centimeters.

Combustion: The process of burning.

Commutator: A series of copper bars connected to an armature winding. The bars are insulated from each other and from the armature.

Compound: A combination of two or more unlike atoms that cannot be broken down into their individual constituents.

Compression: Opposite of tension. Reduction in volume, such as compressing a gas. Also applying pressure to a spring to reduce its length.

Compression gauge: A gauge used to test compression in engine cylinders.

Compression ignition: Principle used in diesel engines to ignite an air-fuel mixture by using heat of compression.

Compression ratio: The ratio of the maximum volume in an engine cylinder with the piston at Bottom Dead Center to the minimum volume in the cylinder with the piston at Top Dead Center.

Compression rings: The top set of piston rings designed to seal between the piston and cylinder to prevent the escape of gases from the combustion chamber.

Compression stroke: The portion of the piston's movement devoted to compressing the air in the engine's cylinder.

Compressor wheel: A fan-like wheel in the turbocharger that forces air into the intake manifold.

Condensation: The process of a vapor becoming a liquid; the reverse of evaporation.

Conduction: The transfer of heat from one object to another by being in direct contact.

Conductor: Any material or device forming a path for the flow of heat or electricity.

Conformability: The ability of a bearing insert to creep or flow slightly so that the shaft and bearing will conform to each other.

Connecting rod: The connecting link between the crankshaft and the piston.

Connecting rod bearing: Inserts which fit into the connecting rods and ride on the crankshaft rod.

Connecting rod cap: The removable lower part of the connecting rod.

Constant mesh gears: Gears that are always in mesh with each other.

Constant speed governors: Governor that maintains engine speed at a constant rpm, regardless of the load. Used to govern engine speed on generators and other stationary applications.

Contraction: Reduction in the size of an object when it cools.

Control rack: A toothed rod inside a mechanical injector pump which rotates the pump plunger to control the quantity of injected fuel.

Control valve: A valve used to control the pressure of fluid in a circuit.

Convection: A transfer of heat from one object to another through a liquid.

Converter: As used in connection with LPG, a device which converts or changes LPG from a liquid to a vapor for use by the engine.

Coolant: Ethylene glycol or other liquid used in a cooling system. The liquid used to cool an engine and contained in the water jacket.

Coolant recovery system: A plastic bottle and hose that is used with a closed cooling system to recover and provide additional coolant when needed.

Coolant Temperature Sensor (CTS): A sensor that monitors engine coolant temperature. Used by the engine control ECM to determine when to enter closed loop and end cold node operation.

Cooling fins: A series of thin metal strips placed between cooling passages to help dissipate heat.

Cooling system: A system that allows air or liquid to circulate to maintain a constant temperature.

Cooling tower: Part of an open cooling system used to store excess water. Because of the cost of operations and environmental regulations, cooling towers are seldom used in modern industrial installations.

Core plugs: *See* Freeze plugs

Corrode: To eat away gradually by chemical action, such as rust.

Corrosion inhibitors: Any chemical added to a substance that prevents the formation of oxidation or corrosion.

Cotter pin: A soft metal pin that is used to secure components in place.

Counterbalance shaft: A shaft designed to cancel out vibrations from the crankshaft and camshaft.

Counterbore: To enlarge a hole to a given depth.

Counterclockwise rotation: Rotation to the left, opposite to that of clock hands.

Countersink: To make a counterbore so that the head of a screw or bolt sets flush or below the surface.

Crankcase: The part of the engine that surrounds the crankshaft.

Crankcase dilution: Condition where fuel enters the crankcase, diluting the engine oil.

Crankpin: The crankshaft section that is attached to the connecting rod.

Crankshaft: The main shaft which supports the connecting rods and turns piston reciprocation into rotary motion.

Crankshaft gear: A gear that is pressed on the crankshaft.

Crankshaft throw: The offset part of the crankshaft where the connecting rods fasten.

Crosshead: Part of the valve train in an engine that actuates two valves per cylinder. Permits two valves in the same cylinder to be opened simultaneously by a single rocker arm.

Crude oil: Liquid petroleum oil that has not been refined.

Cruise control range: Range that is typically limited to the speed range anticipated on the open road. This range is programmable using the low cruise and high cruise limits.

Cu. In.: Abbreviation for cubic inch.

Curb weight: The weight of a unmanned vehicle with fuel, oil, coolant, and all standard equipment installed.

Current: The flow of free electrons through a conductor.

Customer specified parameter: Any parameter value that can be changed whose value is set by the customer. These parameters are protected by customer passwords.

Cycle: A reoccurring period in which a series of actions take place in a definite order.

Cylinder: A hole that has a set depth and contains the piston.

Cylinder head: A detachable portion of an engine which covers the upper end of the cylinder bores and forms part of the combustion chamber. Also includes the valves in the case of overhead valve engines.

Cylinder sleeve: A liner or sleeve interposed between the piston and the cylinder wall or water jacket to provide an easily replaceable surface for the cylinders.

D

Damper: A unit or device used to reduce or eliminate vibration, oscillation, of a moving part, fluid, etc.

Dashboard: Part of vehicle interior containing the instrument cluster, air conditioning controls, switches, etc.

Data link connector: Plastic plug-in terminal with two or more electrical connections used to interface with an engine or vehicle's computers.

Dead center: Extreme upper or lower position of the piston within the cylinder.

Deceleration: The process of slowing down in rotational speed, forward speed, etc.

Deglazer: An abrasive tool used to remove the glaze from cylinder walls before new piston rings are installed.

Degree: A division or part of a circle. One degree is equal to 1/360 of a circle.

Degree wheel: Wheel-like tool attached to the crankshaft. Used to time valves to a high degree of accuracy.

Density: The relative mass per unit volume of an object.

Desiccant: A material used to absorb and remove excess moisture.

Desired rpm: An input to the electronic governor within the ECM. The electronic governor uses inputs from the throttle position sensor, engine speed/timing sensor and customer parameters to determine the desired rpm.

Detergent: An additive used in engine oil to remove and hold in suspension foreign matter that finds its way into the oil.

Diagnosis: The process of analyzing symptoms, test results, etc, to determine the cause of a problem.

Diagnostic event code: This code indicates an event that describes an abnormal engine condition, such as an idle shutdown occurrence. These codes do not typically indicate problems within the electronic system itself.

Diagnostic lamp: Sometimes called the malfunction indicator lamp, it is used to alert the operator to the presence of an active diagnostic code.

Diagnostic Trouble Code (DTC): A code displayed that can be used to determine where a malfunction is located.

Dial gauge: A precision micrometer which indicates exact readings via needle movements on a dial.

Diaphragm: A flexible partition used to separate two different compartments.

Die: A tool for cutting threads.

Diesel engine: An engine that uses diesel oil for fuel. A diesel engine injects fuel oil directly into the cylinders. The compression is so great that the air itself is hot enough to ignite the diesel fuel without a spark.

Diesel injection pump: A pump that develops high pressure to force fuel out of the injectors and into the combustion chambers.

Diesel Oxidizing Catalysts (DOC): Catalyzing elements used to eliminate many of the gases, sulfates, and heavy organic fumes in diesel exhaust.

Digital signal: An electronic signal that uses on and off pulses.

Diode: A semiconductor device that allows current flow in one direction but resists it in the other.

Dipstick: A metal or plastic rod used to determine the quantity of oil or fluid in a reservoir.

Direct combustion: Combustion that takes place in the main chamber of the engine.

Direct current (dc): Electric current that flows steadily in one direction only.

Direct drive: Gear condition in which the crankshaft and driven device turn at the same speed.

Discharge: Process of drawing electric current from a battery.

Displacement: The total volume displaced by the cylinders when moving from BDC to TDC.

Distillation: The process of heating a liquid to the point that its components vaporize and then capturing each separate component as it condenses in a tower or column.

Distribution tubes: Tubes used in the engine cooling system to circulate and direct flow of coolant to hard to reach areas.

Distributor injection pump: A compact pump that uses a single pumping element to meter and delivery pressurized fuel to all cylinders.

Documentation: Repair orders or other means used to record work performed on an engine or vehicle.

DOHC: Abbreviation for a Dual Overhead Camshaft Engine.

Double lap flare: A flare which, when made, utilizes two wall thicknesses.

Dowel pin: A steel pin, usually of circular shape like a cylinder, that is passed through two matching holes to maintain the parts in proper alignment.

Draw: The amount of current required to operate an electrical device.

Draw-filing: Filing by passing the file at right angles to the work.

Driveability: The process of diagnosing, troubleshooting, isolating, and repairing a problem on a vehicle. Generalized term normally used when describing a repair of an engine performance problem.

Drive-fit: A fit between two parts that is so tight that they must be driven together. Also known as a *press-fit*.

Droop: A measure of the change in rpm that occurs from no-load to full-load. Also referred to as *steady state speed regulation*.

Dry air filter: A filter element that requires no oil or other liquid medium to trap dirt particles. Automotive air filter elements are dry filters.

Dry friction: Any resistance to movement between two unlubricated surfaces.

Dry liner: A replacement cylinder liner that does not vome in direct contact with the coolant.

Dry muffler: A type of muffler that does not use water as an additional silencer.

Dual coil vehicle speed sensor: A type of magnetic pickup sensor that senses movement of the teeth on the output shaft of the transmission. It contains two coils, one supplies a signal to the vehicle speedometer, and the other signals the vehicle speed buffer.

Dual fuel tanks: A fuel tank system comprised of two separate fuel tanks. Also referred to as *tandem fuel tank systems*.

Dual voltage control units (DUVAC): Switch that allows a vehicle's starting system to use 24 volts and the vehicle's electrical system 12 volts once the engine is started.

Dust sight window: Inspection port for checking air cleaner condition.

Dye penetrant: A testing material that can be sprayed or painted onto iron and aluminum parts to locate cracks.

Dynamic balance: Condition when the centerline of a revolving object is in the same plane as the object itself.

Dynamic imbalance: Unbalanced condition when the centerline of a revolving object is not in the same plane as the object itself.

Dynamometer: A machine used to measure engine horsepower output. An engine dynamometer measures horsepower at the crankshaft and the chassis dynamometer measures horsepower at the wheels.

E

Ebullient cooling systems: Cooling system that reduces diesel engine temperature by absorbing heat during a phase change from liquid to vapor (steam).

Eccentric: The degree of being off-center. The distance between the center of an eccentric and its axis; the throw. One circle within another circle not having the same center.

ECM: Electronic or Engine Control Module.

EEC: Evaporative Emissions Control.

Electric governors: Governor used in applications where constant speed must be maintained or controlled from a remote location.

Electrode: An electric conductor though which an electric current enters or leaves a medium such as an electrolyte, a nonmetallic solid, a gas, or a vacuum.

Electrolysis: A chemical change that occurs between two dissimilar metals, resulting in a corrosive effect on the metals. Occurs frequently in the presence of sea water.

Electrolyte: A solution of acid and water in a battery.

Electromagnet: A magnet that is produced by placing a coil of wire around a steel or iron bar. When current flows through the wire, the bar becomes magnetized.

Electromagnetic interference (EMI): Electronic noise that is created when two or more wires carrying a strong voltage or signal are allowed to cross. Can also be created by radio waves.

Electromotive force (EMF): Voltage.

Electron: A negatively charged particle that makes up part of the atom.

Electron theory: The accepted theory of electronics that states that electricity flows from negative to positive.

Electronic control module (ECM): Engine control computer that accepts inputs to monitor and provides outputs that control, or change, engine functions.

Electronic control unit (ECU): General term used for any computer that controls a vehicle system.

Electronic engine control: System that monitors and controls engine operation under all conditions.

Electronically controlled unit injector: Mechanically actuated, electronically controlled unit injector which combines pumping, electronic fuel metering, and injecting elements in a single unit.

Electronically erasable programmable read only memory (EEPROM): A type of microprocessor whose programming can be changed by special electronic equipment.

Element: One complete set of positive and negative plates with separators.

Embedability: Refers to a bearing's ability to let small dirt and metal particles embed themselves to avoid scratching the part.

Emissions: Any release of harmful materials into the environment. Gases produced from exhaust, crankcase, and fuel tanks and their contribution to smog.

End play: Amount of lengthwise movement between two parts due to clearance.

Energy: Any capacity for doing work.

Engine: A device that converts energy into useful mechanical motion.

Engine displacement: The volume of space in which the piston moves in a full stroke multiplied by the number of cylinders in the engine.

Engine dynamometer: A device that tests engine output at the crankshaft.

Engine mounts: Pads made of metal, rubber, and plastic. Designed to hold engine to the frame and isolate engine vibrations. May also be liquid-filled.

Engine speed sensor: A sensor that sends information on engine speed and piston location to the ECM.

Engine speed/timing sensor: Sensor that provides a variable amplitude and pulse width modulated signal to the ECM, which translates the input as crankshaft position and engine speed.

EPA: Abbreviation for *Environmental Protection Agency.*

Erasable programmable read only memory (EPROM): A type of microprocessor whose programming can be altered only by erasing it with special equipment and reprogramming.

Ether: A colorless liquid used as a starting fluid in diesel engines.

Ethylene glycol: A liquid chemical used in engine coolant.

Evaporation: The process of a liquid turning into vapor.

Exhaust backpressure: Reverse pressure exerted in the exhaust system by restricted flow.

Exhaust emissions: The products of combustion that are discharged from the engine exhaust.

Exhaust gas analyzer: Electronic calibrated device used to measure the amount of pollutants in exhaust emissions and thereby determine the engine's efficiency.

Exhaust manifold: Connecting pipes between the cylinder head exhaust ports and the exhaust pipes.

Exhaust pipe: Pipe used to connect the exhaust manifold to the muffler.

Exhaust stroke: The portion of the piston's travel that is devoted to expelling the burned fuel charge.

Exhaust system: Components which carry exhaust emissions from the engine. These components include the muffler, exhaust pipes, exhaust manifold, and turbo impeller.

Expansion tank: A plastic tank used to recover excess coolant as part of a closed liquid coolant recovery system.

Explosion doors: Spring-loaded panels installed on large displacement diesel engines that also act as a pressure relief in the event of a crankcase explosion or extreme pressure build-up within the crankcase.

Extreme pressure lubricants (EP): A lubricant compounded to withstand heavy loads such as that imposed between gear teeth.

F

Fahrenheit: Unit of measurement for temperature of which the boiling point of water is 212°F and the freezing point is 32°F.

Failure mode identifier (FMI): Numerical code for failure descriptions. Adapted from SAE standard practice J1587 diagnostics.

Fan: A mechanically or electrically operated device designed to create a moving stream of air, generally for cooling purposes.

Fan clutch: A temperature controlled device mounted to a engine-driven fan. It allows the fan to freewheel when the engine is cold or when the vehicle is at highway speed.

Fast idle: Idle rpm that determines the preset fast idle engine rpm as controlled using the fast idle enable switch. Fast idle operates similar to cruise control, except that it governs engine rpm with the vehicle stationary, or at no-load conditions.

Feathering: Wear pattern that appears and feels like the edges of feathers on a bird.

Feeler gauge: A thin strip of metal ground to a precise thickness used for measuring clearance between two parts.

Ferrography: A type of analysis performed on engine oil samples. Wear metal is separated from other particles in the oil and categorized by size, shape, and the ratio of small-to-large particles.

Ferrous metal: Metal containing iron or steel.

Fiber optic: A path for electricity or data transmission in which light acts as the carrier.

Field: Area surrounded by a magnetic force.

Filter: A device designed to remove foreign substances from air, oil, fuel, etc.

Firewall: Metal bulkhead between the engine and passenger compartment.

Firing order: The order in which engine cylinders must be fired to deliver their power stroke.

Fit: Contact area between two machined surfaces.

Flange: A projecting edge designed to keep a part in place.

Flare: A flange applied to tubing in order to provide a seal and to keep a fitting in place.

Flash codes: In electronically controlled engines, the troubleshooting code numbers that are flashed out on the diagnostic lamp.

Flash point: The point at which a substance will flash or burn when exposed to an open flame.

Flash programming: A method of programming or updating an ECM via the data link connector using an electronic service tool, instead of replacing the components.

Floating piston pin: A piston pin which is not locked in the connecting rod or the piston, but is free to turn or oscillate in both the connecting rod and the piston.

Flywheel: A large heavy wheel that forms the base for the starter ring gear and in which energy is absorbed and stored by means of momentum. Also provides a mounting surface for the torque converter or clutch assembly.

Foot-pound (ft.-lb.): Unit of work equal to the force in pounds multiplied by the distance in feet through which it acts.

Force: The amount of push or pull force exerted upon a body.

Four-stroke cycle engine: A engine design where a power pulse occurs every other revolution of the crankshaft. These strokes are (1) intake stroke; (2) compression stroke; (3) power stroke; and (4) exhaust stroke.

Free electrons: Electrons in outer orbits around the nucleus of an atom. Can be moved out of their orbits relatively easily.

Freeze plugs: A stamped metal disc installed in the holes where the sand core was removed from the casting. Allows for expansion if coolant should freeze, preventing a cracked block. Also called *core plugs.*

Freezing point: The temperature at which a liquid turns into a solid.

Frequency: The rate of change in direction, oscillation, cycles, etc., in a given time span.

Friction bearing: A bearing made of babbitt or bronze with a smooth surface.

Friction: Any resistance to movement between two objects placed in contact with each other.

Fuel: Any substance that will burn and release heat. In the case of a diesel engine, it is usually fuel oil obtained from petroleum.

Fuel injection: A system that sprays diesel fuel directly into the combustion chambers via atomization.

Fuel injector: Valve controlled by an electronic solenoid or spring pressure. Component that actually injectes the fuel into the combustion chamber.

Fuel line freeze: Condition caused by water in fuel turning to ice in cold weather and blocking the fuel lines.

Fuel lines: The portion of the fuel system that carries fuel from the tank to the filter and on to the injection pump and fuel injectors.

Fuel position: On some electronically controlled engines, an internal signal within the ECM, from the electronic governor to the fuel injection control. It is based on desired rpm, FRC fuel position, rated fuel position and engine rpm.

Fuel pump: A device used to draw fuel from the tank and force it into the fuel system.

Fuel ratio control (FRC): On some electronically controlled engines, a limit based on control of the air to fuel ratio. When the ECM senses a higher boost pressure (more air into cylinder), it increases the FRC fuel position limit, allowing more fuel into the cylinder.

Fuel tank: A large tank made of steel, aluminum, or plastic used to store the vehicle's supply of fuel.

Fulcrum: A support on which a lever pivots in raising an object.

Full-flow oil filter: An oil filter that filters all the oil passing through the engine

Full-load setting: On some electronically controlled engines, a number represents the fuel system adjustment made at the factory to "fine-tune" the fuel system for maximum fuel delivery. The correct value for this parameter is stamped on the engine information ratings plate.

Full pressure system: A type of oiling system that draws oil from the sump and forces it through passages in the engine.

Fuse: Protective device that interrupts current flow if an overload condition is present in the circuit.

Fusible link: A special calibrated wire installed in an electric circuit. Will allow an overload condition for short periods. A constant overload will melt the wire and break the circuit.

G

Gallery: A pathway for oil or coolant inside a block, cylinder head, etc.

Galvanize: To coat a metal with a molten alloy mixture of lead and tin. Used to prevent corrosion.

Gas: A nonsolid material that can be compressed. It will expand when heated, and condense when cooled.

Gasket: A material placed between two parts to prevent leakage.

Gassing: Bubbles that rise to the top of battery electrolyte during charging. The bubbles are caused by hydrogen gas produced as a by-product of the charging process.

Gear: A circular object, either flat or cone shaped, upon which a series of teeth are cut.

Gear backlash: The small amount of clearance between the gear teeth of mating gears.

Gear clash: Noise that is heard when gears fail to mesh properly.

Gear down protection: On some systems, a programmable high gear limit used to promote driving in higher gears for increased fuel economy.

Gear pump: A type of pump that uses meshing gears to provide pressure and fluid movement.

Gear ratio: The relationship between the number of turns made by a driving gear to complete one full turn of the driven gear. If the driving gear makes four revolutions in order to turn the driven gear once, the gear ratio would be four to one (4:1).

Gear reduction: Setup in which a small gear is used to drive a larger gear. Produces an increase in torque.

Generator: An electromagnetic device consisting of an armature, field windings, and other parts that produces electricity when turned. Also see *Alternator.*

Glaze: A highly smooth, glassy finish on a cylinder wall. Produced over a long period of time by piston ring friction.

Glaze breaker: An abrasive tool used to remove glaze from a cylinder wall prior to the installation of new piston rings.

Glow plugs: A heating element used to help diesel engines start initially or in cold weather by warming the air in the cylinders.

Governor: A device designed to sense changes in engine speed, then change the amount of fuel delivered to correct for the difference.

GPM: Gallons Per Minute.

Grade markings: Lines placed on the heads of some bolts to indicate tensile strength.

Gram (G): A metric unit of measurement for weight or mass equal to 0.03527 oz.

Grid: Lead screen or plate to which the battery plate active material is affixed.

Grind: To remove metal from an object by means of an abrasive wheel.

Gross horsepower: The brake horsepower of an engine with optimum settings and without allowing for power absorbed by the engine-driven accessories.

Gross torque: The maximum torque produced when measured at the engine's crankshaft. Does not allow for torque consumed by the engine-driven accessories.

Ground: The terminal of the battery connected to the vehicle's frame.

Growler: A device used to test electric motor armatures.

Gum: Any oxidized portions of petroleum products that accumulate in the engine and fuel system.

H

Harmonic balancer: A device that is mounted on one end of a crankshaft. Used to reduce torsional vibration.

Harmonic vibration: A high frequency vibration caused by the crankshaft.

Harness: The wiring bundle (loom) that connects all components of the electronic system.

Hazardous waste: Any chemical or material that has one or more characteristics that make it hazardous to health, life, and/or the environment.

Header: A steel pipe that connects the exhaust manifold to the muffler and tailpipe.

Heat: A form of energy associated with the motion of atoms or molecules and capable of being transmitted by conduction, convection, and radiation.

Heat exchanger: A device, such as a radiator, used to cool or heat by transferring heat from one medium to another. A device which utilizes normally wasted heat for useful purposes.

Heat loss: Heat from burning fuel that is lost in the cylinder without doing useful work.

Heat shield: A sheet of metal or fiberglass used to shelter heat sensitive components, such as wiring, from excessive engine and exhaust heat.

Heat-shrink tubing: Plastic tube used to insulate electrical solder joints that shrinks when heat is applied.

Heat treatment: Application of controlled heating and cooling to a metal that is timed in order to produce certain properties.

Heater core: A radiator-like unit in the blower case in which coolant circulates. Used to heat the vehicle interior.

Heel: The outside or larger part of a gear tooth.

Helical gear: A gear that has teeth cut at an angle to its centerline.

Helicoil: An insert used to repair damaged threads.

Helix: A tapered groove cut into a plunger designed to help control fuel delivery.

Hertz (Hz): A measure of frequency in cycles per second.

Hone: An abrasive tool for correcting small irregularities or differences in diameter in a cylinder.

Honing: Process of removing metal with a fine abrasive stone. Used to achieve close tolerances.

Horsepower (hp): Measurement of an engine's ability to perform work. One horsepower is defined as the ability to move 33,000 pounds one foot in one minute.

Hose clamp: A device used to secure hoses to their fittings.

Hose: A flexible rubber or neoprene tube for carrying water, oil, and other fluids.

Hot tank: A heated chemical cleaning tank that is used to clean parts that are immersed into it. Should not be used with aluminum parts when filled with a strong alkali cleaner.

Hunting: Occurs when there is a rapid oscillation of the governed engine speed. The engine speed moves back and forth around the desired value.

Hydraulic dashpot: This device may be used with the governor to help avoid low idle overrun and stabilize idle speed.

Hydraulic electronic unit injector (HEUI): Unit injector featuring a hydraulically-actuated injection pumping, with an electronically controlled injector. Combines fuel metering and injecting elements into a single unit.

Hydraulic system: An arrangement of tubing and pistons used to transmit force through fluid from one location to another.

Hydraulics: The science of fluids in motion.

Hydrocarbons (HC): A compound composed of hydrogen and carbon atoms. Petroleum is a hydrocarbon. An unwanted exhaust pollutant resulting from unburned fuel.

Hydrometer: A float device used to measure the specific gravity of battery electrolyte to determine its state of charge.

Hygroscopic: The ability to absorb moisture from the air.

I

ID: Abbreviation for *Inside Diameter*.

Idle rpm limit: A programmable parameter that indicates the maximum allowable engine rpm when the rpm is set using the cruise control On/Off switch and the set/resume switch.

Idle shutdown time: A programmable parameter that indicates the length of time (in minutes) that the engine will idle before shutting down.

Idle/PTO bump rpm: A programmable parameter that shows how much the engine rpm will be increased or decreased when the accel/decel switches are briefly toggled.

Idle: Refers to the engine operating at its slowest speed with a machine under no load.

Ignition switch: A key operated switch mounted on the steering column for connecting and disconnecting power to the electrical system.

Impeller: A wheel-like device that has fins cast into it and uses centrifugal force to transfer liquids or air. Also used to refer to parts of a turbocharger.

Indicated horsepower (IHP): The calculated power produced by burning fuel within the cylinders.

Indicator: An instrument for recording the creation of cylinder pressure during the cycle.

Indirect combustion: Takes place in a small prechamber.

Induction: The imparting of electricity to an object by magnetic fields.

Inertia: A physical law which tends to keep a motionless body at rest or also tends to keep a moving body in motion. Effort is required to start a mass moving and to slow or stop it once it is in motion.

Inertia switch: A switch that is designed to operate only if a sudden movement occurs, such as a collision.

Infrared rays: Rays of light that are invisible to the human eye. Used with some dyes to trace fluid leaks.

Inhibitor: A material added to another material to control or prevent some unwanted action, such as corrosion, foaming, etc.

Injection actuation pressure control valve: An electronically controlled dump valve which maintains high pressure in the fuel manifold based on ECM input.

Injection actuation pressure sensor: A device that determines fuel pressure and sends a corresponding electrical signal to the ECM.

Injection pump: A device by which fuel is metered and delivered under pressure to the injectors.

Injector: A valve that is controlled by a solenoid or spring pressure for injecting fuel oil into the combustion chamber of an engine against the pressure of air within the chamber.

Inlet air heater: A device for improving the cold start capability of the engine, reducing smoke by heating the air when the engine is cold.

Inlet port: The opening in the cylinder of a two-cycle engine through which the air enters the cylinder.

Inline: Any group of components in a straight row.

Inline engine: An engine in which all its cylinders are aligned in a straight row.

Input sensors: Any sensor that provides information to an ECM.

Input shaft: The shaft carrying the driving gear.

Insulation: Any material that resists the flow of electrons, heat, or noise.

Intake air temperature sensor (IAT): A thermistor installed in the air cleaner or air intake tube. Sensor used by the ECM to monitor the temperature of the air entering the engine.

Intake manifold: Series of connecting tubes or housing between the air filter and the openings to the intake valves.

Intake stroke: The portion of the piston's movement that is devoted to drawing air into the combustion chamber.

Intake valve: Valve through which air is admitted to the cylinder.

Integral: A device that is formed as part of another unit.

Integrated circuit (IC): A single chip of semiconductor material which contains various electrical components in miniaturized form.

Intercooler: Heat exchanger used to cool the incoming air between the stages of compression.

Intermittent: An event that occurs at different intervals.

Intermittent codes: Computer diagnostic code that does not return immediately after it has been cleared.

Internal combustion engine: An engine that burns fuel within itself as a means of developing power.

Internal gear: A gear with teeth cut on its inward facing surface.

Intravance: An automatic timing device designed to advance engine timing while the engine is running.

Ion: An electrically charged atom or molecule produced by an electrical field, high temperature, etc.

Ionize: To convert partially or completely into ions.

Isochronous: Maintaining constant engine speed independent of the load being carried.

J

Joule: Metric unit of measurement for energy or work equal to a force of one Newton applied through a distance of one meter. One joule is equivalent to 0.737324 ft-lbs.

Journal: The part of a shaft or axle that actually contacts a bearing.

Jump start: Method of starting a vehicle with a weak or dead battery through the use of jumper cables.

Jumper cables: A pair of electrical cables used to start an engine with a weak or dead battery.

Jumper tube: A small pipe used to connect either the inlet or return fuel manifold to the injectors.

Jumper wire: A wire used to make a temporary electrical connection.

K

Keel cooling system: An outboard heat exchanger system which is either attached to or built as part of a ship's hull below the light waterline.

Key: A small metal piece that fits into a groove partially cut into two parts to allow them to turn together.

Keyway or **Keyseat:** A groove or slot cut into a shaft, pulley, or hub that permits the insertion of a key to hold a part on a shaft.

Kickout switch: Refers to the service brake and clutch switches typically used to "kick out" or exit the cruise control mode or PTO/idle set speeds.

Kilometer (km): A metric unit of measurement for distance equivalent to 5/8 of a mile.

Kinetic energy: Any energy associated with motion.

Knock: Engine noise caused by detonation or worn mechanical parts.

Knurl: The process of roughing a piece of metal by pressing a series of cross-hatched lines into the finished surface, which raises the area between the lines.

L

Laminate: To build up or construct something out of a number of thin sheets.

Land: A portion of metal separating the grooves that the piston rings ride against.

Lapping: The process of fitting two surfaces by rubbing them together with an abrasive material between the two surfaces.

Latch mode: In some systems, a programmable parameter option that controls the exhaust brake. The exhaust brake will engage when the service brake pedal is depressed and will remain engaged until this control detects a change in input.

Latent heat: The amount of heat beyond a substance's boiling or melting point required to change a solid to liquid or liquid to vapor.

Lateral runout: Side-to-side movement of a part, shaft, etc.

Lathe: Machine on which a solid piece of material is spun and shaped by a fixed cutting tool.

Lever: A rigid bar or shaft that pivots on a fixed fulcrum. It is used to increase force or to transmit a change in motion.

Lift: Maximum distance a valve head is raised off its seat.

Light emitting diode (LED): A special function diode that lights when forward biased.

Liner: A thin section placed between two parts, such as a replaceable cylinder liner in an engine.

Linkage: Any moveable series of rods, levers, and cables used to transmit motion from one unit to another.

Liquid: A substance in a state that can assume the shape of the vessel in which it is placed without a change in volume.

Liquid withdrawal: A system which draws LPG gas from the bottom of the tank to ensure delivery of liquefied petroleum gas.

Liter: Metric unit of measurement for liquid equal to 2.11 pints. Also used as a measure of volume equal to 61.027 cubic inches.

Load: The equipment or burden that is driven or moved by the engine's power.

Lobes: Egg-shaped projections on a camshaft.

Lock washer: A type of washer used to prevent the accompanying nut from working loose.

Longitudinal: Parallel to the length of the vehicle.

Louver: Slots cut into the hood or body, usually for ventilation.

LPG: Abbreviation for *liquid petroleum gas.*

Lubricant: A material designed to reduce friction between surfaces.

Lug (engine): Condition where the addition of fuel does not cause a corresponding increase in RPM. Any operational state where the engine is operating at or below its maximum torque speed.

M

Magnaflux: A chemical process that is used to check iron or steel components for cracks.

Magnetic field: Area encompassed by magnetic lines of force surrounding a magnet.

Magnetic timing meter: A tachometer that uses magnetism as a triggering device.

Magnetism: Invisible lines of force that attract ferrous metals.

Main bearing: Series of bearings that support the crankshaft in the engine.

Main cap: Metal pieces that bolt to the block. Used to support the crankshaft.

Maintenance-free battery: A sealed battery that requires no additional water or electrolyte during its useful life.

Malfunction indicator light (MIL): Amber-colored light in the instrument cluster used to indicate that a problem exists in a vehicle's computer control system. Also called a *Check Engine* or *Service Engine Soon* light. Generalized term for any instrument cluster light used to indicate a problem in a system.

Manifold: Pipes connecting a series of openings or outlets to a common opening (e.g. exhaust or intake manifold).

Manometer: An instrument used for measuring the pressure of liquids and gases. Usually a tube filled with water or mercury.

Marine technicians: Technicians who specialize in the maintenance and repair of engines on boats and ships.

Master Technician: An experienced technician. Usually used to refer to someone who has passed all of the ASE certification tests in a skill area.

Material Safety Data Sheet (MSDS): Information on a chemical or material that must be provided by the material's manufacturer. Lists potential health risks and proper handling procedures.

Matter: Any substance that makes up anything that occupies space, has weight, and is perceptible to the senses.

Mechanical efficiency: An engine's rating as to how much potential horsepower is lost through friction within its moving parts. The ratio of brake horsepower output of an engine to the indicated horsepower in the cylinders.

Melting point: Temperature at which a solid becomes a liquid.

MEP: Abbreviation for *Mean Effective Pressure.* The difference between compression and combustion pressures.

Meter: Metric unit of measurement of length. One meter is equal to 39.37 inches.

Metric system: A decimal system of measure that is based on 10. Used by most of the world and in some professions worldwide.

Micrometer: A precision instrument for measuring either internal or external dimensions to within thousandths or ten thousandths of an inch or millimeter.

Microprocessor: A small silicon chip that contains elements in a computer. Often referred to as an integrated circuit or IC.

Millimeter (mm): Metric unit of measurement equivalent to .039370 of an inch. One inch is equivalent to 25.4 mm.

Milling: Process of cutting metal with a multitooth rotating cutting wheel.

Misfiring: Failure of one or more cylinders to fire.

Molecule: The smallest portion that matter may be divided into and still retain all of its properties.

Motor: A device which converts electrical or fluid energy to mechanical energy.

MPG: Abbreviation for *miles per gallon.*

MPH: Abbreviation for *miles per hour.*

Muffler: A chamber attached to the end of the exhaust pipe which allows the exhaust gases to expand and cool. It is usually fitted with baffles or porous plates and serves to subdue much of the noise created by the exhaust. A chambered unit attached to a pipe or hose to deaden noise.

Multi-viscosity oil: An engine oil that can exhibit different viscosity characteristics when heated or cooled.

Multifuel engine: An engine designed to start and operate on any one of a variety of fuels without modification. Also known as a *flex fuel engine.*

Multimeter: An electrical test meter that can be used to test for voltage, current, or resistance.

Multiple plunger inline fuel injection pump: A fuel injection pump that uses a separate pumping element for each cylinder in the engine.

Multiplexing: A method of using one communications path to carry two or more signals simultaneously.

MVMA: Motor Vehicle Manufacturers Association.

N

Natural gas: General term for any petroleum-based gas, such as propane and liquefied petroleum gas.

Naturally aspirated: Another name for an engine that has no turbocharger.

NC threads: National Coarse thread sizes.

Needle bearing: An antifriction roller bearing that uses many small diameter rollers in relation to their length.

Negative terminal: A terminal from which current flows on its path to the positive terminal. Usually designated by a minus sign (-).

Neoprene: A synthetic rubber with superior resistance to such chemicals as oil.

Net horsepower: Maximum horsepower at the flywheel with all engine-driven accessories in use.

Net torque: Maximum torque at the flywheel with all engine-driven accessories in use.

Neutral safety switch: A switch that prevents starter engagement if the transmission is in gear.

Neutron: A particle of an atom that has a neutral charge. Forms the central core of an atom along with protons.

Newton-meter (N•m): Metric unit of measurement for torque equivalent to .7376 foot pounds.

NF thread: National fine thread sizes.

NHTSA: National Highway Traffic Safety Administration.

Nonferrous metal: A metal such as aluminum, copper, and brass that contains no iron and is nonmagnetic.

Normally aspirated: An engine without a turbocharger or supercharger which operates at the start of the compression stroke with a cylinder air charge very near or slightly below atmospheric pressure.

North pole: One of the poles of a magnet from which lines of force originate.

Nozzle: An opening through which fuel flows into the combustion chamber. Also referred to as an injection nozzle.

NSC: National Safety Council.

Nut lock: A slotted nut which fits over a standard nut on a wheel spindle. Normally used on nondriving axles.

O

OD: Abbreviation for *Outside Diameter.*

Odometer: A device used to measure and register the number of miles a vehicle has traveled.

OEM: Abbreviation for Original Equipment Manufacturer.

Ohm: A unit of resistance to the flow of electric current. It is equal to that of a conductor in which a current of one ampere is produced by a potential of one volt across the terminals. Unit of measurement for resistance to the flow of electric current in a given unit or circuit.

Ohm's Law: Formula for computing unknown voltage, resistance, or current in a circuit by using two known factors to find the unknown value. The resistance in ohms is equal to the voltage divided by the current in amperes flowing in the circuit.

Ohmmeter: An electrical instrument used to measure the amount of resistance in a given unit or circuit. An electrical instrument designed to measure the resistance of a circuit in ohms.

Oil analysis report: A detailed report that breaks down the amount and types of contaminants in an engine oil sample. Useful for determining if an engine is approaching a major failure.

Oil classification CC, CD, CE: Classification for oil designed for use in diesel engines. The C refers to compression ignition.

Oil cooler: An air or liquid cooled heat exchanger used to remove excess heat from the engine or transmission oil.

Oil filter: A device used to strain abrasive impurities from engine oil.

Oil gallery: Cast or drilled passageway or pipe in an engine. Used to carry oil from one part of the engine to another requiring lubrication or cooling.

Oil nozzle: Nozzle located just under each cylinder in some diesel engines. Used to lubricate the bottom of the piston-connecting rod assemblies.

Oil pickup: Tube and screen that connects to the oil pump and extends to the bottom of the oil pan. Used by oil pump to pick up oil.

Oil pressure gauge: A dash mounted instrument that provides oil pressure readings in pounds per square inch or in kilopascals.

Oil pump: A device that forces oil under pressure to the oil galleries for distribution throughout the engine.

Oil ring: The bottom piston ring which scrapes oil off the engine cylinder walls.

Oil seal: A device used to prevent oil leakage past certain areas, such as around a rotating shaft.

Oil slinger: An impeller or other device attached to a revolving shaft that will throw any oil passing it outward to splash lubricate other components before returning to its point of origin.

Open circuit: Occurs when an electrical wire or connection is broken or a switch is open, preventing the signal or supply voltage from reaching its intended destination.

Open loop: Computer operation mode used when an engine has not reached its optimum operating condition. The ECM operates the engine using a basic set of fixed values.

Optical sensor: A light sensitive device used to determine when to close or open a circuit in relation to its exposure to light.

Orifice: An aperture or restricted opening used to control the flow and delivery of a liquid.

Oscillate: Any back and forth swinging action like that of a pendulum.

Oscilloscope: Test device used to observe voltage by displaying line patterns in relation to time.

Out-of-round: Condition where a cylinder or other round object has greater wear at one diameter than another.

Output device: A computer-controlled device used to change an engine setting.

Output shaft: A shaft that delivers power from within a mechanism, such as a transmission.

Overdrive: An arrangement of transmission gears that results in the driven shaft turning more revolutions than the driving shaft.

Overhead camshaft (OHC): A camshaft mounted above the cylinder head. Usually driven by a timing chain, belt, or a combination of the two.

Overhead valve (OHV): An engine designed with the valves in the cylinder head. Also known as a *valve-in-head engine.*

Overrunning clutch: A clutch mechanism that will drive in one direction and slip in the other.

Oxides of nitrogen (NO$_x$): An undesirable compound of nitrogen and oxygen in exhaust gases. Usually produced when combustion chamber temperatures are excessively high.

Oxidize: To combine an element with oxygen or a catalyst which converts it to its oxide form. This action will form rust in ferrous metals.

Oxidizing catalyst: A chemical or other substance which causes oxygen to combine with the metal to produce rust and scale.

P

Packing: A flexible material used to seal parts that move relative to each other.

Pan: A thin metal cover bolted to the bottom or side of an engine to contain oil.

Parallel circuit: An electrical circuit that has two or more resistance units wired so that current can flow through them at the same time.

Parameter identifier: A two or three digit code which is assigned to each component to identify data via data link to ECM.

Parameter: A programmable value or limit which determines the characteristics or behavior of the engine and/or vehicle.

Parasitic load: Normal electrical load from the ECM, radio, and other electrical components placed on a vehicle's battery when the engine is not operating.

Pascal's Law: A principle of fluids that states that when pressure is applied to a confined fluid, it is transferred undiminished throughout the fluid.

Passive magnetic vehicle speed sensor: A type of vehicle speed sensor that does not require a power and a ground connection. It produces a signal based on the change in magnetic flux of a ferrous metal gear near the sensing tip.

Password: Any group of numeric or alphabetic characters used to restrict access to a given parameter. For example, an electronic system requires a correct password in order to change customer specified parameters (customer passwords) or certain engine specifications (factory passwords). Passwords are also required to clear certain diagnostic codes.

Pedal throttle position sensor: A sensor that measures pedal position and sends a signal to the ECM. The sensor is mounted on a throttle pedal assembly.

Peen: To stretch over by pounding with the round end of a hammer.

Penetrating oil: Special oil used to free corroded parts so that they can be removed.

Permanent magnet: A magnet capable of retaining its magnetic properties over a very long period of time.

Petcock: A valve placed in a tank or line for draining purposes.

Petroleum: Raw oil taken out of the ground to be refined into diesel fuel, gasoline, kerosene, etc.

Phillips head screw: A screw head having a fairly deep cross slot.

Phosgene gas: A toxic, colorless, poisonous gas produced when refrigerant is burned.

Phosphor-bronze: A bearing material composed of lead, tin, and copper.

Photo sensitive diode: A semiconductor device that allows current to flow when exposed to light.

Pilot shaft: A "dummy" shaft temporarily used to align parts for assembly.

Pinion gear: A small gear that is either driven by or driving a larger gear at an angle.

Piston: Cylindrical plug that is closed on one end and attached to a connecting rod at the other. When the fuel charge is fired, the force of the expanding gases will transfer from the closed end through the piston to the connecting rod and crankshaft.

Piston boss: That portion of the piston which supports the piston pin.

Piston collapse: A reduction in the diameter of the piston skirt caused by heat and impact stress.

Piston displacement: The volume of air moved or displaced by movement of the piston from bottom dead center to top dead center.

Piston expansion: An increase in piston diameter due to normal heating.

Piston head: The top of the piston; term used to refer to the portion of the piston above the top ring.

Piston lands: The portion of the piston between the ring grooves.

Piston pin: A cylindrical piece of alloy steel which passes through the piston bosses and upper end of the connecting rod so that movement of the piston is transmitted to the connecting rod. Also called a *wrist pin.*

Piston ring end gap: Distance left between the ends of the rings when installed in the cylinder.

Piston ring expander: A ring placed behind the piston ring. Designed to increase the pressure of the ring against the cylinder wall.

Piston ring grooves: Series of slots in the piston in which the rings are fitted.

Piston rings: A split ring installed in a groove in the piston. Seals the compression chamber from the crankcase.

Piston skirt: The portion of the piston below the rings which is designed to take the side thrust of the piston.

Piston skirt expander: Spring device placed inside a piston skirt to increase its diameter.

Pivot lever: A pin or shaft on which a part rests or turns.

Plastigage: A measuring tool that is compressed between two tightly fitting surfaces, such as bearings, to measure clearance.

Plates: Thin sections of lead peroxide or porous lead used to make up a battery's positive and negative cells.

Platinum (Pt): A precious metal sometimes used as an electrical conductor.

Play: Any movement between two parts.

Polarity: The positive or negative terminals of a battery. Also the north and south poles of a magnet.

Pole piece: Component of motor that keeps the armature rotating.

Pole shoes: Metal pieces around which field coil windings are placed.

Polymer: A chemical chain of simple molecules added to a substance, such as oil, to increase its performance characteristics.

Poppet valve: A disc type valve used to open and close the valve ports, that is opened by the action of the cam and closed by the action of the valve spring.

Port: In diesel engines, the opening in the cylinder head or cylinder that allows air or exhaust gas through.

Port and helix: Consists of a pumping plunger and barrel assembly designed to regulate fuel delivery.

Positive: Terminal to which electrons flow. Usually indicated by a plus (+) sign.

Potential horsepower: A measurement of the maximum amount of horsepower available.

Potentiometer: A variable resistor that can be used to adjust voltage in a circuit.

Pour point: The pour point of an oil is the lowest temperature at which oil will flow under prescribed conditions.

Power: The rate at which work is being done.

Power stroke: The downward movement of the piston that occurs after the air-fuel charge has been ignited.

Power take off (PTO): An auxiliary drive that eliminates the need for a second engine to power any needed accessories.

PPM: Abbreviation for Parts Per Million.

Practical efficiency: The amount of horsepower delivered to the driving wheels.

Precombustion chamber: A portion of the combustion chamber connected to the cylinder through a narrow port.

Preheating: The application of heat in preparation for some further treatment, such as welding.

Preloading: Process of adjusting a bearing so that it has a mild pressure placed upon it.

Press-fit: Condition of fit between two parts that requires pressure to force the two parts together.

Pressure: Any force per unit of area placed on a surface.

Pressure bleeding: A method of bleeding a system by using additional pressure from an external source.

Pressure cap: A cap that is designed to hold a preset amount of pressure.

Pressure relief valve: A valve designed to open at a specific pressure to prevent excessive pressure on a closed system.

Pressure sensor: A sensor that is used to detect excessive high or low pressures in a system.

Pressure-splash system: A system that uses an oil pump to supply oil to the camshaft and crankshaft bearings and the movement of the crankshaft to splash oil onto the cylinder walls and other nearby parts.

Preventive maintenance (PM) program: Organized effort to maintain a documented history of all maintenance and repair work performed on a group of engines, vehicles, etc.

Primary wiring: Small insulated wires which serve the low voltage needs of the vehicle systems.

Printed circuit: An electrical circuit that is made by conductive strips printed on a board or panel.

Programmable read only memory (PROM): A semiconductor chip that contains instructions that are permanently encoded into the chip. Instructions contain base operating information for how components should operate under various conditions.

Progressive shifting: Refers to shifting up through the lower gears quickly by not using excessive engine rpm in each gear. Shifts are made above peak torque but below rated rpm. Using excessive engine (higher) rpm ranges before shifting to the next gear wastes fuel and fails to take advantage of the torque rise of the engine.

Prony brake: A device that utilizes friction to measure the horsepower output of the engine.

Propane: Hydrocarbon-based gas that is mixed with butane and is sometimes used as an engine fuel. A petroleum gas, C_3H_8 which has a boiling point of about -44°F. Also known as *LPG* or *CNG*.

Propylene glycol: A type of liquid chemical used in engine coolant.

Proton: A positively charged particle in the nucleus of an atom.

Prussian blue: A deep blue pigment mixed with a grease-like substance. Used to find low and high spots on a part.

PSI: Abbreviation of pounds per square inch.

Pulse width : A signal consisting of variable width pulses at fixed intervals, whose ratio of time on versus total time off can be varied. Also referred to as *duty cycle.*

Pump: A device that is designed to move coolant, oil, fuel, etc., from one area to another.

Purge: The process of removing air or impurities from a system. *Also see* Bleeding

Push rod: A connecting link in an operating mechanism (e.g. a cam operated rod that operates the rocker arm which opens an engine valve).

Q

Quenching: Process of dipping a heated object into water, oil, or other substance to quickly reduce its temperature.

Quick charge test: Method of determining if a battery's plates are sulfated.

R

R-12 (CFC-12): Refrigerant used in older air conditioning systems. Replaced by R-134a in newer vehicles. Also called *dichlorodifluoromethane (CCl_2F_2).*

R-134a (HFC-134a): Refrigerant used in the air conditioning systems of most vehicles manufactured after 1992. Replaced R-12 due to environmental concerns. Also called *tetrafluroethane (CH_2FCFE).*

Race: The inner and outer portions of an antifriction bearing (either ball or roller type) which provide a surface for the balls or rollers.

Radial: A line at right angles to a shaft, cylinder, etc., center line.

Radial runout: Difference in rotation caused by uneven diameter.

Radiation: Any transfer of heat from one object to another when the hotter object radiates invisible waves of heat that strike surrounding objects, causing them to vibrate and heat. Thermal radiation is energy in the form of wave motion. If it falls on some material object where it is absorbed, it produces an increase in internal energy and results in a rise in temperature.

Radiator: A heat exchanger used to remove heat from engine coolant. It is comprised of a series of finned passageways. As coolant moves through the passageways, heat is given off to the fins, which is then dissipated by passing air through the fins.

Radiator core: A series of finned passages made of copper or aluminum in a radiator through which coolant passes and gives off excess heat.

Radius: Distance in a straight line from the center of a circle to its outer perimeter.

Ram air: Air that is forced through a condenser or radiator or into the engine by vehicle movement.

Random access memory (RAM): A portion of computer memory that serves as temporary storage for data. This data is lost if power to the computer is lost. It is used to store sensor information and any diagnostic trouble codes.

Rated horsepower: Indication of horsepower load that can be safely placed upon an engine for a prolonged period of time.

Ratio: A fixed relationship between things in number, quantity, or degree. Also, the relative size of two quantities expressed as a quotient of one divided by the other. The ratio of 9 to 5 is written 9:5 or 9/5. A fixed relationship between things in number, quantity, or degree.

Raw-water cooling system: An open cooling system that uses an external water source, such as pond, lake, or sea water to cool an engine.

Read only memory (ROM): A type of computer memory that cannot be changed and is not lost if power is cut off. Contains the general information to operate a computer.

Ream: To enlarge or smooth a hole by using a round cutting tool with fluted edges. To finish a hole accurately with a rotating fluted tool.

Reciprocating motion: Any back and forth motion such as the action of pistons in an engine.

Rectified: A term used to describe alternating current (ac) that is changed to direct current (dc).

Reduction gear: A gear that increases torque by reducing the speed of a driven shaft in relation to the driving shaft.

Reed switch: An electronic switch that consists of two metal strips or reeds. The reeds are influenced by a magnetic field, which causes them to open or close, depending on the application.

Reference voltage: A regulated voltage supplied by the ECM to a sensor. The reference voltage is changed by the sensor and used by the ECM to detect a change in operating conditions.

Relative volatility: The point at which a substance or one of its components turns to vapor or "flashes."

Relay: A magnetically operated switch used to make or break current flow in a circuit.

Relieve: The process of removing metal from the valve seat area and cylinder to improve the flow of fuel into the cylinder.

Remote station operation: Refers to engine speed control from a location outside of the vehicle cab, typically for some type of PTO operation for pumping or other application using engine power.

Reserve capacity: The amount of time a battery can produce an acceptable current when not charged by the alternator.

Reservoir: A tank or bottle used to hold a reserve of fluid, such as coolant.

Resistance: The measure of opposition to electrical flow in a circuit.

Resistor: A device placed in a circuit to lower voltage and current flow.

Resonator: A small muffler-like device that is placed in an exhaust system to further reduce exhaust noise.

Retard: To set the timing so that combustion begins later or less degrees before TDC.

Return spring: A spring positioned to close a valve.

Reverse bias: A condition or arrangement where a diode acts as an insulator.

Reverse flow muffler: A muffler that has its internal pipes arranged so that exhaust gases flow in a reverse direction inside the muffler before exiting to the tailpipe.

Reverse flush: Method of cleaning by flushing a cleansing agent through a system in the reverse direction of normal fluid flow. To pump water or a cleaning agent through the cooling system in the opposite direction to normal flow.

Rheostat: A variable resistor used to control current flow.

Ribbed belt: A V-type drive belt that has small ridges added along its length.

Ridge reamer: A device used to remove the metal ridge that forms at the top of a cylinder due to wear prior to piston removal.

Ring expander: A spring device used to hold a ring snugly against the cylinder walls.

Ring gap: Distance between the end of the piston rings when installed in the cylinder.

Rivet: A metal pin used to hold two objects together. One end has a head while the other end must be set or peened over.

Rocker arm: A lever arm used to direct downward motion on a valve stem. An arm used to change upward movement of the cam operated push rod to downward motion to open an engine valve on an overhead valve engine.

Rocker shaft: A shaft upon which rocker arms are mounted in some engines.

Rod cap: The lower removable half of a connecting rod.

Roller bearing: A bearing which contains hardened roller ball bearings between two races.

Roller clutch: A clutch that utilizes a series of rollers placed in ramps. The clutch will provide driving force in one direction, but will slip in the other.

Roller tappets: Type of valve lifter that has a roller attached to the end connected to the cam.

Roller vane pump: A pump which uses spring-loaded vanes that are shaped like rollers to provide the pumping action.

Rollover valve: A valve in the fuel tank or delivery lines that prevents the escape of raw fuel in the event of a vehicle rollover.

Room temperature vulcanizing (RTV): A type of sealant that cures at room temperature.

Roots blower: A blower similar to a gear pump.

Rotary motion: A continuous motion in a circular direction, such as performed by a crankshaft. A circular movement, such as the rotation of a crankshaft.

Rotary pump: A pump that uses a star-shaped rotor.

Rotary valve: A valve that exposes ported holes to allow the entrance and exit of gases.

Rotor housing: A circular metal housing with evenly spaced slots cut into it to provide a housing for rotor vanes.

Rotor pump: A type of pump that uses a central rotor with spring-loaded vanes.

RPM: Abbreviation for revolutions per minute.

Running-fit: A fit in which sufficient clearance has been provided to enable parts to turn freely and to receive lubrication.

Runout: The side-to-side distortion or play of a rotating part.

S

SAE: Society of Automotive Engineers.

SAE data link connector: Normally used to refer to 16-pin data link connector mandated by SAE's on-board diagnostic standards. Also referred to as an *ATA data link connector.*

SAE thread: Commonly referred to as a standard thread. The standard counts the number of threads per inch.

Safety valve: A valve designed to open when internal pressures within a container exceed a predetermined level.

Saybolt test: Test used to determine the viscosity of a fluid.

Saybolt viscosimeter: An instrument used to determine the viscosity of a liquid.

Scale: Accumulation of corrosion and mineral deposits within a cooling system.

Scanner: An electronic tool, usually hand held, that is used to read and interpret diagnostic codes and engine sensor information. Commonly referred to as a *scan tool.*

Scavenging: Refers to the displacement of exhaust gas from a cylinder by incoming fresh air.

Scavenging efficiency: In a two-cycle engine, scavenging efficiency is the ratio of a new air charge trapped in a cylinder to the total volume of air and exhaust gases in the cylinder at port closing position.

Score: A scratch or groove on a finished surface.

SCR: Abbreviation for Silicon Controlled Rectifier.

Screw extractor: A device used to remove broken bolts, screws, etc., from holes.

Scuffing: A roughening of the cylinder wall caused when there is no oil film separating the moving parts and metal-to-metal contact is made.

Seal: A formed device made of plastic, rope, neoprene, or Viton. Used to prevent oil leakage around a moving part, such as a shaft.

Sealant: A liquid or paste material applied to a surface along with or in place of a gasket to prevent oil leaks.

Sealed bearing: A bearing that has been sealed at the factory and cannot be serviced during its useful life.

Seat: A surface upon which another part rests or seats. An example would be a valve face resting on its valve seat.

Sediment: An accumulation of matter or foreign debris, which settles to the bottom of a liquid.

Seize: Condition where two or more parts have forced the lubricant out from between them due to excessive heat and/or friction. When this happens, the parts stick together.

Self-test: A diagnostic test that a computer runs on itself and its associated systems (input sensors and output devices) to ensure that there are no faults.

Self-diagnostics: The ability of a computer to not only check the operation of all of its sensors and output devices, but to check its own internal circuitry and indicate any problems via diagnostic trouble codes.

Self-induction: The creation of voltage in a circuit by varying current in the circuit.

Semi-metallic: A friction material that has metal particles added to an organic compound to increase its useful life.

Semiconductor: A substance, such as silicon, that acts as a conductor or insulator, depending on its operating condition and application.

Sensor: A device used to detect a change in pressure, temperature, or mechanical movement and convert a reference voltage into a usable electrical signal.

Separator: Plastic, rubber, or other insulating material placed between a battery's plates.

Series circuit: A circuit with only one path for current to flow.

Series-parallel circuit: A circuit in which a series and parallel circuits are combined.

Serpentine belt: A single belt used to drive all of the engine-driven accessories.

Service brake clutch: Typically a pressure switch, it is normally closed with the pedal released. Depressing the brake will open the circuit.

Service engine lamp: Often referred to as the "diagnostic lamp", it is used to alert the operator of the presence of an active event, and is used to flash a diagnostic code.

Service program module: A software program on a factory programmable computer chip that is designed to adapt an electronic service tool to a specific application.

Shift point: The points, at either engine rpm or vehicle speed, when a transmission should be shifted to the next gear.

Shim: A thin piece of brass, steel, or plastic inserted between two parts to adjust the distance between them.

Shimmy: A condition where a rotating part shakes from side-to-side.

Short circuit: Occurs when an electrical circuit is connected to an undesirable point. For example, a wire can rub against a vehicle frame until it wears off its insulation and makes electrical contact with the frame itself.

Shrink fit: A fit so tight that one part must be cooled (inner) or heated (outer) to fit on another part.

Shroud: A plastic or metal enclosure around a fan to guide and facilitate air flow.

Shunt: An alternate or bypass portion of a circuit.

Shunt winding: A wire coil that forms an alternate path through which electrical current can flow.

Signal: A voltage or waveform used to transmit information typically from a sensor to the ECM.

Silencer: A device for reducing the noise of intake air or exhaust gases.

Silicon (Si): An element that is neither a good conductor or insulator. By doping silicon with different elements, its characteristics can be changed. It is used to make transistors, integrated circuits, and other semiconductor devices.

Single exhaust system: An exhaust system that has only one pipe leading from the exhaust manifold to the muffler and out to the tailpipe.

Skid plate: A stout metal plate attached to the underside of a vehicle to protect the engine from damage from "grounding out" on rocks, curbs, and road surface.

Skirt: That portion of a piston below the piston pin which takes the side thrust of the piston.

Sleeve: A replaceable pipe-like section that is pressed or pushed into a block.

Slip rings: Metal rings mounted on an alternator drive shaft in which brushes make continuous contact.

Sludge: A composition of oxidized petroleum products along with an emulsion formed by the mixture of oil, water, dirt, and other contaminants. This forms a pasty substance, clogs oil lines and passages, and interferes with engine lubrication.

Smog: Generalized term used to describe air pollution caused by chemical fumes and smoke.

Snap ring: A split ring snapped in a groove to hold a bearing, thrust washer, gear, etc., in place.

Sodium valve: An engine valve that has metallic sodium added to its stem to speed up the transfer of heat from the valve head to the stem and from the guide and block. Sodium-filled valves should never be cut or ground.

SOHC: Abbreviation for *single overhead camshaft.*

Solenoid: An electrically operated magnetic device used to operate some unit. An iron core is placed inside a coil. When electricity is applied to the coil, the iron core centers itself in the coil and, as a result, it will exert some force on anything it is connected to.

Solid state: Any electrical device that has no moving parts, such as a transistor, diode, or resistor.

Solvent: A liquid used to dissolve or thin another material.

South pole: One of the poles of a magnet from which lines of force originate.

Spacer: A piece of metal or other material placed between two parts to provide clearance or thrust force for a fastener.

Specific gravity: A relative weight of a given volume of a specific material as compared to an equal volume of water.

Speed burp: Any sudden, brief, unwanted change in rpm.

Speed control: Method of maintaining a set speed as determined by the driver. Usually referred to as *cruise control.*

Speed droop: Measurement of the change in engine speed from no load to full load.

Speedometer: Instrument used to determine vehicle speed in miles or kilometers per hour.

Speed/timing sensor: A sensor that measures crankshaft position, direction of rotation and engine rpm, and sends a signal to the ECM.

Spindle: A machined shaft on which bearing races rest.

Splash system: An oiling system which supplies oil to moving parts by attaching dippers to the bottom of the connecting rods. These dippers can either dip into shallow trays or into the sump itself. The spinning dippers splash oil over the inside of the engine.

Splines: Metal grooves cut into two mating parts.

Spool valve: A hydraulic control valve shaped like a thread spool.

Sprag clutch: A clutch that will allow rotation in one direction, but not in the other. Commonly referred to as an *overrunning clutch.*

Spring loaded: A device or other component held in place or under tension by one or more springs.

Spring pressure: A component that is held in place by a spring.

Spurt hole: A small hole drilled through the connecting rod that causes oil to spurt and lubricate the cylinder walls.

Sq. ft.: Abbreviation for Square Foot

Sq. in.: Abbreviation for Square Inch.

Stability: Occurs when the governor can maintain desired engine speed without fluctuations.

Staked nut: Type of nut whose edges can be bent downward into a slot to secure it on a shaft.

Stall: To stop rotation or operation.

Starter pinion gear: A small gear on the end of the starter shaft that engages and turns the flywheel ring gear.

Starter: An electric, hydraulic, or pneumatic motor which uses a geardrive to crank (start) the engine.

Static electricity: A charge of electricity generated by friction between two objects.

Stator: A small hub in the torque converter that improves oil flow. Also, the stationary wire field in an alternator.

Straight roller bearing: A bearing which uses straight, non-tapered rollers in its construction.

Stress: The force or strain to which a material is subjected.

Stroke: The distance the piston moves between TDC and BDC.

Stud: A fastener that has threaded rods at both ends.

Sulfated: Condition where the lead in a battery's plates deteriorates and combines with the sulfur from the battery electrolyte to form a sulfate which coats the plates.

Sump: The part of an oil pan that contains the oil.

Supercharger: A blower or pump which forces air into the cylinders at higher-than-atmospheric pressure. The increased pressure forces more air into the cylinder, thus enabling more fuel to be burned and more power to be produced. Superchargers are dependent on engine speed and are most efficient at high engine speeds.

Supply voltage: A constant voltage supplied to a component in order to provide electrical power for its operation. It may be generated by the ECM, or it may be battery voltage supplied by the vehicle wiring.

Swirl: Rotation of a mass of air as it enters the cylinder is known as "swirl." This is one form of turbulence.

Switch: A device to make or break the flow of current through a circuit.

T

T-harness: A test harness is designed to connect into the vehicle or engine harness, allowing normal circuit operation while providing a breakout or "T" to measure signals.

Tachometer: An instrument designed to measure the rotational speed of an engine in revolutions per minute.

Tailpipe: The exhaust pipe running from the muffler to the rear of the vehicle.

Tank gauge unit: A variable resistor float device placed inside the fuel tank to monitor fuel level.

Tap: To cut threads in a hole with a threaded, tapered tool. Also to repair badly damaged threads.

Taper: Wear condition in which a cylinder is worn more at the top than at the bottom.

Tapered roller bearing: A bearing that utilizes a series of tapered steel rollers that operate between an outer and inner race.

Tappet: An adjusting screw for varying the lash or clearance between the valve stem and valve train components. May be built into the valve lifter or rocker arm.

TDC: Abbreviation for Top Dead Center.

Technical service bulletins (TSB): Information published by vehicle manufacturers in response to vehicle conditions, problems, etc., that may not be diagnosed by normal methods.

Temperature gauge: A dash mounted gauge that is used to indicate engine temperature.

Tension: Any pulling or stretching stress placed on an object.

Terminals: The connecting points in an electrical circuit.

Test light: A device that will show the presence of current by lighting a small light.

Thermal efficiency: The percentage of heat developed in a burning fuel charge that is used to develop power.

Thermistor: A device that changes its resistance in relation to heat.

Thermostat: A temperature sensitive device used in cooling systems to control coolant flow in relation to temperature.

Throttle position sensor (TPS): Input sensor used to monitor throttle position.

Throw: Offset portion of the crankshaft designed to accept a connecting rod.

Throw-out bearing: Bearing that is used to minimize pressure between the clutch surface and the throw-out fork.

Thrust bearing: A bearing designed to resist side pressure.

Thrust load: A pushing or shoving force exerted against one body by another.

Thrust washer: A bronze or hardened steel washer placed between two moving parts to prevent longitudinal movement.

Timing chain: Drive chain used to operate the camshaft by turning off the crankshaft.

Timing gear cover: Metal cover placed over the timing gears.

Timing gears: Any group of gears which are driven from the engine crankshaft to cause the valve, ignition, and other engine-driven accessories to operate at the desired time during the engine cycle.

Timing marks: A series of calibrating marks used to set engine timing. The marks can be located on the vibration damper, flywheel, and throughout an engine to check injection and valve opening timing.

Timing sprocket: Chain or belt sprockets on the crankshaft and camshaft.

Tolerance: A permissible variation between two extremes of specifications or dimensions. The amount of variation permitted from an exact size or measurement.

Torque limit: Programmable parameter that limits maximum torque based on PTO configuration and operation.

Torque wrench: A calibrated wrench with a built-in indicator designed to indicate the amount of torque applied to a fastener.

Torque: Any turning or twisting force, usually measured in foot-pounds or Newton-meters.

Total acid number (TAN): Measures the acidity of an oil which can contribute to internal corrosion of parts, oil thickening, and accelerated wear patterns.

Total base number (TBN): A measurement of an oil's ability to neutralize acids formed during combustion.

Transfer pump: A pump used for moving fuel from one tank to another or to deliver fuel to the injection pump.

Transistor: A semiconductor that is used as a switching device.

Transmission: A device that uses gearing and torque conversion to change the ratio between the engine RPM and driving wheel RPM.

Trip recorder: An aftermarket device dedicated to recording vehicle and engine operating parameters during vehicle service. Used to analyze driving habits and produce driver logs.

Troubleshooting: A process of diagnosing or locating the source of the trouble or troubles from observation and testing.

Troubleshooting chart: Diagnostic flow chart that provides step-by-step procedures to test automotive systems.

Turbine: A wheel upon which a series of angled vanes are affixed so that a moving column of liquid or gas will impart rotational movement to the wheel.

Turbocharger: A turbine device that utilizes exhaust pressure to increase the air pressure going into the cylinders. Used particularly in reference to movement of air in the cylinder and combustion chamber.

Turbulence: A violent irregular movement or agitation of a fluid or gas. Violent swirling motion. Fuel injection provides some turbulence. Additional turbulence is provided by the design features of the combustion space.

Two-stroke cycle engine: An engine that requires one complete revolution of the crankshaft to fire each piston once. An engine requiring only one complete revolution of the crankshaft to complete the cycle of events.

U

UAW: Abbreviation for United Auto Workers.

UIC: Abbreviation for Universal Integrated Circuit.

Underground storage tank (UST): Buried tanks used primarily to store fuel but may be used to hold other liquids.

Unit fuel injection systems: A low pressure fuel delivery system. Fuel is delivered to the unit injectors at pressures much lower than other diesel fuel injection systems.

Universal joint: A flexible joint that permits changes in driving angles between a driving and driven shaft.

V

V-belt: A V-shaped belt that is used to turn engine-driven accessories such as the alternator, water pump, and air conditioning compressor.

V-engine: An engine in which the cylinders are arranged in two separate banks.

Vacuum: A pressure in an enclosed area that is lower than ambient pressure.

Vacuum gauge: A test gauge used to determine the degree of vacuum existing in a chamber.

Valve: A device designed to open and close an opening.

Valve clearance: The space between the end of the valve stem and its actuating mechanism (rocker arm or camshaft).

Valve crossheads: Properly positioned on top of the valves, this device forms a bridge that permits the rocker arm to open two valves simultaneously.

Valve duration: The length of time that the valve remains open in degrees of crankshaft movement.

Valve face: The outer edge of the lower part of the valve. Mates with the valve seat.

Valve float: Condition in which valves are forced back open before that have had a chance to seat. Usually occurs at extremely high engine speeds.

Valve guide: A hole machined into the head to support the valve stems as they ride up and down. May be machined oversized to accept a removable guide.

Valve head: The part of the valve below the stem. The valve face is machined on this part.

Valve keeper: Device which snaps into a groove in the valve stem. Used to retain valve and spring assembly in the head.

Valve lash: Valve tappet clearance or total clearance in the valve train with the cam follower on the camshaft's base circle.

Valve lift: The distance that the valves moves from fully closed to fully open.

Valve lifter: A solid or hydraulic plunger that is moved by the camshaft to open the valves.

Valve margin: The width of the edge of the valve between the top of the valve and edge of the face.

Valve overlap: Period in degrees of crankshaft rotation in which both the intake and exhaust valves are partially open.

Valve ports: The openings through the head from the intake and exhaust manifolds to the combustion chamber.

Valve rotator: A device locked to the valve stem which forces the valve to rotate with rocker-arm action.

Valve seal: A seal that is placed over the valve stem to prevent oil leakage between the stem and the guide.

Valve seat: The area on which the face of the valves rest when closed.

Valve spring: A coil spring used to keep valves closed.

Valve spring seat: Machined recess in the head in which the valve spring sits.

Valve spring shim: A precision machined washer used to adjust valve spring tension.

Valve stem: The portion of the valve that is inside the head. The stem rides in the valve guide.

Valve timing: The position of the camshaft to crankshaft position so that the valves will open and close at the proper time.

Valve train: The various parts that make up the valve and operating mechanism.

Vane: Any plate, blade, or the like attached to an axis and moved by, or in air, or a liquid.

Vane pump: A type of pump that uses vanes that throw off or is moved by liquid or air.

Vaporize: A change in state from a liquid to a gas.

Vehicle identification number (VIN): Individual series of letters and numbers assigned to a vehicle by the manufacturer at the factory.

Vehicle speed calibration: A programmable parameter used by the ECM to scale the vehicle speed signal into miles per hour (or kilometers per hour).

Vehicle speed sensor (VSS): An electromagnetic pickup that measures vehicle speed from the rotation of gear teeth in the drive train of the vehicle. Used by the engine's ECM to monitor vehicle speed.

Velocity: Distance traveled in a given amount of time.

Vibration damper: A round metal or rubber weight attached to a crankshaft or other part to minimize vibration.

Viscosimeter: A device used to determine the viscosity of a given liquid. The length of time that it takes a heated liquid to flow through a set orifice determines its viscosity.

Viscosity: A measure of a fluid's ability to flow or its thickness.

Viscosity index: A rating given to oils and other fluids to indicate its resistance to changes in viscosity when heated.

Volatility: The tendency for a fluid to evaporate rapidly or pass off in the form of vapor. For example, gasoline is more volatile than diesel fuel as it evaporates at a lower temperature. The tendency of a fluid to evaporate in relation to temperature.

Volt: Unit of measurement of electrical pressure or force that will move a current of one ampere through a resistance of one ohm.

Voltage drop: A lowering of circuit voltage due to excessive lengths of wire, undersize wire, or through a resistance.

Voltage regulator: A mechanical or electrical device used to control alternator output.

Voltmeter: Instrument used to measure voltage in a given circuit.

Volume: Unit of measurement of space in cubic inches or cubic centimeters.

Volumetric efficiency: A comparison between the actual and ideal volume of gases or air drawn into a cylinder during the intake cycle.

Vortex: A mass of whirling liquid or gas.

W

Wastegate: A valve that vents excess exhaust gas to limit the amount of boost delivered by a turbocharger.

Water detector: A sensor installed in a diesel fuel system which warns the driver of water contamination of the fuel.

Water jacket: The area around the engine cylinders that is left hollow so that coolant may be admitted.

Water pump: The coolant pump; any pump used to circulate coolant through an engine.

Watt: Unit of measurement of electrical power. It is obtained by multiplying volts by amperes.

Watt's law: Formula for computing unknown power, voltage, or current in a circuit by using two known factors to find the unknown value.

Wet friction: The resistance to movement between two lubricated surfaces.

Wet muffler: A muffler that uses water to suppress engine noise and to cool the muffler. Frequently used in marine applications.

Wet sleeve: A thick metal barrel inserted into an engine cylinder. Is constantly in contact with engine coolant.

Wheel speed sensor: Magnetic sensor used in an anti-lock brake system to measure wheel speed.

Wire harness: A group of primary wire encased in a paper or plastic sleeve. Used to ease installation and to prevent wire damage.

Wiring diagram: A detailed drawing showing the location of electronic components and devices that are connected together in a circuit.

Work: A force applied to a body, causing it to move. Measured in foot-pounds, watts, or joules. Mechanical work is done when some object is moved against a resisting force. It is the product of force which acts to produce displacement of body and the distance through which body is displaced in the direction of force.

WOT: Abbreviation for Wide Open Throttle.

Wrist pin: See *Piston pin.*

Z

Zener diode: A silicon diode that serves as a rectifier. It will allow current to flow in one direction only until the applied voltage reaches a certain level. Once it reaches this point, the diode allows current to flow in the opposite direction.

Zinc anodes: Sacrificial metal placed in a marine cooling system to reduce the corrosive effect of electrolysis.

Index